NECK INJURY BIOMECHANICS

PT-141

Edited by
Jeffrey A. Pike

Published by
SAE International
400 Commonwealth Drive
Warrendale, PA 15096-0001
U.S.A.
Phone: (724) 776-4841
Fax: (724) 776-5760
www.sae.org
June 2009

SAE International™

400 Commonwealth Drive
Warrendale, PA 15096-0001 USA
E-mail: CustomerService@sae.org
Phone: 877-606-7323 (inside USA and Canada)
724-776-4970 (outside USA)
Fax: 724-776-1615

ISBN 978-0-7680-2163-9
Library of Congress Catalog Number: 2009926659
SAE Order No. PT-141

For permission and licensing requests, contact SAE Permissions, 400 Commonwealth Drive, Warrendale, PA 15096-0001 USA; e-mail: copyright@sae.org; phone: 724-772-4028; fax: 724-772-9765.

To purchase bulk quantities, please contact SAE Customer Service, e-mail: CustomerService@sae.org, phone: 877-606-7323 (inside USA and Canada), 724-776-4970 (outside USA), fax: 724-776-1615.

Visit the SAE Bookstore at http://store.sae.org

DEDICATION

To Debbie
and the "kids,"
Stacy, Adam, Jessica, Blair, and Emily,

and

to the memory of my sister,
Anne Dolores Pike,

and

to the memory of
Colonel John Paul Stapp
and
Professor Larry Patrick

Other SAE books of interest

Forensic Biomechanics: Using Medical Records to Study Injury Mechanisms
By Jeffrey A. Pike
(Product Code: R-379)

Neck Injury: The Use of X-Rays, CTs, and MRIs to Study Crash-Related Injury Mechanisms
By Jeffrey A. Pike
(Product Code: R-268)

Automotive Safety Handbook, Second Edition
By Ulrich Seiffert and Lothar Wech
(Product Code: R-377)

Vehicle Accident Analysis and Reconstruction Methods
By Raymond M. Brach and R. Matthew Brach
(Product Code: R-311)

Crash Reconstruction Research: 20 Years of Progress (1988-2007)
Edited by Michael S. Varat
(Product Code: PT-138)

For more information, or to order a book, contact SAE at 400 Commonwealth Drive, Warrendale, PA 15096-0001; phone 724-776-4970; fax 724-776-0790; email CustomerService@sae.org.

PREFACE

An earlier SAE International compendium (*Biomechanics of Impact Injury and Injury Tolerances of the Head-Neck Complex*, PT-43, edited by Stanley H. Backaitis) provided a collection of papers from the early 1960s through 1991. Therefore, this compendium will start with the early 1990s and progress through the present. Both the reprints and bibliography sections include entries up to and including 2009.) In addition, the current work includes a Background section that ties together concepts of anatomy, injuries, and imaging to make the included papers useful to a wider range of safety professionals. (For those seeking a more detailed discussion of anatomy, the current work includes a paper devoted entirely to the topic of the anatomy of the cervical spine.) This collection also includes several of the earlier papers to help put current research into a historical perspective.

This compendium includes 40 previously published technical papers, a bibliographic listing of an additional 100 references, and, as mentioned above, a section that provides background information to make the contents more readable and useful to a wide range of safety professionals. The papers are arranged alphabetically by author. Although much of the underlying testing was vehicle-related, the resultant data regarding neck motion and loading are applicable to a broad range of injury scenarios. For example, a study of neck axial compression may offer useful insight into injury associated with sports as well as vehicle rollovers. The papers include a comparison of male versus female injuries (reprint 4) and discussion of child injuries (including reprints 17, 19 and 32). In general, the included works relate to blunt trauma, as opposed to penetrating trauma, but within the blunt trauma category may be applied to a wide range of injury scenarios. To facilitate this, Table I provides a listing of the 40 papers and categorizes the main types of neck motion (e.g., flexion) or neck loading (e.g., tension), and, in the case of vehicle-type impact testing, the primary vehicle impact direction (e.g., rear impact).

The bibliography also is arranged alphabetically by author, and Table II provides a cross reference of those papers arranged chronologically, for those users who wish to focus on the most recent studies, earlier work, or work that was done during a particular period of time.

The 140 papers should offer useful insights, not only for the past and present, but as this ever-changing field continues to evolve, with regard to future applications as well.

ACKNOWLEDGMENTS

First and foremost, it is my pleasure to acknowledge the role of the Advisory Panel. Each of the members listed below participated in several iterations of the paper selection process and, in many instances, took on other tasks as well. Each member of the panel is truly a world-renowned authority, and I thank them all for their willingness to participate and for sharing their time and expertise.

Ultimately, however, injury biomechanics is not about papers and books but about reducing morbidity and mortality from biomechanical trauma. The Advisory Panel members all have "day jobs," in which capacity they have helped save lives and reduce injuries. On behalf of the many individuals they have benefitted, I would like to thank the group members for their contributions to society. I hope this volume will help others to continue these efforts.

ADVISORY PANEL

John M. Cavanaugh, Professor
Department of Biomedical Engineering
Wayne State University
USA

Tom Gibson, Director
Human Impact Engineering
Australia

Guy Nusholtz
Chrysler Group LLC
USA

Koshiro Ono
Japan Automobile Research Institute
Japan

Stephen A. Ridella, Chief, Human Injury Research Division
National Highway Traffic Safety Administration
USA

Gunter P. Siegmund, President
MEA Forensic Engineers & Scientists, Ltd.
Canada

Mats Y. Svensson, Professor
SAFER—Vehicle and Traffic Safety Centre
Chalmers University of Technology
Sweden

David C. Viano
ProBiomechanics LLC
USA

Beth A. Winkelstein, Associate Professor
Bioengineering and Neurosurgery
University of Pennsylvania
USA

Narayan Yoganandan. Professor and Chair, Biomedical Engineering
Department of Neurosurgery
Medical College of Wisconsin
USA

I also would like to thank the many authors whose publications are included, either as reprints or in the Bibliography, as well as the authors of the many fine papers that were not included. Ultimately, this volume was the result of a number of diverse selection criteria, including availability for reprint.

On the dedication page of this book, I have singled out two researchers: Colonel John Stapp and Professor Larry Patrick. Not only were they pioneers but also volunteers --they were test subjects for much of the data they reported. I hope they are looking down from that great deceleration sled in the sky and smiling.

Finally, this is my fourth volume for SAE International—the first three as author and this one as editor. Perhaps the best acknowledgment that I can give to Martha Swiss and the crew at SAE is to mention that there may be a fifth.

Jeffrey A. Pike
SAE Fellow and Series Editor
Biomechanics Consulting, Inc.
USA

INTRODUCTION

The following introduction is intended to provide a brief overview of some aspects of anatomy, injuries, and radiographic imaging that will be useful in the remainder of this book. It is intended to be neither exhaustive nor exhausting, but rather to provide some useful background regarding the anatomy, injuries, and imaging of the cervical spine. Those desiring a more detailed introduction to neck anatomy are referred to Huelke 1979 (reprint included in this volume) or to one of the standard anatomy texts (e.g., *Gray's Anatomy*). Much of the information and illustrations in the following discussion are derived from three publications (fully cited in the Bibliography section): Pike 1997 (a more detailed discussion of general anatomic and injury scaling terminology and concepts and of particular interest to some readers, with the role of biomechanics in federal regulations also addressed), Pike 2002 (neck anatomy, injury, and imaging), and Pike 2008 (using medical records to study injury).

<u>Anatomy</u>

The bony framework of the spinal column is composed of bones of various shapes and sizes. These include a series of ring-like vertebrae stacked on top of each other to form a column-like structure, the vertebral column. The channel bordered by the opening in the center of the vertebrae contains the downward extension of the brain, the brain stem, which leads to the spinal cord.

A shorthand system frequently used to specify spine location designates each of the vertebrae using the first letter of the name of the region in which the vertebra is located. Moving inferiorly (i.e., from the base of the skull toward the tailbone), the regions of the spine are as follows: cervical, thoracic, lumbar, sacral, and coccygeal (Figure 1). Each individual vertebra then is designated by referring to the first letter of the region (e.g., "C" for cervical), and, starting with the uppermost vertebra in that region, numbering the vertebrae consecutively until the bottom vertebra in that region has been named. Thus, the vertebrae of the cervical region are designated as Cl, C2, and so on, down through C7. (Cl is at the head-neck junction, and C7 is at the neck-shoulder junction.) Examples of the use of this nomenclature would be to describe a vertebral fracture "of C5" or a spinal cord injury "at C5."

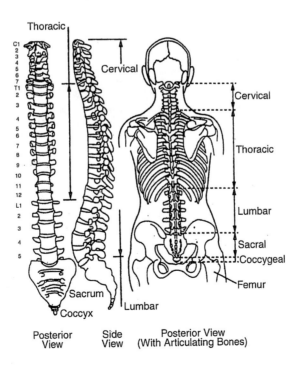

Figure 1 The spine.

The vertebra immediately below C7 is the first vertebra of the thoracic region and therefore is designated as T1. The vertebra immediately beneath this is T2, and so on, until T12. Immediately beneath T12 is the first vertebra of the lumbar region, which is designated as L1. Beneath this are the remaining lumbar vertebrae, which are designated as L2 through L5. Below L5 is the sacrum, and at the tip of the sacrum is the coccyx. Although the constituent sacral vertebrae have fused by adulthood, they are still discernable. Thus, a similar shorthand nomenclature may be applied to the sacral vertebrae, namely, S1 through S5. Similar to the sacral vertebrae, the coccygeal vertebrae, even when fused, remain discernable and may be designated as Co1 through Co4.

When viewed from the side, the normal spine for an individual standing straight, with arms at sides and palms forward, has three major curves (in the mid-sagittal plane) that approximately correspond to the cervical, thoracic, and lumbar regions (Figure 1). The most superior curve, referred to as the cervical curve, is convex ventrally and extends from the top of C2 (the apex of the odontoid process) (Figure 2) to the middle of T2. When the spinal curvature is reduced or exaggerated, it may be indicative of some pathology. Reduced curvature may be referred to as "cervical straightening" and may be an incidental finding or may be indicative of muscle spasm. Three types of spinal curvature are kyphosis, lordosis, and scoliosis. Kyphosis is a spinal curvature that is concave anteriorly (e.g., a hump-type shape); lordosis refers to a curvature that is concave posteriorly (e.g., swayback); and scoliosis refers to a serpentine shape in the coronal plane (e.g.,characterized by uneven shoulders and a prominent shoulder blade).

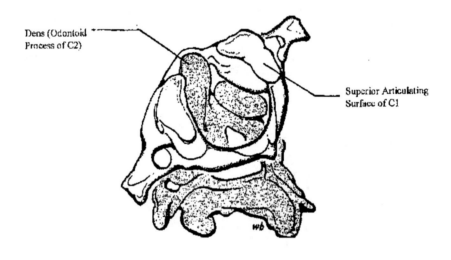

Dens (Odontoid Process of C2)

Superior Articulating Surface of C1

Figure 2 Vertebra C2 (the axis).

The typical cervical vertebra is characterized by an opening in the central section of the bone, and this opening is referred to as the vertebral foramen. The spinal cord passes through this opening (Figure 3). The cord starts at the top of the column and continues down to the lumbar region. Also, there are various ligaments, which are flexible and sinewy tissues that bind the vertebrae together. These ligaments connect two or more adjacent vertebrae or, in some cases, run along the length of the column. The intervertebral discs are located between the bodies of adjacent vertebrae in the cervical, thoracic, and lumbar regions (except for C1 and C2). Branches of the spinal cord, called nerve roots, connect the spinal cord to the face, limbs, and torso (Figure 4).

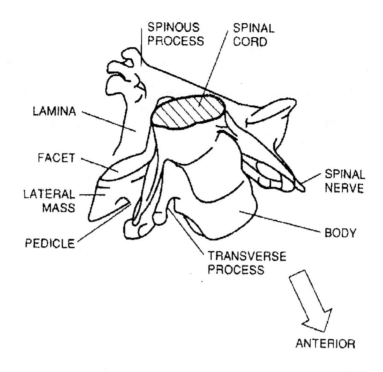

Figure 3 Vertebra and spinal cord.

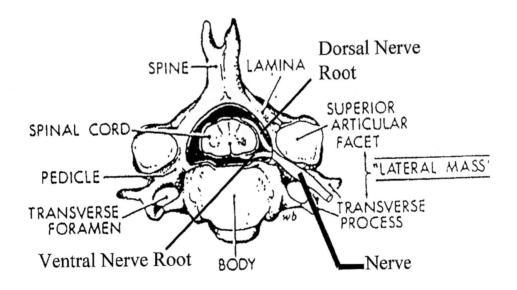

Figure 4 Spinal cord, spinal nerves, and nerve roots.

Injuries

Spinal injuries can run the gamut from very minor to very severe. In fact, at the minor end of the spectrum, there may be disagreement regarding if and/or how an injury occurred. One example of this would be if there is a subjective complaint (e.g., pain) without any objective evidence of injury (e.g., x-ray). Another instance would be if there is an objectively discernable condition (e.g., bulging disc) but disagreement regarding whether or not it was a condition that existed before the incident (a so-called pre-existing condition) or if it's painful. Addressing such topics may provide useful insight with regard to determining tolerance limits and injury mechanisms.

Most spinal injuries can be categorized as a sprain, disc disruption, vertebral fracture, or vertebral dislocation, and any one or a combination of these may or may not involve the spinal cord or the spinal nerve roots. A sprain basically refers to the stretching of soft tissue beyond its elastic limit (i.e., it will not return to its original shape and size, even after the stretching force is discontinued) so that it begins to tear. The soft tissue involved most often is a muscle, ligament, or facet joint capsule (i.e., the fibrous structure enclosing a facet joint). Disc disruption most frequently refers to a tearing or compression of the intervertebral disc, possibly with the extrusion of some of the contents. Vertebral fracture refers to an actual cracking or breaking of one of these bones, and dislocation refers to a relative displacement between adjacent vertebrae.

Spinal injuries frequently are due not to some unique movement, which is exclusively associated with injury, but rather to extremes of the various movements and loads that are part of normal functioning of the spine These include tension, compression, flexion, extension, lateral flexion (lateral bending) rotation, and combinations thereof. In addition to these movements, the cervical spine may be loaded externally (e.g., by impact to the top of the head) in compression or tension, that is, a compressive or stretching force may be exerted along the length of the spine. In a vehicular impact environment, although a particular neck motion may predominate (e.g., someone may sustain what is basically a flexion-type injury), it is unlikely that the neck will undergo a single, pure, two-dimensional motion. For example, one segment of the spine may be undergoing flexion, while another segment is undergoing extension. The motion also may include a time sequence of more than one motion (e.g., a given region of the spine may first undergo flexion and then undergo extension).

Generally, injury to the various non-neural spinal structures (i.e., muscles, ligaments, disc, and vertebra) become much more significant if they affect the neural elements—the spinal nerves and especially the spinal cord. When the neural structures are damaged, loss of sensation or motor function may occur. Thus, vertebral fracture *per se* is not sufficient to cause neurological deficit. (Indeed, a majority of cervical fractures have no associated neurological injury.) Rather, the potential of a fractured or dislocated vertebra to cause injury to the spinal cord is of major concern.

Similarly, a major concern regarding a torn or ruptured disc is whether or not the jelly-like middle of the disc (the nucleus pulposus) is extruded rearward to the extent that it presses on a spinal nerve (a branch of the spinal cord) or possibly even presses on the spinal cord itself (Figure 4). When part of the disc material presses against surrounding soft tissues and possibly injures the spinal cord or spinal nerves, the pain and disability that may be associated with a ruptured disc are experienced. Depending on the extent and nature of the pressure that this material exerts against the neural tissue, some loss of sensation and/or mobility may result.

Similar to a protruding disc, a ligament that normally helps to support the vertebral column may push against the cord. This is particularly true in the elderly, in whom the ligamentum flavum is prone to buckle and, during neck extension, may push against the cord and produce an injury, for instance, to the inner diameter of the cord. When this occurs, it is referred to as the central cord syndrome. Consequently, the overriding concern with regard to injuries of the discs and ligaments is essentially the same as for fractures of the vertebrae—namely, will they cause injury to the spinal cord (or one of the peripheral nerves emerging from either side of the cord)? Thus, a so-called "broken neck" (fracture of one or more cervical vertebrae), although serious in its own right, becomes much more severe if the bone is broken in such a way that the bone then injures the spinal cord (Figure 5).

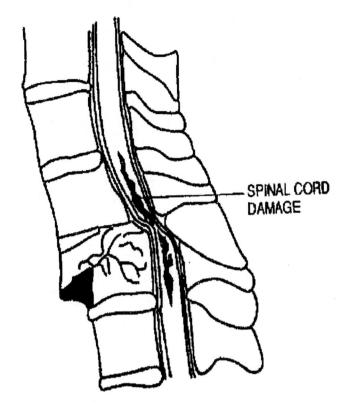

SPINAL CORD
DAMAGE

Figure 5 Vertebral fracture with cord injury.

A complete disruption of the spinal cord, at any particular level, disrupts the functioning of the spinal cord at the site of transection and distally, both sensory and motor. For example, if the spinal cord is severed at the level of C5, all spinal cord function handled by the cord below the C5 level will be lost. Similarly, if the spinal cord is severed at the C4 level, all of the spinal cord functions below the C4 level will be lost. Thus, a spinal cord transection at the C4 level would be equivalent to the loss due to a transection at the C5 level, plus the additional loss of the functioning of the cord segment from C5 to C4. Note that in contrast to spinal cord injury, spinal nerve (or nerve root) (Figure 4) injury generally does not affect the cord, nerves, or nerve roots distal to the injury site and hence does not affect the body regions served by nerves that branch from the cord inferior to the injury site.

As a general guideline, the sacral region of the spinal cord controls defecation, urination, and sexual function and contributes to ambulation. The lumbar region controls the lower limbs and contributes to urination, the thoracic region controls the lower torso, and the cervical region controls the upper torso and upper extremities. Thus, transection of the spinal cord in the lumbar region generally would result in at least a partial loss of lower-limb movement. Typically, the spinal cord *per se* terminates at the L1 level, and the lumbar, sacral, and coccygeal nerve roots emerge from this point and continue down the spinal canal as the cauda equina (horse's tail.)

Generally, if the cord is transected above the L2 level, ambulation, even with traditional mechanical assistance (e.g., a cane or brace) cannot be achieved, and transection anywhere in the cervical region would result in quadriplegia. Sometimes the term "quadraplegia" is used to indicate total paralysis of all four limbs, while the term "quadraparesis" is used to indicate partial loss of motion in all four limbs.

Cervical cord injuries, if in the upper cervical region, can have an additional deleterious effect. The phrenic nerve, which controls the respiratory diaphragm (and hence, the ability to breathe), usually is a combination of nerve branches from the third, fourth, and fifth cervical levels. If the cord is severed above this level, the patient would be unable to breathe, and death would result. An exception to this would be if the patient could be kept alive via artificial respiration until transported to a medical facility. The patient then could be helped to breathe, either by being ventilated mechanically or by being made to breathe "on his or her own" with the assistance of a phrenic nerve pacer (similar to a heart pacemaker).

Note that in practice, most blunt trauma spinal injuries do not involve complete transection of the cord; thus, there may be motor and sensory losses at different levels. Furthermore, blunt trauma may not produce physical interruption of the cord, but rather may set in motion a biochemical cascade that results in tissue death within a few hours. Thus, there is the possibility of medical/surgical intervention to interfere with this process and improve the outcome. A number of systems have been developed for ranking and quantifying injury severity. These include the Abbreviated Injury Scale (AIS), based on the probability of fatality (www.aaam.org), and the American Spinal Injury Association (ASIA) Impairment Scale, based on function (www.asia-spinalinjury.org).

Rather than direct mechanical trauma to the cord *per se* (e.g., impingement on the cord by displaced or fragmented vertebra or disc), neck trauma initially may injure the blood vessels supplying the cord. These types of vascular injuries may be exacerbated if there is already some spinal canal compromise due to degenerative changes (e.g., buckling of the posterior longitudinal ligament) or various diseases (e.g., bony spurs [osteophytes] associated with arthritis), which, in effect, narrows the spinal canal. Pre-existing conditions (e.g., diabetes) that limit circulatory capacity also may predispose individuals to vascular injury.

Note that pre-existing conditions also might limit a person's ability to recover from trauma. For example, a smoker may not heal properly (e.g. arthrodesis). Also, an older person may be more likely to succumb to a given injury than a younger, otherwise healthy individual (Pike 1989).

Radiology

A lateral view usually is the first neck x-ray taken and frequently provides the most information. After trauma, the lateral view often is accomplished with the patient supine and in a cervical collar. In this position, the x-ray source is on one side of the table, and the "film" is on the other side (Figure 6) so that the x-ray beam is essentially horizontal. Lateral views often are supplemented with one or more of the following: A-P (anterior-posterior), odontoid (open mouth), oblique, pillar, swimmer, and flexion-extension views (Pike 2002).

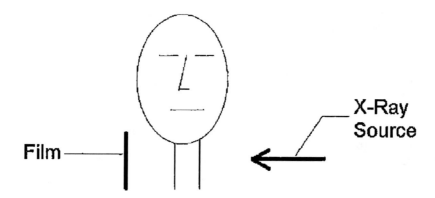

Figure 6 Positioning of x-ray source, body region of interest, and film for plain film radiograph (cross-table lateral view).

The lateral view and other x-rays are especially good at depicting bony structure and thereby provide direct information regarding bone fracture. On the other hand, these images do not depict ligaments and other soft tissue; thus, indirect inferences must be made regarding injuries to structures such as ligaments and discs. One example of a "directly observable injury" is a particular type of vertebral fracture that is referred to as a "wedge fracture" because of the wedge-shape appearance on a lateral view x-ray of the fractured vertebra (Figure 7). This type of injury generally is associated with neck flexion—the fronts of two vertebrae are compressed together until one of them fractures.

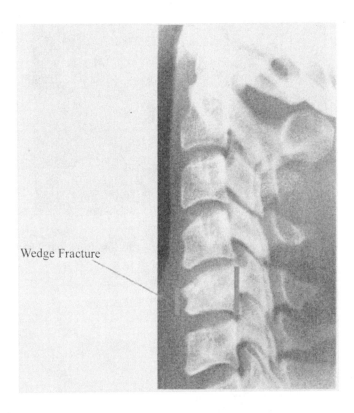

Figure 7 Wedge fracture.

Various clues regarding the location and nature of "indirect-observation injury" are discernible from a lateral view c-spine (cervical spine) plain film x-ray—by drawing a series of curves superimposed on the cervical vertebrae and then checking these curves for parallelness and continuity. As will be discussed, all curves should be smooth and generally parallel, and the lack of these traits can be indicative of various injuries. In effect, this is checking the alignment of the vertebrae and the integrity of the associated ligaments and other soft tissue. For example, although ligaments *per se* are not depicted on x-rays, one function of the ligaments is to maintain vertebral alignment. Therefore, irregularities of alignment can be used to infer information regarding ligamentous injury.

Two of the curves (Figures 8 and 9) that can be drawn on lateral views of the cervical spine are designated as the anterior vertebral body line (Figure 8) and the posterior vertebral body line (Figure 9).

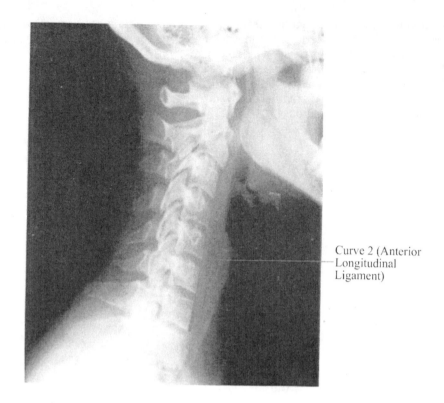

Figure 8 Lateral view—Curve 2 (anterior vertebral body line) (anterior longitudinal ligament [ALL] line).

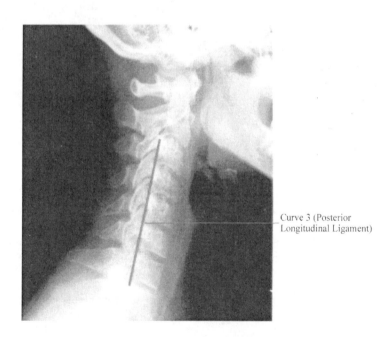

Figure 9 Lateral view—Curve 3 (posterior vertebral body line) (posterior longitudinal ligament [PLL] line).

Misalignment may be indicative of ligament tears, vertebra fracture, or relative displacement of one vertebra with respect to an adjacent vertebra. Any of these may be indicative of neurological injury or the potential for neurological injury.

Alignment

The line labeled "Curve 2," which is the anterior vertebral body line (Figure 8), corresponds to the anterior longitudinal ligament (ALL). A disruption in the smoothness of this curve corresponds to a tear or rupture of the ligament (Figure 10) and an associated vertebral displacement.

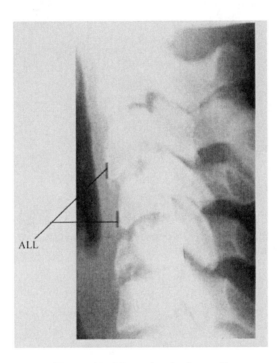

Figure 10 Disruption in Curve 2.

Similarly, the line labeled "Curve 3," the posterior vertebral body line (Figure 9), corresponds to the posterior longitudinal ligament. A disruption in the curve corresponds to a tear or rupture of the ligament and an associated vertebral displacement (Figure 11). Curves 2 and 3 should essentially be parallel. Even a small offset of several millimeters can suggest disruption of either one or both of the ligaments (ALL and PLL).

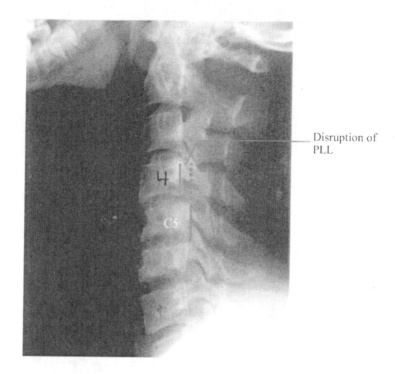

Figure 11 Disruption of Curve 3.

Excessive angulation between adjacent vertebrae is suggestive of ligamentous disruption or vertebral dislocation, Figure 12 depicts approximately 30 degrees of angulation—two to three times the angle at which ligamentous disruption might be suspected.

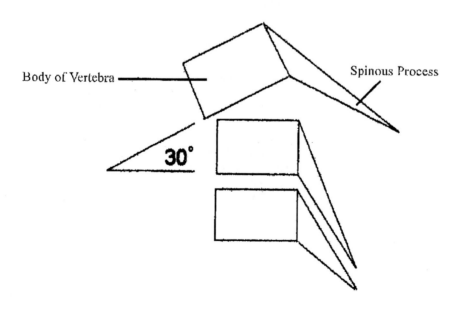

Figure 12 Angulation between adjacent vertebrae.

Also, the height of each vertebral body should be rather constant in the lateral view. For a given vertebra, even a variation in height of several millimeters suggests a compressive fracture of the vertebra.

Cartilage

In this context, the term "cartilage" refers to the intervertebral disc. Spacing between two adjacent vertebrae should be constant. If this spacing is wider anteriorly, it may be indicative of an extension-type injury; if this space is wider posteriorly, it may be indicative of a flexion-type injury. On a plain film x-ray, the disc height can be inferred by noting the separation between adjacent vertebrae, and the disc can be visualized directly on an MRI (Figure 13).

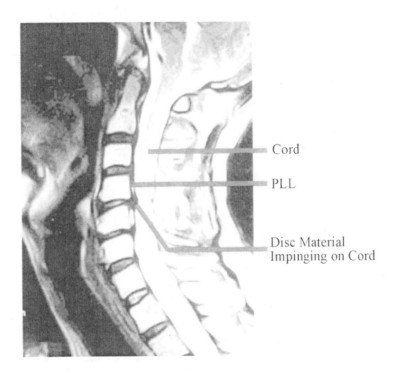

Figure 13 MRI—lateral view.

As mentioned, C1 (the atlas) and C2 (the axis) are shaped quite differently from the other cervical vertebrae and from each other (Figure 14). A special view, the so-called "odontoid view," is used for viewing the odontoid process (dens) of C2 and other structures of the upper cervical spine. The mandible (lower jaw bone) on a "regular" A-P view typically obscures the upper cervical spine region; thus, the odontoid view is used. Basically, the odontoid view is A-P in orientation but with the x-ray beam aimed at the open mouth (Figure 15) to remove the mandible and teeth from the image. The resulting image (Figure 16) can provide much information regarding bone fracture and/or ligament disruption.

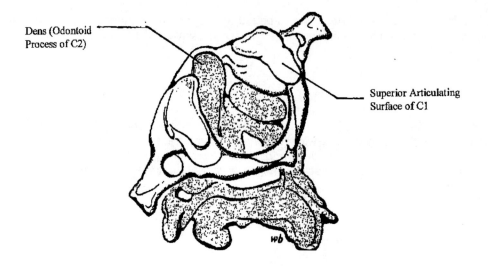

Dens (Odontoid Process of C2)

Superior Articulating Surface of C1

Figure 14 C1 and C2.

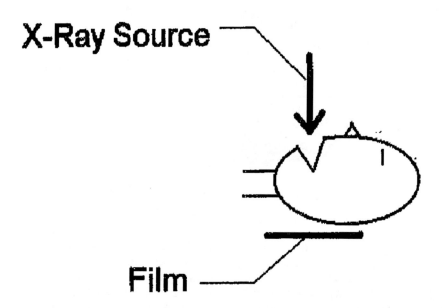

X-Ray Source

Film

Figure 15 Open-mouth odontoid view, beam position.

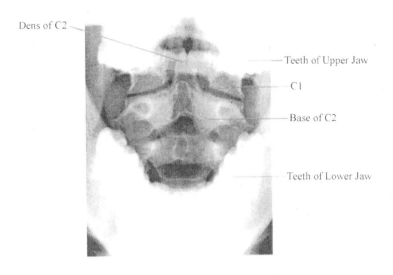

Figure 16 Open-mouth odontoid view, radiograph.

More specifically, the odontoid view is used to check for fracture of either the dens or of the ring of C1 and for ligamentous injury, as evidenced by the improper alignment between C1 and C2. The odontoid process (dens) of C2 usually is held in close approximation to the anterior arch of C1 by the transverse ligament. A separation of more than a few millimeters between the anterior edge of the dens and C1 may indicate disruption of this ligament. The lateral displacement of both halves of C1 is indicative of ligamentous and/or bony injury. Fracture of the bony ring of C1 also may be observed on a CT, which is an x-ray generated image that displays a "slice" of the structure being imaged. As with other x-ray images, bones are readily visualized and soft tissue is not. An example of a bilateral fracture involving both anterior and posterior elements of C2, often referred to as a Jefferson fracture, is provided (Figure 7).

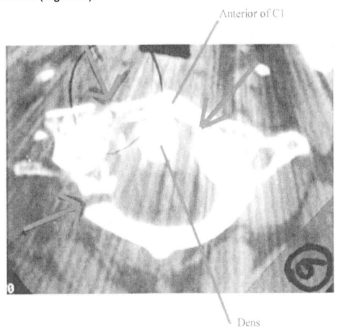

Figure 17 Jefferson fracture (C1), CT (axial view).

Those seeking more information on the topics discussed in this Introduction are referred to the following publications:

- For general terminology:
  Pike, Jeffrey A., *Automotive Safety Anatomy, Injury, Testing, and Regulation* (R-381), SAE International, Warrendale, PA, 1997.

- For more information specifically regarding neck anatomy, injury, or imaging:
  Pike, Jeffrey A., *Neck Injury: The Use of X-Rays, CTs, and MRIs to Study Crash-Related Injury Mechanisms* (R-268), SAE International, Warrendale, PA, 2002.

- For using medical records to study injury mechanisms:
  Pike, Jeffrey A., *Forensic Biomechanics: Using Medical Records to Study Injury Mechanisms* (R-379), SAE International, Warrendale, PA, 2008.

TABLE OF CONTENTS

HYBRID III DUMMY NECK RESPONSE TO AIR BAG LOADING

Venkatesh Agaram, Jian Kang, Guy Nusholtz and Gregory Kostyniuk
DaimlerChrysler
USA
Paper No.469

ABSTRACT

This paper discusses issues related to the Hybrid III dummy head/neck response due to deploying air bags. The primary issue is the occurrence of large moment at the occypital condyles of the dummy, when the head-rotation with respect to the torso is relatively small. The improbability of such an occurrence in humans is discussed in detail based on the available biomechanical data. A secondary issue is the different anthropometric characteristics of the head/neck region of the Hybrid III dummy when compared to humans.

Different modes of interaction between the deploying air bag and the Hybrid III dummy's neck are discussed. Key features of the dummy's response in these interaction modes have been described in light of the laxity of the atlanto-occipital joint and the effect of the neck muscle pairs. Issues for improving the biofidelity of the Hybrid III dummy's neck response due to deploying air bags are discussed.

INTRODUCTION

At present, the occupant response in automotive accidents is estimated by studying the response of the Hybrid III family of dummies in simulated crashes. With the increasing provision of air bags in today's, and possibly future fleet of vehicles, the biofidelity of the Hybrid III dummy neck response to air bag loading has taken on special significance. The characteristics of human response to interaction with deploying air bags however, is not well understood, and consequently, the design of Hybrid III family of dummies may require updating.

The current head-neck design of the Hybrid III dummies may not provide a reliable prediction of the response of human subjects due to a deploying air bag. The major issue, is the occurrence of large moments in the neck of the dummy, with very little rotation of the head relative to the torso. Such a response is unlikely in humans due to different load resisting mechanisms in humans when compared to those in the dummy. Another issue is the significant departure in the anthropometric characteristics of the head/neck region, between the Hybrid III dummy and the humans. The exposed horizontal surface in the chin-jaw region and the near vertical cavity between the jaw and the neck, as well as, the vertical surface behind the chin, provide unrealistic reaction surfaces for loading by an inflating air bag, potentially resulting in unrealistic neck-deformation. Although, the secondary issue of the air bag penetrating the chin-neck-jaw cavity has received some attention in terms of proposals for neck shield design, the fundamental issue of discrepancy in the moment-rotation relationship between that of the Hybrid III dummy and humans has received limited attention.

Melvin et al [1] in their study of air bag interaction with out-of-position drivers used a vinyl-nitrite neck-skin and a Neoprene chin filler on the Hybrid III 5th percentile female dummy, in an attempt to prevent the air bag from entering the neck-chin-jaw cavity. The spine of the dummy was modified in order to allow it to slouch without off from the seat. The authors did not present a detailed analysis to show if their neck shield design successfully prevented the air bag from entering the neck-chin-jaw cavity or not. They did however mention that in their efforts, stiffer neck shields generated alternate load paths to the head, shunting the upper neck load cells, thereby affecting the readouts. The authors did not address the issue of high moments in the neck at low angles of head rotation.

Morris et al [2] have studied three neck shield concepts for Hybrid III 5th percentile female dummy in driver seat, in order to assess their ability to prevent the air bag from entering into the neck-chin-jaw cavity. The concepts were: the standard head skin with a molded foam neck shield, the TMJ head skin (SAE terminology, referred as modified neck-skin in the paper [2]) with a foam neck wrap, and the TMJ head skin with an integrated neck shield. The integrated neck skin was formed by welding vinyl skin to the TMJ head skin such that it wraps around the neck and goes under the jacket. The authors report that the standard head skin with the molded foam neck shield passed the extension calibration test but failed the flexion calibration test. The TMJ head skin with the foam neck wrap passed both the extension and the flexion calibration tests. The TMJ head skin with the integrated neck shield could not be tested for calibration because the gripping under the jacket could not be simulated. Amongst the three neck shield designs considered, the TMJ head skin with the integrated neck shield was the best in preventing the air bag from entering the neck-chin-jaw cavity. However, this design was considered to have inhibited the upper neck load cells from measuring the true loads by restricting the head motion. The authors did not address the improbability of occurrence of high moments at the occipital condyles in humans for very little head rotation, a characteristic exhibited by the Hybrid III dummy.

Kang et al [3] have recently presented their study of the moment-rotation relationship of the Hybrid III dummy head/neck, due to a deploying air bag, by comparing it with the corridors proposed by Mertz and Patrick [4]. The moments at the occipital condyles of the Hybrid III dummy, in certain modes of air bag-neck interaction, go out of the Mertz and Patrick [4] corridors, with very little head rotation relative to the torso. Kang et al [3] also studied two neck shields for their potential to prevent air bags from entering the neck-chin-jaw cavity. One of their neck shield designs consisted in welding a Hybrid II 50th percentile male dummy head skin's chin and neck portion to the underside of the Hybrid III 5th percentile female dummy's standard head skin. The entrance of the air bag into the neck-chin-jaw cavity was successfully prevented. Due to the stiffness of the Hybrid II dummy's chin and neck portion of the head skin, certain neck air bag interaction modes were completely eliminated. It was however not clear if, and to what extent, the neck load cell measurements were affected. Further, the neck shield passed the extension calibration test, while it failed the flexion calibration test due to interference with the jacket. The second neck shield design studied by Kang et al [3] targeted the elimination of a specific air bag neck interaction mode, namely the entrapment of the air bag behind the jaw, in the near vertical jaw-neck cavity. Aluminum patches were used as extensions to the jaw of the dummy's head, preventing the air bag from getting trapped behind the jaw, although not preventing it from getting into the chin-jaw cavity and exhibiting another, less severe, air bag neck interaction mode. This neck shield design successfully passed both the flexion and the extension calibration tests.

From the literature it appears that the attempts to prevent the air bag from entering the chin-neck-jaw cavity of the Hybrid III dummy, have been the major focus. This is not surprising because such efforts do not need fundamental changes to the dummy's head/neck design and only involve the design of neck shields. However, it must be noted that neck shields, which prevent the air bag from entering the chin-neck-jaw cavity effectively, also pass the neck calibration tests, and which do not interfere with the neck load measurements have yet to be designed. Attempts to improve the moment-rotation relationship of the Hybrid III dummy's neck are conspicuous by their absence in reported literature, although this might be a more fundamental way of dealing with the problem of large moment at the occipital condyles at small head rotation angles.

The Hybrid III dummy neck was designed by limiting its flexion and extension responses, described as the relationship between the moment at the occipital condyles and the rotation of the head relative to the torso, to the corridors proposed by Mertz and Patrick [4]. The neck-deformation mode considered by Mertz and Patrick [4] was generated by the motion of the head relative to torso, when the torso was restrained either by the seat belt or the seat back. This essentially results in first bending mode in the dummy's neck. Air bag loading is significantly different from the seatbelt or the seatback loading, because of much larger extent of interaction of the air bag with the head, neck and torso. Further the anthropometric characteristics of the current Hybrid III dummy's neck-jaw complex, is different from that of humans. The neck-chin-jaw cavity in the Hybrid III dummy is easily accessible to the deploying air bag. The large reaction surface offered by the dummy's neck-chin-jaw cavity to the air bag, results in different neck deformation mode when compared to that due to the seatbelt or the seatback loading.

The current Hybrid III dummy's neck response is represented solely through beam-bending, although the laxity of the atlanto-occipital joint and the action of the muscle pairs imply that two separate load paths should be included in the response of the head/neck system. In case of air bag loading, the combination of the beam-like neck structure tuned to first bending mode (based on Mertz and Patrick [3] corridors), and the possibility of entrapment of the air bag within the neck-chin-jaw cavity, can result in second bending mode bending of the dummy's neck. This in turn can lead to high neck moments at very low angles of head rotation relative to torso.

In what follows, is a description of different air bag-neck interaction modes likely with the Hybrid III. More details can be found in Ref.3.

Figure 1. Typical Test Setup

Although the results presented pertain to the Hybrid III 5th percentile female dummy, the arguments are general and apply to the 50th percentile as well as the 95th percentile male. The discussion of the Hybrid III dummy neck bending response in light of the laxity of the atlanto-occypital joint [7,8] and the role of the muscle pairs [9], is then presented. A comparison of the neck response of the Hybrid III dummy, with that of the THOR dummy [10], is also included.

AIR BAG NECK INTERACTION MODES

The results presented here are a summary of a series of static, air bag deployment tests conducted to investigate the head/neck response of the Hybrid III 5th percentile female dummy due to a deploying air bag [3]. The study was limited to frontal passenger air bags, and the seat belts were not used. The test setup is shown in Figure 1. The dummy was placed, leaning towards the instrument panel, in a full-forward passenger seat. The seat was raised two inches from its normal position. The dummy's position was chosen to enhance the probability of air bag entrapment in the neck-chin-jaw cavity.

A standard Hybrid III 5th percentile female dummy, with a TMJ head skin, and a SAE neck shield [2] were used for the baseline tests. The head skin referred to as "modified" in Ref.2 is referred to as TMJ head skin in this paper. The neck shield was a thin, "mouse pad like" material, which was wrapped around the neck. The head skin, in both the chin-jaw area and in the cavity behind the jaw, was painted with chalks of different colors in order to determine whether the air bag was entrapped under the chin or

behind the jaw. High-speed video cameras and film cameras were used to monitor the head/neck and air bag interactions.

The head external loads resulting from air bag impact were calculated using the head accelerations and the upper neck loads. Based on the analysis of the measured response time-histories, the high-speed films, and the colored chalk marks on the air bag, the modes of interaction between the dummy's neck and the deploying air bag were deciphered. Three modes of air bag-dummy interaction: were identified: interaction mode 1, the neck loads are generated primarily from the air bag loading the front of the head, interaction mode 2, the neck loads are generated primarily from the air bag trapped under the chin, and interaction mode 3, the neck loads are generated primarily from the air bag trapped behind the jaw. The three modes of air bag-neck interaction are shown schematically in Figures 2-4. The figures show the configurations at the instant of peak loads (max N_{ij}) in each mode of interaction. Typical time history data are plotted for three typical representative tests in Figure 5-11.

Air bag-Neck Interaction Mode 1

In the first air bag-neck interaction mode, the air bag directly loads the head (Figure 2), leading to a flexion moment at the neck. The neck shear is positive (Figure 5), which implies that the head is pushed rearwards relative to the neck, as a result of membrane tension. The head external shear is in the anterior-posterior direction (Figure 6) confirming the rearward pressure on the head by the air bag. The neck axial force is insignificant in magnitude and changes from compression to tension (Figure 7). The head external axial force is of small magnitude in the superior-inferior direction (Figure 8). The applied external loads are both larger than the corresponding neck forces. This indicates that, in this head/neck loading-pattern, the neck loads are primarily from the air bag loading of the head with only small loads of the air bag directly on the neck. The upper neck moment is in pure flexion (Figure 9), which is an indication that the center of air bag pressure on the dummy's head is beneath the CG of the head. The torso is accelerated rearward by the deploying air bag (Figure 10). The dependence of the upper neck moment on the head rotation relative to the chest is compared to the flexion corridor proposed by Mertz, et al., [6] for Hybrid III 5th percentile female dummy (Figure 12). In this mode of air bag-neck interaction (Case 1, Figure 12), the moment-rotation relationship falls outside the corridor almost immediately after the head rotation starts, indicating that significant neck-moments can occur with little head rotation. The combined upper (Figure 9) and lower neck bending moments (Figure 11) cause the neck to flex into a reflected S-shape in second mode bending.

Figure 2. Air bag Loading the Head Directly

Figure 3. Air bag Trapped in the Chin-Jaw Cavity

Figure 4. Air bag Trapped behind the Jaw

Air bag-Neck Interaction Mode 2

In the second air bag-neck interaction mode, the air bag contacts the head under the chin (Figure 3). The bag is trapped under the chin during the deployment. The neck shear changes to negative (Figure 5), which implies that the head is pulled forward relative to the neck shortly after the initiation of the air bag-dummy

interaction and membrane tension has developed. However, the head external shear is insignificant in magnitude and changes direction from anterior-posterior in the beginning to posterior-anterior in the latter part (Figure 6). This implies that a major portion of neck shear comes from the inertial loading of the head on the neck, the direct loading of the air bag on the neck, or a combination of both. Significant tension load is present in the neck (Figure 7). The head external axial force is in the inferior-superior direction (Figure 8) and causes the head to pull on the neck. The external axial load is close in magnitude to the neck tension. This indicates that the contribution to the neck tension is primarily due to the air bag loads on the head/neck area and not due to the inertial loading of the head. The upper neck moment is pure extension in nature (Figure 9). The chest acceleration is mainly in the rearward direction (Figure 10). The upper neck moment as a function of head rotation relative to the chest is compared to the extension corridor proposed by Mertz, et al., [7] (Figure 12, Case 2). The test data again falls outside the corridor, even at very small head-rotation angles. The forces and moments again cause a second mode bending in the neck. However, in this mode, the deformed shape of the neck is S-shaped as opposed to the reflected S-shape seen in air bag-neck interaction Mode 1 (according to the moments in Figures 9, 11).

Air bag-Neck Interaction Mode 3

In the third air bag-neck interaction mode, the air bag contacts the head below the chin (Figure 4). The fabric is entrapped in the hollow area between the neck and the jaw. As the bag continues to inflate, pressure is built up within the entrapped portion of the air bag and membrane tension develops. The air bag pulls the head forward and upward, possibly pushing on the neck at the same time. The resulting neck shear is negative to a higher degree than in Mode 2 (Figure 5), which implies that the head is pulled forward relative to the neck. The head external shear maintains the posterior-anterior direction (Figure 6) during the whole event confirming the forward pulling of the head. The neck shear is larger than the external shear load. This indicates that a portion of the neck shear comes from the inertial loading of the head on the neck. However, the major contribution to the neck shear is due to the air bag loads on the head/neck area. This could be due, either to the membrane tension in the deploying air bag in front of the dummy, pulling on the air bag material, trapped in the jaw-neck cavity, or the pressure of the trapped air bag material pushing against the neck and the jaw, or both. Tension is evident in the neck (Figure 7)

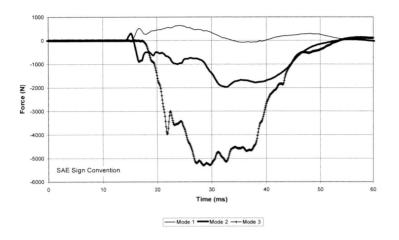

Figure 5. Upper Neck Shear Force.

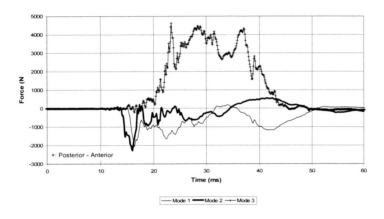

Figure 6. Head External Shear Force.

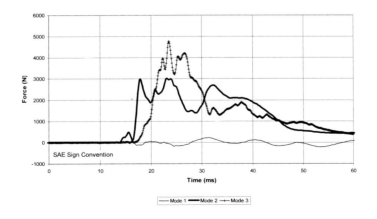

Figure 7. Upper Neck Axial Force.

5

DISCUSSION

The flexion and extension response corridors developed by Mertz and Patrick [3], and Mertz et al [6] are the primary basis for biofidelity of Hybrid III dummy's neck. The human volunteers, whose response was used to devise these extension corridors all have some degree of neck tension. Some tension is due to normal muscle activity holding the head in place. Additional tension could result from anticipation of impact. Under conditions of no neck tension i.e. cadaver response, the head would be expected to translate (due to inertial loading of the head) before showing significant rotation. The Hybrid III dummy neck is designed to deform in first mode bending in order for the moment-rotation curves to remain for the most part within the corridors. The plateau portion represents the maximum moment that the neck muscles can generate in resisting head motion before appreciable head rotation occurs. The initial bending stiffness for the 5th percentile female is 2.06 Nm/degree for flexion and 0.77 Nm/degree for extension.

After reaching a certain point, the neck muscle yields and the head keeps rotating without an increase in the bending moment. When the normal articular voluntary range of motion of the neck is reached, the action of the neck ligaments and/or passive stretch of the neck muscles, increases the bending resistance of the neck. The lower portion of the corridors reflects the elastic behavior of the ligaments and muscles as well as energy dissipation of the muscles during rebound. These corridors represent the neck response in the particular cases of restraint with either the seatbelt or the seatback. However, they were not developed for evaluating air bag loading. By basing the design of the Hybrid III dummy neck on corridors restricted to first mode bending, it is not clear if the dummy's head/neck response in case of loading by the air bag is biofidelic or not.

The three air bag-neck interaction modes observed with the Hybrid III dummy showed second bending mode response. In all the three cases the neck-moment versus head-rotation curves go out of the Mertz et al [6] corridors as soon as the head starts rotating, and never return, indicating that the dummy's neck is undergoing deformations which it is not conceived for.

The occurrence of the second bending mode in the human neck would seem likely in case of compressive loading of the human neck. However, when the air bag applies tensile forces in the neck by applying upward load to the chin-jaw region, a second mode bending seems unlikely because, the tensile load could be resisted only by aligning the muscle pairs with the direction of the load or engaging the ligamentous structure between the head and neck. Such an alignment could happen only after a substantial rotation of the head has occurred by which time the air bag would escape from under the human chin-jaw region.

Further, the human occipital condyle joint appears to have considerable laxity, which allows it to experience significant rotation before it can sustain a substantial moment across the joint [7,8]. Whereas, the current Hybrid III neck exhibits considerable bending resistance at its occipital condyle joint. This lack of compliance may allow large moments to be transmitted to the dummy's neck by the head without significant relative motion. In a human subject, motion and resistance to motion of the neck is accomplished through muscle pairs, which are attached to the skull, the individual vertebra, and the torso. These muscle pairs respond in various group actions to produce the desired movement of the head and neck. The muscle tones are simulated in the dummy through a pair of rubber nodding blocks and four rubber neck-discs. Nightingale, et al [9] studied the effects of upper neck axial and joint rotational stiffness on measured moments in the Hybrid III dummy during air bag loading using MADYMO occupant simulations. They found that decreasing the rotational stiffness had a dramatic effect on the extension moment.

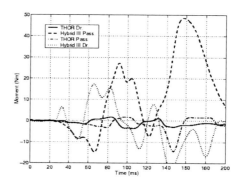

Figure 15. Neck Moment Comparison of THOR Dummy and Hybrid III Dummy [10

The Hybrid III dummy response appears to capture the global moment and head motion correctly in the first bending mode due to the use of the nodding blocks in the head-neck interface. However there is no way of estimating the local moment at the occypital condyles, equivalent to that which would occur in a human. By comparison, the NHTSA advanced dummy, THOR, has a neck system in which the loads on the head are resisted by the combination of a cable system and a beam like neck structure. Consequently, substantial head rotation is possible with relatively low moment in the occipital condyle joint [10]. Comparing the neck response of THOR dummy and Hybrid III dummy in vehicle crashes, the magnitude of the bending moment at the occipital condyle joint in THOR dummy was approximately 1/6 of the Hybrid III for both driver and passenger (Figure 15). This is one possible solution to the neck artifacts seen in the Hybrid III. However, the THOR is a new dummy that has not been evaluated thoroughly.

To eliminate the effect associated with the air bag trapping under the chin or behind the jaw, two experimental neck shield schemes were investigated [3]. The details of these investigations were presented in Ref.3. Essentially it was possible to eliminate the occurrence of both air bag-neck interaction Modes 2 and 3 with one neck shield design, and selectively eliminate only interaction Mode 3 with another neck shield design. In general, it appears that purely with neck shield designs although some level of control can be exercised on the neck-air bag artifact, the problem of passing the neck calibration tests and the problem of interference with the neck load cell measurement are faced.

CONCLUDING REMARKS

The artifact related to the Hybrid III dummy neck-air bag interaction manifests itself as the occurrence of second bending mode of neck deformation, which does not appear to be biofidelic.

Several approaches could be used to solve the problem of the dummy neck artifact. One way would be to design a suitable neck shield, which would prevent the air bag from entering the neck-chin-jaw cavity, and prevent the second bending mode of the neck deformation from occurring. However, in order for the neck shield to be effective, it may require a design that may not pass the calibration tests. In addition it may provide a bypass for a part of the load that should be measured by the load cell. In other words, although a neck shield would seem like a simple solution to the problem, developing a robust one is difficult.

Even if it were possible to come up with an efficient neck shield design, it would need a considerable extent of time before the neck shield would be an accepted part of the testing procedures. A large number of tests will have to be run by several organizations in order to establish confidence in the neck shield. This would be an effort of very significant extent.

In order for a neck shield design to be effective in all situations, it might become imperative to influence the basic response of the dummy head/neck complex by redesigning the basic neck components. In other words, a mixed approach, with a modified neck and a neck shield may become eventually necessary. This would be a very substantial work in itself followed by all the testing by several organizations before it is accepted widely.

A radical solution, in which the head/neck response could be made considerably more biofidelic would be by developing a dual load path system. The system would have a weak central bending structure, representing the vertebrae, and a strong 3D outer cable truss system, representing the muscle pairs. This would be the most time-consuming approach. The Hybrid III dummy has been known for a long time. A lot of experience has been gained with Hybrid III by the safety testing organizations all over the world. With the new system, it will take several years of testing before all the associated problems will be known, and solutions for them could be found.

Further, in order to design a more biofidelic head/neck system, a great deal of research into the human neck-airbag interaction will have to be carried out. At this time there is an acute shortage of biomechanics research results relating air bag deployment and human response. This is an effort, which will have to be carried out before reliable and robust biofidelic dummy neck systems can be produced.

REFERENCES

1. Melvin, J., Horsch, J., McCleary, J., Wideman, L., Jensen, J. and Wolanin, M., "Assessment of Air Bag Deployment Loads with the Small Female Hybrid III Dummy". Proceedings of the 37th Stapp Car Crash Conference, pp. 121-132, SAE Paper 933119, 1993.
2. Morris, C. R., Zuby, D. S. and Lund, A. K., "Measuring Air Bag Injury Risk to Out-of-Position Occupants", paper 98-S5-O-08, Proceedings of the 16th International Technical Conference on the Enhanced Safety of Vehicles, May 31 – June 4, 1998, Windsor, Canada.

3. Kang, J., Agaram, V., Nusholtz, G. and Kostyniuk, G., "Air Bag Loading on In-Position Hybrid III Dummy Neck". SAE paper 2001-01-0179, 2001.

4. Mertz, H. and Patrick, L., "Strength and Response of the Human Neck". Proceedings of the 15th Stapp Car Crash Conference, SAE paper 710855, 1971.

5. Nusholtz, G., Wu, J. and Kaiker, P., "Passenger Air Bag Study Using Geometric Analysis of Rigid Body Motion". Experimental Mechanics, September, 1991.

6. Mertz, H., Irwin, A., Melvin, J., Stalnaker, R. and Beebe, M., "Size, Weight and Biomechanical Impact Response Requirements for Adult Size Small Female and Large Male Dummy". SAE paper 890756, 1989.

7. Snyder, R., Chaffin, D. and Foust, D., "Bioengineering Study of Basic Physical Measurements Related to Susceptibility to Cervical Hyperextension-Hyperflexion Injury". UM-HSRI-BI-75-6 (Research Report), 1975.

8. Goel, V.K., Clark, C.R., Gallaes, K. and Liu, Y.K., "Moment-Rotation Relationships of the Ligamentous Occipital-Atlanto-Axial Complex". J. Biomechanics, Vol.21, No. 8, pp. 673-680, 1988.

9. Nightingale, R., Winkelstein, B. A, Knaub, K.E, and Myers, B. S, "Experiments on the Bending Behavior of Cervical Spine Motion Segments". Proceedings of the 26th International Workshop on Experimental and Computational Biomechanics Workshop, Tempe, Arizona, 1998.

10. Xu, L., et al., "Comparative Performance Evaluation of THOR and Hybrid III". SAE paper 2000-01-0161, 2000.

Effect of Seat Stiffness in Out-of-Position Occupant Response in Rear-End Collisions

Brent R. Benson, Gregory C. Smith, and Richard W. Kent
Collision Safety Engineering

Charles R. Monson
GMH Engineering

ABSTRACT

Accident data suggest that a significant percentage of rear impacts involve occupants seated in other than a "Normal Seated Position". Pre-impact acceleration due to steering, braking or a prior frontal impact may cause the driver to move away from the seat back prior to impact. Nevertheless, virtually all crash testing is conducted with dummies in the optimum "Normal Dummy Seated Position". A series of 7 rear impact sled tests, having a nominal AV of 21 mph, with Hybrid III dummies positioned in the "Normal Dummy Seated Position", "Out of Position" and slightly "Out of Position" is presented. Tests were performed on yielding production Toyota and Mercedes Benz seats as well as on a much stiffer modified Ford Aerostar seat. Available Hybrid III upper and lower neck as well as torso instrumentation was used to analyze and compare injury potential for each set of test parameters. In all cases, neck forces and moments were found to increase when the dummy's torso was leaned forward at impact. For the out-of-position tests, the results showed that the upper neck loads were clearly related to seat stiffness; stiffer seats produced greater neck loading. The lower neck loads, however, showed no definite correlation with seat back stiffness.

INTRODUCTION AND BACKGROUND

Fatalities and serious injuries in rear-end collisions are relatively infrequent; only about four percent of all fatalities result from very severe collisions in this accident mode [James, 199 1]. About seventy percent of the societal harm from rear-end collisions is concentrated in much larger numbers of low-level injuries caused in crashes with ΔV's of less than 20 mph [James, 1996]. Despite the relatively low priority of rear-impact protection compared to other areas of automobile crash protection, and despite the relatively small percentage of harm associated with rear-impact accidents of severity greater than 20 mph, considerable research into improving crash protection in high-severity rear-end impacts is being conducted within the automotive safety community. Proposals that seats should be stiffened to help retain occupants in these high-severity rear-end crashes are countered by concerns that stiffening seat backs will swell the already much larger amassment of harm from the greater number of minor and moderate injuries that occur in lower-severity rear-end impacts [Warner, 199 1].

Historically, crash testing of occupant protection concepts has been performed using anthropomorphic test devices (crash dummies) which are placed and postured in what is designated the Normal Dummy Seated Position (NDSP). In this NDSP, the dummy sits in an erect posture with its buttocks squarely on the seat and its shoulder blades flush against the seat back. For a single-occupant seat, the head and chest are laterally centered on the headrest and seat back, respectively. Any positioning other than the NDSP can be thought of as Out-of-Position (OOP). Except for a 1994 study at Collision Safety Engineering and a few unintentional exceptions [Strother, 1994], virtually all rear-impact experiments reported in the literature have been conducted with the dummies in the NDSP. This is consistent with the need for standardized test protocols to facilitate performance measurement, maximize test repeatability, and improve comparison between tests. Exclusive NDSP testing may be inconsistent, however, with the need to gain a comprehensive understanding of the most effective methods of protecting occupants in rear-end impacts within the real-life environment. The Normal Seated Position (NSP), with the occupant's head in close proximity to the head restraint, is coincidentally the optimal position for occupant injury protection in rear-end impact, an intuitive fact demonstrated by the measured increase in injury levels for the few OOP tests that have been performed [Strother, 1994].

Data from the National Accident Sampling System (NASS) suggest that pre-impact vehicle motions in a significant percentage of rear-impacts may cause many accident victims to not be in the NSP when the impact occurs, even if they happened to be in such a position at the onset of the accident. Many occupants would be involuntarily moved from their initial positions by accelerations caused by abrupt pre-impact braking and/or steering. Rear impacts often happen because the vehicle in front did something unexpected, such as slamming on brakes, hitting the next car in front, or cutting in from another lane. Such actions logically lead to driver responses that tend to move occupants from their initial seated positions.

Other studies have focused on voluntary causes for occupants in rear-impacted automobiles to not be in the NSP. Mackay, et al found that people don't sit in cars the way dummies do, even when driving straight down the road, looking straight ahead [Mackay, 1995]. They found that drivers generally sit further forward in their seats than do crash dummies in the NDSP. This difference was particularly pronounced for small females, who were found to sit with their heads about three-and-a-half inches further forward than the corresponding crash test dummies. Of the population of drivers studied, nearly 50% sat with their heads six inches or more forward of their head restraints while driving straight down the road.

Personal observation and experience suggest several additional reasons for occupants of automobiles to be out-of-position, aside from the simple fact that they may not be comfortable sitting erect in the NDSP, especially for long periods of time. They may be taking advantage of being stopped at an intersection to stretch and readjust their position, or to retrieve something from a purse or bag or from the glove box. They may be reaching for controls, such as the tape player or air conditioner. They may be shifting gears with a manual stick shift. They may be interacting with children in the rear or passenger-side seat. They may have their head turned looking in the rearview mirror or talking with other occupants in the car or someone at the intersection, They may be adjusting their clothing, A study made by the authors at partially obstructed intersections found it very common for people to lean forward several inches to see around the obstructions before proceeding into traffic. There are obviously numerous causes for occupants to not be in the NSP.

That occupants may not be in the optimum restraint position is a fact that was recognized in the mid-1970's during research into passive frontal impact protection. Early air cushion restraint systems that ignored the out-of-position problem were later found to need modification in order to not increase injury levels to out-of-position occupants in near-threshold-deployment levels of frontal impact. Indeed, even some front passenger air bag systems built today may be overly aggressive when deployed to protect an occupant pitched forward by pre-impact braking, or when deployed into a child sitting on the edge of the seat. Hopefully, the lessons learned during these development years will not be ignored by those proposing dramatic seat redesigns.

Neglect of the OOP problem when evaluating seat designs for occupant protection in rear impacts may result in designs that have the potential to increase the overall harm caused by this accident mode. As is shown in the few OOP tests that have been performed, there can be significant differences between the NDSP and the OOP dummy responses and injury measurements [Strother, 1994].

PURPOSE OF STUDY

The purpose of this study was to examine and quantify occupant response data, as a function of seat back stiffness, experienced by a 50th percentile Hybrid III male anthropomorphic test dummy in a rear end collision. Upper and lower neck forces and moments will be examined for two different seated positions: the out-of-position (OOP) where the dummy was pitched forward and his head was 20 inches from the head restraint and also the normal dummy seated position (NDSP).

Automobile seats with markedly different seat back stiffnesses were used in this examination - two production automobile seats and one modified seat that has been proposed as an improvement in rear-impact protection [Saczalski, 1993].

TEST METHODOLOGY

Tests were carried out with the seats mounted to a sled which was impacted into a fixed barrier at a nominal speed of 19 mph, resulting in a AV of approximately 21 mph. As depicted in Figure 1, 86% of AIS O-3 injuries occur at speeds less than 20 mph, while only about 27% of AIS 4-6 (severe or greater) injuries occur in the O-20 mph speed range [James, 1996].

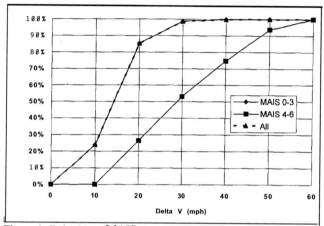

Figure 1: Delta V vs MAIS, NASS 82-86, CDS 88-94; Rear Damaged Light Vehicles; Front Outboard Occupants, Ages 12+

The sled tests were performed at the Collision Safety Engineering crash test facility according to the overall methodology outlined in a paper by Strother in 1994. [Strother, 1994]. A Hybrid III 50th percentile male anthropomorphic test device (ATD) was placed in each seat with the following instrumentation:

Upper neck tension-compression force, F_z ;
Upper neck shear force, F_x ;
Upper neck extension/flexion moment, M_y ;

Lower neck tension-compression force, F_z ;
Lower neck shear force, F_x ;
Lower neck extension/flexion moment, M_y ;

Lumbar tension-compression force, F_z ;
Lumbar shear force, F_x ;
Lumbar extension/flexion moment, M_y ;

Head Acceleration, G_x, G_y, and G_z ;
Thorax Acceleration in the mid-sagittal plane, G_x and G_z .

For the OOP tests the Hybrid III dummy was pitched forward with his buttocks still correctly positioned in the seat cushion, A light weight string tether around the neck helped maintain a distance of 20 inches from the back of the head to the head restraint in all out-of-position tests.

The two production seats selected were driver's seats from a 1989 Toyota Corolla and a 1985 Mercedes 190E. The "limited-yield" seat was a 1988 Ford Aerostar driver's seat modified to render it very stiff as compared with the production seats. The seat in each of the tests was placed in the mid-position and the seat back angle was adjusted to 15 degrees. The inboard as well as the outboard aft edges were measured pre and post impact in order to document residual deformation. The static seat pull characteristics of these seats are presented in Table 1. As can be seen, the modified seat represents a significant increase in seat back stiffness, strength and energy absorption capability over the production seats.

	Stiffness (lbf/in)	Maximum Strength (lbf)	Total Absorbed Energy (ft-lbf)	% of Total Energy Restored
1988 Ford Aerostar (modified)	694.3	3632	1249.4	12.0
1985 Mercedes 190E	128.2	1144	299.7	50.4
1989 Toyota Corolla (4-dr)	100.0	784	400.0	28.2

Table 1: Static Seat Pull Characteristics

TEST RESULTS

A total of seven sled tests were performed, three with the modified Aerostar seat, two with the production Mercedes seat, and two with the production Toyota seat. Typical plots of sled acceleration and velocity during impact are shown in Figure 2. A summary of maximum (peak) values of the instrumented dummy measurements for the seven tests, along with the corresponding maximum 50th percentile male injury tolerance limits, is presented in Table 2.

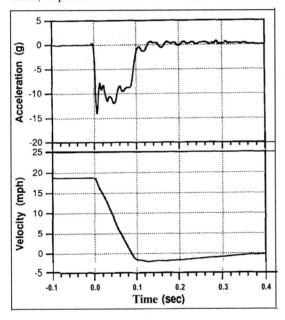

Figure 2: Typical Sled Motion

TESTS WITH MODIFIED AEROSTAR SEAT (S3, S2, S1)

Three modified Aerostar seats were tested, each with the Hybrid-III dummy in a different seated position: (S3) Normal Dummy Seated Position (NDSP), (S2) Torso pitched forward slightly (Slightly OOP), and (S1) Torso pitched forward (OOP).

TEST S3 - Normal Dummy Seated Position: This test was conducted with the back of the dummy against the seat back at impact. In this position, the back of the head was about 4.25 inches forward of the leading surface of the head restraint. Figure 3a through 3f are photographs taken from the high speed film showing the motion of the ATD and seat during the test.

Figure 3a is the photograph taken at time zero, the instant that the sled carriage contacts the fixed barrier. Figure 3b was taken at 46 ms into the event, corresponding to initial rearward seat back deformation, The chest acceleration has reached about 10 g at this point, and the lower neck extension torque is just starting to increase. Note that the head has yet to contact the head restraint. Figure 3c is at about 60 ms, at which time seat back deformation temporarily ceases. The lower neck extension torque is continuing to rise at the same steady rate started at about 46 ms. There is still no head contact with the head restraint at this time.

Figure 3d is taken at about 70 ms into the event. At this time, the dummy's head contacts the head restraint and the seat back deformation resumes. Figure 3e is at 76 ms, at which time the extension torque in the lower neck reaches a peak of about 1226 in-lbf (very nearly reaching the 1240 in-lbf injury limit). Note that at this time the head is in good alignment with the torso.

The last photograph in this series is Figure 3f, taken at 100 ms. The seat back dynamic deformation has reached its maximum value of about 7 deg; the seat back would be reclined a maximum of about 22 deg. From this position the seat back rebounds, pitching the dummy forward into the shoulder belt.

TESTS2 - Torso Pitched Forward Slightly: This test was conducted under the same conditions as Test S3, except that the torso of the dummy was leaned forward in the seat slightly, so that the displacement between the back of the head and the leading edge of the head restraint was about 6 inches.

At about 46 ms into the event, the dummy's shoulders are into the seat back and the first phase of seat back deformation begins. The torso deceleration has reached about 11 or 12 g at this time. At about 60 ms, the seat back deformation is temporarily halted. It is about this time that the dummy's head first contacts the head restraint. At about 80 ms into the crash,

Peak Measurement	units	S1 Lo	S1 Hi	S2 Lo	S2 Hi	S3 Lo	S3 Hi	M1 Lo	M1 Hi	M2 Lo	M2 Hi	T1 Lo	T1 Hi	T2 Lo	T2 Hi	Stat Injury Criteria Lo	Hi
Upper Neck Fx	(lbf)	-233	112	-114	52	-50	23	-56	67	-33	24	-108	31	-18	23	-697	697
Upper Neck Fz	(lbf)	-60	937	-32	450	-33	389	-46	809	-49	332	-106	737	-37	416	-899	742
Upper Neck My	(in-lbf)	-682	596	-293	241	-65	113	-261	451	-172	93	-329	571	-89	92	-505	1684
Lower Neck Fx	(lbf)	-117	196	-51	208	-30	147	-65	483	-35	163	-87	219	-15	197	-697	697
Lower Neck Fz	(lbf)	-99	690	-31	387	-43	353	-81	523	-81	245	-80	408	-61	365	-899	742
Lower Neck My	(in-lbf)	-2938	913	-1720	617	-1226	410	-2984	385	-1269	320	-2959	963	-1474	134	-1240	2200
Lumbar Fx	(lbf)	-129	487	-89	95	-44	88	-211	302	-76	96	-114	389	-63	114	-2401	2401
Lumbar Fz	(lbf)	-156	578	-167	529	-196	399	-322	649	-304	704	-247	450	-126	546	-1571	2850
Lumbar My	(in-lbf)	-2241	701	-559	416	-712	332	-1561	508	-994	502	-1905	101	-1254	230	-3269	10912
Peak Head Accel	(g)	*		50		38		57		26		52		30			
Peak Thorax Accel	(g)	29		27		22		21		17		26		15		60	
HIC (36ms)		*		319		149		451		59		437		61		1000	

Table 2: Summary of Test Results (*Head Z-axis accelerometer failed)

A:t=0msec
Initial contact
Seat back angle 15°

D: t = 90 msec
Dummy's head contacts restraint
Seat back deformation resumes

B: t = 46 msec
Seat back begins rearward deformation
Chest acceleration at 10 g's
Lower neck extension starting to increase

E: t = 76 msec
Extension torque in lower neck peaks at 1226 in-lbf

C: t = 60 msec
Seat back deformation temporarily ceases
Lower neck extension torque continues to rise

F: t = 100 msec
Dynamic seat back deflection ~ 7 °

Figure 3: (TEST S3 ▪ Modified Aerostar Seat) ▪ Normal Dummy Seated Position

the extension moment in the lower neck has reached a maximum of about 1720 in-lbf (exceeding the 1240 in-lbf injury criterion). Again at this time, the head and torso are in reasonably good alignment, with only a slight extension angle evidenced. At 92 ms into the event, the dummy has attained its maximum rearward excursion with little or no seat deformation over that achieved during the 46-60 ms time frame. The seat back recline is estimated from the high speed film at about 22-23 degrees (estimated dynamic seat back deflection of about 7-8 degrees). Finally, at 180 ms, the seat back has sprung forward, recovering a significant portion of its dynamic deformation and propelling the upper body of the dummy forward and slightly upward to re-encounter the shoulder belt. The measured residual seat back deformation was about 4 degrees.

TEST S1 - Torso Pitched Forward: This sled test was conducted under the conditions of Tests S3 and S2, except that the torso of the dummy was pitched forward in a more pronounced attitude. In this position, the back of the head was about 20 inches forward of the head restraint. Figures 4a through 4f document the overall response of the dummy and seat in this test.

Figure 4a is a photograph taken at the onset of the impact. Comparing this photo with Figure 3a, the difference in dummy positioning is clearly seen. Figure 4b is taken at about 62 ms, at which time the seat back yielding begins. The onset of seat deformation occurs later in this test compared with Tests S2 and S3 because of the greater movement required under this condition to place the shoulders of the dummy against the seat back. Figure 4c is at about 72 ms into the crash, at which time the seat back deformation stops. The seat back recline at this instant is estimated from the high speed film at about 23 deg, which would correspond to about 8 degrees of dynamic seat back deflection. As can be seen in this photo, the head is still some distance forward of the head restraint when seat back yield ceases. Figure 4d is at about 100 ms. It is around this time that the tension loads and the neck extension torque in the upper neck, as well as the extension torque in the lower neck begin their climb to values exceeding the injury criteria. Figure 4e is a photograph taken at about 114 ms into the impact - this is the instant where the upper neck tension and the upper and lower neck extension torques reach their above-criteria peak values. Again, the head is in only slight extension at this time. Maximum head extension, in fact, occurs at about 130 ms (Figure 4f) and appears to be less than 45 degrees. From this position, the dummy is thrown forward and slightly upward as energy stored in the seat back is released. About 3 degrees of residual seat back deformation was measured in this test.

SUMMARY - SLED TESTS WITH THE MODIFIED AEROSTAR SEATS

Three sled tests were conducted using the modified Aerostar seats. The dummy position was varied in these tests from the Normal Dummy Seated Position (Test S3) to a position slightly pitched forward (Test S2), and then to a position where the torso is pitched forward to where the

head was about 15-16 inches forward of its NDSP position (Test S1).

In the NDSP test, the dummy lower neck extension moment approached the limiting value for likelihood of injury. Only a slightly different head position in Test S2 relative to that of the NDSP test (Test S3) resulted in about a 50 percent increase in the neck torques, thereby exceeding the injury criterion. With the dummy leaned about 15-16 inches forward of its NDSP at impact (Test S1), the lower neck moment measurement increased markedly, well beyond the injury threshold. In this case, the tension force in the upper neck (937 lbf) also exceeded its respective injury criterion (742 lbf).

TESTS WITH THE MERCEDES SEAT

Two Mercedes seats were tested, one in the Normal Dummy Seated Position (NDSP), and one with the dummy pitched forward (OOP) at impact.

TEST M2 - Normal Dummy Seated Position: Test M2 was conducted with the Mercedes seat with the anthropomorphic dummy in the Normal Dummy Seated Position (NDSP). In this position, the dummy's shoulders are firmly against the seat back, and the back of the dummy's head is about 2 inches forward and upward from the front surface of the head restraint. Figures 5a through 5f document the response of the dummy and the seat during this test.

Figure 5a is at the onset of the deceleration, i.e. at time zero. The seat back is positioned in the erect position so that the initial seat back recline is about 15 deg. In Figure 5b, at time 36 ms, the seat back is just beginning to yield rearward. The chest A-P acceleration is climbing up to a plateau of about 7.5 g, which will be reached at about 48 ms into the event. In Figure 5c, at 124 ms, the head just contacts the head restraint as seat back yield continues in the upper portion of the seat back frame. In Figure 5d, at 132 ms, the lower neck extension moment enters a period of relatively rapid increase, reaching a peak of 1269 in-lbf at about 142 rns, Figure 5e. This torque level is just above the criterion value of 1240 in-lbf.

Finally, at about 180 ms, the yielding of the upper portion of the seat back ceases and the dummy begins its rebound phase of motion (Figure 5f). The maximum head extension angle occurs at about this time. As measured from the high-speed film, the seat back attains a maximum angle of about 50 degrees (outboard side) to 57 degrees (inboard side) for a maximum dynamic deflection of about 35-42 degrees. A residual seat back angle of about 50.6 degrees was measured on the inboard side of the seat post-impact indicating a recovery of approximately 7 degrees.

TEST M1 - Torso Pitched Forward: Test M1 was conducted under the same conditions as Test S1 except that the Mercedes seat was employed rather than the modified Aerostar seat. That is, the dummy torso was pitched forward so that, at impact, the back of the head was about 20 inches forward of the head restraint and about 18 inches

A: t = 0 msec
Initial contact
Seat back angle 15"

D: t = 100 msec
Upper neck tension and extension begin climbing

B: t = 62 msec
Seat back begins to yield

E: t = 114 msec
Injury criteria values are exceeded in the upper neck tension and upper and lower neck extension torques

C: t = 72 msec
Seat back deformation stops
Seat back angle at ~ 23 °

F: t = 130 msec
Maximum head extension

Figure 4: (TEST S1 ▪ Modified Aerostar Seat) ▪ Out-of-Position

A: t=0msec
Initial contact
Seat back angle 15°

D: t = 132 msec
Lower neck extension values increase rapidly

B: t = 36 msec
Seat back begins to yield
Chest acceleration 7.5 g's

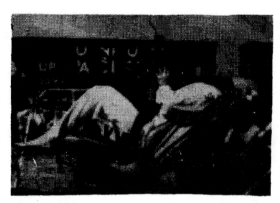

E: t = 142 msec
Lower neck extension peaks at 1269 in-lbf

C: t = 124 msec
Head contacts head restraint
Seat back yield continues

F: t = 180 msec
Seat yielding ceases, dummy begins rebound
Maximum head extension
Dynamic seat back deflection to ~ 35-42°

Figure 5: (TEST M2 • Mercedes Seat) • Normal Dummy Seated Position

A:t=Omsec
Initial contact
Seat back angle 15°

D: t = 108 msec
Dummy's head rotated into contact with head restraint
Second phase of seat back yield begins

B: t = 70 msec
Shoulders have reached seat back
Seat back begins to yield

E:t= 150msec
Seat back yielding ceases
Maximum dynamic seat back deformation at -45 °

C:t= 100 msec
Seat back yield ceases

F: t = 400 msec
Rebound phase

Figure 6: (TEST M 1 - Mercedes Seat) - Out-of-Position

forward of its NDSP. Figures 6a through 6f illustrate the dummy kinematics and seat response during this test.

Figure 6a is taken at the initiation of the impact. Comparing this photo with Figure 5a, one can see the similarity in dummy pre-impact positioning. Figure 6b is at about 70 ms into the event, at which time the shoulder area of the dummy torso has reached the seat back and the seat back starts its initial phase of rearward yield. During this first phase of seat back yield, the upper neck tension force and the lower neck extension moment start increasing (i.e. at about 75-80 ms). Figure 6c is taken about 100 ms into the impact, the time at which this initial phase of seat back yield is over. At about 108 ms (Figure 6d), the dummy's head has rotated into contact with the head restraint and the second phase of seat back yield starts. This second phase of seat back deformation lasts until about 150 ms (Figure 6e), at which time the seat attains its maximum dynamic seat back angle of about 45 degrees (for a maximum dynamic seat back yield of about 30 deg). During this time period, i.e. around 125 ms, the upper neck tension force and the lower neck extension moment reach peaks that exceed their respective injury criteria. Figure 6f is taken at about 400 ms into the event, at which time .the dummy is reacting to the release of energy stored in the seat back by rebounding upward and forward. About 17 degrees of permanent seat back deformation was measured after this test.

SUMMARY OF TESTS WITH THE MERCEDES SEATS

Two tests were conducted with the Mercedes seats; one test with the ATD in the Normal Dummy Seated Position (Test M2), and the other test with the dummy torso pitched forward at impact. In both tests, the lower neck extension moment exceeded the injury criterion. In addition, during the OOP test, the upper neck tension reached a peak of just over 800 lbf, exceeding the injury criterion of 742 lbf.

A comparison of the neck forces and torques achieved with the Mercedes seat and those achieved by the modified Aerostar seat, indicates that the forces and torques are comparable, despite the fact that the permanent deformations of the Mercedes seats were significantly greater than those of the modified Aerostar seats. Interestingly, the residual seat deformations were greater in the tests with the NDSP dummies as compared with those experienced in the OOP tests.

TESTS WITH THE TOYOTA SEAT

Two Toyota seats were tested, one in the Normal Dummy Seated Position (NDSP), and one with the dummy pitched forward (OOP) at impact.

TEST T2 - Normal Dummy Seated Position: Test T2 was conducted using the Toyota seat, with the anthropomorphic dummy in the Normal Dummy Seated Position. With the shoulder area of the dummy pressed against the seat back, this puts the rear surface of the head about 1 ½ inches forward and slightly up from the leading surface of the head restraint.

Figures 7a through 7f document the response of the dummy and seat in this test. Figure 7a is at the initiation of the impact. Figure 7b is taken at 40 ms into the event, at which time the seat back begins yielding. The A-P chest deceleration has increased to about 10 g at this point in time, and the onset of seat back deformation is accompanied by a temporary drop in chest deceleration, which returns to the 10 g level at about 50 ms, climbing slightly thereafter. Figure 7c is at 110 ms, at which time the head contacts the head restraint. It is at this time that the lower neck extension moment begins to rise more rapidly to its peak of 1474 in-lbf at about 130 ms (Figure 7d). At about 150 ms the seat back yield ceases as the seat back attains its peak deflection. From film analysis, it is estimated that the maximum seat back angle attained was approximately 60 deg (Figure 7e). With an initial recline of 15 deg, this would correspond to about 45 degrees of dynamic seat back deformation. Post impact seat back recline angle was measured at 60 deg, indicating little seat back recovery.

TEST T1 - Torso Pitched Forward: Test T1 was conducted using the Toyota seat under the same conditions as Tests S1 and M1, with the dummy torso pitched forward at impact. Figures 8a through 8f document the dummy kinematics and seat response for this test.

Figure 8a is at time zero and Figure 8b is at about 60 ms when the seal back deformation begins. In Figure 8c, at 130 ms, the head first contacts the head rest. During the period between Figures 8c and 8d the lower neck extension torque begins to increase. In Figure 8d, at 140 ms, this extension torque reaches a maximum of about 2959 in-lbf, well above the 1240 in-lbf criterion and at a level comparable to that in the other OOP tests with the modified Aerostar seat and the Mercedes seat. At 170 ms into the event, Figure 8e, the seat back attains its maximum deformation, with the backrest tilted rearward at about a 45 degree angle, corresponding to about 30 degrees of dynamic seat back deformation. As can be seen in Figures 8d and 8e, the head and neck extension is not excessive in this test. Finally, in Figure 8f, the dummy is undergoing its mild rebound motion during seat back unloading. The residual seat back recline measured after the test was about 40 deg, indicating about 5 degrees of seat back angle recovery.

SUMMARY OF TESTS CONDUCTED WITH THE TOYOTA SEATS

Two tests were conducted with the Toyota seats, one with the dummy in the Normal Dummy Seated Position and the other with the dummy torso pitched forward at impact. Looking first at the test in the NDSP, the resulting neck forces and torques in this test (T2) are comparable to those obtained in the earlier tests with the modified Aerostar seat and the Mercedes seat. Examining the tests conducted with the dummy torso pitched forward at impact, the neck loads sustained by the dummy in the Toyota seat are lower than those sustained by the dummy seated in the other two seats. As was the case with the modified Aerostar seat and the Mercedes seat, the test with the NDSP occupant produced

A:t=Omsec
Initial contact
Seat back angle 15°

D: t = 130 msec
Lower neck moment peaks

B: t = 40 msec
Seat back yield begins
Chest acceleration 10 g's

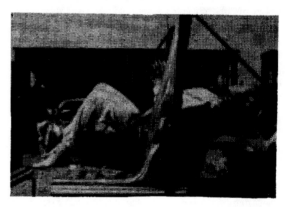

E: t = 150 msec
Seat back yield ceases
Maximum dynamic deflection at -60"

C: t = 110 msec
Head contacts the head rest
Lower neck extension moment begins to rise

Figure 7: (TEST T2 - Toyota Seat) - Normal Dummy Seated Position

A: t = 0 msec
Initial contact
Seat back angle 15°

D:t= 140msec
Extension torque reaches a maximum

B: t = 60 msec
Seat back deformation begins

E: t = 170 msec
Dynamic seat back deflection maximum at -45 °

C: t = 130 msec
Head contacts head restraint

F: t = 210 msec
Dummy is rebounding

Figure 8: (TEST T1 · Toyota Seat) · Out-of-Position

21

more seat back deformation than the test with the dummy torso pitched forward at impact.

DISCUSSION OF TEST RESULTS

On the basis of the seven tests conducted during this program it appears that, if one ignores the slightly excessive lower neck extension torques (relative to the accepted injury criteria [Mertz, 1968]), these three seat designs are capable of providing reasonable protection to a belted 50th percentile male occupant in a nominally 21 mph AV rear impact where that occupant is in the Normal Dummy Seated Position. When the occupant's upper torso is pitched forward at the instant of impact, however, neck forces and moments increase. This increase is relatively more dramatic as seat stiffness increases, with greater loads produced by stiffer seats. In fact, when the dummy head was simply moved forward less than 2 inches relative to the NDSP, the dummy in the modified Aerostar seat sustained an increase in the lower neck extension torque of almost 50 percent (from 1226 in-lbf to 1720 in-lbf).

Figure 9 shows the measured upper-neck compression and tension loads as functions of the time duration for which the measurements exceeded the plotted values. The time duration-dependant injury criteria for tension and compression of the 50th percentile male upper neck are also plotted for comparison. Tests S 1 (Modified Aerostar Seat), Ml (Mercedes Seat) and T1 (Toyota Seat) have been plotted such that occupant response as a function of the seat stiffness could be examined. A similar plot is presented in Figure 10 which shows the measured upper-neck fore-aft shear loads and corresponding 50th percentile male injury

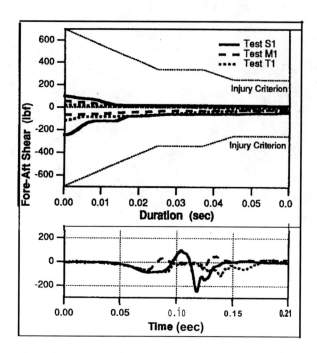

Figure 10: Upper Neck Shear Loads for Tests S 1, M 1 & T 1

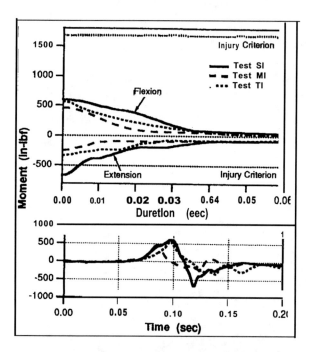

Figure 11: Upper Neck Moment Loads for Tests S 1, M 1 & T 1

criteria for the same three out-of-position tests [Melvin, 1985]. Finally, Figure 11 plots the upper neck moment loads and corresponding injury criteria. It is seen that the stiff modified Aerostar seat produced the greatest upper neck loads for each of the measured modes. This difference is especially remarkable in the neck extension measurements, where the modified Aerostar values are approximately twice those of the two production seats.

Figures 12, 13 and 14 are data observed in the lower neck for the axial loads, shear loads, and moment loads

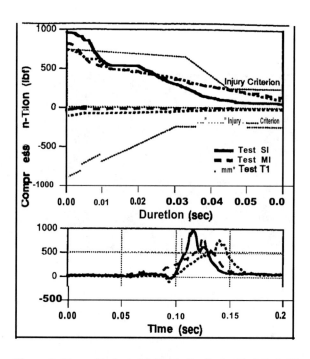

Figure 9: Upper Neck Axial Loads for Tests S1, Ml, & Tl

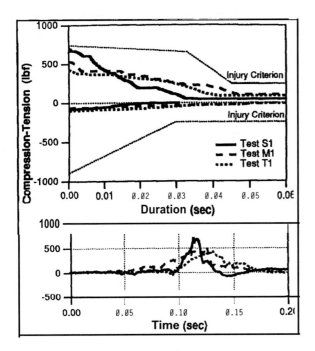

Figure 12: Lower Neck Axial Loads for Tests S 1, M 1 & T 1

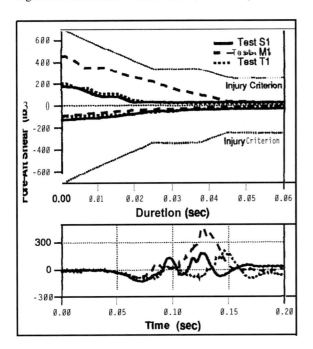

Figure 13: Lower Neck Shear Loads for Tests S 1, M 1 & T1

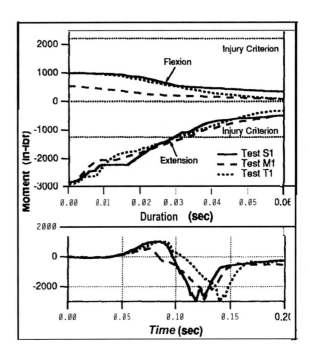

Figure 14: Lower Neck Moment Loads for Tests S 1, M 1 & T1

respectively. In contrast to the upper neck loads, the lower neck loads show no clear correlation with seat stiffness.

As can be seen in Figure 14, in all the OOP tests the lower neck extension torques exceeded the injury criterion by a substantial margin, including those in the test with the relatively less stiff Toyota seat. Since this criterion was exceeded by such a wide margin, either these values suggest very real injury potential, or the injury criterion is too conservative. Moreover, given the fact that these seats have significantly different stiffness properties, it seems that the magnitude of these lower neck torques are somewhat independent of seat properties; they are certainly not a function of the magnitude of seat deformation.

Interestingly, none of the test dummies underwent *any* extreme hyperextension in any of these tests and the timing of the peak of the lower neck extension moment did not correspond to either the instant of maximum neck extension or the instant of maximum dummy and seat displacement. This demonstrates the inadequacy of evaluating injury potential by simple visual examination of crash-test films without adequate proper dummy neck instrumentation.

To the extent that research efforts are focused on the relatively low-priority area of rear impact protection, further research may be warranted to determine a more appropriate lower neck torque injury criterion and to determine whether or not occupants pitched forward at the instant of rear impact are exposed to any significant lower neck trauma. If it is determined that they are, then perhaps a novel seat design concept like the "catcher's mitt" seat developed by General Motors will show some advantage in this regard.

For all of the seats tested, there was more seat deformation in the tests conducted with the dummy in the Normal Dummy Seated Position as compared to that sustained in tests with the dummy torso pitched forward at impact. This seems counter-intuitive, since relative velocity between the upper body and seat back is allowed to build up in the OOP tests before contact, seemly requiring additional seat back energy absorption. What appears to happen, however, is that during the period of time wherein the torso is rotating into contact with the seat back, ride down energy is being absorbed by the dummy without causing seat back deformation, since the loads into the seat back are occurring relatively close to the seat back anchorage system.

REFERENCES

[James, 1996]

James, Michael B., et. al; "Severe and Fatal Injuries in Rear-End Impacts"; to be published.

[James, 1991]

James, Michael B., et. al; "Occupant Protection in Rear-End Collisions: I. Safety Priorities and Seat Belt Effectiveness"; SAE #9 129 13; 35th Stapp Conference; pg. 369; November 199 1.

[Mackay, 1995]

Mackay, G.M., et. al; "Rear End Collisions and Seat Performance - To Yield or Not To Yield." 39th Annual AAAM Proceedings, p. 23 1. Chicago, IL; October 16-18, 1995.

[Melvin, 1985]

Melvin, John W.; "Advanced Anthropomorphic Test Device (AATD) Development Program, Phase 1 Reports: Concept Definition, UMTRI for US Department of Transportation, NHTSA, p. 101; 1985.

[Mertz, 1968]

Mertz, Harold J.; "Car Occupant Response to Rear-End Collision - A Mathematical Model," 1st International Conference on Vehicle Mechanics, p. 588, Detroit, MI, July 16-18 1968.

[Saczalski, 1993]

Saczalski, Kenneth J., et. al; "Field Accident Evaluations and Experimental Study of Seat Back Performance Relative to Rear-Impact Occupant Protection"; SAE #930346; Seat System Comfort and Safety; pg. 151; March 1993.

[Strother, 1994]

Strother, Charles E., et. al; "Response of Out-of-Position Dummies in Rear Impact", SAE #94 1055; In-Depth Accident Investigations: Trauma Team Findings in Late Model Vehicle Collisions, pg. 65; February 28, 1994.

[Warner, 1991]

Warner, Charles Y., et. al; "Occupant Protection in Rear-End Collisions: II. The Role of Seat Back Deformation in Injury Reduction", SAE #912914; 35th Stapp Conference; pg. 379; March 199 1.

A SLED TEST PROCEDURE PROPOSAL TO EVALUATE THE RISK OF NECK INJURY IN LOW SPEED REAR IMPACTS USING A NEW NECK INJURY CRITERION (NIC)

Ola Boström
Yngve Håland
Rikard Fredriksson
Autoliv Research
Mats Y. Svensson
Hugo Mellander
Chalmers University of Technology
Sweden
Paper Number 98-S7-O-07

ABSTRACT

Today's cars do not sufficiently prevent neck injuries in rear end impacts. So called whiplash injuries are often sustained at low velocities. According to Swedish road casualty statistics, the risk for whiplash injuries increases dramatically with the velocity change (Δv) of the impacted car in the interval between 10-20 km/h. During recent years, much progress has been made in research concerning this issue. This includes new findings from injury statistics, better knowledge of injury mechanisms (even if they are not yet fully understood) and development of suitable rear impact dummies.

This paper describes a new sled test procedure involving two levels of rear impact severity. In the proposed procedure, a new neck injury criterion (NIC) which is a measure of the effect of violence to the neck, is used to evaluate the level of neck protection.

Seats, from two cars with different neck injury-risk rating (according to Swedish statistics), have been tested according to the new procedure and compared with a new seat concept. The results indicate that a seat back with a low yielding limit has a lower risk of neck injury, which is reflected in lower NIC-values.

INTRODUCTION

When designing car seats to prevent injuries in high Δv rear-end collisions (Δv above 25 km/h and 10 g in crash pulse), there already exist sled test procedures including risk evaluation criteria (Viano, 1994). For this level of severity most researchers agree that neck hyper extension and occupant ramping up the seat back (with the potential for secondary impact of the occupant with the rear seat and the rear window) must be avoided. By improving the head rest and stiffening the seat, the occupant may be protected from life threatening injuries. On the other hand, AIS 1 classified neck injuries, sustained mostly at low speed rear-end collisions (Eichberger et al. 1996, Parkin et al. 1995) have been given increased attention over the last ten years. According to Nygren et al. (1984), Lundell et al. (1998) and v Kock et al. (1995, 1996), these injuries are by far the most common injury type in rear end impacts and cause long term disability in 1 out of 10 injury cases (Nygren, 1984). Despite these facts, there are no established test methods nor evaluation criteria for low speed rear impacts. A reasonable requirement for a test procedure, evaluating disabling neck injuries in these impacts, would be the ability to discriminate between circumstances with different injury risk (Jakobsson et al., 1994).

According to an in-depth study of neck injuries by Olsson et al. (1990) the shape of the crash pulse has a greater influence on the severity of the neck injury than the amount of transferred energy. Recent work by Krafft (1998) shows that the existence of a tow bar as well as being hit by a car with a transversely mounted engine significantly increases the risk of long term disability in rear impacts. It is tempting to believe that, in a rear impact these two factors influence the mean or peak struck car acceleration.

Boström et al. (1996) proposed a new neck injury criterion (NIC) based on a hypothesis of Aldman (1986) and the findings of Svensson (1993). The idea of NIC is to measure the effect of the violence to the neck (normally not life threatening) during the initial retraction phase, phase 1 in Figure 1.

The scientific basis for the NIC-criterion has been further substantiated in recent work, where NIC-values in simulated real-life rear-end collisions have been compared with the actual injury outcome (Boström and Krafft et al., 1997a). The NIC has been found to be sensitive to the seat structure characteristics, the car Δv, and the car crash pulse.

The aim of this paper was to propose and evaluate a new sled test procedure to characterize a car seat from neck injury risk point of view. The design of the test method is based on real-life crash data and research in biomechanics as well as experience from various sled tests and full-scale car tests.

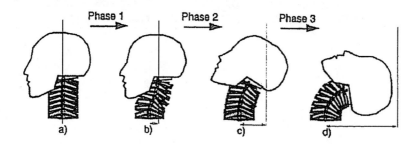

Figure 1 - Schematic view of four parts of the head-neck motion during a rear-end collision: a) initial posture, b) maximum retraction, c) maximum rearward angular velocity of the head is reached, d) hyper extension. The vertical line represents a reference plane in rest. (from Svensson, 1993)

PROPOSED SLED TEST PROCEDURE

The most appropriate crash pulse to use in a sled test with a car seat to simulate a rear impact, should be based on a large set of full scale crash tests with the particular car model. However, the purpose of this study was to evaluate the properties of a seat independently of the corresponding car structure. That is, the ambition was that a seat performing well in this study should perform well in any car regardless of the car structure properties.

In the proposed sled test procedure the seats are exposed to acceleration pulses giving the same Δv but with different acceleration time-history. The difference between the chosen acceleration levels represents the difference between the striking/struck car having a stiff or soft frontal/rear structure, or the difference between a struck car with or without a tow bar. The influence on occupant loading due to such differences have been investigated by Håland et al. (1996) and Boström et al. (1998) by means of full scale crash tests. According to Krafft (1998) these factors indicate a difference in disability risk.

The expected effect of the violence to the neck of a human occupant is measured by the NIC response. The NIC and the tolerance level are defined according to equations 1 - 4.

$$NIC = a_{relative} * 0.2 + v_{relative}^2 \qquad [m^2/s^2] \qquad (1.)$$
calculated at maximal retraction (posture b in Figure 1)

$$a_{relative} = a_{T1} - a_{C1} \qquad [m/s^2] \qquad (2.)$$
local x-acceleration, T1=lower neck, C1=upper neck

$$v_{relative} = \text{time integral of } a_{relative} \qquad [m/s] \qquad (3.)$$

$$\text{Tolerance level of NIC} = 15 \ m^2/s^2 \qquad (4.)$$

In eq. 1, 0.2 [m] is a length parameter. Depending on the biofidelity of the dummy response, these equations may have to be changed, for example by making assumptions about the upper neck (C1) acceleration.

The hypothesis is that a seat which is tolerant to different rear impact crash pulses and has low NIC values, up to maximal retraction, is a good seat with low risk of neck injury.

METHOD

Two standard production seats, seat B ("Bad") and G ("Good"), and an anti whiplash seat, AWS, were tested with a Hybrid III (HIII) 50th percentile male dummy. The AWS has a force controlled yielding of the seat back to give the neck a gentle acceleration until maximum retraction is passed. According to real-life disability data analysed by Krafft (1998), in rear impacts, the seat G car model is much safer than the seat B car model. This agrees with the ranking based on police reported accidents presented by Boström and Krafft et al. (1997a).

The chosen Δv in the sled tests was 15 km/h representing an impact speed of approximately 25 km/h (for equal masses of the target/bullet cars). Two pulses, from now on called the 4g and the 8g pulse, were used in the tests (Figure 2).

The seat back angle was measured by the use of an SAE H-point machine (dummy). It was placed in each seat model and the seat back angle was adjusted so the torso-line was 25 degrees to the vertical. The resulting seat back angle for each seat model was measured and used in the sled tests. The H-point of the HIII was positioned according to the H-point machine and the upper torso was pushed into the seat back with the same force as with the H-point machine. Finally the baseline of the head was placed in a horizontal position.

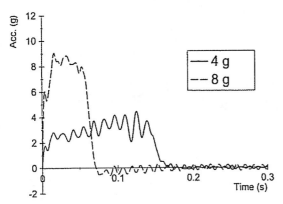

Figure 2 - For the evaluation of the proposed test concept, these pulses were chosen; the 4g and the 8g pulse.

For seat G, the test was repeated with a 5th percentile HIII female dummy seated on a child cushion. The purpose was to evaluate the weight influence on the test results. The reason for the child cushion was to prevent the 5th percentile female dummy from sinking into the seat below the transverse upper seat back beam.

The neck injury criterion (equations 1-4), was configured for the use of the HIII dummy. The neck (as well as the complete spine) of the HIII dummy is far from biofidelic regarding the initial retraction phase. Therefore, the relative acceleration in equation 2 is not applicable for a HIII dummy. On the other hand, the upper neck (C1) acceleration of an unaware human occupant is relatively low until the moment of maximal retraction (eq. 1 and posture b in Figure 1). This is true as long as the head is not accelerated by the head rest during the retraction phase. In order to evaluate the risk of injury/level of protection for a given seat, NIC50 as defined in eq. 5 - 9 was used as a criterion in the current evaluation.

$$NIC50 = a_{lower\ neck} * 0.2 + v_{lower\ neck}^2 \quad [m^2/s^2] \quad (5.)$$
$$\text{calculated when}$$

$$d_{lower\ neck} = 50\ mm \quad (6.)$$

$$a_{lower\ neck} =$$
(local) x-component of the lower neck acc. $[m/s^2]$ (7.)

$$v_{lower\ neck},\ d_{lower\ neck} = \text{time integral and double time}$$
$$\text{integral of } a_{lower\ neck} \quad (8.)$$

$$\text{Tolerance level, NIC50} = 15\ m^2/s^2 \quad (9.)$$

Equations 5-9 are the conformed alternative to eq. 1-4 with the assumption of zero upper neck (C1) acceleration during the initial retraction phase (phase 1 in Figure 1) and the occurrence of maximal retraction after 50 mm of lower neck displacement relative to a non accelerating head.

In addition to NIC50, the upper neck extension moment and shear force (My and Fx) were also measured. To evaluate the rebound effect of the seats, the relative upper torso rebound was calculated as follows:

$$\text{Relative upper torso rebound} =$$
$$= (\text{max. lower neck speed} - \Delta v)/\Delta v \quad (10.)$$

If for example the interaction between occupant and seat-back in a rear impact is totally plastic (non-elastic), the maximum neck speed in eq. 10 becomes approximately Δv and the relative upper torso rebound becomes zero. If on the other hand, the interaction is totally elastic, the maximum neck speed becomes approximately $2\Delta v$, with a relative upper torso rebound close to 1 (100%).

RESULTS

The performance of seat G and of seat AWS compared to seat B were quite different. The lower neck acceleration of the HIII in seat B was considerably affected by the difference in pulse, which was not the case for seat G and the seat AWS (Figures 3-5). The resulting NIC50 values for the production seats were in agreement with the disability analysis made by Krafft (1998). It was found that the level of the pulse influenced the NIC value significantly for seat B, but not for seat G (Figure 6). Actually, only the 8g pulse for seat B resulted in NIC values well above the injury threshold of 15 m^2/s^2.

There was no correlation found between the relative upper torso rebound values and the expected injury outcome (Figure 7). Actually it seemed as seat G was even more elastic than seat B.

For all tests, the traditional neck criteria, upper neck extension moment and shear force, were well below the AIS2+ tolerance levels (57 Nm/1100 N) proposed by Backaitis and Mertz (1994) (Figures 8-9). However, for seats G and B, the shear force (Fx) values were lower in the 4g pulse tests compared to the 8g pulse tests. For the 8g pulse, the seat B Fx value was higher than the corresponding values for seat G and seat AWS.

The results of the test with the elevated HIII 5th percentile female dummy were comparable with the results with the HIII 50th percentile male dummy. There was no substantial difference regarding the NIC response for the two pulses (Figure 10). The lighter dummy experienced, however, slightly higher NIC50 values.

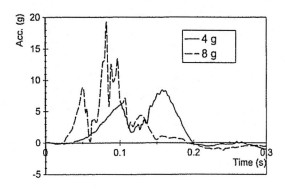

Figure 3 - Lower neck acceleration for seat B for the two crash pulses.

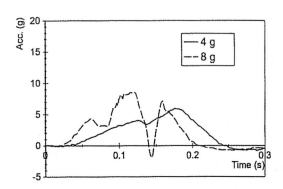

Figure 4 - Lower neck acceleration for seat G for the two crash pulses.

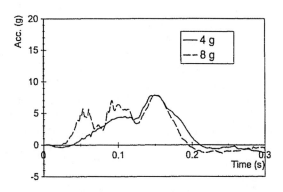

Figure 5 - Lower neck acceleration for seat AWS for the two crash pulses.

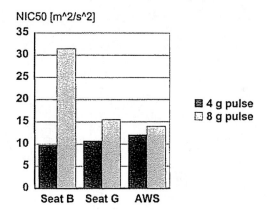

Figure 6 - NIC50 values for the HIII 50th percentile male dummy for the two pulses for seat B, G and AWS. The tolerance level is 15 m²/s².

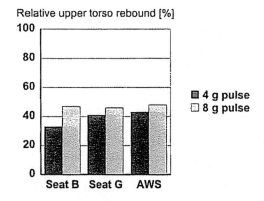

Figure 7 - Relative upper torso rebound, defined in eq. 10, for the HIII 50th percentile male dummy for the two pulses for seat B, G and AWS.

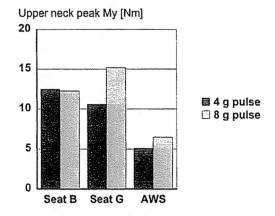

Figure 8 - Peak upper neck torque, My, for the HIII 50th percentile male dummy for the two pulses for seat B, G and AWS.

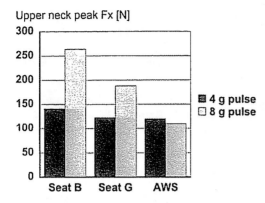

Figure 9 - Peak upper neck shear force, Fx, for the HIII 50th percentile male dummy for the two pulses for seat B, G and AWS.

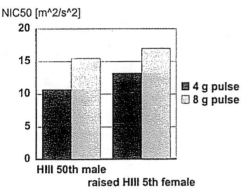

Figure 10 - NIC50 values for the HIII 50th percentile male dummy compared to the values for the raised HIII 5th percentile female dummy for the two pulses for seat G.

DISCUSSION

The pulses (the 4g and 8g pulses) and the Δv (15 km/h) used in this study were chosen on the basis of full scale rear impacts, where the impact speed was 25 km/h. If it is determined that the 8g pulse is not an accurate representation of an average injurious impact, the level of pulse and/or the Δv will have to be changed in the sled test procedure.

The explanation for the lower NIC50 values and the insensitivity to the shape of the acceleration pulse for seat G and seat AWS is clearly the "softer" performance indicated by the lower neck accelerations shown in Figure 3-5. In addition, the "softer" performance of seat G and seat AWS also resulted in decreased upper neck shear forces. This is in agreement with the analysis of a series of

sled tests with a HIII dummy equipped with a Rear Impact Dummy (RID) neck developed by Svensson and Lövsund (1992), where the *initial* upper neck torque and shear force maxima were shown to correlate with the NIC values (Boström et al., 1997b). However, in contrast to the NIC values, for all sled tests in this study as well as in the study by Boström et al. (1997b), the peak upper neck moment and shear force were well below the tolerance levels (57 Nm/1100 N; Backaitis and Mertz, 1994). In this study only the NIC50 values, in agreement with disability data, prove seat B being worse than seat G.

The major limitations of the seat test procedure (performed) seem to be the disregard of the seat geometry, the restriction to low velocity and the focus on AIS 1 injuries. The motivation for a test with these limitations is the fact that head rests of car seats have a low efficiency (Nygren 1984, Brault et al. 1998) and that high velocity rear impacts and AIS2+ injuries are rare (Otte et al., 1997). The proposed sled test procedure seems to evaluate the risk of neck injury and the level of protection in an elementary way. That is, the test is able to discriminate between cars (seats) with rather different disability rankings. In order to evaluate a seat more precisely, taking the seat geometry into account, a dummy with more human like properties regarding spinal motion is needed. Such a dummy is under development in Sweden and will be presented later (Davidsson et al., 1998).

The use of a dummy representing an average female instead of an average male would be more appropriate since females are at higher risk (Krafft et al., 1996). In this study, the 5th percentile female dummy was elevated with a cushion in order to simulate a light (compared to an average) female with an average seating height. As a result, the NIC values were slightly higher. However, no *more* information was gained. It appears, on the basis of this limited dummy weight study, that a seat that accelerates a dummy representing an average male in a gentle way (low NIC values), regardless of the acceleration profile, will accelerate a 50th percentile female dummy in a similar manner.

CONCLUSION

The proposed seat test procedure evaluates the risk of neck injury and level of protection in typical low speed rear impacts. It includes two different acceleration pulses and uses the NIC as the main injury risk indicator. To conclude the findings of this study:

• The proposed sled test procedure with the HIII dummy appears to be relevant for an elementary evaluation of car seats regarding the risk of neck injury in low velocity rear-end impacts.

• Seats (seat backs) with low yielding limit are tolerant to different rear impact crash pulses and have low NIC50 values.

• A gentle neck acceleration, until maximal neck retraction is passed (posture b in Figure 1), could prevent neck injuries with a risk of permanent disability from occurring, as the NIC50 value would be below the tolerance level.

ACKNOWLEDGEMENT

The authors wish to thank Maria Krafft at Folksam Research for valuable help with the choice of seats.

REFERENCES

Aldman B (1986): An Analytical Approach to the Impact Biomechanics of Head and Neck. Proc. 30th Annual AAAM Conf., LC 64-1965, pp. 439-454.

Backaitis SH, Hertz HJ (1994): Hybrid III - the first human-like crash test dummy, SAE PT-44, ISBN 1-56091, pp. 407-422.

Boström O, Svensson MY, Aldman B, Hansson HA, Håland Y, Lövsund P, Seeman T, Suneson A, Säljö A, Örtengren T (1996): A new neck injury criterion candidate - based on injury findings in the cervical spinal ganglia after experimental neck extension trauma. Proc. 1996 Int. Conf. on the Biomechanics of Impact (IRCOBI), Dublin, Ireland, pp. 123-136.

Boström O, Krafft M, Aldman B, Eichberger A, Fredriksson R, Håland Y, Lövsund P, Steffan H, Svensson M Y, Tingvall C (1997a): Prediction of neck injuries in rear impacts based on accident data and simulations, Proc. 1997 Int. Conf. on the Biomechanics of Impact (IRCOBI), Hannover, Germany, pp. 251-264.

Boström O, Håland Y, Lövsund P, Mellander H, Svensson MY (1997b): A comparison between NIC-values and upper neck moment during the early phase of neck motion in low speed rear impacts, Twenty-Fifth Int. Workshop on Human Subjects for Biomechanical Research, Lake Buena Vista, Florida, in press.

Boström O, Håland Y, Fredriksson R, Eriksson L, Krafft M, Lövsund P, Svensson M Y (1998): manuscript in preparation 1998 IRCOBI, Göteborg, Sweden.

Brault JR, Wheeler JB, Siegmund GP, Brault EJ (1998): Clinical response of human subjects to rear end automobile collisions, Archives of physical medicine and rehabilitation, Vol 79, January 1998, pp. 72-80.

Davidsson J, Linder A, Svensson MY, Flogård A, Håland Y, Jakobsson L, Lövsund P, Wiklund K (1998): manuscript in preparation 1998 IRCOBI, Göteborg, Sweden.

Eichberger A, Geigl BC, Moser A, Fachbach B, Steffan H, Hell W, Langwieder K (1996): Comparison of different car seats regarding head-neck kinematics of volunteers during rear-end impact, Proc. 1996 Int. Conf. on the Biomechanics of Impact (IRCOBI), Dublin, Ireland, pp. 153-164.

Håland Y, Lindh F, Fredriksson R, Svensson MY (1996): The influence of the car body and the seat on the loading of the front seat occupant's neck in low speed rear impacts, Proc. 29th ISATA Conf., Florence, Italy, pp. 21-30.

Jakobsson L, Norin H, Jernström C, Svensson S, Johnsen P, Isaksson-Hellman I, Svensson MY (1994): Analysis of different head and neck responses in rear-end car collisions using a new humanlike mathematical model. Proc. 1994 Int. Conf. on the Biomechanics of Impact (IRCOBI), Lyon, France, pp. 109-126.

v Koch M, Nygren Å, Tingvall C (1994): Impairment Pattern In Passenger Car Crashes, a Follow-up of Injuries Resulting in Longterm Consequences. ESV conference, Munich. 94-S5-O-02, pp. 776-781.

v Koch M, Kullgren A, Lie A, Nygren Å, Tingvall C (1995): Soft tissue injury of the cervical spine in rear-end and frontal car collisions. Proc. Int. Conf. on the Biomech. of Impacts (IRCOBI), pp. 273-283 and J. of Traffic Medicine Vol. 25 No. 3-4, pp. 89-96, 1997.

Krafft M, Kullgren A, Lie A, Nygren Å, Tingvall C (1996): Whiplash associated disorder - Factors influencing the incidence in rear-end collisions. Proc. 18th ESV Conf. in Melbourne, Australia, paper 96-S9-O-09.

Krafft M (1998): A comparison of short and long term consequences of AIS 1 Neck Injuries, in rear impacts, manuscript in preparation 1998 IRCOBI, Göteborg, Sweden.

Lundell B, Jakobsson L, Alfredsson B, Jernström C, Isaksson-Hellman I (1998): Guidelines for and the design of a car seat concept for improved protection against neck injuries in rear end collisions, SAE 980301, Int. Congress and Exposition Detroit, Michigan Feb. 23-26, 1998.

Nygren Å (1984): Injuries to Car Occupants - Some Aspects of the Interior Safety of Cars. Akta Oto-Laryngologica, Supplement 395, Almqvist & Wiksell, Stockholm, Sweden, ISSN 0365-5237.

Olsson I, Bunketorp O, Carlsson G, Gustafsson C, Planath I, Norin H, Ysander L (1990): An In-Depth Study of Neck Injuries in Rear End Collisions. Proc. 1990 Int. Conf. on the Biomech. of Impact, (IRCOBI), Bron, Lyon, France, pp. 269-282.

Otte D, Pohlemann T, Blauth M (1997): Significance of soft tissue neck injuries AIS 1 in the accident scene and deformation characteristics of cars with delta-v up to 10 km/h, Proc. 1997 Int. Conf. on the Biomechanics of Impact (IRCOBI), Hannover, Germany, pp. 265-283.

Parkin S, Mackay GM, Hassan AM, Graham R (1995): Rear end collisions and seat performance -To yield or not to yield, 39th annual proceedings-AAAM, pp. 231-244.

Svensson MY, Lövsund P (1992): A dummy for rear-end collisions - development and validation of a new dummy-neck, Proc. 1992 Int. Conf. on the Biomechanics of Impact (IRCOBI), Verona, Italy, pp. 299-310.

Svensson MY (1993): Neck Injuries in Rear-End Car Collisions - Sites and Biomechanical Causes of the Injuries, Test Methods and Preventive Measures, Doctoral thesis, Dept. of Injury Prevention, Chalmers Univ. of Techn., S-412 96 Göteborg, Sweden, ISBN 91-7032-878-1.

Viano DC (1992): Influence of seatback angle on occupant dynamics in simulated rear-end impacts, SAE 922521.

FEMALE VOLUNTEER MOTION IN REAR IMPACT SLED TESTS IN COMPARISON TO RESULTS FROM EARLIER MALE VOLUNTEER TESTS

Anna Carlsson[1,2], Astrid Linder[1], Mats Svensson[2], Johan Davidsson[2], Sylvia Schick[3], Stefan Horion[3], Wolfram Hell[3]

[1] VTI, Swedish National Road and Transport Research Institute, Sweden
[2] Chalmers University of Technology, Sweden
[3] LMU, Ludwig-Maximilians-Universitaet Muenchen, Germany

ABSTRACT

Vehicle related crashes causing neck injuries (whiplash) are costly and common, and injury statistic data shows a larger risk of neck injuries for females than for males. This study aims at investigating differences between female and male dynamic response in rear impacts. Rear impact sled tests with female volunteers were carried out and the results were compared with previously performed tests with males in matching test conditions. The volunteer tests were performed at a change of velocity of 7 km/h. The comparison of the average response of the males and the females and their response corridors showed several differences. The horizontal head acceleration peak value was on average 40% higher and occurred on average 18% earlier for the female volunteers compared to the male volunteers. The NIC value was 45% lower and 30% earlier for the females, probably due to a 27% smaller initial head-to-head restraint distance and thereby a 24% earlier head restraint contact. The results provide characteristic differences between dynamic responses of females and males in low speed rear impacts. These results contribute to the understanding of human dynamic response in rear impacts. In addition, they can be used in the process of future development if numerical and/or mechanical human models for crash testing.

Keywords: WHIPLASH, VOLUNTEERS, KINEMATICS, REAR IMPACTS, SLED TESTS

IN THE EUROPEAN UNION it is estimated that more than 300 000 citizens suffer Whiplash Associated Disorders (WAD) from vehicle collisions every year. The associated socio-economic impact of these injuries is in the order of 4 billion Euros per year (Whiplashkommisionen, 2005). The injuries are found in all impact directions with rear impacts being the most common in terms of cause of injury in accident statistics. Since the end of the 1960's, epidemiological data has shown that females have 40-100% larger risk of sustaining WAD than males (among others Krafft et al., 1997 and Jakobsson et al., 2004). However, little is known about the differences between females and males in terms of dynamic response. Such response data is primarily established by performing volunteer tests. These test results are a necessary input for the development of improved occupant models (crash dummies and computational models).

METHODS AND MATERIALS

Volunteer rear impact sled tests were performed at the change of velocity of 7 km/h. The volunteers were fitted with accelerometers and film markers and placed in a laboratory seat. The laboratory seat was designed to resemble a car seat of the late 1990s and consisted of panels which enable the motion of the different parts of the seat back to be monitored during the impact. The film marker and accelerometer positions, sled system and seat were the same as those used in Davidsson et al. (1999).

Eight female volunteers with an average weight of 60 kg, height of 1.66 m, seating height of 0.88 m, neck circumference of 0.33 m, and an average age of 24 years were run on a sled that was rear impacted by a second, bullet-sled. The volunteers were healthy with no history of neck pain and were chosen to represent an average female. The test series were ethically approved by an ethical committee at LMU (Klinikum der Universität München, Ethikkommission), ref number 319-07, and before the tests the volunteers were examined by a medical doctor. The volunteers were instructed to sit in their own preferred position in the seat, look in the forward direction and relax prior to the test. Response corridors were generated as the average value ± one standard deviation (SD) for each time sample. These responses were compared to responses obtained from (Davidsson *et al.* 1999) which contains the average response of five males with an average weight of 77 kg, height of 1.82 m, seating height of 0.92 m, neck circumference of 0.37 m, and an average age of 30 years.

SLED - The sled used for the test series was a target-bullet sled in which the acceleration pulse (Fig. 1) was generated by a bending bar. The coordinate system was defined according to the SAE J211 standard.

INSTRUMENTATION - Tri-axial accelerometers were placed on the left side of the volunteers' head, approximately at the centre of gravity of the head. Two linear accelerometers, in x- and z-direction, were placed on a holder attached to the skin close to T1 (the upper thoracic vertebra). Linear accelerometers were placed on the bullet sled and on the target sled. A tape switch with a release force of 5N was attached to the steel bar on the target sled and the signal from this switch defined the start of the impact, T=0. The motion of the volunteers was monitored with high-speed video at 1000 frames per second.

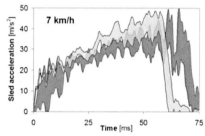

Fig. 1 - *The sled acceleration.*
The dark corridor represents tests with males and the light corridor represents tests with females.

The upper torso displacement angle was defined as the angle between the film mark placed at the T1 and the film mark placed at the clavicle. The horizontal displacement of the head and T1 were expressed relative to the sled coordinate system with the x-axis forward (in the sled travelling direction). The head-to-head restraint distance was measured before the test and the contact time was identified from film analysis. The Neck Injury Criterion (NIC) value was calculated according to Boström *et al.* (1996), and the accelerations for the NIC calculations were filtered in CFC60.

RESULTS

The comparison of the average response and the corresponding corridors of the males and the females showed several differences.

The horizontal head acceleration data (Fig. 2a) showed that the increase of the horizontal head acceleration started earlier for the females and that the peak acceleration for the females occurred earlier, and had a higher peak value, than the peak acceleration for the males. On average, the peak value for females was 72 m/s^2 at 119 ms compared to 51 m/s^2 at 145 ms for the males.

The increase of the horizontal T1 acceleration started earlier for the females compared to the males (Fig 2b). On average, it took 61 ms for the females and 74 ms for the males to reach the T1 acceleration 20 m/s^2. The peak values were of the same magnitude for the males and the females.

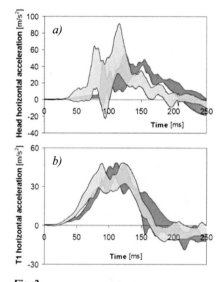

Fig. 2
a) The horizontal head acceleration.
b) The horizontal T1 acceleration.
The dark corridors represent tests with males and the light corridors represent tests with females.

The rearward horizontal displacement peaks of the head and T1 for the females were smaller and earlier than that of the males in the negative direction (Fig. 3a,b). The average value for the males was 149 mm at 144 ms and for the females it was 111 mm at 126 ms. The corresponding numbers for the horizontal T1 negative displacement were 100 mm at 122 ms for the males and 87 mm at 108 ms for the females.

The negative peak of the head relative T1 horizontal displacement (Fig. 3c) was smaller and earlier for the average female (40 mm at 163 ms) compared to the average male (79 mm at 186 ms).

The positive peak of the head angle (Fig. 3d) was larger for the males (24°) compared to the females (18°) while the positive peak of the upper torso angle (Fig. 3e) was larger for the females (19°) compared to the males (14°).

The first (negative) peak of the head relative upper torso angle (Fig. 3f) was smaller and earlier for the average male (6° at 100 ms) than the average female (9° at 117 ms). The second (positive) peak was larger and earlier for the males (12° at 180 ms) than the females (7° at 208 ms).

The average NIC value for the females was lower and earlier (3.4 m^2/s^2 at 62 ms) compared to the males (6.3 m^2/s^2 at 88 ms).

The initial head-to-head restraint distance was on average smaller for the females (62 mm) compared to the males (86 mm) and the head-to-head restraint contact occurred on average earlier for the females (71 ms) compared to the males (94 ms).

DISCUSSION

The horizontal head acceleration peak value was on average 40% higher and occurred on average 18% earlier for the female volunteers compared to the male volunteers. Similar results have been reported by Hell *et al.* (1999) and Carlsson *et al.* (2008). The earlier occurrence of the head acceleration peak for the female volunteers in the present study may be due to smaller initial head-to-head restrain distance for the females (62 mm) compared to the males (86 mm). Smaller head-to-head restraint distance for females has also been reported by Jonsson *et al.* (2007), for all seating positions in a Volvo V70.

The smaller initial head-to-head restraint distance for the females resulted in an earlier head-to-head restraint contact time, a smaller horizontal head displacement and head extension angle, and lower NIC values. Since the head-to-head restraint contact occurred earlier for the females, and the contact force between the upper torso and seat structure peaked after the head-to-head restraint contact, the relative acceleration between the head and T1 was smaller for the females, leading to a lower NIC value for females than for males. The NIC value was on average 45% lower and 30% earlier for the female volunteers compared to the male volunteers. Similar differences in the NIC value between

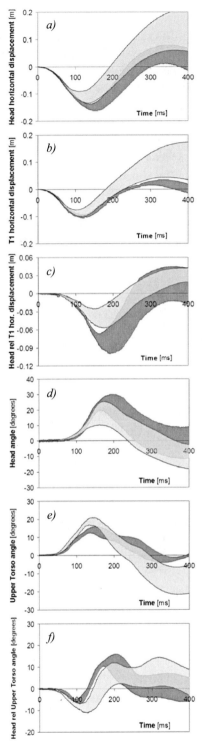

Fig. 3
a) *The horizontal head displacement.*
b) *The horizontal T1 displacement.*
c) *The horizontal head rel. T1 displacement.*
d) *The head angular displacement.*
e) *The upper torso angular displacement.*
f) *The head rel. upper torso angular displacement.*
The dark corridors represent tests with males and the light corridors represent tests with females.

female and male volunteers have been reported by Carlsson *et al.* (2008).

The increase of the T1 horizontal acceleration started earlier for the females compared to the males which is consistent with what has been reported by Hell *et al.* (1999) and Ono *et al.* (2006).

Further studies are needed of both male and female volunteers in car seats in order to establish their dynamic response in rear impacts. Such data is fundamental in order to develop mathematical and mechanical models of both males and females for rear impact tests. These models are essential in order to develop and evaluate the performance of new anti-whiplash systems for all adult occupants.

CONCLUSIONS

This paper quantifies the differences between female and male volunteer response in a rear impact. The data showed that the head x-acceleration peak value was 40% higher and 18% earlier for the females compared to the males. The NIC value was 45% lower and 30% earlier for the females, probably due to a 27% smaller initial head-to-head restraint distance and thereby a 24% earlier head restraint contact. These results contribute to the understanding of human dynamic response in rear impacts. In addition, they can be used in the process of future development of numerical and/or mechanical human models for crash testing.

ACKNOWLEDGEMENTS

This study was funded by the Swedish Governmental Agency for Innovation Systems (VINNOVA). Special thanks for valuable help with data collection and processing to Kristin Thorsteinsdottir at LMU and Carsten Reinkemeyer at Allianz.

REFERENCES

Boström, O., Krafft, M., Aldman B., Eichberger, A., Fredriksson, R., Håland Y., Lövsund P., Steffan, H., Svensson, M.Y., Tingvall, C. (1997): *Prediction of Neck Injury Based on Accident Data and Simulations.* Proc. IRCOBI conf., Hannover, Germany, pp. 251-283.

Carlsson, A., Linder, A., Svensson, M., Siegmund, G. (2008): *Dynamic Responses of Female and Male Volunteers in Rear Impacts.* World Congress on Neck Pain, Los Angeles, USA

Davidsson, J., Flogård, A., Lövsund, P., and Svensson, M.Y. (1999): *BioRID P3 - Design and performance compared to Hybrid III and volunteers in rear impacts at ΔV=7 km/h.* (99SC16). Proc. 43rd Stapp Car Crash Conference, pp. 253-265. Society of Automotive Engineers, Warrendale, PA.

Hell, W., Langwieder, K., Walz, F., Muser, M., Kramer, M., Hartwig, E. (1999): *Consequences for Seat Design due to Rear End Accident Analysis, Sled Tests and Possible Test Criteria for Reducing Cervical Spine Injuries after Rear-End Collision.* Proc. IRCOBI conf., Sitges, Spain, pp. 243- 259.

Jakobsson L. Norin, H., Svensson, M.Y. (2004): *Parameters Influencing AIS 1 Neck Injury Outcome in Frontal Impacts.* Traffic Inj. Prev., Vol. 5, No 2, pp. 156–163.

Jonsson B., Stenlund H., Svensson M.Y. and Björnstig U. (2007): *Backset and Cervical Retraction Capacity among Occupants in a Modern Car.* Traffic Inj. Prev., Vol. 8, No 1, pp. 87–93.

Krafft M, Kullgren A, Lie A, Nygren Å, Tingvall C. (1997): *Soft Tissue Injury of the Cervical Spine in Rear-end End Car Collisions.* Traffic Medicine, 25(3-4).

Ono K, Ejima S, Suzuki Y, Kaneoka K, Fukushima M, Ujihashi S (2006): *Prediction of neck injury risk based on the analysis of localized cervical vertebral motion of human volunteers during low-speed rear impacts.* Proc. IRCOBI conf., Madrid, Spain, pp. 103-113.

Whiplashkommissionen (2005): http://www.whiplashkommissionen.se/pdf/WK_finalreport.pdf

2006-01-0067

The RID2 Biofidelic Rear Impact Dummy: A Validation Study Using Human Subjects in Low Speed Rear Impact Full Scale Crash Tests. Neck Injury Criterion (NIC)

Arthur C. Croft
Spine Research Institute of San Diego; Southern California University of Health Sciences

Mathieu M.G.M. Philippens
TNO Science and Industry

ABSTRACT

Human subjects and the recently developed RID2 rear impact crash test dummy were exposed to a series of full scale, vehicle-to-vehicle crash tests to evaluate the biofidelity of the RID2 anthropometric test dummy on the basis of calculated neck injury criterion (NIC) values. Volunteer subjects, including a 50th percentile male, a 95th percentile male, and a 50th percentile female, were placed in the driver's seat of a vehicle and subjected to a series of three low speed rear impact crashes each. Both subjects and dummy were fully instrumented and acceleration-time histories were recorded. From this data, velocities of the heads and torsos were integrated and used to calculate the NIC values for both crash test subjects and the RID2. The RID2 dummy is designed to represent a 50th male. The overall performance and biofidelity of the RID2 compared most favorably to the human subject who was, himself, a 50th percentile male. Although the number of tests was small, the biofidelity of the RID2, in the context of the smaller female and larger male, was limited. The overall performance and biofidelity of the RID2 was reasonable when compared to the 50th percentile male volunteer. It is possible that under real world crash conditions, in which the occupant of the target vehicle is exposed to an unexpected impact, that their NIC values might be more comparable to those of the RID2, suggesting that its biofidelity could have been underestimated as a result of the alerted status of the crash test volunteers.

INTRODUCTION

Whiplash injury has become recognized as a significant public health problem in recent years (Spitzer et al., 1995). Some authors describe the minor neck or cervical spine injury resulting from any motor vehicle crash as *whiplash*. However, the risk for injury from the rear impact vector crash has been widely reported as being higher than for other vector crashes (Bylund and Bjornstig, 1998, Borchgrevink et al., 1996, Borchgrevink et al., 1997, Krafft, 1998, Richter et al., 2000). The outcomes in rear impact crashes at low speeds have also been reported to be less favorable than those of frontal or other crash vectors (Krafft, 1998), and long-term disability, a term which has not been operationally defined in most studies, has been variously reported to be 2% (Gargan et al., 1997), 5% (Borchgrevink et al., 1996), 7% (Radanov et al., 1993, Gozzard et al., 2001), 8% (Pettersson et al., 1997), 10% (Nygren, 1984), 12% (Gargan and Bannister, 1990, Kasch et al., 2001a, Kasch et al., 2001b), 16% (Bylund and Bjornstig, 1998), and 24% (Ettlin et al., 1992). The incidence of whiplash injury and disability have been increasing in recent years (Richter et al., 2000, Holm et al., 1999, Richter et al., 1999, Galasko et al., 2000).

Although rear impact crashes represent a minority of crash types, accounting for only about 25% of all crashes, they represent a disproportionate risk for injury (Holm et al., 1999). This differential risk may be explained through human subject crash testing. In one study, the subjects' head linear accelerations were found to be markedly higher in rear impacts vs. frontal impact crashes with crash speeds and other variables held constant, and subjects rated these crashes markedly less tolerable than frontal crashes (Croft et al., 2002a). The fact that the largest group exposed to this form of trauma are persons between the ages of 20-40 years of age and disability in this group results in a high loss of productive years of life, and the fact that this is potentially a preventable injury (or at least one in which the risk can be greatly reduced) make research in this area a high public health priority.

In order to better understand the forces imposed during low speed rear impact crashes (LOSRIC), human subjects have been placed in vehicles under full scale crash conditions (Croft et al., 2002a, Severy et al., 1955,

West et al., 1993, Szabo et al., 1994, McConnell et al., 1993b, McConnell et al., 1995b, Szabo and Welcher, 1996, Siegmund et al., 1997, Brault et al., 1998, Croft et al., 2002b, Croft et al., 2002c). As in the case of higher speed crash tests, crash test dummies would be the preferable test subjects in low speed crashes. Unfortunately, the modern Hybrid III anthropometric test device (ATD) lacks adequate biofidelity to serve as a valid proxy for human subjects in the special application of LOSRIC and this has lead to the development of a series of specialized rear impact dummies (RID) (Svensson et al., 1993, Davidsson et al., 2001, Philippens et al., 2002, Cappon et al., 2000). Early attempts were simply to modify existing Hybrid III dummies by substituting a more supple neck, but the Hybrid III's rigid thoracic spine rendered such configurations impractical. It is clear from observation of human subject crash testing that, under direct loading from the seat back, the thoracic curve, which is normally kyphotic in humans, will be flattened. This was first observed by McConnell et al. (McConnell et al., 1993a, McConnell et al., 1995a) who reported the resulting vertical motion of the head (which is partially contributed to by a ramping up the seat back) to be as much as 3.5 inches. This flattening of the thoracic spine and vertical rise of the head is also associated with compression of the spine (Bertholon et al., 2000). It has been postulated that this type of loading can fracture cervical vertebral end plates and may be factor in some cases of chronic pain (Freeman et al., 2001). Vertical motion of the head will also alter the relative head restraint geometry by increasing the topset, **Figure 1**. Thus, a flexible thoracic spine is necessary in the rear impact dummy in order to improve its biofidelity by allowing some degree of torso flattening.

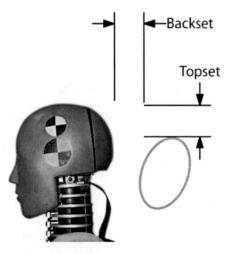

Figure 1. Critical head restraint geometry are described in terms of the horizontal distance between the head and restraint, *backset*, and the vertical distance from the top of the head to the top of the restraint, *topset*.

Svensson and Lövsund (Svensson and Lovsund, 1992) developed a new neck for use with the Hybrid III dummy—the RID neck—and validated it with previous human studies and impact testing. They tested three different neck stiffnesses and compared these to tests with the original Hybrid III neck, which was markedly more stiff. The purpose of this work by TNO (TNO Crash-Safety Research Centre, The Netherlands) was to develop a neck for the 50th percentile Hybrid III dummy to be used in rear impact simulations of up to 25.8 km/h. The TNO Rear Impact Dummy (TRID) Neck was an improvement, but still lacked a flexible thoracic spine, as the TRID neck was a replacement for the Hybrid III neck only.

The RID2 (First Technology Safety Systems, Plymouth, Michigan) has a fully mobile neck in all ranges, and a flexible thoracic spine with motion possible at one joint. Tests comparing the Hybrid III, BioRID, and RID2 have been conducted (Philippens et al., 2002, Zellmer et al., 2002, Cappon et al., 2000) and have generally found the BioRID and RID2 dummies to be comparable to each other and significantly more biofidelic than the Hybrid III for testing under LOSRIC conditions. However, these studies have not been conducted under real world boundary conditions (i.e., full scale low velocity vehicle-to-vehicle crash tests) with living human subjects and RID2 side-by-side.

A potential limitation to some of the validation studies that have been completed to date is that, in many case, seats and seat backs are sometimes purpose-built for the tests and might not be representative of real production car seats in terms of seat back stiffness, compliance, and overall restitutional behavior. One of the ultimate goals in crash testing is to develop a surrogate or proxy for human subjects which will allow testing under conditions which are not suitable for human subject testing (e.g., a high frequency of tests and/or high acceleration pulses) because of the health risks imposed. In order to develop such a device, it is necessary to validate the ATD against living human subjects within the boundary conditions for which the ATD is intended, and that was the goal of this study. Main Section

MATERIALS AND METHODS

All human subjects and the RID2 were instrumented for every test. Crash test vehicles were also instrumented. The subject's headgear array consisted of three triaxial blocks of IC Sensors 3031-050 (50 g) accelerometers tightly affixed to the head via a lightweight headband. Peripheral head acceleration measurements were resolved to the approximate head static center of gravity via an algorithm which utilized the locations of each triaxial block relative to known anatomical landmarks. A low profile triaxial block of thoracic accelerometers was constructed using two Entran EGAXT-50 (50 g) accelerometers and one IC Sensors 3031-050 (50 g) accelerometer. The accelerometers were affixed to the subjects with medical adhesive and tightly fitted straps at the approximate level of C7-T1 on the anterior torso. For the lumbar measurements, a lightweight uniaxial IC Sensors 3031-050 (50 g) accelerometer was affixed with medical adhesive to the base of the subjects' lumbar spines at approximately the level of L5-S1. Target and bullet vehicle accelerometers consisted of a triaxial block of 3031-050 (gain adjusted to ± 15 g full scale) accelerometers affixed with sheet metal screws to each vehicle's chassis at the approximate static center of gravity. Analog to digital conversion was performed by a 12-bit A/D converter operating with a maximum conversion rate of 330,000 samples per second. All data were collected following the general theory of Society of Automotive Engineers (SAE) Recommended Practice: Instrumentation for Impact Test—Part 1—Electronic Instrumentation—J211/1 Mar95. (SAE, 1996). All accelerometer data was collected at 1000 Hz. Vehicle accelerations were filtered using an SAE Class 60 filter. Vehicle changes in velocity were calculated from vehicle acceleration data filtered with an SAE Class 180 filter. Occupant accelerometer data was filtered with and SAE Class 60 filter. Vehicle speeds were also measured using an MEA 5th wheel (MacInnis Engineering Associates, Richmond, BC Canada) attached to each vehicle. Data were acquired at 128 Hz simultaneously for both vehicles for the period 1 sec before to 4 sec after impacts. Time traps for recording vehicle impact speed consisted of custom built Timer Interval Meter with internal clock calibrated to an NIST traceable source. The pressure sensitive tape switches were Tape Switch Corporation Type 102A, requiring 40 ounce pressure for activation. RID2 instrumentation consisted of triaxial head cg linear accelerometers, skull cap force transducer, a 6 component upper neck load cell, a 6 component lower neck load cell, T1-level triaxial linear accelerometers, T12-level triaxial accelerometers, triaxial pelvis linear accelerometers, and 6 inclinometers used for static positioning.

A total of 9 tests, consisting of three tests each with the RID2 as front seat passenger and a human volunteer as driver, were performed, **Table 1**. In each of the three tests series, accurate rear impact test speeds were facilitated using a trunk lid-mounted speedometer on the bullet vehicle which was fed by an MEA 5th wheel. The bullet vehicle was pushed by a practiced push team capable of speeds in excess of 16.2 km/h with reproducibility of +/- 0.3 km/h. In all 9 crash tests the bullet vehicle was a 1994 Ford Crown Victoria (1727 kg) and the target vehicle was a 1989 Chrysler Le Baron (1290 kg). Both vehicles were inspected for damage prior to and after each test. Neither vehicle sustained any significant residual structural damage in these 9 tests and no repairs were necessary to guarantee repeatability or reproducibility of crash conditions. In all tests, the human subject was instructed to place his/her foot on the brake using the same force as he/she would normally use in traffic, and were all allowed to assume their relaxed, normal seating posture. They were also instructed to keep their eyes open and to place their hands lightly on the steering wheel and not to grip it.

Summary of crashes			
Crash #	Subj-ects *	V_c (km/h)	delta V (km/h)
1	DV	8.4	6.0
2	DV	12.4	8.7
3	DV	16.3	11.0
4	AF	9.0	6.6
5	AF	11.9	8.4
6	AF	15.0	10.3
7	RC	8.7	5.5
8	RC	11.1	7.9
9	RC	15.3	9.0

* Human subjects were seated in driver's seat. RID2 ATD was in passenger seat in all tests.
Subject DV: 27-year-old male, 1.8 m in height, 81 kg (50th percentile male)
Subject AF: 24-year-old female, 1.6 m in height, 56 kg (50th percentile female)
Subject RC: 19-year-old male, 1.9 m in height, 109 kg (95th percentile male)
In all cases the bullet vehicle was a 1994 Ford Crown Victoria and the target vehicle was a 1989 Chrysler Le Baron.
V_c: closing velocity
delta V: change in velocity

Table 1.

The stationary target vehicle was placed in neutral with the motor turned off.

Selection/exclusion criteria for human subjects included a willingness to participate in low speed crash tests, no history of significant spinal pain or headaches, and no prior significant injuries to the spine. Each volunteer also was examined by a licensed physician to ensure their fitness for participation, and cervical spine range of motion was measured using a CROM device and recorded before and after the tests were completed. Radiographic studies were undertaken before and after all crash tests to insure volunteer safety.

Institutional review board approval was obtained. In all cases, participants were fully informed of the potential risks of crash testing, and consent for participation was obtained in full accordance with the Office for Protection from Research Risks (OPRR) of the Department of Health and Human Services, and the recommended *Belmont Report*. Subjects were interviewed after each test and were given the opportunity to terminate their participation at any time without penalty. The RID2 ATD was calibrated and repositioned prior to each test in accordance with the manufacturer's recommendations. In this paper, the SAE right hand coordinate system is used to represent vectors and motion paths. The proposed neck injury criterion considers the relative acceleration and velocity between the top and bottom of the spine and is given as (Croft et al., 2002c):

$$NIC = a_{rel}x0.2 + v_{rel}^2 \qquad \text{Eq. (1)}$$

where a_{rel} and v_{rel} are the relative horizontal acceleration and velocity between the bottom (T1) and top (C1) of the cervical spine. The constant, 0.2, represents the approximate length of the neck in meters. This equation accounts for what is now widely held to be one of the most important risk factors in LOSRIC injury—the retraction of the head (head lag) during the first 100 or so milliseconds of the crash sequence (Siegmund et al., 1997, Brault et al., 1998). The equation for NIC is calculated as follows:

$$NIC(t) = a_{rel}(t)x0.2 + \left[V_{rel}(t)\right]^2$$

$$\text{where } a_{rel}(t) = a_x^{T1}(t) - a_x^{Head}(t), \qquad \text{Eq. (2-3)}$$

$$\text{and } V_{rel}(t) = \int a_x^{T1}(t)dt - \int a_x^{Head}(t)dt$$

where a_x^{T1} = the acceleration-time history measured in the antero-posterior (x) direction at the level of the first thoracic vertebra in units of g. Likewise, a_x^{Head} = the acceleration-time history measured in the antero-posterior (x) direction at the location of the center of gravity of the head in units of g.

The integration of the acceleration (converted to m/s^2) at the level of head center of gravity in the time domain, giving the velocity in the x-direction (resulting in units of m/s), is expressed by:

$$\int a_x^{Head}(t)dt \qquad \text{Eq. (4)}$$

The integration of the acceleration (converted to m/s^2) of the first thoracic vertebra in the time domain, giving the velocity in the x-direction (resulting in units of m/s), is expressed by:

$$\int a_x^{T1}(t)dt \qquad \text{Eq. (5)}$$

RESULTS

No serious injuries were reported by the subjects and all subjects completed all three of their crash tests and all post-crash examinations. One female subject, however, did report mild neck discomfort and headache following the third test.

Overall, the comparison of the RID2 and 50th percentile human subjects' head linear (x) acceleration-time histories are quite good, **Figure 2**. In all cases, the RID2's acceleration was somewhat greater in both phases, but the morphology of the acceleration pulses was found to be generally good in this study. The NIC values of the RID2 and human subjects of the current study are provided in **Table 2**. These values are also plotted in **Figure 3**.

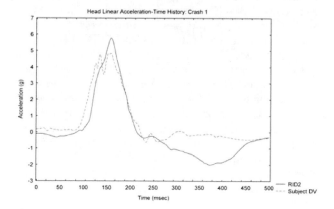

Figure 2. Exemplar acceleration time history comparison between RID2 and human subject. Both are 50th percentile males. The multiple peaks seen in the curve of the human subject are likely minor artifacts induced during the time of head contact with the head restraint resulting from relative motion between head and headgear. The lower negative acceleration of the human subject during the period between 250 and 450 msec is the result of muscle activation.

Neck Injury Criterion (NIC) (m^2/s^2)			
Crash #	RID2	Human subject	Variation (%)
1	5.6	6.6	17.9
2	6.8	10.3	51.5
3	8.0	8.2	2.5
4	5.4	0.8	-85.2
5	9.1	3.2	-64.8
6	7.6	1.6	-79.0
7	5.2	3.0	-42.3
8	6.3	3.6	-42.9
9	8.1	4.7	-42.0

Table 2.

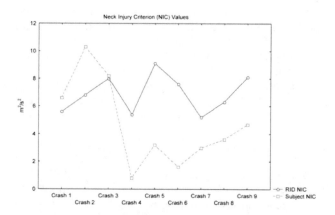

Figure 3. Plot of NIC values for RID2 and human subjects for all crash tests.

DISCUSION

The higher NIC values seen in subject DV in crashes 1-3, which averaged 24% higher than those of the RID2, may reflect the difference in compliance between the human and RID2 cervical spines. In crashes 4-6, subject AF had markedly lower NIC values than the RID2 (averaging 76.3% lower), probably as a function of her initially very low backset, which reduced the differential motion between the torso and head and which demonstrates how optimal head restraint geometry is a critical factor in reducing NIC values. In crashes 7-9, the 95th percentile male subject's (RC) NIC values parallel those of the RID2 but averaged 42.4 % lower, probably as a result of his larger mass. Generally, higher RID2 NIC values coincided with higher speed changes, with one exception. This is likely the result of subtle positioning variation of the RID2 between tests.

Overall, the RID2 performed adequately under crash conditions that are representative of real world crashes at low speeds. Peak linear x acceleration of the RID2 head was always higher than that of the human subjects, averaging 51% higher. The average RID2 head linear x acceleration was 31% higher than those of the 50th percentile male and female subjects. There are potential limitations with this kind of crash testing in terms of its external validity. Recent research has demonstrated that subject awareness alters the responses to staged crashes significantly, with later muscular activation recorded in both surprised male and female subject groups, higher amplitude muscular contraction in the male surprised group, and greater head retraction ranges in the female group compared to groups who were alerted to the impending event (Siegmund et al., 2003).

Thus, aware subjects of crash tests cannot be considered fully representative of the subgroup of real world crash victims who are unaware of the impending rear impact crash. Siegmund et al. (Siegmund et al., 2003) speculated that previous reports of clinical symptoms generated in crash test experiments may underestimate the risk of whiplash in real crashes. The scientific limitations of extrapolating risk estimates from these studies has been reported previously (Freeman et al., 1999). Taking these facts into consideration, the RID2's overall biofidelity might have been underestimated by the current study. It is noteworthy that persons who report having been caught unaware of the impending rear impact crash are at greater risk of injury (Dolinis, 1997, Sturzenegger et al., 1994) and have been reported to have a significantly worse prognosis (Ryan et al., 1994).

As a result of the small number of tests that volunteers can be exposed to for safety reasons, and as a result of the practical limitation on the total number of tests that can be run and the small number of volunteers used, it is not possible to apply meaningful statistical analysis to this set of data. Other factors also limit the external validity of these tests. Many of the variables present in real world crashes, such as offset crash conditions, bumper over- or under-ride, variations between relative masses of crashing vehicles, and the wide variety of seat back and head restraint designs and stiffnesses cannot be accounted for in small-scale crash test studies of this kind. Moreover, subjects were placed into ideal positions, with head restraints adjusted in their upright position, and all were healthy subjects who were both medically screened for known risk factors and were aware of the impending crash. Extrapolations regarding the risk for injury from this study to real world crashes cannot be made with any degree of reliability and should be discouraged..

CONCLUSION

The RID2 dummy is designed to represent a 50th male. The overall performance and biofidelity of the RID2 compared most favorably to the human subject who was, himself, a 50th percentile male. Its overall higher ranges of head acceleration and calculated NIC values compared to the human subjects were generally consistent and potentially explainable on the basis of pre-crash head restraint geometry and differences in body size between the RID2 and the three volunteers. It is possible that under real world crash conditions, in which the occupant of the target vehicle is exposed to an unexpected impact, their excursions and accelerations might be more comparable to those of the RID2, suggesting that its biofidelity could have been underestimated as a result of the alerted status of the crash test volunteers. This is a variable we cannot easily evaluate for practical and ethical reasons. Although the number of tests was small, the biofidelity of the RID2, in the context of the 50th percentile female and 95th percentile male, was limited.

ACKNOWLEDGMENTS

This research was made possible, in part, by a contribution from the Center for Research into Automotive Safety and Health (CRASH). We would like to thank First Technology Safety Systems, Plymouth, Michigan, for the use of the RID2 dummy.

REFERENCES

Bertholon, N., Robin, S., Le-Coz, J.-Y., Potier, P., Lassaue, J.-P., & Skalli, W. (2000, September 20-22). *Human head and cervical spine behaviour during low-speed rear end impacts: PMHS sled tests with a rigid seat.* Paper presented at the Proceedings of the International IRCOBI Conference on Biomechanics of Impacts, Montpelier France.

Borchgrevink, G. E., Lereim, I., Royneland, L., Bjorndal, A., & Haraldseth, O. (1996). National health insurance consumption and chronic symptoms following mild neck sprain injuries in car collisions. *Scand J Soc Med, 24*(4), 264-271.

Borchgrevink, G. E., Stiles, T. C., Borchgrevink, P. C., & Lereim, I. (1997). Personality profile among symptomatic and recovered patients with neck sprain injury, measured by MCMI-I acutely and 6 months after car accidents. *J Psychosom Res, 42*(4), 357-367.

Brault, J. R., Wheeler, J. B., Siegmund, G. P., & Brault, E. J. (1998). Clinical response of human subjects to rear-end automobile collisions. *Arch Phys Med Rehabil, 79*(1), 72-80.

Bylund, P. O., & Bjornstig, U. (1998). Sick leave and disability pension among passenger car occupants injured in urban traffic. *Spine, 23*(9), 1023-1028.

Cappon, H. J., Philippens, M. M. G. M., van Ratingen, M. R., & Wismans, J. S. H. M. (2000, Sept 20-22). *Evaluation of dummy behaviour during low severity rear impact.* . Paper presented at the International Research Council on the Biomechanics of Impact (IRCOBI) Conference, Montpellier, France.

Croft, A., Haneline, M., & Freeman, M. (2002a). Differential Occupant Kinematics and Forces Between Frontal and Rear Automobile Impacts at Low Speed: Evidence for a Differential Injury Risk. *International Research Council on the Biomechanics of Impact (IRCOBI), International Conference, Munich, German, September 18-20,* 365-366.

Croft, A., Haneline, M., & Freeman, M. (2002b, September 18-20). *Differential occupant kinematics and forces between frontal And rear automobile impacts at low speed: evidence for a differential injury risk.* . Paper presented at the International IRCOBI Conference on the Biomechanics of Impact, Munich, Germany.

Croft, A. C., Herring, P., Freeman, M. D., & Haneline, M. T. (2002). The neck injury criterion: future considerations. *Accid Anal Prev, 34*(2), 247-255.

Davidsson, J., Lovsund, P., Ono, K., Svensson, M., & Inami, S. (2001). A comparison of volunteere, BioRID P3, and Hybrid III performance in rear impacts. *J Crash Prev Inj Control, 2*(3), 203-220.

Dolinis. (1997). Risk factors for 'whiplash' in drivers: a cohort study of rear-end traffic crashes. *Injury, 28*(3), 173-179.

Ettlin, T. M., Kischka, U., Reichmann, S., Radii, E. W., Heim, S., Wengen, D., et al. (1992). Cerebral symptoms after whiplash injury of the neck: a prospective clinical and neuropsychological study of whiplash injury. *J Neurol Neurosurg Psychiatry, 55*(10), 943-948.

Freeman, M. D., Croft, A. C., Rossignol, A. M., Weaver, D. S., & Reiser, M. (1999). A review and methodologic critique of the literature refuting whiplash syndrome. *Spine, 24*(1), 86-96.

Freeman, M. D., Sapir, D., Boutselis, A., Gorup, J., Tuckman, G., Croft, A. C., et al. (2001, Nov 29-Dec 1). *Whiplash injury and occult vertebral fracture: a case series of bone SPECT imaging of patients with persisting spine pain following a motor vehicle crash.* Paper presented at the Cervical Spine Research Society 29th Annual Meeting, Monterey, California.

Galasko, C. S. B., Murray, P. A., & Pitcher, M. (2000). Prevalence and long-term disability following whiplash-associated disorder. . *Journal of Musculoskeletal Pain, 8*, 15-27.

Gargan, M., Bannister, G., Main, C., & Hollis, S. (1997). The behavioural response to whiplash injury. *J Bone Joint Surg Br, 79*(4), 523-526.

Gargan, M. F., & Bannister, G. C. (1990). Long-term prognosis of soft-tissue injuries of the neck. *J Bone Joint Surg Br, 72*(5), 901-903.

Gozzard, C., Bannister, G., Langkamer, G., Khan, S., Gargan, M., & Foy, C. (2001). Factors affecting employment after whiplash injury. *J Bone Joint Surg Br, 83*(4), 506-509.

Holm, L., Cassidy, J. D., Sjogren, Y., & Nygren, A. (1999). Impairment and work disability due to whiplash injury following traffic collisions. An analysis of insurance material from the Swedish Road Traffic Injury Commission. *Scand J Public Health, 27*(2), 116-123.

Kasch, H., Bach, F. W., & Jensen, T. S. (2001). Handicap after acute whiplash injury: a 1-year prospective study of risk factors. *Neurology, 56*(12), 1637-1643.

Kasch, H., Stengaard-Pedersen, K., Arendt-Nielsen, L., & Staehelin Jensen, T. (2001). Pain thresholds and tenderness in neck and head following acute whiplash injury: a prospective study. *Cephalalgia, 21*(3), 189-197.

Krafft, M. (1998, September 16-18). *A comparison of short- and long-term consequences of AIS 1 neck injuries, in rear impacts.* Paper presented at the International IRCOBI Conference on the Biomechanics of Impact, Goteborg, Sweden.

McConnell, W., Howard, R., & Guzman, H. (1993). Analysis of human test subject kinematic responses to low velocity rear end impacts. *SAE Tech Paper Series, 930889,* 21-30.

McConnell, W., Howard, R., & Poppel, J. (1995). *Human head and neck kinematic after low velocity rear-end impacts: understanding "whiplash."* 952724. Paper presented at the Proceedings of the 39th Stapp Car Crash Conference Proceedings.

McConnell, W. E., Howard, R. P., & Guzman, H. M. (1993). Analysis of human test subject kinematic responses to low velocity rear end impacts. *SAE Tech Paper Series, 930889,* 21-30.

McConnell, W. E., Howard, R. P., Poppel, J. V., & al., e. (1995). *Human head and neck kinematic after low velocity rear-end impacts: understanding "whiplash."* Paper presented at the 39th Stapp Car Crash Conference Proceedings.

Nygren, A. (1984). Injuries to car occupants--some aspects of the interior safety of cars. A study of a five-year material from an insurance company. *Acta Otolaryngol Suppl 395,* 1-164.

Pettersson, K., Hildingsson, C., Toolanen, G., Fagerlund, M., & Bjornebrink, J. (1997). Disc pathology after whiplash injury. A prospective magnetic resonance imaging and clinical investigation. *Spine, 22*(3), 283-287; discussion 288.

Philippens, M., Cappon, H., van Ratingen, M., Wismans, J., Svensson, M., & Sirey, F. (2002). Comparison of the Rear Impact Biofidelity of BioRID II and RID 2. *Stapp Car Crash Journal, 46,* 461-476.

Radanov, B. P., Di Stefano, G., Schnidrig, A., & Sturzenegger, M. (1993). Psychosocial stress, cognitive performance and disability after common whiplash. *J Psychosom Res, 37*(1), 1-10.

Richter, M., Otte, D., & Blauth, M. (1999). Acceleration injuries of the cervical spine in seat-belted automobile drivers. Determination of the trauma mechanism and severity of injury. *Orthopade, 28*(5), 414-423.

Richter, M., Otte, D., Pohlemann, T., Krettek, C., & Blauth, M. (2000). Whiplash-type neck distortion in restrained car drivers: frequency, causes and long-term results. *Eur Spine J, 9*(2), 109-117.

Ryan, G. A., Taylor, G. W., Moore, V. M., & Dolinis, J. (1994). Neck strain in car occupants: injury status after 6 months and crash-related factors. *Injury, 25*(8), 533-537.

Severy, D., Matthewson, J., & Bechtol, C. (1955). *Controlled automobile rear-end collisions -- an investigation of related engineering and medical phenomena.* Paper presented at the In Medical Aspects of Traffic Accidents, Proceedings of the Montreal Conference.

Siegmund, G., King, D., Lawrence, J., Wheeler, J., Brault, J., & Smith, T. (1997). Head/neck kinematic responses of human subjects in low-speed rear-end collisions. *SAE Technical Paper 973341,* 357-385.

Siegmund, G. P., Sanderson, D. J., Myers, B. S., & Inglis, J. T. (2003). Awareness affects the response of human subjects exposed to a single whiplash-like perturbation. *Spine, 28*(7), 671-679.

Spitzer, W. O., Skovron, M. L., Salmi, L. R., Cassidy, J. D., Duranceau, J., Suissa, S., et al. (1995). Scientific monograph of the Quebec Task Force on Whiplash-Associated Disorders: redefining "whiplash" and its management. *Spine, 20*(8 Suppl), 1S-73S.

Sturzenegger, M., DiStefano, G., Radanov, B. P., & Schnidrig, A. (1994). Presenting symptoms and signs after whiplash injury: the influence of accident mechanisms. *Neurology, 44*(4), 688-693.

Svensson, M., & Lovsund, P. (1992, September 9-11). *A dummy for rear end collisions: development and validation of a new dummy neck.* Paper presented at the International IRCOBI Conference on the Biomechanics of Impacts, Verona, Italy.

Svensson, M., Lovsund, P., Haland, Y., & Larsson, S. (1993). Rear-end collisions: a study of the influence of backrest properties on head-neck motion using a new dummy. *SAE Tech Paper Series, 930343,* 129-142.

Szabo, T., JB, W., & Anderson, R. (1994). Human occupant kinematic response to low speed rear-end impacts. *SAE Tech Paper Series, 940532,* 23-35.

Szabo, T., & Welcher, J. (1996). Human subject kinematics and electromyographic activity during low speed rear impacts. *SAE paper 962432,* 295-315.

West, D., Gough, J., & Harper, T. (1993). Low speed collision testing using human subjects. *Accid Reconstruct J, 5*(3), 22-26.

Zellmer, H., Muser, M., Stamm, M., Walz, F., Hell, W., Langweider, K., et al. (2002). *Performance comparison of rear impact dummies: Hybrid III (TRID), BioRID and RID 2.* Paper presented at the International IRCOBI Conference on the Biomechanics of Impact, Munich, Germany.

CONTACT

Dr. Arthur C. Croft the Spine Research Institute of San Diego. 2371 Via Orange Way, Suite 105, Spring Valley, California, 91978, USA. E-mail: drcroft@srisd.com.

983156

Head-Neck Kinematics in Dynamic Forward Flexion

Bing Deng
Computer Engineering Technologies

John W. Melvin and Stephen W. Rouhana
General Motors Research and Development Center

ABSTRACT

Two-dimensional film analysis was conducted to study the kinematics of the head and neck of 17 restrained human volunteers in 24 frontal impacts for acceleration levels from 6g to 15g. The trajectory of the head center of gravity relative to upper torso reference points and the rotation of head and neck relative to the lower torso during the forward motion phase were of particular interest. The purpose of the study was to analyze the head-neck kinematics in the mid-sagittal plane for a variety of human volunteer frontal sled tests from different laboratories using a common analysis method for all tests, and to define a common response corridor for the trajectory of the head center-of-gravity from those tests.

INTRODUCTION

As the structural linkage between the head and chest of an anthropomorphic dummy, the neck plays a vital role in determining the dynamics of the head during a crash and in determining whether or not a head impact will occur. Dummy necks must have humanlike response characteristics to be useful for evaluating automotive occupant protection systems.

Performance requirements of neck structures were proposed by Mertz and Patrick in the early seventies [1, 2]. They conducted a series of sled tests using a single male volunteer and six embalmed cadavers. A rigid seat, a lap belt and a criss-crossed pair of shoulder straps were used to restrain the subjects. Neck response was defined by the relationship between the moment at the occipital condyles and the rotation of the head relative to the torso. These requirements are necessary but not sufficient to describe humanlike neck response.

Three head motion requirements were suggested by Melvin et al. in 1973 [4] for dummy neck performance: 1) the range of angular motion of the head relative to the torso, 2) the trajectory of the center of gravity of the head relative to the torso, and 3) the resistance of the neck to motion of the head during loading and rebound.

In 1984, Wismans and Spenny [5] analyzed human head-neck motion in frontal flexion based on eleven tests conducted by the Naval Biodynamics Laboratory (NBDL), with two human volunteers. They introduced a linkage system with two ball and socket joints. The head motions were defined relative to the first thoracic vertebral body (T1), while the rotations of the T1 reference were neglected. Thunnissen et al. [6] improved the analysis by correcting the errors due to T1 instrumentation mount rotation.

Realistic head center of gravity trajectories are critical to improve dummy neck designs. Because of the difficulties in determining the head trajectory relative to the torso, due to the lack of a well-defined reference point, the recommendations of Melvin et al. have not, in general, been realized. There have also been differing opinions as to the extent to which the upright position of the Navy volunteers produces a different range of motion from the range that might occur in a real automotive crash where the occupants are typically more reclined. An analysis of the human head-neck response of a variety of volunteer tests using a single analysis method was needed to investigate and resolve these issues.

The objective of this study is to analyze the human head-neck kinematics in dynamic frontal flexion with primary emphasis on forward motion in the mid-sagittal plane. A variety of human volunteer frontal sled tests from different laboratories were analyzed. Data are presented with attention to the position of the head center of gravity and the head-neck rotation relative to the torso. Comparative measures of response were estimated for all tests, and a single neck response specification is introduced for the automotive environment.

HUMAN VOLUNTEER DATABASE

Three different databases of frontal impacts with male subjects were analyzed:

1. Eight tests with seven military volunteers done at the Naval Biodynamics Laboratory (NBDL) at up to 15-g acceleration levels.

2. Nine tests with nine military volunteers done at Air Force Research Laboratory (AFRL) at up to 10-g acceleration levels.

3. Seven tests with two student volunteers done at Wayne State University (WSU) at up to 10-g acceleration levels.

All three databases were run on horizontal HYGE-type accelerators. The volunteers were fairly representative in height and mass of the 50th percentile adult male. The general protocols used for the experiments conducted in these three studies have been published in the open literature (see the specific references below). The AFRL and WSU tests have never been analyzed for the purpose of defining head-neck kinematics, however.

In the NBDL tests, the subjects were restrained in an upright position by left and right shoulder straps; a lap belt and an inverted V-pelvic strap tied to the lap belt (Figure 1). Upper arm and wrist restraints were used to prevent flailing. The tests were designed as the "neck up, chin up" configuration [7].

Figure 1. NBDL test setup with "neck up, chin up" configuration. Test 3890.

In the AFRL tests, a conventional double shoulder strap and lap belt harness restrained the subjects (Figure 2). A generic rigid seat with the seat back reclined 13° from vertical and the seat pan inclined 6° upward from horizontal was used [8].

In the WSU tests, a 1978-production passenger car seat and three-point lap-shoulder belt restraint system with inertial reel were used to contain the occupant. The car toe board was simulated to position the feet and legs of

the test subject similar to that of a car occupant. The lap belt was worn snugly and the shoulder belt was slack on the inertial reel (Figure 3). A redundant torso belt was used as one of the safety back up features [9]. Table 1 shows the experiment numbers, the peak sled acceleration and maximum velocity, and the subject I.D. numbers, height and weight. Twenty-four experiments from the three different research institutes were analyzed.

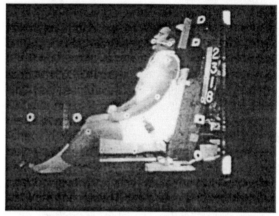

Figure 2. AFRL test setup. Test 2318.

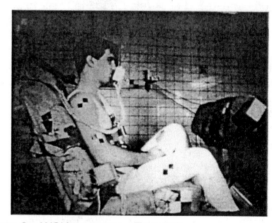

Figure 3. WSU test setup. Test 142.

ANALYSIS METHOD

High-speed 16 mm movie films containing the aforementioned tests were obtained from the University of Michigan Transportation Research Institute (NBDL tests), the Air Force Research Laboratory (AFRL tests), and Wayne State University (WSU tests). Films were digitized using a NAC 160F motion analyzer. Onboard lateral camera views were analyzed using planar photogrammetric methods. Two points a known distance apart were digitized to provide a scale factor to convert the digitized units into physical displacement units. All films had a timing mark that operated at 100 Hz. The film speeds of NBDL, AFRL, and WSU databases were 500 frames per second, 480 and 500 frames per second, and 350 frames per second, respectively. Two targets on the sled were digitized to account for film translational and rotational jitters. Only motions in the mid-sagittal plane were considered in this study.

Since the various human volunteer tests were obtained from three different laboratories, each group of tests had particular objectives. Photo targets were placed differently for the three databases. A common analysis method, which was suitable to measure the head-neck kinematics relative to the torso, was developed to accommodate the differences. This method used the configuration in the time zero frame as a template, locating the target positions on this template, and identifying corresponding marker locations in the subsequent motion. Digitized targets included two markers on the head instrumentation mount (H1 and H2); two approximated head anthropometric markers: the occipital condyles (O.C.) and the tragion (T.G.); two markers on the neck anterior contour: the highest visible point (NF2) and the lowest visible point (NF1); two markers on the neck posterior contour: the highest visible point (NB2) and the lowest visible point (NB1); and the shoulder belt to upper torso contact point (S.H.). Two markers (Torso1 and Torso2) along the lower torso contour (parallel to the seatback orientation initially) provided the lower torso rotation. Also digitized were two markers on the sled (Sled1 and Sled2) as the fixed translational and rotational references. Figure 4 illustrates the thirteen targets on a schematic template.

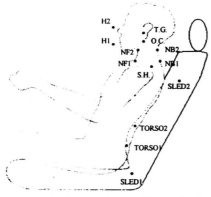

Figure 4. A schematic template to illustrate the digitized targets for the head-neck kinematics relative to the torso in frontal impacts.

The targets were digitized at a rate of every 2 film frames. The accuracy of the target position location was determined to be 0.4 percent (0.4 mm over 102 mm gage length). Some of the targets disappeared momentarily due to obstructions to vision during the motion. Such small data gaps were filled by cubic curve fitting. Position data was smoothed with a low pass filter at 20 Hz cutoff frequency. This filtering level was chosen to eliminate discontinuities in the data while maintaining the overall quality of the position information. Data were expressed in the sled coordinate system, where positive X is the posterior-anterior direction and positive Z is the inferior-superior direction.

Table 1.   Human Volunteer Frontal Impacts

Naval Biodynamics Laboratory					
Test No.	Peak Acc	Max Vel	Sub. ID	Height	Weight
	g	m/s		cm	kg
LX 3890	8.2	11.97	134	178.3	74.8
LX 3983	15.6	17.55	134	178.3	74.8
LX 3889	8.2	12.02	130	180.1	72.6
LX 3894	8.4	12.07	131	167.0	67.6
LX 3997	8.1	12.12	132	172.9	79.8
LX 3895	8.2	12.00	133	161.7	61.2
LX 3898	8.3	12.15	135	171.6	68.9
LX 3901	7.9	11.88	136	185.4	88.5

Air Force Research Laboratory					
Test No.	Peak Acc	Max Vel	Sub. ID	Height	Weight
	g	m/s		cm	kg
A2340	9.82	13.61	M-15	166.1	63.5
A2318	9.79	13.66	B-6	177.8	77.6
A2252	9.59	13.58	T-1	167.9	75.8
A2251	9.57	13.56	L-4	182.9	83.9
A2320	9.79	13.61	S-3	176.8	79.8
A2263	9.58	13.61	T-2	180.1	73.0
A2062	9.47	13.33	J-3	179.1	75.8
A2071	9.45	13.36	P-4	162.1	57.6
A2075	7.83	11.45	B-4	175.3	90.7

Wayne State Univ. Bioengineering Center					
Test No.	Peak Acc	Max Vel	Sub. ID	Height	Weight
	g	m/s		cm	kg
W81	6.1	7.06	JL	178.5	74.8
W140	6.0	6.84	TC	177.8	68.0
W141	6.0	6.84	TC	177.8	68.0
W142	6.8	7.69	TC	177.8	68.0
W143	7.5	8.31	TC	177.8	68.0
W144	8.1	9.16	TC	177.8	68.0
W145	9.6	10.06	TC	177.8	68.0

An anatomical head coordinate system was established to obtain the head center of gravity location from the digitized tragion position. This body fixed frame originated at the tragion with the X-axis pointing to the infraorbitale and the Z-axis perpendicular to the X-axis (Figure 5). Based on 50th percentile male anatomy, the head center of gravity is located 32 mm upward and 6 mm forward relative to the tragion in this moving frame [10]. A coordinate transformation was used to obtain the instantaneous head center of gravity location relative to the tragion in the fixed sled coordinate system.

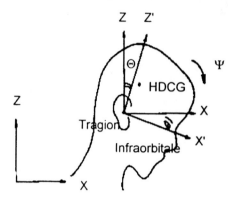

sled coordinate system

Figure 5. Coordinate transformation between the head anatomical coordinate system and the sled coordinate system.

$$X = 6.0 \times \cos(\theta+\Psi) + 32.0 \times \sin(\theta+\Psi)$$

$$Z = -6.0 \times \sin(\theta+\Psi) + 32.0 \times \cos(\theta+\Psi)$$

Where θ is the initial angle between the head anatomical coordinate system and the sled coordinate system, and Ψ is the head rotation in the subsequent motion (positive value for flexion and negative value for extension). The head center of gravity position in the sled reference system is then obtained as:

$$X_{CG} = X + X_{TRAGION}$$

$$Z_{CG} = Z + Z_{TRAGION}$$

Head rotation was calculated from the two markers on the head instrumentation mount (H1 and H2). Two markers along the lower torso contour (Torso1 and Torso2) provided the lower torso rotation. A mid-point of neck thickness (N.M.) was calculated by averaging the four neck targets: NF1, NF2, NB1, and NB2. The vector from the approximate occipital condyles target (O.C.) to this neck mid-point (N.M.) represented the neck rotation. C++ programs were developed to carry out the calculations. Time-history data of the head rotation, the neck rotation, the lower torso rotation, and the absolute head center of gravity positions in the sled reference frame were obtained (Figure 6).

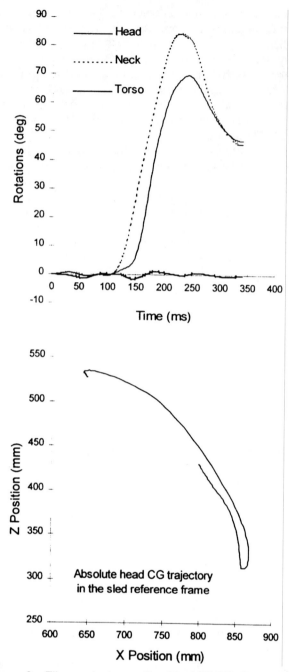

Figure 6. Film analysis results of test N3889. Data plots included the head rotation, neck rotation, lower torso rotation, and the absolute head CG trajectory relative to the sled reference frame.

RESULTS

HEAD CG TRAJECTORY RELATIVE TO THE TORSO – The trajectory of the head center of gravity relative to the torso is important for defining dummy neck performance requirements. A common torso reference for a variety of tests needs to be determined to represent realistic head trajectory during head-neck dynamic flexion. There were some difficulties in uniquely determining the head center of gravity trajectory relative to the torso. If the torso has any translation and/or rotation, then a torso reference must be identified in order to obtain the head trajectory associated with head-neck motion only. All tests were reviewed and several approaches were developed to separate the head-neck trajectory from the whole body trajectory, that is, to examine that motion of the head CG solely controlled by the neck.

Approach 1 – The shoulder belt to upper torso contact point was chosen as the reference for the displacement of the head center of gravity. This torso reference followed the upper torso translational motion. As the torso became restrained by the shoulder belt(s), significant head-neck forward rotation began to occur. Accordingly, the contact point between the shoulder(s) and the shoulder belt(s) became fixed relative to the sled. The head center of gravity trajectory during head-neck motion could then be obtained by eliminating the translational motion of the upper torso. Figure 7 presents the paths of the head center of gravity relative to the shoulder reference point during the loading phase of twenty-four volunteer tests.

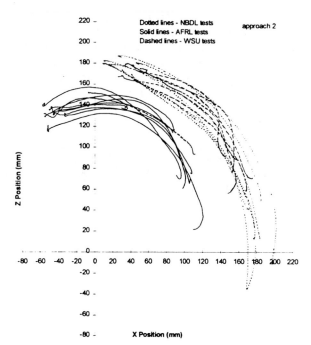

Figure 8. Head CG trajectories relative to the lower neck mid-point.

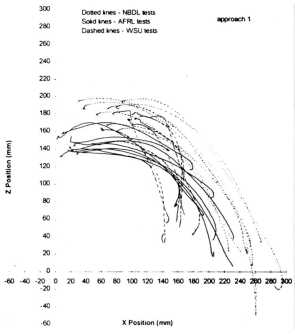

Figure 7. Head CG trajectories relative to the shoulder belt to upper torso contact point.

Approach 2 – A lower neck mid-point was generated as the torso reference by averaging the lower neck markers NF1 and NB1. This reference represented the visible base of the cervical spine. The head center of gravity trajectory expressed relative to this mid-point would represent the head-neck kinematics relative to the torso during pure dynamic flexion. However, this reference location was affected by changes in the neck shape, due to muscle activity, during the motion. The trajectories of the head center of gravity relative to the lower neck mid-point during dynamic forward flexion are shown in Figure 8.

Approach 3 – Because of the problems associated with neck shape change at the base of the neck, a third approach used an alternative method to determine a neck base point. For each test, an initial neck length was calculated as the distance between the occipital condyles and the mid-point of NF1 and NB1 in the first film frame (Figure 9). Each length was considered a constant for the rest of the analysis of each particular test. The neck mid-point, N.M., was obtained by averaging the four neck targets: NF1, NF2, NB1, and NB2. For each subsequent frame, a vector from the approximate occipital condyles target (OC) to the neck mid-point, (N.M.) was extended using the initial neck length to obtain the neck base point. This reference point reflected the motion of the estimated neck-torso joint and the reference location was not affected by the neck shape change. However, this reference position was dependent on the initial visibility of the base of the neck and was found to be affected by the different test setups. The relative head center of gravity trajectories obtained by this method are presented in Figure 10.

With the exception of a single test, Figures 8 and 10 show that the NBDL and WSU data fall into a single trajectory corridor using either Approach 2 or 3. The AFRL data appears to be a different corridor, however. As shown in Figure 2, the neck base in the AFRL tests was obscured to some extent by the clothing worn by the subjects. This obscuration led to an estimated neck base point that was closer to the head CG than in the other groups. Although an arbitrary estimate of the actual position of the neck base point about 50 mm lower than those used would appear to be supported by the separation of the data in both Figure 8 and 10, such a shift was not used, in part because of the obvious shift in starting positions for those tests. However, the important result from these two methods is that the trajectories of the head CG, when separated from the translational and rotational motion of the thoracic spine and other whole body torso motions, follow arcs that appear to fall into a single corridor.

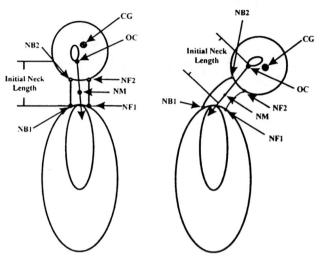

Figure 9. Schematic representation of Approach 3.

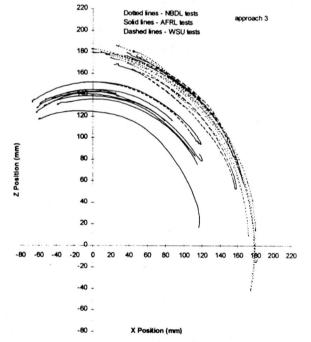

Figure 10. Head CG trajectories relative to the neck base point (using the initial neck length).

To obtain comparable measurements of head-neck kinematics relative to the torso and to resolve a single response specification for a variety of tests, the absolute head center of gravity trajectory in the sled coordinate system was studied. The instantaneous centers of curvature and curvature radii were calculated for each test (Figure 11). Based on the head center of gravity positions at frame i-1, i, and i+1 (AI-1, AI, AI+1), the instantaneous center of curvature was obtained by solving the intersection of two lines: one was the mid-length perpendicular line defined by (AI-1, AI,), and the other was the mid-length perpendicular line defined by (AI, AI+1). The curvature radius was defined as the distance between the position Ai and the corresponding center of curvature. The occupant head kinematics during dynamic forward flexion could be described as initial translation followed by rotation. Average values of the instantaneous centers of curvature and curvature radii were calculated for each test during the rotational phase. An averaged circular arc appeared to be a quite reasonable approximation for each absolute head trajectory during the rotational motion. Figure 12 illustrates the curvature analysis results for each experiment, where the solid line is the absolute trajectory of the head center of gravity during the loading phase and the shaded line is the fitted averaged circular arc.

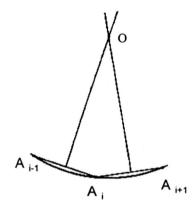

Figure 11. Center of curvature analysis.

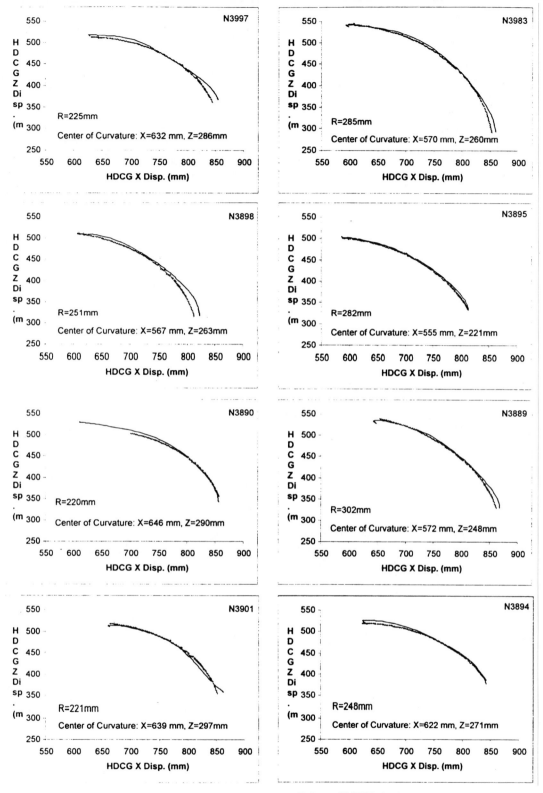

Figure 12a. Circular arc curve fitting of NBDL tests.

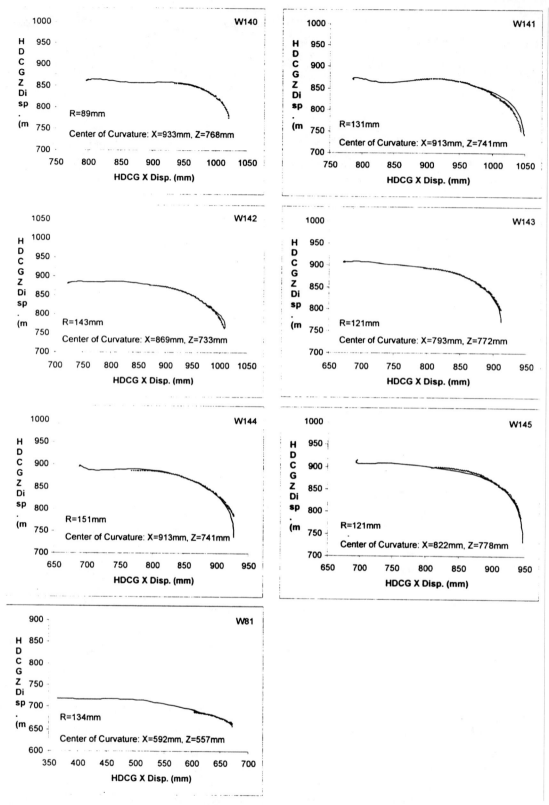

Figure 12b. Circular arc curve fitting of WVU tests.

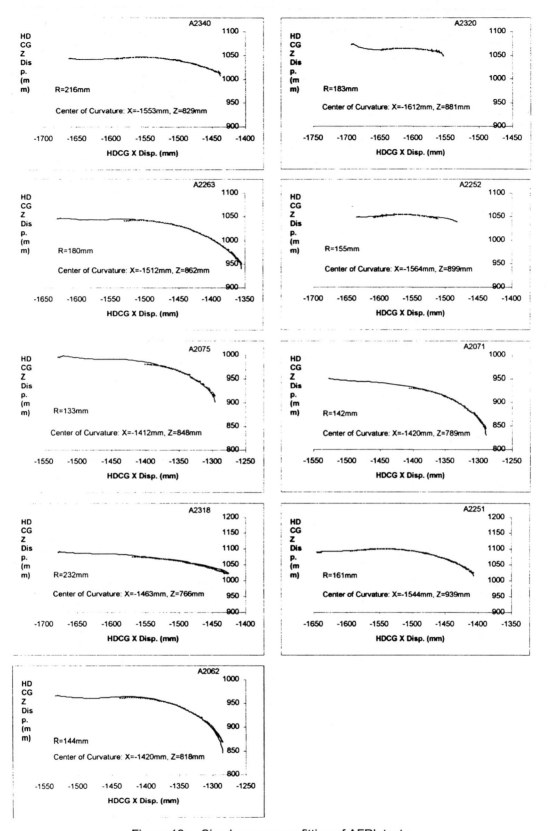

Figure 12c. Circular arc curve fitting of AFRL tests.

53

The average radius for the WSU subjects was 127 mm (89-151 mm range, 20 mm S.D.), for the AFRL subjects was 172 mm (133-247 mm range, 34 mm S.D.), and for the NBDL subjects was 254 mm (221-302 mm range, 32 mm S.D.). An analysis of variance of the radii indicated that the group means are significantly different (P < 0.05) and a Newman-Keuls multiple contrasts analysis indicated all of the means were different from one another. The mean value of the combined radii from all three groups is 186 mm (60 mm S.D.).

The distance from the head CG to the C7/T1 joint for the mid-sized male is 177 mm (UMTRI, 1983). Since the average radii for the WSU and AFRL subjects is less than that value, it appears that those subjects had head CG trajectories that were controlled by the cervical spine and its musculature alone. Conversely, the average radius of the NBDL subjects was well above that value, which suggests that their head motions were produced, in part, by motion in the thoracic spine below the cervical spine.

Approach 4 – Since the goal of this project was to define a common response corridor for the trajectory of the average male head center of gravity from the volunteer tests, it was decided to use an average male neck length as the initial neck length. This was defined as the distance between the occipital condyles and the cervical spine base (C7/T1) and was chosen to be 125 mm for all subjects. This 125 mm length is approximately the active length of the Hybrid III dummy neck and is similar to the initial neck length used by Thunnissen et al. [6] in their analysis of the NBDL data. It is also consistent with the mean radius of curvature of the three groups (186 mm). Thus, using the method of the third approach, but changing the constant from the initial neck length to 125 mm, a new neck base point was generated to serve as the torso reference point in Approach 4. Figure 13 shows the relative head CG trajectories of twenty-four tests using this approach (Approach 4).

The human cervical spine has lordotic curvature in its normal state. The seatback angle and the test configuration affect the initial head-neck orientation. If the head is to remain approximately level while the torso is inclined along the seat, the neck angle should represent the natural curvature of the human cervical spine. The typical head-neck configurations of the three databases were "head/neck flexed forward" for WSU volunteers, "head/neck extended backward" for AFRL subjects, and "head extended/neck flexed" for NBDL volunteers. Angular adjustment of the head and cervical spine is necessary to obtain the trajectory range for comparable test configurations. The head/neck angle is defined as the angle between the vector from the neck base point (Approach 4) to the head center of gravity and the vertical direction. The occupant head/neck inclined 5° forward from vertical was chosen as the standard configuration for this analysis. The starting positions of the relative head center of gravity trajectories obtained using the fourth approach were shifted (rotated along the trajectory) based on the differences between the subjects' initial head/neck angular positions and the standard head/neck orientation. Table 2 shows the rotation angles subtracted or added for each test, where positive values represent flexion angle subtracted and negative values represent extension angle added.

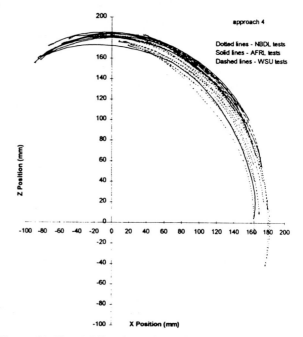

Figure 13. Head CG trajectories relative to the neck base point (using 125 mm).

The initial seatback angle also plays an important role in determining the range of the head trajectory. An upright seating position could produce more forward angular head/neck motion. Among the twenty-four analyzed tests, the WSU database used a standard car seat with the seat back inclined 24° rearward from vertical. The NBDL and the AFRL databases used rigid seats with 0° and 13° seatback angles respectively. The typical automotive test configuration of a seat back inclined 24° rearward from vertical (WSU seatback angle) was chosen as the standard configuration for this analysis. The ending positions of the relative head center of gravity trajectories obtained using the fourth approach were reduced by 24° of angular rotation (along the trajectories) for the NBDL tests and 11° angular rotation (along the trajectories) for the AFRL experiments. Figure 14 presents the head CG trajectories relative to the cervical spine base point (obtained using the fourth approach, for a typical automotive test configuration (the head/neck inclined 5° forward and a seat back inclined 24° rearward), for twenty-four human volunteer frontal impacts.

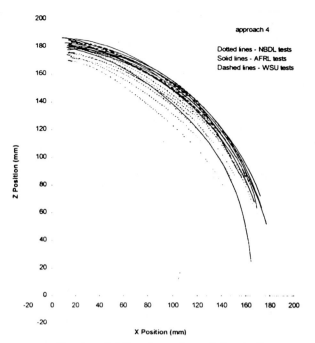

approach 4

Dotted lines - NBDL tests
Solid lines - AFRL tests
Dashed lines - WSU tests

Figure 14. The head CG trajectories relative to the cervical spine base point (using 125 mm) for a typical automotive test configuration (the head/neck inclined 5° forward and a seat back inclined 24° rearward).

Table 2. Initial head/neck angles used to adjust starting points of HDCG trajectories ("+" flexion, "-" extension).

NDL Test No.	H/N ∠ deg	AFRL Test No.	H/N ∠ deg	WSU Test No.	H/N ∠ deg
LX 3890	5	A2340	26	W81	9
LX 3983	5	A2318	35	W140	-18
LX 3889	-8	A2252	32	W141	-7
LX 3894	-3	A2251	33	W142	-3
LX 3997	2	A2320	31	W143	-7
LX 3895	1	A2263	26	W144	-10
LX 3898	-1	A2062	30	W145	-11
LX 3901	-8	A2071	22		
		A2075	20		
Mean	-1		28		-7
S.D.	5		5		8

The data shown in Figure 14 were analyzed to define a common response corridor for the trajectory of the center of gravity of the average male head, statistically. For each X position, ranging from 15 mm to 200 mm with 5 mm increments, the mean values and standard deviations of Z were obtained from the twenty-four tests. For each averaged Z position, the standard deviations of the X values were calculated. The averaged head CG trajectory,

with X and Z deviations, is shown in Figure 15. The corridor determined by Thunnissen et al. [6] for the NBDL data only is also shown.

Approach 5 – The individual absolute head CG trajectory plots shown in Figure 12 exhibit a characteristic in which the final stage of the trajectory of most of the subjects can be fitted with a constant, but different, radius of curvature. The potential influences of the seat back angle, restraint system type, initial head/neck posture and the degree of muscle activity may have their greatest effect on the initial stages of the head motion and lesser effects as the limit of forward motion is reached. Accordingly, it was decided to normalize the trajectories by aligning the final endpoints of each subject as a technique to define the average absolute trajectory of all the subjects with no assumptions of a body reference point. The adjusted trajectories are shown in Figure 16. This transformation of the data appears to produce an overlapping of the responses that is less sensitive to the initial effects mentioned above. The data shown in Figure 16 were analyzed statistically. For each X position, ranging from -240 mm to 0 mm with 5 mm increments, the mean values and standard deviations of Z were obtained from the twenty-four tests. For each averaged Z position, the standard deviations of the X values were calculated. The averaged head CG trajectory with X and Z deviations are shown in Figure 17. The Approach 5 technique was applied to the head O.C. data (Figure 18) and produced similar results to that of the head CG data. The trajectories shown in Figure 18 were analyzed statistically also. The averaged head O.C. trajectory with X and Z deviations are shown in Figure 19.

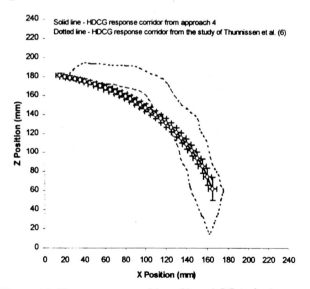

Solid line - HDCG response corridor from approach 4
Dotted line - HDCG response corridor from the study of Thunnissen et al. (6)

Figure 15. The average corridor of head CG trajectory relative to the cervical spine base point (using 125 mm) from twenty-four volunteer frontal impacts.

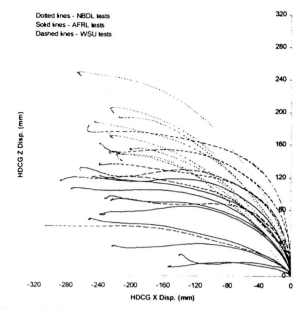

Dotted lines - NBDL tests
Solid lines - AFRL tests
Dashed lines - WSU tests

Figure 16. The absolute head CG trajectories relative to each individual CG ending point of loading phase.

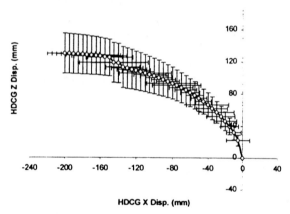

Figure 17. The average corridor of absolute head CG trajectories relative to each individual ending point of the loading phase.

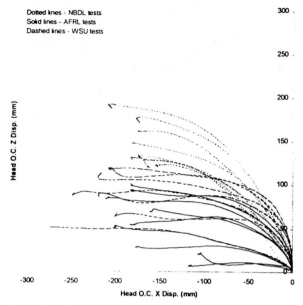

Dotted lines - NBDL tests
Solid lines - AFRL tests
Dashed lines - WSU tests

Figure 18. The absolute head OC trajectories relative to each individual OC ending point of loading phase.

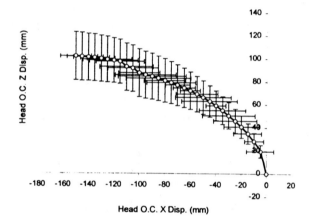

Figure 19. The average corridor of absolute head O.C. trajectories relative to each individual ending point of the loading phase.

A comparison of the average responses for the head CG trajectories found by Approach 4 and Approach 5 was made by shifting the response shown in Figure 15 using the ending point as a reference. The result is shown in Figure 20 along with the shifted O.C. data from Figure 19. The different ending positions between head CG and head O.C. were calculated for each of the twenty-four tests. The averaged value, which is x = -39 mm and Z = −29 mm, were used to shift the ending point of head O.C. trajectory. The two methods for determining the head CG trajectory produced similar results during the rotation phase of the head CG motion.

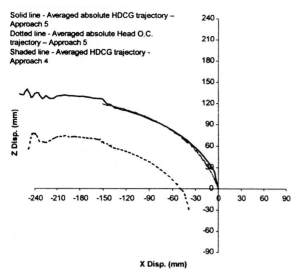

Figure 20. Comparison of Approaches 4 and 5.

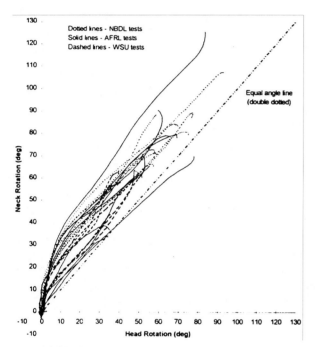

Figure 21. Neck rotation vs. head rotation.

HEAD-NECK ROTATIONS RELATIVE TO THE TORSO – Head rotation was calculated from the two markers on the head instrumentation mount (H1 and H2). Two markers along the lower torso contour (Torso1 and Torso2) provided the lower torso rotation. A mid-point of neck thickness (N.M.) was calculated by averaging the four neck targets: NF1, NF2, NB1, and NB2. The vector from the approximate occipital condyles target (O.C.) to this neck mid-point (N.M.) represented the neck rotation.

The neck rotation is plotted as a function of the head rotation in Figure 21. Figure 22 presents the relation between the relative head rotation and the relative neck rotation using the lower torso as a common reference angle. The two figures are quite similar due to little rotation of the restrained lower torso. Note that, with one exception, the neck rotation is always greater than the head rotation. The data below 40 degrees tends to separate into those curves, which tend to follow the equal angle line, and those that rise rapidly in neck angle and then run parallel to the equal angle line.

To define a common response corridor for the rotational response, the data shown in Figure 21 were calculated statistically. For each head rotation, ranging from 0° to 90° with 2° increments, the mean values and standard deviations of neck rotation were obtained from the twenty-four tests. For each averaged neck rotation, the standard deviations of the head rotation were calculated. The averaged neck rotation as a function of head rotation with standard deviations is shown as the solid line in Figure 23.

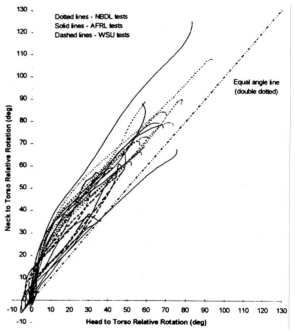

Figure 22. Neck-to-lower torso rotation vs. head-to-lower torso rotation.

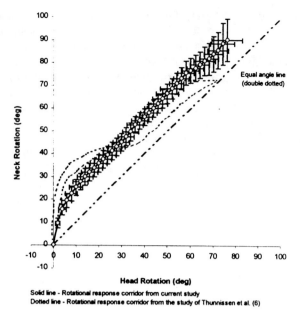

Solid line - Rotational response corridor from current study
Dotted line - Rotational response corridor from the study of Thunnissen et al. (6)

Figure 23. The average corridor of neck rotation as a function of head rotation.

DISCUSSION

Absolute head CG trajectory in the sled reference frame shows that after translational motion ends, the motion of the head is predominantly a circular arc. The translational components to the motion, which were separated out by Approaches 4 and 5, were caused by a number of factors. The primary factor appears to have been the type of restraint system, with the three-point belt system allowing for the most upper torso motion. In contrast, the double shoulder harnesses in the NBDL and AFRL tests held onto the shoulders more tightly and resulted in the initiation of head rotation sooner than in the WSU tests.

In addition, judging from the radii of curvature, the NBDL subjects had more upper torso flexion than the AFRL and WSU subjects. The shorter radius of curvature of the WSU subjects is understandable, since a three-point shoulder belt crosses the chest at mid-sternum which should reduce the ability of the thoracic spine to bend forward. That is, the motion would be constrained to occur higher in the spine than in the thoracic segment. The shoulder belts of the AFRL subjects were connected to the lap belts at the centerline of the subject. As such they made a "V" shape from the shoulders to just below the waist. This would have restricted the ability of the thoracic spine to bend forward because the shoulder belts covered a large part of the mid-sagittal section of the rib cage. In contrast, the shoulder belts of the NBDL subjects were parallel from the shoulders down to the lap belts. The parallel shoulder belts restrained the shoulders and lateral rib cage, but appear to have allowed the central portion of the rib cage to move forward somewhat before being stopped. This likely accounts for the increased mobility of the thoracic spine relative to the other two test series and for the significantly larger radii of curvature observed in the absolute trajectory analysis.

Additional factors influencing the trajectory differences may be the upright seating position and the experience of the NBDL volunteers due to repeated testing over time. In any case, the NBDL volunteers exhibited radii of curvature of head motion that project well down into the thoracic spine, while the other two data sources produced motions based in the cervical spine only.

The method of analysis chosen here (Approach 4) to normalize the data to the average male neck length allows the trajectory data from the NBDL tests to be referenced to a neck point that provides data that can be compared to the other data sources (Figure 13). Finally, adjustment of the resulting trajectories from all the tests to a standard position that accounts for initial head/neck orientation and torso inclination has produced a set of responses that can be statistically analyzed (Figure 14). The resulting corridor in Figure 15 defines, for the mid-sized male, the head center of gravity trajectory under the test conditions of the volunteer experiments normalized to the automotive test condition. This corridor lies near that found by Thunnissen et al. [6] for higher severity NBDL tests. Approach 5, using the absolute trajectory data normalized for end point position, adds support to the results of Approach 4. As shown in Figure 20, the two approaches produce very similar results in terms of an average radius of curvature of the final, rotational phase of the head CG trajectory.

The relationship between neck angle and head angle is another feature of the NBDL tests that has provided much discussion in its application to dummy neck performance specifications. As seen in Figures 21 and 22 there can be a variety of responses to this relationship, sometimes referred to as "head lag". The averaged data shown in Figure 23 indicates that, on average, the neck angle increases more rapidly than the head angle for the first 25 degrees of neck rotation. In contrast, the NBDL data analyzed by Thunnissen et al. [6] (Figure 23) shows a much more rapid rise in neck angle that levels off to an equal angle slope after 40 degrees of neck rotation. Because this relationship may be dependent on the initial attitude of the subjects' head and neck and the state of muscle activity prior to the impact, it is difficult to specify an appropriate head/neck angle response corridor for an automotive test dummy neck. Common practice in setting up and positioning a crash test dummy is based on an average seated position of a normally seated occupant with no awareness of an impending crash. As such, the inclusion of any features that involve voluntary action on the part of the occupant, either due to muscle pre-tensing or head pre-positioning, is not consistent with current practice. Dummy necks do include postural neck muscle activity in their bending response, which tends to produce a neck angle versus head angle near the line of equal angle, as do some of the volunteers.

CONCLUSIONS

The head motions in twenty-four tests with seventeen belt-restrained human volunteers in frontal impacts under three different sets of initial conditions in three different laboratories were analyzed using a common method. The head-neck kinematics in the mid-sagittal plane was studied. The head CG trajectories were studied in an absolute (sled) reference frame and using four different body reference points. The apparent differences in trajectory and head attitude were accounted for and normalized to produce a single average response specification for head trajectory was produced using the fourth approach. The data generated from this study provide new information on the mid-sized male head center of gravity trajectory as controlled by the neck and head/neck angular motion relative to the torso in frontal impacts with a belt restrained torso.

ACKNOWLEDGEMENTS

This study is financed by GM pursuant to an agreement between GM and the U.S. Department of Transportation. The authors wish to express their special thanks to Mr. John Buhrman at Air Force Research Laboratory, Dr. Sal Guccione at the Naval Biodynamics Laboratory, and Dr. Paul Begeman at Wayne State University for providing detailed experimental information. The authors would also like to acknowledge the suggestions and advice on methods to analyze the experimental data offered by the members of the SAE Human Mechanical Response and Injury Criteria Subcommittee (HMRICS).

REFERENCES

1. H.J. Mertz and L.M. Patrick "Investigation of the Kinematic and Kinetics of Whiplash". Proceedings of the 11th Stapp Car Crash Conference, SAE 670919, 1967.
2. H.J. Mertz and L.M. Patrick "Strength and Response of the Human Neck". Proceedings of the 15th Stapp Car Crash Conference, SAE 710855, 1971.
3. H.J. Mertz, R.F. Neathery, and C.C. Culver "Performance Requirements and Characteristics of Mechanical Necks". Human Impact Response, PP 263-288, 1973.
4. J.W. Melvin, J.H. McElhaney, and V.L. Roberts "Evaluation of Dummy Neck Performance". Human Impact Response, PP 247-261, 1973.
5. J. Wismans and C.H. Spenny "Head-Neck Response in Frontal Flexion". Proceedings of the 28th Stapp Car Crash Conference, SAE 841666, 1984.
6. J. Thunnissen, J. Wismans, C.L. Ewing, and D.J. Thomas "Human Volunteer Head-Neck Response in Frontal Flexion: A New Analysis". Proceedings of the 39th Stapp Car Crash Conference, SAE 952721, 1995.
7. C.L. Ewing and D.J. Thomas "Torque versus Angular Displacement Response of Human Head to -Gx Impact Acceleration". Proceedings of the 17th Stapp Car Crash Conference, SAE 730976, 1973.
8. D. Ma, L.A. Obergefell, and A.L. Rizer "Development of Human Articulating Joint Model Parameters for Crash Dynamics Simulations". Proceedings of the 39th Stapp Car Crash Conference, SAE 952726, 1995.
9. P.C. Begeman, R.S. Levine, and A.I. King "Belt Slip Measurements on Human Volunteers and the Part 572 Dummy in Low -Gx Impact Acceleration". Proceedings of the 27th Stapp Car Crash Conference, SAE 831635, 1983.
10. Anthropometry of Motor Vehicle Occupants, UMTRI, 1983.

Traffic Injury Prevention, 7:61–69, 2006
Copyright © 2006 Taylor & Francis Group, LLC
ISSN: 1538-9588 print / 1538-957X online
DOI: 10.1080/15389580500413000

Taylor & Francis
Taylor & Francis Group

Influence of Seat Geometry and Seating Posture on NIC$_{max}$ Long-Term AIS 1 Neck Injury Predictability

LINDA ERIKSSON

Autoliv Sverige AB, Vårgårda, Sweden; Applied Mechanics, Chalmers University of Technology, Göteborg, Sweden

ANDERS KULLGREN

Folksam Research, Stockholm, Sweden

Objective. *Validated injury criteria are essential when developing restraints for AIS 1 neck injuries, which should protect occupants in a variety of crash situations. Such criteria have been proposed and attempts have been made to validate or disprove these. However, no criterion has yet been fully validated. The objective of this study is to evaluate the influence of seat geometry and seating posture on the NIC$_{max}$ long-term AIS 1 neck injury predictability by making parameter analyses on reconstructed real-life rear-end crashes with known injury outcomes.*

Methods. *Mathematical models of the BioRID II and three car seats were used to reconstruct 79 rear-end crashes involving 110 occupants with known injury outcomes. Correlations between the NIC$_{max}$ values and the duration of AIS 1 neck injuries were evaluated for variations in seat geometry and seating posture. Sensitivities, specificities, positive predictive values, and negative predictive values were also calculated to evaluate the NIC$_{max}$ predictability.*

Results. *Correlations between the NIC$_{max}$ values and the duration of AIS 1 neck injuries were found and these relations were used to establish injury risk curves for variations in seat geometry and seating posture. Sensitivities, specificities, positive predictive values, and negative predictive values showed that the NIC$_{max}$ predicts long-term AIS 1 neck injuries also for variations in seat geometry and seating postures.*

Conclusion. *The NIC$_{max}$ can be used to predict long-term AIS 1 neck injuries.*

Keywords   Rear Impacts; Neck Injury Criteria; Injury Probability; Mathematical Modelling; NIC$_{max}$

Soft-tissue neck injuries sustained in rear-end crashes are rising in both number and frequency (von Koch et al., 1994; Morris & Thomas, 1996; Kullgren et al., 2002). Most occupants sustaining these types of injuries recover within a month, but one out of ten of the injured suffer from symptoms for longer than one month (Nygren et al., 1985). A way to reduce the number of AIS 1 neck injuries, or at least to shorten the duration of the symptoms, is to equip cars with restraints that decrease the risk of sustaining these injuries. Development of restraints requires accurate models of the human, the car, and the crash; validated criteria identifying harmful and harmless crash situations are essential. Knowledge about the injuries, and the factors influencing the injury risk, facilitates the development of restraints that protect occupants in a variety of crash situations.

Our knowledge of AIS 1 neck injuries is insufficient: the injury mechanisms are still not fully understood, however epidemiological studies have identified factors that influence the in-

Received 28 November 2004; accepted 7 July 2005.
Address correspondence to Linda Eriksson, Applied Mechanics, Chalmers University of Technology, SE-412 96 Göteborg, Sweden. linda. E-mail: linda.eriksson@me.chalmers.se

jury risk. Among these factors, the crash pulse may be the most influential. An analysis of crashes with recorded crash pulses and known injury outcomes showed that the mean acceleration of the crash pulses correlated well with the injury risk (Krafft et al., 2002). Variations between car models greatly influence the injury risk, and this influence remains after compensation for car body weight (Eichberger et al., 1996; Boström et al., 1997). Seat characteristics have been shown to influence the injury risk (Parkin et al., 1995; Morris & Thomas, 1996) and the occupant kinematics (Svensson et al., 1996; Prasad et al., 1997; Watanabe et al., 2000), in rear-end impacts.

Injury criteria for AIS 1 neck injuries have been proposed and attempts have been made to validate or disprove these. Many methods have been used, and their common link is to compare test data with real-life data to find a robust relationship between the criterion values and the corresponding injury outcomes. Among the proposed criteria, the NIC (Boström et al., 1996) is the most frequently evaluated. The NIC is given in Equation 1; a_{rel} is the relative acceleration and v_{rel} is the relative velocity between the T1 and head local x-coordinates. To transform the NIC into a scalar value, the NIC$_{max}$, defined as the maximum value during the first 150 ms of a crash, was presented by Boström et al. (2000). NIC_{max} seems to predict AIS 1 neck injuries (Boström et al., 1997; Eichberger et al., 1998; Wheeler et al., 1998; Zuby et al., 1999; Boström et al., 2000; Eriksson & Boström, 2002; Kullgren et al., 2003).

To date, the evaluation studies have not been extensive enough to fully validate the NIC_{max}: the tests used have not been realistic enough, the real-life data too general, or both. An evaluation method that can combine realistic models with injury outcomes is the reconstruction of real-life crashes by the use of mechanical and mathematical models. These crashes must be well-documented in terms of known risk factors, and the models used for the reconstruction must respond properly to these risk factors. Moreover, the data on the occupants' injuries must be accurate. This method was used by Kullgren et al. (2003) to correlate the NIC_{max}, estimated in mathematical models, with injury outcome for 110 occupants in 79 crashes. The data describing these crashes were recorded crash pulses, the car models, and the occupants' gender, age, previous neck injuries, and present injury outcomes. Risk factors known to influence the AIS 1 neck injury risk were omitted from that study.

The objective of this study is to evaluate the influence of seat geometry and seating posture on the NIC_{max} long-term AIS 1 neck injury predictability by making parameter analyses of reconstructed real-life rear-end crashes with known injury outcomes by the aid of mathematical modelling.

$$NIC(t) = 0.2 \times a_{rel}(t) + v_{rel}(t)^2 \qquad (1)$$

METHOD

Real-Life Crashes

In this study, 79 crashes with 110 occupants were analyzed. These crashes were selected from a database at Folksam Insurance Company, Sweden, which contains all crashes, irrespective of repair cost or injury outcome, with cars fitted with crash pulse recorders that were insured by Folksam. The inclusion criteria in this study were single rear-end crashes with recorded crash pulse and front seat occupant with no previous long-term AIS 1 neck injury. Among these, the crashes with the three most frequently represented car models were selected; these three models were of the same brand. The crash pulse recorders measured acceleration versus time at a sampling frequency of 1000 Hz, for rear-end crashes with the principle direction of force within $\pm 30°$, and the trigging threshold was approximately 3g.

The occupant injury outcomes were established from medical notes and interviews with the occupants. All medical information was used in the injury classification and a follow-up of possible medical symptoms was carried out at least six months after the crash. The questionnaire of symptoms and the process of defining injury severity were structured in cooperation with a medical doctor. The symptoms noted were those associated with pain, stiffness, and musculoskeletal signs, and with neurological symptoms, such as numbness. The injury outcomes were divided into three categories depending on the duration of the neck injury symptoms: no symptoms, symptoms lasting less than one month (initial symptoms), and symptoms lasting more than one month (long-term symptoms). Fourteen of the 110 occupants reported long-term symptoms.

Madymo Models

To analyze the influence of seat geometry and seating posture on the neck injury risk, Madymo models were used; the one simulating the occupant was the Madymo BioRID II (release date March 1, 2003). This is an upgrade of the validated Madymo model of the BioRID I developed by Eriksson (2002). The crashes selected from the Folksam database included three car models. Since these cars were equipped with different kinds of seats, models of the passenger seats were made in Madymo. These Madymo seat models were based on those used in Kullgren et al. (2003).

Sled tests were carried out for the three seats (Seat Type 1, 2, and 3) and the BioRID II (manufactured in Sweden) at Δv 23 km/h with a mean acceleration of 4.5g, a peak acceleration of 8.9g, and a pulse duration of 145 ms, and at Δv 8.5 km/h with a mean acceleration of 2.4g, a peak acceleration of 5.0g, and a pulse duration of 100 ms. All seats used in the sled tests were second-hand, but they were screened for damage. Four sled tests were carried out for Seat Type 1 fitted with two types of head restraints and the 4.5g crash pulse; one sled test with Seat Type 1 was carried out with the 2.4g crash pulse. Two sled tests with the 4.5g crash pulse and one test with 2.4g crash pulse were made for each of Seat Types 2 and 3.

The test matrix for the sled tests is given in Table I. In all sled tests, the dummy was seated in a normal posture and no seat belts were used. Seat-back inclinations were adjusted to a torso angle of $25°$ with an H-point tool and the head restraints were placed in their lowest position. None of the head restraints could be inclined, and none of the seats were equipped with adjustable lumbar support. The seats were placed in the middle of their front-aft rails. The dummy responses were measured and filtered according to Davidsson (1999). For each type of seat, the spreads in dummy responses were calculated and used to establish corridors for the local x and z accelerations in the dummy head, C4, T1, T8, L1, and pelvis. All sled tests were filmed at 1000 Hz and motions were digitized.

For each seat, the geometries of the seat cushion, the seat back, and the head restraint contours were implemented in

Table I Test matrix for the sled tests

Sled test no.	Seat	Head restraint	Crash pulse
# 1	Type 1	Standard I	4.5g
# 2	Type 1	Standard I	4.5g
# 3	Type 1	Standard II	4.5g
# 4	Type 1	Standard II	4.5g
# 5	Type 1	Standard I	2.4g
# 6	Type 2	Standard	4.5g
# 7	Type 2	Standard	4.5g
# 8	Type 2	Standard	2.4g
# 9	Type 3	Standard	4.5g
# 10	Type 3	Standard	4.5g
# 11	Type 3	Standard	2.4g

Madymo to achieve correct contact areas between the seat and the dummy. Also, parts of the seat structure that may influence the dummy kinematics during the crashes were implemented. Each seat cushion was modelled as a plane with contact characteristics acting on the dummy pelvis, hips, and femurs, and with a cylinder representing the seat front edge restricting the motions of the lower legs. Each seat was connected to a floor by a joint that allowed the seat to move somewhat back and down during the crashes. The floors were horizontally oriented planes that controlled the motion of the dummy feet. The seat backs were connected to the seat cushions by joints simulating the recliners. For two of the seats, additional joints were placed in the upper part of the recliners, since these seats broke during the sled tests with the 4.5g crash pulse.

The frame and the recliner of the third seat were stiff enough not to break in the sled tests, hence no joint at the upper part of the recliner was necessary for that seat type. The Madymo models of the head restraints were connected by joints to the upper beams of seat back; the heights of the head restraints could be adjusted analogous to the adjustment of the mechanical counterparts. All joints connecting frame structures in the seats were given the characteristics of an elastic component with hysteresis and a damping component.

The head restraints were modelled as cylinders. For each seat, three planes built up the seat-back padding, and each of these planes was connected to the seat-back frame by Kelvin spring-damper elements corresponding to the mattress of springs behind the padding in the mechanical seat backs. Ellipsoids were placed along the seat-back sides to interact with the arms during the crashes. Finally, the estimated stiffness characteristics of the Madymo seat models were tuned with the aim to fit the responses from the Madymo models into the response corridors established from the sled tests carried out with the 2.4g and 4.5g crash pulses.

Variations in Seat Geometry and Seating Posture

To assess the influence of seat geometry and seating posture on the NIC_{max}, a simulation test matrix was defined and applied to the Madymo model of the 79 crashes selected from the Folksam database. This test matrix consists of four variables, A to D, which defines 100 combinations of seat geometry and seating posture. The Madymo BioRID II pelvis, legs, and feet were placed in their normal positions, while the spine curvature, seat-back inclination, and head restraint position were varied for the 100 combinations.

Variable A defines the seat-back inclination. This variable was distributed normally with mean 0° and a standard deviation 1.33°. A negative A value corresponds to a more upright seat back than the normal inclination, while a positive A value corresponds to a more inclined seat back than the normal. To maintain initial contact between the dummy lower back and seat back independent of seat-back inclination, the A variable also defined the initial angular displacement between the sacrum and the last lumbar vertebra, and between adjacent lumbar vertebrae. The five angular displacements were of the same magnitude, and

were linearly related to the A variable. These angular displacements were thereby also normally distributed: mean 0° and a standard deviation of 0.67° per joint. To maintain the head base in its normal horizontal position, the angular displacements of six adjacent thoracic vertebra joints were initially changed. The B variable defined which six joints were changed; this variable was uniformly distributed. The lowermost and uppermost of these joints were rotated half the rotation magnitude of one lumbar joint, but in the opposite direction from the lumbar joints; the four intermediate joints were rotated the same amount as the lumbar joints, but in the opposite direction.

The initial lumbar and thoracic spine curvatures are also influenced by variable C. Negative C values correspond to flexion of the lumbar joints and the lower thoracic joints and extension of the upper thoracic joints; positive C values to extension of the lumbar joints and the lower thoracic joints and flexion of the upper thoracic joints. The magnitude of the angular displacement of each joint corresponds to the amplitude of a sinus function with a period of 17 joints; the maximum amplitude is the absolute value of the C variable. The period begins at the joint between the two lowest lumbar vertebrae and ends at the joint between the uppermost thoracic vertebra and the lowermost cervical vertebra. The summation of the angular displacements along the spine defined by variable C is zero; hence the head base remains horizontal. The C variable is normally distributed with mean value −1.0° and a standard deviation of 0.67°.

The fourth variable, D, defines the height of the head restraint. All seats in this study were equipped with head restraints with the same adjustable range at three discrete heights separated by 1.85 cm. The lowest position was used for 69 of the 100 simulations, the middle position for 28 simulations, and the highest position for three simulations.

The variables A to D were randomly chosen within their distributions to form the 100 combinations of seat geometry and seating posture. Due to differences in seat geometry, the initial spread of the head-to-head-restraint distances differed for the three seats. The horizontal head-to-head-restraint distance was measured between the most rearward point of the head and the most forward point of the head restraint; for Seat Type 1 the distance was between 26.2 and 95.7 mm (median value 59.9 mm), for Seat Type 2 between 35.1 and 104.6 mm (median value 76.6 mm), and for Seat Type 3 between 54.4 and 124 mm (median value 8.62 mm). The vertical distance was measured between the highest point of the head and the highest point of the head restraint; for Seat Type 1 the distance was between 74.7 and 121.8 mm (median value 117.5 mm), for Seat Type 2 between 63.8 and 112.4 mm (median value 108.7 mm), and for Seat Type 3 between 24.1 and 72.6 mm (median value 68.7 mm).

Since variables A to D changed the initial seat geometry and dummy seating posture, the seat-back plane positions also had to be changed to obtain equilibrium of the seat-dummy system when the simulated crash started. This was solved independently for each simulation by keeping the dummy and the seat-back frame in their chosen positions, while simultaneously letting the seat-back planes move before the crash started.

Injury Risk Curves

Injury risk curves relate the risk of sustaining an injury to criterion values. In this study risk curves for long-term AIS 1 neck injuries versus the NIC_{max} were established by relating NIC_{max} values, estimated in Madymo simulations of real-life crashes, to occupant injury outcome. The crashes selected from the Folksam database included 110 occupants: the risk curves were established on the injury outcome for all occupants by using the crashes including two occupants twice in the calculations. The simulation test matrix containing 100 combinations of seat geometries and seating postures was applied to all crashes, which resulted in a spread of possible NIC_{max} values for each occupant. In each simulation, the Madymo seat model used was that corresponding to the real-life crash in which the crash pulse was recorded. For every occupant the median, minimum, and maximum NIC_{max} values were determined from the 100 cases. Thus, risk curves for long-term AIS 1 neck injuries could be established. Each risk curve was calculated by dividing the NIC_{max} values into five intervals, calculating the ratio of injured occupants in each interval, and connecting these ratios to a curve by linear interpolation.

NIC_{max} Predictability

To analyze the NIC_{max} predictability of long-term symptoms diagnostic tests were used: the sensitivities, specificities, positive predictive values, and negative predictive values, and their 95% confidence intervals. In this part of the study each of the 100 combinations of seat geometry and seating posture were analyzed to determine the influence on the NIC_{max} of variations in seat geometry and seating posture for different crash situations. For each analyzed crash situation, the Madymo seat model used was that corresponding to the real-life crash in which the crash pulse was recorded.

The definitions of the diagnostic tests are shown in Table II (Altman et al., 2000). Confidence intervals were calculated using Wilson's equation, which has better statistical properties than the traditional method for proportions close to zero and close to one, since the confidence interval is asymmetrical (Altman et al., 2000):

$$\frac{2pn + z^2 - z(z^2 + 4pn\sqrt{1-p})}{2(n+z^2)} \quad \text{to}$$

$$\frac{2pn + z^2 + z(z^2 + 4pn\sqrt{1-p})}{2(n+z^2)} \tag{2}$$

where p is the analyzed proportion, n is the sample size, and z the $100(1 - \alpha/2)$ percentile from the standard normal distribution. The 95% confidence interval is given by $z = 1.96$.

First, the specificities, positive predictive values, and negative predictive values, and their 95% confidence intervals, were calculated for two fixed sensitivities, 0.79 and 0.93, for the 100 combinations of seat geometry and seating posture. Also, the resulting NIC_{max} threshold levels were calculated for these sensitivities. The fixed sensitivity of 0.79 corresponds to the assumption that 11 occupants of the 14 with long-term symptoms in the real-life crashes have estimated NIC_{max} values above a given threshold; the fixed sensitivity of 0.93 assumes that 13 of the 14 have estimated NIC_{max} values above a given threshold. These analyses showed the spreads in NIC_{max} values for fixed sensitivity levels.

Then, the diagnostic tests were applied for three fixed threshold levels of the NIC_{max} in order to analyse the NIC_{max} predictability for selected threshold levels. The lowest of the three fixed NIC_{max} thresholds was approximately the lower decile for the sensitivity of 0.93, while the highest of the three fixed NIC_{max} thresholds was approximately the upper decile for the sensitivity of 0.79. The middle value was defined as the average of the fixed high and low levels. These analyses showed the spread in sensitivities, specificities, positive predictive values, and negative predictive values, for selected NIC_{max} thresholds, which can be used to assess the NIC_{max} predictability: a perfect criterion with an accurate threshold results in all four diagnostic test variables equal to one.

Table II Definitions of sensitivity, specificity, positive predictive value, and negative predictive value

| | | AIS 1 Neck Injury Symptoms | | Total |
		Long-term	None or initial	
NIC_{max} value	Above threshold	a	b	a + b
	Below threshold	c	d	c + d
	Total	a + c	b + d	

- Sensitivity = a/(a+c)
  The proportion of occupants with long-term symptoms who have NIC_{max} values above a fixed threshold
- Specificity = d/(b+d)
  The proportion of occupants with no or initial symptoms who have NIC_{max} values below a fixed threshold
- Positive predictive value = a/(a+b)
  The proportion of occupants with NIC_{max} values above a fixed threshold who have long-term symptoms
- Negative predictive value = d/(c+d)
  The proportion of occupants with NIC_{max} values below a fixed threshold who have no or initial symptoms

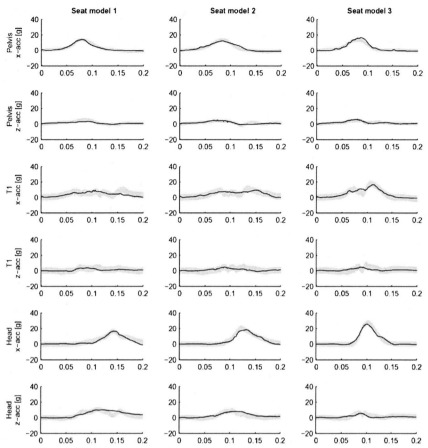

Figure 1 Comparison between sled tests and Madymo simulations for the 4.5g crash pulse. All curves are given versus time in seconds. Black lines are responses from Madymo simulations; grey areas are response corridors established from sled tests.

RESULTS

Madymo Models

Head, T1, and pelvis accelerations generally fit into the response corridors for the 4.5g crash pulse, with the exceptions of some responses during the rebound phase of the crashes (Figure 1). Generally, the Madymo models imitated the responses from the sled tests better for the 4.5g crash pulse than for the 2.4g crash pulse. The NIC_{max} values from the Madymo simulations and from the sled tests exposed to the 4.5g and the 2.4g crash pulses are shown in Table III.

Table III Comparison of Madymo simulations and sled tests in term of the NIC_{max} for the 4.5g crash pulse and the 2.4g crash pulse, respectively

		NIC_{max} [m²/s²]		
		Madymo	Sled tests	
4.5g crash pulse	Seat model 1	26	24	20
	Seat model 2	21	23	25
	Seat model 3	20	23	27
2.4g crash pulse	Seat model 1	6.0	14	
	Seat model 2	8.3	13	
	Seat model 3	9.7	13	

Injury Risk Curves

The distributions of NIC_{max} values as a result of the variation in seat geometry and seating posture are shown in Figure 2. Each curve in these graphs corresponds to one occupant: those with long-term symptoms are shown in the upper graph and those with initial or no symptoms in the lower graph. The NIC_{max} values for occupants with long-term symptoms and the NIC_{max} values for those with initial or no symptoms overlap. Nevertheless, most of the NIC_{max} values related to occupants with initial or no symptoms are in ranges of relatively low values, while the ranges are greater for NIC_{max} values corresponding to those occupants that sustained long-term symptoms.

Correlation was found between NIC_{max} values and the duration of AIS 1 neck injury symptoms. Thus, an injury risk curve could be established (Figure 3). At a NIC_{max} value of approximately 15 m²/s² the injury risk was 18% when the median NIC_{max} value were used, and the injury risk was 55% when the minimum NIC_{max} values were used.

NIC_{max} Predictability

For the 100 combinations of seat geometries and seating postures, the spreads in NIC_{max} values, sensitivities, specificities, positive predictive values, and negative predictive values, and

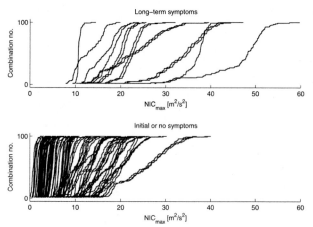

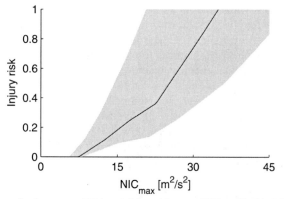

Figure 2 Spreads in NIC_{max} as a result of varied seat geometry and seating posture. Each curve corresponds to one occupant in one crash; the spread for each occupant is a result of the 100 combinations of seat geometry and seating posture. For those crashes involving two occupants, the curve corresponding to one of the occupants was moved 0.6 m²/s².

their 95% confidence intervals, are shown in Figures 4 and 5. The confidence intervals were wider for the sensitivities and positive predictive values than for the specificities and negative predictive values. This can be attributed mainly to the smaller number of occupants with long-term AIS 1 neck injury symptoms in relation to occupants with no or initial symptoms.

Figure 4 shows that the NIC_{max} threshold ranged between 10 m²/s² and 23 m²/s² for the fixed sensitivity of 0.79, and between 8.0 m²/s² and 14 m²/s² for the fixed sensitivity of 0.93. As can be seen in Figure 2, for most of the occupants the NIC_{max} values were distributed like a normal distribution with a tail of comparatively small NIC_{max} values. These tails were the reason for the relatively wide ranged of NIC_{max} threshold in Figure 4; as well as the relatively low sensitivities, the relatively high specificities, and the relatively high positive predictive values, for a few of the cases in Figure 5.

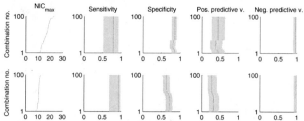

Figure 4 NIC_{max} values, sensitivities, specificities, positive predictive values, and negative predictive values for the fixed sensitivities of 0.79 (above) and 0.93 (below). In the left column, the NIC_{max} values are given. Each of these graphs contains 100 NIC_{max} values, as a result of the 100 combinations of seat geometry and seating posture. The other graphs show the result from the diagnostic tests. Each of these graphs contains the diagnostic test values in black and their 95% confidence intervals in grey; thus the black lines contains 100 diagnostic tests as a result of the 100 combinations of seat geometry and seating posture. The diagnostic tests quantify how well NIC_{max} predicts long-term AIS 1 neck injuries; for a perfect criterion, all diagnostic test variables are equal to one.

DISCUSSION

Mathematical Models

Mathematical modelling is a useful tool in conducting studies with a large number of parameters or parameter levels. Furthermore, mathematical modelling can be used to reconstruct real-life crashes. In this study, 79 real-life crashes involving 110 occupants were reconstructed, and a parameter analysis was made for each crash. Altogether 7900 mathematical simulations were performed to evaluate the influence of seat geometry and seating posture on the NIC_{max}.

The real-life crashes analyzed in this study were reconstructed by using Madymo models. The seat models were tuned to sled tests and two crash pulses were used: mean acceleration

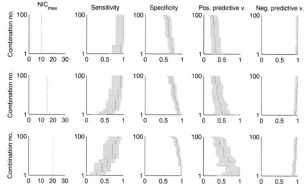

Figure 5 NIC_{max} values, sensitivities, specificities, positive predictive values, and negative predictive values for the fixed NIC_{max} thresholds of 10 (above), 15 (middle), and 20 (below). In the left column, the NIC_{max} values are given. Each of these graphs contains 100 NIC_{max} values, as a result of the 100 combinations of seat geometry and seating posture. In this figure the diagnostic tests were applied on fixed NIC_{max} thresholds, therefore all NIC_{max} values in the same graph were fixed. The other graphs show the result from the diagnostic tests. Each of these graphs contains the diagnostic test values in black and their 95% confidence intervals in grey; thus the black lines contains 100 diagnostic tests as a result of the 100 combinations of seat geometry and seating posture. The diagnostic tests quantify how well NIC_{max} predicts long-term AIS 1 neck injuries; for a perfect criterion, all diagnostic test variables are equal to one.

Figure 3 Long-term AIS 1 neck injury risk versus NIC_{max}. The black line is the risk curve established from the median NIC_{max} values; the grey area represents the risks bounded by the minimum and the maximum NIC_{max} values, associated with the combinations of seat geometry and seating posture.

4.5g and Δv 23 km/h, and mean acceleration 2.4g and Δv 8.5 km/h. This selection of crash pulses was based on a study by Krafft et al. (2002) showing that most occupants who sustain long-term symptoms were exposed to a crash pulse with a mean acceleration higher than 4.5g and Δv higher than 15 km/h; occupants with no long-term symptoms had on average been exposed to a crash pulse with a mean acceleration lower than 3.9g and Δv lower than 10 km/h. The crash pulses used in this study thereby corresponded to one crash with a high probability of long-term symptoms and another crash with a low probability. Thus, the Madymo models used in this study are in accord with relevant crash situations, which indicates that the conclusions drawn from this study are accurate.

Although the T1 and head x-accelerations fitted into the corridors for the 4.5g crash pulse, the NIC_{max} estimated in the Madymo simulations did not match exactly the values measured in the mechanical tests for any of the seats (Table III). However, the trends of the NIC_{max} as a result of varied seat geometry and seating posture are most likely accurate, since the discrepancies in the NIC_{max} between the simulations and the sled tests were caused mainly by incorrect timing of the head to head restraint contact or by the absence of peaks in the T1 accelerations within the range of the corridors. As can be seen in Table III, the NIC_{max} values estimated in the Madymo simulations are much lower than the values measured in the corresponding mechanical tests for the 2.4g crash pulse. These discrepancies most likely made the injury risk seem greater for low NIC_{max} values in Figure 3. Moreover, in Figure 5, the sensitivities are probably somewhat low and the specificities somewhat high, for the fixed NIC_{max} of 10 m^2/s^2.

The Madymo BioRID II (release date March 1, 2003) was used to model the occupants in this study. The BioRID II better mimic the head rotation and neck loads than does the BioRID I. However, the cables acting as neck muscle substitutes on the mechanical BioRID II have not yet been modelled on the Madymo dummy, which resulted in somewhat inadequate head rotations, especially for extensive rotations of the head and neck. Inadequate head rotation can influence the time for head-to-head-restraint contact and the NIC_{max} value; this most likely occurred for the crashes with the most powerful crash pulses in this study.

To achieve realistic distributions of relatively normal seat geometries and seating postures for the three seats modelled in this study, a voluntary study was carried out. The volunteers were asked to incline the seat back to a driving position; the seat-back inclinations were measured and the seating postures estimated. It was found that the placement of the pelvis on the seat was almost the same independent of seat-back inclination, therefore the same placement of BioRID II pelvis, legs, and feet were used in all simulations. The seat-back inclination was assumed to be normally distributed, since the majority of the volunteers inclined the seat back to approximately the same degree. Nevertheless, there were too few volunteers to show statistically that the inclinations were normally distributed or to estimate the mean value and the standard deviation. Several studies have shown that most head restraints are left in their lowest position (Nygren et al., 1985; Viano & Gargan, 1996; Cullen et al., 1996), therefore 69 of 100 head restraints were placed in the lowest position, while only three were in the highest position, in the simulation test matrix used in this study.

Evaluation of NIC_{max}

Many methods can be used to validate or disprove injury criteria. Kullgren et al. (2003) reconstructed a broad set of real-life crashes including both injured and uninjured occupants. Their study showed that the NIC_{max} is applicable to predict AIS 1 neck injury risk. All reconstructed crashes in their study were performed with the mid-sized male dummy BioRID II seated in a normal posture and with the seat in a normal position. Since there are no validated mathematical and mechanical occupant models with variable size, age, or gender properties available, the present study focused on the influence of seat geometry and seating posture on the AIS 1 neck injury risk. As can be seen in Figure 2, the seat geometry and seating posture highly influence the NIC_{max} values. Therefore, it is important to maintain the same seat geometry and dummy seating posture within test series in order to obtain comparable results, as well as vary those factors for different crash pulses to avoid sub-optimization when developing safety systems.

This study shows that although the variations in seat geometry and seating posture resulted in spreads of the NIC_{max}, a correlation to injury risk was found and a risk curve was established (Figure 3). An essential criterion characteristic when developing safety system is that the injury risk is zero for some criterion values, otherwise it should not be possible to develop a safety system that undoubtedly prevents from injuries. Figure 3 shows that the neck injury risk is zero for small NIC_{max} values. Hence, for a safety system that only produces low NIC_{max} values the risk of sustaining a neck injury is infinitesimal. For moderate and high NIC_{max} values the spread in risk is wide. This supports earlier studies that have shown that the NIC_{max} is sensitive to seat geometry and seating posture. Nevertheless, since the risk is zero for low NIC_{max} values and then continuously raises, this supports the results from the diagnostic tests that NIC_{max} is applicable to predict long-term AIS 1 neck injury risk also when seat geometry and seating posture are varied. Since individual seating posture, seat adjustment, and medical factors in the real-life crashes were unknown in this study, and since only three car models from the same brand were analyzed, Figure 3 should not be used to propose an NIC_{max} threshold for long-term AIS 1 neck injuries.

The diagnostic tests were applied in order to evaluate the NIC_{max} predictability of long-term AIS 1 neck injuries. A high sensitivity shows that the threshold is selected so that almost all occupants with long-term symptoms have criterion values above the threshold. This study showed that very few combinations of seat geometry and seating posture resulted in relatively low sensitivities (Figure 5). It is not likely that all occupants with long-term symptoms were seated in identical positions; therefore these low sensitivities should be used as low-probability worst-case situations. The positive predictive value measures how well the criteria can differentiate occupants with long-term injuries from occupants with initial or no symptoms, among those with estimated criterion values above a selected threshold. As can be seen in Figures 4 and 5, the positive predictive values were generally low, indicating that high proportions of occupants with NIC_{max} values above the chosen thresholds have initial or no symptoms. The negative predictive values were close to one for all cases, indicating that the proportions of occupants with initial or no injuries are high among those with estimated NIC_{max} values below the selected threshold. Hence, this study shows that the NIC_{max} can be used to predict long-term AIS 1 neck injuries.

CONCLUSIONS

The sensitivities and the negative predictive values were high for almost all combinations of seat geometry and seating posture analyzed; hence the NIC_{max} is a robust criterion and can be used for estimating the risk of long-term AIS 1 neck injuries.

A risk curve for long-term AIS 1 neck injuries as a function of the NIC_{max} was established. By the aid of mathematical modelling, which has the advantage of absolute repeatability in comparison with mechanical modelling, the spread in the NIC_{max} values, as a result of the combinations of seat geometry and seating posture, was evaluated. This spread indicates that the seat geometry and seating posture should be taken into account when estimating NIC_{max} values, and that the seat geometry and dummy seating posture should be identical within test series to obtain comparable results.

The spread in the risk curve is bounded by the maximum and minimum NIC_{max} values estimated for each occupant in a simulated crash when seat geometry and seating posture were varied. The outer limits of the spread are a worst-case scenario; consequently they are not usually represented in real-life crashes.

REFERENCES

*Altman DG, Machin D, Bryant TN, Gardner MJ. (2000) Statistics with confidence, 2nd Edition. BMJ Books.

Boström O, Fredriksson R, Håland Y, Jakobsson L, Krafft M, Lövsund P, Muser MH, Svensson MY. (2000) Comparison of car seats in low speed rear-end impacts using the BioRID dummy and the new neck injury criterion (NIC). *Accident Analysis and Prevention*, Vol. 32, No. 2, pp. 321–328.

Boström O, Krafft M, Aldman B, Eichberger A, Fredriksson R, Håland H, Lövsund P, Steffan H, Svensson, MY, Tingvall C. (1997) Prediction of neck injuries in rear impacts based on accident data and simulations. *Proceedings of the 1997 International IRCOBI Conference on the Biomechanics of Impacts*, September 24–26, Hannover, Germany, pp. 251–264.

Boström O, Svensson MY, Aldman B, Hansson H, Håland Y, Lövsund P, Seeman T, Suneson A, Säljö A, Örtengren T. (1996) A new neck injury criterion candidate—based on injury findings in the cervical spinal ganglia after experimental neck extension trauma. *Proceedings of the 1996 IRCOBI International Conference on the Biomechanics of Impacts*, September 11–13, Dublin, Ireland, pp. 123–136.

Cullen E, Stabler K, Mackay GM, Parkin S. (1996) Head restraint positioning and occupant safety in rear impacts: The case for smart restraints. *Proceedings of the 1996 International IRCOBI Conference on the Biomechanics of Impacts*, September 11–13, Dublin, Ireland, pp. 137–152.

Davidsson J. (1999) BioRID II final report. Report December 30th, Department of Machine and Vehicle Design, Chalmers University of Technology, Göteborg, Sweden.

Eichberger A, Geigl BC, Moser A, Fachbach B, Steffan H, Hell W, Langwieder K. (1996) Comparison of different car seats regarding head-neck kinematics of volunteers during rear end impact. *Proceedings of the 1996 International IRCOBI Conference on the Biomechanics of Impacts*, September 11–13, Dublin, Ireland, pp. 153–164.

Eichberger A, Steffan H, Geigl M, Svensson MY, Boström O, Leinzinger PE, Darok M. (1998) Evaluation of the applicability of the neck injury criterion (NIC) in rear end impacts on the basis of human subject tests. *Proceedings of the 1998 International IRCOBI Conference on the Biomechanics of Impacts*, September 16–18, Göteborg, Sweden, pp. 321–333.

Eriksson L. (2002) Three-dimensional mathematical models of the BioRID I and car seats, for low-speed rear-end impacts. *Traffic Injury Prevention*, Vol. 3, No. 1, pp. 75–87.

Eriksson L, Boström O. (2002) Assessing the relevance of NIC and its correlation with crash-pulse parameters: Using the mathematical BioRID I in rear-end impacts. *Traffic Injury Prevention*, Vol. 3, No. 2, pp. 175–182.

Krafft M, Kullgren A, Ydenius A, Tingvall C. (2002) Influences of crash pulse characteristics on whiplash associated disorders in rear impacts—crash recording in real life crashes. *Traffic Injury Prevention*, Vol. 3, No. 2, pp. 141–149.

Kullgren A, Eriksson L, Boström O, Krafft M. (2003) Validation of neck injury criteria using reconstructed real-life rear-end crashes with recorded crash pulses. *Proceedings of the 18th International Technical Conference on the Enhanced Safety of Vehicles*, May 19–22, Nagoya, Japan, Paper No. 344–O.

Kullgren A, Krafft M, Ydenius A, Lie A, Tingvall C. (2002) Developments in car safety with respect to disability—Injury distributions for car occupants in cars from the 80s and 90s. *Proceedings of the 2002 International IRCOBI Conference on the Biomechanics of Impacts*, September 18–20, Munich, Germany, pp. 145–154.

Morris A, Thomas P. (1996) A study of soft tissue neck injuries in the UK. *Proceedings of the 15th International Technical Conference on the Enhanced Safety of Vehicles*, May 13–16, Melbourne, Australia, Paper No. 96-S9-O-08, pp. 1412–1425.

Nygren Å, Gustavsson H, Tingvall C. (1985) Effects of different types of headrests in rear-end collisions. *Proceedings of the 10th International Technical Conference on Experimental Safety Vehicles*, July 1–4, Oxford, UK, pp. 85–90.

Parkin S, Mackay GM, Hassan AM, Graham R. (1995) Rear end collisions and seat performance—to yield or not to yield. *Proceedings of the 39th Annual Association for the Advancement of Automotive Medicine Conference*, October 16–18, Chicago, Illinois, USA, pp. 231–244.

Prasad P, Kim A, Weerappuli DPV. (1997) Biofidelity of anthropomorphic test devices for rear impacts. *Proceedings of the 41st Stapp Car Crash Conference*, November 13–14, Orlando, Florida, USA, Paper No. 973342, pp. 387–415.

Svensson MY, Lövsund P, Håland Y, Larsson S. (1996) The influence of seat-back and head-restraint properties on the head-neck motion during rear-impact. *Accident Analysis and Prevention*, Vol. 28, No. 2, pp. 221–227.

Watanabe Y, Ichikawa H, Kayama O, Ono K, Kaneoka K, Inami S. (2000) Influence of seat characteristics on occupant motion in low-speed rear impacts. *Accident Analysis and Prevention*, Vol. 32, No. 2, pp. 243–250.

Wheeler J, Smith T, Siegmund G, Brault J, King D. (1998) Validation of the neck injury criterion (NIC) using kinematic and clinical results from human subjects in rear-end collisions. *Proceedings of the 1998 International IRCOBI Conference on the Biomechanics of Impacts*, September 16–18, Göteborg, Sweden, pp. 335–348.

Viano D, Gargan M. (1996) Headrest position during normal driving: Implication to neck injury risk in rear crashes. *Accident Analysis and Prevention*, Vol. 28, No. 6, pp. 665–674.

von Koch M, Nygren Å, Tingvall C. (1994) Impairment pattern in passenger car crashes, a follow-up of injuries resulting in long-term consequences. *Proceedings of the 14th International Technical Conference on the Enhanced Safety of Vehicles*, May 23–26, Munich, Germany, Paper No. 94-S5-O-02, pp. 776–781.

Zuby D, Vann T, Lund A, Morris C. (1999) Crash test evaluation of whiplash injury risk. *Proceedings of the 43rd Stapp Car Crash Conference*, October 25–27, San Diego, California, USA, Paper No. 99SC17, pp. 267–278.

THE MECHANISMS OF EARLY ONSET C5/C6 SOFT-TISSUE NECK INJURY IN REAR IMPACTS

Tom Gibson[1], Ola Bostrom[2], Anders Kullgren[3], Bruce Milthorpe[4]
[1]Human Impact Engineering, Australia, [2]Autoliv Research, Sweden, [3]Folksam Research, Sweden, [4]Graduate School of Biomedical Engineering, UNSW, Australia

ABSTRACT

An anatomically based, multi-body model of the C5/C6 motion segment was developed to study soft-tissue neck injury mechanisms in rear impacts. This was integrated into the MADYMO-based van der Horst head and neck model. Responses were compared with volunteer test data up to head restraint impact. Soft-tissue

(n=78) and known long-term pain outcomes. Facet capsule shear and impingement injury mechanisms at C5/C6 were demonstrated. Facet capsule loading correlated well with NIC_{max} and was able to predict the risk of AIS1 neck injuries with persisting pain.

Keywords: Whiplash, rear impact, soft tissue, neck, injury, models.

THE DEVELOPMENT OF SOFT-TISSUE NECK INJURY CRITERIA for rear impacts has been hampered by the difficulty of relating the injury outcome of an occupant as a result of a rear impact on a vehicle to specific injury mechanisms. In turn, the lack of knowledge regarding the injury mechanism and of accepted injury criteria has slowed the development of appropriate test methodologies for the design of vehicle systems to mitigate rear impact injury. The optimisation of the design of vehicle safety systems for the minimisation of whiplash needs a better understanding of human tolerance to these injuries.

A significant amount of data regarding soft tissue neck injury already exists from the testing of volunteers and post-mortem humans as well as field and clinical studies. A methodology is required to link the available data together and guide the investigation into those areas requiring further study. Mathematical models of the neck have now reached a level of development allowing them to be used to make these links. The investigation of these injuries requires a model possessing a high level of dynamic biofidelity at various levels: as a part of a human body model, as a head and neck model and as a neck motion segment model. Possibly one of the most advanced models in terms of validation and capabilities is the MADYMO-based Human Model. The biofidelity and capabilities of this model have been proven when used to model dynamic rear impact tests with volunteers, Meijer et al. (2001). The detailed multi-body neck model was developed and validated by van der Horst (2002) and is illustrated in Figure 1. It includes active muscle capability.

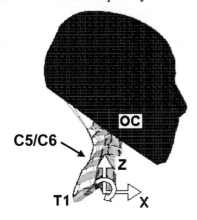

Fig. 1 – The van der Horst (2002) human head and neck model with T1 axes.

The use of the van der Horst neck model to investigate soft-tissue neck injury causation required improvements in its ability to sense injury at the motion segment level. Gibson (2005) developed a compatible C5/C6 motion segment model, Figure 2(b), which had the van der Horst (2002) head and neck model as its basis. A new motion segment model including the cervical disc structure and better injury sensing for the disc and facet capsule was developed. The model was validated with static *in-vitro* experimental data.

This paper describes the application of this detailed C5/C6 motion segment model to link together available early AIS1 neck injury data to the motion of the neck in real crashes, to demonstrate its use as a tool for investigating soft-tissue neck injury.

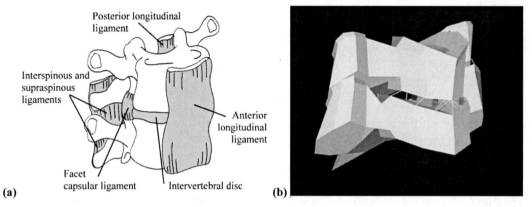

Fig. 2 – (a) Sketch of the C5/C6 motion segment, showing the vertebral bodies and major ligaments, and (b) the C5/C6 motion segment model, Gibson (2005).

METHODS AND MATERIALS

The stages in the study are shown in Figure 3. The C5/C6 motion segment model (Study A) was re-integrated into the MADYMO-based multi-body human head and neck model developed by van der Horst (2002) (Study B), to allow realistic dynamic neck loads to be applied. The responses and injury-

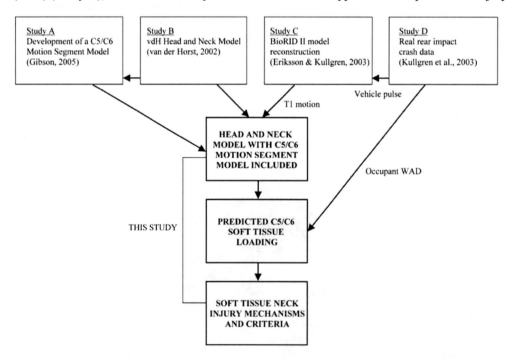

Fig. 3 – Diagram showing the sections of the study.

sensing capability of this new head and neck model were compared with the response data for the original model by van der Horst to ensure that the validation of this model was unaffected. The new head and neck model was then applied to the investigation of a group of real crashes investigated by Kullgren et al. (2003) (Study D). The crash sample, see below, was of rear impacts to vehicles equipped with a crash-pulse recorder and with known post-crash injury outcomes for the occupants. Kullgren et al. (2003) had reconstructed the crashes with a MADYMO-based BioRID II dummy and seat model in Study C. The BioRID II dummy and seat model responses were in turn, validated with sled tests. The motions of the dummy thorax, T1, from these reconstructions were used to drive the head and neck model. Responses obtained from the C5/C6 model form the basis for the analysis of the soft-tissue injury causation in rear impacts presented here.

STUDY A - C5/C6 MOTION SEGMENT MODEL The C5/C6 motion segment model was developed in MADYMO by Gibson (2005) from the geometry of the van der Horst (2002) head and neck model (Study B below). The C5/C6 motion segment included changes to give improved biofidelity and injury-sensing capability of the disc and facet capsule, while maintaining the biofidelity. A disc structure of ligaments and point restraints was implemented in the model matching the disc anatomy described by Mercer & Bogduk (1999). The anulus fibrosis has a crescent shape with the bulk of the ligament fibres anteriorly and inclined at about 45° to the neck axis, Figure 4. The disposition of the anulus fibrosis fibres was demonstrated to have an effect on the motion segment kinematics, Gibson (2005). The model responses have been validated quasi-statically with respect to *in-vitro* test data.

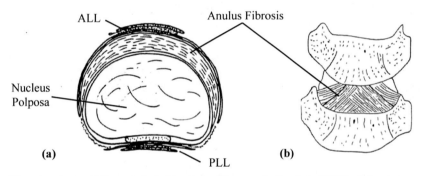

Fig. 4 - The structure of the cervical disc from Mercer & Bogduk (1999): (a) a transverse view of the C5/C6 cervical spine motion segment disc with the major ligaments and (b) the anterior view of the anulus fibrosis with the ligament fibres inclined at 45°.

STUDY B - THE VAN DER HORST HUMAN HEAD AND NECK MODEL The van der Horst head and neck model is a detailed multi-body neck model implemented in MADYMO (Figure 1). It consists of a rigid head, rigid vertebrae (C1, C2, C3, C4, C5, C6, C7 and T1), non-linear visco-elastic discs, frictionless facet joints, non-linear visco-elastic ligaments and controllable segmented contractile muscles, (van der Horst, 2002). The model vertebral shapes are based on the scanned vertebra of an individual's neck. Each disc is represented as a simple 6-degree of freedom joint and the facet capsule as a simple single pivot sliding joint with four spring/damper units representing the capsular ligaments. The muscles follow the curvature of the neck, with realistic lines of action of the muscle forces, and can be actuated as active muscles. The model has been validated quasi-statically with respect to *in-vitro* test data as well as dynamically with volunteer test data. This validation by van der Horst was accepted for the purposes of this study.

STUDY C - CRASH RECONSTRUCTION WITH THE MADYMO BIORID II AND SEAT MODEL The individual occupant motion from rear impact crash data in Study D was reconstructed in Eriksson & Kullgren (2003) and Kullgren et al. (2003). The seats in the three car models selected for reconstruction differed in geometry and stiffness characteristics. The seats were measured and the cushion, the seat back, and the head restraint contours as well as the seat structures were developed to give correct dummy kinematics during the crash for the specific vehicle.

The BioRID II dummy used as a basis for the simulation was the first version manufactured by Chalmers University and Autoliv, and not Denton. Kullgren et al. validated each seat model by means of two sled tests with the BioRID II (at a ΔV=23km/h and a mean acceleration of 4.5g). All the selected crashes were reconstructed in MADYMO using the recorded crash pulses from Study D (below). The BioRID II was seated in a standard seated posture with no seat belts. The seat back inclination corresponded to a torso angle of 25° and the head restraints were placed in their lowest positions.

STUDY D - REAL WORLD CRASH DATA From 1996, the Folksam Insurance Company, Sweden, fitted more than 40,000 cars with crash pulse recorders to measure real world acceleration-time histories in rear end crashes, Kullgren et al. (2003). For this project the three most highly represented car models were selected: where the vehicles were involved in a single rear-end crash with a recorded crash pulse; and, front seat occupants with no previous history of long-term AIS1 neck injury were involved. This gave 78 crashes with a total of 110 front seat occupants for the analysis. The average crash ΔV was 10km/h with a mean acceleration of 3.5g. The maximum change of velocity was 33.2km/h and the maximum mean acceleration was 10.2g. The whiplash injury symptoms of the vehicle occupants were collected by telephone interviews. The symptoms were divided into four categories by duration: no pain (Group 0); pain duration for less than one month (Group 1); pain for more than one month but less than 6 months (Group 2); and, pain for greater than 6 months (Group 3).

IN THIS STUDY local T1 accelerations predicted by the BioRID II dummy and seat models were used for the 78 reconstructions of the Swedish crashes (Kullgren et al., 2003). These T1 accelerations were used to drive the T1 segment of modified head and neck model, Figure 1(a). Hence the neck motion used in this study includes the effect of the impact severity and the vehicle parameters, including the seat characteristic responses. The head and neck model was used in its standard posture of an erect 50th-percentile male volunteer (van der Horst, 2002). Only the passive muscle responses of the head and neck model were used. This was consistent with the early onset injury being studied and the 150ms response time found by Siegmund, Brault & Chimich (2000a) for the initiation of muscle response in volunteers subject to rear impacts. The simulation for the driver (n=78) and a front seat passenger (n=32) in the same crash are identical, but the actual injury outcome from the crash is used.

The motions and forces resulting from the applied T1 accelerations and predicted for the C5/C6 motion segment were analysed for comparison with the actual whiplash injuries to the vehicle occupants. Table 1 summarises the component loading investigated. The strains were calculated from the elongation predicted by the model for each ligament divided by the average ligament lengths from several sources. The ALL (18.3 mm) and FC (6.72 mm) ligament lengths were as measured by Yoganandan, Kumaresan & Pintar (2000). The length for the anulus fibrosis ligaments was derived from the average seperation of the C5 and C6 vertebral bodies in the anterior region of the disc divided by sin 45°, to account for the inclination of the fibres. The NIC_{max} (Bostrom et al., 1996) is the only current neck criteria designed to correlate with early-onset injury in rear impacts (Kullgren at al., 2003), and this was calculated for each case.

Table 1 – The C5/C6 motion segment soft tissue components investigated.

Symbol	Component	Units
ALL	Anterior longitudinal ligament	Strain %
FC	Anterior facet capsule ligament	Strain %
AF	Anterior anulus fibrosus ligament fibre	Strain %
BFz	Back of facet surface	Force, N
Fx	Facet capsule	Motion in x direction, m
Fz	Facet capsule	Motion in z direction, m
AFx	Anulus fibrosus	Motion in x direction, m
AFz	Anulus fibrosus	Motion in z direction, m
NPx	Nucleus pulposus	Motion in x direction, m
NPz	Nucleus pulposus	Motion in z direction, m
FPP	Facet surface	Separation, m

PREDICTION OF SYMPTOM DURATION The ability of the C5/C6 motion segment model loading to predict the persisting (greater than 1 month) whiplash symptoms of vehicle occupants was investigated by means of diagnostic tests based on the concept of positive and negative predictive values (Altman et al., 2000). A criterion value was selected based on a threshold for the component loading, such that if the loading was greater than the threshold, then it was predicted the subject was injured. These thresholds were chosen to give a sensitivity of 85%. The confidence intervals for the thresholds were calculated. Based on these criterion thresholds and the actual persisting whiplash symptoms, a 2-by-2 contingency table was constructed, Table 2.

Table 2 – The 2x2 contingency table

Neck Injury Symptoms		Long-term	No or initial	Total
Criterion Value	Above threshold	a	b	a + b
	Below threshold	c	d	c + d
	Total	a + c	b + d	a + b + c + d

Based on the contingency table above, the following indexes are defined:

- **Sensitivity = a/(a + c):** The proportion of occupants with long-term symptoms that have estimated criterion values above a fixed level.
- **Specificity = d/(b + d):** The proportion of occupants with no symptoms or initial symptoms that have estimated criterion values below a fixed level.
- **Positive predictive value = a/(a + b):** The proportion of occupants with estimated criterion values above a fixed level that have long-term symptoms.
- **Negative predictive value = d/(c + d):** The proportion of occupants with estimated criterion values below a fixed level that have no symptoms or initial symptoms.

LIMITATIONS OF THIS STUDY are related to the use of the linked simulations performed for other purposes. To reconstruct the possible early injury causation in the C5/C6 motion segment of the neck in real accidents required two major simplifying assumptions. Firstly, the C5/C6 level of the neck was assumed to be a significant source of injury in rear impacts. Support for this comes from various sources including: accident studies (Gibson et al., 2000); clinical investigation (Aprill & Bogduk, 1992); and the testing of both volunteers (Kaneoka & Ono, 1998) and cadavers (Deng et al., 2000a). Secondly, it was assumed the injury was most likely caused during the initial forward motion of the thorax with respect to the head, before the head impacts with the head restraint. Support for this comes from accident studies demonstrating the partial effectiveness of head restraints in reducing soft-tissue neck injury, such as that of Morris and Thomas (1996) and again the testing of both volunteers and cadavers with the high-speed radiography in Kaneoka & Ono (1998) and Deng et al. (2000a). Current injury hypotheses based on volunteer testing (Ono et al., 1997) and cadaver testing (Yoganandan et al., 1998; Deng et al., 2000b) have emphasised the early-onset injury mechanism. In this study, the soft-tissue loading at C5/C6 of the head and neck model was analysed only during the early stage of the impact, prior to contact with the head restraint.

Each area of the study had limitations with respect to the data used. In summary:

- The use of the in-vehicle crash pulse recording improves the accuracy of the vehicle crash parameters, but occupant factors such as seating position and stature were not available;
- The BioRID II 50th-percentile male dummy and seat model with no seatbelt for the crash reconstruction has limitations regarding T1 motion biofidelity of the dummy model, the effect of ramping up the seat back, occupant posture and anthropometry.
- The reconstruction of the vertebral motion by the head and neck model has limitations with regard to biofidelity, position of the neck, fixed anthropometry, passive muscle responses and lack of head contact with the head restraint.

RESULTS

VERIFICATION OF THE HEAD AND NECK MODEL responses was by comparison of the new C5/C6 motion segment with the original head and neck model and volunteer corridors derived from sled test with a rigid seat (ΔV=9.3 km/h) by Davidsson et al. (1999), Figure 5. The results obtained were found to be very similar to the validation of head and neck model with passive muscle responses in rear impacts by van der Horst (2002). The overall motion of the head was found to be the same in the time period up to 150 ms. Therefore, the original validation of the model by van der Horst was accepted for the purposes of this study. The results obtained by Meijer et al. (2001) using the whole MADYMO human body model including the van der Horst head and neck model are included for comparison. The models fail to predict the slight flexion motion of the neck that occurs in the initial 100 ms of the T1 motion. The effect of this motion is evident in the slightly flexed upper cervical spine in the high-speed radiograph taken at t=44 ms for the volunteer in Figure 6.

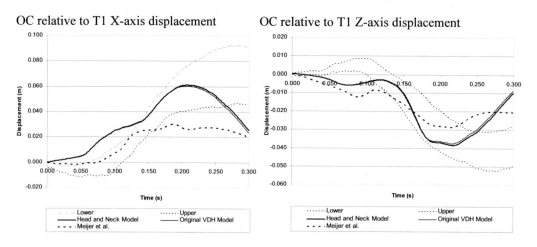

Fig. 5 – Comparison of the neck responses predicted by the head and neck model, the original van der Horst model and the Meijer et al. (2001) human body model, compared with the corridors (mean ± SD) from the volunteer sled test with a rigid seat (ΔV=9.3 km/h) by Davidsson et al. (1999).

The head and neck motion predicted by the model are compared in Figure 5 to the vertebral motion of the Ono et al. (1997) volunteer (S6) from the high-speed radiographs recorded during the test. The neck link moves from upright and neutral to full extension with similar timing. Detail comparisons of the vertebral motion from high-speed radiography of Ono et al. (1997) volunteer and Deng et al. (2000b) cadaver testing were made. These showed large variations in the individual test subject vertebral motion during the test. It was difficult to use the angular motion of the vertebra as a verification tool without significantly more test data being available.

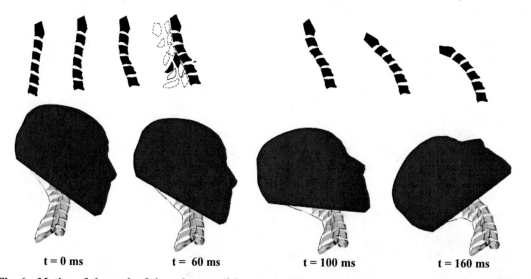

Fig. 6 - Motion of the neck of the volunteer (S6) up to t=160 ms from high-speed radiography (top row), Ono et al. (1997), and of the head and neck model (lower row).

COMPARISON OF REAL CRASH CASES Two of the crash cases, one of high and one of low severity are presented as examples of the responses predicted by the C5/C6 model. The local T1 accelerations for three cases are compared with an Ono et al. (1997) volunteer test in Figure 7.

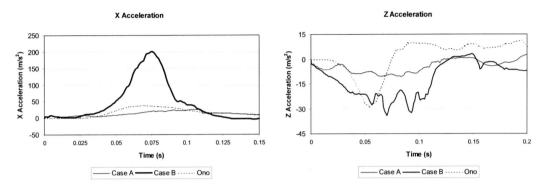

Fig. 7 – Local T1 accelerations predicted by the BioRID II and seat model for Case A, Case B and for the Ono et al. (1997) 8.0 km/h volunteer tests used to drive the head and neck model.

Case A is a low-severity rear impact of 2.5g. The driver was uninjured and the passenger had pain for 1-month duration (Injury Group 1). Case B is one of the more severe crashes in the crash data, with a T1 x-acceleration of 20g. The driver had pain persisting for longer than 6 months (Injury Group 3) post impact. The Ono volunteer tests had no head restraints, no recorded injuries and a measured T1 x-acceleration of 3.5g. Due to the test set-up the T1 z-acceleration of 2.9g for the volunteer is about 80% of that for the x-direction acceleration, where, on average for the crash reconstructions, the z-acceleration is only 40% of the x-acceleration.

The NIC_{max} was predicted by the head and neck model for the two cases and the Ono volunteer test. The low-severity case, Case A, had a NIC_{max} of 4 (at t=110 ms) the severe impact, Case B, a NIC_{max} of 31 (at t=75 ms), and the Ono volunteer test had a NIC_{max} of 8 (at t=70 ms). If the positive NIC_{max} is greater than 15, then the impact is likely to result in longer-term injury, Bostrom et al. (1996).

Due to the combined extension and retraction motion of the head and neck due to the rear impact, the predicted elongations were greater for ligaments in the anterior portions of the neck structures; the predicted anterior longitudinal ligament [ALL] and the anterior facet capsule [FC] ligament elongations for the three examples are in Figure 8. Elongations resulting from the severe impact, Case B, were significantly higher than the others.

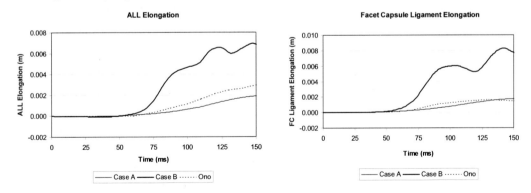

Fig. 8 – The C5/C6 anterior longitudinal ligament elongation and anterior FC ligament elongation predicted for Case A, Case B and the Ono et al. (1997) volunteer test.

The C5/C6 model predicted the critical motion segment bearing-surface displacements for the three cases. The variation from the rest position of the rear edge of the facet surface [FPP] using C6 as the reference was predicted, Figure 9. When FPP has a negative value, as in Case B and for a limited time for the Ono volunteer, it is an indication of the facet surfaces coming together and the possibility of facet surface impingement exists. In Case A the motion is dominated by rearward motion of the head, the facet surfaces close slightly and then separate.

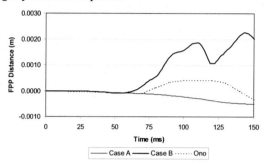

Fig. 9 – The separation of the posterior facet surfaces [FPP] using C6 as the reference. A negative value is an indication of possible facet surface impingement.

THE CRASH RECONSTRUCTIONS Four of the cases could not be used due to instability leading to asymmetrical responses. The 74 remaining rear impact crashes with associated whiplash injury outcomes were classified into the 4 neck injury groups by duration of the resultant pain. In Figure 10, the separation distance of the posterior facet surfaces for the C5/C6 motion segment were plotted against time for the four injury groups (defined in the summary of Study D). The cases can be seen to have one of two types of early C5/C6 motion segment trajectory. The first is connected with the more severe impacts, as represented by Case B. The motion is dominated by the shearing load due to head inertia and leads to the facet surfaces separating with little chance of impingement. In the second type of motion, the rear facet surfaces move together during the early motion (up to t=150 ms), as represented by the low severity Case A, and there is possible impingement or interaction of the facet capsular surfaces and other structures such as the meniscus.

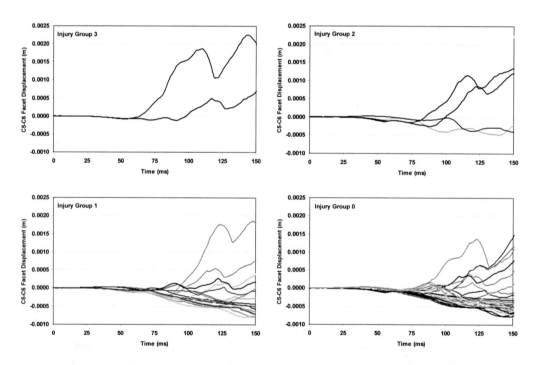

Fig. 10 – The separation of the facet surfaces [FPP] predicted by the C5/C6 motion segment model with time during the rear impact crash reconstructions (n=74).

PREDICTION OF SYMPTOM DURATION The ability of the C5/C6 peak loading data for the 74 drivers and 29 passengers remaining to predict the longer duration whiplash symptoms was analysed. Seven of the drivers and 3 passengers had longer duration whiplash symptoms (defined as greater than 1 month). The predictive capability of each load type from Table 1 was analysed for high and low cut-offs for drivers only and for drivers and front seat passengers combined. The thresholds aimed to give a sensitivity of 85%, the actual low cut off selected giving 86% sensitivity and the high 71%. The contingency tables were constructed for each criterion, allowing the sensitivity, specificity, positive predictive value, negative predictive value and the 95% confidence intervals to be calculated. The results for both the high and low cut-off values are given in Table 3 for the drivers only and Table 4 for the drivers and passengers combined.

The strains of the anterior aspects of the major C5/C6 motion segment ligaments, ALL, FC and AF were all good predictors of longer-duration pain outcomes. The x-displacements of the bearing components of the motion segment, AFx, NPx and the facet surfaces, Fx, were also good predictors of the occurrence of longer duration pain, as was the displacement in the z-direction of the posterior nucleus pulposa, NPz. NIC_{max} was also shown to be a good predictor of injury.

Table 3 – The loading of the C5/C6 motion segment components for the drivers only.

(a) For the low cut-off.

Component	Drivers Only				
	Cut-off	Sensitivity	Specificity	Positive Predictive Value (±Confidence Interval)	Negative Predictive Value (±Confidence Interval)
ALL	0.13	0.86	0.72	0.24 (±0.17)	0.98 (±0.04)
FC	0.40	0.86	0.74	0.25 (±0.17)	0.98 (±0.04)
AF	0.33	0.86	0.74	0.25 (±0.17)	0.98 (±0.04)
BF z	-36.3	0.86	0.00	0.08 (±0.06)	0.00 (±0.00)
F x	-0.003	0.86	0.75	0.26 (±0.18)	0.98 (±0.04)
F z	-.00004	0.86	0.00	0.08 (±0.06)	0.00 (±0.00)
AF x	-0.002	0.86	0.75	0.26 (±0.18)	0.98 (±0.04)
AF z	-.00006	0.86	0.13	0.09 (±0.07)	0.90 (±0.19)
NP x	-0.0025	0.86	0.75	0.26 (±0.18)	0.98 (±0.04)
NP z	-.00034	0.86	0.72	0.24 (±0.17)	0.98 (±0.04)
FPP	0.00044	0.86	0.04	0.09 (±0.07)	0.75 (±0.42)
NICmax	10.86	0.86	0.72	0.24 (±0.17)	0.98 (±0.04)

(b) For the high cut-off.

Component	Drivers Only				
	Cut-off	Sensitivity	Specificity	Positive Predictive Value (±Confidence Interval)	Negative Predictive Value (±Confidence Interval)
ALL	0.17	0.71	0.90	0.42 (±0.28)	0.97 (±0.04)
FC	0.50	0.71	0.85	0.33 (±0.24)	0.97 (±0.05)
AF	0.23	0.71	0.88	0.38 (±0.26)	0.97 (±0.04)
BF z	-41.8	0.71	0.06	0.07 (±0.06)	0.67 (±0.38)
F x	-0.004	0.71	0.85	0.33 (±0.24)	0.97 (±0.05)
F z	-0.00005	0.71	0.01	0.07 (±0.06)	0.33 (±0.53)
AF x	-0.003	0.71	0.85	0.33 (±0.24)	0.97 (±0.05)
AF z	-0.00007	0.71	0.40	0.11 (±0.09)	0.93 (±0.09)
NP x	-0.003	0.71	0.85	0.33 (±0.24)	0.97 (±0.05)
NP z	-0.0004	0.71	0.75	0.23 (±0.18)	0.96 (±0.05)
FPP	0.0004	0.71	0.06	0.07 (±0.06)	0.67 (±0.38)
NICmax	12.88	0.71	0.82	0.29 (±0.22)	0.96 (±0.05)

Table 4 – The loading of the C5/C6 motion segment components for the drivers and passengers only.

(a) For the low cut-off.

Component	Drivers and Passengers				
	Cut-off	Sensitivity	Specificity	Positive Predictive Value (±Confidence Interval)	Negative Predictive Value (±Confidence Interval)
ALL	0.13	0.89	0.73	0.24 (±0.15)	0.99 (±0.030)
FC	0.40	0.89	0.76	0.26 (±0.15)	0.99 (±0.03)
AF	0.33	0.89	0.76	0.26 (±0.15)	0.99 (±0.03)
BF z	-36.3	0.89	0.00	0.08 (±0.05)	0.00 (±0.00)
F x	-0.003	0.89	0.77	0.27 (±0.16)	0.99 (±0.03)
F z	-0.00004	0.89	0.00	0.08 (±0.05)	0.00 (±0.00)
AF x	-0.002	0.89	0.77	0.27 (±0.16)	0.99 (±0.03)
AF z	-0.00004	0.89	0.04	0.08 (±0.05)	0.80 (±0.35)
NP x	-0.002	0.89	0.77	0.27 (±0.16)	0.99 (±0.03)
NP z	-0.0003	0.89	0.72	0.24 (±0.14)	0.99 (±0.03)
FPP	0.0004	0.89	0.03	0.08 (±0.05)	0.75 (±0.42)
NICmax	10.85	0.89	0.72	0.24 (±0.14)	0.99 (±0.03)

(b) For the high cut-off.

Component	Drivers and Passengers				
	Cut-off	Sensitivity	Specificity	Positive Predictive Value (±Confidence Interval)	Negative Predictive Value (±Confidence Interval)
ALL	0.17	0.78	0.87	0.37 (±0.22)	0.98 (±0.03)
FC	0.50	0.78	0.85	0.33 (±0.20)	0.98 (±0.03)
AF	0.46	0.78	0.87	0.37 (±0.22)	0.98 (±0.03)
BF z	-39.90	0.78	0.02	0.07 (±0.05)	0.50 (±0.49)
F x	-0.004	0.78	0.85	0.33 (±0.20)	0.98 (±0.03)
F z	-0.00005	0.78	0.01	0.06 (±0.05)	0.25 (±0.42)
AF x	-0.003	0.78	0.85	0.33 (±0.20)	0.98 (±0.03)
AF z	-0.00006	0.78	0.14	0.08 (±0.06)	0.87 (±0.17)
NP x	-0.003	0.78	0.85	0.33 (±0.20)	0.98 (±0.03)
NP z	-0.0004	0.78	0.76	0.23 (±0.15)	0.97 (±0.03)
FPP	0.00041	0.78	0.043	0.07 (±0.05)	0.67 (±0.38)
NICmax	12.88	0.78	0.84	0.32 (±0.19)	0.98 (±0.03)

An injury criterion for use in designing safety equipment needs to predict a non-injurious event with absolute accuracy (100%), but only requires reasonable accuracy (10-50%) when predicting an injurious event. For the 85% sensitivity (the proportion of injured occupants above the chosen cut-off/threshold used here), the negative predictive values for the good predictive parameters (ALL, FC and AF ligament elongations; AFx, NPx, Fx and NPz displacements) were close to 100%. The proportion of all occupants below the threshold that were uninjured was close to 100%, or stated in a different way, this was the probability of correctly predicting a non-injurious event. The positive predictive value is approximately 30%. This is the proportion of all occupants above the threshold that were uninjured and is the probability of correctly predicting an injurious event. The corresponding thresholds were found to be 17% relative elongation of the ALL, 50% relative elongation of the FC, and 23% relative elongation of AF – considering drivers only and the high cut-off point. The predictive capabilities were slightly improved when the both drivers and passengers were considered.

DISCUSSION

In a multi-body model, such as the neck and head model, the ligaments are represented by discrete spring/damper units with attachments to the superior and inferior vertebra, Figure 1B. The ALL was shown to have the highest predicted elongation and as it is also the longest ligament in the motion segment the lowest predicted strain in a rear impact, followed by the AF and the FC, which are each progressively shorter and have higher strains.

In the more severe cases the acceleration pulse causes significantly higher strains of all the ligaments. For Case B, the ALL predicted maximum ligament strain was 36% and for the FC ligaments it was 90%, Table 5. Failure strains of 31% (±5.9%) for the ALL and 116% (±19.6%) for the FC ligaments were measured quasi-statically by Yoganandan, Kumaresan and Pintar (2000). As allowable dynamic ligament failure loads have been found to be rate dependent (Yoganandan et al., 1998), these predicted ligament strains are most probably in the area of sub-catastrophic failure. The strain levels for Case A and the Ono volunteer test are significantly lower and by this criteria are unlikely to be injurious. The NIC_{max} predictions agree with this.

In the severe impact example Case B in Table 5, the predicted maximum strain in the C5/C6 anterior ligaments of the facet capsule [FC] is in the region of sub-catastrophic failure suggested by Siegmund et al. (2000b). The onset of the shear loading to the motion segment in Case B appears to be so rapid that the facet surfaces are simply pulled apart. When all the crash data is investigated, the predicted average strains for the cases with pain duration of greater than 1 month (n=7), the average peak strain is 64% (±20%). For the cases with pain duration less than 1 month (n=19), the predicted average strain is 37% (±17%) and for the remainder of cases with no pain (n=48) the predicted strain is 31% (±15%). All this indicates a correlation between the predicted anterior facet ligament strain and the duration of pain in this sample of crashes. This gives support to the facet shear mechanism of whiplash injury suggested by Deng et al. (2000b).

Table 5 – The predicted C5/C6 motion segment maximum strain for Case A and Case B, and the Ono et al. (1997) volunteer test compared with *in-vitro* measured failure strains from various sources.

Ligament	Case A		Case B		Ono Volunteer		Failure Strain % (±SD)
	Strain %	Injury	Strain %	Injury	Strain %	Injury	
ALL	10	–	36	+	18	–	31 (±5.9) [a.]
AF fibre	13	–	60	+	23	–	60 [b.]
FC	25	–	90	+	26		116 (±19.6) [a.] 94 (±85) [c.] 35 (±21) [d.]

Sources: a. Yoganandan, Kumaresan & Pintar (2000), b. Pintar et al. (1986), c. Siegmund et al. (2000b), d. Sub-catastrophic failure, Siegmund et al. (2000b).

When the facet impingement for the three cases is considered, the predicted dynamic motion of C5/C6 facet surfaces for Case A and the Ono volunteer appear to make such an event possible. Figure 8 indicates a decrease in the facet surface separation occurs between 40 and 75ms, supporting the Ono et al. (1997) facet impingement hypothesis. However, the separation distance of the facet surfaces [FPP] and the force on the posterior region of the facet surfaces [BFz] were not found to be good predictors of long duration pain. These parameters with poor predictive capabilities had positive predictive values close to zero (considering the 95% confidence interval). When the predicted FC ligament strain is plotted against the total rear facet surface compression force, an interesting pattern is seen in Figure 10. The two Group 3 injury cases, the most severely injured vehicle occupants in the study, have shear-dominated motion, which includes positive separation of the facet surfaces. Three of the five cases from Group 2 also have separation of the facet surfaces. The remaining two cases have contact of the facet surfaces.

A similar split in the motion types is found in the lower-severity injury groups with initial pain, and no injury. It appears that at the higher impact severities, the shear displacement velocity of C5 with respect to C6 becomes great enough that the facet capsular surfaces separate before the extension rotation of C5 with respect to C6 occurs (shear injury to the facet capsule). It is the timing and magnitude of the extension rotation of C5 with respect to C6, which causes the facet surfaces to impinge. Figure 10 shows the two regions clearly: the shear injury region due to facet ligament strain greater than 40% in the upper right, and the possible facet impingement region related to force on the rear facet surface of greater than 35 N. The predicted force on the facet surface in compression is based on the assumption in the van der Horst (2002) head and neck model that the facet is twice as stiff as the disc. This assumption requires further laboratory-based testing.

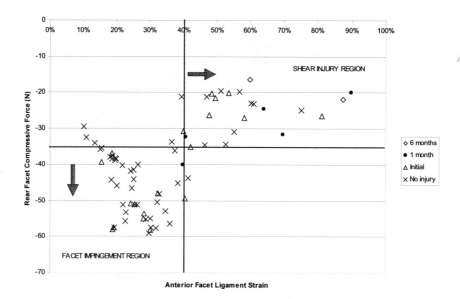

Fig. 11 – Predicted C5/C6 anterior facet ligament strain plotted with the rear facet surface compression force for the drivers only (n=74) in the crash reconstructions with the different injury outcomes. The dotted lines and arrows represent the greater than 40% strain and 35N force regions.

SUMMARY

A detailed C5/C6 motion segment model was used to investigate early AIS1 soft-tissue neck injury in rear impacts. This required linking together the results obtained from the four independent studies. A C5/C6 motion segment model developed by Gibson (2005) was included in the MADYMO-based human head and neck model, van der Horst (2002). The head and neck model was driven by T1 accelerations derived from the reconstruction of a group of real life crashes using a MADYMO BioRID II dummy and seat model, where the injury outcome was known (Kullgren et al., 2003). Injuries to the occupants of the crashed vehicles were available in the form of pain duration following the impact.

The predicted early C5/C6 motion segment component strains resulting from the crashes were correlated with the real life AIS1 neck injuries to the vehicle occupants. Good correlations were obtained with predicted strain of the model C5/C6 motion segment joint ligaments, with critical levels occurring in the anterior facet capsule ligaments. The thresholds predicted for long-term pain duration as a result of a rear impact was found to be 40% strain of the anterior FC ligaments and a 35N load on the rear of the facet surfaces. The model also predicted good correlation between NIC_{max} and the duration of pain following the impact.

The C5/C6 model predicted two distinct motions resulting from the crash loads and therefore supports the two main hypotheses of early whiplash-associated injury causation: a shear dominated motion, Deng et al. (2000b); and, a rotation dominated motion with possible facet impingement, Ono et al. (1997).

REFERENCES

Altman, D, Machin, D, Bryant, T, & Gardner, M (2000), *Statistics with Confidence*, 2nd Edition, British Medical Journal Books, UK.

Aprill, C & Bogduk, N (1992), 'The prevalence of cervical zygapophyseal joint pain: A first approximation', *SPINE*, vol. 17, pp. 744-7.

Bostrom, O, Svensson, MY, Aldman, B, Hansson, H, Håland, Y, Lövsund, P, Seeman, T, Suneson, A, Säljö, A, & Ortengren, T (1996), 'A new neck injury criterion candidate – based on injury findings in the cervical spinal ganglia after experimental neck extension trauma', *Proceedings of the 1996 IRCOBI International Conference on the Biomechanics of Impacts.*

Davidsson, J, Flogard, A, Lövsund, P & Svensson, MY (1999), 'BioRID P3 – design and performance compared to Hybrid III and volunteers in rear impacts at Delta V = 7 km/h', *Proceedings of the 43rd STAPP Car Crash Conference.*

Deng, B, Begeman, PC, Yang, KH, Tashman, S & King, AI (2000a), 'Kinematics of human cadaver cervical spine during low speed rear-end impacts', *44th STAPP Car Crash Conference*, Paper no. 2000-01-SC13.

Deng B, Luan F, Begeman P, Yang K, King, A & Tashman, S (2000b), 'Testing the shear hypothesis of whiplash injury using experimental and analytical approaches', in N Yoganandan & F Pintar (eds.), *Frontiers in Whiplash Trauma*, IOS Press: Netherlands.

Eriksson, L & Kullgren, A (2003), 'Influence of seat geometry and seating posture on NIC_{MAX} and Nkm AIS 1 neck injury predictability', *Proceedings of the International Conference on the Biomechanics of Impacts.*

Gibson, T, Bogduk, N, MacPherson, J & McIntosh, A (2000), 'Crash characteristics of whiplash associated chronic neck pain', *J Musculoskeletal Pain*, vol. 8, pp. 87-95.

Gibson, T (2005), 'Development and Validation of a C5/C6 Motion Segment Model', PhD Thesis submitted to the University of NSW, Australia.

Kaneoka, K & Ono, K (1998), 'Human volunteer studies in whiplash injury mechanisms', in N Yoganandan, F Pintar, S Larson & A Sances (eds.), *Frontiers in Head and Neck Trauma*, IOS Press, Netherlands, pp. 313-325.

Kullgren, A, Eriksson, L, Bostrom, O & Krafft, M (2003), 'Validation of neck injury criteria using reconstructed real-life rear-end crashes with recorded crash pulses', *Proceedings of the Enhanced Safety of Vehicles Conference*, Japan.

Meijer, R, Ono, K, van Hoof, J, Kaneoka, K (2001), 'Analysis of rear end impact response using mathematical human modelling and volunteer tests (volunteer data & simulation)', *Journal of SAE, Spring Convention*, 20015357.

Mercer, S & Bogduk, N (1999), 'The ligaments and anulus fibrosus of the human adult cervical intervertebral disc', *SPINE*, vol. 24 (7), pp. 619-28.

Morris, A & Thomas, P (1996), 'A study of soft tissue neck injuries', *Proceedings of the 15th International Conference on Experimental Safety Vehicles*, Melbourne.

Ono, K, Kaneoka, K, Wittek, A & Kajzer, J (1997), 'Cervical injury mechanism based on the analysis of human cervical vertebral motion and head-neck-torso kinematics during low speed rear impacts', *41st STAPP Car Crash Conference*, Paper no. 973340, pp. 339-356.

Pintar, F, Mykelburst, J, Sances, A & Yoganandan, N (1986), 'Biomechanical Properties of the human intervertebral disc in tension', in SA Lantz & AI King (eds.), *Proceedings of the Advances in Bioengineering Conference,* USA.

Siegmund, G, Brault, JR & Chimich, DD (2000a), 'Do cervical muscles play a role in whiplash injury?', *Journal of Whiplash and Related Disorders*, vol. 1(1).

Siegmund, GP, Myers, BS, Davis, MB, Bohnet, HF & Winkelstein, BA (2000b), 'Human cervical motion segment flexibility and facet capsular ligament strain under combined posterior shear, extension and axial compression', *Proceedings of the 44th STAPP Car Crash Conference*, Paper No. 2000-01-SC12.

Stemper, B, Yoganandan, N & Pintar, F (2004), 'Validation of a head-neck computer model for whiplash simulation', *Med. Biol. Eng. Comput.*, vol. 42, pp. 333-8.

van der Horst, M (2002), 'Human head neck response in frontal, lateral and rear end impact loading - modelling and validation', PhD Thesis, the Technical University, Eindhoven, Netherlands.

Yang, KH & King AI (1984), 'Mechanism of facet load transfer as a hypothesis for low back pain. *Spine,* 1984;9, pp557-565.

Yoganandan, N, Kumaresan, S & Pintar, FA (2000), 'Geometric and mechanical properties of human cervical spine ligaments', *Journal of Biomechanical Engineering*, vol. 122, pp. 623-9.

Yoganandan, N, Pintar, FA, Cusick, JF, Sun, E & Eppinger, R (1998), 'Whiplash Injury Mechanisms', Whiplash '98 Symposium, p. 23.

2000-01-0166

An Energy Based Analysis for Neck Forces in Frontal Impacts

Joseph E. Hassan and Guy Nusholtz
Daimlerchrysler

ABSTRACT

In 1997 Mertz, Prasad and Irwin [1] have described a technique for the development of injury risk curves for measurements made with the CRABI and Hybrid III family of biofidelic child and adult dummies that are used to evaluate restraint systems in frontal collision simulations. They further developed normalized injury risk curves for neck tension, neck extension moment, combined neck tension and extension moment for adults and children.

The approach described by Mertz et al [1], is based on lines of equivalent stress and uses the maximum normal stress theory of failure to impose limits of the risk of injuries. In this paper a complementary approach is described based on the maximum energy of failure and lines of constant energy. A special case of this approach in 1D is used to develop the assessment values obtained by Mertz et al [1]. Limitations and advantages of the energy based approach are described, with especial emphasis on future implementation.

INTRODUCTION

Mertz et al [8] and Prasad and Daniel [9] have conducted tests to assess the effects of deploying passenger airbag interactions with animals that were chosen to represent the size, weight and state of tissue development of 3-year old children. In their studies a series of matched tests was conducted using the 3-year old "airbag" dummy. This allowed the various injury severities experienced by the pig to be correlated with corresponding dummy response measurements. Both studies concluded that neck tension was the best indicator of the onset of AIS ≥3 neck injury with no AIS ≥ 3 neck injury occurring below a neck tension of 1160 N. Mertz and Prasad [1] developed injury risk curves for extension moment and the combination of tension and extension moment based on the combined data published indepedently in [8] and [9]. Since both of these data sets are estimates of the tolerance of a 3-year old child, their analytical treatment used analytical normalization techniques for size and strength in order to give estimates of injury risk curves for any ages of children or sizes of adults.

Mertz and Prasad's [1] treatment of the available test data is based on lines of constant stress to develop combined neck injury risk. However, it does not refer to muscle effect in the overall treatment. While force/stress treatment is a reliable analytical approach, the effect of muscles is not taken into consideration due to the fact that allowable forces and stresses in the muscle cannot be assessed in a normalized manner. The approach outlined in this paper can include the muscle effect as part of modeling neck injury risk. While it is being demonstrated in this work in the 1-D domain, it can be easily extended to 2 and 3D analyses.

ANALYTICAL TREATMENT

The following is a discussion for the development of energy based formulation for injury risk curves for AIS ≥ 3 neck injury for tension/extension loading of the neck based on peak neck tension, peak neck extension moment which can be used for the CRABI and HYBRID III family of child and adult dummies.

Table 1. Neck Circumferences for Various Ages of Children and Sizes of Adults

Dummy	Circum.(mm)	A_c
CRABI 6	221	0.906
CRABI 12	226	0.918
CRABI 18	226	0.926
H III - 3 Yr.	244	1.000
H III - 6 Yr.	264	1.082
H III - Sm. Fem.	304	1.246
H III - Mid-Male	383	1.57
H III - Lg. Male	421	1.725

SIZE SCALE FACTORS – Several anthropometry studies [11-14] have found that the neck size can be characterized by neck circumference. A typical relationship that describes such dependency is given in table 1 and in [1]. Since the reference published data in [1] applies to 3-year old, a neck size scale factor, A_c, can be defined as,

$$A_c = \frac{(\text{Neck Circumference})}{(\text{Neck Circumference of a 3-years old})} \quad (1)$$

ENERGY BASED FAILURE CONSIDERATIONS – Historically the maximum tissue stress was chosen as the failure mechanism for determining risk injury criterion and the failure stress level was assumed to be independent of age. These assumptions allow the elastic modulus of the ligament to vary with age. However, it does not explicitly explain the deterioration of the tissue and bones ultimate strength with age. *Currey and Butler [22]* showed that the modulus of elasticity and bending strength both increase with age until the age of 30 years, but decrease thereafter. Similar results were found by Goldstein, et al [23] and Zhou, et al [24] . All studies reviewed indicate that human injury tolerance reduces as age increases because of the aging of human bones and tissues. The mechanical properties of the human bones change significantly with age mainly due to increased porosity and the change of mineralization of bones. It was also reported that the reduction in ultimate strength and fracture toughness of human bones is about 25% to 55% within the entire life span. Human soft tissues have exhibited similar reductions in their strength.

Table 2. Modulus dependency on age

Subject	Modulus	E/E_3 ratio	λ_w
6-Months	0.282	0.596	1.991
12-Months	0.322	0.681	1.863
18-Months	0.362	0.765	1.757
3-Year Old	0.473	1.000	1.537
6-Year Old	0.667	1.410	1.294
Sm. Adult Fem.	1	2.114	1.057
Mid. Size Male	1	2.114	1.057
Lg Adult Male	1	2.114	1.057

The distortion energy theory of failure predicts that failure will occur whenever the distortion energy in a unit volume equals the distortion energy in the same volume when uniaxially stressed to its failure strength. This approach enables the treatment of the complex stress cases with ease, by considering adding different energies from different configurations. Therefore, a complex geometry such as the neck can be handled easily by considering the energy dissipated by each component of the structure. Treatment in this regard does not have to use a 1D simplifications, as in the case of constant stress theory, to derive neck injury risk.

Figure (4) depicts the complex geometry of the human neck. Representation of the muscle effect in the maximum stress theory of failure requires assessment of this complex geometry in the form of cross section area, and other inertia properties in the spatial domain. The distortion energy theory however, does not require complex geometry definitions. In fact several researchers [26,27] have already quantified the muscle effect in the form of mass, spring damper idealization, which is easily adapted to the energy approach.

Figure 4.   Neck muscles

It should also be noted that the three neck criteria; maximum neck tension, maximum neck extension moment and maximum value of the combination, are measures of "macro" structural load carrying capacities of the neck structure. As such their failure values will be dependent not only on the ligament load, but also on the corresponding load being transmitted by the muscle groups. Therefore, an analytical treatment that can simplify taking into account different models available for muscle behavior is extremely attractive. The maximum distortion energy theory as described below, is a viable tool in this regard.

As a verification to the distortion energy failure approach, a special case of 1D- uniform stress is used to derive the same parameters obtained by Mertz et al. [1]. In this 1D case the strain energy reaches a threshold failure level defined by $U_d = ((1+\nu)^*\sigma^2)/3E$ Where ν and E are the Poisson's ratio and the elasticity Modulus respectively. Therefore, it accounts for variations of properties with age as reflected on E and ν.

NECK TENSION – The relationship between the ratios of neck tension forces, A_F, the sizes of the necks A_A, and the average stresses, A_σ, can be expressed as,

$$A_\sigma = A_F /A_A \qquad (2)$$

The distortion energy theory of failure for 1D- uniform stress indicates that the strain energy reaches a threshold failure level defined by

$$U_d = ((1+\nu)^*\sigma^2)/3E \qquad (3)$$

Where ν and E are the Poisson's ratio and the elasticity modulus respectively.

Further define $(1+\nu)/3E = 1/ (E_3 \lambda^2_w) \qquad (4)$

Where E_3 is the elasticity Modulus of the reference 3 year old child. Researchers [22,23,24] point out that the variation of Poisson's ratio with respect to the range of age considered here is very limited. Therefore, the assumption is made to take an average value of ν to be 0.27 for all the ages considered in this approach. Therefore, one can further define

$$\lambda^2_w = 2.37 (E/E_3) \tag{5}$$

$$\text{Or for a 3 year old} \quad \lambda_w = 1.54 \tag{6}$$

Therefore λ_w will represent the dependency of energy to failure on age as a result of changing the Elastic modulus as shown in table 2.

Substitution from equations 4, 5 and 6 in 3 yields,

$$U = A^2_\sigma /(E_3 \lambda^2_w) \tag{7}$$

For children and adults, the average strain energy based on the cross-sectional areas of their necks will be taken as equal for equal injury severity. The ratio of their cross-sectional areas, A_A, will be taken as the ratio of the square of their neck circumferences, A_c^2. Therefore, from Equation 2, 6 and 7, the ratio of the tensile forces that corresponding to equal injury severity can be shown to be,

$$(A_F/ A_{F3}) = (\lambda_w/1.54) \cdot A_c^2 \tag{8}$$

where A_{F3} is the tensile force for the reference 3-year old child.

Since data for neck tension forces and corresponding neck injury severities for the 3-year old child exist, the neck tension forces causing the same injury severity in other size occupants can be determined using equation 8.

The neck tension force and corresponding neck injury severity data of Mertz et al [8] and Prasad and Daniel [9] were analyzed in [1], in order to determine the injury risk curve for AIS $\geq$ 3 neck injury based on neck tension forces experienced by a 3-year old child. The results given in [1] lists the neck forces normalized by the tension force, F_1, that corresponds to a 1 percent risk of AIS $\geq$ 3 neck injury. For the 3-year old, this force is 1070 N. For any size person, the corresponding value A_F can be determined from Equation 8, or

$$A_F = (\lambda_w/1.54) \cdot A_c^2 \cdot 1070N \tag{9}$$

Where A_c is defined by equation 1.

The normalized injury risk curve for the 3-year old child is identical to the normalized risk curve for any size occupant provided the normalized force is computed by the relationship by Equation 4. Figure 1 gives the risk curve for AIS $\geq$ 3 neck injury based on normalized neck tension for any size person. Normalizing values for various child and adult dummies are given in table 3. These values were calculated using Equation 8 and the values of A_c

given in Table 1. Note that the normalized values are the neck tensions that produce a 1 percent risk of AIS $\geq$ 3 neck injury for the corresponding dummy. Further, this curve gives an estimate of the injury risk when the neck is being loaded in tension and extension which is the loading mode experienced by the pigs and the child dummy in the biomechanical tests.

Table 3. Tension

Subject	A_C	λw	$A_F(N)^+$ Stress based	$A_F(N)$ Energy based
6-Months	0.906	1.187	690	678
12-Months	0.918	1.268	740	744
18-Months	0.926	1.345	810	803
3-Year Old	1.000	1.537	1070	1070
6-Year Old	1.082	1.825	1420	1487
Sm. Adult Fem.	1.246	2.235	1960	2415
Mid. Size Male	1.57	2.235	3110	3835
Lg Adult Male	1.725	2.235	3760	4629

+ From Ref. [25]

NECK EXTENSION MOMENT – For structures whose cross-sectional areas can be characterized by a single length scale factor, A_L, Mertz et al [13] have shown that the relationship between the ratios of the internal bending moments, A_M, and the internal bending stresses, A_σ, can be expressed as,

$$A_\sigma = A_M / A_L^3 \tag{10}$$

Again we specify equal strain energy to failure for equal injury severity. From Equation 10, the ratio of neck extension stresses associated with neck injury is,

$$A_\sigma/ A_{\sigma3} = (A_M / A_{M3}) \cdot (A_{L3}^3/ A_L^3) \tag{11}$$

Where A_{M3} and A_{L3} are the values corresponding to the 3-year old child reference

$$\text{Recall that } A_L/A_{L3} = A_c^3 \tag{12}$$

And using equations 3,4,5 and 6 one can arrive at

$$(A_M/ A_{M3}) = (\lambda_w/1.54) \cdot A_c^3 \tag{13}$$

Since data for neck extension moments and corresponding neck injury severities for the 3-year old child exists, the neck extension moments causing the same injury severity in other size occupants can be estimated from Equation 13.

The neck extension moment and corresponding neck injury severity data of Mertz et al [8] and Prasad and Daniel [9] were analyzed using the Mertz/Weber Method [6]. The injury risk for AIS $\geq$ 3 neck injury based on the neck extension moments experienced by a 3-year old child were reported in [1]. The neck extension moments are normalized by the moment, M_1, that corresponds to a 1 percent risk of AIS $\geq$ 3 neck injury. For the 3-year old, this moment was reported to be 13 Nm. For any size person, the corresponding value M_1 can be determined from Equation 13, or

$$M_1 = (\lambda_w/1.54) \cdot A_c^3 \; 13.0 \; Nm \qquad (14)$$

Where A_c is defined by Equation 1.

The normalized injury risk curve for the 3-year old child is identical to the normalized risk curve for any size occupant provided the normalized moment is computed for the relationship given by Equation 14. Figure 2 gives the injury risk curve for AIS $\geq$ 3 neck injury based on normalized neck extension moment for any size person. Normalizing values for various child and adult dummies are given in table 4. These values were calculated using Equation 14 and the values of A_c given in Table 1 and λ_w given in table (2) Note that the normalized values are the neck extension moments that produce a 1 percent risk of AIS $\geq$ 3 neck injury for the corresponding dummy. Further, this curve gives an estimate of the injury risk when the neck is being loaded in tension and extension, which is the loading mode experienced by the pigs and the child dummy in the biomechanical tests.

Table 4. Normalized extension

Subject	A_c	λw	M1(Nm) Stress Based[+]	M1(Nm) Energy based
6-Months	0.906	1.187	7.6	7.5
12-Months	0.918	1.268	8.2	8.3
18-Months	0.926	1.345	9.1	9.0
3-Year Old	1.000	1.537	13.0	13.0
6-Year Old	1.082	1.825	18.6	19.6
Sm. Adult Fem.	1.246	2.235	29.7	36.6
Mid. Size Male	1.57	2.235	59.4	73.1
Lg Adult Male	1.725	2.235	78.7	97.0

+ From Ref [25]

COMBINED TENSION AND EXTENSION MOMENT – The following is an approach to combining the tension and extension moment loading. Let A be the cross-sectional area of the membrane and D be the distance from the anterior surface of the atlas to its posterior surface. Assume that one half of the measured tensile force is carried by the membrane and that the membrane tensile force produced by the extension moment is equal to the measured extension moment divided by D. With these assumptions, the total force in the anterior membrane is,

$$P = M_E / D + F_T / 2 \qquad (15)$$

And the stress is,

$$\sigma = P/A = (AD)^{-1} [M_E + D F_T/2] \qquad (16)$$

In terms of neck scale factor, A_c, defined by Equation 1, Equation 9 can be written as,

$$\sigma = (A_c^3 A_3 D_3)^{-1} [M_E + A_c D_3 F_T/2] \qquad (17)$$

$$\text{or } \sigma = K \cdot (A_c^3 A_3 D_3)^{-1}$$

where A_3 and D_3 are values corresponding to the 3-year old dimensions, and $\quad K = [M_E + A_c D_3 F_T/2]$

As previously defined, the distortion energy theory of failure for 1D- uniform stress indicates that the strain energy reach a threshold failure level defined by

$$U_d = ((1+ \nu)^* \sigma^2)/3E \qquad (18)$$

Where ν and E are the Poisson's ratio and the elasticity Modulus respectively.

$$\text{Further define} \quad (1+\nu)/3E = 1/ (E_3 \lambda_w^2) \qquad (19)$$

Where E_3 is the elasticity Modulus of the reference 3 year old child. Researchers [23] stated that the variation of Poisson's ratio with respect to the range of age considered here is very limited. Therefore, the assumption is made to take an average value of ν to be 0.27 for all the ages considered in this approach. Therefore, one can further define

$$\lambda_w^2 = 2.37 \; (E/E_3) \qquad (20)$$

$$\text{Or for a 3 year old} \quad \lambda_w = 1.54 \qquad (21)$$

Therefore λ_w will represent the dependency of energy to failure on age as a result of changing the Elastic modulus.

Substitute of values from equations (17),(19) and (20) in equation 18, one can obtain

$$U_d = (A_c^3 A_3 D_3)^{-2} [M_E + A_c D_3 F_T/2]^2 / (E_3 \lambda_w^2) \qquad (22)$$

Table 7 gives values of K calculated from the data of Mertz et al [8] and Prasad and Daniel [9]. Since these data are representative of a 3-year old child, $A_c = 1$, $\lambda_w = 1.54$ and $D_3 = 25.2$ mm were used in the calculations. These data were analyzed by the Mertz / Weber Method [6] for AIS $\geq$ 3 neck injury based on values for a 3-year old child. For a 1 percent risk of AIS $\geq$ 3 neck injury, $K = 20.0$[1]. The corresponding energy, U_{d1}, can be calculated from equation 22 as,

$$U_{d1} = 400.0 /E_3 (1.54.A_3 D_3)^2 \qquad (23)$$

$$\text{Or } U_{d1} = 169/(E_3) \cdot (A_3 D_3)^2$$

The energy level given by equation 22 can be normalized by U_{d1}, therefore, the energy level corresponding to a 1 percent risk of AIS $\geq$ 3 neck injury can be written as,

$$U_d/U_{d1} = K^2/(169 \cdot \lambda_w^2 \cdot A_c^6) \qquad (24)$$

It is convenient in energy computations to define the normalized root mean energy in the form of the square root of the energy ratio. In such a case, it is convenient to define

$$R_{TE} = (U_d/U_{d1})^{1/2} \qquad (25)$$

Or

$$R_{TE} = (13 \cdot \lambda_w \cdot A_c^3)^{-1} [M_E + A_c D_3 F_T/2] \qquad (26)$$

---

[1] This value was later revised in ref [25]

88

The normalized energy ratio can be expressed in terms of the ordinate value, M_C, and the abscissa value, F_C, of the constant energy line corresponding to 1 percent risk of AIS ≥ 3 neck injury for any size occupant, or,

$$R_{TE} = M_E / M_C + F_T / F_C \qquad (27)$$

Where from inspection of equation 26.

$$M_C = 13.0 \cdot A_c^3 \lambda_w \qquad Nm \qquad (28)$$

$$F_C = 1032 A_c^2 \lambda_w \qquad N \qquad (29)$$

The injury risk curve for AIS ≥ 3 neck injury for combined normalized neck tension and extension moment for the 3-year old child can be obtained by dividing the tension and extension moments, by the corresponding values of M_C and F_C given by Equation 28 and 29 noting that $A_c = 1$ for the 3-year old child. The resulting curve is identical to the injury risk curve for any size occupant provided M_C and F_C are calculated using Equations 28 and 29.

Figure 3 gives the injury risk curve for AIS ≥ 3 neck injury based on the normalized root energy ratio, R_{TE}, produced by combined neck tension and extension moment for any size person. Values of M_C and F_C for various sizes of child and adult dummies are given in the table 5. These values were calculated using Equations 28 and 29 and the values of A_c and λ_w given in Table (1) and table [2] respectively.

Table 5.   Combined tension and moment

Subject	Fc(N) Energy Based	Mc(N.m) Energy Based	Fc(N) Stress+ Based	Mc(N.m) Stress Based+
6-Months	1005	11	1110	12.7
12-Months	1103	13	1180	13.8
18-Months	1190	14	1290	15.1
3-Year Old	1586	20	1710	21.6
6-Year Old	2205	30	2260	30.1
Sm. Adult Fem.	3581	56	3130	49.3
Mid. Size Male	5685	112	4970	98.6
Lg Adult Male	6863	149	6000	131.0

+ From Ref [25]

DISCUSSIONS AND CONCLUSION

The efficacies of the various injury risk curves are the best fit when used to assess risks for subjects of the same age and size as that of the original test subjects. The normalization for size and material strength consideration is based on the laws of classical structural mechanics. The size scale factors have excellent efficacy since they are based on average dimensions taken from anthropometry studies. The strength scale factors lack rigorous supporting data since tissue failure stress data based on variations of age and dynamic loading rate are not found in the technical literature. If found, it can be accounted for naturally using the above approach. The

distortion energy theory utilized in this paper, allows for variation of the elastic modulus with age and consequently variations in the strain energy level at failure as a function of age. For materials with time-dependent properties, this is an essential requirement.

For the neck, the load it can carry prior to injury will be dependent, not only on the strain energy to failure level of the ligament, but more importantly on the degree of muscle tensing that has occurred prior to and during loading. It is the neck muscles that protect the neck ligaments from being overloaded. The animals whose data were used to develop the neck injury risk curves were anesthetized. They had some passive muscle reaction, which was well below maximum active muscle tension. Upper bounds of neck injury risk curves corresponding to maximum active muscle tension can be obtained by adding such levels to the critical values listed in the legends of each risk curve. Based on analysis of human volunteer tests, Mertz et al [19, 20] noted that, statically, the average man could resist 1100 N (225 lb) in pure tension and 23.7 Nm (17.5 ft-lb) moment when resisting neck extension. Using Equations 3 and 6 and the neck circumference data given in Table 1, corresponding static neck muscle strength values were calculated for the other sizes of people and are given in Table 6. These static strength values were added to the critical values for neck injury risk curves (Figs. 1 – 3) to give critical values for maximum muscle tensing which are given in Table 7.

The above energy treatment however, simplifies the mathematical modeling of the effect of muscles and their influence on injury risk. To assess the potential for neck injury based on measured internal reactions between the head and the neck, the degree of muscle tone must be specified as an added energy term in equation 15. This is especially true in cases where the muscle effect is easier to measure in the form of stiffness and damping energy rather than forces which require some assessment as to how the internal neck load is distributed among the muscle groups and ligaments.

Table 6.   Static Muscle strength

Maximum Static Muscle Strengths		Critical Values for Risk Curves Based on Maximum Static Muscle Strengths [1]			
Tension (N)	Ext.Mom. (Nm)	F1 (N)	M1 (Nm)	Fc (N)	Mc (Nm)
366	4.6	1246	14.3	1676	19.5
383	4.9	1303	15.2	1723	20.4
383	4.9	1303	15.2	1743	20
446	6.1	1516	19.1	2036	26.1
522	7.8	1772	25.6	2382	33.1
693	11.9	2353	37	3163	50.6
1100	23.7	3740	74	5020	101.1
1330	31.5	4520	98.3	6060	134.2

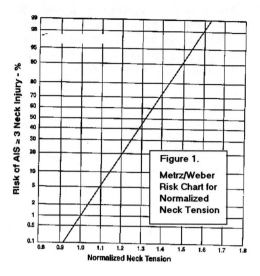

Figure 1.
Mertz/Weber Risk Chart for Normalized Neck Tension

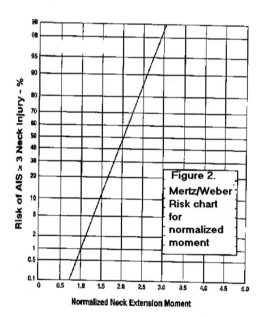

Figure 2.
Mertz/Weber Risk chart for normalized moment

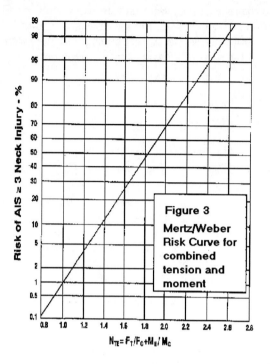

Figure 3
Mertz/Weber Risk Curve for combined tension and moment

$N_{TE} = F_T/F_C + M_E/M_C$

Table 7.

	Pk. Neck Tension			Pk. Neck Moment		
Tension (N)	AIS	Tension (N)	AIS	Ext. Mom. (Nm)	AIS	Ext. Mom. (Nm)	AIS
400	0	1410	0	11.3	0	26.0	0
525	0	1430	6	14.0	0	26.0	0
525	0	1445	3	15.0	4	26.0	0
525	0	1460	0	15.8	0	29.4	6
560	0	1480	4	17.0	0	30.0	0
574	0	1490	4	18.0	1	33.9	6
588	0	1500	3	18.0	3	37.3	4
625	0	1500	4	18.0	0	37.3	0
635	0	1530	3	20.0	4	42.4	4
680	0	1570	0	20.0	3	46.0	5
805	1	1920	4	20.0	4	46.3	4
813	0	1925	4	20.0	0	46.3	6
813	0	1925	6	20.0	0	47.5	6
855	0	2270	4	20.0	0	63.0	3
938	0	2270	6	23.0	0	64.0	4
943	2	2270	6	23.0	0	64.0	6
960	0	2680	3	23.2	0	64.0	6
1050	0	2820	6	24.0	5	66.0	4
1150	0	2960	4	25.0	0	66.0	3
1160	6	3040	5	25.0	0	67.0	2
1250	4	4100	5	25.4	0	80.0	5
1260	5			25.4	0		

REFERENCES

1. Mertz, H. , Prasad,P. and Irwin,A.L. "Injury Risk Curves for Children and Adults in Frontal and Rear Collisions", SAE973318

2. Prasad, P. And Mertz, H. J., "The Position of the United States Delegates to the ISO Working Group 6 on the Use of HIC in the Automotive Environment", SAE 851246, 1985.

3. Mertz, H. J., Prasad, P. And Nusholtz, G., "Head Injury Risk Assessment for Forehead Impacts", SAE 960099, February, 1996.

4. Mertz, H. J., Prasad, P. And Nusholtz, G., "Head Injury Risk Assessments Based on 15 ms HIC and Peak Head Acceleration Criteria", Proceeding of AGARD Meeting on Impact Head Injury, November 7-9, 1996.

5. Mertz, H. J., Horsck, J. D., Horn, G., and Lowne, R. W., "Hybrid III Sternal Deflection Associated with Thoracic Injury Severities of Occupants Restrained with Force – Limiting Shoulder Belts", SAE 910812, February 1991.

6. Viano, D. V. And Lau, I. V., "Thoracic Impact: A Viscous Tolerance Criterion", Proceeding of the Tenth Experimental Safety Vehicle Conference, July, 1985.

7. Mertz, H. J. And Weber, D. A., "Interpretations of the Impact Responses of a 3-Year Old Child Dummy Relative to Child Injury Potential", Proceedings of the Ninth International Technical Conference on Experimental Safety Vehicles, Kyoto, Japan, November 1-4, 1982. (Also Published in SAE 826048, SP-736 Automatic Occupant Protection Systems, February, 1988).

8. Wolanin, M. J., Mertz, H. J., Nyznyk, R. S., and Vincent, J. H., "Description and Basis of a Three-Year-Old Child Dummy for Evaluation Passenger Inflatable Restraint Concepts", Proceedings of the Ninth International Technical Conference on Experimental Safety Vehicles, Kyoto, Japan, November 1-4, 1982 (Also published in SAE 826040, SP-736, February, 1988).

9. Mertz, H. J., Driscoll, G. D., Lenox, J. B., Nyquist, G. W., and Weber, D. A., "Responses of Animals Exposed to Deployment of Various Passenger Inflatable Restraint System Concepts for a Variety of Collision Severities and Animal Positions", Proceedings of the Ninth International Technical Conference on Experimental Safety Vehicles, Kyoto, Japan, November 1-4, 1982. (Also published in SAE 826047, PT31).

10. Prasad, P. And Daniel, R. P., "A Biomechanical Analysis of Head, Neck and Torso Injuries to Child Surrogates Due to Sudden Torso Acceleration", Twenty-Eighth Stapp Car Crash Conference, SAE 841656, November, 1984.

11. Neather, R. F., Kroell, C. K. And Mertz, H. J., "Prediction of Thoracic Injury from Dummy Responses", Nineteenth Stapp Car Crash Conference, SAE 751151, November, 1975.

12. Weber, K. And Lehman, R. J., "Child Anthropometry for Restraint System Design", UMTRI-85-23m June, 1985.

13. Schneider, L. W., Robbins, D. H., Pflug, M. A., and Snyder, R. G., "Development of Anthropometrically Based Design Specification for an Advanced Adult Anthropomorphic Dummy Family", Volume 1, UMTRI-83-53-1, December, 1983.

14. Mertz, H. J., Irwin, A. L., Melvin, J. W., Stalnaker, R. L., and Beebe, M. S., "Size, Weight, and Biomechanical Impact Response Requirements for Adult Size Small Female and Large Male Dummies", SAE 890756, March, 1989.

15. Mertz, H. J., and Irwin, A. L., "Biomechanical Basis for the CRABI and Hybrid III Child Dummies", Forty-First Stapp Car Crash Conference, November, 1997.

16. Mertz., H. J., "Anthropomorphic Test Divices", Accident Injury – Biomechanics and Prevention, Springer-Verlag, N. Y., 1993.

17. "Anthropomorphic Dummies for Crash and Escape System Testing", AGARD-AR-330, July, 1996.

18. Melvin, J. W., "Injury Assessment Reference Values for the CRABI 6-month Infant Dummy in a Rear-Facing Infant Restraint with Airbag Deployment", SAE 950872, February, 1995.

19. Rouhana, S. W., Jedrzejczak, E. A. And McCleary, J. P., "Assessing Submarining and Abdominal Injury Risk in the Hybrid III Family of Dummies: Part II – Development of the Small Female Frangible Abdomen", Thirty-Fourth Stapp Car Crash Conference, SAE 902317, November, 1990.

20. Mertz, H. J. And Patrick, L. M., "Investigation of the Kinematics and Kinetics of Whiplash", Eleventh Stapp Car Crash Conference, October, 1967.

21. Mertz, H. J. And Patrick, L. M., "Strength and Response of the Human Neck", SAE 710855, Fifteenth Stapp Car Crash Conference, November 1971.

22. Currey, J. D. And Butler, G.,"The mathematical properties of bone tissue in children", Journal of Bone and Joint Surgery, 57A, 810-814, 1975.

23. Goldstein,S. Et. Al. Biomechanics of bone", Accidental Injury, ed. Nahum, A. M. And Melvin, J.W., Springer-Verlag, 1993.

24. Qing Zhou, et. Al.,"Age Effects on Thoracic Injury Tolerance", SAE962421.

25. Mertz, H., G. Nusholtz and P. Prasad,"Proposal for Dummy Response Limits for FMVSS 208 Compliance Testing", Biomechanics Task Group, AAMA, Dec. 1998.

26. De Jager,M., Sauren, A., Thunnissen and Wismans J."A Global and a detailed mathematical model for Head-Neck Dynamics",SAE962430.

27. Yogananadan, N., Pintar,Maiman,Cusick,Sances and Walsh, "Human head-neck biomechanics under axial tension", Med. Eng. Phys. V. 18, No. 1,1996.

Anatomy of the Human Cervical Spine and Associated Structures

D. F. Huelke
Univ. of Michigan Medical School

GENERAL CONCEPTS

THE CERVICAL REGION OF THE BODY, "the neck", extends above, from the base of the skull and a line drawn along the lower border of the mandible (jaw), to below, the level of the first rib. This lower boundry, the first rib, extends from the first thoracic vertebra to approximately the top of the sternum (breast bone) (Fig. 1).

The human neck has two basic functions: 1) to give support and mobility to the head, and 2) to act as a conduit, interconnecting structures of the head and face with those of the thorax and abdomen.

Located anteriorally are the visceral components of the neck. These include small muscles of the anterior neck, the esophagus to the stomach, the larynx and upper trachea to the lungs the thyroid gland and arteries and veins for the blood supply and drainage of the head, brain and facial areas. Major nerves from the brain and from the cervical spinal cord also pass vertically through this region. Posterially are the cervical vertebrae, their associated muscles, and nerves.

SUPERFICIAL LANDMARKS

Superficially, certain cervical landmarks can easily be identified. Dividing the neck into its anterior and posterior regions are two muscles, one on each side, the sternocleidomastoid muscles (Fig. 1). These muscles pass diagonally downward from their attachments behind the ears to the upper portion of the sternum. The sternocleidomastoid muscle is noticeable when the head is turned to the opposite side. It stands out as a vertical cord at the side of the neck.

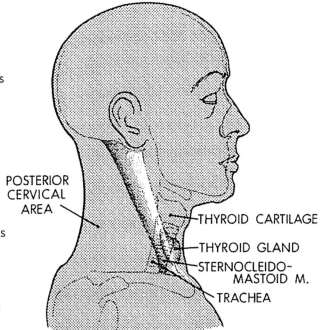

Fig. 1 - Lateral view of the neck showing the anterior visceral compartment and the posterior cervical area that includes vertebrae and associated musculature

On the anterior midline the thyroid cartilage of the larynx (voice box) is readily seen. Its most forward projection or bump is the laryngeal prominence (Adams apple) (Fig. 2).

Posteriorly the spines of the cervical vertebrae are difficult to palpate, for they are very

──────── ABSTRACT ────────

The anatomy of the major structures of the neck are presented including the anterior throat structures and the components of the cervical spine and associated structures. Basic neck movements and generalized muscle actions are described.

Anatomical relationships of the bones, ligaments, muscles and joints of the cervical spine are emphasized as a foundation for a clear understanding of the structural elements involved in neck fractures and dislocations.

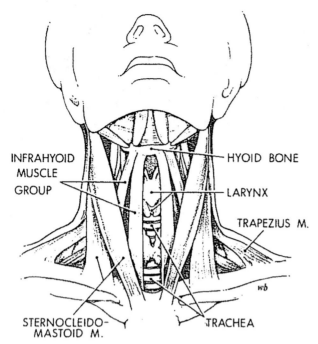

Fig. 2 - The anterior visceral compartment bounded on the side by the sternocleidomastoid muscles. The larnyx and trachea are found on the midline

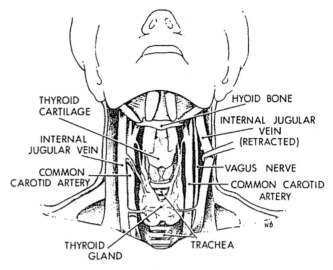

Fig. 3 - Deep anatomy of the anterior compartment showing the thyroid gland, major nerves and blood vessels, and the thyroid cartilage

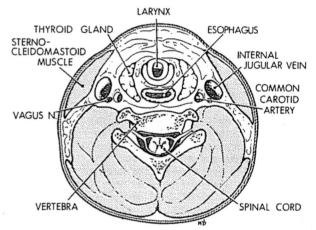

Fig. 4 - A schematic cross-section of the neck showing a cervical vertebra and associated muscles with visural structures located anteriorly

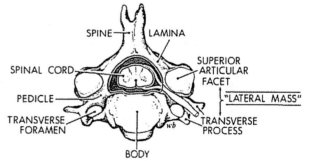

Fig. 5 - The components of a cervical vertebra

short, not like those in the thoracic area that markedly project rearward and are easily seen through the skin. Of the cervical vertebral spines, that of C7 is usually the first one to be easily palpated and thus is referred to as <u>vertebra</u> prominens (prominent vertebra).

ANTERIOR COMPARTMENT

As stated before, the larynx is anterior in the neck (Fig. 3). It serves as a fairly rigid non-bony housing for the passageway of air to and from the lungs and for sound and voice production. It consists of a cartilaginous shell made up of several units which does not have the strength of bone, and thus is susceptible to trauma. Obviously, collapse of the cartilaginous larynx will occlude the airway, thereby preventing adequate respiratory exchange.

Located deep beneath the sternocleidomastoid muscles are the <u>carotid sheaths</u> (Fig. 3). Included in this bundle of nerves and blood vessels on each side is the carotid artery, internal jugular vein, and major nerves from the brain, as well as some cervical nerves to thoracic and abdominal organs. In general, these structures are only vulnerable to serious injury by deep penetrating neck wounds. The carotid arteries supply the face, nose and mouth, and are the major arterial supply of the brain. Likewise, the major venous return of the brain and deep facial structures is carried by the internal jugular vein.

POSTERIOR COMPARTMENT

The posterior compartment of the neck includes a central core of cervical vertebrae, surrounded on their lateral and posterior sides

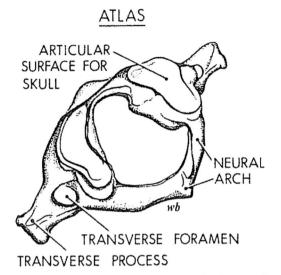

ATLAS

ARTICULAR SURFACE FOR SKULL

NEURAL ARCH

TRANSVERSE FORAMEN

TRANSVERSE PROCESS

wb

Fig. 6 - The atlas-the first cervical vertebra

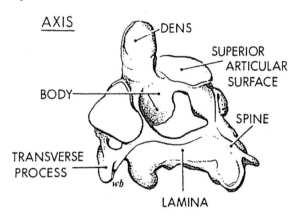

AXIS

DENS

SUPERIOR ARTICULAR SURFACE

BODY

SPINE

TRANSVERSE PROCESS

wb

LAMINA

Fig. 7 - The axis-the second cervical vertebra

by neck muscles (Fig. 4). Anteriorly the verte- brae are covered by small flat muscles and by the visceral structures of the neck. In addition, the nerves which arise from the cervical spinal cord pass into and between the lateral muscles in this posterior compartment. The cervical vertebrae are seven in number and the lower five are identical in structure to each other. Only the first and second cervical vertebrae, the atlas and axis, are uniquely different.

Typically, a cervical vertebra is much small- er than those in the thoracic or lumbar areas (Fig. 5). All vertebrae of the entire vertebral column, with the exception of the first cervical vertebra, have a large cube of bone, the body, as the front portion of the vertebra. The body is a honeycomb block of interconnecting spincules of bone, the cortex, surrounded by a thin shell of harder compact bone.

Behind the body of the vertebrae is the neural arch, a bony ring that ends posteriorly in the spine (Fig. 5). The spines of the cervical vertebrae are very short and blunted in comparison to those in the thoracic or lumbar areas. Only in the lower cervical vertebrae can the spines be

more easily palpated with the spine of C7 being most prominent. However, by deep palpation on the posterior midline of the neck most all of the spines of the vertebrae can be identified. The neural arch extends from the posterior lateral sides of the body, surrounds an open area, the vertebral foramen. When the vertebrae are stacked one on top of the other the individual vertebral foramina then form the vertebral canal through which the spinal cord passes. The various parts of this bony ring have specific names. On each side of the ring of bone are the lateral masses, including the superior and inferior articular processes and a lateral extension on each side, the transverse process that has a hole in it, the transverse foramen through which the vertebral artery will pass to the base of the brain. Between the body and the lateral mass is a very short segment of bone, the pedicle; behind the lateral mass, between it and the spine, is the bony lamina. On the surface of the superior articular process is a plain smooth area, the superior articular facet. This is the joint sur- face for the articulation with the vertebra imme- diately above it, contacting the inferior artic- ular facet of the inferior articular process, located beneath the transverse process.

As indicated previously, the first and sec- ond cervical vertebrae are noticeably different. The first cervical vertebrae, the atlas, is a ring of bone (Fig. 6); it does not have a body. Its spinous process is absent, but it does have all the other components of a vertebra including the lateral mass, pedicle, and lamina and articular facets. The second cervical vertebra, the axis, is also different. Although it has all the parts and characteristics of the lower cervical vertebrae it has, extending upward from the body of C2, a small finger-like process of bone, the dens (odontoid process) (Fig. 7). This is the body of the first cervical vertebra, which during de- velopment fuses to the body of the second vertebra and not to that of the atlas. When they are joined together the dens fits into the front part of C1 (Fig. 8).

CERVICAL ARTICULATIONS

When all of the cervical vertebrae are stack- ed one upon another it can be seen that the superior articular facets of a lower vertebra join with the inferior articular facets of the bone immediately above it. Similarily, the bodies of one vertebra are stacked on those above and below. All of the cervical vertebrae are united one with another by means of ligaments and discs. A fibro- cartilaginous intervertebral disc is found between the bodies of each cervical vertebra and are in- separable from the bone itself, i.e., a disc can- not be easily removed without tearing away some of the bone of adjacent bodies (Fig. 9). The disc allows a certain amount of movement between the individual bodies; it is also a shock absorbing mechanism.

In front of the bodies of the vertebrae is the anterior longitudinal ligament which runs the

ARTICULATION OF C1 & C2

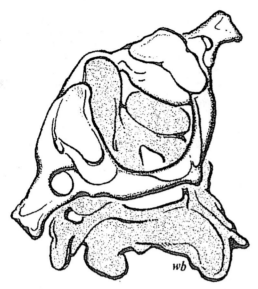

Fig. 8 - The articulation between the atlas and axis

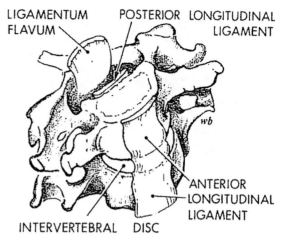

Fig. 9 - Some of the ligaments of the cervical vertebra

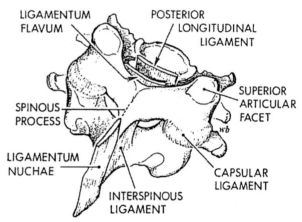

Fig. 10 - Ligaments of the cervical vertebrae as seen in posterior view

full length of the vertebral column, from the base of the skull to the tip of the coccyx (tailbone). The anterior longitudinal ligament interconnects each body with the adjacent vertebral bodies, and limits the range of extension movement. Immediately behind the body is another continous ligament that extends from the upper part of the vertebral chain downward to the base of the coccyx; being located behind the bodies of the vertebrae it is named the posterior longitudinal ligament. Within the vertebral canal, interconnecting adjacent laminae, is a serially arranged group of ligaments called the ligamentum flavum. These ligaments being found at each and every space between the adjacent laminae are not continuous as are the anterior and posterior longitudinal ligaments.

Laterally, each of the adjacent superior and inferior articular facets are completely enclosed by a ligamentous structure, the capsular ligaments (Fig. 10). Beneath the capsular ligaments are all the characteristic structures of a moveable joint including joint fluid (synovial fluid), articular cartilage covering the surface of each facet, and a synovial membrane which produces the synovial fluid. It is through these joints that much of the neck movement is permitted.

Posteriorly, between each and every spine are the interspinous ligaments that are found not only in the cervical region but also through the length of the vertebral column. In addition, there is a ligament superimposed on the tips of the spinous processes, the superspinous ligament. This ligament in the cervical region is called the ligamentum nuchae; it interconnects the spines and also extends from the tips of the spines to just beneath the skin of the back of the neck and is

an attachment area of many of the deep neck muscles.

The articulations between the first cervical vertebra and the base of the skull are the atlanto-occipital joints (Fig. 8). These are oval joint surfaces on the upper part of the lateral masses of the first cervical vertebra with mirror image counterparts on the base of the skull. These joints are also surrounded by all the typical capsular structures as are the other lateral mass articulations. Because of the oval surface, which is longer from front to back than it is from side to side, only a forward and backward head rocking motion is allowable at these joints. Nodding of the head in a "yes" manner characterizes the approximate range of motion at these joints. Completely surrounding the circular shaped first cervical vertebra are ligaments that span between the upper surface of this bony circle to the base of the skull. Specifically, this is called atlanto-occipital membrane, which tightly secures the atlas to the base of the skull.

Of all the cervical joints those between the first and second cervical vertebra are the most

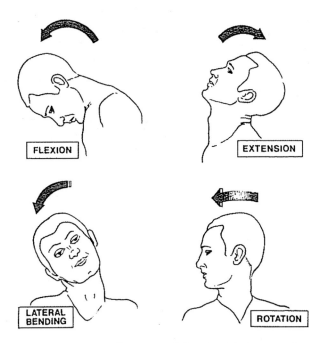

FLEXION

EXTENSION

LATERAL BENDING

ROTATION

Fig. 11 - General types of head-neck movements

unique and different. As elsewhere, the lateral mass articular surfaces interconnect with one another between C1 and C2 but the dens of the C2 vertebra sticks up vertically just inside the front portion of the ring of the atlas. This dens is held in place by strong ligaments that extend horizontally to attach to the inner aspect of the body of C1. Additional ligaments extending upward from the tip of the dens to the lower portion of the base of the skull. These ligaments then appear in a cross-shape and thus are called cruciate ligaments. Thus, the atlas and head can now pivot about the dens, allowing a slight side-to-side movement in this area as in the act of shaking the head when indicating "no".

MOVEMENTS OF THE CERVICAL SPINE

Movements of the cervical spine are limited by the type of articulations, normal stretch of the ligaments that surround the vertebrae, and by the muscles that attach to the skull or to the cervical vertebrae. We have already discussed that the nodding of the head in a "yes" manner is a movement between the skull and the first cervical vertebra, however, all other motions of the head and neck, such as forward bending (flexion), rearward neck and head bending (extension), tilting the head and neck to one side (lateral bending) or looking over ones shoulder (lateral rotation), generally occur between each and every one of the cervical vertebra (Fig. 11). However, at the first and second cervical vertebra the only action that was permitted was that of the rotating the head in the "no" manner.

Most of the actions of the vertebral spine are through the lateral articulations. Cervical

movement is limited by the joint capsules surrounding all of these articulations in the cervical spine, by the minimal amount of compression of the discs and limited stretch of the interspinous ligaments, the ligamentum flavum, the ligamentum nuchae and the longitudinal ligaments. Excessive motion, in any direction, will necessitate either stretching of the ligaments, compression of the disc and/or bony elements of the cervical spine, associated possibly with tearing of certain of the ligaments and bone fractures.

NECK MUSCULATURE

The muscles that are found in the cervical area are all of the voluntary type, that is, under the direct control and will of the individual. These muscles are basically of three types; muscles crossing through the neck region without having a direct attachment to the cervical vertebrae, those which pass through the area and have some attachments in the cervical region, and those that attach from one cervical vertebra to another.

The sternocleidomastiod muscle is a prime example of a muscle passing through the area, having direct action on the head, but not attaching to neck structures. With its attachments to the skull and to the upper sternum, contraction of the sternocleidomastoid muscle will bend the head toward that side and rotate it so that the chin points to the opposite side. When both sternocleidomastoid muscles act together the chin is elevated. Posteriorly are the trapezius muscles that attach at the base of the skull and spread out onto the shoulders giving the characteristic outline of the neck-shoulder curvature. This muscle also attaches in the posterior cervical midline area to the ligamentum nuchae. Bilateral actions of the trapezius draws the head backwards.

Deep around the posterior and lateral sides of the neck are muscles that attach to some of the individual cervical vertebrae, and lower down to ribs. Contractions of these muscles either cause rib elevation, or if the ribs are stabilized, neck bending to the side on which the muscle is contracting. Similarly, muscles that are continuous with other fibers along the vertebral column extend upward from the thoracic region, inserting into individual vertebra. These muscles can cause lateral bending, or individual rotation of the vertebra. Contracting bilaterally they produce extension or stabilize the neck. Deep cervical muscle fibers attach from one cervical vertebra to another, or may attach one or two segments higher. These are basically rotary muscles.

All of these deep cervical muscles, however, act as stabilizers of the cervical vertebrae so that when they contact there is a mass of muscles which have not only a protective function, but hold the neck and head in an upright position. In order to get neck flexion there needs to be relaxation of the posterior cervical muscles with bilateral contraction of the sternocleidomastoid and other muscles. Partial relaxation of the posterior cervical muscles allows the head to partially flex.

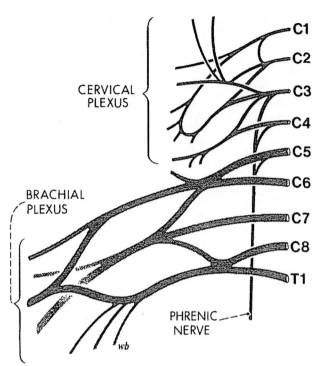

Fig. 12 - Diagrammatic representation of the cervical and brachial plexuses. Note specifically the origin of the phrenic nerve from C3, 4, 5

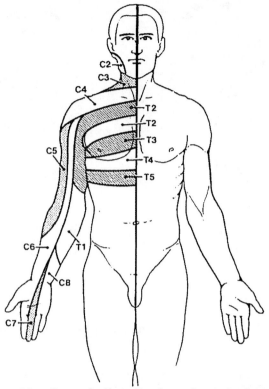

Fig. 13 - General distribution of cervical and upper thoracic nerves

Such a movement is characterized by someone in the seated position starting to fall asleep; the head starts to bob forward and in a jerky motion due to these muscles relaxing intermittently. Neck extension from a flexed or neutral position is due to the contraction of the deep cervical muscles and the trapezius. Thus the flexion action is basically a relaxation of the deep cervical neck muscles with some active contraction of the lateral and anterior muscles. Anterior to the upper cervical vertebrae are several thin flat muscles, attaching to the vertebrae and to the base of the skull. These muscles aid in neck flexion and in lateral neck bending.

CERVICAL NERVES

The nerves that arise from the cervical portion of the spinal cord pass outward between the pedicles and immediately join with those either above or below them (Fig. 12). Thus, this combination of nerves, a plexus, allows for reorganization and redistribution of individual nerve fibers to various areas. The cervical plexus supplies the skin of the neck and side of the face with deeper branches going directly into the muscles that surround the lateral and posterior sides of the cervical vertebrae. The cervical plexus is formed from cervical nerves 1-4. The nerves that arise from the lower cervical levels (C5-C8)

likewise emerge between the pedicles and immediately form the large brachial plexus for the innervation of the upper extremity (Fig. 13, 14).

In that the majority of neck fractures and spinal cord involvements occur in the area of the lower cervical vertebrae, the cervical plexus nerves are spared. However, depending on the level of the cord injury in the lower cervical area there may be incomplete nerve deficite and thus sparing of all or part of the upper extremity nerves may occur (Fig. 14). The nerves of the brachial plexus are both motor and sensory, thus, the neck injured individual may have sensory loss of the upper extremity in association with motor loss as well. Of all the nerves in the cervical area the phrenic nerve is extremely important (Fig. 12). This nerve arises from components of cervical nerves C3, 4, and 5 which join together to pass downward through the thorax, outside of the vertebral canal, to the respiratory diaphram. Thus a paralyzing spinal cord injury at or below the C5-C6 level will not affect the phrenic nerve and the respiratory diaphragm will remain functional. However, if severe spinal cord damage occurs above the C3 level, death usually results, for the roots of the phrenic nerve are still within the spinal cord and paralysis of the respiratory diaphragm results.

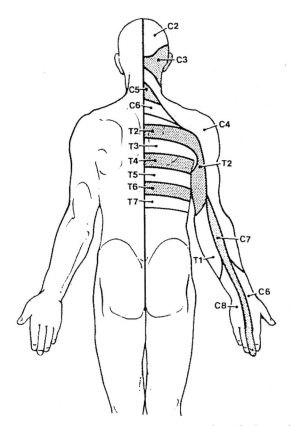

Fig. 14 - Posterior view of cervical and thoracic
nerve distribution

RESEARCH STUDY ON NECK INJURY LESSENING WITH ACTIVE HEAD RESTRAINT USING HUMAN BODY FE MODEL

Yuichi Kitagawa*, Tsuyoshi Yasuki*, Junji Hasegawa*
*Toyota Motor Corporation, Japan

ABSTRACT

According to the traffic accident data in Japan, the number of rear impacts is the largest among the various car crash scenarios on the road. Neck injury due to low speed rear impacts continues to be a significant issue in terms of social cost. Although the precise injury mechanism is still not determined clearly, it is commonly recognized that a relative motion between the head and torso could be one of the factors causing the injury. NIC and some other indicators have been developed to assess injury risk and some are used in laboratory tests. Recent studies focus on the joint capsules as a potential site of neck pain. Experiments were conducted to evaluate Joint Capsule Strain under simulated loading conditions. Various seat systems have been developed aiming to help reduce the injury risk. There have been two approaches used in reducing the relative motion between the head and torso. One is to allow the occupant torso to sink into the seatback, and the other is to support the head as early as possible. The second approach has led to the development of active head restraint systems. The active head restraint system is designed to move the head restraint forward (closer to the occupant's head) when activated in rear impacts. The effectiveness of such systems is verified evaluating the dummy readings (indicator values) and sometimes investigating the field data. The authors previously conducted a study using a human body finite element (FE) model to estimate the effectiveness of a fixed head restraint system in terms of both NIC and Joint Capsule Strain. The results confirmed that seat design factors such as the forward location of the head restraint could help lower the indicator values. This study investigates the effectiveness of the active head restraint system in reducing both NIC and Joint Capsule Strain. A human FE model is used to simulate the occupant head and neck motion with the different head restraint systems. The study also analyzes the cervical vertebral motions to investigate the function of the active head restraint system in correlation with the strain growth in the joint capsules

Keywords: Rear impacts, Neck injury, Seats, Active Head Restraint

JAPANESE TRAFFIC ACCIDENT DATA shows an increasing trend of rear impacts and that the number of rear impacts is the highest among the various car crash scenarios in Japan followed by collisions at intersections. Most of them are relatively low speed impacts and injury outcome is generally less severe, in terms of AIS coding. However, the insurance companies report that the claims of neck pain such as whiplash injury are the majority of total automotive insurance costs. Despite the frequency of neck injury in the field, its mechanism is not clearly determined yet. A common understanding is that a relative motion between the head and the torso could contribute to the cause of injury. Hypotheses were made focusing on hyperextension of the neck, stretch of the cervical muscles and the pressure gradient in the spinal canal. Recent studies assume that the cervical facet joint is a potential site of neck pain. Deng et al. (2000) analyzed the kinematics of the cervical vertebrae of a Post Mortem Human Subject (PMHS) using a high speed X-ray camera, while Ono et al. (1997) conducted a cineradiographic analysis on living human subjects (volunteers) in simulated rear impacts. They hypothesized that impingement in the facet joint could be a possible cause of pain in rear impacts. Yang et al. (1996) noted that shear motion in the cervical joint could attribute the pain to the joint capsules. Yoganandan et al. (2001) also supported the facet joint impingement mechanism as a possible mechanism of neck injuries. Winkelstein et al. (1999) examined the deformation of the joint capsules related to the relative motion between adjacent vertebrae. They observed an increase of the principal strain in the capsules when pretorque was applied, while bending itself did not raise the strain up to a harmful level. There are some other studies that analyzed the facet joint motion (Sundararajan et al., 2004, Lee et al., 2004). The authors of this paper have conducted a study using a human body FE model to simulate head and neck kinematics of an occupant during a rear impact and to estimate tissue strain in the cervical joint capsules (Kitagawa et al., 2006). The study showed that the relative acceleration and displacement are generated after the impact and the joint capsule strain more correlated to the relative displacement. Another study on seat design factors found that the stiffness of the reclining joint, the location of the head restraint and the head restraint supporting stiffness are relatively important to help lower NIC and Joint Capsule Strain (Kitagawa et al., 2007).

On the other hand, research efforts have been made in developing criteria to assess neck injury risk. Boström et al. (1996) proposed a criterion called NIC based on their assumption that the pressure gradient in the spinal fluid could be a cause of injury. The indicator evaluates the amplitude of relative acceleration and velocity between the head and torso in its formula, and it has become a popular indicator in laboratory tests and assessment tests. There are some other indicators proposed for injury assessment, neck shear/tension force, moment, Nkm LNL, Rebound Velocity, Head Restraint Contact Time and T1 acceleration (Muser, 2000, Schmitt, 2001, Heitplatz, 2003). Because the injury mechanism is not clearly determined yet, multiple indicators are evaluated in assessment tests. Product seat systems have been developed aiming to help lower the neck injury risk in terms of the proposed criteria. A common approach is to locate the head restraint forward, expecting a shorter Head Restraint Contact Time, in order to reduce the relative motion between the head and torso. Other challenges are, for example, to soften the upper part of the seatback allowing the torso to sink (Sekizuka, 1998), and to design the reclining joint to yield absorbing energy (Jakobsson, 2004). An advanced technology in this area is the active head restraint system, where the head restraint is moved forward (closer to the head) when a rear impact occurs. Various systems have been developed by car makers. A popular approach is to activate the head restraint with mechanical linkages. Such product seat systems are often called "anti-whiplash" systems. Kullgren et al (2007) conducted a study investigating the effectiveness of such systems based on the Swedish accident data. The results showed that the cars fitted with the advanced anti-whiplash systems had 50% lower risk of neck injuries leading to long-term symptoms.

This study examined the effectiveness of the active head restraint system using a human body FE model. Considering the recent studies on the neck pain mechanism, Joint Capsule Strain was evaluated as well as NIC. The strain growth was attributed to the vertebral displacement, and the magnitude of the displacement was correlated with the interaction between the head and the head restraint. The study also focused on the alignment of the head restraint motion with respect to the facet joint surface, assuming its contribution to the strain growth.

METHOD

HUMAN BODY FE MODEL: The study uses a human body FE model named the Total Human Model for Safety (THUMS), which was jointly developed by Toyota Motor Corporation and Toyota Central Research and Development Laboratory. THUMS represents an average-sized adult male person (175 cm and 77 kg) aiming to simulate body kinematics and responses during car crashes. The model includes all the bony parts and major ligaments so that the model can simulate a real human subject. Joint modeling is important for simulating body kinematics during crashes. The ligaments were modeled connecting bones to bones and contacts were defined between them. This modeling method simulates realistic joint motions and calculates force transmission through the joint. Each body part also imitates the human tissue material in terms of mechanical response against external loading. Such material properties are defined based on the literature data (Yamada, 1970, Yoganandan et al., 1998), while the geometry of the body parts refers to human body databases. The model also includes skin, fat, muscle, brain and internal organs, and most of them are simplified as solid parts in THUMS. The neck muscles are modeled with 1D discrete elements to simulate their passive responses against stretch. The cervical facet joint capsules are newly introduced into the model for this study in order to calculate Joint Capsule Strain. The model includes approximately 60,000 nodes and 80,000 elements, and it runs on a commercial finite element code LS-DYNA^{TM}. Figure 1 is a close-up view of the neck part with the joint capsules added. The major ligaments included are: the anterior longitudinal ligament (ALL), the posterior longitudinal ligament (PLL), the ligamentum flavum (LF), the interspinous ligament (ITL), the supraspinous ligaments (SSL), and the intertransverse ligament (ISL). Relative motion between adjacent vertebrae generally occurs around the facet joints located on the right and left sides of the neural arch. The joints are covered with the joint capsules, and the capsule tissues were modeled with membrane elements. The model also includes the synovial fluid (represented with solid elements) which may affect the magnitude of joint capsule strain due to its uncompressive property. Internal contacts in the cervical region such as vertebra-to-vertebra or vertebra-to-ligament are also defined as treated in other joints. The surface-to-surface and/or node-to-surface algorisms in LS-DYNA™ are adopted to handle such contacts.

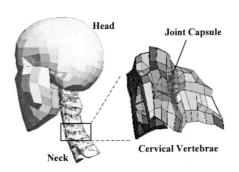

Figure 1. THUMS Occupant Model and Joint Capsule Modeling.

MODEL VALIDATION: The model has been validated against human impact responses reported in the literature (Iwamoto et al., 2002). The mechanical responses of the model against external loading were verified part by part such as impacts to the head, thorax, abdomen, pelvis and lower extremity. The responses were compared to those of Post Mortem Human Subject (PMHS) in the literatures. The model responses were found to be within the PMHS corridors. For rear impact simulations, the authors previously validated the model in its whole body motion and at the neck component level (Kitagawa et al., 2006). Head and neck kinematics in a whole body motion were verified comparing to the literature data with human volunteer subjects (Ono et al., 1997). An inclined sled system was used to generate low velocity rear impact, and head and neck kinematics were captured with a high speed X-ray (Figure 2). A numerical simulation was conducted using the THUMS model duplicating the test condition. Comparisons of rotations of the vertebral bodies found that the model well simulated the human head and neck responses, although it did not represent muscular effects working to reduce the head-neck motion after 100 ms (Figure 3). The calculated time history curve of the head rotation was shifted earlier compared to that of the test. This is possibly because of a difference in their initial head orientations. Another validation examined the validity of the cervical joint model in predicting Joint Capsule Strain (Kitagawa et al., 2008). Siegmund et al. (2000) conducted quasi-static loading tests where C3-C4 components segmented out from PMHS were subjected to shear and compressive loadings (Figure 4). The anterior-posterior displacement and the sagittal rotation of C3 with respect to C4 were monitored. Joint Capsule Strain was estimated from relative displacement among photo markers posted to the capsule tissue. A corresponding part was extracted from the THUMS neck, and then equivalent boundary conditions were applied to the model. The anterior-posterior displacement and sagittal rotation were calculated from nodal displacement in the model and Joint Capsule Strain was directly output from elements representing the capsule tissue. Figure 5 compares the test data and the simulation results, where the test data were plotted as corridors and the simulation results were plotted as curves. A comparison confirmed that the calculated Joint Capsule Strain (curves) were within the test corridors. These validation results indicate that the model can simulate vertebral motion and estimate joint capsule strain.

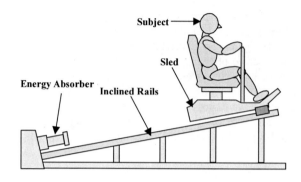

Figure 2. Inclined Sled Test System (Ono et al., 1997).

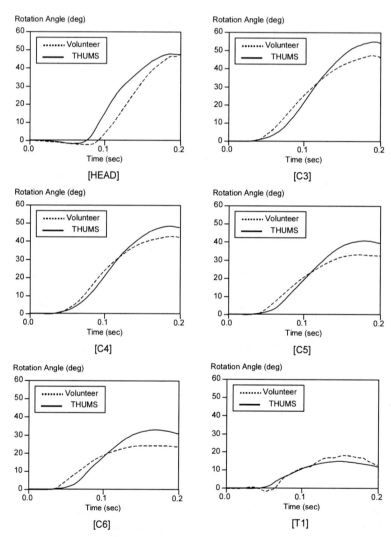

Figure 3. Comparison of Head and Vertebral Rotations between THUMS and Volunteer.

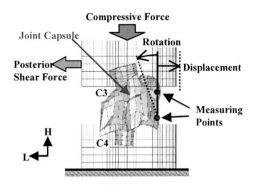

Figure 4. Loading Test on C3-C4 Component.(Siegmund et al., 2000)

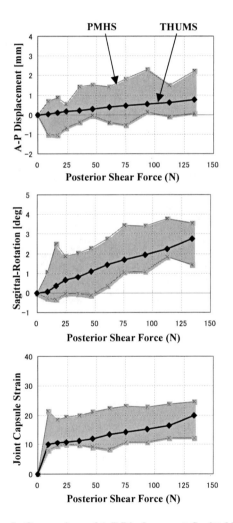

Figure 5. Comparison of A-P Displacement, Sagittal Rotation and Joint Capsule Strain between PMHS and THUMS.

SEAT MODEL: Two seat models were prepared for rear impact simulations. The first one is a prototype seat model with a fixed head restraint. Its design concept is called "WIL" introduced by Sekizuka et al. (1998), which reduces the relative motion between the head and torso allowing the torso to sink into the seatback. The effectiveness of this concept was previously examined in laboratory tests and with FE simulations (Sawada et al., 2005). The other one is a prototype seat equipped with an active head restraint system. The design concept is based on the WIL but is enhanced introducing an activation system explained later. In both models, the geometrical features of the seat structure were incorporated, including the construction of components and their mechanical properties. The material properties of the seat frames and cushion foams were also incorporated. Contacts between the parts were handled by the algorism implemented in the LS-DYNA™. Figure 6 shows overall views of the seat models. Each model was validated in advance so that its mechanical response against external loading simulated that in a physical experiment. A typical external load in rear impacts is a rearward force from the occupant torso to the seatback. Rotational stiffness of the reclining joint is thought to be one of the major factors. The joint stiffness property in the seat model was carefully adjusted to simulate actual seatback response. The active head restraint system in this study works to move the head restraint forward and upward. Such a motion is generated along guiding tubes where the head restraint supporting poles are inserted. The system is activated when the occupant pelvis loads the lower unit implemented in the seatback during a rear impact. The displacement of the lower unit is transmitted to the upper unit by attached cables, and the upper unit lifts the supporting poles (Figure 7). The model precisely simulates the linkage mechanism, and its reaction against loading was carefully verified.

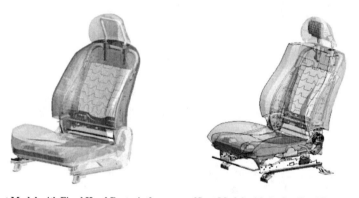

[Seat Model with Fixed Head Restraint]   [Seat Model with Active Head Restraint]

Figure 6. Prototype Seat Models.

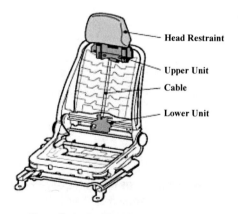

Figure 7. Active Head Restraint System.

REAR IMPACT SIMULATION: Rear impact simulations were conducted using the THUMS occupant model and the seat models. Figure 8 shows the simulation model. The posture of THUMS was adjusted so that it takes a natural sitting posture on the seat model, basically following the seating procedure in neck assessment tests. The adjustment was performed applying gravity in the downward and rearward direction, taking into account contacts between the THUMS body parts and the seat cushion surface. A small gap was allowed between the head and the head restraint, which was basically equal to the amount resulting with the 3D mannequin or the BioRID dummy. After the posture was adjusted, rear impact simulations were conducted. Each seat model was mounted on a rigid floor plate and the floor plate was accelerated assuming a rear impact. According to the research on rear collision data by the Ministry of Land, Infrastructure and Transport of Japan (2002), the average delta-V for rear impacts was around 16 km/h, which is more severe than roughly 60 percent of all rear collisions. A delta-V of 25 km/h covers approximately 90 percent of all rear collisions. This study focuses on the higher delta-V condition as a challenge to examine the benefit of the latest head restraint systems, although the authors have conducted simulation studies at delta-V's of both 16 and 25 km/h. In the following rear impact simulations, a triangular acceleration pulse representing an impact at a delta-V of 25 km/h was applied to the model (Figure 9). In the FE input data, the impact condition was defined as a velocity boundary condition applied to the floor plate, converting the acceleration pulse to the velocity time history. An explicit integration scheme was used for the rear impact simulations, with an integration time-step determined from the material density and the stiffness (without mass scaling), and was terminated 200 ms after impact.

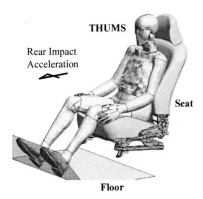

Figure 8. Model for Rear Impact Simulation (Fixed Head Restraint).

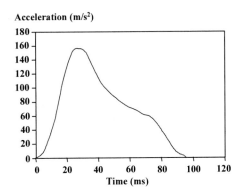

Figure 9. Triangle Acceleration Pulse for Input (delta-V = 25 kmh).

Contacts between the occupant body and the seat were again handled with the algorism in the LS-DYNA™. The contact search area was divided into four regions for the analysis. One was between the head and the head restraint, the second one was between the upper torso and the seatback surface, the third one was between the pelvis and the seatback surface, and the last one was between the lower extremities (buttock, thighs and calves) and the seat cushion surface. A set of nodes and elements was defined in advance, where time history responses such as displacement, velocity, acceleration and strain were output for analysis. The surface-to-surface algorism in the LS-DYNA™ was used for contact calculations between the occupant body and the seat surface. The algorism automatically outputs the contact force as an integrated amount for each defined region. The calculated data were output to ASCII files at a time interval of 0.1 ms, and then processed with Post-Processing Software for plotting. The authors previously conducted a parametric study varying seat design factors to investigate the correlation among neck injury indicators (Kitagawa et al., 2007). The result showed that NIC represented the amplitude of relative acceleration between the head and the torso, while Joint Capsule Strain grew with the magnitude of neck extension. This study uses these two indicators in estimating neck injury risk. NIC is calculated using the following equation:

$$NIC = 0.2 * (A_{T1} - A_{Head}) + (V_{T1} - V_{Head})^2$$

where A_{Head} and A_{T1} are the accelerations measured at the head (center of gravity) and T1 respectively, and V_{Head} and V_{T1} are the velocities at the head and T1. In general, the maximum indicator values are taken for evaluation in injury assessment. Joint Capsule Strain is directly output from the element forming the capsule tissues. The maximum peaks in time history curves from all the capsule elements are compared and the largest strain value is used for evaluation.

RESULTS

HEAD AND NECK KINEMATICS: The head and neck kinematics of the THUMS occupant model were analyzed in the case of the seat equipped with a fixed head restraint. Figure 10 shows the entire motion of the occupant model in frames at 0, 50, 100 and 130 ms after the initiation of impact. Although the simulation was conducted applying a forward acceleration pulse reproducing the actual rear impact scene, the paper describes the occupant motion in the vehicle coordinate system (on the rigid floor) for a better understanding. There is a gap between the head and the head restraint at the initial state (0 ms), while the lower torso (buttock) contacts the seatback in the initial seating position. At the beginning of impact, the torso and the head start moving rearward, but the torso is immediately supported by the seatback, while the head moves free until it contacts the head restraint. In this simulation case, the head to head restraint contact occurred around 70 ms. The seatback frame deforms rearward as the occupant body loads into it. The magnitude of deformation reaches its maximum peak around 100 ms. The torso of the model moves in the forward direction at this point, which is called a 'rebound' motion. The head still moves back for a while as the head restraint deformation reaches its maximum peak at around 130 ms.

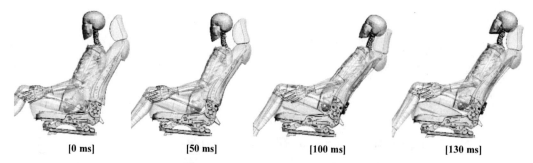

[0 ms]   [50 ms]   [100 ms]   [130 ms]

Figure 10. Occupant (THUMS) Motion in Rear Impact with Fixed Head Restraint.

Then the head turns to move forward. Contact forces between the occupant body and the seat were examined. Figure 11 plots the contact forces calculated at the pelvis, torso and head portions of the model. The contact force at the pelvis rises first, followed by the torso and the head rises last. The magnitude of each peak force may be related to the mass of the corresponding body part and the contact area. The contact force at the pelvis is generated by the inertia of the lower body and the posterior side of the buttock contacts the lower part of the seatback. The torso also has a large contact area but the upper torso does not contact the seatback at the initial state (0 ms), resulting in a delay in the latter rise after 50 ms. The magnitude of the peak is relatively lower than that at the pelvis because of relatively smaller mass of the torso. The contact force at the head is much smaller due to its lighter mass. The peak for the head appears around 130 ms, corresponding to the timing of the maximum deformation of the head restraint, which is later than the other peaks. Figure 12 shows the horizontal accelerations calculated at the pelvis, T1 and head. The pelvis is accelerated first, as also observed in the contact force, because the lower torso contacts the seatback from beginning. The T1 acceleration rises almost at the same timing, but its increasing trend is not continuous, possibly due to a gap behind the upper torso and penetration into the seatback. Note that T1 is located at the upper end of the torso. The head is accelerated last. There is a small rise from 30 to 70 ms possibly due to a transmitted motion through the cervical spine. The acceleration rise is more prominent after 70 ms where the head contacts the head restraint surface. The T1 acceleration rises again after the head to head restraint contact. The timings of the acceleration peaks at the head and T1 are close to those of the contact forces, while the timing of the head acceleration peak appears around 100 ms which is earlier than that of the contact force around 130 ms. The amplitudes of the acceleration peaks are much closer to each other compared to the maximum magnitudes in the contact

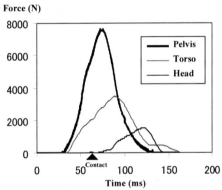

Figure 11. Contact Forces at Pelvis, Torso and Head with Fixed Head Restraint.

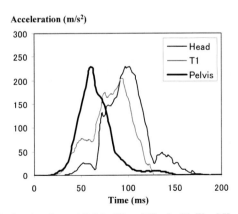

Figure 12. Accelerations at Pelvis, T1 and Head with Fixed Head Restraint.

forces. Figure 13 plots the time history curves of NIC and Joint Capsule Strain calculated from the simulation. The strain was calculated at all the cervical joints from OC-C1 to C7-T1. The maximum strain value was always found at the C6-C7 joint under the impact condition assumed in this study. A comparison of the time history curves indicates the difference in timing of their peaks. NIC reaches its maximum peak around 70 ms when the head contacts the head restraint. Joint Capsule Strain has its maximum peak around 140 ms, which is slightly later than the timing of the maximum deformation of the head restraint.

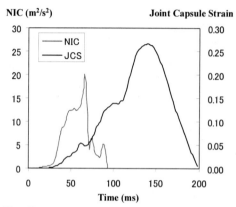

Figure 13. Time History Curves of NIC and Joint Capsule Strain (C6-C7) with Fixed Head Restraint.

ACTIVE HEAD RESTRAINT: A rear impact simulation was also conducted using the seat model equipped with the active head restraint system. Figure 14 shows a close-up view of the head and neck with the motion of the active head restraint during the rear impact. The frames were selected from the initial state (0 ms), 60 ms and 100 ms. There is a gap between the head and the head restraint at the initial state, as observed in the previous case. The seatback deforms as the occupant leans on it, and the head restraint tends to move away as a result. As the deformation mostly grows around the reclining joint located at the bottom of the seatback, the head restraint moves backward and downward. The active system moves the head restraint forward and upward. It contacts the occupant head around 60 ms cancelling the frame deformation in moving away. The system keeps supporting the head after the contact. The relative motion between the head and torso is not significant at 100 ms compared to that at 60 ms. The horizontal accelerations calculated at the pelvis, T1 and head were plotted in Figure 15. Similar to the previous case, the pelvis acceleration rises first, followed by the torso and the head. Note that the seat configuration and the mechanical property of the seatback structure are different from the model with the fixed head restraint. Likewise, the acceleration responses are different from the previous case. The prominent difference is that the amplitude of the head acceleration is relatively higher than the other two curves, because the active head restraint more greatly supported the occupant's head. There is a small rise in the head acceleration right after 50 ms, which is possibly due to the T1 acceleration rise as similarly observed in the previous case. The T1 acceleration rises earlier but has a small stop and a decrease, then it increases again synchronizing with the head acceleration. A similar trend was observed in the previous case although the amplitude and timing were different. Figure 16 and 17 compare NIC and Joint Capsule Strain respectively between the two cases. There are two peaks in the NIC curve with the active head restraint system while the other case has only one. The maximum value is lower in the case with the active system. The difference in Joint Capsule Strain is relatively larger. The maximum peak appears around 100 ms and there is another but smaller peak around 140 ms, while only one peak was observed in the previous case.

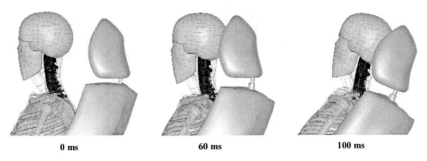

0 ms   60 ms   100 ms

Figure 14. Close-Up View of Head and Neck Motion with Active Head Restraint.

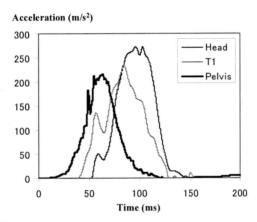

Figure 15. Accelerations at Pelvis, T1 and Head with Active Head Restraint.

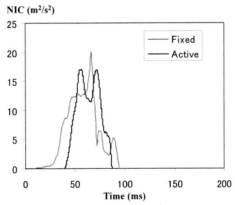

Figure 16. Comparison of NIC between Fixed and Active Head Restraints.

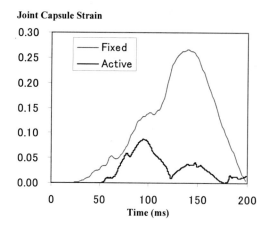

Joint Capsule Strain

Figure 17. Comparison of Joint Capsule Strain (C6-C7) between Fixed and Active Head Restraints.

DISCUSSION

The relative acceleration term in the NIC formulation greatly affects its maximum value. Assuming that the T1 acceleration has an increasing trend, the NIC value becomes lower as the contact timing is earlier. This is why the maximum (first) NIC value was lower in the case with the active head restraint system. The second NIC peak observed in that case possibly came from the head acceleration rise which was relatively milder than that in the other case with the fixed head restraint. The head restraint support structure of the active system could be relatively less stiff when compared to that in the fixed structure. The milder rise of the head acceleration resulted in another peak in the NIC curve. However, the amplitude of both peaks are lower than that of the single peak in the first case.

The mechanism of strain growth in the joint capsule was examined focusing on the vertebral motion. The study selected the C6-C7 unit as it showed the maximum strain value among the cervical joints from C1 to C7. The C6 motion was analyzed with respect to C7, converting the global X and Z-displacements to a local coordinate system defined along the facet joint surface. Figure 18 illustrates the C6-C7 unit showing the local coordinate system used for analysis. Z-displacement means a stretch in the joint while X-displacement corresponds to a shear motion in the defined coordinate system. While the contribution from these displacement components to the neck pain is not known, the study monitored them to better understand the vertebral motion. The calculated local X and Z-displacements with the fixed head restraint are shown in Figure 19 as time history curves, superimposed with the Joint Capsule Strain. The left axis indicates the magnitude of displacement while the right one means the magnitude of strain. The timing of the maximum peak of Joint Capsule Strain was close to that of X-displacement, while the first rise of strain synchronized with Z-displacement. Figure 20 shows the calculated displacements in the second case with the active head restraint system. The magnitudes of both displacements are smaller in this case compared to that in the previous case. A similar correlation was found between the peaks of strain and displacements, but the maximum strain peak appeared when Z-displacement reached its maximum peak. There is another strain peak synchronizing with that of X-displacement, but the magnitude is much smaller. It suggests that the magnitude of Joint Capsule Strain in the second case was lower mostly because of small X-displacement. Note that the facet joint surface is inclined around 45 degrees against the spinal column. The shear deformation (X-displacement) in the joint causes a posterior-inferior motion of C6 with respect to C7. The forward and upward motion of the active head restraint aligns with the direction of the shear deformation. The magnitude of Joint Capsule Strain can be reduced by supporting the head from the posterior-inferior side, that is, moving the head restraint forward and upward. The active

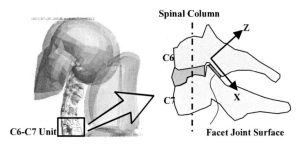

Figure 18. Local Coordinate System for Vertebral Motion Analysis.

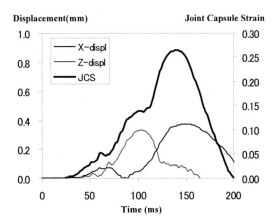

Figure 19. Local X and Z Displacements and Joint Capsule Strain (C6-C7) with Fixed Head Restraint.

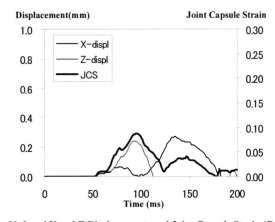

Figure 20. Local X and Z Displacements and Joint Capsule Strain (C6-C7) with Active Head Restraint.

head restraint system begins to activate when the pelvis pushes against the loading unit. The early rise and the relatively larger magnitude of contact force at the pelvis indicate that the pelvis force could be an improved trigger mechanism to activate the system. The timing of the head to head restraint contact became shorter; around 60 ms with the active system while 70 ms with the fixed head restraint. The head restraint keeps supporting the occupant head as long as the lower unit is loaded by the pelvis. Although the strain can grow past 100 ms in some circumstances, the active head restraint system presented in this study could support the head for a sufficient period of time.

Toyota had developed a seat design concept "WIL" to help reduce loading to the occupant neck during a rear impact. This technology has been applied to product seat designs since 1997, employing energy absorbing design features and head restraint located relatively forward compared to the seatback. The effectiveness of the seat system was examined through dummy tests and THUMS simulations, confirming that both NIC and Joint Capsule Strain indicated lower values (Sawada et al., 2005). According to the field study conducted by Kllugren et al. (2007), investigating actual rear collision database in Sweden, the WIL seats showed lower risk of neck injury compared to those without an anti-whiplash system. The active head restraint system presented in this study is an extension of the WIL Concept seat. The results of this study indicated that the active head restraint system could further help reduce the neck injury risk in low speed rear impacts.

LIMITATIONS

The study has limitations, as in many studies with human body FE models. Although the head and neck kinematics of the THUMS model were validated against the literature data, the number of data samples and loading conditions were limited. The study assumes that the model could simulate the human head and neck response even when loading conditions and seat models are changed. The lack of active muscle function brings another limitation in simulating living human responses. The model cannot mimic muscular contraction due to spinal reflex, which could resist against a sudden stretch of the muscle. The model possibly overestimates the magnitude of body movement in the later phase of impact. The mechanism of neck pain in the form of Joint Capsule Strain needs further studies from a neurophysiology point of view. It should be also studied how vertebral displacements in X and Z directions contribute to the neck pain. Based on all such limitations, the calculated Joint Capsule Strain as well as the NIC value should be treated comparatively rather than absolutely. The results and findings were mostly discussed in terms of the difference (increase or decrease) among the study cases. Future study will focus on more advanced modeling including active muscular responses and also incorporating the latest research findings.

CONCLUSIONS

Rear impact simulations were conducted using a human body FE model, THUMS, representing an average size male occupant. The cervical system including the facet joint capsules was incorporated to the model. The validity of the model was examined comparing its mechanical responses to those in the literature such as the whole body motion of the volunteer subject and the vertebral motion in the PMHS tests. Rear impact simulations were conducted using the validated THUMS model and two prototype seat models, one had a fixed head restraint and the other one was equipped with an active head restraint system. The active head restraint system works moving the head restraint forward and upward when the lower unit is loaded by the pelvis. The head and neck kinematics and responses were analyzed from the simulation results. The force and acceleration rose at the pelvis first, followed by T1 and the head. The early timing of force rise and its magnitude indicated that the pelvis force was a good trigger for the active head restraint system. The results showed that the head was supported earlier in a case with the active head restraint system, and both NIC and Joint Capsule Strain were lowered. The study also analyzed the mechanism of strain growth in the joint capsules. Relatively greater strain was observed in the direction of the facet joint surface, which was around 45 deg inclined to the spinal column. The forward and upward motion of the active head restraint was aligned with the direction of the joint deformation, and contributed to lower strain in the joint capsules. The results indicated that the active head restraint could help reduce the neck injury risk, not only by supporting the head at an early timing but also through its trajectory stopping the joint deformation.

ACKNOWLEDGMENTS

THUMS has been developed in collaboration with the Toyota Central Research and Development Laboratory. The authors would like to thank Toyota Technical Development Corporation for its assistance with the modeling and simulation work.

REFERENCES

Boström, O., M. Svensson, B. Aldman, H. Hansson, Y. Haland, P. Lovsund, T. Seeman, A. Suneson, A. Saljo and T. Ortengen. 1996. "A New Neck Injury Criterion Candidate Based on Injury Findings in the Cervical Spinala Ganglia after Experimental Neck Extension Trauma." Proc. International Conference on the Biomechanics of Impacts. 123-136.

Deng, B., P. Begeman, K. Yang, S. Tashman and A. King. 2000. "Kinematics of Human Cadaver Cervical Spine During Low Speed Rear-End Impacts." Stapp Car Crash Journal 44: 171-188.

Hasegawa, J. 2004. "A Study of Neck Soft Tissue Injury Mechanisms During Whiplash Using Human FE Model." Proc. International Conference on the Biomechanics of Impacts. 321-322.

Iwamoto, M., Y. Kisanuki, I. Watanabe, K. Furusu, K. Miki and J. Hasegawa. 2002. "Development of a Finite Element Model of the Total Human Model for Safety (THUMS) and Application to Injury Reconstruction." Proc. International Conference on the Biomechanics of Impacts. 31-42.

Jakobsson L. 2004. "Field analysis of AIS1 neck injuries in rear end car impacts- injury reducing effect of WHIPS." Thesis Chalmers University of Technology, Göteborg, Sweden.

Japanese National Police Agency. 2004. "Traffic Green Paper 2003" (in Japanese). 1-23.

Kitagawa, Y., T. Yasuki and J. Hasegawa. 2006. "A Study of Cervical Spine Kinematics and Joint Capsule Strain in Rear Impacts Using a Human FE Model." Stapp Car Crash Journal 50: 545-566.

Kitagawa, Y., T. Yasuki and J. Hasegawa. 2007. "Consideration of Possible Indicators for Whiplash Injury Assessment and Examination of Seat Design Parameters using Human FE Model." 07-0093. 20th ESV Conference.

Kitagawa, Y., T. Yasuki and J. Hasegawa. 2008. "A Study of Head and Neck Kinematics in Rear Impact and Whiplash Injury Predictors using Human FE Model." World Congress on Neck Pain.

Kraft, R., A. Kullugren, A. Ydenius, O. Bostrom, Y. Haland and C. Tingvall. "Rear Impact Neck Protection by Reducing Occupant Forward Acceleration – A Study of Cars on Swedish Roads Equipped with Crash Recorders and a New Anti-Whiplash Device." Proc. International Conference on the Biomechanics of Impacts. 221-231.

Lee, K., Davis, M., Mejilla, R. and Winkelstein, B. 2004. "*In Vivo* Cervical Facet Capsule Distraction: Mechanical Implications for Whiplash and Neck Pain." Stapp Car Crash Journal 48: 373-395.

Lu, Y., C. Chen, S. Kallakuri, A. Patwardhan and J. Cavanaugh. 2005. "Neural Response of Cervical Facet Joint Capsule to Stretch: A Study of Whiplash Pain Mechanism." Stapp Car Crash Journal 49: 49-66.

Ministry of Land, Infrastructure and Transport. 2002. The 3rd Car Safety Symposium.

Kullgren, A., Krafft, M., Lie, A. Tingvall, C. 2007. "The Effect of Whiplash Protection Systems in real-Life Crashes and Their Correlation to Consumer Crash Test Programmes." 07-0468. 20th ESV Conference.

Ono, K. and K. Kaneoka. 1997. "Motion Analysis of Human Cervical Vertebrae During Low Speed Rear Impacts by the Simulated Sled." Proc. International Conference on the Biomechanics of Impacts. 223-237.

Sawada, M., Hasegawa, J. 2005. "Development of New Whiplash Prevention Seat." 05-0288 19th ESV Conference.

Sekizuka, M. 1998." Seat Designs for Whiplash Injury Lessening. " 98-S7-O-06 16th ESV Conference.

Siegmund, G. P., B. S. Myers, M. B. Davis, H. F. Bohnet and B. A. Winkelstein. 2000. "Human Cervical Motion Segment Flexibility and Facet Capsular Ligament Strain under Combined Posterior Shear, Extension and Axial Compression." Stapp Car Crash Journal 44: 159-170.

Svensson, Y., B. Aldman, P. Lövsund et al. 1993. "Pressure Effects in the Spinal Canal During Whiplash Extension Motion: A Possible Cause of Injury to the Cervical Spinal Ganglia." Proc. International Conference on the Biomechanics of Impacts. 189-200.

Yamada, H. 1970. "Strength of Biological Materials." Evans, F. G. (Ed). Williams & Wilkins Company, Baltimore.

Yoganandan, N., A. Pintar, S. Kumaresan and A. Elhagediab. 1998. "Biomechanical Assessment of Human Cervical Spine Ligaments." 42nd Stapp Car Crash Conference.

Winkelstein, B., R. Nightingale, W. Richardson and B. Myers. 1999. "Cervical Facet Joint Mechanics: Its Application to Whiplash Injury." Proc. 43rd Stapp Car Crash Conference. 243-265.

THE EFFECT OF WHIPLASH PROTECTION SYSTEMS IN REAL-LIFE CRASHES AND THEIR CORRELATION TO CONSUMER CRASH TEST PROGRAMMES

Anders Kullgren
Maria Krafft
Folksam Research and Karolinska Institutet, Sweden
Anders Lie
Swedish Road Administration and Karolinska Institutet, Sweden
Claes Tingvall
Swedish Road Administration, Sweden and Monash University, Australia
Paper number: 07-0468

ABSTRACT

The objective was to study the influence of various types of car seats aimed at protecting whiplash injuries on real-life injury outcome. Furthermore, the aim was to study correlation between whiplash consumer crash tests and real-life injury outcome. In both cases the influence on long-term whiplash symptoms were studied.

Since 1997 various seats aimed at lowering the risk of whiplash injuries have been introduced in cars. The cars were divided into groups according to the safety technology used. Since 2003 consumer crash test programmes have been running. The correlation on group level between whiplash injury outcome in real-life crashes and the test results of consumer crash tests both in Sweden by Folksam and the Swedish Road Administration and by IIWPG were studied.

The results show that cars fitted with more advanced whiplash protection systems had 50% lower risk of whiplash injuries leading to long-term symptoms than cars launched since 1997 without whiplash systems. All three whiplash preventive technologies studied, RHR (Reactive Head Restraints), WhiPS (Whiplash Prevention System), and WIL (Whiplash Lessening System), showed lower risk of whiplash injury leading to long-term symptoms than cars fitted with standard seats.

A correlation was found between consumer whiplash crash tests and real-life outcome. It was found that cars rated in the worst group in the IIWPG and Folksam/SRA ratings had 43% and 60% higher risk of long-term symptoms in real-life crashes, respectively, than cars rated in the best group.

A limitation with the tests is that the consumer crash test programmes are conducted with the seat only, while the real-life injury outcome concerns the performance of the whole car.

It can be concluded that seats aimed at preventing whiplash injuries in general also lower the risk in real-life crashes. Furthermore it can be concluded that results from existing consumer crash test programmes for whiplash correlate with real-life injury outcome.

INTRODUCTION

In October 1997 the Swedish parliament decided upon the new road traffic safety policy in Sweden, the so-called Vision-Zero (Kommunikations-departementet 1997). An important part in the policy is to minimise health losses and not accidents or injuries in general. Health losses include fatal injuries and severe injuries where the person not is recovering within reasonable time, i.e. the focus is set on the public health problem.

Apart from fatalities, injuries leading to disability reported by insurance companies are a good indication of the number of serious road traffic injuries. They also give a good picture of both the typical injuries and the type of crashes that primarily should be in focus for road traffic safety actions. In Sweden more than 3,500 permanently disabled car occupants are reported every year (with a disability of at least 10% according to the classification used by Swedish insurance companies) (Försäkringsförbundet 1996). More than 50% of those are whiplash injuries. It is therefore important that the society focuses on reducing whiplash injuries.

In modern cars on the Swedish market, whiplash injuries account for approximately 70% of all injuries leading to disability (Folksam 2005). Most occupants reporting whiplash injuries recovers within a week, while between 5% and 10% will get more or less life lasting problems (Nygren 1984, Krafft 1998, Whiplashkommissionen 2003).

Whiplash prevention initiatives

Whiplash preventive measures have so far been focussed on developments of the seat. Since the 70s head restraints have been implemented more and more frequently. To date all seating positions in most car models are fitted with head restraints. The whiplash injury reducing effects of head restraints have been shown to be relatively low, between 5% and 15% (Nygren et al 1985, Morris and Thomas 1996). In order to increase the vehicle crashworthiness in high-speed rear end crashes, vehicle seats have become stiffer since the late 80s (Krafft 1998). Stiffer seats have probably increased the whiplash injury risks in low-speed rear-end crashes.

Based on this knowledge more advanced whiplash protection devices have been introduced on the market. The better protection is achieved through improved geometry and dynamic properties of the head restraint or by active devices that move in a crash as the body loads the seat. The main ways to lower the whiplash injury risk are to minimise the relative motion between head and torso, to control energy transfer between the seat and the body and to absorb energy in the seat back.

To date several systems exist, for example RHR or AHR (Reactive Head Restraint or Active Head Restraint) in several car models, WhiPS (Whiplash Prevention System) in Volvo and Jaguar, WIL (Whiplash Injury Lessening) in Toyota. RHR was firstly introduced in Saab cars in 1998 (SAHR) (Wiklund and Larsson 1997), and is today the most common whiplash protection concept on the market. It exists in several models from for example Audi, Ford, Mercedes, Nissan, Opel, Skoda, Seat and VW. RHR is a mechanical system that actively moves the head restraint up and closer to the head and in a crash. Saab has apart from the head restraint also designed the seat back structure to better support the torso in a rear end crash. WhiPS was first introduced in Volvo cars in 1999 (Lundell et al 1998, Jakobsson 1998). The seat back is in a crash moved rearwards and yields in a controlled way to absorb energy. The Toyota system WIL (Sekizuka 1998) has no active parts and is only working with improved geometry and softer seat back. Ford has also introduced seats without active or reactive parts in the headrest, but with an improved design aimed at preventing whiplash injury.

Studies have been presented showing the effect of the Saab RHR and Volvo WhiPS indicating an injury reducing effect of approximately 40-50% (Viano and Olssén 2001, Insurance Institute for Highway Safety (IIHS) 2002, Jakobsson 2004, Krafft et al 2003). Apart from that the information of real-life performance of different systems is limited.

In recent years some consumer rating programs have been developed and introduced. In 2003 Folksam and the Swedish Road Administration (SRA) started crash testing of car seats, where each seat is exposed to three different tests. Also the German ADAC started crash testing of car seats using multiple tests for each seat (ADAC website). In 2004 the insurance initiative IIWPG (International Insurance Whiplash Prevention Group) started consumer crash testing in Europe and in the USA (IIHS and Thatcham websites). In those tests each seat was exposed to one test. Studies of the correlation between crash test results and real-life performance is rare.

Objectives of the study

The objective was to study the influence of various types of car seats aimed at protecting whiplash injuries on real-life injury outcome. Furthermore, the objective was to study correlation between whiplash consumer crash tests and real-life injury outcome. In both cases the influence on long-term whiplash symptoms were studied.

METHOD/MATERIAL

The study was based on two different data sources. To calculate the proportion of injuries leading to long-term symptoms all whiplash injuries in rear-end crashes reported to the insurance company Folksam between 1998 and 2006 were used. In total 6383 reported whiplash injuries were included. To calculate relative risk of an injury in rear-end crashes all two-car crashes reported by the police between 1998 and 2006 were used, in total 15587 crashes.

Injury classification

Claims reports including possible medical journals for all crashes with injured occupants between 1998 and 2006 were examined. Whiplash injuries reported in rear-end crashes within a range between +/-30 degrees from straight rear-end were noted.

Insurance claims were used to verify if the reported whiplash injuries led to long-term symptoms. Occupants with long-term symptoms were defined as those where a medical doctor examined the occupant and the occupant claimed injury symptoms for more than 4 weeks, which corresponds to a payment of at least 2000 SEK in the claims handling process used by Folksam. Out of the 6383 persons reporting a whiplash injury, 912 (13%) led to long-term symptoms according to that definition.

Calculation of relative injury risk

According to Evans (1986), when two cars collide with each other, the injury risk for Car 1 in relation to Car 2 can be expressed as the number of injured occupants in Car 1 in relation to the number in Car 2. This is equal to the risk of injury in car 1 in relation to the risk of injury in Car 2, which can be denoted as p_1 / p_2. Assuming that the probabilities p_1 and p_2 are independent, and that the injury risk in Car 2 can be expressed as the injury risk in Car 1 multiplied by a constant, four cases can be summed: x_1, x_2, x_3 and x_4. The relative injury risk in the whole range of impact severity is equal to equation (1). In this study the relative injury risk for the sum of all cars in each group studied was calculated.

In a similar way the relative risk of injury in rear-end crashes can be calculated with the same technique, where the number of crashes with injured drivers in the struck car in rear-end crashes in relation to the number of crashes with injured drivers in the striking car are summed, see Table 1. The method used in this study to calculate relative injury risk has been further described by Hägg et al. (1992) and Hägg et al (1999).

The initially presented method is relevant for cars of similar mass. If Car 1 and Car 2 have unequal mass, the exposure to impact severity will be unequal as well. While crashworthiness rating based on real-life experience should preferably show the benefit or dis-benefit of mass, the current method would give too much attention to mass, as it would also include the benefit or dis-benefit for the colliding partner. When calculating the injury risk for car models relative to the average car, it is important that the relative injury risk for all car models can be compared with the identical average car. This is not the case if the influence of mass differences on the exposure for the collision partner is not compensated. The initial estimate, equation (1), must therefore be modified to take mass relations into account. The factor m was calculated for the car models in each group under study, and thus used to compensate the relative injury risk for the models in each group, see equation (2).

Table 1. Classification of combinations of injured drivers in the struck and striking car in rear-end crashes.

		Drivers in the striking car		Total
		driver injured	driver not injured	
Drivers in the struck car	driver injured	x_1	x_2	$x_1 + x_2$
	driver not injured	x_3	x_4	
Total		$x_1 + x_3$		

x_1 = number of crashes with injured drivers in both cars
x_2 = number of crashes with injured drivers in struck car and not in the striking car
x_3 = number of crashes with injured drivers in striking car and not in struck car
x_4 = number of crashes without injured drivers in both cars

$$R = (x_1 + x_2) / (x_1 + x_3) \qquad (1).$$

$$R_{modified} = R * m^{((M-Maverage)/100)} =$$
$$= (x_1 + x_2) / (x_1 + x_3) * m^{((M-Maverage)/100)} \qquad (2).$$

M is the mass of the studied vehicle and $M_{average}$ is the average mass of all vehicles. In these calculations the factor m was set as 1.035, see Hägg et al. (1992), which means that the mass effect used to control for the exposure on impact severity was 3.5% per 100 kg. The relative risk of sustaining an injury with long-term symptom was calculated as the product of the relative injury risk and the proportion of occupants with long-term symptoms in relation to the number of reported whiplash injuries.

Categories of cars studied

The whiplash injury and disability risks were calculated for some different categories;

- If the car was fitted with a specially designed whiplash protection system. Those not fitted with whiplash protection system were divided in cars launched before and after 1997.

- Kind of whiplash protection system in cars launched after 1997.
- Performance in the IIWPG ratings.
- Performance in the Folksam/SRA ratings.

The whiplash protections systems defined are RHR-Reactive Head Restraint, WhiPS (Volvo) and WIL (Toyota). Cars with seats fitted with RHR were divided into Saab RHR and RHR in the other manufacturers. Standard seats were defined as those not fitted with any of the systems mentioned above. A group with standard seats tested in consumer ratings was also compared.

RESULTS

A summary of the results is presented in Table 2. Detailed number of crashes and injured for the calculation of relative injury risk is presented in Table 3 in the Appendix.

Cars fitted with more advanced whiplash protection systems had approximately 50% lower proportion of whiplash injuries leading to long-term symptoms as cars with standard seats launched after 1997. Also, the relative risk of a sustaining a whiplash injury leading to long-term symptoms was approximately 50% lower in cars fitted with more advanced whiplash protection systems than in cars with standard seats launched after 1997. Compared with cars launched before 1997 with standard seats the difference was even higher.

It was also found that cars with RHR, WhiPS or WIL, all had lower risk of whiplash injuries leading to long-term symptoms compared with cars with standard seats. Saab cars with RHR showed lower whiplash injury risk than the group of cars with RHR seats from other manufacturers.

Standard seats tested in consumer ratings had lower whiplash injury risk than other standard seats.

A correlation was found between both IIWPG and Folksam/SRA ratings and proportion of injuries leading to long-term symptoms as well as for relative risk of sustaining a whiplash injury leading to long-term symptoms. Car seats rated in the worst group (Red) in the Folksam/SRA crash tests had 60% higher risk of long-term whiplash injury risk than car seats rated in the best group (Green+). Cars rated in the worst group (Poor) in the IIWPG crash tests had 43% higher risk compared with cars seats rated in the best group (Good).

Table 2. Proportions of injuries with long-term symptoms, relative injury risk in rear-end crashes and relative risk of a whiplash injury with long-term symptoms.

Type of study		Whiplash injuries leading to long-term symptoms			Relative injury risk in rear-end crashes		Relative risk of long-term symptoms
		Reported whiplash injuries (n)	Injuries leading to disability (n)	Proportion of injuries leading to disability (p_{dis})	Number of crashes	Relative injury risk (R)	Relative risk of disability (R* p_{dis})
Special whiplash protection system	Cars with a system	534	40	7,5%	1216	0,977	0,073
	Standard seats 97-	1571	213	13,6%	2488	1,051	0,143
	Standard seats -97	4109	635	15,5%	11883	0,970	0,150
Kind of whiplash protection system (Car models from model year 1997)	RHR	165	10	6,1%	433	1,11	0,067
	Saab RHR	114	6	5,3%	341	0,98	0,052
	Other RHR	51	4	7,8%	92	1,04	0,081
	WhiPS	89	6	6,7%	631	0,95	0,064
	WIL	264	20	7,6%	125	1,10	0,083
	Std seats tested in consumer ratings	196	20	10,2%	368	1,06	0,108
	Other std seats	1366	194	14,2%	2125	1,04	0,148
IIWPG rating	Good	253	17	6,7%	1083	0,95	0,064
	Acceptable	52	3	5,8%	49	1,24	0,071
	Marginal	86	5	5,8%	105	1,21	0,070
	Poor	205	18	8,8%	235	1,04	0,092
	Not tested seats	5615	836	14,9%	14107	0,98	0,146
Folksam/SRA rating	Green+	140	8	5,7%	729	0,98	0,056
	Green	314	21	6,7%	1089	0,98	0,066
	Yellow	77	4	5,2%	60	1,30	0,068
	Red	23	2	8,7%	40	1,03	0,089
	Not tested seats	5798	857	14,9%	14392	0,99	0,147

DISCUSSION

Whiplash injuries leading to permanent disability are serious and account for the vast majority of injuries leading to permanent disability (Nygren 1984, Krafft 1998). Many initiatives to reduce the problem have been taken, where most car manufacturers also include whiplash protection in their designs of new models (Lundell et al 1998, Wiklund and Larsson 1998). Many are also introducing more advanced whiplash protection systems in their models. Measuring the performance of recent introduced whiplash

prevention technology is very important for future activities in legislation and consumer testing, such as EuroNCAP. In recent years many initiatives of consumer rating system aimed at measuring neck injury risk in rear-end crashes have been launched. But the correlation between real-life whiplash injury outcome and results from these consumer rating programmes has to date not been presented.

Existing consumer crash testing is focussed the seat performance since the seat plays a major role in protecting the occupants from whiplash injury. This approach is probably relevant in today's situation, where the seat plays a major role for the whiplash injury risk, but since real-life outcome concerns the performance of the whole car, the results could be influenced by the difference.

The definition of long-term symptoms used in this study was chosen because it takes several years, sometimes up to 6 years, until a degree of permanent disability can be finally set and verified according to the system used by the insurance companies in Sweden (Försäkringsförbundet 1996). To be able to use this definition crashes older than 6 years can only be used, which is not applicable to study whiplash preventive systems introduced the latest 6 years.

Due to the limited number of crashes and injured it was not possible to study the performance of single car models, only groups of cars. All various car models fitted with reactive head restraints (RHR) may have different performance in real-life crashes. In this study it was only possible to study the difference between Saab RHR and RHR for other manufacturers, such as Audi, Ford, Nissan, Opel and VW. No major difference between these could be verified.

The results from this study is very positive and show that efforts made by car manufacturers to reduce whiplash injury risks has been successful, although there are still potential improvements to make. It is also positive that test results from consumer test programmes correlate with real-life performance. Also in this case there are still potential improvements to make to better mirror real-life injury risks. There is always a need to verify crash test results with results from real-world crashes.

Results from existing consumer crash test programmes indicate a large variation in protection. Some seats perform well even without more advanced whiplash protection systems, while some seats fitted with for example RHR received poor rating results. Identifying that a seat has a whiplash protection device is not enough. It stresses the need for consumer test programmes to be used as guidance for consumers in picking the best cars and it also stresses the need for validation of their performance in real-life crashes.

Finally, it is important to stress that further efforts should be made to improve car seats and also other safety technology to reduce whiplash injuries leading to permanent disability. Although the attempts made so far reduces the whiplash injury risk a lot, there is still a long way to go. In modern cars, whiplash injury accounts for approximately 70 % of all injuries leading to disability (Folksam 2005). Even if half of the whiplash injuries in rear-end crashes could be avoided, whiplash is still the most dominating injury leading to permanent disability.

CONCLUSIONS

- Cars fitted with advanced whiplash protection systems had 50% lower risk of whiplash injuries leading to long-term symptoms compared with standard seats launched after 1997.
- The whiplash prevention systems, RHR (Reactive Head Restrains), WhiPS or WIL, had lower risk of whiplash injuries leading to long-term symptoms compared with standard seats launched after 1997.
- A correlation was found between consumer crash test programmes and real-life whiplash injury outcome. Cars with seats rated as good in the consumer crash tests had lower risk of whiplash injuries leading to long-term symptoms compared with seats with poor results.

REFERENCES

ADAC website: www.adac.de

Evans L (1986) Double pair comparison – a new method to determine how occupant characteristics affect fatality risk in traffic crashes. *Accid. Anal. and Prev.*, Vol. 18, No. 3:pp 217-227.

Folksam (2005) How safe is your car? 2005. www.folksam.se or Folksam Research 10660 Stockholm Sweden.

Försäkringsförbundet (1996) Gadering av medicinsk invaliditet –96 (only in Swedish). IFU Utbildning AB, Stockholm Sweden.

Hägg A, v Koch M, Kullgren A, Lie A, Nygren Å, Tingvall C (1992) Folksam Car Model Safety Rating 1991-92, Folksam Research 106 60, Stockholm, Sweden.

Hägg A, Krafft M, Kullgren A, Lie A, Malm S, Tingvall C, Ydenius A (1999) Folksam Car Model Safety Ratings 1999, Folksam Research 106 60, Stockholm, Sweden.

IIHS website: www.iihs.org

Jakobsson L (1998) Automobile Design and Whiplash Prevention. In: Whiplash Injuries: Current Concepts in Prevention, Diagnoses and Treatment of the Cervical Whiplash Syndrome. Edited by Robert Gunzberg, Maerk Szpalski, Lippincott-Raven Publishers, pp. 299-306, Philadelphia.

Jakobsson L. Whiplash Associated Disorders in Frontal and Rear-End Car Impacts. Biomechanical. Thesis for the degree o doctor of philosophy. Crash Safety Dividson , Dep of Machine and Vehicle Systems. Chalmers University of Technology, Sweden 2004.

Kommunkationsdepartementet (1997) På väg mot det trafiksäkra samhället, DS 1997:13, (English short version: En Route to a Society with Safe Road Traffic, Selected extract from Memorandum prepared by the Ministry of Transport and Communications, DS 1997:13), Stockholm.

Krafft M (1998) Non-Fatal Injuries to Car Occupants - Injury assessment and analysis of impacts causing short- and long-term consequences with special reference to neck injuries, Thesis, Karolinska Institutet, Stockholm, Sweden.

Krafft M, Kullgren A, Lie A, Tingvall C (2003) Utvärdering av whiplashskydd vid påkörning bakifrån – verkliga olyckor och krockprov.

("Evaluation of whiplshprotection systems in rear-end collisions – real-life crashes and crash tests"). Folksam and SRA, Folksam research 10660 Stockholm, Sweden.

Lundell B, Jakobson L, Alfredsson B, Lindström M, Simonsson L (1998) The WHIPS seat – A car seat for improved protection against neck injuries in rear end impacts. Paper No 98-S7-O-08, Proc. 16th ESV Conf, 1998, pp. 1586-1596.

Nygren Å (1984) Injuries to car occupants – some aspects of the interior safety of cars. Akta Oto-Laryngologica, Supplement 395. Almqvist & Wiksell, Stockholm. Sweden. ISSN 0365-5237.

Nygren Å, Gustavsson H, Tingvall C (1985) Effects of Different Types of Headrests in Rear-end Collisions. Proc. of the 9th ESV conf. Oxford, UK. pp85-90

Sekizuka M (1998) Seat Designs for Whiplash Injury Lessening, Proc. 16th Int. Techn. Conf. on ESV, Windsor, Canada.

Thatcham website: www.thatcham.org

Viano D and Olsen S (2001) The Effectiveness of Active Head Restraint in Preventing Whiplash. The Journal of TRAUMA Vol 51:pp959-969.

Whiplashkommissionen (2003) (Swedish whiplash commission) www.whiplashkommissionen.se.

Wiklund K, Larsson H (1997) SAAB Active Head Restraint (SAHR) - Seat Design to Reduce the Risk of Neck Injuries in Rear Impacts, SAE Paper 980297, Warrendale.

APPENDIX

Table 3. Numbers of crashes with different combinations of injured occupants and relative injury risks in rear-end crashes.

		No. crashes	X_1	X_2	X_3	R	m	$R_{modified}$
Cars with and without whiplash protection	Seats with whiplash system	1216	351	461	501	0,95	1,03	0,98
	Standard seats from MY 1997	2488	711	1075	952	1,07	0,98	1,05
	Standard seats until MY 1997	11883	3093	5013	4986	1,00	0,97	0,97
Type of whiplash device	RHR	433	140	157	172	0,95	1,17	1,11
	Saab RHR	341	117	116	133	0,93	1,05	0,98
	Other RHR	92	23	41	39	1,03	1,00	1,04
	WHIPS	631	160	238	273	0,92	1,03	0,95
	WIL	125	43	53	39	1,17	0,94	1,10
	Standard seats tested 97-	368	96	170	149	1,09	0,98	1,06
	Other standard seats 97-	2125	618	912	807	1,07	0,97	1,04
IIWPG rating	Good	1083	306	400	459	0,92	1,03	0,95
	Acceptable	49	13	25	17	1,27	0,97	1,24
	Marginal	105	24	56	41	1,23	0,98	1,21
	Poor	235	76	97	86	1,07	0,98	1,04
	Good+Acc	1132	319	425	476	0,94	1,03	0,97
	Marg+Poor	340	100	153	127	1,11	0,98	1,09
	Not tested	14107	3724	5969	5830	1,02	0,97	0,98
Folksam/SRA rating	Green+	729	181	193	226	0,92	1,06	0,98
	Green	1089	293	279	332	0,92	1,07	0,98
	Yellow	60	17	22	12	1,34	0,97	1,30
	Red	40	16	9	9	1,00	1,03	1,03
	Above average	1099	295	285	333	0,92	1,03	0,95
	Below average	90	31	25	19	1,12	0,95	1,07
	Not tested	14392	3800	6115	5942	1,02	0,97	0,99
Total		15587	4155	6549	6438	1,01	0,99	1,00

2009-01-0386

Initial Assessment of the Next-Generation USA Frontal NCAP: Fidelity of Various Risk Curves for Estimating Field Injury Rates of Belted Drivers

Tony R. Laituri, Scott Henry, Brian Kachnowski and Kaye Sullivan
Ford Motor Company

ABSTRACT

Various frontal impact risk curves were assessed for the next-generation USA New Car Assessment Program (NCAP). Specifically, the "NCAP risk curves" — those chosen by the government for the 2011 model year NCAP — as well as other published risk curves were used to estimate theoretically the injury rates of belted drivers in real-world frontal crashes. Two perspectives were considered: (1) a "point" estimate of NCAP-type events from NCAP fleet tests, and (2) an "aggregate" estimate of $0 \leq \Delta V \leq 56$ km/h crashes from a modeled theoretical vehicle whose NCAP performance approximated the average of the studied fleet. Four body regions were considered: head, neck, chest, and knee-thigh-hip complex (KTH). The curve-based injury rates for each body region were compared with those of real-world frontal crashes involving properly-belted adult drivers in airbag-equipped light passenger vehicles.

The assessment yielded mixed results. For the head, all of the studied curves yielded injury rates within the confidence limits for both the point and aggregate estimates. For the neck, the NCAP risk curve for the combined-loading metric (N_{ij}) yielded results well outside the confidence limits for both the point and aggregate estimates. For the chest, the NCAP risk curve yielded results within the confidence limits for the point estimate, but the result was somewhat lower than the mean. For the KTH, all of the studied curves yielded results outside the confidence limits for both the point and aggregate estimates. Accordingly, different available risk curves were recommended for the neck and chest, and the need for research to improve the KTH risk estimates was demonstrated.

INTRODUCTION

Risk curves provide the foundation for the United States NCAP by converting test dummy responses into human occupant injury probabilities. For example, consider the multi-step process in the present frontal NCAP. The head response from the mid-sized male test dummy is substituted into a severe-to-fatal injury risk curve, yielding a head injury probability; the same process is applied for the chest response; the two resulting probabilities are used to compute a combined probability of injury; that single probability is evaluated subject to a grading scale; and a star rating is assigned.

The National Highway Traffic Safety Administration (NHTSA) recently published numerous significant changes to the NCAP pertaining to 2011+ model year vehicles (NHTSA, 2008 and 2008a), and risk curves will again play a major role. Some of the changes for the frontal NCAP will include revised head and chest metrics, additional risk computations for the upper neck and the knee-thigh-hip complex, and assessment of injuries that are easier to sustain. To affect these changes, NHTSA selected a risk curve for each considered body region from the available literature. *For easy reference in the present study, the risk curves selected by NHTSA will be called "NCAP risk curves."*

SAE Customer Service: Tel: 877-606-7323 (inside USA and Canada)
                        Tel: 724-776-4970 (outside USA)
                        Fax: 724-776-0790
                        Email: CustomerService@sae.org
SAE Web Address: http://www.sae.org
Printed in USA

9-2009-01-0386

SAE *International*

One advantage of using risk curves is that injury risks can be estimated without having to wait years for field data to amass; one disadvantage is that risk curves are updated frequently as more information becomes available. Therefore, a comprehensive assessment of the NCAP risk curves would be valued by safety advocates.

In the present study, the NCAP risk curves and others will be assessed by comparing field- and curve-based injury rates of belted adult drivers in frontal crashes in a four-step process, by which the paper is organized:

(1) Collect <u>field data</u> to compute injury rates, by body region, pertaining to
- a. NCAP-type crashes which provide a point-estimate,
- b. More common crashes (that include NCAP-type crashes) which provide an aggregate estimate,

(2) Collect <u>test dummy responses</u> from
- a. NCAP tests of the fleet, and
- b. A theoretical field model of a vehicle whose NCAP performance approximated the average of the studied fleet.

(3) Collect available risk curves, by body region, and

(4) Compare field- and various curve-based injury rates, by body region.

All available published risk curves known to the authors will be considered in the present study. Different analysts may prefer different risk curves for many possible reasons. Therefore, no risk curves will be excluded based on analyst experience, choice of statistical method, and so forth. Rather, the available risk curves will be shown, their underlying assumptions will be reported, and their performance for estimating injury rates relative to those from USA field data will be assessed. Finally, *for this study, a risk curve will be deemed "acceptable" if it yields estimated injury rates within the confidence limits of the field data; if both the NCAP risk curve and other risk curves for a particular body region yield acceptable results, the NCAP risk curve will be "recommended."*

PART 1: USA FIELD DATA

Field data from the NASS/CDS were selected subject to the following criteria:

- Calendar years of NASS = 1993-2006
- Vehicles = 1994-2006 MY VIN-identified light passenger vehicles, fitted with driver airbags
- Occupants = Adult (age 15 years and older) drivers
- Restraint = Properly-belted
- Crashes = Towaway, 11-1 o'clock, full-engagement frontals (non-rollovers).

The resulting field data is summarized in Table 1; therein

- "ΔV" is the longitudinal speed change bin, with a 9.6 km/h interval. For example, "ΔV=8 km/h" represented "0 km/h ≤ ΔV ≤ 9.6 km/h."

Table 1. Field Data: Belted Adult Drivers in 11-1 o'clock, Full Engagement Frontal Crashes

ΔV (km/h)	Severity	Crashes n	Crashes N	Crashes % of All	Head/Face n	Head/Face N	Head/Face AIS3+ Risk (%)	Neck/Spine n	Neck/Spine N	Neck/Spine AIS3+ Risk (%)	Chest n	Chest N	Chest AIS3+ Risk (%)	KTH n	KTH N	KTH AIS2+ Risk (%)	Any of the Four n	Any of the Four N	Any of the Four Risk (%)
8	Both	25	12,869	1.1	0			0			0			1	20	0.2	1	20	0.2
16	Both	914	476,282	40.3	1	377	0.1	1	38	0.01	1	108	0.0	7	1,440	0.3	10	1,963	0.4
24	Both	1,034	385,214	32.6	5	262	0.1	1	27	0.01	8	1,035	0.3	20	3,929	1.0	32	5,136	1.3
32	Both	705	204,999	17.4	3	85	0.0	4	247	0.12	13	920	0.5	23	2,396	1.2	36	3,206	1.6
40	Both	344	63,877	5.4	4	1,046	1.6	0			12	747	1.2	27	1,416	2.2	39	2,614	4.1
48	Both	168	22,532	1.9	6	381	1.7	2	58	0.26	13	752	3.3	20	1,552	6.9	34	2,496	11.1
56	Both	101	6,617	0.6	6	701	10.6	2	94	1.43	19	1,670	25.2	16	1,301	19.7	28	2,128	32.2
64	Both	46	3,805	0.3	2	261	6.9	1	3	0.07	6	348	9.2	12	1,007	26.5	14	1,295	34.1
72	Both	25	2,608	0.2	2	443	17.0	0			8	636	24.4	9	1,311	50.3	15	1,504	57.7
80	Both	13	819	0.1	3	95	11.6	1	40	4.84	2	68	8.3	1	171	20.9	6	334	40.8
88	Both	18	945	0.1	2	45	4.8	0			10	535	56.6	12	714	75.6	13	737	78.0
Total	Both	3,393	1,180,567	100.0	34	3,696	0.3	12	507	0.04	92	6,820	0.6	148	15,257	1.3	228	21,434	1.8
8	NPBLE	22	11,732	1.0	0			0			0			1	20	0.2	1	20	0.2
16	NPBLE	631	366,276	31.0	0			0			0			6	1,428	0.4	6	1,428	0.4
24	NPBLE	423	184,995	15.7	2	123	0.1	0			3	351	0.2	7	1,398	0.8	11	1,781	1.0
32	NPBLE	533	177,880	15.1	2	80	0.0	0			5	576	0.3	12	1,721	1.0	18	2,304	1.3
40	NPBLE	260	53,366	4.5	3	546	1.0	0			9	522	1.0	18	694	1.3	28	1,668	3.1
48	NPBLE	126	18,334	1.6	2	40	0.2	1	8	0.05	8	556	3.0	12	986	5.4	21	1,531	8.4
56	NPBLE	84	4,529	0.4	6	701	15.5	1	53	1.18	15	783	17.3	9	735	16.2	18	984	21.7
64	NPBLE	28	2,872	0.2	1	251	8.8	0			1	42	1.5	5	680	23.7	6	931	32.4
72	NPBLE	15	1,682	0.1	1	429	25.5	0			3	478	28.4	6	625	37.2	8	674	40.1
80	NPBLE	4	213	0.0	0			0			1	12	5.9	0			1	12	5.9
88	NPBLE	7	426	0.0	0			0			4	185	43.5	5	361	84.9	6	384	90.2
Total	NPBLE	2,133	822,305	69.7	17	2,170	0.3	2	61	0.01	49	3,505	0.4	81	8,648	1.1	124	11,717	1.4
8	PBLE	3	1,137	0.1	0			0			0			0			0		
16	PBLE	283	110,005	9.3	1	377	0.3	1	38	0.03	1	108	0.1	1	12	0.0	4	535	0.5
24	PBLE	611	200,219	17.0	3	139	0.1	1	27	0.01	5	684	0.3	13	2,531	1.3	21	3,355	1.7
32	PBLE	172	27,119	2.3	1	5	0.0	4	247	0.91	8	344	1.3	11	675	2.5	18	902	3.3
40	PBLE	84	10,511	0.9	1	500	4.8	0			3	225	2.1	9	722	6.9	11	946	9.0
48	PBLE	42	4,199	0.4	4	341	8.1	1	49	1.18	5	197	4.7	8	566	13.5	13	965	23.0
56	PBLE	17	2,088	0.2	0			1	41	1.97	4	887	42.5	7	566	27.1	10	1,144	54.8
64	PBLE	18	933	0.1	1	10	1.1	1	3	0.27	5	306	32.9	7	327	35.1	8	364	39.1
72	PBLE	10	926	0.1	1	14	1.5	0			5	159	17.1	3	686	74.0	7	831	89.9
80	PBLE	9	606	0.1	3	95	15.6	1	40	6.54	1	56	9.2	1	171	28.3	5	322	53.1
88	PBLE	11	520	0.0	2	45	8.6	0			6	350	67.3	7	353	68.0	7	353	68.0
Total	PBLE	1,260	358,263	30.4	17	1,526	0.4	10	445	0.12	43	3,316	0.9	67	6,609	1.8	104	9,717	2.7

- "Severity" has two possible levels: potentially barrier-like events (PBLE) or not (NPBLE). The details for this categorization have been discussed in Laituri et al. (2003a).

- "Crashes" pertains to drivers in towaway crashes irrespective of injury outcomes; "n" is the sample size, and "N" is the weighted estimate.

- To reduce the effect of potentially overly influential NASS weighting factors, weighting factor truncation was used: 500 for serious-to-fatal injured drivers (AIS3+), 1000 for moderate injured drivers (AIS2), and 5,000 for all other drivers. The abbreviated injury scale, AIS, will be discussed in Part 3.

- The injury data are considered by body region: "head/face," "neck/spine," "chest," "KTH," and "Any of the Four" body regions. Each "n" entry represents a driver with at least one injury in that body region at the chosen AIS level. Therefore, if a driver had one AIS3 head injury, one AIS3 chest injury, and two AIS2 KTH injuries, the n's for the head, chest, and KTH would increment by one; the KTH would not increment by two.

- "Neck/spine" pertains to injuries for the entire spine: cervical, thoracic, and lumbar. However, the NCAP Notice (NHTSA, 2008) called for consideration of only the cervical spine (upper neck). Therefore, the neck/spine injury rates computed from these field data were expected to be higher than those considered in the NCAP Notice, as there are more considered injuries for the same number of crashes.

- In the present study, "neck" is used for the "neck/spine" field data, for brevity.

- NHTSA bins some AIS2-level neck injuries as AIS3+ injuries (NHTSA, 1999 and 2008). That practice seemed inconsistent with the present study. However, it will be investigated in future studies.

- "Risks" (injury rates) are computed from the weighted data: that is, N injured drivers / N crash-involved drivers.

PART 1(A): NCAP-TYPE CRASHES IN THE FIELD

Table 1 shows that, even with 14 years of NASS data, NCAP-type events were relatively rare (i.e., of the 3,393 towaway frontal crashes in the sample, only 17 involved belted drivers in PBLE with ΔV=56 km/h and an airbag deployment). To address the small-sample size issue, NCAP event risks were estimated two different ways:

(1) The data from one lower and one higher bin were combined with the 56 km/h bin to produce a 48-64 km/h bin. The results of this analysis will be termed "collapsed-data" estimates. (The underlying data are highlighted in Table 1.)

(2) The injury rates for the PBLE were fitted with Gompertz curves, as a function of speed, and the resulting fits were then evaluated at 56 km/h. The results of this analysis will be termed "Gompertz-fit" estimates. Appendix 1 contains more information about the fitting procedure and the resulting fits.

These two approaches were used to estimate the injury rates for the aforementioned body regions. As shown in Table 2, the two approaches yielded somewhat different results. However, there were no observed trends between the two: sometimes the Gompertz-fit-based injury rates were lower; sometimes they were higher.

The same conclusions were drawn from these two approaches:

In NCAP-type real-world crashes,

- **The KTH and chest injury rates were <u>much</u> higher than those of the head and neck.**

- **KTH injury rates > Chest injury rates > Head injury rates > Neck injury rates.**

The lower and upper 95th%ile confidence limits for the collapsed-data estimate were also computed with statistical software (SAS, 2002); they are shown in Figure 1.

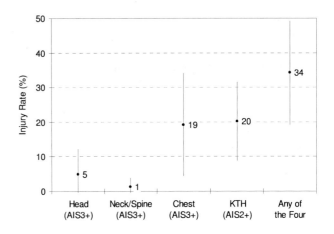

Figure 1. Injury Rates and 95% Confidence Limits for Approximated NCAP-type Events (Collapsed)

Table 2. Injury Rates in Approximated NCAP-type Events based on Two Approaches

48-64 km/h, PBLE Crashes		AIS3+ Head		AIS3+ Neck		AIS3+ Chest		AIS2+ KTH		Any of the Four	
n	N	n	N	n	N	n	N	n	N	n	N
77	7,220	5	351	3	93	14	1,390	22	1,459	31	2,473
Collapsed Data: 48-64 km/h		Injury Rate ≈ 5%		Injury Rate ≈ 1%		Injury Rate ≈ 19%		Injury Rate ≈ 20%		Injury Rate ≈ 34%	

| **Gompertz Fit @ 56 km/h** | | Injury Rate ≈ 3% | | Injury Rate ≈ 0% | | Injury Rate ≈ 20% | | Injury Rate ≈ 28% | | Injury Rate ≈ 42% | |

Driver age was also considered, and the results are in Table 3. Note that the average age of drivers in the NCAP-type events was 46 years old. However, the average age of drivers with chest and/or KTH injuries was over 50 years old. Figure 2 shows that the ages of the crash-involved drivers did not follow a normal distribution.

Table 3. Average Age of Drivers in NCAP-type Events (Collapsed), for Various Outcomes

Belted Driver, Airbag-Equipped Vehicle	Average Age
Crash-Involved	46
w/ AIS3+ Head Injury	26
w/ AIS3+ Neck Injury	48
w/ AIS3+ Chest Injury	59
w/ AIS2+ KTH Injury	55
w/ Any of the Four Injuries	50

injury sources in the field data were atypical of NCAP crash tests (e.g., left side interior). Consequently, *the injury rates estimated from the NCAP test data were expected to be lower than those observed in the field data.*

PART 1(B): AGGREGATE RISK ESTIMATES FROM FIELD DATA

Table 4 shows the injury rates for the aggregate perspective (crashes with $0 \leq \Delta V \leq 56$ km/h, including both PBLE and NPBLE). Note that the attendant counts — both crashes and injured-drivers — were <u>much</u> larger than those for the NCAP-type events (cf. Table 2, for example, n=77 crashes vs. n=3,291 crashes).

Consistent with one of the conclusions for the point estimates in Part 1A,

For crashes with $0 \leq \Delta V \leq 56$ km/h:

- **KTH injury rates > Chest injury rates > Head injury rates > Neck injury rates.**

Figure 3 shows the lower and upper 95th%ile confidence limits for the aggregate injury rates.

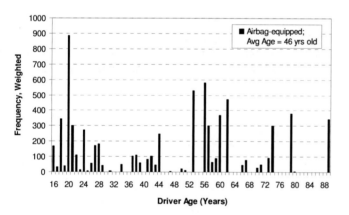

Figure 2. Distribution of Ages of Drivers in Approximated NCAP-type Events (Collapsed)

The reported "injury source" was also studied for the relevant injuries recorded for drivers in these crashes. Those data are in Appendix 2. Some of the reported

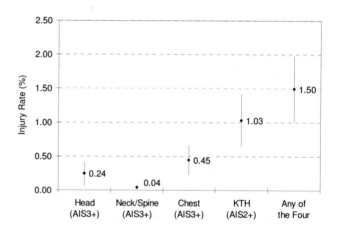

Figure 3. Injury Rates and Confidence Limits for $0 \leq \Delta V \leq 56$ km/h Crashes

Table 4. Injury Rates for $0 \leq \Delta V \leq 56$ km/h Crashes

Crashes, ΔV≤56 km/h		AIS3+ Head		AIS3+ Neck		AIS3+ Chest		AIS2+ KTH		Any of the Four	
n	N	n	N	n	N	n	N	n	N	n	N
3,291	1,172,390	25	2,852	10	464	66	5,232	114	12,054	180	17,563
		Inj Rate ≈ 0.24%		Inj Rate ≈ 0.04%		Inj Rate ≈ 0.45%		Inj Rate ≈ 1.03%		Inj Rate ≈ 1.50%	

Table 5 shows the average ages of the attendant drivers. Note that all of the mean ages were less than 50 years old.

Table 5. Mean Ages of Belted Drivers in
$0 \leq \Delta V \leq 56$ km/h Crashes

Belted Driver, Airbag-Equipped	Average Age
Crash-Involved	36
w/ AIS3+ Head Injury	34
w/ AIS3+ Neck Injury	46
w/ AIS3+ Chest Injury	48
w/ AIS2+ KTH Injury	34
w/ Any of the Four Injuries	37

PART 1(C): SENSITIVITY ANALYSIS

To address further the concerns about small sample sizes of the field dataset, a sensitivity analysis was conducted for the injury rates by body region.

Two datasets were formed by using different screening filters for belted drivers in 11-1 o'clock full-engagement, non-rollover, towaway frontal crashes in the 1993-2006 calendar years of NASS:

1. "Original" = 1994-2006 MY light vehicles fitted with airbags, that is, the dataset studied to this point, and

2. "Broader" = All MY light vehicles, irrespective of airbag fitment, that is, a less homogeneous dataset.

Appendix 3 contains more of the details of this sensitivity analysis. It shows that the conclusions drawn from the original dataset applied to the broader dataset as well. Specifically, for both the point estimates as well as the aggregate estimates,

- KTH injury rates > Chest injury rates > Head injury rates > Neck injury rates.

- The neck injury rates were consistently much lower than those of the other body regions.

Therefore, while the ordinary cautions regarding small sample sizes apply, we concluded that (a) the original dataset yielded injury rates consistent with a much larger and less homogeneous dataset, and (b) those injury rates and their rank ordering provided useful direction for assessing the fidelity of the aforementioned risk curves. Accordingly, the original dataset was used for the subsequent analyses.

PART 2(A): DUMMY RESPONSES IN NCAP TESTS

Test dummy data were collected from USA NCAP tests involving HIII50 drivers in 2001-2005 MY light vehicles (n=183). The distributions of these occupant responses are shown in Figure 4. Therein, normal distributions were fitted, and the means and standard deviations are also shown.

For later reference, the peak combined loading metric for the upper neck (N_{ij}) was distributed as follows: 58% were from N_{TF}, 32% from N_{TE}, 2% from N_{CF}, and 8% from N_{CE}, where the subscripts are as follows: T=tension, C=compression, E=extension, and F=flexion. Therefore, tension was involved in 90% of the N_{ij}'s (i.e., 58% + 32%).

In Part 4, this NCAP dataset will be used to evaluate the fidelity of the various risk curves for estimating injury rates by body region for NCAP-type crashes.

PART 2(B): DUMMY RESPONSES IN THEORETICAL FIELD MODEL

Theoretical field modeling was also conducted to assess the fidelity of the various risk curves for estimating injury rates from an aggregate perspective, thus providing a more comprehensive theoretical assessment of the various risk curves. The field model will first be conceptualized; a more detailed description then follows.

Theoretical Field Model:

For each body region, the aggregate injury rate was computed by applying the general field model from Laituri et al. (2003b). Specifically, 42 math models were used, based on the following matrix for belt+bag drivers:

- Three occupant sizes = {HIII05, HIII50, and HIII95},
- Seven ΔV's = {8, 16, 24, 32, 40, 48, 56 km/h}, and
- Two severity types = {PBLE, NPBLE},

with real-world effects such as seat positioning and initial belt tensions, as discussed in Laituri et al. (2003b).

The aggregate injury rate was computed from the field model as follows:

$$\text{Injury Rate} \equiv \sum_{i=1}^{n=7 \text{ speeds}} \sum_{j=1}^{n=2 \text{ severities}} \sum_{k=1}^{n=3 \text{ sizes}} W_i(\Delta V) * \left[R_{ijk}(\Delta V) \right]. \quad (1)$$

In Equation 1, the "Injury Rate" is the estimated number of injured drivers per 100 crashes through 56 km/h — the aforementioned aggregate injury rate metric of Table 4, but evaluated with the theoretical field model.

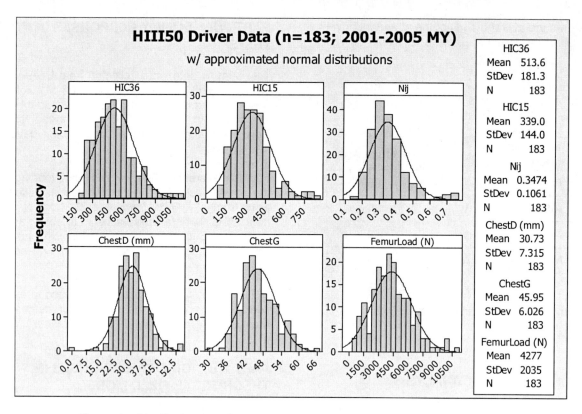

Figure 4. Distributions of HIII50 Occupant Responses in NCAP tests (n=183)

Also, in Equation 1, W_i is a speed-dependent real-world weighting factor for each event, defined as follows:

$$W_i(\Delta V) \equiv W_{size_i}(\Delta V) * W_{speed_i}(\Delta V) * W_{severity_i}(\Delta V) \quad (2)$$

i = speed change indicator, for example, i =1 (ΔV=8 km/h) to i=7 (ΔV=56 km/h) .

Each weighting factor was computed from field data crashes, irrespective of outcome. These factors were important in the computation, as they established the event probabilities (e.g., lower-speed events occurred more frequently than higher-speed events).

$\left[R_{ijk}(\Delta V) \right]$ is the injury probability for the studied body region for a particular crash event, denoted by the subscript "ijk." This injury probability was computed by substituting the pertinent occupant response into the attendant risk curve. Biomechanical scaling was used to convert risk curves for the HIII50 into curves for the other dummy sizes, as discussed in Part 3. Moreover, an injury threshold was considered: if the event risk was less than 1%, the event risk was set to zero, as discussed in Laituri et al. (2003b, 2004, and 2005.)

Finally, if the age and gender of the driver were part of a particular risk curve assessment, Equations 1-2 were first evaluated for the age-gender subgroup. Then, the resulting injury rates were weighted by sub-group crash involvement as in Laituri et al. (2004, 2005, and 2006).

Appendix 4 contains the real-world weighting factors and the occupant responses from the simulations. A spreadsheet computer program was created to evaluate Equation 1.

Details:

The aim was to model a vehicle whose occupant responses approximated those of an "average" vehicle. This was accomplished in a multi-step process.

The modeling started with a mid-sized passenger car whose driver restraint system consisted of a prototype dual-stage airbag and a 4-kN load-limiting retractor.

Next, test results for belted HIII05, HIII50, and HIII95 in the mid-sized passenger car — or sled-test versions of it — were compared with corresponding Madymo models. Validations have been discussed in Laituri et al. (2001, 2005). Therein, the models were deemed to have acceptable correlation.

Next, the occupant responses of the HIII50 in the NCAP condition from this model were compared with the fleet test results. As shown in Table 6, all of the occupant responses for the average passenger car were close to the fleet average, except for the femur load. Accordingly, the knee bolster in the model was stiffened to yield a closer match to the fleet-average femur load; those results are presented in the final column of Table 6.

Table 6. Comparison of NCAP Results: Fleet Averages versus Average Passenger Car (Original and Revised)

Occupant Response	Fleet Avg. (n=183)	Avg. Pass Car Model (Original)	Avg. Pass Car Model (Stiffer Bolster)
HIC36	514	513	517
HIC15	339	363	353
Upper Neck, N_{ij}	0.35	0.34	0.41
Chest Deflection (mm)	30.7	29.7	31.6
Chest Acceleration (g's)	46.0	46.5	48.7
Max Femur Load (N)	4,277	2,070	4,270

The average passenger car model with the stiffer knee bolster was chosen to represent the airbag-equipped fleet.

Finally, the occupant model was generalized for 42 modeled crashes involving the average passenger car with the stiffer knee bolster.

The results of the modeling will be discussed in Part 4.

PART 3: RISK CURVES FOR FRONTAL IMPACTS

The Abbreviated Injury Scale was used in the present study. It categorizes injuries on a threat-to-life scale, defined as follows:

AIS0: No Injury
AIS1: Minor
AIS2: Moderate
AIS3: Serious
AIS4: Severe
AIS5: Critical
AIS6: Maximum .

Subject to a specified level of injury, researchers use available data to derive risk curves.

Recall that a risk curve converts a test-dummy occupant response into the probability of a human occupant sustaining an injury at a specified level for a specified body region. A risk curve could be derived directly if, for every real-world crash, the following human occupant information were known: bone strength, degree of muscle activation, all of the attendant human occupant responses, all of the applied forces, torques, and their levels of distribution. None of this information is available from the NASS database or any other database of real-world crashes. Consequently, researchers are relegated to using inferential methods which are often based on limited data, surrogates, and approximations.

One of those approximations is the use of post mortem human subjects (PMHS). Researchers often use the injury outcomes of tests involving PMHS to derive a risk curve and its corresponding statistical goodness-of-fit between the test outcomes and the resulting probability function. Another approximation is the use of biomechanical scaling — the method by which responses for one size of occupant are converted into those for other occupant sizes. Moreover, different analysts often use different statistical methods, sometimes yielding very different risk curves for the same dataset (NHTSA, 1999). Finally, for an adult population, some body regions demonstrate pronounced age dependence (e.g., chest), while for other body regions, there is less evidence of age effects (e.g., head). Consequently, risk curves for a particular body region are updated frequently as the science of biomechanics matures and as more information becomes available.

The present NCAP uses HIC36 and chest acceleration from the instrumented mid-sized dummy to estimate head and chest AIS4+ risks. Recall that, in the 2008 NCAP Notice, NHTSA made numerous changes, by

- changing the head and chest metrics to HIC15 and chest deflection, respectively,
- computing risks for more body regions (viz., the upper neck and the KTH), and
- considering lower levels of injury (viz., AIS3+ for the head, neck, and chest; AIS2+ for the KTH).

Various attendant risk curves are discussed in the next section, and they are considered by body region.

Head:

Figure 5 depicts various published risk curves for the head. All of these risk curves were derived from analyses based on the same dataset of PMHS head drop tests, discussed in Mertz et al. (2003). The dataset consisted of various injury outcomes. In Figure 5, the NCAP risk curve is labeled "AIS3+, NHTSA (2008)." As explained below, the curves differed because of the underlying statistical methods.

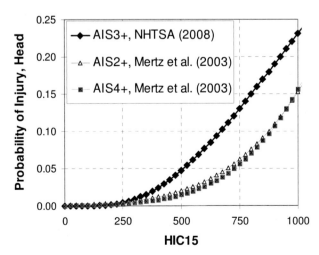

Figure 5. Various Risk Curves for the Head

NHTSA's risk curve was based on extending the work of Hertz (1993). Therein, injury thresholds were fitted with a lognormal distribution. Also, the data for the injured were assumed to be left-censored; the data for the uninjured were assumed to be right-censored. Originally, the curve was derived for AIS2+ injuries. However, according to the NCAP Notice (NHTSA, 2008), the AIS3+ curve was derived by "using real-world data to determine the relative incidence of different severity brain injuries." No further explanation was provided.

Mertz et al. (2003) used a statistical method which became known as the Mertz/Weber method. Therein, injury threshold levels are assumed to be normally distributed. The injured specimen with the lowest applied risk variable — in this case, the injured specimen with the lowest HIC — is defined as the "weakest"; the uninjured specimen with the highest applied risk variable is defined as the "strongest." Recall that two statistics define a normal distribution: the mean and the standard deviation. For the Mertz/Weber method, the mean is the average of the risk variables associated with the weakest and strongest specimens. The standard deviation is estimated by using a median rank table for the data between the weakest and strongest. NHTSA (1999) reported concerns about this statistical method, including too much reliance on two data points, and no diagnostics on goodness of fit. Accordingly, *NHTSA has not adopted risk curves based on the Mertz/Weber method.*

These head risk curves will be assessed in Part 4.

Upper Neck:

The NCAP Notice defines the upper neck risk by the maximum risk from three different loading conditions and related metrics: combined loading (N_{ij}), tension (F_T), and compression (F_C).

Combined Loading:

The AIS3+ upper neck risk curve used in the NCAP Notice was from NHTSA's analysis for advanced airbags (NHTSA, 1999). Therein, data from matched tests of out-of-position HIII 3-year old test dummies and anesthetized porcine subjects were analyzed. The dummies provided the occupant responses; the porcine subjects provided the injury outcomes. These tests were conducted in the 1980's by Mertz et al. (1982) and Prasad and Daniel (1984). Due to the nature of the testing, only tension, extension, and their combination were studied with these data. Based on some assumptions discussed later, NHTSA extended the tension-extension combined loading metric (N_{TE}) to the more general metric (N_{ij}).

The issues and concerns related to the risk curve development for the upper neck are summarized as follows:

(1) In the underlying data from the 1980's studies, the time-dependent nature of N_{TE} for the 3-year old was not considered. As such, in 1999, NHTSA could only

use the <u>peak</u> forces and moments. The use of peak data, however, is inconsistent with the contemporary computation of N_{TE}. The extent of the difference is discussed in the next two items.

(2) In 2000, Mertz and Prasad computed the peak stress from the combined loading in the upper neck of the HIII 3-year old. The authors referenced the stress equation as the "kernel." Specifically, the authors digitized the data from the aforementioned 1980's studies, and from the resulting time histories, they derived an equation for estimating the corresponding forces and moments that produced the peak-stress condition. Therefore, Mertz and Prasad were more faithful to today's time-synched N_{TE} calculation. They subsequently derived a corresponding risk curve by applying the Mertz-Weber method to the peak kernel data. In 2003, they republished the curve.

(3) The risk curves from Mertz et al. (2000, 2003) and NHTSA (1999, 2008) were significantly different. Recall that they were based on different data. Moreover, they were derived via different statistical methods. As demonstrated in Appendix 5, the curves differed primarily because of the choice of statistical method.

(4) Mertz and Prasad (2000) identified two tests as being potentially overly influential in their development of an N_{TE} risk curve for the HIII 3-year old. Those tests were designated "F2" and "F8."

(5) The porcine subjects had no muscle tone. More muscle tone reduces the attendant risk. Accordingly, Mertz and Prasad (2000) derived an equation for the 3-year old which accounted for various degrees of muscle tone (m=0 to 1): $N_{TE} = N_{TE_{max}} - 0.177\,m$.

To address these issues, the matched-test dataset was revisited. Figure 6 summarizes the findings.

The first curve in Figure 6 is the NCAP curve. It produced 3.8% risk for $N_{TE}=0$. The second curve is from Nakahira et al. (2000), derived by applying a modified maximum likelihood (MML) method to the dataset from the Alliance of Automobile Manufacturers (1999); the equation was provided by Mullen (2008). The third curve is from Mertz and Prasad (2000); it was republished in Mertz et al. (2003). The fourth curve was derived by applying the statistical method typically used by NHTSA (binary logistic regression analysis) to the kernel dataset of Mertz and Prasad (2000). It produced 4.4% risk for $N_{TE}=0$. The fifth curve was derived by removing the potentially-overly influential tests (F2 and F8) from the dataset. The sixth curve resulted from the addition of a (0,0) point to the dataset, while the seventh curve pertained to 50% of maximum muscle tone. The final curve resulted from the use of the modified maximum likelihood method on the kernel dataset from Mertz and Prasad.

The goodness-of-fit statistics were computed for these models, where applicable, subject to the NHTSA N_{TE} computation. The results are summarized in Table 7.

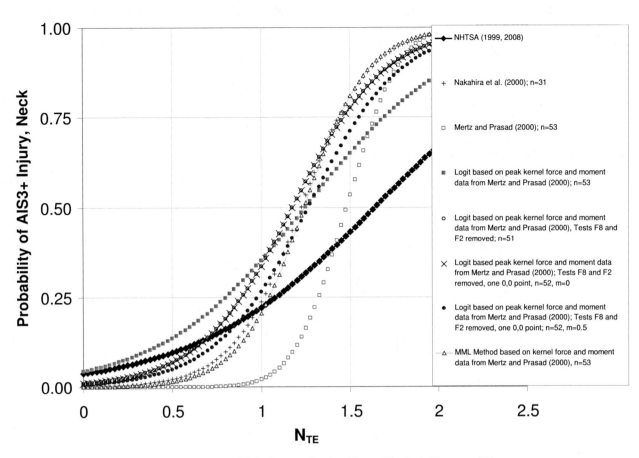

Figure 6. Various Risk Curves for the Upper Neck, in Terms of N_{TE}

Table 7. Goodness-of-Fit Data for Various Logit-based Risk Curves for N_{TE}

Logit-based Risk Curve: ($P = \frac{1}{1+e^{-Logit}}$)	Sample size	Log Likelihood	Hosmer-Lemeshow Goodness of Fit, P'	Goodman-Kruskal Gamma
NHTSA (1999, 2008) Logit = -3.227 + 1.969 N_{TE}	53	-35.115	0.352	0.48
Nakahira et al. (2000) Logit = -6.403 + 5.231 N_{TE}	31	No Estimate	No Estimate	No Estimate
Present: Logit based on kernel data from Mertz and Prasad (2000), n=53 Logit = -3.08750 + 2.4721 N_{TE}	53	-31.959	0.878	0.48
Present: Logit based on kernel data from Mertz and Prasad (2000), Tests F8 and F2 removed Logit = -4.5165 + 3.855 N_{TE}	51	-28.068	0.894	0.55
Present: Logit based on kernel data from Mertz and Prasad (2000), Tests F8 and F2 removed, (0,0) added Logit = -4.54094 + 3.8758 N_{TE}	52	-28.079	0.940	0.57
Present: MML based on kernel data from Mertz and Prasad (2000) Logit = -6.94098 + 5.5933 N_{TE}	53	-35.829	0.003	0.46

Examination of Table 7 indicates the following:

- The NCAP curve showed relatively weak goodness-of-fit and gamma values: 0.352 and 0.48, respectively.

- The goodness-of-fit of the models improved as the peak stress-state kernel data from Mertz and Prasad (2000) were used. Specifically, P' increased from 0.352 to 0.878; however, the gamma did not change.

- Both the goodness-of-fit and the gamma improved by removing the two potentially-overly influential tests, and by adding a (0,0) point to the dataset.

- The goodness-of-fit for the MML method was unacceptably weak (i.e., P' = 0.003).

All of the p-values for the constant and coefficient in the conventional logits were less than 0.05. Therefore, that statistic was not reported. The statistical method used for the Mertz and Prasad curve does not easily lend itself to goodness-of-fit computations — one of the reasons why NHTSA prefers the logistic method. As such, no fit statistics were reported for that curve. Also, it was inappropriate to assess the statistical model for the 50% muscle-tone, as the porcine subjects had negligible muscle tone. Therefore, no fit statistics were generated for it.

These upper neck risk curves will be assessed in Part 4.

Tension:

Figure 7 depicts various published risk curves for the upper neck in terms of tensile load.

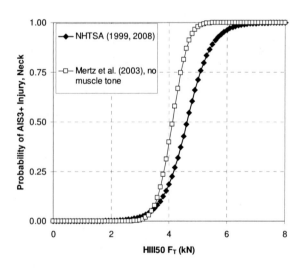

Figure 7. Various Risk Curves for the Upper Neck, in Terms of Tensile Load for the HIII50

The aforementioned data from the matched tests involving porcine and HIII 3-year olds were used by Mertz et al. (2003) and NHTSA (2008) to derive their tension-based risk curves. Because the outcomes were based on anesthetized porcine subjects, both risk curves did not account for muscle tone. Recall that risk decreases as muscle tone increases. Accordingly, the curve from Mertz should be considered an as upper-limit risk curve. (Mertz et al. did, however, provide a theory and a formula to account for the effect of muscle tone.)

NHTSA's AIS3+ risk curves for the HIII50 and HIII05 are as follows:

HIII50: $$P_{neck} = \frac{1}{1 + e^{-(-10.9745 + 2.375\,F_T)}} \qquad (3)$$

HIII05: $$P_{neck} = \frac{1}{1 + e^{-(-10.958 + 3.770\,F_T)}} \qquad (4)$$

from p. 95 of 120 in the 2008 NCAP Notice (NHTSA, 2008), where F_T is in kN. Note that both the constants and the force coefficients of the logit differ. Biomechanical scaling is generally applied to the predictor (force in this case), not the constants. Therefore, it was unclear how scaling was incorporated.

The risk curves for the HIII05 are in Figure 8.

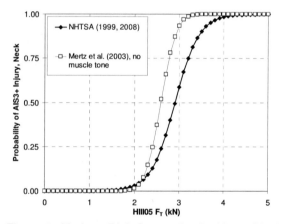

Figure 8. Various Risk Curves for the Upper Neck, in Terms of Tensile Load for the HIII05

These upper neck risk curves will be assessed in Part 4.

Compression:

Mertz (1993) reported the injury assessment reference values (IARVs) for the neck tension and compression. As shown in Figure 9, these IARVs were duration-dependent. Mertz advised that these curves should be interpreted as boundaries. As an example, data above the tension boundary represented an outcome with "potential for significant neck injury due to axial neck tension loading"; data below the boundary represented "injury unlikely" outcomes. Note that the longer-duration IARVs for tension and compression were equal. However, the shorter-duration IARVs were not equal. At that time, 1993, Mertz did not report the corresponding

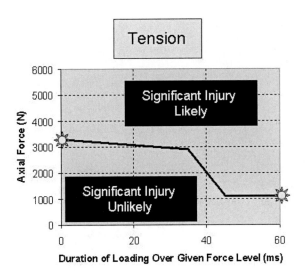

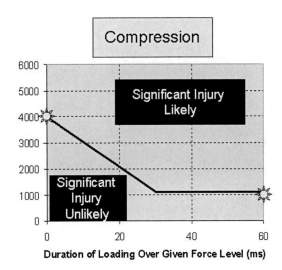

Figure 9. Injury Assessment Curves from Mertz (1993) for the Upper Neck Force, as Measured in the HIII50 Test Dummy

risk curves. Later, Mertz et al. (1997, 2000, 2003) reported a risk curve for tension, based on analysis of the aforementioned matched tests. The authors never published a risk curve for compression.

Figure 10 shows the NCAP risk curve for the upper neck in terms of compressive load. It is identical to NHTSA's risk curve for tensile loading. NHTSA likely assumed that the equal, long-duration loading IARVs would manifest in the risk curves.

These upper neck risk curves will be assessed in Part 4.

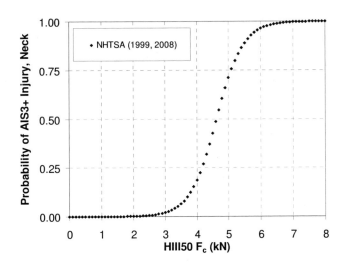

Figure 10. NCAP Risk Curve for the Upper Neck, in Terms of Compressive Load on the HIII50

Chest:

Figure 11 depicts various published risk curves for the chest.

The NCAP risk curve was selected from the set of age-dependent risk curves from Laituri et al. (2005); NHTSA chose to evaluate the set of risk curves at 35 years old to approximate the average crash-involved driver. The age-dependent risk curves were derived by (a) applying the MML method to a PMHS dataset, subject to two covariates: age of the PMHS and its maximum normalized sternum compression, and (b) applying a closure equation to relate PMHS normalized sternum compression to HIII50 chest compression. The curve for 30-year olds from Laituri et al. (2003b) was derived by applying an empirical method to account for age. The resulting risk curve was nearly identical to that from Laituri et al. (2005), evaluated at 35 years. The other risk curves did not account for age. Note that the curves from Mertz et al. (1991) and Laituri et al. (2005), evaluated at 50 years, were nearly identical. Consequently, Laituri et al. (2005) surmised that the drivers in the dataset used by Mertz et al. were likely older individuals. The risk curves from Prasad et al. (2004) and Laituri et al. (2004) were derived by cross-correlating injury rates in the field with model-estimated chest deflections for field events involving belt-only drivers. In both studies the drivers were categorized by age-gender groups:

> "YM"   15-49 year old men,
> "YF"   15-49 year old women,
> "OM"   50+ year old men, and
> "OF"   50+ year old women.

As reported in Prasad et al. (2004), the age effect was more significant than the gender effect. Note that the OM risk curves from Prasad et al. and Laituri et al. were nearly identical. The cross correlations from Prasad et al. involved only HIII50 occupant responses, whereas the cross correlation from Laituri et al. involved the responses of more occupants.

These thoracic risk curves will be assessed in Part 4.

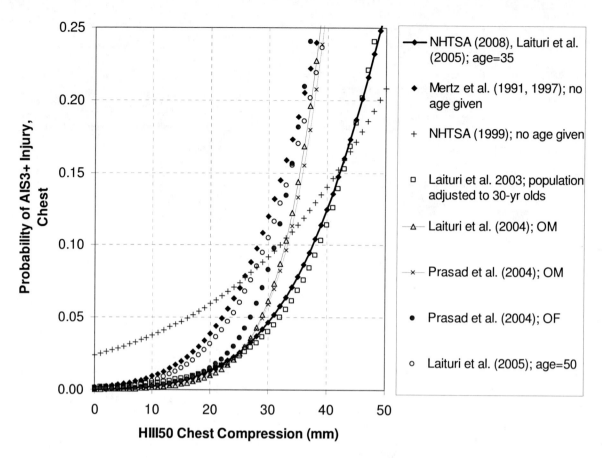

Figure 11. Various Risk Curves for the Chest

Knee-Thigh-Hip Complex:

All of the available risk curves for the KTH were derived by analyzing knee-loaded PMHS and by assuming that the load applied to the knee of the PMHS approximated the load applied to the knee of the PMHS approximated the HIII50 femur load. NHTSA (1999, 2008) used the PMHS data summarized in Morgan et al. (1989), and used binary logistic regression methods to derive the risk curve. The International Standards Organization (ISO, 2003) applied the Mertz-Weber method to the dataset from Morgan et al. to derive their risk curve. Laituri (2006) included more PMHS data, accounted for gender and age, used bioscaling to normalize the loads, applied arbitrary censoring, and produced a set of Weibull curves for different ages and gender. Laituri et al. reported that that their risk curves understated high-speed injury rates for unbelted/unrestrained drivers in NASS.

Rupp (2006) derived a risk curve for the hip. However, that risk curve was not cast in terms of the HIII50 dummy. As such, it was not assessed herein.

Figure 12 shows the various available femur-load risk curves. Therein, only the most risk-producing risk curve from the set of Laituri et al. is shown. Note that the curve from NHTSA (1999, 2008) is nearly identical to the curve

from Laituri et al. for older men and women for femur loads less than 8 kN. However, the curve from ISO differs from the other two.

All of these KTH risk curves will be assessed in Part 4.

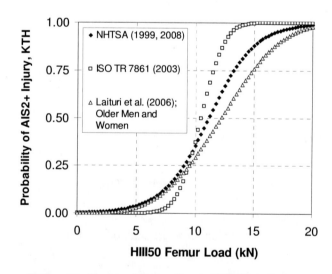

Figure 12. Various Femur-Force KTH Risk Curves

PART 4: COMPARISON OF FIELD- AND CURVE-BASED RISKS

For each body region, the various curve-based injury rates were compared to the aforementioned NASS-based injury rate. Recall that the curve-based point and aggregate injury rate estimates were expected to be lower than the injury rates from field data, as the field data included some contacts atypical of laboratory testing. This expectation applied for all four body regions, for the two crash domains.

The curve-based injury rates for the two crash domains were computed as follows:

- Point Estimate (NCAP-type events): The NCAP dataset for drivers was evaluated with the aforementioned risk curves to assess their fidelity. Specifically, the peak occupant responses for 183 driver tests in NCAP were evaluated with the various risk curves; average risks for each body region were then computed. For example, 183 HIC15's yielded 183 AIS3+ probabilities using NHTSA's risk curve; the average probability from those 183 was 0.02, or 2%.

- Aggregate Estimate ($0 \leq \Delta V \leq 56$ km/h events): The field model for the average passenger car was used. The aggregate injury rate was estimated for the aforementioned 42 simulations representing real-world frontal impacts involving belted drivers with supplemental airbag restraint. Each event risk was estimated from the attendant risk curve; each event risk was weighted by its real-world frequency, irrespective of outcome; the event risks were summed over the studied speeds.

If body region had an available set of age- dependent or age/gender-dependent risk curves — in the present study, those body regions were the chest and KTH — the aggregate injury rate was computed by using all of the risk curves in the set. However, to remain consistent with the NCAP Notice (NHTSA, 2008), *only one risk curve will be used for the point estimate.* Therefore, if during the assessment, a <u>set</u> of risk curves demonstrates better fidelity for the aggregate estimate, only <u>one</u> risk curve from that set will be recommended to estimate that body-region risk in the NCAP.

Recall that, in the present study, a risk curve whose estimated injury rate fell within the confidence limits of the field data was deemed "acceptable." *For this study, if both the NCAP risk curve and other risk curves for a particular body region yielded acceptable results, the NCAP risk curve was recommended over the others.*

The following discussion presents the injury rate comparisons by body region and by crash domain. At the end of each section, there will be conclusions and recommendations.

Head: The point estimates are in Table 8; the aggregate estimates are in Table 9. Therein, "LCL" is the lower 95% confidence limit; "UCL" is the upper 95% confidence limit.

The NCAP risk curve produced results within the confidence limits for both the point and aggregate-risk estimates. The same conclusion was also drawn for the studied risk curves from Mertz et al. (2003).

Conclusion (Head): The NCAP risk curve for the head demonstrated acceptable fidelity, as it yielded injury rates within the confidence limits for both the point and aggregate estimates.

Table 8. Comparison of AIS3+ Injury Rates for the <u>Head</u>; <u>Point-Estimate</u> (NCAP-type Events)

Risk Curve	NCAP Tests (n=183), average	NASS NCAP-type Events
NHTSA (1999, 2008)	2%	Collapsed-Data Estimate ≈ 5%
Mertz et al. (2003); AIS2+	1%	[LCL=0% UCL=12%]
Mertz et al. (2003); AIS4+	1%	Gompertz-Fit Estimate ≈ 3%

Table 9. Comparison of AIS3+ Injury Rates for the <u>Head</u>; <u>Aggregate Estimate</u> (Field Model: $0 \leq \Delta V \leq 56$ km/h)

Risk Curve	Field Model w/o a Threshold	Field Model w/ Threshold = 1%	NASS $0 \leq \Delta V \leq 56$ km/h
NHTSA (1999, 2008)	0.08 %	0.07 %	0.24 %
Mertz et al. (2003); AIS2+	0.17 %	0.03 %	[LCL=0.07% UCL=0.42%]
Mertz et al. (2003); AIS4+	0.09 %	0.03 %	

Upper Neck: The point estimates are in Table 10; the aggregate estimates are in Table 11.

The NCAP risk curve for N_{ij} demonstrated unacceptable fidelity, as it yielded injury rates <u>well</u> <u>outside</u> the confidence limits for both the point and aggregate estimates. Specifically, for the aggregate estimate, the NCAP risk curve yielded an injury rate two orders of magnitude greater than the field result.

Recall that by convention the aforementioned field data for the "neck" pertained to injuries for the entire spine (cervical, thoracic, and lumbar), whereas the NCAP called for consideration of only the cervical spine. Therefore, the fidelity of the NCAP risk curve for N_{ij} was even worse than the aforementioned results demonstrate. Specifically, the injury rate for $0 \leq \Delta V \leq 56$ km/h crashes was 0.04% (see Figure 3); it would have been 0.02% if the thoracic and lumbar injuries were excluded.

Tables 10 and 11 also showed that when the logistic statistical method was applied to the kernel data, the point and aggregate estimates demonstrated worse fidelity than the NCAP N_{ij} risk curve. When the MML method was applied to the kernel data for N_{TE}, the fidelities improved. Specifically, the point estimate was within the confidence limits, and the aggregate estimate approached zero, although the estimate was outside of the near-zero-magnitude confidence limits. Recall, however, that the goodness of fit of the matched-test dataset degraded to an unacceptable degree for the MML method when applied to the kernel data, as shown in Table 7. Therefore, that risk curve could be discounted for that reason.

Some of the risk curves for the upper neck did demonstrate acceptable fidelities; they included the N_{TE} risk curve from Mertz et al. (2003), the tension-only curves from NHTSA and Mertz et al. (2003), and the compression-only curves from NHTSA. (Recall that the compression-only curve from NHTSA was based on an assumption.)

One notion is to consider <u>only</u> tension, as it was represented in 90% of the N_{ij}'s for the point estimate (Part 2A) and in 81% of the N_{ij}'s for the aggregate estimate (Table A8). Moreover, Nusholtz et al. (2003) analyzed 45 of the aforementioned matched tests for the neck curve derivation, including the attendant time histories. After conducting various statistical analyses, those authors concluded that tension alone best classified the outcomes, and that the inclusion of extension reduced the predictive strength of the statistical model. However, in the present study, the very limited field data in Table A3 did not provide clear indication of the potential injury mechanisms. Therefore, while tension is predominant in both the testing and the modeling, NHTSA's call for N_{ij} – where all four combined loading states are considered – is more comprehensive.

<u>Conclusion (Neck):</u> The NCAP risk curve for N_{ij} demonstrated unacceptable fidelity. To evaluate the point estimate, the upper neck risk could be more faithfully represented by the maximum risk computed from three risk curves: NCAP tension-only, NCAP compression-only, and the Mertz combined-loading N_{TE} risk curve. For the studied data, this proposal was equivalent to excluding the Mertz combined-loading N_{TE}. (The tension-only curve from Mertz et al. (2003) could also be used.)

Table 10. Comparison of AIS3+ Injury Rates for the <u>Neck</u>; <u>Point-Estimate</u> (NCAP-type Events)

Risk Curve	Metric	NCAP Tests[*] (n=183), avg.	NASS "NCAP-type" Events
NHTSA (1999, 2008)	N_{ij}	7.4 %	
NHTSA (1999, 2008)	N_{TE}	6.6 %	
Nakahira (2000)	N_{TE}	0.9 %	
Mertz et al. (2003)	N_{TE}	0.3 %	
Present: Logit based on kernel data from Mertz and Prasad (2000), n=53	N_{TE}	8.6 %	
Present: Logit based on kernel data from Mertz and Prasad (2000), Tests F8 and F2 removed, (0,0) added, m=0.5, n=52	N_{TE}	2.5 %	Collapsed-Data Estimate ≈ 1% [LCL = 0% UCL = 4%] Gompertz-Fit Estimate ≈ 0%
Present: MML based on kernel data from Mertz and Prasad (2000), n=53	N_{TE}	0.9 %	
NHTSA (1999, 2008)	F_T	0.7 %	
Mertz et al. (2003)	F_T	0.6 %	
NHTSA (1999, 2008)	F_C	0.5 %	
NHTSA (1999, 2008)	Max Risk (F_T, F_C)	0.7 %	

[*] No consideration of thresholds

Table 11. Comparison of AIS3+ Injury Rates for the Neck; Aggregate Estimate (Field Model: $0 \leq \Delta V \leq 56$ km/h)

Risk Curve	Metric	Field Model w/o Threshold	Field Model w/ Threshold = 1%	NASS $0 \leq \Delta V \leq 56$ km/h
NHTSA (1999, 2008)	N_{ij}	5.87 %	5.87 %	
NHTSA (1999, 2008)	N_{TE}	5.61 %	5.61 %	
Nakahira et al. (2000)	N_{TE}	0.57 %	0.31 %	
Mertz et al. (2003)	N_{TE}	0.00 %	0.00 %	
Present: Logit based on kernel data from Mertz and Prasad (2000), n=53	N_{TE}	7.10 %	7.10 %	
Present: Logit based on kernel data from Mertz and Prasad (2000), Tests F8 and F2 removed, (0,0) added, m=0.5, n=52	N_{TE}	2.46 %	2.46 %	0.04 % [LCL = 0.01% UCL = 0.07%]
Present: MML based on kernel data from Mertz and Prasad (2000), n=53	N_{TE}	0.57 %	0.31 %	
NHTSA (1999, 2008)	F_T	0.01 %	0.00 %	
Mertz et al. (2003)	F_T	0.00 %	0.00 %	
NHTSA (1999, 2008)	F_c	0.00 %	0.00 %	
NHTSA (1999, 2008)	Max Risk (F_T, F_C)	0.01 %	0.01 %	

Chest: The point estimates are in Table 12; the aggregate estimates are in Table 13.

The NCAP risk curve from Laituri et al. (2005), based on analysis of PMHS, yielded results within the confidence limits for the point estimate when evaluated for 35-year olds, but the result was somewhat lower than the mean (curve-based: 6%, NASS-based: 19%). Moreover, the set of age-dependent risk curves from Laituri et al. (2005), when evaluated as a six-age-group set in the aggregate risk assessment, yielded results outside the confidence limits (curve-based: 1.39%, NASS-based: 0.45%). However, a set of four age-gender curves from Prasad et al. (2004) — derived by cross-correlating field risks with model-estimated chest deflections for field events involving belt-only drivers — yielded better results from both perspectives. This was likely due to the heavy demands placed on the closure model derived in Laituri et al. (2005). The closure model involved a transform between the PMHS normalized sternum compression and the HIII50 chest deflection, based on a small sample of matched tests. The curves from Prasad et al. (2004) and Laituri (2004) did not require this transform. Of those two sets of age-gender risk curves, those from Prasad et al. better captured the contributions to the aggregate population injury rate. Specifically, without providing all of the underlying data, both sets of age-gender risk curves yielded injury rates for the YM, YF,

and OM subgroups that were similar and within the confidence limits of the related field data. However, the OF risk curve from Laituri yielded an injury rate that was outside the field confidence limits for the OF drivers; the injury rate from the OF risk curve from Prasad yielded a result within the confidence limits. Consequently, the set of curves from Prasad et al. (2004) were preferred to the set of curves from Laituri et al. (2005) and Laituri et al. (2004) for estimating the aggregate injury rates.

To evaluate the thoracic risks for the point estimate, NHTSA called for the NCAP risk curve to be evaluated at 35 years — the average crash-involved age from their analyses of field data. In the present analysis, the average age of drivers in $0 \leq \Delta V \leq 56$ km/h crashes was indeed 35 years. However, recall that the average age in NCAP-type events was somewhat higher (46 years old), with a non-normal age distribution. Zhou et al. (1996) showed that the chest risks are age-dependent. As such, it was difficult for any one risk curve to approximate all field events. However, if one curve were to be used to quantify the chest risk for NCAP-type events, the OM curve from Prasad et al. (2004) is recommended. As shown in Table 12, it yielded an injury rate for the NCAP-type events that better agreed with the field data than the NCAP risk curve (i.e., relative to the NASS result of 19%, the recommended curve yielded 11% whereas the NCAP risk curve yielded 6%.)

Table 12. Comparison of AIS3+ Injury Rates for the Chest; Point-Estimate (NCAP-type Events)

Risk Curve	NCAP Tests (n=183), average	NASS "NCAP-type" Events
NHTSA (2008)=Laituri et al. (2005); age=35	6 %	Collapsed-Data Estimate ≈ **19%** [LCL = 4% UCL = 34%] Gompertz-Fit Estimate ≈ **20%**
Mertz et al. (1997); no age given	15 %	
NHTSA (1999); no age given	10 %	
Laituri et al. (2003); population empirically adjusted to represent 30-year olds	6 %	
Prasad et al. (2004); OM curve only	11 %	

Table 13. Comparison of AIS3+ Injury Rates for the Chest; Aggregate Estimate (0 ≤ ΔV ≤ 56 km/h)

Risk Curve(s)	Field Model w/o Threshold	Field Model w/ Threshold = 1%	NASS 0 ≤ ΔV ≤ 56 km/h
NHTSA (2008)=Laituri et al. (2005); age=35	1.36 %	1.18 %	**0.45 %** [LCL = 0.23, UCL = 0.66]
Mertz et al. (1997) ; no age given	3.92 %	3.91 %	
NHTSA (1999); no age given	5.87 %	5.87 %	
Laituri et al. (2003); population empirically adjusted to represent 30-year olds	Thresholds were called for in that paper	1.14 %	
Laituri et al. (2005); ages = 20, 30, 40, 50, 60, 70	Thresholds were called for in that paper	1.39 %	
Laituri et al. (2004); using four curves: YM, YF, OM, OF	Thresholds were called for in that paper	0.63 % (Thresholds = 0, 0, 1, 1)	
Prasad et al. (2004); using four curves: YM, YF, OM, OF	Thresholds were called for in that paper	0.49 % (Thresholds = 0, 0, 1, 1)	

Figure 13 demonstrates the reason for the improved fidelity for the point estimate: For deflections more than 26 mm, the recommended risk curve yielded higher injury probabilities than the NCAP risk curve.

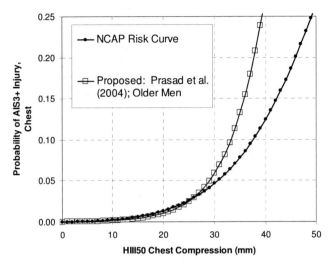

Figure 13. Comparison of Thoracic Risk Curves

approximately one order of magnitude lower than the NASS-based injury rates. More research will be needed to reconcile these differences.

Conclusion (KTH): For evaluation of the point estimate, the NCAP risk curve should be used, but only until a better risk curve is available.

Table 14. Comparison of Point-Estimate Risks: AIS2+ KTH Risks in NCAP-type Events

Risk Curve	NCAP Tests (n=183), average	NASS "NCAP-type" Events
NHTSA (1999, 2008)	5 %	Collapsed-Data Estimate ≈ **20 %**
ISO (2003)	0 %	[LCL = 8, UCL = 32]
Laituri et al. (2006), Older Drivers	4 %	Gompertz-Fit Estimate ≈ **28 %**

Conclusion (Chest): The NCAP risk curve from Laituri et al. (2005) yielded an injury rate within the confidence limits for the point estimate, although it was somewhat lower than the mean. For the aggregate injury rate assessment, the set of age-dependent risk curves from Laituri et al. (2005) yielded results outside the confidence limits. The set of age-gender curves from Prasad et al. (2004), however, yielded better results from both perspectives, subject to a 1% threshold. Of that set, the risk curve for older men (age=50+) was recommended for the next-generation NCAP, in order to recover more faithfully the injury rate in the field for such events.

KTH: The point estimates are in Table 14; the aggregate estimates are in Table 15.

All of the studied curves yielded results outside of the confidence limits for both the point and aggregate-risk estimates. The curve-based injury rates were

Overall: The set of recommended risk curves for the frontal component of the next-generation USA NCAP is in Appendix 6.

Figure 14 illustrates one of the principle findings of the present study: If the NCAP risk curves were applied to the 183 tests in NCAP dataset, 61% of those tests would have had the neck assigned the highest risk of the four body regions. Yet, this was inconsistent with the aforementioned field data analyses. Therein, the neck injury rate was consistently the lowest of the four. Figure 14 also shows that the recommended set of risk curves would address this overemphasis of the upper neck (i.e., the neck would never have been assigned the highest risk).

Table 15. Comparison of AIS2+ Injury Rates for the KTH; Aggregate Estimate (0 ≤ ΔV ≤ 56 km/h)

Risk Curve	Field Model w/o Threshold	Field Model w/ Threshold = 1%	NASS 0 ≤ ΔV ≤ 56 km/h
NHTSA (1999,2008)	0.41 %	0.01 %	**1.03 %**
ISO (2003)	0.00 %	0.00 %	[LCL = 0.64, UCL = 1.41]
Laituri et al. (2006), Older Drivers	< 0.41 % by inspection	< 0.41 % by inspection	

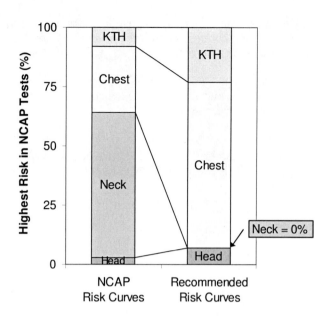

Figure 14. Survey of NCAP Database (n=183): Body Region with Highest Risk, by Risk Curve Set

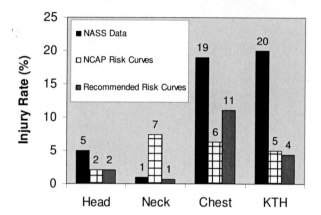

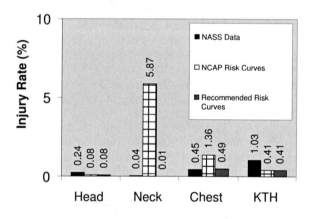

Figure 15. Assessment of NCAP Risk Curves versus Recommended Risk Curves for Drivers: AIS3+ Injury Rates for Head, Neck, and Chest; AIS2+ for KTH

Conclusions (Overall):

Figure 15 shows the injury rates estimated from the NASS field data, the NCAP risk curves, and the recommended curves. The top plot pertains to the point estimates of the NCAP-type crashes; the bottom plot pertains to the aggregate estimates for $0 \leq \Delta V \leq 56$ km/h crashes.

Fundamentally, the NCAP risk curves should rank order the injury rates by body region consistent with the injury rates in the field. Presently, the NCAP risk curves do not accomplish that aim (see Figure 15: "Point Estimate," white/thatched bars vs. black bars). The recommendation for the neck would immediately address the misranking of the neck injury rate — the neck would then appropriately have the lowest injury rate, leaving the chest and KTH as the two higher injury rates. The recommendation for the chest would also improve the chest risk estimates. However, the understated KTH risks are necessarily a topic for additional research.

LIMITATIONS AND FUTURE STUDIES

The present study had limitations which should be considered in future studies. Some related comments follow.

The NASS dataset used in the present study involved a small sample size. Concerns about sample size are commonplace when considering NASS field data, as approximately 4,500 towaway crashes are investigated each year. These crashes include vehicles of various model years, containing various restraint systems, in all crash modes (front, side, rear, rollover, other, and unknown), with both belted and unbelted drivers.

Therefore only a small portion of the NASS data would ever involve the circumstances of interest in the present study: a belted driver in an airbag-equipped vehicle in a non-rollover, full-engagement frontal crash. This sample of cases is further reduced if a particular ΔV range is required, especially if the lower frequency, higher speed crashes are of interest. Although a sensitivity analysis was conducted showing that the injury rates and rank-order of those rates by body region did not vary much as different field filters were introduced, the present study should be revisited as more field data becomes available to assess its robustness.

Error estimates were not computed for the field models due to a lack of statistical data to estimate those effects. For example, statistics of variation for all of the variables in the field model would be needed. Some of those variables include airbags (sizes, sensor timings, etc.), belts (load levels, routings, etc.), and occupants (ages, sizes, seating positions, etc.). For a more balanced comparison with the field data, error estimates and robustness should be considered in future studies.

The underlying PMHS datasets used for deriving some of the risk curves also had relatively small sample sizes. As more tests are conducted, injury thresholds for the population should be quantified more precisely. Those refinements should be represented in the attendant risk curves.

The thoracic risk curves from Laituri et al. (2005) involved a closure model that transformed PMHS normalized sternum compression to HIII50 chest deflection, based on a small sample of matched tests. More lower-severity matched tests should be conducted to refine the low-deflection yet-high-frequency portion of that transform.

The thoracic risk curves from Prasad et al. (2004) were derived by using fitted field data. In the fit, the gender effect was relatively weaker statistically than the age effect. This fit should be revisited as more field data becomes available.

NHTSA binned some AIS2 neck injuries with the AIS3+ neck injuries (NHTSA, 1999 and 2008). That practice seemed inconsistent in the present study. However, it will be investigated in future studies.

The field model was based on Madymo models of drivers involving a belt+airbag restraint system. As more laboratory tests become available, related model correlations should be conducted to refine the field model.

Finally, only risk curves with parametric forms were considered in the present study.

CONCLUSIONS

In 2008, NHTSA published a Notice to update the USA NCAP. Therein, risk curves convert occupant responses from frontal and side impact test dummies into human occupant injury probabilities. In the present study, we assessed the fidelity of the underlying risk curves for estimating injury rates in real-world frontal impacts. Specifically, the NCAP risk curves were used to estimate the injury rates of belted drivers in airbag-equipped light vehicles in real-world crashes. The injury rates were considered from two perspectives: (1) a "point" estimate pertaining to NCAP-type events, involving NCAP fleet tests of vehicles and (2) an "aggregate" estimate pertaining to 0-56 km/h towaway crashes, involving modeling of a theoretical vehicle whose NCAP performance approximated the average of the studied fleet. In both assessments, four different body regions were considered: head, neck, chest, and KTH. Finally, other published risk curves for these four body regions were also evaluated.

The bases for the comparisons were injury rates computed from the NASS field data. A risk curve whose estimated injury rates fell within the confidence limits of the field data was deemed "acceptable." For this study, if both the NCAP risk curve and other risk curves for a particular body region yielded acceptable results, the NCAP risk curve was recommended over the others.

The assessment yielded mixed results. For the head, all of the studied curves yielded results within the confidence limits for both the point and aggregate risk estimates. For the neck, the NCAP risk curve for the combined-loading metric N_{ij} yielded results well outside the confidence limits for both the point and aggregate risk estimates. For the chest, the NCAP risk curve yielded results within the confidence limits for the point estimate, but the result was somewhat lower than the mean. For the KTH, all of the studied curves yielded results outside of the confidence limits for both the point and aggregate risk estimates. From this analysis, we recommended different available risk curves for the neck and chest, and we demonstrated the need for more research to improve KTH risk estimation. The recommended set of risk curves for the next-generation frontal NCAP is in Appendix 6. The non-NCAP risk curves for the head and KTH demonstrated no substantial improvements over the NCAP curves, whereas the proposed risk curves for the neck and chest yielded improved fidelity with real-world injury rates. The recommended risk curve for combined loading for the neck was based on a statistical method generally unaccepted by NHTSA.

ACKNOWLEDGEMENTS

We thank Kevin Siasoco and Marvin Nutt for reviewing the manuscript and providing numerous suggestions. Finally, this paper is dedicated in memory of Joan C. Laituri; R.I.P.

REFERENCES

Alliance of Automobile Manufacturers (1999) "Supplemental Notice of Proposed Rulemaking, FMVSS 208, Occupant Crash Protection," Docket No. NHTSA-1999-6407-0040.

Hertz, E. (1993) "A Note on the Head Injury Criteria (HIC) as a Predictor of the Risk of Skull Fracture," Proceedings of the Association for the Advancement of Automotive Medicine.

International Standards Organization, Technical Report, ISO/TR 7861:2003(E), "Road Vehicles — Injury Risk Curves for Evaluation of Occupant Protection in Frontal Impact," 2003.

Laituri, T., Sriram, N., Kachnowski, B., Scheidel, B., and Prasad, P. (2001) "Theoretical Evaluation of the Requirements of the 1999 Advanced Airbag SNPRM— Part One: Design Space Constraint Analysis," SAE-2001-01-0165.

Laituri, T., Prasad P., Kachnowski, B., Sullivan, K., and Przybylo P. (2003a) "Predictions of AIS3+ Thoracic Risks for Belted Occupants in Full-Engagement, Real-World Frontal Impacts: Sensitivity to Various Theoretical Risk Curves," SAE-2003-01-1355.

Laituri, T., Kachnowski, B., Prasad, P., Sullivan, K., and Przybylo, P. (2003b) "A Theoretical, Risk Assessment Procedure for In-Position Drivers Involved in Full-Engagement Frontal Impacts," SAE-2003-01-1354.

Laituri, T., Sullivan, D., Sullivan, K., and Prasad, P. (2004) "A Theoretical Math Model for Projecting AIS3+ Thoracic Injury for Belted Occupants in Frontal Impacts," 2004-22-0020, Stapp Car Crash Journal, Vol. 48.

Laituri, T., Prasad, P., Sullivan, K., Frankstein, M., and Thomas, R. (2005) "Derivation and Evaluation of a Provisional, Age-Dependent, AIS3+ Thoracic Risk Curve for Belted Adults in Frontal Impacts," SAE-2005-01-0297.

Laituri, T., Henry, S., Sullivan, K., and Prasad, P. (2006) "Derivation and Theoretical Assessment of a Set of Biomechanics-Based, AIS2+ Risk Equations for the Knee-Thigh-Hip Complex," 2006-22-0054, Stapp Car Crash Journal, Vol. 50.

MADYMO, User's Manual 3D, V5.4.1, TNO Road-Vehicles Research Institute, 1999.

Mertz, H., Driscoll G., Lenox J., Nyquist G., and Weber D. (1982) "Responses of Animals Exposed to Deployment of Various Passenger Inflatable Restraint System Concepts for a Variety of Collision Severities and Animal Positions," Proceedings of the Ninth International Technical Conference on Experimental Safety Vehicles, pp. 352-368.

Mertz, H., Horsh, J., Horn, G., and Lowe, R. (1991) "Hybrid III Sternal Deflection Associated with Thoracic Injury Severities of Occupants Restrained with Force-Limiting Shoulder Belts," SAE910812.

Mertz, H. (1993) "Anthropomorphic Test Devices," In Accidental Injury: Biomechanics and Prevention, ed. A. Nahum and J. Melvin, pp. 66-84. Springer-Verlag, New York.

Mertz, H., Prasad, P., and Irwin, A. (1997) "Injury Risk Curves for Children and Adults in Frontal and Rear Collisions," Paper No. 973318, 41st Stapp Car Crash Conference.

Mertz, H. and Prasad, P. (2000) "Improved Neck Injury Risk Curves for Tension and Extension Moment Measurements of Crash Test Dummies," Stapp Car Crash Journal 44: 59-75.

Mertz, H., Irwin, A., and Prasad, P. (2003) "Biomechanical and Scaling Bases for Frontal and Side Impact Injury Assessment Reference Values," Stapp Car Crash Journal, Vol. 47, pp. 155-188.

Morgan, R., Eppinger, R., and Marcus, J. (1989) "Human Cadaver Patella-Femur-Pelvis Injury Due to Dynamic Frontal Impact to the Patella," Proceedings of the 12th International Technical Conference on Experimental Safety Vehicles, Sweden.

Mullen, Chris (2008). Personal communication.

Nakahira, Y., Furukawa, K., Niimi, H., Ishihara, T., Miki, K., and Matsuoka, F. (2000) "A Combined Evaluation Method and a Modified Maximum Likelihood Method for Injury Risk Curves," Conference of the International Research Council on the Biomechanics of Impact (IRCOBI).

NHTSA (1999) "Development of Improved Injury Criteria for the Assessment of Advanced Automotive Restraint Systems – II," Docket No. NHTSA-2000-7013-0003.

NHTSA (2008) "New Car Assessment Program," Docket No. NHTSA-2006-26555, 120 pp., July 11, 2008.

NHTSA (2008a) "Notice of Postponement of Enhancements to the New Car Assessment Program (NCAP)," Docket No. NHTSA-2006-26555, 4 pp., December 19, 2008.

Nusholtz, G., Domenico, L. D., Shi, Y., and Eagle, P. (2003) "Studies of Neck Injury Criteria Based on Existing Biomechanical Test Data," Accident Analysis and Prevention, Vol. 35.

Prasad, P. and Daniel, R. (1984) "A Biomechanical Analysis of Head, Neck, and Torso Injuries to Child Surrogates Due to Sudden Torso Acceleration," SAE Paper No. 841656.

Prasad, P., Laituri, T., and Sullivan, K. (2004) "Estimation of AIS3+ Thoracic Risks of Belted Drivers in NASS Frontal Crashes," Journal of Automobile Engineering, Paper No. D15603.

Rupp, J., (2006) "Biomechanics of Hip Fractures in Frontal Motor-Vehicle Crashes," Doctoral dissertation, The University of Michigan.

SAS (2002), v 9.1, SAS Institute Inc., Cary, NC.

Zhou, Q., Rouhana, S., and Melvin, J. (1996) "Age Effects on Thoracic Injury Tolerance," SAE962421, 40th Stapp Car Crash Conference.

NOTATIONS

SYMBOLS

δ	Chest deflection of test dummy
ΔV	Vehicle speed change (longitudinal)
μ	Mean
σ	Standard deviation

SUBSCRIPTS

CE	Compression and Extension, combined loading
CF	Compression and Flexion, combined loading
E	Extension
HIII50	Hybrid III, mid-sized male test dummy
ij	Four possible evaluation of combined loading: i=tension or compression; j=flexion or extension
neck	Upper neck
T	Tension
TE	Tension and Extension, combined loading
TF	Tension and Flexion, combined loading

DEFINITIONS, ACRONYMS, ABBREVIATIONS

DEFINITIONS

Risk	Probability of sustaining an injury at a specified level
A,B,C	Dummy variables

ACRONYMS

AIS	Abbreviated Injury Scale
AIS2+	Abbreviated Injury Scale, Level 2 or higher
AIS3+	Abbreviated Injury Scale, Level 3 or higher
CDS	Crashworthiness Data System
CDF	Cumulative Density Function
ChestD	Chest deflection in test dummy
ChestG	Chest acceleration in test dummy
DOF	Direction of Principal Force

F	Force, upper neck
FMVSS	Federal Motor Vehicle Safety Standard
HIC	Head Injury Criterion
HIC15	Head Injury Criterion, 15 ms-based
HIC36	Head Injury Criterion, 36 ms-based
HIII05	Hybrid III, small female test dummy
HIII50	Hybrid III, mid-sized male test dummy
HIII95	Hybrid III, large-sized male test dummy
IARV	Injury Assessment Reference Value
KTH	Knee-Thigh-Hip
L	Left
LCL	Lower 95th%ile Confidence Limit
M	Moment, upper neck
MML	Modified Maximum Likelihood
MY	Model Year
n	Sample size
N	Weighted NASS estimate
N_{ij}	Combined loading metric for the upper neck, maximum of the four possible evaluations
NASS	National Automotive Sampling System
NCAP	New Car Assessment Program
NFS	Not Further Specified
NHTSA	National Highway Traffic Safety Administration
NPBLE	Not a Potentially Barrier-Like Event
OF	Older Women (ages, 50+)
OM	Older Men (ages, 50+)
P, p	Probability
PBLE	Potentially Barrier-Like Event
PMHS	Post-Mortem Human Subject(s)
SNPRM	Supplemental Notice of Proposed Rule Making
UCL	Upper 95th%ile Confidence Limit
USA	United States of America
VIN	Vehicle Identification Number
W	Real-world weighting factor
YF	Younger Women (ages, 15-49)
YM	Younger Men (ages, 15-49)

APPENDIX 1: GOMPERTZ FITS OF INJURY RATES IN PBLE

Various parametric forms were considered to fit the PBLE injury rates by body region across the speed range. Logit, Weibull, and Gompertz fits were assessed. Excel Solver was used to change the parameters of the fit equations to minimize the total squared error between actual and field injury rates, summed over all of the ΔV bins. Only bins with 10 or more crashes were considered, to reduce the effect of potentially overly influential data. Additionally, a zero-zero point was included: (ΔV=0, Injury Rate=0). The Gompertz form yielded the best fits, that is:

$$P\,(\%) = 100\,e^{-\alpha\,e^{-\beta(\Delta V)}}$$

where ΔV is in km/h.

The resulting Gompertz parameters are reported in Table A1, and the corresponding fits are shown in Figures A1-A5. Note the differences in the y-axis scale for Figures A1-A5.

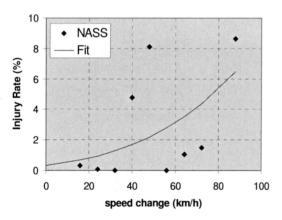

Figure A1. AIS3+ Head Injury Rates in PBLE

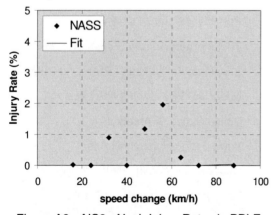

Figure A2. AIS3+ Neck Injury Rates in PBLE

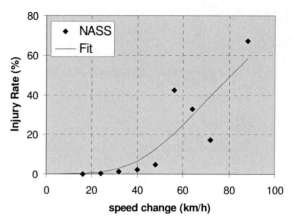

Figure A3. AIS3+ Chest Injury Rates in PBLE

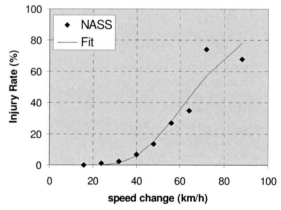

Figure A4. AIS2+ KTH Injury Rates in PBLE

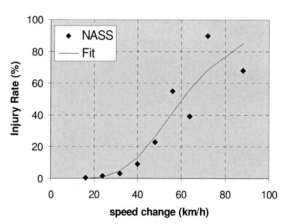

Figure A5. Injury Rates (Any of the 4 Body Regions) in PBLE

Table A1. Gompertz Parameters, from Fits of Field Data (PBLE)

Body Region	Injury Level	α	β
Head	AIS3+	5.69954	0.00832
Neck	AIS3+	25.32161	0.01366
Chest	AIS3+	10.63809	0.03390
KTH	AIS2+	21.75530	0.05081
Any of the Four	AIS2+, 3+	16.35891	0.05225

APPENDIX 2: REPORTED INJURY SOURCES IN APPROXIMATED NCAP-TYPE EVENTS

The reported "injury source" from NASS was also studied for the relevant injuries recorded for drivers of airbag-equipped passenger cars in approximated NCAP-type events, collapsed over the longitudinal speed change range of 48-64 km/h. Those data are shown in Tables A2-A5. Therein, "DOF" stands for "direction of force" and "NFS" stands for "Not Further Specified."

For the KTH-injured drivers in the NCAP-type events, there were 22 drivers with 36 KTH AIS2+ injuries. Table A6 shows the distribution of these 36 injuries by KTH complex subregion; femoral shaft fracture was the most frequent injury.

Table A2. AIS3+ Head-Injured Drivers (n=5)

ΔV (km/h)	DOF	Reported Injury Source	Body Part
48	12	steer wheel	Cerebrum NFS
48	12	other	Cerebrum NFS
48	12	other	Cerebrum NFS
64	12	steer wheel	Cerebrum NFS
48	12	windshield	Cerebrum NFS

Table A3. AIS3+ Neck-Injured Drivers (n=3)

ΔV (km/h)	DOF	Reported Injury Source	Body Part
48	12	other	Fracture odontoid
56	12	seat/head rest	Disc fracture
64	12	steer wheel	Cord contusion

Table A4. AIS3+ Chest-Injured Drivers (n=14)

ΔV (km/h)	DOF	Reported Injury Source	Body Part
48	12	air bag	Lung NFS
64	12	steer wheel	Lung NFS
48	11	seat belt	Thoracic cavity NFS
64	11	steer wheel	Rib cage
56	11	steer wheel	Thoracic cavity NFS
56	12	air bag	Lung NFS
56	12	steer wheel	Lung NFS
64	12	steer wheel	Lung NFS
48	12	L side interior	Rib cage
48	12	seat belt	Rib cage
56	12	seat belt	Rib cage
64	1	L side interior	Rib cage
48	12	L arm/hardware	Diaphragm
64	12	steer wheel	Rib cage

Table A5. AIS2+ KTH-Injured Drivers (n=22)

ΔV (km/h)	DOF	Reported Injury Source	Body Part
48	1	knee bolster	Femur
48	12	inst panel	Femur
48	12	inst panel	Femur
64	12	knee bolster	Femur
56	12	knee bolster	Knee
64	11	knee bolster	Femur
56	11	knee bolster	Femur
56	12	L arm/hardware	Femur
56	12	knee bolster	Femur
64	11	knee bolster	Femur
48	12	knee bolster	Femur
48	12	knee bolster	Femur
64	12	seat/head rest	Pelvis
56	11	knee bolster	Pelvis
56	12	knee bolster	Pelvis
48	12	knee bolster	Knee
64	1	knee bolster	Femur
48	12	L arm/hardware	Pelvis/Femur
56	12	knee bolster	Femur
48	12	knee bolster	Pelvis
64	12	knee bolster	Pelvis/Hip
64	12	knee bolster	Femur

Table A6. Distribution of Injuries* in the KTH Region

Part	Injury Description	AIS	Frequency
Knee	Patella Tendon Laceration	2	0
	Knee Dislocation; NFS	2	1
	Knee Dislocation; no articular cartilage	2	0
	Knee Laceration into Joint	2	0
	Knee Meniscus Tear	2	1
	Knee Sprain	2	1
Thigh	Femur fracture but NFS as to site	3	2
	Femur fracture; open/displaced/comminuted	3	2
	Femur fracture; condylar	3	1
	Femur fracture; head	3	0
	Femur fracture; intertrochanteric	3	1
	Femur fracture; neck	3	0
	Femur fracture; shaft	3	10
	Femur fracture; subtrochanteric	3	1
	Femur fracture; supracondylar	3	2
Hip	Hip Dislocation NFS	2	0
	Hip Dislocation; no articular cartilage	2	4
	Pelvis fracture; NFS	2	1
	Pelvis fracture; closed	2	5
	Pelvis fracture; open/displaced/comminuted	3	4
		Total =	36

* Some of the possible AIS2+ KTH injuries were not observed in the studied sample; they were listed for completeness.

APPENDIX 3: SENSITIVITY ANALYSIS

To address concerns about small sample sizes of the field dataset, a sensitivity analysis was conducted for the injury rates by body region.

Two datasets were formed by using different screening filters for belted drivers in 11-1 o'clock full-engagement, non-rollover, towaway crashes in the 1993-2006 calendar years of NASS:

1. "Original" = 1994-2006 MY light vehicles fitted with airbags, and

2. "Broader" = <u>All</u> MY light vehicles, <u>irrespective</u> of airbag fitment.

Note that the "Original" dataset was the dataset studied earlier, and that it was more homogeneous than the broader dataset.

Figure A6 contains the results of the comparison. The left column pertains to the original dataset; the right column pertains to the broader dataset. The top row pertains to the point estimate; the bottom row pertains to the aggregate estimate. The crash counts, both sample (n) and weighted (N), are also reported.

Therefore, the broader dataset involved 1.7 or 1.9 times more crashes than the original dataset, depending on the reference.

Figure A6 shows that the two main conclusions drawn previously from the original dataset also applied to the broader dataset. Specifically, for both the point estimates as well as the aggregate estimates,

- KTH injury rates > Chest injury rates > Head injury rates > Neck injury rates.

- The neck injury rates were consistently <u>much lower</u> than those of the other body regions.

The magnitudes of the injury rates for the point estimate were somewhat different. There was less difference for the aggregate estimate. Therefore, while the ordinary cautions regarding small sample sizes apply, we concluded that (a) the original dataset yielded injury rates consistent with a much larger and more non-homogeneous dataset, and (b) those injury rates and their rank ordering provided useful direction for assessing the fidelity of the aforementioned risk curves.

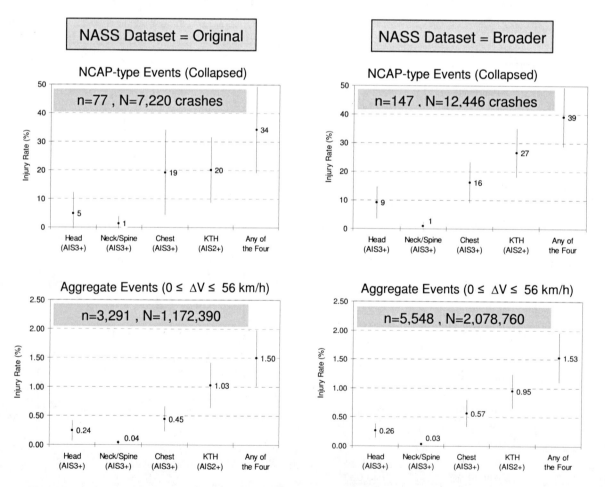

Figure A6. Comparison of Injury Rates by Body Region, for Two Datasets: Original versus Broader

APPENDIX 4: THEORETICAL FIELD MODEL DETAILS

Tables A7-A9 contain the real-world weighting factors for the various field models considered in the present study. All of these factors were computed from the aforementioned field data irrespective of outcome. Table A7 pertains to the field models which involve age-dependence. Table A8 pertains to those with age/gender dependence. Table A9 pertains to those which are irrespective of age or gender. Note that in each model, these factors necessarily sum to unity. The weighting factor for occupant size was from Laituri et al. (2003b).

The simulation results are in Table A10. Therein, a deployment schedule for the dual-stage driver airbag was prescribed. Real-world effects (e.g., seating position) were included in these models.

Table A7. Real-World Weighting Factors for Field Models involving Age Effects

	W_age		W_speed		W_severity		W_size	
Age	Towaways (N)	weight	ΔV (km/h)	weight	type	weight	occupant	weight
15-25	416,558	0.355	8	0.015	PBLE	0.146	HIII05	0.22
			16	0.362	NPBLE	0.854	HII50	0.59
			24	0.324			HIII95	0.19
			32	0.210				
			40	0.068				
			48	0.015				
			56	0.007				
26-35	258,170	0.220	8	0.006	PBLE	0.072	HIII05	0.22
			16	0.373	NPBLE	0.928	HII50	0.59
			24	0.337			HIII95	0.19
			32	0.206				
			40	0.047				
			48	0.029				
			56	0.003				
36-45	184,173	0.157	8	0.010	PBLE	0.243	HIII05	0.22
			16	0.396	NPBLE	0.757	HII50	0.59
			24	0.389			HIII95	0.19
			32	0.133				
			40	0.058				
			48	0.011				
			56	0.003				
46-55	166,916	0.142	8	0.017	PBLE	0.131	HIII05	0.22
			16	0.487	NPBLE	0.869	HII50	0.59
			24	0.296			HIII95	0.19
			32	0.140				
			40	0.044				
			48	0.012				
			56	0.003				
56-65	66,485	0.057	8	0.007	PBLE	0.288	HIII05	0.22
			16	0.524	NPBLE	0.712	HII50	0.59
			24	0.268			HIII95	0.19
			32	0.139				
			40	0.020				
			48	0.027				
			56	0.016				
66+	80,089	0.068	8	0.000	PBLE	0.092	HIII05	0.22
			16	0.504	NPBLE	0.908	HII50	0.59
			24	0.301			HIII95	0.19
			32	0.093				
			40	0.052				
			48	0.040				
			56	0.011				

Table A8. Real-World Weighting Factors for Field Models involving Age/Gender Effects

Age/Gender	W age/gender		W speed		W severity		W size	
	Towaways (N)	weight	ΔV (km/h)	weight	type	weight	occupant	weight
YM	432,642	0.369	8	0.011	PBLE	0.126	HIII05	0.01
			16	0.406	NPBLE	0.874	HII50	0.80
			24	0.329			HIII95	0.19
			32	0.175				
			40	0.054				
			48	0.019				
			56	0.006				
YF	481,915	0.411	8	0.016	PBLE	0.159	HIII05	0.22
			16	0.372	NPBLE	0.841	HII50	0.77
			24	0.335			HIII95	0.01
			32	0.192				
			40	0.053				
			48	0.027				
			56	0.005				
OM	113,810	0.097	8	0.029	PBLE	0.244	HIII05	0.01
			16	0.465	NPBLE	0.756	HII50	0.80
			24	0.330			HIII95	0.19
			32	0.104				
			40	0.032				
			48	0.030				
			56	0.010				
OF	144,023	0.123	8	0.000	PBLE	0.116	HIII05	0.22
			16	0.534	NPBLE	0.884	HII50	0.77
			24	0.274			HIII95	0.01
			32	0.128				
			40	0.038				
			48	0.020				
			56	0.006				

Table A9. Real-World Weighting Factors for Field Models; Irrespective of Age or Gender

Age	W age		W speed		W severity		W size	
	Towaways (N)	weight	ΔV (km/h)	weight	type	weight	occupant	weight
All	1,172,391	1.000	8	0.011	PBLE	0.142	HIII05	0.22
			16	0.406	NPBLE	0.858	HII50	0.59
			24	0.329			HIII95	0.19
			32	0.175				
			40	0.054				
			48	0.019				
			56	0.006				

Dummy	ΔV (km/h)	Severity	Head		Upper Neck							Chest		Femur	
			HIC36	HIC15	Fz+ (kN)	Fz- (kN)	Nij	Nte	Ntf	Nce	Ncf	ChestG	ChestDef (mm)	Load, L (kN)	Load, R (kN)
HIII05	8	PBLE	15	10	0.73	-0.1	0.28	0.28	0.08	0.06	0.14	18.1	-16.7	-0.06	-0.05
HIII05	16	PBLE	44	20	0.73	-0.1	0.41	0.41	0.17	0.03	0.17	20.9	-17.6	-0.14	-0.08
HIII05	24	PBLE	62	34	0.71	-0.1	0.41	0.41	0.18	0.04	0.15	23.3	-18.5	-0.16	-0.06
HIII05	32	PBLE	113	81	0.74	-0.09	0.38	0.38	0.21	0.04	0.15	26.4	-18.9	-0.15	-0.05
HIII05	40	PBLE	138	105	0.88	-0.09	0.45	0.45	0.25	0.04	0.18	29.7	-19.5	-0.03	-0.10
HIII05	48	PBLE	258	180	1.08	-0.09	0.47	0.47	0.27	0.04	0.03	33.4	-23.3	-0.07	-0.04
HIII05	56	PBLE	384	227	1.14	-0.09	0.51	0.51	0.27	0.04	0.32	37.1	-24.4	-0.06	-0.12
HIII50	8	PBLE	13	7	0.34	-0.13	0.1	0.09	0.1	0.05	0.1	10.0	-14.4	-0.46	-0.28
HIII50	16	PBLE	51	30	0.74	-0.15	0.17	0.14	0.15	0.05	0.17	19.0	-19.9	-0.47	-0.30
HIII50	24	PBLE	99	57	0.93	-0.15	0.23	0.18	0.23	0.05	0.22	23.6	-22.1	-0.52	-0.35
HIII50	32	PBLE	164	79	1.02	-0.14	0.29	0.18	0.29	0.05	0.03	29.2	-23.5	-1.01	-0.32
HIII50	40	PBLE	295	232	1.11	-0.14	0.21	0.18	0.21	0.05	0.02	31.4	-27.1	-1.41	-0.33
HIII50	48	PBLE	796	796	1.21	-0.18	0.24	0.24	0.19	0.05	0.02	35.0	-31.2	-3.07	-0.34
HIII50	56	PBLE	431	241	1.49	-0.13	0.28	0.16	0.28	0.05	0.02	42.2	-31.0	-3.23	-0.34
HIII95	8	PBLE	12	6	0.44	-0.16	0.11	0.1	0.08	0.02	0.11	13.1	-17.4	-1.04	-0.58
HIII95	16	PBLE	57	27	0.73	-0.18	0.18	0.14	0.13	0.03	0.18	20.4	-25.5	-0.52	-0.37
HIII95	24	PBLE	103	49	0.87	-0.15	0.19	0.16	0.19	0.04	0.02	22.4	-24.8	-0.46	-0.81
HIII95	32	PBLE	256	118	1.19	-0.44	0.26	0.16	0.25	0.02	0.26	23.6	-27.4	-0.58	-1.70
HIII95	40	PBLE	527	527	1.33	-0.55	0.26	0.18	0.09	0.26	0.02	28.5	-28.5	-0.19	-1.67
HIII95	48	PBLE	811	811	1.91	-0.13	0.29	0.29	0.15	0.02	0.02	35.2	-30.8	-1.51	-1.20
HIII95	56	PBLE	540	478	1.78	-0.13	0.28	0.28	0.26	0.02	0.02	42.2	-36.7	-1.92	-1.08
HIII05	8	NPBLE	9	7	0.61	-0.09	0.25	0.25	0.09	0.05	0.09	10.4	-12.3	-0.14	-0.26
HIII05	16	NPBLE	21	10	0.8	-0.1	0.35	0.35	0.1	0.02	0.14	15.6	-16.1	-0.11	-0.13
HIII05	24	NPBLE	31	16	0.94	-0.09	0.46	0.46	0.12	0.08	0.15	19.6	-17.2	-0.09	-0.09
HIII05	32	NPBLE	42	22	0.84	-0.09	0.43	0.43	0.2	0.02	0.18	22.7	-18.2	-0.09	-0.08
HIII05	40	NPBLE	84	56	0.8	-0.09	0.43	0.43	0.2	0.02	0.03	26.6	-20.0	-0.04	-0.02
HIII05	48	NPBLE	127	87	0.95	-0.1	0.5	0.5	0.24	0.03	0.03	30.9	-22.6	-0.05	-0.02
HIII05	56	NPBLE	211	166	1.11	-0.09	0.51	0.51	0.29	0.03	0.03	35.5	-24.4	-0.07	-0.02
HIII50	8	NPBLE	4	2	0.17	-0.08	0.08	0.05	0.08	0.02	0.04	5.7	-10.8	-0.47	-0.29
HIII50	16	NPBLE	18	9	0.37	-0.1	0.12	0.09	0.12	0.04	0.11	10.1	-15.2	-0.46	-0.15
HIII50	24	NPBLE	50	28	0.69	-0.11	0.15	0.14	0.15	0.05	0.13	16.2	-18.7	-0.49	-0.26
HIII50	32	NPBLE	95	55	0.93	-0.13	0.21	0.18	0.21	0.05	0.02	22.4	-22.1	-0.50	-0.27
HIII50	40	NPBLE	148	67	1.06	-0.22	0.28	0.13	0.28	0.05	0.15	28.4	-24.1	-0.83	-0.40
HIII50	48	NPBLE	308	227	1.09	-0.14	0.22	0.17	0.22	0.05	0.02	32.9	-29.1	-1.68	-0.30
HIII50	56	NPBLE	517	517	1.17	-0.14	0.19	0.19	0.18	0.05	0.02	36.6	-32.5	-2.41	-0.35
HIII95	8	NPBLE	4	2	0.17	-0.08	0.08	0.05	0.08	0.02	0.04	5.7	-10.8	-0.47	-0.29
HIII95	16	NPBLE	16	8	0.44	-0.14	0.12	0.1	0.08	0.03	0.12	10.9	-17.2	-1.09	-0.76
HIII95	24	NPBLE	52	26	0.67	-0.15	0.15	0.13	0.15	0.06	0.15	17.5	-22.1	-1.49	-1.54
HIII95	32	NPBLE	127	59	0.78	-0.16	0.22	0.16	0.22	0.02	0.02	24.5	-23.8	-1.23	-0.73
HIII95	40	NPBLE	209	107	1	-0.16	0.27	0.17	0.27	0.02	0.02	23.7	-26.7	-1.37	-0.58
HIII95	48	NPBLE	342	316	1.31	-0.16	0.2	0.16	0.2	0.01	0.06	27.2	-26.4	-0.73	-2.52
HIII95	56	NPBLE	662	662	1.49	-0.15	0.24	0.24	0.12	0.16	0.02	31.3	-32.6	-0.61	-2.68

APPENDIX 5: COMBINED LOADING EQUATIONS

The AIS3+ upper neck risk curve used in the NCAP Notice was presented in the 1999 SNPRM analysis for advanced airbags (NHTSA, 1999). Therein, data from matched tests of out-of-position HIII 3-year old test dummies and anesthetized porcine subjects were analyzed. The dummies provided the occupant responses and the porcine subjects provided the injury outcomes. These tests were conducted in the 1980's by Mertz et al. (1982) and Prasad and Daniel (1984). Due to the nature of the testing, only tension and extension (and their combination) were studied with these data.

In 1999, NHTSA provided their tension-extension equation for the HIII 3-year old:

$$N_{TE} \text{ [3-year old]} = \frac{F_T}{2120} + \frac{M_E}{27} . \qquad (A1)$$

In 2000, Mertz and Prasad provided their equation:

$$N_{TE} \text{ [3-year old]} = \frac{F_T}{2130} + \frac{M_E}{26.8} . \qquad (A2)$$

Therefore, the foundations for the subsequent analyses were nearly identical.

Mertz and Prasad (2000) subsequently derived a risk curve from the aforementioned matched tests. Their risk curve was based on an estimated peak stress state for the upper neck of the 3-year old test dummy; the injury outcomes (AIS3+ injury or not) pertain to the porcine subjects with negligible muscle tone. Later, Mertz et al. (2003) adjusted their N_{TE} equation to represent in-position HIII50's with 80% of their maximum muscle tone. The resulting N_{TE} equation is shown below in Equation A3. In the NCAP Notice, the N_{TE} for the HIII50 was computed according to Equation A4.

151

Mertz et al. (2003): $\quad N_{TE} = \dfrac{F_T}{6780} + \dfrac{M_E}{133} \qquad$ (A3)

NHTSA (2008): $\qquad N_{TE} = \dfrac{F_T}{6806} + \dfrac{M_E}{135} \qquad$ (A4)

Thus, the N_{TE} computations are slightly different. Moreover, NHTSA and Mertz et al. used different statistical methods to derive their respective risk curves for these N_{TE} inputs. Specifically, NHTSA used binary logistic regression, while Mertz used a modified median rank (Mertz-Weber) method.

Of these two differences in the risk curve derivation, the more significant difference was the applied statistical method. This may be illustrated by means of an example. Suppose that a test yields the following force and moment inputs: F_T=3300 N and M_E=57 N-m. Those data were evaluated using the risk curves from NHTSA and Mertz et al. The results are in Table A11. Therein, the difference between the two N_{TE}'s was slight (0.907 vs. 0.915), but the difference between the risk estimates was substantial (19.1% vs. 0.8%).

Table A11. Example of Effects: Different N_{TE} Calculations and Different Risk Curves

Inputs	N_{TE} Computation		AIS3+ Risk Estimate	
F_T=3,300 N; M_E = 57 N-m	NHTSA (2008)	Mertz et al. (2003)	NHTSA (2008)	Mertz et al. (2003)
	0.907	0.915	0.191	0.008

APPENDIX 6: SET OF RECOMMENDED RISK CURVES FOR USA FRONTAL NCAP

The set of recommended risk curves for the revised frontal NCAP (for HIII50 drivers) is as follows:

Head:

$$P_{head} = \text{lognormal CDF } (\mu_{HIC15} = 7.45231;\ \sigma_{HIC15} = 0.73998)$$

$$= \text{NORMSDIST}((\text{LN(HIC15)}-7.45231)/0.73998),\ \text{in EXCEL}$$

Upper Neck:

$$P_{neck} = \max\left[A, B, C\right]$$

where

$$A \equiv P_{tension} = \frac{1}{1 + e^{-(-10.9745 + 2.375\,F_T)}}$$

$$B \equiv P_{compression} = \frac{1}{1 + e^{-(-10.9745 + 2.375\,F_C)}}$$

$$C \equiv P_{combined}$$

$$= \text{normal CDF } (\mu_{N_{TE}} = 1.480;\ \sigma_{N_{TE}} = 0.235)$$

$$= \text{NORMDIST}(N_{TE},\ 1.480,\ 0.235,\ \text{TRUE}),\ \text{in EXCEL}$$

where F is in kN and N_{TE} is either computed subject to Equation A3 or approximated by Equation A4 (recall Table A11).

Chest:

$$P_{OM} = \frac{1}{1 + e^{-(-8.141 + 0.179\,\delta_{HIII50})}}$$

where δ_{HIII50} is in mm, and where any risk less than or equal to 1% is set to zero.

KTH:

$$P_{KTH} = \frac{1}{1 + e^{-(-5.7949 + 0.5196\,F)}}$$

where F is the HIII50 femur load in kN.

This equation should only be a placeholder; it did not demonstrate acceptable fidelity.

CHANGE OF VELOCITY AND PULSE CHARACTERISTICS IN REAR IMPACTS: REAL WORLD AND VEHICLE TESTS DATA

Astrid Linder
Matthew Avery
The Motor Insurance Repair Research Centre
Thatcham, United Kingdom
Maria Krafft
Anders Kullgren
Folksam Research, Sweden
Paper No. 285

ABSTRACT

Impact severity in collisions that can cause soft tissue neck injuries are most commonly specified in terms of change of velocity. However, it has been shown from real-world collisions that mean acceleration influences the risk of these injuries. For a given change of velocity this means an increased risk for shorter duration of the crash pulse. Furthermore, dummy response in crash tests has shown to vary depending on the duration of the crash pulse for a given change of velocity. The range of duration for change of velocities suggested for sled tests that evaluate the protection of the seat from soft tissue neck injuries are still to be established. The aim of this study was to quantify the variation of duration of the crash pulse for vehicles impacted from the rear at change of velocities suggested in test methods that evaluate the protection from soft tissue neck injuries. Crash pulses from the same vehicle models from different generations in real-world collisions producing a similar change of velocity were also analysed.

The results from the crash tests show that similar changes of velocity can be generated with various durations of crash pulses for a given change of velocity in rear impacts. The results from real-world collisions showed that a similar change of velocity was generated with various durations and shapes of crash pulses for the same vehicle model.

INTRODUCTION

Rear impacts causing AIS 1 (AAAM 1990) neck injuries most frequently occur at delta-Vs (changes of velocity) below 30 km/h in the struck vehicle (Parkin et al., 1995, Hell et al., 1999, Temming and Zobel, 2000). Furthermore, it has been shown that mean acceleration (i.e. the duration of the crash pulse for a given delta-V) influences the risk of AIS 1 neck injuries (Krafft et al., 2002). It has also been shown that the shape of the crash pulse influences.

risk of AIS 1 neck injuries in frontal impacts (Kullgren et al., 1999). Acceleration pulses from rear impacts shows that the same delta-V can cause a large variation in acceleration pulse shapes in the struck vehicle (Krafft, 1998, Zuby et al., 1999, Heitplatz et al., 2002). From real-world collisions it has been shown that the acceleration pulse also can vary in shape (i.e. duration of crash pulse, maximum magnitude of acceleration, onset rate etc) in impacts of similar delta-Vs (Krafft, 1998).

Dummy response in crash tests has been shown to vary depending not only on the delta-V but also on the duration of the crash pulse for a given delta-V (Linder et al., 2001a). The range of the duration of the crash pulse that corresponds to a specific delta-V in rear impacts has been shown to cover a wide range for vehicles impacted at the rear at a delta-V of up to 11 km/h (Linder et al., 2001b). The range of the duration of the crash pulse that corresponds to a specific delta-V in rear impacts that can cause AIS 1 neck injuries remains to be established. The range of the duration of the crash pulse for a specific delta-V is necessary to establish when designing impact severities for sled test methods that evaluate the safety performance of a seat in rear impacts, particularly in respect of AIS 1 neck injuries. Such test methods are at the moment under development Cappon et al. (2001), Muser et al. (2001), Langwieder and Hell (2002) and Linder (2002) and under discussion in groups like IIWPG (International Insurance Whiplash Prevention Group), EuroNCAP (European New Car Assessment Program), EEVC (European Enhanced Vehicle Safety Committee) Working group 12 and ISO (International Organization for Standardization) TC22/SC10/WG1. The delta-V suggested in sled test in these methods that represent the delta-V where the majority of rear impacts are reported is 15 or 16 km/h (Cappon et al., 2001, Muser et al., 2001 and Langwieder and Hell, 2002).

The first aim of this study was from laboratory crash tests to quantify the variety of mean acceleration monitored in different vehicles impacted in the same way. The second aim was to demonstrate the variety of the duration and shape of the crash pulse in the same vehicle model from real-world crashes producing similar delta-V.

MATERIALS AND METHODS

Laboratory Crash Tests

Sixteen vehicles (Table 1) were impacted at the rear either with a barrier or with a vehicle of the same make and model as the impacted vehicle. The barrier used in the OW test had a weight of 1000 kg (Figure 1). The barrier used in the CR tests had a weight of 1800 kg. The vehicles were impacted at the rear with 100 % overlap. The test where a vehicle was impacted by another vehicle (test OW3739, CR01001 and CR01002), the same make and model of vehicle was used as the bullet vehicle. The mass of the cars used were from 1010 kg - 1966 kg. The OW9999 vehicle was from 1983 series car (the actual vehicle was a used vehicle new in 1993 and with no structural corrosion) and the other vehicles were from the mid 1999.

Table 1.
The weight of the impacted vehicles and the impact velocity of the barrier or the impacting vehicle in the rear impacts.

Impact No.	Vehicle mass (kg)	Impact velocity (km/h)
OW9999	1190	18.3
OW3660	1450	30.0
OW3737	1965	52.4
OW3749	1445	36.9
OW3763	1347	35.7
OW3759	1493	32.1
OW3760	1493	43.8
OW3718	1010	40.0
OW3539	1384	24.9
OW3594	1405	35.2
OW3500	1339	18.5
CR98001	1450	24.0
CR98002	1800	24.0
CR98006	1750	24.0
CR01001	1439	32.0
CR01002	1461	32.0

Figure 1.

A rigid barrier impacting the rear of the vehicle.

One vehicle model was impacted both with a barrier and with another vehicle in order to compare the crash pulse from a rigid barrier to that generated by an impacting vehicle. The crash pulses from the laboratory tests were from previously performed tests at the Motor Insurance Repair Research Centre in the UK and at the Insurance Institute for Highway Safety in the US. The accelerometer was mounted at the base of the B-pillar on the left hand side in the vehicles in the OW tests. The vehicles in the OW tests were right-hand drive vehicles for the UK market. The vehicles in the CR tests were left-hand drive vehicles. The accelerometer was mounted on a steel bar pinned between the front door hinge-pillar and the b-pillar on the left hand side in the vehicles in the CR98001 and CR98002 tests. The accelerometer was mounted to the floor in the vehicle centreline just behind the front row of the seats in the vehicles in the CR98006 and the CR01 tests. The CR01 tests were performed with vehicles of the same make and model for the US and European market. These vehicles were structurally identical except from the bumper system. All vehicles were a conventional monocoque construction.

Real-World Rear Impacts

Since 1995, Folksam in Sweden have been equipping various new car models with one-dimensional crash-pulse recorders, mounted under the driver or passenger seat to record the crash pulse obtained during real-world impacts. The crash-pulse recorder is based on a spring mass system where the movement of the mass is registered on photographic film. When a vehicle equipped with a crash recorder has been involved in a collision the crash pulse is analysed by Folksam and the outcome for the occupants in terms of injuries is analysed by Folksam. In this study, crash pulse from rear impacts with two generation of the same vehicle model, T1 and T2 from 1993 and 1998, were presented.

Data Acquisition and Analysis

The crash pulse measured as the acceleration signals of the vehicle were filtered in accordance with SAE CFC 60 and the velocity was calculated by integrating the acceleration. The duration of the crash pulse (T_p) and the delta-V were identified from the filtered acceleration curves and the velocity curves.

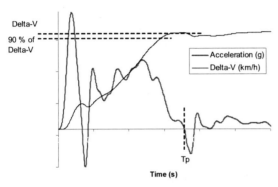

Figure 2.

Schematic drawing showing how the duration of the crash pulse T_p were identified from the graphs.

The T_p was defined as the time when the acceleration changed from positive to negative after 90 % of delta-V had occurred. Mean acceleration was calculated, defined as delta-V(at Tp)/Tp. For a given change of velocity a higher mean acceleration thus correspond to this a shorter duration of the crash pulse. For the graphic presentation of the crash pulses in this study the pulses were all adjusted so that the acceleration of 1 g occurred at time zero. Furthermore the crash pulses were filtered with CFC 36 (corresponding to a cut of frequency of 60 Hz) since oscillations in the crash pulses were found (Figure 3).

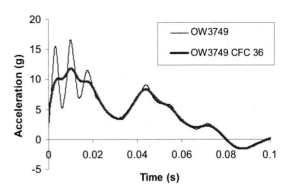

Figure 3.

Example of the oscillations filtered out by the CFC 36 filtering compared to the SAE standard filtering (CFC 60).

RESULTS

The results showed a considerable variation of the duration of the crash pulse for a similar delta-V both for different vehicles impacted the same way and for the same vehicle model impacted in various ways in real-world collisions. Furthermore, various pulse shapes were registered in the same vehicle from impacts which generated a similar delta-V.

Laboratory Crash Tests

The crash pulses from sixteen vehicles rear impacted with delta-Vs from 10.2 km/h to 19.4 km/h were examined. The duration of the crash pulse were between 65 ms and 130 ms. This resulted in mean accelerations between 3 g and 7.9 g (Figure 4 and 5 and Table 2).

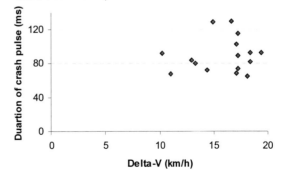

Figure 4.

The duration of the crash pulse versus delta-V from vehicles impacted at the rear with a rigid barrier or with another vehicle.

Table 2.
The delta-V, mean acceleration and duration of the crash pulse, T_p, from the vehicles impacted at the rear with a rigid barrier or another vehicle.

Impact No.	Delta-V (km/h)	T_p (ms)	a_{mean} (g)
OW9999	10.2	92	3.0
OW3660	17.1	69	7.0
OW3737	17.1	103	4.7
OW3749	18.4	82	6.4
OW3763	17.2	89	5.5
OW3759	18.4	93	5.6
OW3760	17.2	74	6.6
OW3718	18.1	65	7.9
OW3539	13.3	80	4.9
OW3594	13.0	84	4.4
OW3500	11.0	68	4.4
CR98001	16.6	130	3.6
CR98002	14.4	72	5.7
CR98006	14.9	129	3.3
CR01001	19.4	93	5.6
CR01002	17.2	116	4.1

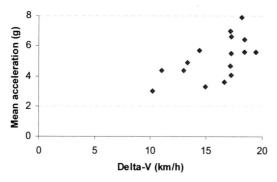

Figure 5.

The mean acceleration versus delta-V from vehicles impacted at the rear with a rigid barrier or with another vehicle.

The crash pulses from thirteen vehicles impacted with a rigid barrier (all OW tests except OW3759 and the CR98 tests) showed a range of mean acceleration from 3 g to 7.9 g. The crash pulses from the three vehicles impacted with another vehicle (OW3759 and CR01 tests) showed a range of mean acceleration from 4.1 g to 5.6 g.

The crash pulses recorded in the laboratorial tests are displayed in Figures 6-12.

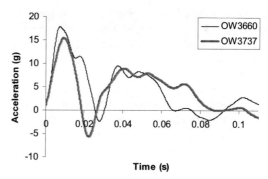

Figure 6.

The crash pulses from the OW3660 and OW3737, in tests at 100 % overlap with an impacting barrier generating a delta-V of 17.1 km/h.

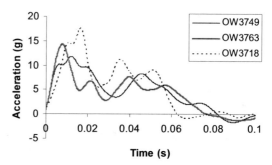

Figure 7.

The crash pulses from the OW3749, OW3763 and OW3718, in tests at 100 % overlap with an impacting barrier generating a delta-V of 17.2 km/h to 18.4 km/h.

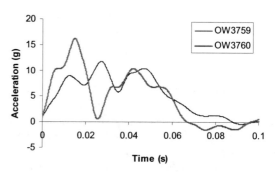

Figure 8.

The crash pulses from the OW3759 (car-to-car), and OW3760 (barrier to car), in tests at 100 % overlap with a delta-V of 17.2 km/h and 18.4 km/h.

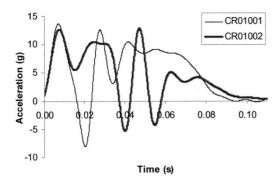

Figure 9.

The crash pulses from the car-to-car test with the same vehicle model for the US and European market, in tests at 100 % overlap with a delta-V of 17.2 km/h and 19.4 km/h.

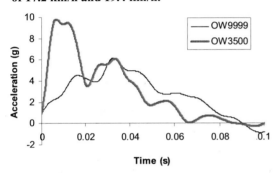

Figure 10.

The crash pulses from the OW3500 and OW9999, in tests at 100 % overlap with an impacting barrier generating a delta-V of 10.2 km/h to 11 km/h.

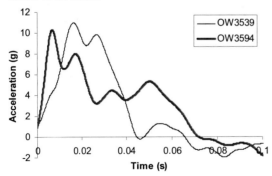

Figure 11.

The crash pulses from the OW3539 and OW3594, in tests at 100 % overlap with an impacting barrier generating a delta-V of 13 km/h to 13.3 km/h.

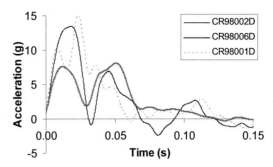

Figure 12.

The crash pulses from theCR98001, CR98002, and CR98006, in tests at 100 % overlap with an impacting rigid barrier generating a delta-V of 14.4 km/h to 16.9 km/h.

Real-World Rear Impacts

A large range of durations of crash pulses were found in the same type of vehicle where a similar delta-V was generated. Figure 13 and 14 and Table 3 shows the duration of the crash pulse, the mean acceleration and the delta-V from the real-world impacts from the vehicles T1 and T2. Furthermore, a considerable difference in shape of the crash pulse was registered in these cases (Figure 15 and 16). The duration of the pulses ranged from 77 ms - 134 ms. For vehicle T1, a change of velocity between 12.0 - 14.7 km/h and duration of the crash pulse from 77 ms to 109 ms was registered. For vehicle T2, a change of velocity between 17.1 - 20.4 km/h and duration of the crash pulse from 100 ms to 134 ms was registered.

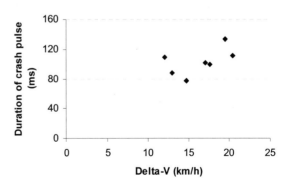

Figure 13.

The duration of the crash pulse and the delta-V from the crash recorder data from two different vehicle models of the same make.

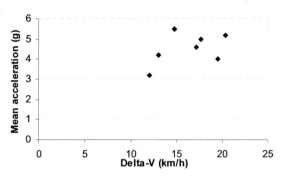

Figure 14.

The mean acceleration and the delta-V from the crash recorder data from two different year models of the same make and model of vehicle.

Table 3.
The duration of the crash pulse and the delta-V from the crash recorder data from two different year models of the same make and model of vehicle.

Car	CPR Number	Delta-V (km/h)	T_p (ms)	a_{mean} (g)
T1	C29521	14.7	77	5.5
T1	C30044	13.0	88	4.2
T1	C29614	12.0	109	3.2
T2	C30032	20.4	111	5.2
T2	C29732	19.5	134	4.0
T2	C29876	17.6	100	5.0
T2	C29739	17.1	102	4.6

Figures 15 and 16 shows the acceleration pulses from the real-world impacts from the vehicles T1 and T2.

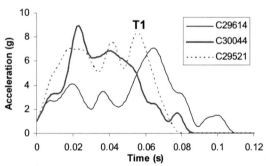

Figure 15.

The crash pulse measured in vehicle T1 in collisions with a change of velocity between 12.0 - 14.7 km/h.

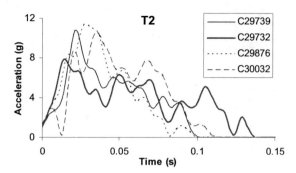

Figure 16.

The crash pulse measured in vehicle T2 in collisions with a change of velocity between 17.1 - 20.4 km/h.

DISCUSSION

A large variation in duration of crash pulse for a given delta-V and pulse shape can be produced in vehicles manufactured in the mid 1990s in rear impacts (Figure 6-12). Both delta-V and mean acceleration (i.e. duration of the crash pulse for a given delta-V) have been shown to influence the risk of AIS 1 neck injuries (Krafft et al., 2002). For a given delta-V a longer pulse will result in a lower mean acceleration and a lower risk of neck injuries (Krafft et al., 2002). The variation in durations of crash pulse for a given delta-V revealed in this study implies that vehicle seats aimed at reducing the risk of an AIS 1 neck injury should be designed in such a way that they provide the optimum protection in rear impacts in crashes where a great variation in duration of the crash pulse for a given delta-V might occur. These findings emphasise the importance of mean acceleration or the duration of crash pulse for a specific delta-V to be specified, in addition to delta-V, for sled tests that evaluate the protection from AIS 1 neck injuries of the seat, as suggested by Linder (2002).

A large variety of durations of crash pulse for a specific delta-V will be produced in the same car model, as exemplified by the real-world crash pulses collected from two year models of the same vehicle make and model (Figure 15 and 16). Therefore it can be expected that any vehicle will in real-world collisions be exposed for a large variety of durations of crash pulses for a specific delta-V. This might indicate that the design of the seat would have the largest potential to reduce the risk of AIS 1 neck injury in a rear impact since a huge variety of pulse shapes will be generated in the same vehicle model due to the various configurations of the collisions.

In this study the duration of the crash pulse (T_p) was defined as the time when the acceleration changed from positive to negative after 90 % of delta-V had occurred. This definition was used to ensure that the main part of the energy was transferred into the impacted vehicle at T_p. From the crash pulses analysed for this study it was found to be a robust definition of the duration of the crash pulse.

The crash pulses were filtered with CFC 36 due to oscillations found in the crash pulses. It has been surmised that these oscillations may be due to the mounting methods used to attach the accelerometers to the vehicles. For the real-world data the oscillations could be due to the design of the crash recorder. The filtering of CFC 36 was chosen instead of the CFC 60 and did not influence the delta-V from any of the pulses (as exemplified in Figure 3). The benefit of the CFC 36 filtering was that it highlighted the main characteristics of the crash pulses and was thus the rational of the choice.

The two vehicles of the same make and model for the US and European market which were tested in this study had different bumper systems. The European bumper system (crush cans, bottom, Figure 17) was designed for the NCAR damageability test and required replacement after a test. The US bumper system (hydraulic shock absorbers, top Figure 17) resulted in no damages in both rear-into-flat barrier and rear-into-pole impact test at five mile per hour.

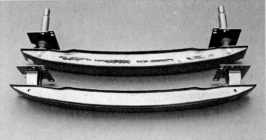

Figure 17.

The US bumper (upper) and the European bumper (lower) from the vehicle tested in test SL01001 and SL01002.

The US and European bumper systems resulted in similar shape of the crash pulse for the first 10 ms (Figure 9). After that the first peak acceleration was reach the shape of the two pulses developed

somewhat differently in terms of when maximum and minimum magnitude of the pulses was reached.

The range of delta-V explored in this study cover the range where rear impacts causing AIS 1 neck injuries most frequently occur (Parkin et al., 1995, Hell et al., 1999, Temming and Zobel, 2000). The main part of the crash tests and real-world data were from delta-Vs at or close to those suggested as delta-Vs for sled tests that evaluate the protection from neck injuries in rear impacts. The delta-V for these sled tests has been proposed to 15 km/h or 16 km/h (Cappon et al., 2001, Muser et al., 2001 and Langwieder and Hell, 2002). For each vehicle in the crash tests a range of durations of the crash pulse for a specific delta-V according to various crash configurations as for the real-world data can be expected. The range of durations of crash pulses for delta-Vs at 14.9 km/h or 17.1 km/h would, according to the results shown in Figure 4 and 13, be at least 69 ms to 130 ms which correspond to a range of mean acceleration of 3.3 g to 7 g. The range of duration of the crash pulses published by Heitplatz et al. (2002) were for the delta-V of 15.7 km/h to 16.9 reported to be approximately 90 ms to 110 ms. These findings are within the range of what has been found in this study. And not surprisingly, with a larger number of vehicle tested the range of duration for a specific delta-V widens, as show in this study.

Mean acceleration has for frontal collisions been shown to influence the risk of injuries (Ydenius, 2002). In that study it was shown that increased mean acceleration increased the risk of MAIS 1 injuries. Of the MAIS 1 injuries in Ydenius (2002) neck injuries are approximately 30 % of these. As a consequence, Ydenius findings emphasises the findings in this study of the importance of mean acceleration with respect to neck injuries.

Recently, attention has been focused on the need to define an acceleration pulse for standardised rear impact testing to evaluate the risk of AIS 1 neck injuries. In sled test proposals (Cappon et al., 2001, Muser et al., 2001 and Langwieder and Hell, 2002) corridors for the crash pulses with a wide range of durations of the crash pulse for a specific delta-V has been suggested to be used. From the results of this study it is not possible to identify a typical mean acceleration (which correspond to a duration of the crash pulse) for a specific delta-V either in the laboratory crash tests or from the real-world data. It might be the case that in rear impacts with a risk of AIS 1 neck injuries there is not one typical pulse or impact severity to be found. Rather a range of duration of crash pulses and delta-Vs that influence the risk of injury. Therefore it is suggested that duration of the crash pulse or mean acceleration, in addition to delta-V, should be specified for impact severity of sled test that evaluate the protection from the seat in rear impacts. This should be taken into consideration in such tests to minimizing the risk of sub optimization of seat protective performances.

CONCLUSIONS

From laboratorial tests with various vehicles impacted at the rear, a range of crash pulse durations between 65 ms to 130 ms was found for delta-Vs from 10.2 km/h to 19.4 km/h. Furthermore, from real-world rear collisions of the same vehicle make, a range of duration of crash pulse between 77 ms to 134 ms was found for delta-Vs from 12 km/h to 20.4 km/h.

This study shows that a similar delta-V can be generated by a variety of mean accelerations. Since mean acceleration have been found to be the main factor influencing the risk of AIS1 neck injuries, both delta-V and the duration of the crash pulse for a specific delta-V (i.e. mean acceleration) should be taken into consideration when defining impact severities in sled test procedure for vehicle seat safety performance assessment. In a sled test procedure a specification of a delta-V is therefore suggested to be accompanied with a specification of the mean acceleration or the duration of the crash pulse and the range of duration for a given delta-V of crash pulses that the seat could be exposed to, be taken into consideration in such tests.

ACKNOWLEDGEMENT

We thank Dr Andreas Moser and Mr Magnus Kock for practical and theoretical support regarding filtering of crash pulses. Thanks are also given to the Insurance Institute of Highway Safety for the supply of the CR test data.

REFERNCES

AAAM. (1990) The Abbreviated Injury Scale – 1990 Revision. American Association for Automotive Medicine, Des Plaines IL.

Cappon, H., Philippens, M., and Wismans, J. (2001) A new test method for the assessment of neck injuries in rear-end impacts, *Proc. 17th ESV Conf.*, Amsterdam, The Netherlands, Paper No. 242.

Hell, W., Langwieder, K., Walz, F., Muser, M., Kramer, M., and Hartwig, E. (1999) Consequences for seat design due to rear end accident analysis, sled tests and possible test criteria for reducing cervical spine injuries after rear-end collisions, *Proc. IRCOBI Conf.*, Sitges, Spain, pp. 243-259.

Heitplatz, F., Raimondo, S., Fay, P., Reim., J., and de Vogle, D. (2002) Development of a generic low speed rear impact pulse for assessing soft tissue neck injury risk, *Proc. IRCOBI Conf.*, Munich, Germany, pp. 249-260.

Krafft, M., Kullgren, A., Ydenius, A., and Tingvall, C. (2002) Influence of Crash Pulse Characteristics on Whiplash Assossiated Disorders in Rear Impacts – Crash Recording in Real-Life Impacts, *Traffic Injury Prevention*, Vol. 3 (2), pp 141-149.

Krafft, M. (1998) *Non-fatal injuries to car occupants – Injury assessment and analysis of impact causing short- and long-term consequences with special reference to neck injuries*, Ph.D. Thesis, Karolinska Institute, Stockholm, ISBN 91-628-3196-8.

Kullgren, A., Thomson, R., and Krafft, M. (1999) The effect of crash pulse shape on AIS1 neck injuries in frontal impacts. *Proc. of the IRCOBI Conference*, Sitges, Spain, pp. 231-242.

Langwieder, K., and Hell, W. (2002) Proposal of an International Harmonized Dynamic Test Standard for Seat/Head Restraint, *Traffic Injury Prevention*, Vol. 3 (2), pp 150-158.

Linder, A. (2002) *Neck Injuries in Rear Impacts: Dummy Neck Development, Dummy Evaluation and Test Condition Specifications*, Ph.D. Thesis, Chalmers University of Technology, Göteborg, Sweden, ISBN 91-7291-106-9.

Linder, A., Olsson, T., Truedsson, N., Morris, A., Fildes, B., and Sparke, L. (2001a) Dynamic Performances of Different Seat Designs for Low and Medium Velocity Rear Impact, *Proceedings of the 45th Annual AAAM Conference*, San Antonio, USA.

Linder, A., Avery, M., Krafft, M., Kullgren, A., and Svensson, M.Y. (2001b) Acceleration Pulses and Crash Severity in Low Velocity Rear Impacts - Real World Data and Barrier Test, *Proceedings of the 17th ESV Conference*, Amsterdam, The Nederlands, Paper No. 216.

Muser M., Zellmer, H., Walz. F., Hell, W., and Langwieder, K. (2001) Test procedure for the evaluation of the injury risk to the cervical spine in a low speed rear end impact, Proposal for the ISO/TC22 N 2071 /ISO/TC22/SC10, www.agu.ch/pdf/iso.v5.pdf.

Parkin S., Mackay G.M., Hassan A.M., and Graham R. (1995) Rear End Collisions And Seat Performance- To Yield Or Not To Yield, *Proc. Of the 39th AAAM Conference*, Chicago, Illinois, pp. 231-244.

Temming, J., and Zobel, R. (2000) Neck Distortion Injuries in Road Traffic Crashes (Analysis of the Volkswagen Database), In: *Frontiers in Whiplash Trauma: Clinical & Biomechanical*, Yoganandan, N. and Pintar, F.A. (Eds.), ISO Press, Amsterdam, The Netherlands, ISBN 1 58603 912 4, pp. 118-133.

Ydenius (2002) Influence of Crash Pulse Duration on Injury Risk in Frontal Impacts Based on Real Life Crashes, *Proc. IRCOBI Conf.*, Munich, Germany, pp. 155-166.

Zuby, D.S., Troy Vann, D., Lund, A.K., and Morris, C.R. (1999) Crash Test Evaluation of Whiplash Injury Risk, *Proc. of the 43rd STAPP Car Crash Conference*, San Diego, USA, pp. 267-278.

Human Head and Neck Kinematics After Low Velocity Rear-End Impacts - Understanding "Whiplash"

Whitman E. McConnell, Richard P. Howard, Jon Van Poppel, Robin Krause, Herbert M. Guzman, John B. Bomar, James H. Raddin, James V. Benedict, and Charles P. Hatsell

Biodynamic Research Corp.

ABSTRACT

A second series of low speed rear end crash tests with seven volunteer test subjects have delineated human head/neck dynamics for velocity changes up to 10.9 kph (6.8 mph). Angular and linear sensor data from biteblock arrays were used to compute acceleration resultants for multiple points on the head's sagittal plane. By combining these acceleration fields with film based instantaneous rotation centers, translational and rotational accelerations were defined to form a sequential acceleration history for points on the head. Our findings suggest a mechanism to explain why cervical motion beyond the test subjects' measured voluntary range of motion was never observed in any of a total of 28 human test exposures. Probable "whiplash" injury mechanisms are discussed.

INTRODUCTION

Our first study, done in February 1991 and reported[31] to the SAE in March 1993, involved four test subjects who were exposed to a series of ten low delta-V (4 - 8 kph or 2.5 - 5.0 mph) rear end impacts between standard road vehicles in an approved human test study protocol. Several other studies published since our first presentation have reported similar findings[5,52,55]. Drawing from the practical lessons learned during our first test series, a second test series was conducted in July 1993 utilizing seven test subjects (including three return volunteers from the first series) who represented a somewhat broader age and anthropometric variation range. This paper reports findings from our analysis of their kinematic responses to a test series of fourteen rear end impacts including a higher range of impact related ΔV (5.8 - 10.9 kph or 3.6 - 6.8 mph) than was previously studied. The resulting head, neck and torso kinematics from a total of eighteen human and four Hybrid III anthropometric test device (ATD) exposures were recorded using a variety of improved electronic and high speed film based data collection methods. The resulting data were analyzed with visual graphic methods and with specially developed mathematical techniques to interpret and combine both angular and translational biteblock accelerometer information with digitized high speed film data. With the resulting clearer picture, all but one of our observations from the first article were confirmed, some of our earlier observations and unanswered questions were able to be refined or corrected and a biomechanically rational explanation of human head, neck, torso, seatback and head restraint interaction during rear end collisions can now be offered along with comments suggesting a proposed mechanism of injury related to the often referred to, but ill defined "whiplash" syndrome.

APPARATUS AND TEST SUBJECTS

VEHICLES AND TEST SITE - The total of 14 test-collisions were performed using three vehicles; one vehicle was used primarily as the striking or "bullet" vehicle, leaving the two other vehicles as primary "target" or struck vehicles. The bullet vehicle was a 1984 GMC C-1500 pick-up truck, while the target vehicles consisted of a 1986 Dodge 600 Convertible and a 1984 Buick Regal Limited Coupe. The test site was an unused section of paved road characterized by a very mild slope running downward (approx. 2% grade) in the direction of motion of the vehicles. A ramp inclined at approximately 14° was constructed and used to launch the bullet vehicle to the test protocol's desired impact speed (Figure 1).

Some modifications were made to the test-vehicles for practical and safety reasons. The GMC's front bumper was replaced by a steel reinforced and wood faced structure which better withstood the horizontal impact forces during repeated impacts with the struck cars. Since previous testing had shown that the struck vehicle had acquired most of its velocity change prior to any kinematic response of the occupant, the acceleration pulse shape variation expected because of this bumper modification did not appear to affect our occupant kinematics. Modifications to the struck vehicles included the addition of steel braces which were placed behind the front seats to safeguard against seatback failure. Since safety brace to seat back contact never

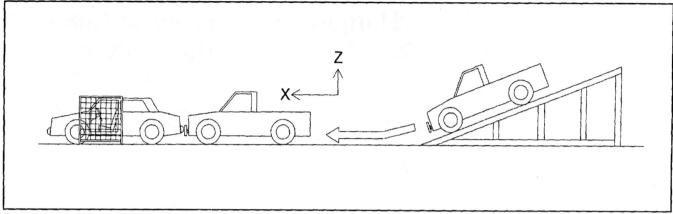

Figure 1 - Test Area and Ramp Setup

occurred during any of the test runs, our safety braces did not influence seat performance or occupant kinematics. Another modification was to remove the front doors of each test vehicle, allowing for better visibility of both the occupant and the seat dynamics. These doors were replaced with frame structures which maintained the rigidity of the vehicles. Factory standard head restraints in both the target vehicles were normally kept in their most fully raised position. In the sedan, where most of the head and neck kinematic data was obtained, the top of the test subjects' head was from 16 to 20 cm. (6.3 to 7.9 inches) above the top of the head restraint and the back of their heads were between 5.1 to 11.7 cm. (2 to 4.6 inches) in front of the head restraint's forward surface. The factory standard 3-point restraint systems were used for each test run with the intentional exception for one test run when both human driver and right front seat passenger were asked not to wear their restraint systems. Each vehicle was inspected after each test run for any safety related problems, as well as any mechanical or impact related changes that might influence test outcomes. Repairs to the vehicles were made as necessary prior to each test run.

TEST SUBJECTS - The test protocol for the current test series was reviewed by the University of Texas Health Science Center Institutional Review Board and IRB Protocol #9010099006 of the University of Texas health Science Center, under DHHS Regulation 46.110(3), approved the use of eight human test subjects selected from the staff of Biodynamic Research Corporation. Seven healthy fully informed volunteer male test subjects, ages from 32 to 59 years, ranging in height and weight from 173 - 188 cm and 76 - 118 kg. (68 - 74 in. and 167 - 260 lb.), completed a pre-testing medical history and physical evaluation, including cervical spine radiographic studies and measurement of each subject's voluntary maximum neck range of motion (extension and flexion) and normal upright head carriage angle.

Test subject marking for photographic analysis included photographically visible marks placed (a) on each individually fitted biteblock and accelerometer assembly, (b) just behind and above the orbital angle of the outboard eye on the skin of the lateral forehead/face, (c) below and slightly behind the left external auditory

canal over the mastoid prominence as an approximation of the lateral projection of the upper end of the cervical spine and (d) a "flag" type target affixed to the skin over the spinous prominence of the first thoracic vertebra (T-1), meant to be visible from the side. (Figure 2) Same sized target marks were placed laterally over the outboard glenohumeral joint and elbow on a tight fitting garment worn over the torso and arms. A similar target mounted on a light plastic strip was affixed with Velcro to a tightly fitted corset-like garment worn over the hips and lower abdomen to approximate the lateral projection of the outboard hip joint position. Bilateral foam rubber ear plugs were utilized as additional visual targets and to further mask the already low operating noises from the distant high speed cameras that might alert the test subjects to the exact time of the impact. A Hybrid III anthropomorphic test dummy (ATD) was fitted with a biteblock type accelerometer assembly and had similar right side anatomical reference point markings applied, with the exception of the corset-target strip assembly, ear plugs and the T-1 flag.

INSTRUMENTATION - Fourteen of the eighteen total human test exposures were accomplished with our test subjects instrumented with biteblock accelerometers. (Figure 3) The individually fitted biteblocks required only a minimal amount of bite pressure in order to remain tightly in place during impact. A biteblock typically housed an array of three translational accelerometers (Endevco 7290-10/30) mounted to record x-, y-, and z-accelerations. The x (fore-aft) and z (up-down) accelerations were recorded using 30-G accelerometers, while a 10-G accelerometer measured y (left-right) accelerations. In most cases, biteblocks also accommodated an array of up to three angular accelerometers (ATA ARS-01) sensitive to head angular motions in the pitch, roll and yaw directions (i.e., sagittal, coronal and transverse planes, respectively.) The ATD was, for the most part, instrumented the same as our human test subjects. Four of the human test exposure runs were accomplished with the data acquisition biteblock assembly replaced by specially constructed light weight dental appliances (split biteblocks) that carried two "flag" type photographic targets which were fixed to the subject's upper (maxillary) and lower (mandibular) teeth to assess any

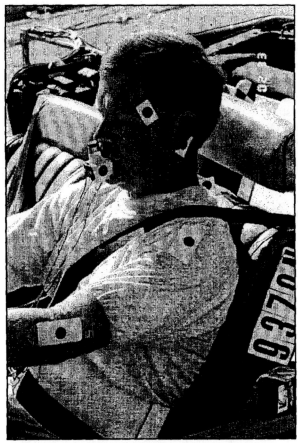

Figure 2 - Test Subject Instrumentation and Marking

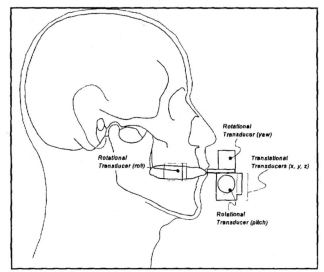

Figure 3 - Transducer Locations

jaw motion.

The electronic instrumentation (ideally 6 transducers per subject) was usually implemented for the drivers and, during the dual position tests, the passenger in the struck car. Additional pairs of accelerometers measuring x- and z-accelerations were mounted onto the central consoles (i.e., near the vehicle's center of gravity) of both the striking and struck vehicles. All transducers were connected to a Bridge Conditioner and Amplifier System (BCAS, Endevco, 68207-6) which was interfaced with a personal computer via an IEEE-488 general purpose bus. All signals were digitized with a data acquisition system board (RC Electronics IS-16E/CR) in the personal computer and were recorded at 2000 samples per second. The data acquisition system was manually triggered a short time before impact. After triggering, data was collected for approximately 1½ seconds and stored in the mass storage media of the computer.

PERIPHERAL DATA ACQUISITION EQUIPMENT - Detailed photographic documentation of test runs was accomplished with four 16-mm high-speed motion picture cameras (two Redlake Fastax and two Photosonics 1B) running at approximately 250 frames per second and equipped with a 100 Hz LED timing light. These long focal length lens cameras were placed during most of the test runs at two fixed positions about 7.6-10.1 m (30-40 ft) from each struck vehicle side and focused to record about the first 183-244 cm. (6-8 ft.) of the occupant's movement. Both the vehicle and the occupants carried target markings which were later digitized frame by frame using a motion analyzer (NAC MOVIAS 160F) to produce position vs. time data. Four standard 8-mm video cameras (Sony EVO-9100) were also used for qualitative documentation of the events. For this test series two free standing photographic grids were constructed consisting of large rectangular frames strung with white cords creating a 15.25 cm (6 inch) square grid system to provide a photographically visible earth fixed reference system. These grids were placed for each test as close as practically possible to each side of the target vehicle.

The closing velocity of the striking vehicle was determined using a speed trap which consisted of a succession of tape switch contacts. In each collision, one wheel of the striking vehicle rolled over the speed trap just prior to making contact with the target vehicle. The electric pulses recorded by the data acquisition system then facilitated the calculation of the closing velocity of the striking vehicle.

In order to coordinate film and transducer data, triggered electronic strobe lights were placed in the field of view of the cameras. These flashed at impact time when the tape switches mounted on the rear bumper of the struck vehicle were compressed. The change in voltage induced by the tape switches was also monitored by a channel on the data acquisition system.

TEST PROCEDURES - Each test run was conducted according to a pre-established test protocol which had scheduled the three day series of human subject test runs planned so each human test subject would have no more than a single daily exposure to a struck vehicle delta-V of 8 kph (5.0 mph) or greater and that each test subject would be exposed to increasing delta-Vs. It had been determined from the first test series, and reconfirmed during the second, that driving the bullet vehicle, even for the higher speed collisions, was not particularly uncomfortable and was unlikely to be injurious. As a result, occupants of the 8 kph and over struck vehicle runs typically might have functioned as a driver of the striking vehicle during other higher speed runs on the same day. For each test run the striking vehicle was backed up the ramp to a marked position and released with the transmission in neutral and the engine running. It then rolled down the ramp, over a short section of roadway and through the impact point speed trap where the velocity was recorded. This procedure was repeated until the desired pre-determined impact speed was reproducibly achieved. When all was ready, the vehicle to be struck was placed into its stationary position at the impact point and the previously prepared test subjects and/or the ATD were situated in the vehicle and their sensors connected to the recording hardware. The striking vehicle then rolled down the ramp to the impact point. As before, no vehicle control inputs, except for minimal bullet vehicle steering to ensure centerline contact between vehicles, were made during all but one test run. For this particular test the target vehicle driver was instructed to keep the brakes firmly applied during the test. After every test collision, each test subject's physical condition and subjective symptom experience was checked, a post test assessment of vehicle damage was completed, and electronic/photographic test result data storage was accomplished. Daily informal test subject checks were conducted for approximately one month after the test series and then periodically thereafter for long term subjective symptom assessment. All test subjects have remained accessible to follow-up. Data from fourteen manned vehicle to vehicle test collisions were recorded.

COMPUTATIONAL APPROACH

The data output of the second test series yielded two types of computationally useful data; acceleration data obtained from transducers and position/angulation data based on digitized high-speed film.

KINEMATIC ANALYSIS - Since a substantial amount of head rotation and translation in the sagittal plane occurred simultaneously in every test, accelerations at various points on the head of an occupant were not expected to equal accelerations at the biteblock. A goal of this work was therefore to use the information obtained from the biteblock transducers and from digitized high speed film to determine accelerations experienced at other locations on the head of a test subject.

Naming B as an arbitrary location on the head of a subject (Figure 4), an expression for its acceleration was written as follows,

$$\ddot{\vec{R}}_B = \ddot{\vec{R}}_{A/B} + \ddot{\vec{R}}_A \qquad [1]$$

in which the bold type face denotes vector quantities.

In terms of components given in the reference frame of the vehicle we then had,

$$(\ddot{\vec{R}}_B)_x = -R_{A/B}\,(\alpha \sin\,\theta + \omega^2 \cos\,\theta) + (\ddot{\vec{R}}_A)_x \qquad [2]$$

$$(\ddot{\vec{R}}_B)_z = R_{A/B}\,(\alpha \cos\,\theta - \omega^2 \sin\,\theta) + (\ddot{\vec{R}}_A)_z \qquad [3]$$

in which ω and α are the angular velocity and acceleration, respectively, of the link AB (i.e., of the occupant.) The parameter α was obtained directly from angular transducers, ω was obtained after numeric integration of α, and θ was directly derived from film. The second terms in both equations [2] and [3] represented measured accelerations and were thus readily available from translational transducers after the application of a coordinate transformation discussed below.

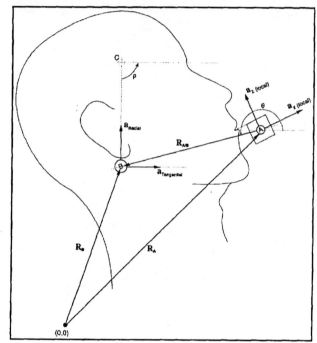

Figure 4 - Calculation Landmarks (nomenclature)

As previously mentioned, measured accelerations (at the biteblock) and computed accelerations (at any other location on the head) encompassed simultaneous translational and rotational effects. Film data became very helpful in the quantification of rotational contributions because it allowed for the computation of the centers of rotation, C, associated with the rotational motion of link AB (i.e., rotational motion of the head). The knowledge of C, whose method of computation is described further below, gave an immediate insight as to the approximate magnitudes of pure rotational effects. Indeed, tangential and radial accelerations for any point were thus available as $r\alpha$ and $r\omega^2$, respectively, where r was the distance from the computed center of rotation to the point of interest.

PREPARATION OF FILM DATA - Film data required considerable preparation before being computationally useful. Film data were digitized using a motion analyzer to yield position data (in terms of digitized units) versus motion picture frame numbers. For every camera view, two points a known distance apart (12" generally) were digitized to provide a scale factor that allowed for the conversion of digitized units into physical displacement units. The 100 Hz LED timing marks appearing alongside the actual footage were digitized as well. Computations based on these digitized timing marks provided more accurate frame rates than the stated nominal 250 frames/second rate characteristic to each of our high speed cameras.

The list of position targets that were digitized for each camera view, at a rate of every 2 to 5 film frames, encompassed up to 17 targets (the number of digitized targets varied somewhat depending on the field of view of each camera.) Targets of most interest for this analysis were the Grid Point (*GD*, a fixed reference point digitized in order to account for film jitter), the Vehicle Point (*VH*), and, on the head of a test subject, the temple (*TM*), the earplug (*EP*), the mastoid (*MS*), and the biteblock (*BB*). Also digitized were an upper and lower marker on the seat backrest.

As a result of the head's kinematic response, some of the head targets disappeared momentarily due to obstructions to vision during the motion of the head. *MS* and *EP* would typically be briefly concealed by the overlying shoulder harness, for example. Unmanageable high contrast sunlight/shadows would occasionally be a

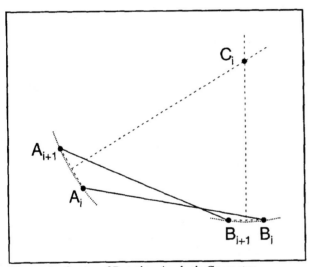

Figure 5 - Center of Rotation Analysis Geometry

problem as well. When such gaps of data were encountered, existing data before and after these gaps were used to create cubic curve fits which, in turn, allowed for the gaps of data to be filled. Once all film data was obtained for every target of interest, the film data was then smoothed with a 2nd order Butterworth filter at 80 Hz break frequency, and splined (with cubics) to transducer data time steps (5 ms.).

PREPARATION OF ACCELEROMETER DATA - Unlike the film data, accelerometer data did not require much post-test computational manipulation aside from filtering and subtracting the gravitational 1 G component. Both rotational and translational accelerometer data were filtered using a low-pass Butterworth filter with cut-off frequency of 80 Hz.

COMPUTATIONS BASED ON FILM DATA - Several parameters were calculated from position data. The angle θ and its numerically differentiated 1st and 2nd derivatives were computed yielding θ_{film}, ω_{film}, and α_{film} values versus time. Secondly, based on the

coordinates of A_i, A_{i+1}, B_i, and B_{+1}, instantaneous centers of rotation C_i were obtained for the motion of link AB (Figure 5). Coordinates of C_i were determined by solving for the intersection of two lines; one line being the mid-length perpendicular of the line defined by [A_i, A_{i+1}], and the other line being the mid-length perpendicular of the line defined by [B_i, B_{i+1}]. Clearly, the locating of C_i was a task that was very sensitive to small angle variations in the two lines expected to intersect. This is the reason why position data were fitted with smooth and continuous cubic functions.

$$a_X = a_x \cos \phi - a_z \sin \phi \qquad [4]$$

$$a_Z = a_x \sin \phi + a_z \cos \phi \qquad [5]$$

COORDINATE FRAMES OF REFERENCE - Transducer accelerations were naturally obtained in the reference frame of the moving head. For consistency, all acceleration data were eventually transformed to an earth orthogonal reference frame using the coordinate transformation calculation given below, where φ is an angle describing the orientation of a body with respect to the horizontal plane and lower case subscripts represent a local reference frame. In this work φ would generally be replaced by θ if dealing with total acceleration or by ρ if dealing with centrifugal or tangential accelerations.

FILTERING - Transducer acceleration data (both angular and translational) were filtered using a 2nd order Butterworth filter with an 80 Hz cut-off frequency. Position data were also smoothed using a 2nd order Butterworth filter. Another type of filter that proved to be useful was the ATM (Alpha Trimmed Mean) filter. It was helpful because derivatives of angular film data frequently produced jagged results which were efficiently smoothed using the ATM filter without affecting existing smooth sections.

CUSTOM SOFTWARE AND SYNCHRONIZATION OF DATA - A C++ program which included an ATM filter and a polynomial curve fitting procedure, was written to help in carrying out the present analysis. After entering data from test run files holding digitized timing marks, filtered position data *vs.* frame numbers, and the filtered accelerometer data *vs.* time, this program provided many of the results that were sought and allowed for the proper synchronization of film and transducer data.

Every run of the program handled acceleration and position data for two targets at a time, providing, as a part of the output, acceleration data associated with those two targets. Most commonly, out of the two targets analyzed, one target would be the one at which the transducers were located (i.e., point A in Figure 4). Based on the knowledge of the accelerations that had occurred at the biteblock target (point A), accelerations at the other target were then automatically computed. An additional option in the program allowed the user to request acceleration data at any point on the sagittal plane of the head other than the two original targets. The location of this new point simply had to be defined with respect to the two original targets in terms of a distance and an angle.

RESULTS AND FINDINGS

Data from each run consisted of medical history and observations obtained before and after each test exposure, the raw electronic biteblock and vehicle accelerometer data and the high-speed film/video record. The use of angular accelerometers on the biteblock arrays increased the information available and considerably facilitated the accuracy of the photographically visible test analysis. The photographically visible test subject marking system and high speed photography was greatly improved from the first test series and the addition of the earth fixed visible grid system permitted better qualitative and quantitative assessment of test subject displacement-time information down to 4 millisecond intervals.

COMPUTATIONAL RESULTS - Listed in Appendix A are sample angular data (Figs. A1 and A2) and acceleration data (Figs. A3, A4 and A5). Transducer and film angular data were compared in order to assess the synchronization of the two sources of data. Proper synchronization was important because both the angle θ, and center of rotation computation were based on film data.

The computed time varied centers of rotation were found to match well with the simultaneous apparent centers of rotation observed on film. Examples of these are shown in the sample vector plots included in Appendix Figs. A4 and A5. This match was valid from impact time to approximately 150 msec after impact, at which time the then substantial translation of the head/neck complex lessened the accuracy of the center of rotation analysis. Up to approximate 150 msec after impact it was possible to evaluate contributions of rotational effects at any point in the sagittal plane of the head, particularly at the Mastoid point which typically experienced the higher accelerations. The tangential acceleration was found to typically reach values exceeding 10 G's during the period prior to 150 msec after impact. This appeared to account for a significant proportion of the acceleration values experienced at the Mastoid point.

TEST RELATED CLINICAL FINDINGS - The human exposure portion of this test series was accomplished during a consecutive three day period. (See Table 1.) Four test subjects were exposed to three test runs each, one test subject had four test exposures and two subjects were exposed only once. The single test exposure for the two test subjects was due to their limited time availability. The same limitation prevented the eighth test candidate's participation in the test series. Four individuals on the approved test panel had participated in the first series of rear end collision tests completed almost 2.5 years before the current test series. These individuals reported having had no testing related interval symptoms. Three of these prior test participants continued to have normal cervical radiographic studies and the fourth, who had minimal cervical degenerative changes on his first cervical radiographs was found to have had no perceptible progression of those findings on his second set of x-rays. All of the new test participants had normal cervical spine x-rays and reported no underlying health problems. One of the new test subjects did report in his pre-test interview being slightly prone to heightened muscle tension associated with vigorous physical activity or stress. Test related symptoms reported by the test subjects are summarized in Table 1. All test subjects reported some test related "awareness" or discomfort symptoms, however slight and/or fleeting. Fleeting headaches reported by Szabo[52], et al, were also noted by most of our test subjects. Those test subjects who were multiply exposed during the three day period noted that their early, just noticeable symptoms, became more pronounced as their test exposures continued. Test subject #2 was the individual who reported increased muscle tension with stress and injury and was the one who participated in the two leftward turned head (30 and 45 degrees) off center line test runs (Table 1). After his third test run and his second turned head test run, he developed somewhat more significant discomfort symptoms than his peers and was not permitted to participate further. His first exposure was with a 30 degree left head turn at a delta-V of 5.8 kph (3.6 mph) and his assessment the day after was that he had not been made uncomfortable at all by this exposure. The next day he underwent a 10.0 kph (6.2 mph) head straight exposure early in the day and later reported the onset of a mild frontal headache. His third exposure and his second test run late on day 2 was with a 45 degree left head turn and a delta-V of 8.2 kph (5.1 mph). This was reported to be a subjectively much more stressfulexposure. He developed an uncomfortable, predominantly right sided, anterior and posterior lower neck muscle strain later that evening and was asked not to participate on day 3 of the test series. Test subject #2 was clearly the most symptomatic of the test panel subjects. His discomfort symptoms steadily cleared over 3-4 days after the test series. All other test subjects reported themselves completely symptom free before or during the third day after the test series was completed. In general, our test subjects reported two types of neck discomfort symptom patterns. Most subjects, 4 of 7, reported either upper or lower anterior neck strap muscle discomfort symptoms. Two subjects reported their discomfort symptoms occurred at the posterior base of their necks. The test subject exposed to the off centerline testing reported both types of discomfort. In the two years that have passed between the tests and the writing of this report none of the subjects have reported any symptoms referable to their test exposures.

CHRONOLOGY OF KINEMATIC EVENTS - Our first paper[31] established a convenient five phase description of the occupant kinematics of the low velocity (range 4-8 kph) rear end collisions that were investigated. These phases were termed:

Table 1. Struck Vehicle Test Subject Clinical Symptoms

Test #	Delta-V kph (mph)	Day #	Test Subject Driver	Symptoms (Driver)	Test Subject Passenger	Symptoms (Passenger)
1 Cnv	10.3 (6.4)	1	#1 *	Supernuchal (occipital) H/A, onset 30 min., lasted 45 min.	—	
2 Sed	5.8 (3.6)	1	#2 (Head left at 30°)	Slight "twinge" upper right trapezius muscle, onset about 45 min., lasted 5 min.	—	
3 Cnv	8.0 (5.0)	1	#3 *	"Sensation" at base of neck post impact, left immediately, minor H/A onset 708 hours, lasted until aspirin after 10 hours.		
4 Cnv	8.0 (5.0)	1	#4 *	Transient H/A after head restraint strike, lasted 10 min.	—	
5 Sed	7.7 (4.8)	1	#5 (no restr)	"Awareness" of posterior neck base, onset few minutes, lasted 12 hours.	#6 (no restr)	Mild neck "awareness", onset few min., lasted few hours.
6 Sed	10.0 (6.2)	2	#2	Mild frontal H/A, onset few minutes, lasted overnight.	ATD	
7 Sed	9.2 (5.7)	2	#7	Mild neck awareness, onset few min., lasted few hours.	#3	Dull H/A from head restraint strike, gone in few seconds, 1-2 hours later, mild frontal H/A, lasted few hours.
8 Sed	10.0 (6.2)	2	#1	Lower anterior strap muscle soreness, onset noted just before test, lasted after next test.	ATD	
9 Sed	8.9 (5.5)	2	#5	Increased low posterior neck (C-7) mild discomfort, onset few min., lasted until next test. Later in evening neck was less stiff and discomfort decreased.	#4	H/A, occipital & Rt. retro-orbital, onset at impact, lasted 4 hours, soreness upper SCM muscles, onset 16 hours, lasted 2-3 days.
10 Sed	8.2 (5.1)	2	#2 (Head left at 45°)	Left frontal H/A & occipital soreness, onset in few minutes, discomfort in anterior strap & lower posterior muscles, onset 15 hours, lasted 3-4 days.	—	
11 Cnv	8.0 (5.0)	2	#3 (Brakes set)	H/A & residual neck extension soreness in lateral posterior muscles, onset 15 hours, lasted one day.	—	
12 Sed	8.7 (5.4)	3	#5	Frontal H/A, onset few min., lasted 5 min. Continued low posterior neck mild discomfort, at (C6-7), pre-existing, lasted about 3 days.	#4	Soreness anterior strap muscles, pre-existing, lasted 1-2 more days.
13 Sed	10.9 (6.8)	3	#3	None reported.	ATD	
14 Sed	10.9 (6.8)	3	#1	Lower anterior strap muscle soreness, lasted approximately 2 days after last test.	ATD	

Legend: Cnv = Convertible Sed = Sedan (no restr) = no restraint system used * = split upper & lower bite blocks

Phase 1 - Initial Response (0 to 100 milliseconds)

Phase 2 - Principal Forward Acceleration (100 to 200 milliseconds)

Phase 3 - Head Overspeed/Torso Recovery (200 to 300 milliseconds)

Phase 4 - Head Deceleration/Torso Rest (300 to 400 milliseconds)

Phase 5 - Restitution Phase (400 to 600 milliseconds)

Our experience during the current test series found the five phase description still valid and convenient. However, one of the principal objectives of the second test series was to extend the experimental range of delta-V above 8 kph (5 mph) to the 11.0 kph (7 mph) level, as well as to investigate in more detail the biomechanically important second and third phases. The improvement in data collected and the higher collision velocities experienced during these tests have further clarified some important areas necessary to understanding a very complex biomechanical event as it evolves over an increasing energy spectrum.

There are several observations that became clear after analyzing the new data. The test subjects' absolute vertical motion with relation to the earth that was reported to have occurred during the first test series was not observed with the higher speeds and better earth fixed reference point visualization of the second test series. At delta-Vs over 8 kph (5 mph), there was very little or no absolute upward body translation with respect to the earth. The relative vertical motion of the test subject's upper trunk and torso ("ramping") with respect to the vehicle seat back, however, was clearly accentuated at the greater speeds of the second test series. This pronounced upward ramping over the seatback of each subject greatly affected their subsequent head and neck kinematics. When the calculated data was reviewed in 2.5 or 4 millisecond intervals there were considerable variations in the resultant acceleration directions of the top of the cervical spine from instant-to-instant between test subjects. However, when all of the test subject's kinematic responses were reviewed as a whole, there was a commonality of response that is summarized as follows. Figures 6 and 7 are provided with trace lines from TM, MS and BB to highlight the observed sequencing of rotary and translational motions for a typical test (Test Run #9). Figure 8 diagramatically represents the principle features of the typical kinematic sequences we observed.

PHASE 1 (0-100 milliseconds) - For all test subjects there was no detectable body motion for at least 50 to 60 milliseconds after the bumper sensors indicated contact between striking and struck vehicles. The struck test vehicle, seat and base of the seat back moved forward about 6.4-7.6 cm. (2.5-3 in.) in this interval and the seat back deflected rearward an average

of 3-5 degrees. Due to the rearward slope of the forward moving seat bottom there was an apparent illusion of hip and thigh elevation with respect to the seat. When hip and pelvis are tracked with respect to the earth fixed grid it was clear that hip and thigh stay stationary while the sloped seat moved out from under, causing the low back area to contact and compress the seatback at a point somewhat higher than if the individual were more slowly slid rearward over the sloping seat bottom.. As this happens, the test subject's pelvis and low back area compressed, and became braced by, the low seat back cushion and started deflecting the entire seat back rearward and relatively away from the still stationary torso. Between 60 and 80 milliseconds the vehicle and seat base moved to about 10.2 cm. (4 in) and the seat back deflected rearward a total of about 6-7 degrees. During this period the initially stationary upper body was pulled forward from below by the pelvis and the lower portion of the trunk intersected the increasing rearward slope of the seat back. The net result of the torso staying vertically stationary as it laid back on the forward moving, rearward deflecting, and therefore vertically lowering seat back was a 5-8 cm. (2-3 in.) ramping up of the shoulders and upper thoracic spine (T-1) over the seat back incline. (Figures 6 & 7.) The mid-back then began intersecting the forward moving seat back surface. As it was pushed forward the thoracic curve became straightened out, further increasing the ramping effect. By 80 milliseconds the test subject's T-1 target flag had moved forward about 2.54 cm. (1 in.).

During this period the head had not moved at all, although the base of the neck had also moved forward along with T-1. The normal neck muscle tone that keeps the head erect was now exerting a forward pull to the top of the neck and thus the neck's attachment to the head. (Figure 9) From about 80 milliseconds onward the forward accelerative forces at the top of the neck built rapidly. By 100 milliseconds T-1 and the base of the neck moved forward another 1-2 cm. and the top of the neck had just begun to move.

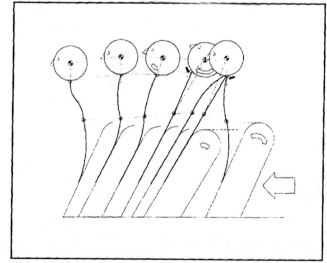

Figure 8 - Kinematic Response Diagram

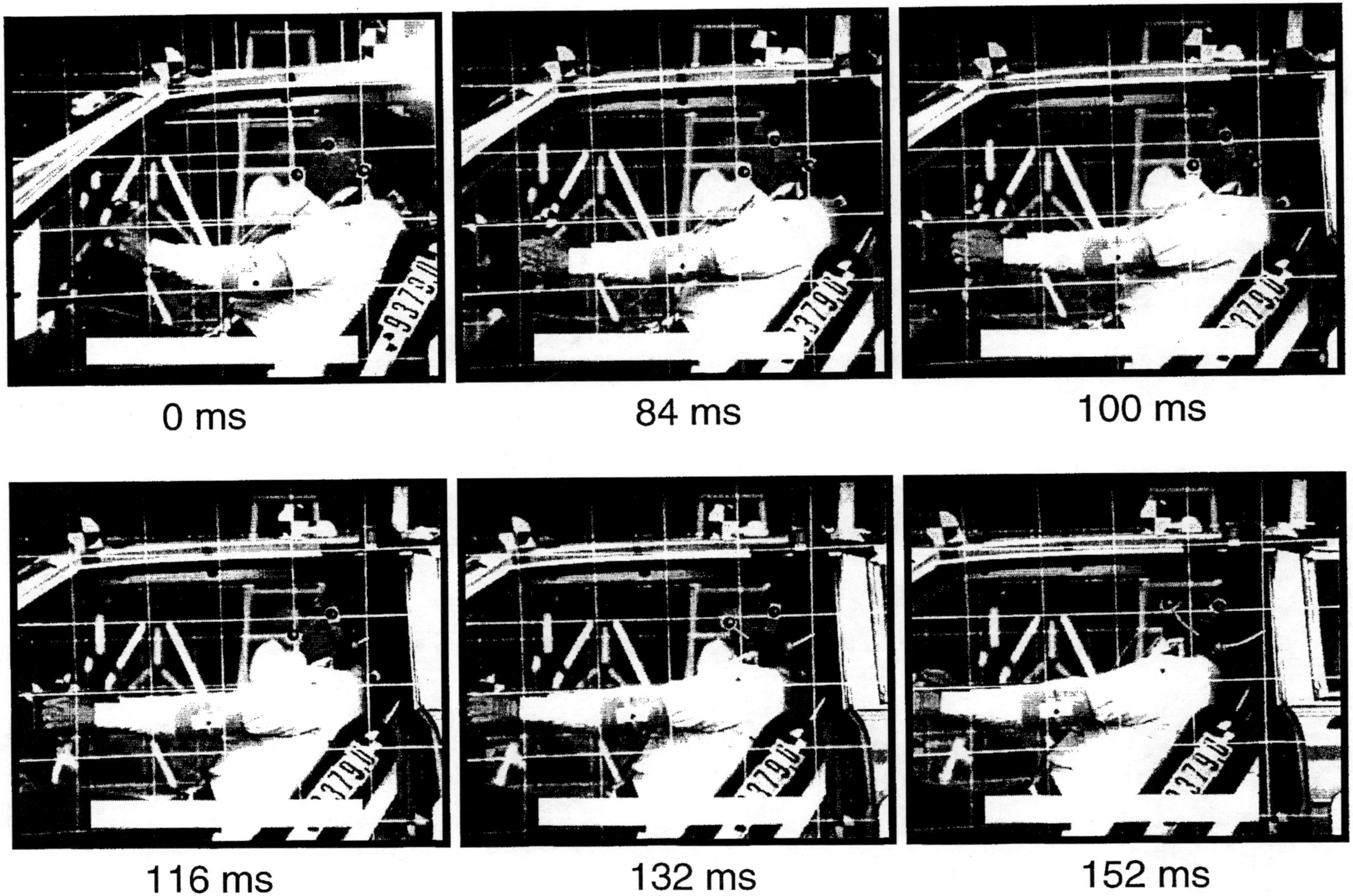

Figure 6: Typical Head/Neck Kinematics

0 ms   84 ms   100 ms

116 ms   132 ms   152 ms

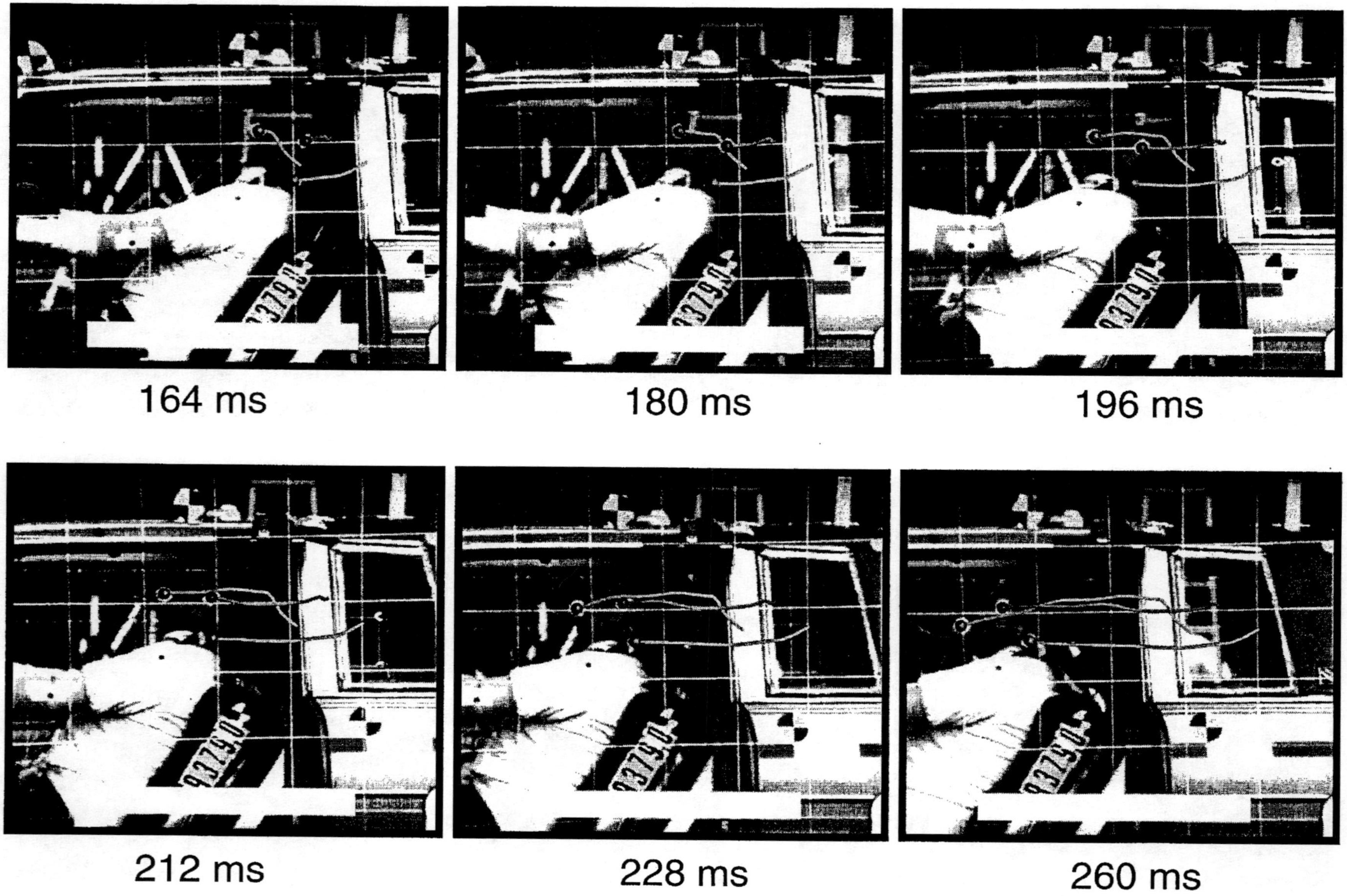

Figure 7: Typical Head/Neck Kinematics

PHASE 2 (100-200 milliseconds) - By 110 - 120 milliseconds the seat back had reached its maximum rearward deflection of 10-14 degrees. (Note. The convertible had a more compliant seat back than the sedan and had a higher maximum deflection by about 2 degrees) and the typical test subject reached a maximum top of the neck (mastoid point) acceleration (5 - 15 G). The vehicle had traveled forward about 15.2 - 17.8 cm. (6 - 7 in.) and T-1 moved forward about 8 - 10.2 cm. (3 - 4 in.) The neck still appeared oriented almost vertically at this point. Instances were clearly observable on the high speed film of several test subject's sternocleidomastoid (SCM) muscles visibly bulging under their skin in response to the abrupt loading of a functioning muscle group. This early head motion appeared to be entirely rotational, about a point within the head which was probably close to the head and neck complex's center of gravity. The most apt analogy to the human head and neck at this point is the response of a heavy ball and chain to a pull on the chain tangential to the chain's attachment to the ball. Between about 110 and 170 milliseconds the head rotated about 10 - 15 degrees and then started to translate forward under the impetus of the neck top's continued forward motion. Around 180 - 200 milliseconds the typical test subject's head reached its maximum rearward rotation and maximum neck extension of 18 - 51 degrees. Taller subjects tended to peak 10 - 30 milliseconds later. All of test subject 3, contacted the fully elevated head restraint on the top surface with their lower occipital scalp in a downward direction. This resulted in the head restraints being driven downward on their adjustable mountings on every run over 8 kph. (Note. Test subject 3, who had the shortest seated height of the test panel appeared to have more quickly contacted the upper portion of the front surface of the head restraint and

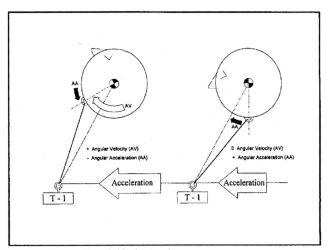

Figure 10 - Ball and Chain Analogy

higher on his occipital scalp than did his peers.) At 180 - 200 milliseconds the seat back deflection and therefore the torso and T-1 angle had decreased to about 5-6 degrees as its work of accelerating the torso was being accomplished. The maximum head angle achieved minus the increase in torso rearward extension gave a good approximation of the actual maximum neck extension during the test exposure. All subjects had less neck extension than their maximum voluntary neck extension as recorded prior to testing by about 10-40 degrees. (Note. Test subject 3 intersected the head restraint prior to achieving the same rotation as his peers and achieved a higher margin of unused extension capability. The rest of the test subjects had a 10-14 degree margin.) For test subjects other than number 3, the head restraint by itself did not appear to significantly arrest the rearward rotation of the test subject's head since these particular head restraints moved relatively easily downward in their adjustment pathways. The ball and chain analogy (Figure 10) offers a likely mechanism for the observed self limitation of the head's rearward rotation and neck extension at this point. As the base of the neck continued to be accelerated forward there was traction of the top of the neck and at the point of the neck's attachment to the head. The further the head rotated the occipital condyles forward and up, the greater the resistance from C-1 which was being pulled down and forward.

PHASE 3 (200-300 milliseconds) - Early during this period the seat back began to return to its pre-impact angle and the seat cushion to its normal configuration. The torso had just about achieved the vehicle's velocity or slightly greater, regained its normal forward curve and moved relatively downward with respect to the seat back. The head, early in this periodhad achieved a somewhat greater velocity than the torso and, with measured decelerations of 1.5-2.5 G, was actively being slowed down by the neck. During this period active control of the head's position appeared to be regained as the head was starting to approach the "over the top" position at about the 280-320 millisecond point.

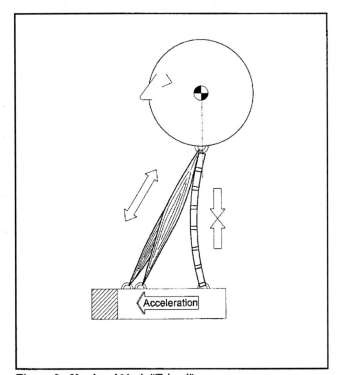

Figure 9 - Head and Neck "Tripod"

PHASE 4 (300-400 milliseconds) - The head continued forward relatively faster than the shoulders but was being decelerated in a level, eyes fixed and focused fashion. Previously described in the first report was a "level head bob" maneuver in which the upper portion of the neck extends and the lower neck flexes while slowing the head. Torso and lower body began to achieve their post impact rest positions during this period.

PHASE 5 (400-600 milliseconds) - By this time the test subject had just about totally achieved the vehicle's impact related velocity change and was returning to his pre-impact position. The slightly higher positioning of the trunk and hips from the pretest position was an unanswered question from the first paper that may be explained by the seat slope derived "landing" of the pelvis and low back area higher up the seat back as described in the above Phase 1 kinematics description.

DISCUSSION

The unique kinematic patterns associated with human test subjects exposed to low velocity rear end collisions shown by this study, while difficult to completely describe in print, are readily apparent when shown in a slow motion format. This is particularly true when the whole photographic record of the test series is viewed back to back as many times as necessary and the associated moment-to-moment G forces of selected anatomical points are available. Key points of this pattern begin with the initial absolutely fixed body position while the vehicle and seat assembly move forward. The seat back impinges forward onto the lower body and responds to the load of accelerating the lower body by deflecting rearward away from the upper torso. When the seat back swings rearward in the vehicle, the seat back top and head restraint are lowered with respect to the still stationary torso, which is met by the forward moving (in earth based coordinates) seat back at a point on the seat back higher than was true in the pre-collision rest position. Added to this ramping motion is the acceleration related uncoiling of the thoracic curve, resulting in the T-1 mounting point for the neck rising quite a bit higher up the plane of the seat back. The head remains stationary during this period, but, due to the falling position of the head restraint, appears to "loop over" the advancing and lowering head restraint. Seated height appears important with taller people contacting the head restraint more on the top surface and shorter people striking more on the forward face of the head restraint. These stature differences result in qualitative differences readily visible to the eye, but difficult to point out and characterize in the calculated data. Figures 11 and 12 graphically show the time history of calculated instantaneous acceleration vectors of a family of points located over the sagital plane of the head. The first major acceleration of the head is angular with the initial 10 - 15 degrees about a nearly stationary point of rotation. Here at about 110 - 120 milliseconds, typically lasting for a period less than 15 - 20 milliseconds, is where the highest mastoid point accelerations (5 - 15 G) were computed. The neck is still almost fully erect at this time and is beginning to show indications of muscular resistance to what is then a transverse acceleration for the still "stiff" head and neck column complex.

Head rotation begins to become apparent after about 100 to 130 milliseconds and then blends with the onset of an increasing forward translation of the entire head. The head's rearward rotation begins to slow as first the ball and chain effect operates and then the overspeed of the head in relation to the top of the torso reverses the head rotation at about 180 - 200 milliseconds. A total rearward head rotation of up to approximately 50 degrees occurs over 60 - 90 milliseconds and then reverses over a longer rebound period. The rearward deflecting seat and the uncurving of the thoracic spine allows considerable rearward angulation of T-1, lessening the extension of the neck required by the head/neck's apparently self-limited rotation/extension. Much of the forward translational acceleration of the head takes place during the subsequent period between 120 to 200 milliseconds, although there are portions of the head's mass below the apparent center of rotation that become accelerated in the forward direction during the rotational period.

Because the head and neck have continued to rotate during this period the translational acceleration vector becomes more in line with the neck's rearward curving longitudinal axis and at much lower (perhaps 2 - 4 G) average force levels than the initially experienced mastoid point accelerations of 5 - 15 G's produced during the primarily rotational phase of head motion.

It must be pointed out that these 5 - 15 G values are calculated for a single reference point undergoing an arcing motion (hence a mostly tangential acceleration) about a center of rotation that is located higher and more forward than the mastoid point and the head's center of gravity. Although calculated G-loadings at the mastoid were noticeably higher during the primarily rotational motion of the head, it is important to note that the early rotational kinetic energies ($\frac{1}{2}I\omega^2$) imparted to the head invariably reached levels about five to six times lower than translational kinetic energies imparted to the head ($\frac{1}{2}mv^2$). Therefore, from the standpoint of actual neck loading (and injury potential), the force needed to produce the observed mastoid tangential acceleration during the early period required less neck loading than would occur if the same magnitude of translational acceleration were produced at the head's center of gravity.

Our preliminary assessment of the translational G forces experienced at the head's center of gravity show a range of about 3 - 6 peak forward G's. Subsequent to these events the rebounding seat back and recoiling torso result in a rapid retreat of the shoulders and T-1 down parallel with the seat back surface, resulting in a vigorous motion of the torso downward and forward with respect to the vehicle.

It was noted earlier that several special test variations were instituted to answer some of the

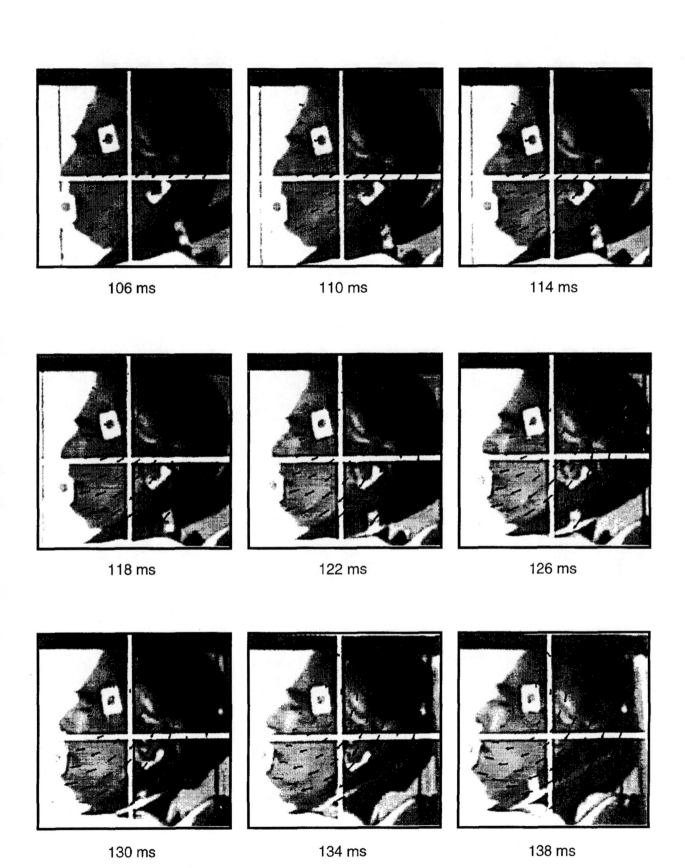

106 ms   110 ms   114 ms

118 ms   122 ms   126 ms

130 ms   134 ms   138 ms

Figure 11: Test 7 Acceleration Vector Fields
(See appendix figure A.5 for vector scale)

175

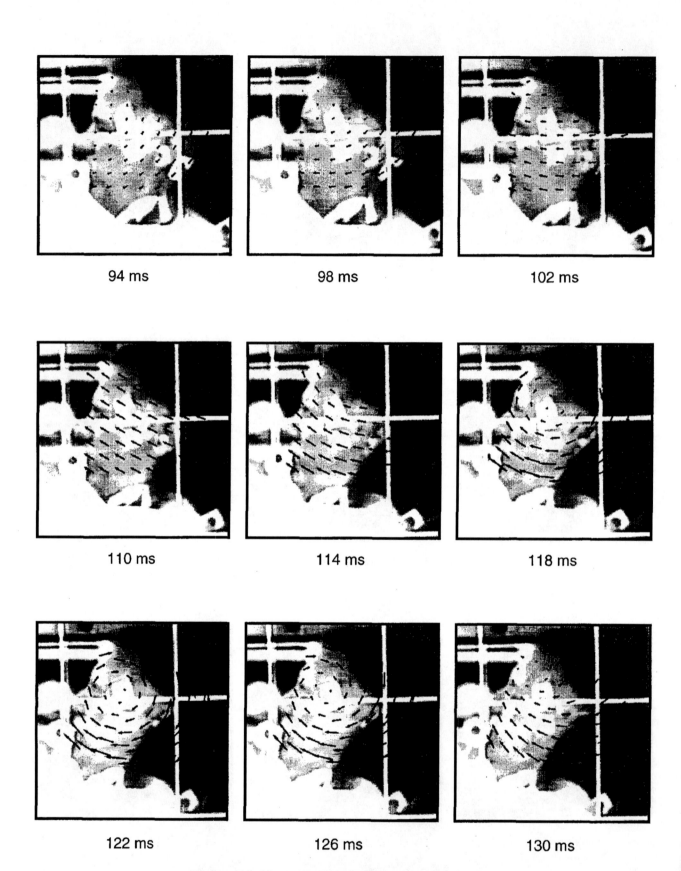

94 ms

98 ms

102 ms

110 ms

114 ms

118 ms

122 ms

126 ms

130 ms

Figure 12: Test 12 Acceleration Vector Fields
(See appendix figure A.4 for vector scale)

questions raised by our earlier test series. Two test subjects on one test run did not use their restraint system. There were no apparent significant qualitative or quantitative differences in their kinematic responses when compared to other test runs. It appears that restraint system use may not play a significant role in this type of low velocity event. The results of the off center line tests have been detailed above. It appears that muscular strain injury potential may be moderately increased for individuals in some low velocity rear end collisions if their heads are significantly turned. The split biteblock tests for tracking jaw motion are the subject of another investigation to be reported later.

Based on these observations, noting that the greatest accelerations occurred while the neck was upright, that muscular resistance was demonstrably occurring during this time period and the fact that our repetitively exposed test subject's symptoms support the assumption of either an acute, mild strain injury to anterior and posterior neck musculature or perhaps mild compressive irritation to vertebral structures in the base of the neck, it seems reasonable to believe our test related neck injuries must have occurred very early in our test subject's kinematic response to the experimental collisions. If this is so, a proposed mechanism for injury is illustrated by Figure 9. At rest in driving position, the head might be considered to be a ball-like mass which must be actively controlled by neck muscles with some degree of constant contraction to keep the head stable against gravity and whatever miscellaneous movements that might be occurring. The anterior neck strap muscles and the two SCM muscles can be thought of as forming the front two legs of a three legged structure, with the cervical spine forming the rear leg. At 90 - 110 milliseconds the base of this system has moved forward while the mass of the head lags behind. To accelerate the head the front two legs of the tripod (composed of partially contracted "toned" muscle) must be in tension and the rear leg (the cervical vertebrae) must be in some compression. The analog of this situation is the individual who, while carrying a moderate weight in his arms, unexpectedly has a very heavy weight added suddenly to his load. Acute biceps muscle strain is the likely result. It is also conceivable that mild acute compressive irritation of some of the more sensitive structures could occur in the posterior parts of the compressed and partially extended spinal leg of the tripod, most likely in the posterolateral facet joint columns and/or spinous processes.

CONCLUSIONS

The better detailed and more extensive data from this second test series permits refinement of some of the conclusions about occupant kinematics presented previously for a broader, more energetic range of rear-end collisions and a greater number and anthropomorphic variety of test subjects. The absolute Gz accelerations with reference to the earth's orthogonal frame of reference were not as significant as expected from our previous work, however, the Gz forces experienced by our test subjects with reference to their ramping interaction with the vehicle seats and seat backs were perhaps more significant than previously anticipated. The relative lack of human test subject low back differential motion, as it becomes quickly braced by the advancing seat back, makes any injury to this area quite unlikely as a result of a low velocity rear end collision. The early first motion of the human head is primarily rotational, blending into rotational and increasingly translational motion with a mechanism that seems to self-limit and then correct the rearward head rotation and neck extension. Not only was there no cervical hyperextension occurring in any test subject, there was a substantial margin of physiologic neck extension left at the point of maximum rotation achieved during each test, well before any of the test subjects would have reached their normal voluntary maximum extension limits. Since all of our test subjects, particularly the multiply exposed ones, developed some form of typical "whiplash" symptoms, it seems reasonable to finally conclude that hyperextension was not the cause of their symptoms. The observed early peak accelerations at 110 - 120 milliseconds, while the neck is still mostly erect and resisting principally transverse forces suggests that this is most likely when muscle strain and possibly compressive irritation may occur. After this period, forces on the neck begin to return to those experienced during day to day activity levels. If that is true, then standard design automobile head restraints, while still very valuable, are in the wrong position to mitigate low velocity neck injuries of this type. These acute neck strain injuries would have already occurred before a passive "head catcher" could intervene. Flexible seat backs appear to be a more effective mechanism to decrease the magnitude of kinematic neck extension during these low velocity events, but still can't be fully protective for this type of injury and also must remain stiff enough to handle the more serious high speed rear end collision. After our present experience with a higher energy test series, 8 kph (5 mph) still seems to be a convenient ΔV threshold for assessing injury potential. For events progressively below this level, the acute muscle strain injury likelihood decreases, probably quite rapidly, even for "thin necked" people. For single events above this level the likelihood of transient acute neck and shoulder muscle strain injury and possible mild compressive irritation of the posterior neck may increase. Several of our test subjects have accumulated 6 - 8 significant (above 8 kph or 5 mph) rear-end collision exposures, without the subsequent development of any persistent cervical symptoms of any kind. As this is written it has been more than four years since the first test series and over two years since the second test series. After analyzing both of our test series there were no observed biomechanical events that could have resulted in permanent cervical injury, and there have been no subsequent indications of any persistent "soft tissue injury" symptoms by any of our test subjects.

BIBLIOGRAPHY

1. Algers, G.; Petersson, K.; Hildingsson, C.; Toolanen, G. Surgery for chronic symptoms after whiplash injury. Follow-up of 20 cases. Acta Orthop Scand 1993 Dec; 6496):654-656

2. Allen, M.E.; Weir-Jones, I.; Motiuk, D.R.; Flewin, K.R.; Goring, R.D.; Kobetitch, R.; Broadhurst, A. Acceleration Perturbations of daily living. A comparison to "whiplash". Spine 1994;19(11):1285-1290

3. The anatomy of the cervical region with particular reference to the symptomology of the "whiplash" lesion. J Bone Joint Surg [Am] 1956 Jan; 38(1):438

4. Awerbuch, M.S. Whiplash in Australia: illness or injury? Med J Aust 1992 Aug 3;156(3):193-196

5. Bailey, M. N.; Wong, B. C.; Lawrence, J. M. Data and methods for estimating the severity of minor impacts. IN: Accident Reconstruction: Technology and Animation V (SP-1083). Warrendale, PA, Society of Automotive Engineers, 1995, pp. 139-175 (SAE Paper #950352)

6. Bring, G.; Wetman, G. Chronic posttraumatic syndrome after whiplash injury. Scand J Prim Health Care 1991;9:135-141

7. Broman, H.; Pope, M.H.; Benda, M.; Svensson, M.; Ottosson, C.; Hansson, T. The impact response of the seated subject. J Orthop Res 1991 Jan;9(1):150-154

8. Burgess, J. Symptom characteristics in TMD patients reporting blunt trauma and/or whiplash injury. J Craniomandib Disord 1991 Fall;5(4):251-257

9. Burgess, J.A.; Dworkin, S.F. Litigation and post-traumatic TMD: how patients report treatment outcome. J Am Dent Assoc 1993 Jun;124(6):105-110

10. Carlsson, G.; Nilsson, S.; Nilsson-Ehle, A.; Norin, H.; Ysander, L.; Ortengren, R. Neck injuries in rear end collisions: biomechanical considerations to improve head restraints. Proceedings of the 1985 International IRCOBI/AAAM Conference on the Biomechanics of Impacts, Gothenborg, Sweden, June 24-26, 1988. Stockholm, Sweden, FOLKSAM Group, 1985, pp. 277-289

11. Cesari, D.; Neilson, I.D. Further consideration of the European side impact test procedure. Proceedings of the 11th International Technical Conference on Experimental Safety Vehicles, Washington, D.C., May 12-15, 1987. Washington, D.C., U.S. Department of Transportation, National Highway Traffic Safety Administration, 1987, pp. 133-136

12. Clemens, H.J.; Burow, K. Experimental investigation on injury mechanisms of cervical spine at frontal and rear-front vehicle impacts. Proceedings of the 16th Stapp Car Crash Conference, Detroit, MI, November 8-10, 1972 (Society of Automotive Engineers Proceedings No. P-45). New York, NY, Society of Automotive Engineers, 1972, pp. 76-104 (SAE Paper #720960)

13. de Jager, M.; Sauren, A.; Thunnissen, J.; Wismans, J. A three-dimensional head-neck model: validation for frontal and lateral impacts. Proceedings of the 38th Stapp Car Crash Conference, Fort Lauderdale, FL, October 31-November 2, 1994 (P-279). Warrendale, PA, Society of Automotive Engineers, 1994, pp. 93-110 (SAE Paper #942211)

14. Evans, R.W. Some observations on whiplash injuries. Neurol Clin 1992 Nov;10(4):975-997

15. Friedman, R.; Harris, J.P.; Sitzer, M.; Schaff, H.B.; Marshall, L.; Shackford, S. Injuries related to all-terrain vehicular accidents: a closer look at head and neck trauma. Laryngoscope 1988 Nov; 98(11):1251-1254

16. Galasko, C.S.B.; Murray, P.M.; Pitcher, M.; Chambers, H.; Mansfield, S.; Madden, M.; Jordon, C.; Kinsella, A.; Hodson, M. Neck sprains after road traffic accidents: a modern epidemic. Injury 1993 Mar;24(3):155-157

17. Gay, J.R.; Abbott, K.H. Common whiplash injuries of the neck. JAMA 1953 Aug 29;152(18):1698-1704

18. Geigl, B. C.; Steffan, H.; Leinzinger, P.; Roll; Muhlbauer, M.; Bauer, G. The movement of head and cervical spine during rearend impact. Proceedings of the 1994 International Conference on the Biomechanics of Impacts, Lyon, France, September 21-23, 1994. Bron, France, IRCOBI Secretariat, 1994, pp. 127-138

19. Goldman, S.; Ahlskog, J.E. Posttraumatic cervical dystonia. Mayo CLin Proc 1993 May;68(5):443-448

20. Hamer, A.J.; Gargan, M.F.; Bannister, G.C.; Nelson, R.J. Whiplash injury and surgically treated cervical disc disease. Injury 1993 Sep;24(8):549-550

21. Hendler, E.; O'Rourke, J.; Schulman, M.; Katzeff, M.; Domzalski, L.; Rodgers, S. Effect of head and body position and muscular tensing on response to impact. Proceedings of the 18th Stapp Car Crash Conference, Ann Arbor, MI, December 4-5, 1974 (Society of Automotive Engineers Proceedings No. P-56). Warrendale, PA, Society of Automotive Engineers, 1974, pp. 303-337 (SAE Paper #741184)

22. Howard, R.P.; Benedict, J.V.; Raddin, J.H., Jr.; Smith, H.L. Assessing neck extension-flexion as a basis for temporomandibular joint dysfunction. J Oral Maxillofac Surg 1991 Nov;49(11):1210-1213

23. Jakobsson, L.; Norin, H.; Jernstrom, C.; Svensson, S. E.; Johnsen, P.; Isaksson-Hellman, I.; Svensson, M. Y. Analysis of different head and neck responses in rear-end car collisions using a new humanlike mathematical model. Proceedings of the 1994 International Conference on the Biomechanics of Impacts, Lyon, France, September 21-23, 1994. Bron, France, IRCOBI Secretariat 1994 pp. 109-125

24. Kallieris, D.; Mattern, R.; Miltner, E.; Schmidt, G.; Stein, K. Considerations for a neck injury criterion. Proceedings of the 35th Stapp Car Crash Conference, San Diego, CA, November 18-20, 1991 (Society of Automotive Engineers Proceedings No. P-251). Warrendale, PA, Society of Automotive Engineers, 1991, pp. 401-417 (SAE Paper #912916)

25. King, D.J.; Siegmund, G.P.; Bailey, M.N. Automobile bumper behavior in low speed impacts. Presented at the International Congress and Exposition, Detroit, MI, March 1-5, 1993. Warrendale, PA, Society of Automotive Engineers, 1993, pp. 1-18. SAE Paper #930211

26. Lee, J.; Giles, K.; Drummond, P.D. Psychological disturbances and an exaggerated response to pain in patients with whiplash injury. J Psychosom Res 1993;37(2):105-110

27. Livingston, M. Whiplash injury: misconceptions and remedies. Aust Fam Physician 1992 Nov;21(11):1642-1643,1646-1647

28. Luo, Z.P.; Goldsmith, W. Reaction of a human head/neck/torso system to shock. J Biomech 1991;24(7):499-510

29. Marshall, S.; Langley, J.; Phillips, D. Severity and medical costs associated with the road casualty problem in New Zealand. Proceedings of the 36th Annual Conference of the Association for the Advancement of Automotive Medicine, Portland, OR, October 5-7, 1992. Torrance, CA, Nissan North America, 1992, pp. 397-412

30. Matsushita, T.; Sato, T. B.; Hirabayashi, K.; Fujimura, S.; Asazuma, T.; Takatori, T. X-ray study of the human neck motion due to head inertia loading. Proceedings of the 38th Stapp Car Crash Conference, Fort Lauderdale, FL, October 31-November 2, 1994 (P-279). Warrendale, PA, Society of Automotive Engineers, 1994, pp. 55-64 (SAE Paper #942208)

31. McConnell, W.E.; Howard, R.P.; Guzman, H.M.; Bomar, J.B.; Raddin, J.H., Jr.; Benedict, J.V.; Smith, H.L.; and Hatsell, C.P. Analysis of human test subject kinematic responses to low velocity rear end impacts. Presentation to Society of Automotive Engineers, Inc., 1993 SAE International Congress & Exposition, Detroit, MI, March 1993.

32. McKenzie, J.A.; Williams, J.F. The dynamic behaviour of the head and cervical spine during "whiplash". J Biomech 1971 Dec;4(6):577-590

33. Mertz, H.J. Neck injury. Presented at "Biomechanics and Its Application to Automotive Design", a continuing education course developed by the Society of Automotive Engineering's Automotive Body Act Passenger Protection Committee. New York, NY, Society of Automotive Engineers, 1973, pp. 1-29

34. Nilson, G.; Svensson, M. Y.; Lovsund, P.; Haland, Y.; Wiklund, K. Rear-end collisions - the effect of the seat-belt and the crash pulse on occupant motion. Proceedings of the 14th International Technical Conference on Enhanced Safety of Vehicles, Munich, Germany, May 23-26, 1994. Washington, D.C., U.S. Department of Transportation, National Highway Traffic Safety Administration, 1994, pp. 1630-1638

35. Ono, K.; Kanno, M. Influences of the physical parameters on the risk to neck injuries in low impact speed rear end collisions. Proceedings of the 1993 International Conference on the Biomechanics of Impacts, Eindhoven, Netherlands, September 8-10, 1993. Cedex, France, International Research Council on Biokinetics of Impacts (IRCOBI), 1993, pp. 201-212

36. Orner, P.A. A physician-engineer's view of low velocity rearend collisions. IN: Automobile Safety: Present and Future Technology (Society of Automotive Engineers SP-925). Warrendale, PA, Society of Automotive Engineers, 1992, pp. 11-18 (SAE Paper #921574)

37. Parmar, H.V.; Raymakers, R. Neck injuries from rear impact road traffic accidents: prognosis in persons seeking compensation. Injury 1993 Feb;24(2):75-78

38. Pennie, B.; Agambar, L. Patterns of injury and recovery in whiplash. Injury 1991;22(1):57-59

39. Robinson, D.D.; Cassar-Pullicino, V.N. Acute neck sprain after road traffic accident: a long-term clinical and radiological review. Injury 1993 Feb;24(2):79-82

40. Ryan, G.A.; Taylor, G.W.; Moore, V.M.; Dolinis, J. Neck strain in car occupants: the influence of crash-related factors on initial severity. Med J Aust 1993 Nov 15;159:651-656

41. Scott, M.W.; McConnell, W.E.; Guzman, H.M.; Howard, R.P.; Bomar, J.B.; Smith, H.L.; Benedict, J.V.; Raddin, J.H., Jr.; and Hatsell, C.P. Comparison of human and ATD head kinematics during low-speed rearend impacts. Presentation to Society of Automotive Engineers, Inc., 1993 SAE International Congress Exposition, Detroit, MI, March 1993.

42. Sehmer, J.M. Rear-end accident victims. Importance of understanding the accident. Can Fam Physician 1993 Apr; 39:828-831

43. Severy, D.M.; Mathewson, J.H.; Bechtol, C.O. Controlled automobile rear-end collisions: an investigation of related engineering and medical phenomena. Can Serv Med J 1955;11:757-759

44. Shea, M.; Wittenberg, R.H.; Edwards, W.T.; White AA 3d; Hayes, W.C. In vitro hyperextension injuries in the human cadaveric cervical spine. J Orthop Res 1992 Nov;10(6):911-916

45. Siegmund, G. P.; Bailey, M. N.; King, D. J. Characteristics of specific automobile bumpers in low-velocity impacts. IN: Accident Reconstruction: Technology and Animation IV (SP-1030). Warrendale, PA, Society of Automotive Engineers, 1994, pp. 333-372 (SAE Paper #940916)

46. Spitzer, W. O.; Skovron, M. L.; Salmi, L. R.; Cassidy, J. D.; Duranceau, J.; Suissa, S.; Zeiss, E. Scientific Monograph of the Quebec Task Force on Whiplash-Associated Disorders: Redefining "Whiplash" and Its Management. Spine 1995 Apr 15; 20 Suppl(8):1-73

47. Svensson, M.Y.; Lovsund, P. A dummy for rear end collisions. Proceedings of the 1992 International Conference on the Biomechanics of Impacts, Verona, Italy, September 9-11, 1992. Cedex, France, International Research Council on Biokinetics of Impacts (IRCOBI), 1992, pp. 299-310

48. Svensson, M.Y.; Lovsund, P.; Haland, Y.; Larsson, S. The influence of seatback and head restraint properties on the head-neck motion during rear impact. Proceedings of the 1993 International Conference on the Biomechanics of Impacts, Eindhoven, Netherlands, September 8-10, 1993. Cedex, France, International Research Council on Biokinetics of Impacts (IRCOBI), 1993, pp. 395-406

49. Svensson, M.Y.; Lovsund, P.; Haland, Y.; Larsson, S. Rear-end collisions: a study of the influence of backrest properties on head-neck motion using a new dummy neck. IN: Seat System Comfort and Safety (SP-963). Warrendale, PA, Society of Automotive Engineers, 1993, pp. 129-142 (SAE Paper #930343)

50. Svensson, M. Y.; Aldman, B.; Hansson, H. A.; Lovsund, P.; Seeman, T.; Suneson, A.; Ortengren, T. Pressure effects in the spinal canal during whiplash motion: a possible cause of injury to the cervical spine ganglia. Proceedings of the 1993 International Conference on the Biomechanics of Impacts, Eindhoven, Netherlands, September 8-10, 1993. Cedex, France, International Research Council on Biokinetics of Impacts (IRCOBI), pp. 1993, pp. 189-200.

51. Szabo, T.J.; Welcher, J. Dynamics of low speed crash tests with energy absorbing bumpers. IN: Automobile Safety: Present and Future Technology (Society of Automotive Engineers SP-925). Warrendale, PA, Society of Automotive Engineers, 1992, pp. 1-9 (SAE Paper #921573)

52. Szabo, T. J.; Welcher, J. B.; Anderson, R. D.; Rice, M. M.; Ward, J. A.; Paulo, L. R.; Carpenter, N. J. Human occupant kinematic response to low speed rear-end impacts. IN: Occupant Containment and Methods of Assessing Occupant Protection in the Crash Environment (SP-1045). Warrendale, PA, Society of Automotive Engineers, 1994, pp. 23-36 (SAE Paper #940532)

53. Taylor, J.R.; Finch, P. Acute injury of the neck: anatomical and pathological basis of pain. Ann Acad Med Singapore 1993 Mar; 22(2):187-192

54. Taylor, J.R.; Twomey, L.T. Acute injuries to cervical joints: an autopsy study of neck sprain. Spine 1993;18(9):1115-1122

55. West, D.H.; Gough, J.P.; Harper, G.T.K. Low Speed Rear-End Collision Testing Using Human Subjects. Accident Reconstruction Journal 1993 May/June; pp. 22-26

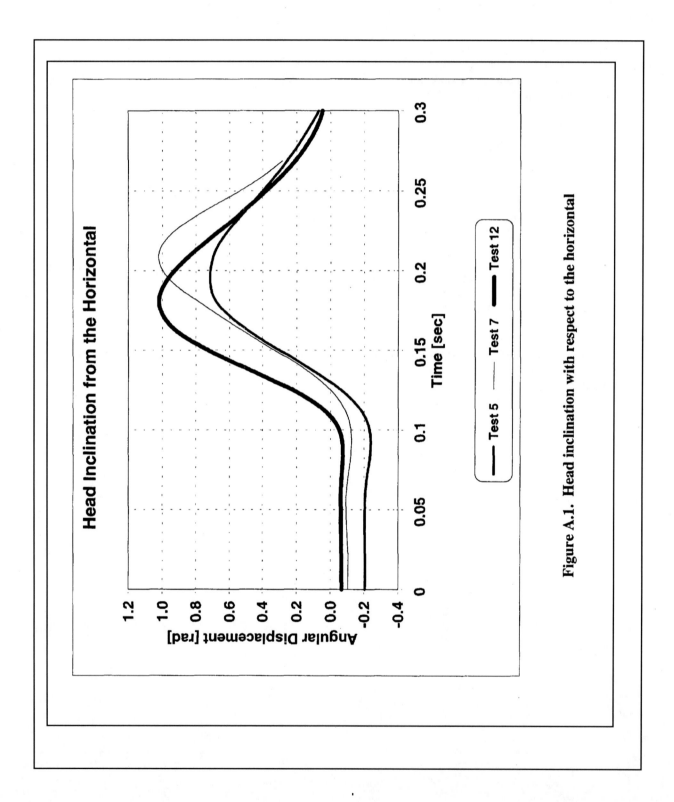

Figure A.1. Head inclination with respect to the horizontal

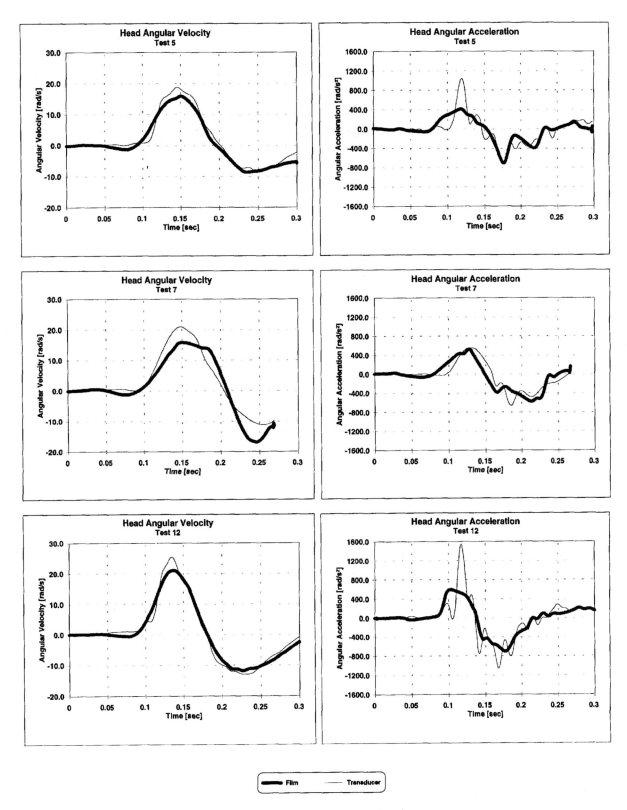

Figure A.2. Head angular velocities and accelerations as
measured with transducers and from digitized film.

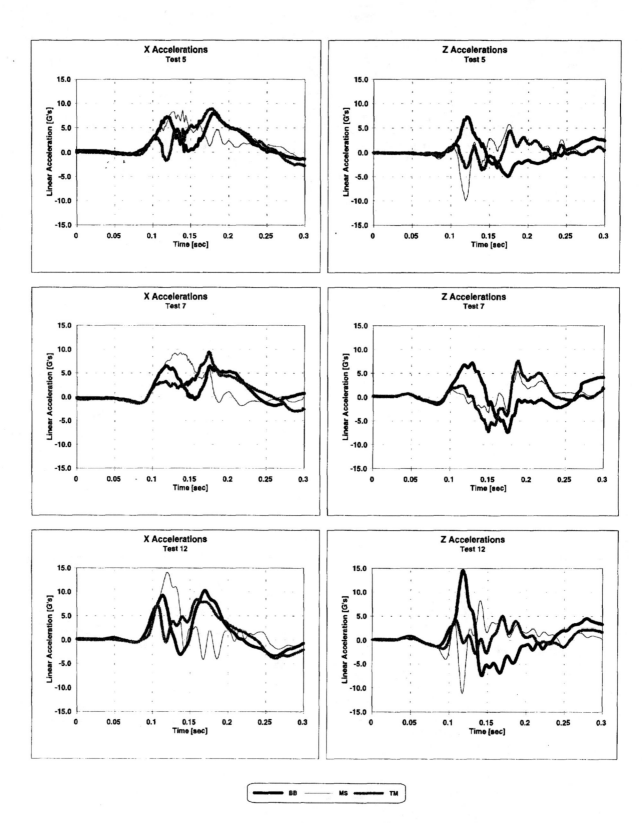

Figure A.3. Acceleration at the digitized locations in the head sagittal plane.

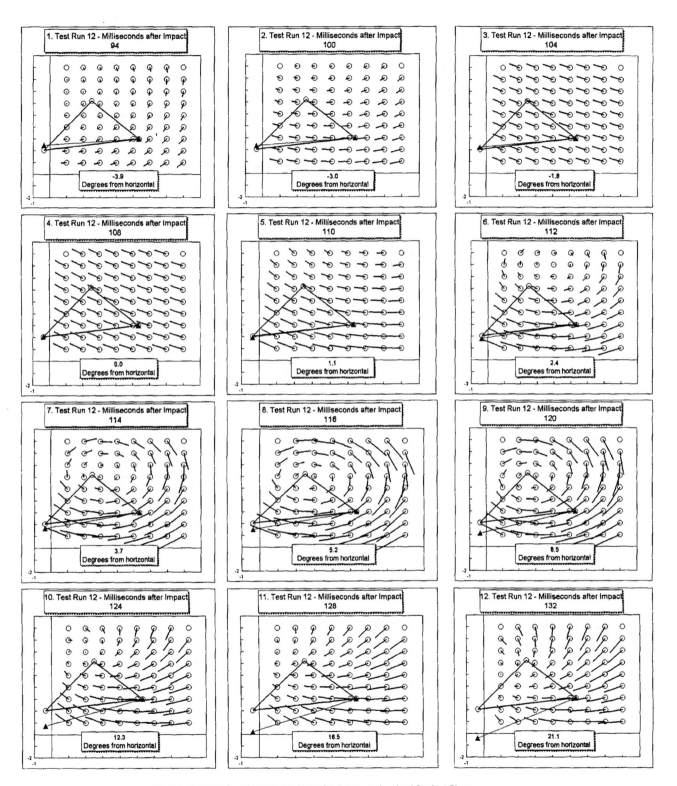

Fig. A4. Acceleration Vector Fields Using 70 Points on the Head Sagittal Plane
Test 12 - Solid Triangles connected by light line is the horizontal plane.
- Dark Lines Connect BB, TM and MS points [open circles].
- Circles around points of origin represent about 2 "G" vector length.

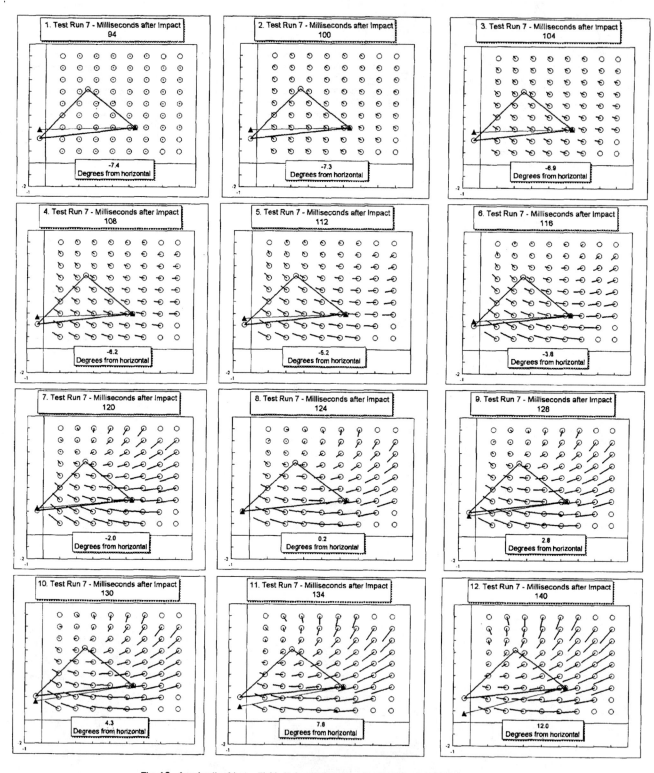

Fig. A5. Acceleration Vector Fields Using 66 Points on the Head Sagittal Plane
Test 7 - Solid Triangles connected by light line is the horizontal plane.
- Dark Lines Connect BB, TM and MS points [open circles].
- Circles around points of origin represent about 2 "G" vector length.

710855

Strength and Response
of the Human Neck*

H. J. Mertz
Research Laboratories, General Motors Corp.

L. M. Patrick
Wayne State University

SEVERAL RESEARCHERS, Mertz and Patrick (1)**, Mertz (2), Ewing, et al. (3, 4), and Tarriere and Sapin (5) have published papers dealing with the response and tolerance of the human neck in an impact environment. Ewing, et al. (3, 4) have presented two papers based on hyperflexion data obtained from human volunteer tests conducted at Wayne State University. In these tests, the subjects, restrained by a full harness, were exposed to increasing levels of sled decelerations which produced increasing severities of hyperflexion. Resultant accelerations measured by accelerometers secured to the occiput, mouth, and first thoracic vertebra, were given as functions of time as well as the angular velocity of the head. No attempt was made in these papers to analyze the acceleration data and determine the forces which acted on the head and produced these accelerations.

---

*Work sponsored in part by U.S. Army Natick Laboratories, Natick, Mass.

**Numbers in parentheses designate References at end of paper.

Tarriere and Sapin (5) presented hyperextension response data obtained on four human volunteers. These data consisted of the angular displacements of the head relative to the torso which were measured from high-speed movies of the experiments. These data were double-differentiated to obtain head acceleration values. No attempt was made to analyze the data for neck forces.

A detailed method for analyzing the forces and moments generated by the neck on the head during hyperextension was given by Mertz and Patrick (1). They used an instrumented human volunteer and cadavers as test subjects and showed that the magnitude of the torque developed by the neck on the head at the occipital condyles was an excellent indicator of neck trauma for hyperextension while the resultant shear and axial forces acting at the occipital condyles did not correlate with the degree of trauma. For a 50th percentile adult male, they recommended a noninjurious tolerance limit of 35 ft lb for the torque developed at the occipital condyles during hyperextension.

The objectives of this paper are to:

1. Describe the structure of the neck.

ABSTRACT————————————

Human volunteers were subjected to static and dynamic environments which produced noninjurious neck responses for neck extension and flexion. Cadavers were used to extend this data into the injury region. Analysis of the data from volunteer and cadaver experiments indicates that equivalent moment at the occipital condyles is the critical injury parameter in extension and in flexion. Static voluntary levels of 17.5 ft lb in extension and 26 ft lb in flexion were attained. A maximum dynamic value of 35 ft lb in extension was reached without injury. In hyperflexion, the chin-chest reaction changes the loading condition at the occipital condyles which resulted in a maximum equivalent moment of 65 ft lb without injury. Noninjurious neck shear and axial forces of 190 lb and 250 lb are recommended based on the static strength data obtained on the volunteers. Neck response envelopes for performance of mechanical necks are given for the extension and flexion modes of the neck.

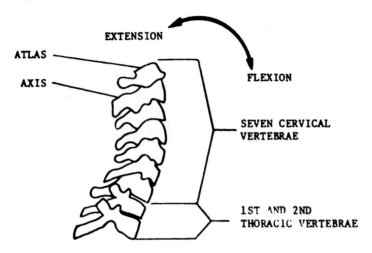

EXTENSION

ATLAS

AXIS

FLEXION

SEVEN CERVICAL
VERTEBRAE

1ST AND 2ND
THORACIC VERTEBRAE

Fig. 1 - Bony structure of the neck showing the seven cervical vertebrae

2. Discuss data pertaining to the range of motion of the head relative to the torso in the sagittal plane.

3. Present data relative to the static strength of the neck in flexion and extension.

4. Present dynamic response and strength data for the human neck in flexion and extension.

5. Recommend noninjurious tolerance values for hyperextension and flexion of the neck.

STRUCTURE OF THE NECK

The bony structure of the neck consists of seven cervical vertebrae as shown in Fig. 1. No two cervical vertebrae of a given neck are identical, and very marked differences occur in the first and second vertebrae. The first cervical vertebra, the atlas, provides direct support for the head with the superior articular surface of the atlas bearing against the occipital condyles of the skull forming a synovial joint. The atlas does not have a body or a spinous process which are characteristic of the other cervical vertebrae.

The second cervical vertebra, the axis, forms a pivot around which the atlas, carrying the head, rotates from left to right. This pivot, the odontoid process, articulates through a synovial joint with the posterior portion of the anterior arch of the atlas and is prevented from moving posteriorly and compressing the spinal cord by the transverse ligament of the atlas.

The third through the seventh cervical vertebrae are similar in shape and function, except that the spinous process of the individual vertebra increases in length as one proceeds down the cervical spine. Each vertebra articulates with the adjacent ones through synovial joints. These joints are held in place by highly inextensible, fibrous ligaments. Separating the adjacent bodies of the vertebrae, but not including the union of the atlas and axis, are fibrocartilaginous intervertebral discs. The integrity of the cervical spine is maintained by numerous ligamentous connections. The anterior and posterior longitudinal ligaments are continuous over the length of the vertebral column. The anterior longitudinal ligament is attached to the base of the occipital bone and descends in front of the vertebral bodies to the sacrum. In between, it is firmly attached to the intervertebral discs and margins of the vertebrae. The posterior longitudinal ligament is attached superiorly to the membrana tectoria which is affixed to the occipital bone, extends posteriorly to the vertebral bodies to the sacrum, and is attached to the vertebral bodies and discs between.

Articulation of the neck is accomplished through muscle pairs which are attached to the skull, the individual vertebra, and the torso. These muscle pairs, which are symmetric with respect to the midsagittal plane, respond in various group actions to produce the desired movement of the head and neck. The muscle pairs which provide resistance to extension of the neck and rearward rotation of the head are the longus capitis, longus colli, rectus capitis anterior, scalenus anterior, the hyoids, sternothyroid, and sternocleidomastoid. Because of their attachment to the mastoid process which is slightly below and behind the occipital condyles, the sternocleidomastoids increase the extension of the neck when the head is rotated rearward and the neck is already extended. However, if the neck is flexed and head rotated forward, the action of this muscle pair flexes the head and resists extension.

The muscles which prevent flexion of the neck and forward rotation of the head are located posteriorly of the vertebral bodies and are the trapezius, levator scapulae, splenius capitis, longissimus capitis, splenius cervicis, semispinalis capitis, semispinalis cervicis, obliquus capitis inferior, obliquus capitis superior, rectus capitis posterior major, and rectus capitis posterior minor.

The relative position of attachment and cross-sectional size of the muscles which are inserted in the base of the skull are shown in Fig. 2. The total area of attachment of the postvertebral muscles is much greater than that of the prevertebral group. Also, the centroid of the area of the postvertebral muscles is further from the occipital condyles than is the centroid of the area of the prevertebral muscle. This implies that the head should be able to resist a larger applied flexing moment than extending moment. An additional resisting moment is developed in flexion when the chin contacts the torso. The contact force produces a moment about the occipital condyles which further limits flexion.

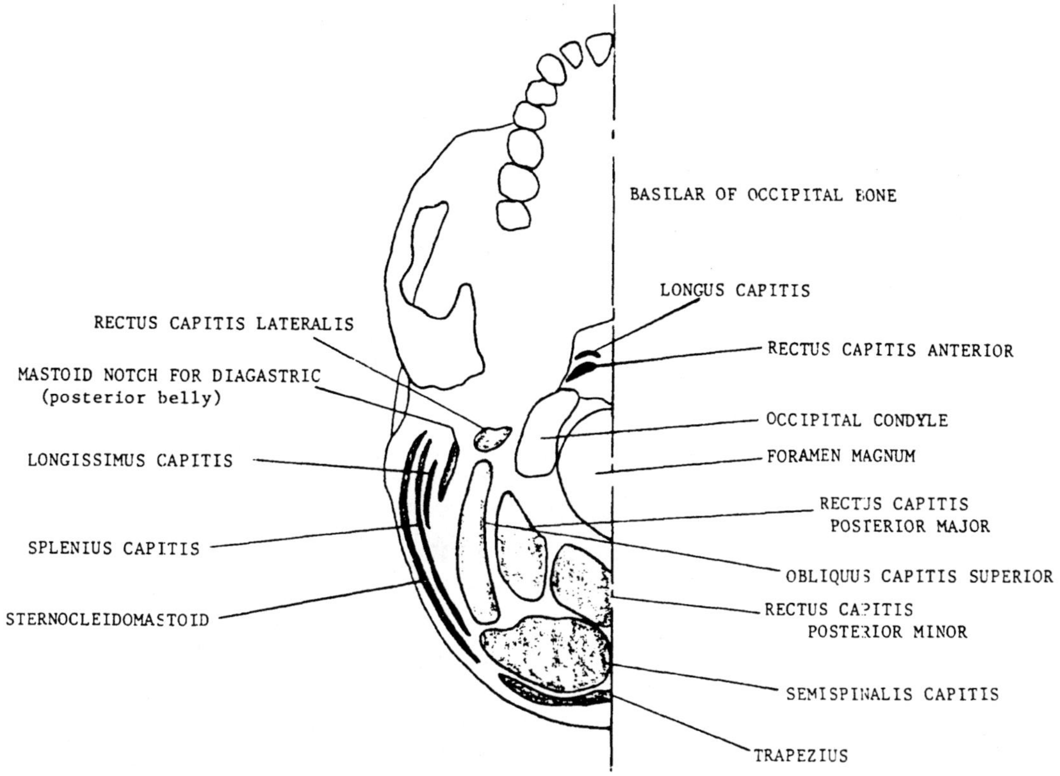
BASILAR OF OCCIPITAL BONE
LONGUS CAPITIS
RECTUS CAPITIS LATERALIS
RECTUS CAPITIS ANTERIOR
MASTOID NOTCH FOR DIAGASTRIC
(posterior belly)
OCCIPITAL CONDYLE
FORAMEN MAGNUM
LONGISSIMUS CAPITIS
RECTUS CAPITIS
POSTERIOR MAJOR
SPLENIUS CAPITIS
OBLIQUUS CAPITIS SUPERIOR
STERNOCLEIDOMASTOID
RECTUS CAPITIS
POSTERIOR MINOR
SEMISPINALIS CAPITIS
TRAPEZIUS

Fig. 2 - Relative position of attachment of the muscles which are inserted in the base of the skull

RANGE OF MOTIONS OF THE HEAD

The degree of articulation of the cervical spine varies from person to person. A person with a long, thin neck will have a greater range of head motion than a person with a short, thick neck. An athletic individual will have a greater range of head motion than his sedentary counterpart. Age and sex will also influence the range of neck articulation. Several investigators have studied the range of motion of the cervical spine for various classes of individuals.

Buck, et al. (6) measured the voluntary head motions of 100 individuals, 53 females and 47 males, whose ages ranged from 18-23 years. For the male subjects, the flexion range was 50-90 deg with a mean of 66 deg and a standard deviation of 8 deg. The extension range was 51-92 deg with a mean of 73 deg and a standard deviation of 9 deg. Combining these figures gives a total excursion range of 101-182 deg with a mean of 139 deg. The female subjects had a greater degree of flexibility in the cervical spine than the male subjects. The female means for flexion, extension, and total excursions were 69, 81, and 150 deg as compared to 66, 73, 139 deg for the male subjects. All of these measurements were taken with the subjects producing the head rotations with their neck muscles with no external forces being applied to the head.

Granville and Kreezer (7) made measurements on 10 male subjects between 20-40 years of age. The average height of the group was 5 ft 9 in, with a range of 5 ft 4 in-6 ft 1 in. The average weight was 142 lb with a range of 120-172 lb. Head movements were recorded for both voluntary and forced head-neck responses. The mean angulation for flexion was 59.8 deg with a standard deviation, σ of 11.7 deg for the voluntary response and 76.4 deg with a $\sigma = 9.2$ deg for the forced response. For extension, the mean voluntary response was 61.2 deg with a $\sigma = 26.8$ deg and the mean force response was 77.2 deg with a $\sigma = 25.1$ deg. Based on these means, average total excursions of 121 deg voluntary and 153.6 deg forced were computed. A questionable point of this data is the large spread of 76.4 deg in the forced angulation as compared to the voluntary 59.8 deg for flexion. In the flexed position, the chin is against the chest so the large spread between the forced and voluntary angulation is not expected.

Ferlic (8) made measurements of the motion of the cervical spines on subjects of various age groups. He did not classify his results in terms of flexion and extensions, but gave only total excursions all of which were voluntary and not forced. No attempt was made to immobilize the torso except that the subjects were instructed to move only the heads and necks. His results for various age groups which included both male and and female subjects were as follows: 15-24 years old, 139 ± 19 deg; 25-34 years old, 127 ± 22 deg; 35-44 years old, 120 ± 19 deg; 45-54 years old, 120 ± 15 deg; and 55-64 years old, 116 ± 22 deg. The average range of motion for all age groups was 127 ± 19.5 deg.

The volunteer, LMP, who was the subject of the dynamic hyperflexion and extension tests which are to be discussed in this paper, is quite representative of a 50th percentile adult male. He is 68 in tall and weighs 160 lb, and was 49 years old at the time of testing which is, according to the 1960 census,

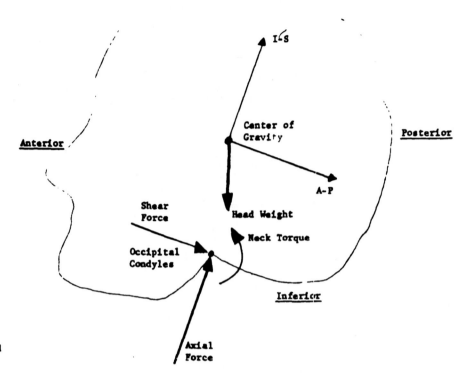

Fig. 3 - Free body diagram of the head
for hyperextension

seven years older than the median age of 42 years for the age
group of males over 18 years old. The volunteer LMP is
employed in a sedentary position and does not have unusual
muscle tone as a result of his occupation. His neck circumference is 15 in and length is approximately 5-1/2 in. Measurements were made of his normal range of motion of his cervical
spine. The volunteer LMP was seated in a rigid chair with his
torso inclined 15 deg rearward and his head in a normal upright
position. The volunteer was capable of rotating his head
relative to his torso 62 deg forward and 60 deg rearward using
hand pressure on his head to achieve these extremes. The
total excursion of 122 deg compares well with the average of
120 ± 15 deg given by Ferlic for the 45-54 year old group.
Consequently, the neck articulation of the volunteer LMP will
be used as a standard reference in the remainder of this paper.

METHOD OF ANALYZING NECK RESPONSE

Since the neck consists of seven cervical vertebrae which
can move relative to each other, the neck cannot be considered
a rigid body and, consequently, its motion cannot be analyzed
by simple rigid body mechanics. To consider the motion of
each vertebra requires the knowledge of the forces acting on
each vertebra and resulting accelerations. In the case of human
volunteer studies, these measurements would be impossible to
make with any reasonable degree of accuracy.

An alternate approach is to analyze the kinematics and
kinetics of the head, since in hyperextension and flexion the
motion of the head is controlled by the forces generated by
the neck. In the time domain of hyperextension and flexion

response of the neck produced by torso acceleration, the head
can be considered a rigid body and, consequently, Newton's
Laws of rigid body mechanics can be used, as was noted by
Mertz and Patrick (1).

For hyperextension resulting from torso acceleration, the
forces acting on the head are produced entirely by the neck
structure. For analytical purposes, these forces can be resolved
into a force acting at the occipital condyles, the bony mating
surface between the skull and the atlas, and a resultant torque.
The only other external force acting on the head is the force
of gravity. The resultant force is resolved into two components, an axial force directed along the axis of the vertebral
column parallel to the odontoid process, and a mutually
perpendicular shear force which produces a distributed bending moment along the cervical spine.

Applying Newton's Laws of rigid body motion to the head
(Fig. 3), the following equations are obtained:

$$\overline{T}_O + \overline{r}_{G/O} \times \overline{W}_H = I_G \overline{\alpha} + m_H \overline{r}_{G/O} \times \overline{a}_G \qquad (1)$$

$$\overline{F}_O + \overline{W}_H = m_H \overline{a}_G \qquad (2)$$

The inertia properties of the head, I_G and m_H, and the
geometric length $\overline{r}_{G/O}$ can be obtained or approximated by
the methods described by Mertz (2). Experimentally, the
acceleration of the head, $\overline{a}_G$ and $\overline{\alpha}$, during hyperextension can
be measured using accelerometers affixed to the head. The
corresponding torque, shear, and axial forces can be calculated

190

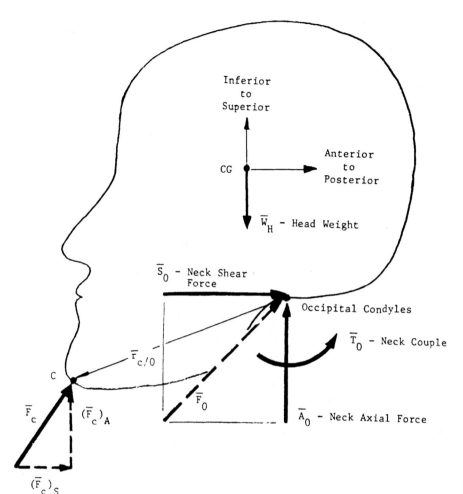

Inferior
to
Superior

CG

Anterior
to
Posterior

$\overline{W}_H$ - Head Weight

$\overline{S}_O$ - Neck Shear Force

Occipital Condyles

$\overline{T}_O$ - Neck Couple

$\overline{r}_{c/O}$

C

$\overline{F}_c$

$(\overline{F}_c)_A$

$\overline{F}_O$

$\overline{A}_O$ - Neck Axial Force

$(\overline{F}_c)_S$

Fig. 4 - Free body diagram of the head for hyperflexion including chin contact with the chest

by Eqs. 1 and 2. Because of the number of force carrying elements, no attempt was made to distribute the loads throughout the neck structure.

Hyperflexion analysis is similar to that described for hyperextension except that the force developed by the chin when it contacts the chest must be taken into account (Fig. 4). The dynamic equations for hyperflexion are:

$$\overline{F}_O + \overline{F}_C + \overline{W}_H = m_H \overline{a}_G \qquad (3)$$

$$\overline{T}_O + \overline{r}_{G/O} \times \overline{W}_H + \overline{r}_{C/O} \times \overline{F}_C = I_G \overline{\alpha} + m_H \overline{r}_{G/O} \times \overline{a}_G \quad (4)$$

Letting

$$\overline{T}_R = \overline{T}_O + \overline{r}_{C/O} \times \overline{F}_C$$

Eq. 4 becomes,

$$\overline{T}_R + \overline{r}_{G/O} \times \overline{W}_H = I_G \overline{\alpha} + m_H \overline{r}_{G/O} \times \overline{a}_G \qquad (5)$$

Again, knowing the inertial and geometrical properties of the head, the resultant forces and moments can be computed from Eqs. 3 and 5 by measuring the head accelerations. The

resultant force, $\overline{F}_R$, acting on the head produced by the neck structure and chin contact force can be resolved into A-P and S-I components which are denoted as $\overline{F}_{A-P}$ and $\overline{F}_{S-I}$, respectively. The resulting torque, T_R, includes the moment of the chin contact force about the occipital condyles. This technique was used to analyze the static and dynamic hyperflexion data presented in this paper. For the static analysis, the acceleration terms in the equations are set equal to zero.

STATIC STRENGTH OF THE HUMAN NECK

The strength of the human neck under static loading systems has been studied by several investigators. Morehouse (9) reported on static strength tests performed on a 30-year old male, who was 5 ft 10 in tall, weighed 152 lb, and had a neck circumference of 13-1/2 in. This volunteer withstood A-P and P-A forces of 18 and 50 lb applied at a level slightly above the ear. Assuming a distance of 3 in from the line of action of the force to the occipital condyles, the torques developed by the neck structure to withstand extension and flexion of the neck were 4.4 and 12.5 ft lb, respectively.

Mertz (2) and Mertz and Patrick (1) describe static strength tests for determining values of maximum torque developed at

191

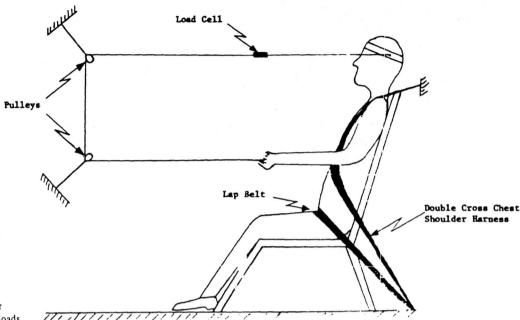

<figure>

Fig. 5 - Test setup for applying static head loads

</figure>

Table 1 - Summary of Voluntary Static Human Neck Torque Levels Developed at The Occipital Condyles for Various Neck Curvatures

Neck Position	Neck Torque Developed at Occipital Condyles, ft-lb	
	Resist Neck Flexion	Resist Neck Extension
Normal	23.5	10.5
Extended	25.0	17.5
Flexed	26.0	12.5

Notes:
1. Extension values obtained from paper by Mertz and Patrick (1).
2. Torque levels given to nearest 1/2 ft-lb.

the occipital condyles for resisting neck extension. They showed that the strength of the neck is dependent on its curvature and on the direction of the applied load. Mertz and Patrick combined their results with those of Carroll, et al. (10) and presented a summary of voluntary static human tolerance values for the neck based on the reactions developed at the occipital condyles. Part of these results are given in Table 1.

Additional tests have been conducted to determine the static strength of the human neck in resisting flexion. The test set-up and data analysis was similar to that employed by Mertz and Patrick (1). For each test, the volunteer was restrained in a rigid chair and applied a load to his head via a system of pulleys as shown in Fig. 5. A load cell was placed in series with the cord through which the load was applied. Films of the loading sequence, synchronized with load cell response, provided necessary information about the direction of the applied load from which the torque developed at the occipital condyles and the shear and axial forces generated by the neck structure were computed. Tests were conducted for three different initial curvatures of the neck: normal, extended, and flexed. For each of these initial positions, the volunteer applied a load which was directed anteriorly (P-A) relative to his head with the line of action of the force passing through the center of gravity of his head. With this loading condition, the neck structure generated forces which were similar to those required to prevent hyperflexion of the neck caused by sudden torso acceleration. The magnitude of the load was increased slowly until the volunteer could not maintain his head position or until he terminated the loading because of fear of injury and/or pain.

A total of 90 static neck strength tests were conducted on 10 volunteers. The maximum static reactions developed at the occipital condyles for the three neck positions of each of the ten volunteers are given in Table 2. The sign convention for these reactions is depicted on Fig. 4, the free-body diagram of the head. To resist the forward rotation of the head, a positive shear force and negative torque must be applied to the head at the occipital condyles. The axial force can be either positive or negative depending upon the position of the head and direction of the applied load. With the neck flexed, the maximum neck torques ranged from 14.8-25.9 ft lb. With the neck holding the head upright, the torque ranged from 16.1-23.4 ft lb. The listed values of the shear and axial forces are not maximum tolerable loads, since the resistive neck torque limited the application of higher loads.

Values for the maximum occipital torque for resisting neck

Table 2 - Maximum Static Neck Reactions Developed at The Occipital Condyles for Various Neck Positions,
Load Applied Essentially in The P-A Direction

Volunteer	Maximum Occipital Torque, ft-lb			Maximum Shear Force, lb			Maximum Axial Force, lb		
	Extended	Normal	Flexed	Extended	Normal	Flexed	Extended	Normal	Flexed
LMP	−18.0	−23.4	−25.9	50.9	63.9	71.0	50.8	15.8	− 61.4
GDG	−16.8	−16.3	−15.5	39.8	41.9	73.9	31.5	19.3	−100.5
RAE	−16.9	−22.7	−14.8	33.4	57.3	45.1	50.5	16.1	− 17.2
DJV	−25.0	−23.1	−18.8	58.8	57.1	50.5	39.8	15.0	− 5.4
HA	−11.4	−20.5	−21.2	44.0	60.6	55.6	44.3	12.1	− 63.3
CJM	−13.1	−20.6	−24.1	45.0	49.5	57.2	40.2	11.4	− 44.9
KCH	−11.0	−16.1	−16.1	25.2	39.4	41.1	20.9	14.7	− 18.7
JVG	−10.9	−16.9	−22.5	33.6	48.0	65.0	86.1	16.8	− 74.9
FTD	−20.1	−21.6	−25.2	68.3	62.0	59.1	85.1	25.9	− 63.1
DAW	−12.6	−21.8	−17.6	36.2	43.6	40.7	59.6	13.5	− 11.4

Table 3 - Maximum Static Force Reactions
Acting on The Head at The
Occipital Condyles

Shear Force, lb		Axial Force, lb	
A-P	P-A	Tension	Compression
190	190	255	250

flexion are presented in Table 1 along with the values given by Mertz and Patrick for resisting neck extension. In its normal upright position, the neck is stronger in resisting flexion, 23.5 ft lb, than it is in resisting extension, 10.5 ft lb, because the major muscles of the neck are located posterior of the occipital condyles and can generate a larger resistive couple to limit neck flexion.

According to Mertz and Patrick (1), the maximum voluntary static neck reaction of one volunteer was 255 lb in tension and 250 lb in compression. For neck shear, the neck can withstand a force of 190 lb acting posteriorly (A-P) relative to the head. A lower bound for an anteriorly (P-A) directed shear force can be obtained by equating it to the A-P reaction of 190 lb because the neck structure is inherently stronger in this direction. For the neck to produce a P-A shear force on the head at the occipital condyles, the odontoid process bears against the bony anterior arch of the atlas, while to develop an A-P shear force the action tends to separate the joint. The values for the maximum tolerable static shear and axial forces are summarized in Table 3.

The static strength values listed in Tables 1 and 3 would apply to neck response characteristics in low g environments where the viscous resisting forces produced by the muscles are not a major portion of the resisting torque. In high g environments, the viscous contribution of the muscle reaction would be comparable to its static strength component, producing a higher resistive torque than that predicted by the static strength analysis. Consequently, the static strength values are considered as lower bounds for the neck strength in a dynamic environment.

Fig. 6 - Overall view of WHAM I with subject restraint in rigid chair

DYNAMIC NECK RESPONSE AND TOLERANCE LEVELS FOR HYPERFLEXION

DESCRIPTION OF EXPERIMENTS - The subjects used for the dynamic flexion tests were a human volunteer, LMP, and four human cadavers. The subjects were restrained in a rigid chair which was mounted on an impact sled, WHAM I, shown in Fig. 6. The sled travels on two horizontal rails and is accelerated pneumatically over a distance of 6 ft to the prescribed velocity. During the acceleration stroke, a headrest is used to maintain the head in an upright position. After reaching its prescribed velocity, the sled coasts for a short distance during which the sled velocity is measured with a magnetic pickup. The sled is stopped with a specially designed hydraulic cylinder which produces a repeatable deceleration pulse. The stroke of the stopping cylinder is continuously variable up to 22 in. Consequently, for a given sled velocity the magnitude and duration of the deceleration pulse is determined by the length of the stopping distance.

The seat, shown mounted on the sled in Fig. 6, was rigidly constructed using steel angles for the main structural components and plywood coverings for the seat back and bottom.

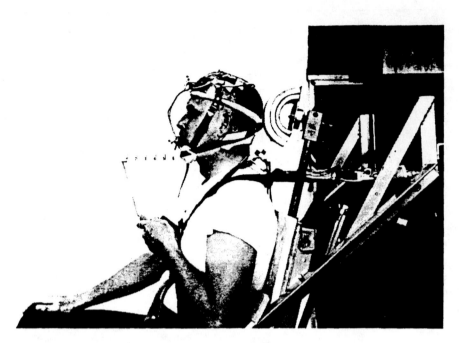

Fig. 7 - Close-up view of the head and neck transducer attached to volunteer LMP. Two accelerometers mounted to bite plate, front of helmet, and first thoracic vertebrae

These coverings were padded with a layer 5/8-in thick Rubatex to distribute the load to the subject's back and buttocks. A rigidly mounted headrest covered with three layers of Rubatex was used to maintain head position during sled acceleration.

The restraint system consisted of a lap belt and two individual shoulder harnesses which crisscrossed the chest at midsternum. Each belt was made of nominally 2-in wide, standard automotive webbing material fastened with a standard automotive seat belt buckle. Load cells were fastened to the ends of the belts to monitor the belt loads. The upper shoulder harness mounts were adjustable so that this portion of the harness could be kept horizontal, independent of the size of the subject. In addition to these restraints, the subject's feet were securely fastened to the foot support and his wrists were strapped to the armrests to prevent flailing of these appendages during the run.

The kinematics of the head of each subject were obtained from accelerometers which were attached to their heads (Fig. 7). For the volunteer, two uniaxial accelerometers were mounted on an acrylic bite plate which was molded to conform to his teeth. The accelerometers were oriented with their sensitive axes orthogonal and lying in the midsagittal plane of the head. For the cadaver tests, the accelerometers were attached to its mouth using dental acrylic, molded to the contour of the oral cavity and teeth as a mounting base. A second pair of accelerometers was attached to a lightweight fiber glass helmet which was securely fastened to the subject's head. The sensitive axes of these accelerometers were orthogonal in the midsagittal plane. From the outputs of these two pairs of accelerometers, the resultant acceleration of the center of mass of the head and the angular acceleration of the head were calculated as a function of time by the method outlined by Mertz (2). For the volunteer runs, a pair of accelerometers were mounted on the subject's back in the vicinity of the spinous process of the first thoracic vertebra, as shown in

Fig. 8 - Close-up view of head of the volunteer showing the additional head weight located above the center of mass of the head

Fig. 7. The output of these accelerometers yields the translation acceleration of the first thoracic vertebra at the base of the neck.

The objective of the research project for which these hyperflexion tests were conducted was to determine the kinetic, kinematic, and physiological effects produced by varying the mass, center of gravity, and mass moment of inertia of the head by the addition of a helmet. Consequently, tests were conducted with lead weights attached to the fiber glass helmet. Four configurations shown in Figs. 7-10 were evaluated:

1. No additional weight except the lightweight fiber glass helmet.

2. Three pounds of additional weight located above the center of the mass of the head.

3. Three pounds of additional weight located approximately at the center of mass of the head.

194

Table 4 - Geometric and Inertial Properties of the Heads of the Subjects

| | Characteristic Lengths | | | | | Inertial Properties | |
Subject	Circum. in	Width in	A-P Length in	Radius of Gyration in	Vol., in³	Wt., lb	Mass Moment of Inertia, lb-in-s-s
Volunteer-LMP	22.3	5.90	7.50	2.68	259	10.8	0.200
Cadaver–1404	22.3	6.19	7.88	2.80	244	10.1	0.205
Cadaver–1538	22.3	5.94	7.50	2.68	239	9.94	0.185
Cadaver–1548	22.6	6.56	7.75	2.75	249	10.35	0.203
Cadaver–1530	24.9	7.19	7.75	2.75	289	11.00	0.215

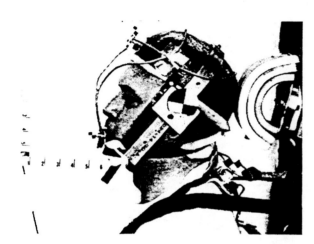

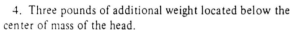

Fig. 9 - Close-up view of the head of the volunteer showing the additional head weight located approximately at the center of mass of the head

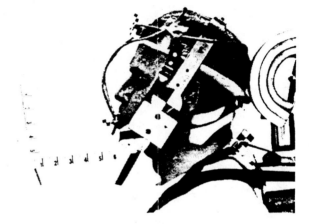

Fig. 10 - Close-up view of the head of the volunteer showing the additional head weight located below the center of mass of the head

4. Three pounds of additional weight located below the center of mass of the head.

The geometric and inertial properties of the heads of the subjects are given in Table 4.

This paper will not address the question of varying the inertial properties of the head, but will utilize the data generated during these tests to describe response and tolerance limits for the neck in flexion.

The transducers used to monitor the various parameters were:

1. Four Statham strain gage accelerometers, Model A-52, to measure head acceleration.

2. Two Statham strain gage accelerometers, Model A-52, to measure the acceleration of the first thoracic vertebra.

3. Four strain gage load cells to measure the restraint harness loads.

4. A Statham strain gage accelerometer, Model A-6, to monitor sled deceleration.

5. A magnetic pickup which recorded the time required for the sled to traverse a known distance at a constant velocity to calculate sled velocity.

The outputs of all transducers were appropriately conditioned and recorded by a light-beam oscillograph.

Two high-speed, 16 mm cameras were used to photograph each run. A close-up view of head and neck motion was taken with a Photosonics rotating prism camera operating at 600 frames/s. An overall view of the subject's motion was obtained with a Milliken framing camera operating at 500 frames/s. A timing generator was used to mark the edges of the films and the oscillograph record and synchronization was obtained by switching the frequency of the timing generator from 1000 Hz to 100 Hz during each test.

VOLUNTEER EXPOSURES - The volunteer LMP was subjected to 46 sled runs of various degrees of severity for the four configurations of additional head weight. During these tests, the volunteer attempted to achieve two different degrees of initial muscle tenseness: relaxed and tense. For the relaxed condition, the volunteer relaxed all of his muscles insofar as he was able to do so while still maintaining an upright head position. For the runs with his muscles tensed, the volunteer tensed his muscles as completely as possible during the entire run.

With his muscles tensed, the volunteer was subjected to sled rides ranging from 1.9-6.8 g for the additional head weight configurations of Helmet Only, Weight Centered, and Weight Low (Table 5). With the weight placed above the center of mass of his head, the volunteer rode the sled at 9.6 g level— his most severe exposure. This run (Run 79) resulted in a pain in the neck and back which lasted for several days.

A summary of the impact conditions and restraint system

Table 5 - Volunteer Sled Runs Conducted at Various Levels of Sled Deceleration
for Different Configurations of Additional Head Weights and for
Different Degrees of Neck Muscle Tone

Additional Head Weight	Neck Muscle Tone	Sled Plateau Deceleration Level, g
Helmet only	Tensed	2.0, 2.1, 2.9, 3.3, 3.3, 3.7, 3.9, 4.4, 4.7, 5.3, 6.8
	Relaxed	2.0, 2.1, 2.9, 3.3, 3.8, 4.1, 4.2, 4.4
Weight high	Tensed	1.5, 2.2, 3.0, 3.1, 4.6, 5.1, 6.9, 8.0, 9.6
	Relaxed	None
Weight centered	Tensed	2.0, 2.9, 4.1, 4.1, 5.1, 6.2
	Relaxed	2.1, 2.8, 4.1
Weight low	Tensed	1.9, 2.9, 3.9, 4.1, 5.8, 6.6
	Relaxed	2.1, 2.7, 4.2

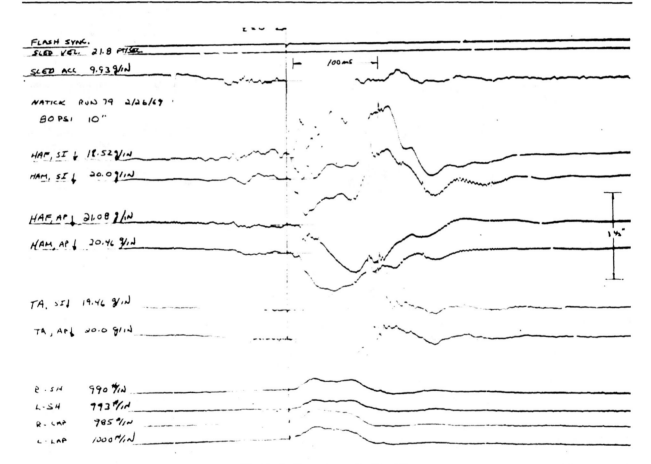

Fig. 11 - Oscillograph record of Run 79

load for all of the volunteer runs is given in Table 6. There does not appear to be a correlation between the magnitude of the onset of sled deceleration and the severity of the volunteer's exposure. Run 79, which was the most severe impact condition as noted by the pain experienced by the volunteer, had an onset of 6200 g/s with a plateau sled deceleration level of 9.6 g. Three other runs (Run 105, 108, and 109) had higher onset levels (6500 g/s, 6300 g/s, and 6400 g/s, respectively) without producing any adverse effects. The plateau sled deceleration level appears to be a good indicator of the volunteer's exposure since Run 79 had the highest level of all the volunteer runs. A more direct indicator of the severity of the exposure is the onset and magnitude of the shoulder harness loads. These two parameters give the "coupling" between the volunteer and the sled. The highest onsets for the shoulder harness loading (9700 and 8300 lb/s) and the largest restraint system loads (280 and 210 lb for the shoulder belts and 310 and 330 lb for the lap belt) occurred during Run 79.

An oscillograph record of Run 79 is shown in Fig. 11. The data read-out of the head acceleration traces of this record at

Table 6 - Summary of Impact Conditions and Restraint System Loads for Volunteer Runs

			Sled Kinematics			Restraint System								
			Deceleration Pulse			Onset		Maximum Shoulder Load*				Add.**	Neck†	
Run No.	Vel., fps	Stop. Dist. in	Onset g/s	Peak, g	Plat., g	R. Sh, lb/s	L. Sh, lb/s	R. Sh, lb	L. Sh, lb	R. Lap, lb	L. Lap, lb	Head Wt.	Mus-cles	Comments
22	10.9	15	–	2.9	2.0	–	–	20	30	30	40	W/O	T	
23	10.8	15	–	3.1	2.0	–	–	40	30	40	40	W/O	R	
24	16.1	15	–	5.3	3.7	–	–	60	60	60	60	W/O	T	
25	16.1	15	–	5.9	3.3	–	–	100	100	80	80	W/O	T	Subject involuntarily tensed neck muscles
26	15.9	15	–	5.7	3.8	–	–	80	80	80	80	W/O	R	
27	17.7	15	–	6.4	4.4	–	–	80	90	80	90	W/O	T	
28	17.7	15	–	7.2	4.4	–	–	120	130	110	120	W/O	R	
73	12.7	20	2450	4.2	1.5	400	800	60	80	90	100	H	T	
74	12.7	10	2400	5.7	3.1	1500	2100	100	100	110	110	H	T	
75	14.7	10	2400	7.0	4.6	4900	2800	120	110	170	140	H	T	
76	17.0	10	3600	8.8	5.1	4300	3900	130	150	200	200	H	T	
77	18.9	10	5600	10.2	6.9	6000	5500	190	190	260	230	H	T	
78	20.4	10	5500	12.2	8.0	9300	7500	250	220	310	310	H	T	Felt neck pain during run.
79	21.8	10	6200	14.0	9.6	9700	8300	280	210	310	330	H	T	Felt pain extending from neck into back. Does not desire to go higher
80	14.9	20	3000	5.6	2.2	1200	1200	90	80	100	100	H	T	
81	17.0	20	5000	6.9	3.0	1500	1700	90	110	110	110	H	T	
82	14.9	20	3500	5.8	2.1	800	1000	90	80	100	90	W/O	T	
83	14.9	10	2700	5.8	4.7	2300	2500	110	110	140	130	W/O	T	
84	17.0	20	4300	5.8	2.9	1000	1100	100	90	120	110	W/O	T	
85	14.9	20	3200	5.5	2.1	1000	1000	100	90	90	90	W/O	R	
86	14.7	10	3100	7.6	4.1	3400	2100	150	120	180	170	W/O	R	
87	17.0	20	4600	7.1	2.9	1700	1200	100	90	110	110	W/O	R	
88	14.9	20	3300	5.7	1.9	500	500	80	80	100	100	L	T	
89	14.9	10	3300	7.1	4.1	2900	2400	110	120	150	130	L	T	
90	17.0	20	3500	6.9	2.9	1000	1100	80	100	100	100	L	T	
91	14.8	20	3600	5.7	2.1	900	900	90	90	90	90	L	R	
92	14.8	10	3300	7.4	4.2	3400	3100	150	140	150	160	L	R	
93	17.0	20	3800	6.9	2.7	1800	1800	110	100	100	100	L	R	
94	14.9	20	3000	5.8	2.0	600	700	90	70	80	90	C	T	
95	14.9	10	3600	7.0	4.1	2700	1900	140	120	110	130	C	T	
96	16.8	20	4400	6.8	2.9	900	900	100	90	100	100	C	T	
97	14.7	20	3300	5.7	2.1	700	700	70	80	100	80	C	R	
98	14.7	10	3200	7.4	4.1	3500	2800	150	130	190	180	C	R	
99	16.8	20	4900	6.6	2.8	1500	1200	110	100	120	110	C	R	
100	17.0	10	2700	6.9	5.1	3200	2300	120	120	160	170	C	T	
101	18.9	10	4700	10.8	6.2	5000	4000	160	150	210	220	C	T	
102	20.3	20	6100	8.2	4.1	1800	1100	150	120	120	110	C	T	
103	17.0	10	4500	8.8	5.8	4300	3200	140	130	180	180	L	T	
104	18.9	10	5200	10.6	6.6	5100	3600	180	130	200	220	L	T	
105	20.4	20	6500	8.2	3.9	2100	1400	120	100	130	130	L	T	
106	16.8	10	3500	8.9	5.3	4000	3000	150	120	200	190	W/O	T	
107	18.7	10	5000	10.7	6.8	5300	4400	190	160	250	250	W/O	T	
108	20.6	20	6300	7.8	3.9	2500	1200	120	110	130	130	W/O	T	
109	18.9	20	6400	7.7	3.3	1400	1000	110	90	105	105	W/O	T	
110	18.9	20	4300	7.8	3.3	1100	1200	120	100	110	110	W/O	R	
111	20.8	20	6200	8.2	4.2	2400	1800	150	110	140	130	W/O	R	

Notes:

*Lap Loadings are combined loadings of the lower shoulder harness and lap belt corresponding to the maximum upper shoulder harness load.

**W/O - Helmet only; H - Added weight high; L - Added weight low; C - Added weight centered.

†T - Neck muscles tensed; R - Neck muscles relaxed.

Table 7 - Natick Project Oscillograph Read-Out of Run 79 Subject LMP

Time	Bite Plate Acc.		Forehead Acc.	
	AP	SI	AP	SI
ms	G	G	G	G
0.0	0.0	0.0	0.0	0.0
7.0	0.5	2.1	0.1	1.5
16.7	2.4	8.4	3.4	3.3
20.1	6.1	11.7	5.1	6.5
24.7	7.5	11.9	3.8	2.9
31.4	10.3	10.7	4.8	6.6
34.8	11.4	9.5	5.4	6.7
42.1	12.4	7.3	7.2	4.4
53.4	13.2	6.7	11.2	2.3
59.8	13.1	5.9	13.4	3.0
67.1	11.8	5.4	15.2	3.3
71.6	11.6	6.6	16.3	2.9
83.5	9.5	6.3	17.4	2.6
99.3	5.6	-10.3	12.5	-13.0
105.0	6.9	-7.7	14.0	-13.7
121.1	5.0	-9.7	10.2	-15.8
131.2	5.1	-5.1	6.9	-8.6
138.2	2.5	-5.0	5.7	-2.5
148.4	1.1	-2.6	4.0	-2.6
170.2	2.6	6.0	4.2	6.1
188.8	1.5	3.1	0.6	3.2
202.5	-0.1	0.8	-1.4	3.6
242.3	-0.7	2.2	-0.8	1.9
265.1	-1.0	0.1	-1.0	0.1
299.9	-0.7	0.8	0.3	-0.2
333.9	-0.7	-1.2	-0.5	-0.3

Table 8 - Natick Project Data Analysis of Run 79 Volunteer LMP

	Kinematics of the Head				Reactions at Occ. Cond.		
	Acc. of C.G.		Angular	Rel.	Equiv.	A-P	Axial
Time,	AP	IS	Acc.	Angle,	Moment,	Force,	Force,
ms	G	G	RA/S/S	deg	ft-lb	lb	lb
0.0	0.0	0.0	0.0	-13.0	0.7	-3.3	14.1
7.0	-0.2	-1.0	86.0	-13.0	3.8	-5.6	-0.1
16.7	0.0	-7.4	147.0	-10.0	5.5	-2.5	-92.6
20.1	3.2	-7.0	356.0	-8.0	-1.0	43.3	-87.2
24.7	1.4	-7.0	459.0	-6.0	8.4	18.5	-86.6
31.4	4.6	-3.1	573.0	-0.0	-1.4	65.6	-29.9
34.8	4.1	-2.6	573.0	3.0	0.2	59.4	-23.8
42.1	5.5	-2.1	451.0	10.0	-9.1	81.8	-16.6
53.4	8.4	-3.9	239.0	21.0	-26.5	126.3	-43.4
59.8	7.9	-5.8	97.0	30.0	-28.5	120.2	-71.4
67.1	7.6	-7.9	-78.0	38.0	-32.7	118.7	-103.0
71.6	9.1	-9.0	-125.0	43.0	-39.8	141.5	-119.2
83.5	8.9	-10.7	-301.0	58.0	-44.3	141.1	-145.6
99.3	9.5	1.0	-760.0	68.0	-60.8	150.5	21.7
105.0	11.3	-0.6	-729.0	70 0	-66.9	177.5	-1.6
121.1	7.9	1.1	-685.0	66.0	-52.6	127.1	22.9
131.2	6.3	1.8	-318.0	58 0	-36.2	102.6	34.8
138.2	0.6	-0.7	-317.0	52 0	-13.7	19.0	-0.8
148.4	1.5	-0.8	-250.0	42 0	-15.0	30.6	-0.8
170.2	-0.1	-5.6	98.0	20.0	2.4	3.2	-67.4
188.8	-0.6	-1.6	144.0	-1.0	6.3	-8.5	-8.5
202.5	-3.3	-0.5	129.0	-4.0	16.6	-49.5	7.6
242.3	-1.5	-1.6	72.0	-13.0	9.0	-26.1	-8.5
265.1	-1.4	-0.4	6.0	-15.0	6.4	-23.6	8.7
299.9	-0.9	-1.8	-43.0	-13.0	3.4	-17.0	-12.7
333.9	-0.7	0.4	-39.0	-13.0	2.7	-14.8	20.0

various time increments is given in Table 7. From this acceleration-time data and the movie data of head and torso position as a function of time, the acceleration of the center of gravity of the head, the angular acceleration of the head, the relative position of the head with respect to the torso, and the reactions of the neck and chin forces transferred to the occipital condyles were computed as described previously. A sample of the output from the computer program used to generate this data is given in Table 8 for Run 79.

The torque calculated at the occipital condyles as a function of the angular position of the volunteer's head relative to his torso for several sled deceleration levels with no additional head weight are shown in Figs. 12 and 13 for the conditions of muscles relaxed and tensed, respectively. There is a marked difference between these two families of response curves. For example, the neck response for the 3.3 g sled run with muscles relaxed is characterized by an initial overshoot of 10 ft lb followed by a plateau of 7 ft lb extending to 25 deg of relative head rotation with a maximum relative angular head position of 32.5 deg. On the other hand, the comparative 3.3 g run with muscles tensed, produced a response curve which rose to a peak of 9.5 ft lb followed by a decreasing torque-angle relation which extended to a maximum relative angular displacement of 8.5 deg, followed by a rapid return of the head toward its initial position.

The maximum responses which occurred during the various volunteer runs are presented in Tables 9-12 for the conditions of Helmet Only, Weight High, Weight Centered, and Weight Low, respectively. For each run, the tables give the plateau sled deceleration level; the maximum torque and the time at which it occurred with reference to the beginning of the sled deceleration pulse; the maximum A-P force acting on the head and its corresponding time to peak; the initial head position relative to a normal upright seating position (torso inclined 15 deg to vertical with A-P axis of the head horizontal); the maximum change in position of the head relative to its initial position and the time at which the maximum angulation occurs; and the degree of initial neck muscle tone.

The maximum neck response occurred during the sled sequence where the volunteer wore the additional head weight above the center of gravity of his head and tensed his neck muscles (Fig. 14). The curves represent the maximum neck response for a series of increasing sled deceleration levels and will be taken as the basis for the development of a maximum response envelope for the neck in flexion. It is evident that neck responses of lesser magnitude can be achieved by the same person, depending on his muscle tone as demonstrated by the curves in Figs. 12 and 13; however, specifying anything other than a maximum response condition would hinder the optimization of restraint system design. It should be emphasized that the torque calculated for the ordinate of these

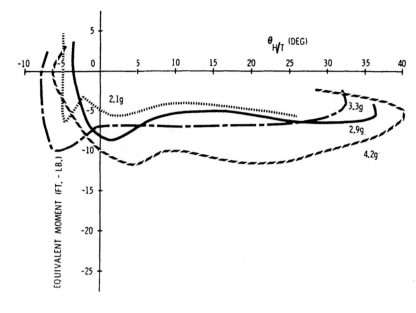

Fig. 12 - Equivalent moment about the occipital condyles as a function of the angular position of the head relative to the torso for various impact severities, volunteer LMP, neck muscles relaxed, no additional head weight

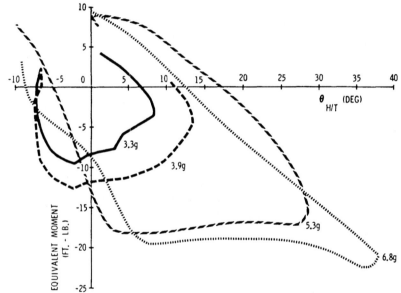

Fig. 13 - Equivalent moment about the occipital condyles as a function of the angular position of the head relative to the torso for various impact severities, volunteer LMP, neck muscles tensed, no additional head weight

curves includes the moment produced by the chin reaction on the chest as well as the moments produced by the muscles and ligaments of the neck structure.

The threshold of pain for the volunteer LMP occurred in Run 78 where the additional weight was positioned above the center of mass of the head. The torque level for this run was 43.5 ft lb with a maximum relative head position of 57 deg. His voluntary tolerance level was reached in Run 79 when he experienced a pain from the back of his neck, extending down into the middle of his back. This exposure produced a stiff neck which lasted several days. The maximum torque was 65 ft lb with a maximum relative angular displacement between his head and torso of 70 deg.

CADAVER EXPOSURES - The purpose of the cadaver runs was to obtain a comparison of the cadaver and volunteer responses to the same sled deceleration pulses for the various

configurations of additional head weight and then to subject the cadaver to more severe conditions in order to extend the data into the injury region. A total of 132 runs were conducted using four human cadavers. These cadavers were identified as Cadaver 1404, Cadaver 1530, Cadaver 1538, and Cadaver 1548.

The transducers and instrumentation used for the cadaver runs were identical to those used for the volunteer runs, except that the thoracic acceleration was not recorded. X-rays were taken of the cadaver's neck to determine whether a particular sled ride caused any observable neck damage. The data obtained from the cadaver runs were analyzed in the same manner as the data obtained from the volunteer runs. A summary of the more germane findings follows.

Cadavers 1530, 1538, and 1548 were subjected to a sequence of high-severity sled rides at plateau accelerations up to

Table 9 - Various Maximum Responses for Volunteer LMP, Helmet Only

Run No.	Plat. Sled, g	Max. Moment		Max. A-P Force		Head Position			Initial Neck Muscle Tension
		T_R, ft-lb	Time, ms	F_{A-P}, lb	Time, ms	Initial $(\theta_{H/T})_0$, deg	$\Delta\theta_{H/T}$, deg	Max. Rotation Time, ms	
23	2.0	- 4.8	105	19	115	-13	26	204	Relaxed
85	2.1	- 6.3	54	25	125	- 4	36	305	Relaxed
87	2.9	- 8.6	92	36	92	- 4	40	250	Relaxed
110	3.3	- 9.8	75	39	75	- 4	36	246	Relaxed
26	3.8	- 7.5	75	33	95	-10	43	215	Relaxed
86	4.1	-15.3	96	63	107	- 6	42	147	Relaxed
111	4.2	-11.8	73	58	110	- 4	44	223	Relaxed
28	4.4	-13.0	99	50	99	-14	41	199	Relaxed
22	2.0	- 7.2	50	25	50	-14	1	69	Tensed
82	2.1	- 7.0	61	27	61	- 7	7	187	Tensed
84	2.9	- 9.0	70	36	82	- 6	6	227	Tensed
109	3.3	- 9.5	65	38	100	- 5	14	211	Tensed
25	3.3	-11.0	85	45	100	- 1	14	161	Tensed
24	3.7	-13.0	57	48	90	-19	11	118	Tensed
108	3.9	-12.2	80	56	100	- 6	19	213	Tensed
27	4.4	-13.5	87	60	87	-15	10	97	Tensed
83	4.7	-15.5	66	56	66	- 8	11	125	Tensed
106	5.3	-18.3	65	78	92	- 7	36	123	Tensed
107	6.8	-22.5	110	92	90	- 7	44	123	Tensed

Table 10 - Various Maximum Responses for Volunteer LMP, Additional Weight Above C. G. of Head

Run No.	Plat. Sled, g	Max. Moment		Max. A-P Force		Head Position			Initial Neck Muscle Tension
		T_R, ft-lb	Time, ms	F_{A-P}, lb	Time, ms	Initial $(\theta_{H/T})_0$, deg	$\Delta\theta_{H/T}$, deg	Max. Rotation Time, ms	
73	1.5	- 7.4	76	24	64	-12	4	92	Tensed
80	2.2	-10.9	185	36	175	-13	20	210	Tensed
81	3.0	-12.0	150	43	155	-11	25	202	Tensed
74	3.1	-17.0	94	59	94	-14	20	122	Tensed
75	4.6	-21.8	103	77	99	-13	33	138	Tensed
76	5.1	-28.7	87	104	87	-13	50	143	Tensed
77	6.9	-33.3	87	120	82	-13	59	130	Tensed
78	8.0	-43.5	80	150	80	-14	71	124	Tensed
79	9.6	-65.0	105	177	105	-13	83	105	Tensed

Notes:
1. The time listed represents the time interval from the initiation of sled deceleration to peak response.

2. No runs were conducted with the neck muscles initially relaxed.

14 g with the additional weight placed at the center of mass of the head. The corresponding neck response curves are shown in Figs. 15-17, respectively. These graphs are plots of the summation of the moments of the neck and chin forces with respect to the occipital condyles as a function of the angular position of the head with respect to the torso for various plateau sled deceleration levels. The response curves for the three cadavers are quite different. Cadaver 1530 had a very stiff neck and, consequently, for a given plateau deceleration level, his relative head angulation was quite small; it did not exceed 20 deg during the 12.0 g sled ride. The neck of this cadaver responded as a linear spring. The neck of Cadaver 1548 was not quite as stiff as Cadaver 1530 as noted by comparing their response curves. Cadaver 1538 had a loose neck with little neck resistance occurring until an appreciable relative head rotation was achieved. The main resistance developed by the neck of Cadaver 1538 was due to ligamentous straining caused by chin contact with the chest as noted by the rapid increase in slope, 12.5 ft lb/deg, at approximately 60 deg of relative head angulation.

Table 11 - Various Maximum Responses for Volunteer LMP, Additional Weight at CG of Head

Run No.	Plat. Sled, g	Max. Moment		Max. A-P Force		Head Position			Initial Neck Muscle Tension
		T_R, ft-lb	Time, ms	F_{A-P}, lb	Time, ms	Initial $(\theta_{H/T})_0$, deg	Max. Rotation		
							$\Delta\theta_{H/T}$, deg	Time, ms	
97	2.1	− 9.0	64	33.0	100	− 5	−2/14	64/263	Relaxed
99	2.8	− 9.4	95	51.3	117	− 2	−6/27	39/276	Relaxed
98	4.1	−15.0	85	74.6	112	− 6	−3/47	36/164	Relaxed
94	2.0	− 8.7	58	32.9	122	− 6	−4	191	Tensed
96	2.9	− 9.7	63	39.4	63	− 4	−4/2	82/238	Tensed
95	4.1	−15.0	82	69.8	82	− 5	−4/18	43/144	Tensed
102	4.1	−12.2	74	65.9	94	−11	−6/22	40/212	Tensed
100	5.1	−20.7	74	95.6	97	−13	44.4	123	Tensed
101	6.2	−26.8	113	112.6	87	− 9	51.1	132	Tensed

Note:

1. The time listed represents the time interval from the initiation of sled deceleration to peak response.

Table 12 - Various Maximum Responses for Volunteer LMP Additional Weight Below CG of Head

Run No.	Plat. Sled, g	Max. Moment		Max. A-P Force		Head Position			Initial Neck Muscle Tension
		T_R, ft-lb	Time, ms	F_{A-P}, lb	Time, ms	Initial $(\theta_{H/T})_0$, deg	Max. Rotation		
							$\Delta\theta_{H/T}$, deg	Time, ms	
91	2.1	− 5.7	100	28	135	− 4	− 2/38	41/282	Relaxed
93	2.7	− 7.8	100	44	100	− 6	− 6/33	63/273	Relaxed
92	4.2	−14.2	114	67	96	− 7	− 3/30	47/166	Relaxed
88	1.9	− 7.2	74	33	74	− 7	− 7/1	111/311	Tensed
90	2.9	−10.0	92	45	92	− 4	− 5/3	60/252	Tensed
105	3.9	−12.0	132	65	102	− 4	−13/11	47/238	Tensed
89	4.1	−14.0	95	72	95	− 8	− 4/20	34/146	Tensed
103	5.8	−16.4	103	92	94	− 5	− 8/38	27/155	Tensed
104	6.6	−23.0	98	124	85	−10	− 7/51	37/131	Tensed

Note:

1. The time listed represents the time interval from the initiation of sled deceleration to peak response.

Cadavers 1404, 1530, and 1548 were subjected to a sequence of high-severity sled runs with the additional weight placed above the center of mass of their heads. The corresponding neck response curves are given in Figs. 18-20. As before, the response curves for these cadavers are different due to the various degrees of neck stiffnesses. Cadaver 1404 had the loosest neck. The main resistance developed by the neck of this cadaver was due to the ligamentous straining caused by contact of the chin with the chest. The slope of the torque-head angle curve for Cadaver 1404 was 12.5 ft lb/deg which was the same as noted for Cadaver 1538 (Fig. 16). The main difference in the response curves for Cadaver 1404 (Fig. 18) and Cadaver 1538 (Fig. 16) is that Cadaver 1404 achieved a larger relative head angulation, 95 deg as compared to 73 deg. This difference is largely due to differences in neck geometry between the two cadavers.

The maximum responses of Cadavers 1404, 1538, 1548, and 1530 for their various sled rides are given in Tables 13-16 respectively. For each run, the maximum values of the summation of the moments about the occipital condyles and the summation of the A-P forces acting on the head are given. Also, the initial position of the head relative to its normal position with the subject seated (A-P axis of the head horizontal and torso inclined rearward at 15 deg to vertical) and the maximum change in the angular position of the head referenced to the torso are presented. The times which are listed represent the times required to reach the various peak responses referenced to the beginning of sled deceleration.

Maximum torque levels of 130 and 140 ft lb were achieved by Cadavers 1538 and 1404 without producing any damage to the cervical spine as noted by x-ray analysis. A maximum A-P force of 470 lb was developed on the head of Cadaver 1538 by

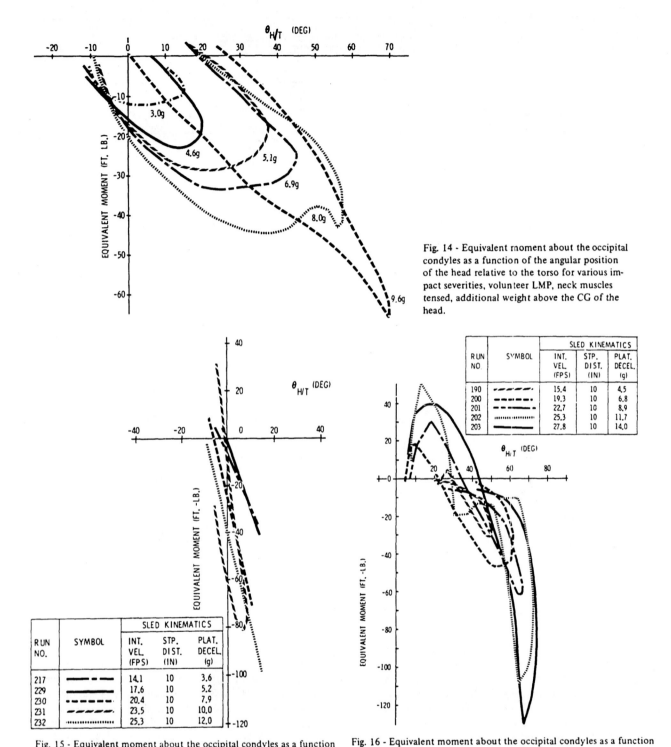

Fig. 14 - Equivalent moment about the occipital condyles as a function of the angular position of the head relative to the torso for various impact severities, volunteer LMP, neck muscles tensed, additional weight above the CG of the head.

RUN NO.	SYMBOL	SLED KINEMATICS		
		INT. VEL. (FPS)	STP. DIST. (IN)	PLAT. DECEL. (g)
190		15.4	10	4.5
200		19.3	10	6.8
201		22.7	10	8.9
202		25.3	10	11.7
203		27.8	10	14.0

RUN NO.	SYMBOL	SLED KINEMATICS		
		INT. VEL. (FPS)	STP. DIST. (IN)	PLAT. DECEL. (g)
217		14.1	10	3.6
229		17.6	10	5.2
230		20.4	10	7.9
231		23.5	10	10.0
232		25.3	10	12.0

Fig. 15 - Equivalent moment about the occipital condyles as a function of the angular position of the head relative to the torso for various high-severity runs, Cadaver 1530, additional weight at the CG of the head

Fig. 16 - Equivalent moment about the occipital condyles as a function of the angular position of the head relative to the torso for various high-severity runs, Cadaver 1538, additional weight at the CG of the head

the neck structure and chin contact with the chest without producing any discernible neck trauma.

RESPONSE CHARACTERISTICS OF THE NECK IN FLEXION - As noted previously, the volunteer LMP is quite representative of a 50th percentile adult male from a weight,

height, age, and occupation standpoint. Consequently, the neck response of the volunteer LMP is considered representative of a 50th percentile adult male.

With the added weight placed above the center of mass of his head, the volunteer was quite apprehensive about the possi-

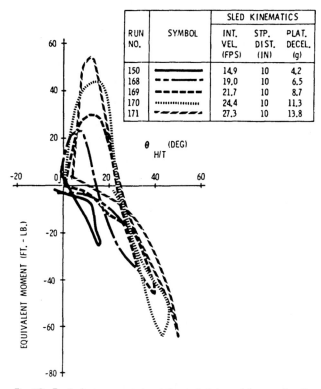

RUN NO.	SYMBOL	SLED KINEMATICS		
		INT. VEL. (FPS)	STP. DIST. (IN)	PLAT. DECEL. (g)
150	——	14.9	10	4.2
168	– – –	19.0	10	6.5
169	— — —	21.7	10	8.7
170	··············	24.4	10	11.3
171	—·—·—	27.3	10	13.8

Fig. 17 - Equivalent moment about the occipital condyles as a function of the angular position of the head relative to the torso for various high-severity runs, Cadaver 1548, additional weight at the CG of the head

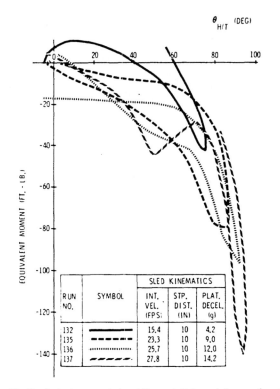

RUN NO.	SYMBOL	SLED KINEMATICS		
		INT. VEL. (FPS)	STP. DIST. (IN)	PLAT. DECEL. (g)
132	——	15.4	10	4.2
135	– – –	23.3	10	9.0
136	··············	25.7	10	12.0
137	—·—·—	27.8	10	14.2

Fig. 18 - Equivalent moment about the occipital condyles as a function of the angular position of the head relative to the torso for various high-severity runs, Cadaver 1404, additional weight above the CG of the head

bility of incurring a neck injury and, therefore, exerted more effort to maintain control over his head motion than he did for any of the other configurations of additional head weight. Consequently, the family of torque-angle curves (Fig. 14), obtained with the added weight placed above the center of mass of the head with neck muscles tensed, will be taken as the basis for describing a response envelope for various degrees of neck flexion for the 50th percentile adult male.

This family of curves and a proposed response envelope are depicted in Fig. 21. The ordinate of the graph is the summation of the moments of the neck and chin forces with respect to the occipital condyles with angular position of the head relative to the torso as the abscissa. The zero angle is referenced to the position where the A-P axis of the head is horizontal and the torso is inclined rearward 15 deg to the vertical. The initial slope of the envelope curve is taken parallel to the initial slopes of the volunteer curves and is 3 ft lb/deg. This portion of the envelope is extended to the 45 ft lb level where the envelope assumes a constant torque level to 45 deg of relative head rotation which agrees with the trend of the volunteer data. From the point (45, -45) the envelope is extended on a straight line to the point (65, -65) which is the level of the maximum torque generated by the volunteer. The unloading portion of the envelope consists of two straight line segments which closely parallel the unloading portion of the volunteer curves. The bottom segment starts at the point (75, -65) and extends to (55, -20) from which the second segment starts and ends at (35, 0). The choice of the starting angle of 75 deg for the bottom segment of the unloading curve was arbitrarily taken as 10 deg greater than the end point of 65 deg for the bottom segment of the loading curve and places the maximum angle experienced by the volunteer midway between these two curves.

Ratios of the areas contained within the individual volunteer loading and unloading curves to the areas between the loading curves and the abscissa were computed and are 0.90, 0.83, 0.87, 0.80, 0.76, and 0.51 for the increasing degrees of severities as expressed by the maximum torque levels. This shows that as the rate of loading increases, the ratio of the energy dissipated by the neck muscles to that stored in the muscles decreases.

In order to extend the response envelope to higher torque levels, the cadaver response data were considered. The neck responses of cadavers used in the test program varied greatly as can be noted by comparing the curves in Figs. 15-20. The differences in response characteristics noted from the above comparisons are due primarily to differences in neck stiffnesses between the cadavers. Another interesting comparison is the change in neck stiffness which occurs after a number of tests have been conducted on the same cadaver, as shown in Figs. 22-25 for Cadavers 1404, 1530, 1538, and 1548, respectively. For each cadaver, there is a progressive loosening of the neck with the increase in the number and degree of severity of the runs conducted with the cadaver. Consequently, the neck response of the cadavers is constantly changing. The largest changes occurred for Cadavers 1538 and 1548, while the smallest change occurred for Cadaver 1404. Because of the consis-

Table 13 - Maximum Responses of Cadaver 1404 for Different Inertial Properties of the Head

Run No.	Plat. Sled, g	Max. Moment		Max. A-P Force		Head Position			Weight Config.
		T_R, ft-lb	Time, ms	F_{A-P}, lb	Time, ms	Initial $(\theta_{H/T})_0$, deg	Max. Rot. $\Delta\theta_{H/T}$, deg	Time, ms	
133	2.2	− 14.7	173	58.0	170	−3	61	176	W/O
112	2.4	− 11.6	198	45.0	172	−5	46	198	W/O
113	2.4	− 12.2	170	50.0	170	−7	49	200	W/O
114	2.5	− 14.3	168	57.0	168	−6	50	188	W/O
128	2.9	− 22.0	175	86.5	157	−5	63	180	W/O
115	3.0	− 18.5	148	73.0	142	−3	57	142	W/O
122	3.0	− 19.5	167	71.0	167	−5	58	197	W/O
123	3.0	− 19.5	162	73.0	152	−5	60	182	W/O
116	3.4	− 23.8	142	92.0	132	−3	66	160	W/O
117	4.6	− 28.0	125	93.0	125	−5	71	150	W/O
118	2.0	− 16.8	175	80.0	170	−5	53	175	L
119	3.0	− 25.0	155	105.0	155	−7	58	155	L
120	3.5	− 28.4	138	125.0	128	−7	61	150	L
121	4.3	− 27.0	141	112.0	131	2	64	158	L
124	2.0	− 23.0	172	90.0	170	−2	63	172	C
125	3.0	− 32.0	153	117.0	153	−3	66	153	C
126	3.6	− 34.5	124	135.0	124	−5	62	124	C
127	4.3	− 43.0	135	147.0	135	−3	69	135	C
129	2.3	− 32.5	198	105.0	198	−4	68	198	H
130	3.0	− 41.0	174	135.0	174	−7	73	174	H
131	3.6	− 44.0	147	140.0	147	−8	74	155	H
134	3.7	− 45.0	153	144.0	153	−7	78	153	H
132	4.2	− 42.5	140	141.0	140	−5	75	140	H
135	9.0	− 80.0	100	264.0	100	−4	86	110	H
136	12.0	− 97.0	105	330.0	105	−5	92	105	H
137	14.2	−140.0	93	357.0	93	−7	95	100	H

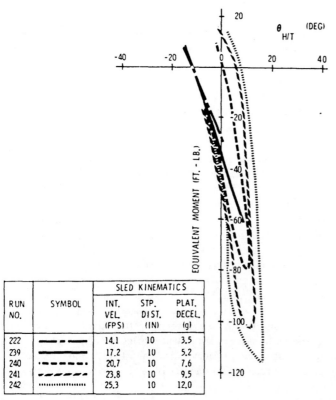

RUN NO.	SYMBOL	SLED KINEMATICS		
		INT. VEL. (FPS)	STP. DIST. (IN)	PLAT. DECEL. (g)
222	— — —	14.1	10	3.5
239	————	17.2	10	5.2
240	·—·—·—	20.7	10	7.6
241	⁄ ⁄ ⁄ ⁄	23.8	10	9.5
242	··············	25.3	10	12.0

Fig. 19 - Equivalent moment about the occipital condyles as a function of the angular position of the head relative to the torso for various high-severity runs, Cadaver 1530, additional weight above the CG of the head

Table 14 - Maximum Responses of Cadaver 1538 for Different
Inertial Properties of the Head

| Run No. | Plat. Sled, g | Max. Moment | | Max. A-P Force | | Head Position | | | |
| | | T_R, ft-lb | Time, ms | F_{A-P}, lb | Time, ms | Initial $(\theta_{H/T})_0$, deg | Max. Rot. | | Weight Config. |
							$\Delta\theta_{H/T}$, deg	Time, ms	
178	2.2	− 11.3	150	49.0	131	8	35	160	W/O
179	3.0	− 17.0	100	71.5	115	10	41	115	W/O
186	3.0	− 18.5	100	75.0	115	10	44	118	W/O
191	3.0	− 18.0	94	75.0	115	10	44	115	W/O
197	3.0	−	−	−	no film	−	−	−	W/O
198	3.0	− 16.5	130	70.0	123	10	49	142	W/O
204	3.0	− 28.5	154	100.0	149	−1	68	176	W/O
180	3.7	− 22.0	90	86.0	109	10	41	109	W/O
181	4.0	− 25.8	83	78.0	90	16	44	100	W/O
182	2.2	− 13.0	100	61.0	130	13	35	120	L
183	3.1	− 16.9	95	70.5	95	10	40	110	L
184	3.4	− 24.0	95	83.5	100	10	43	110	L
185	4.2	− 26.3	93	82.0	75	18	46	105	L
187	2.1	− 17.9	110	73.0	130	13	44	130	C
188	3.0	− 21.5	100	91.0	118	11	47	120	C
189	3.7	− 25.2	95	107.0	110	10	51	120	C
190	4.5	− 30.5	80	106.0	85	20	50	90	C
200	6.8	− 47.0	85	184.0	90	3	61	110	C
201	8.9	− 62.0	83	225.0	80	5	67	90	C
202	11.7	−109.0	73	360.0	73	7	72	90	C
203	14.0	−130.0	68	473.0	68	4	73	96	C
192	2.4	− 18.0	135	70.0	127	8	43	132	H
199	2.6	− 20.5	149	75.0	130	6	53	150	H
193	3.0	− 23.5	123	88.0	117	12	52	134	H
194	3.0	− 23.0	150	83.0	150	5	50	135	H
195	3.7	− 25.0	130	99.0	135	5	53	135	H
196	4.2	− 30.0	100	110.0	108	5	57	130	H

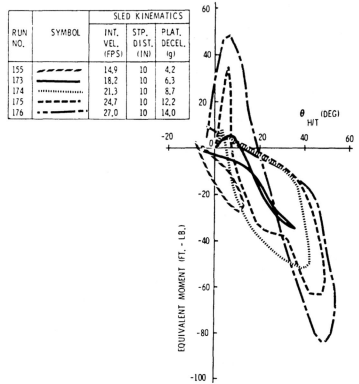

| RUN NO. | SYMBOL | SLED KINEMATICS | | |
		INT. VEL. (FPS)	STP. DIST. (IN)	PLAT. DECEL. (g)
155	―――――	14.9	10	4.2
173	─────	18.2	10	6.3
174	··············	21.3	10	8.7
175	‐ ‐ ‐ ‐	24.7	10	12.2
176	‐·‐·‐·	27.0	10	14.0

Fig. 20 - Equivalent moment about the occipital condyles as a function of the angular position of the head relative to the torso for various high-severity runs, Cadaver 1548, additional weight above the CG of the head

Table 15 - Maximum Response of Cadaver 1548 for Different
Inertial Properties of the Head

Run No.	Plat. Sled, g	Max. Moment		Max. A-P Force		Head Position			
		T_R, ft-lb	Time, ms	F_{A-P}, lb	Time, ms	Initial $(\theta_{H/T})_0$, deg	Max. Rot.		Weight Config.
							$\Delta\theta_{H/T}$, deg	Time, ms	
138	2.3	-11.6	60	37.5	43	-0.2	4.7	70	W/O
139	2.8	-13.2	70	47.5	50	-0.2	5.8	80	W/O
156	3.0	-13.7	65	51.5	117	-2.2	9.5	95	W/O
162	3.0	-13.8	130	62.0	120	-2.8	17.4	140	W/O
177	3.0	-13.9	120	70.0	120	5.0	32.6	135	W/O
140	3.5	-15.5	75	62.0	110	0.0	7.7	75	W/O
141	4.1	-22.5	70	75.0	60	-0.7	11.5	80	W/O
157	6.3	-32.0	75	116.0	70	0.4	19.1	85	W/O
158	8.7	-39.5	65	163.0	70	-0.5	22.8	80	W/O
159	12.0	-51.7	70	204.0	70	-0.5	27.3	85	W/O
160	14.0	-45.0	55	223.0	55	0.2	31.7	67	W/O
142	2.0	-10.0	80	41.5	120	-0.9	4.9	90	L
143	3.0	-11.8	80	64.0	110	-0.5	8.7	90	L
144	3.6	-14.5	85	86.0	107	-0.6	10.0	85	L
145	4.1	-21.5	73	80.0	110	-2.6	13.5	80	L
163	6.3	-27.0	75	141.0	95	-1.7	24.8	85	L
164	9.3	-37.0	70	199.0	82	-1.1	29.3	100	L
165	11.7	-39.0	80	212.0	60	1.3	34.2	90	L
166	13.9	-52.0	80	222.0	70	-0.8	37.7	90	L
147	2.1	-11.1	120	56.5	132	-1.8	8.5	120	C
148	3.0	-13.5	80	70.0	120	-2.2	11.2	130	C
149	3.3	-16.5	80	90.0	110	-1.4	13.0	110	C
150	4.2	-25.0	80	106.0	90	-1.5	16.7	90	C
168	6.5	-36.0	95	174.0	95	-1.0	35.2	95	C
169	8.7	-46.2	90	223.0	80	1.7	40.2	100	C
170	11.3	-63.0	80	300.0	75	0.0	45.8	88	C
171	13.8	-64.0	85	280.0	70	2.8	50.0	85	C
152	2.0	-12.0	110	50.0	100	-8.2	4.0	130	H
153	2.9	-17.2	117	75.0	117	-8.6	6.5	117	H
154	3.5	-18.6	110	89.0	110	-9.0	9.4	120	H
155	4.2	-27.0	80	106.0	100	-9.0	12.7	110	H
173	6.3	-36.0	100	154.0	100	-0.3	34.7	100	H
174	8.7	-52.0	100	220.0	85	0.9	42.1	107	H
175	12.2	-64.0	90	245.0	80	-0.1	49.0	100	H
176	14.0	-85.0	80	320.0	80	-3.5	54.0	90	H

tency of the response curves, the data obtained from Cadaver 1404 was used to extend the volunteer response envelope shown in Fig. 21.

The slope of the torque-angle curve for Cadaver 1404 (Fig. 18), in the region of hyperflexion of the neck is 12.5 ft lb/deg and is due primarily to the moment produced by the chin reacting against the chest and the straining of the posterior neck ligaments. This loading rate was used to extend the voluntary response envelope from 65 ft lb to the level of 140 ft lb (Fig. 26). The unloading rate was taken parallel to the loading curve since the trend is for the loading and unloading curves to approach each other at the higher angles as indicated for Cadaver 1404 and the volunteer responses in Figs. 18 and 21.

The response envelope places the following conditions on the response of any mechanical simulations of the human neck:

1. The evaluation of the neck structure must be made with the neck mounted in a structure which includes a head and a chest so the chin can contact the torso.

2. The torque measured must include the moment produced by the contact force between the chin and chest, if any.

3. The torque-angle relationship must be determined dynamically and both the loading and unloading portions of the curve must lie within the response envelope of Fig. 26.

4. To insure adequate damping, the ratio of the area between the loading and unloading curves to the area between the loading curve and the abscissa should not be less than 0.5 for that portion of the response curve below the 45 ft lb moment level.

These conditions apply to the flexion response of the 50th percentile adult male and are necessary, but not sufficient, conditions to describe a unique head-neck response to flexion. The geometric and inertial properties of the head, neck, and chest must be defined as well as the linear displacement range

Run No.	Plat. Sled, g	Max. Moment		Max. A-P Force		Head Position			Weight Config.
		T_R, ft-lb	Time, ms	F_{A-P}, lb	Times, ms	Initial $(\theta_{H/T})_0$, deg	Max. Rot. $\Delta\theta_{H/T}$, deg	Time, ms	
205	2.1	− 8.3	80	26.0	63	− 6.0	−1.0	80	W/O
223	2.9	− 12.2	60	38.0	70	− 6.0	−0.2	80	W/O
243	2.9	− 13.4	85	49.0	85	− 3.0	−5.2	85	W/O
206	3.0	− 12.4	73	27.0	73	− 8.0	−3.2	85	W/O
228	3.0	− 12.5	80	43.0	80	− 6.0	3.8	88	W/O
207	3.3	− 18.0	73	57.0	73	− 8.0	−1.0	73	W/O
208	3.8	− 30.0	70	90.0	70	− 8.0	3.0	80	W/O
234	5.7	− 43.0	70	148.0	70	− 4.0	13.4	70	W/O
235	7.3	− 52.0	65	188.0	65	− 3.0	11.9	65	W/O
236	10.0	− 62.0	65	248.0	65	− 3.0	13.0	65	W/O
237	12.0	− 74.0	63	310.0	63	− 4.0	11.2	53	W/O
209	2.1	− 5.8	90	34.0	130	− 7.0	−2.1	100	L
210	2.7	− 12.9	80	48.0	90	− 7.0	0.0	110	L
211	3.2	− 15.8	78	60.0	85	− 6.0	1.0	78	L
212	3.6	− 27.4	68	84.0	68	− 7.0	3.3	88	L
224	5.5	− 34.0	70	132.0	60	− 4.0	7.4	75	L
225	7.0	− 45.0	68	200.0	78	− 4.0	8.3	68	L
226	9.5	− 62.0	68	275.0	68	− 1.0	9.4	64	L
227	11.3	− 80.0	65	350.0	65	− 1.0	10.0	60	L
214	2.1	− 9.4	75	33.0	90	− 6.0	0.3	75	C
215	2.9	− 13.4	60	46.0	60	− 6.0	0.8	70	C
216	3.3	− 17.0	75	66.0	63	− 6.0	1.9	75	C
217	3.6	− 27.2	70	99.0	80	− 4.0	7.9	70	C
229	5.2	− 49.0	75	188.0	75	− 4.0	12.0	70	C
230	7.9	− 73.0	68	274.0	68	− 5.0	9.2	63	C
231	10.0	− 84.0	70	330.0	70	− 6.0	8.5	65	C
232	12.0	−110.0	58	375.0	65	− 3.0	13.8	65	C
219	2.2	− 12.0	65	35.0	110	−13.0	−4.4	110	H
220	2.9	− 16.6	75	49.0	115	−13.0	−3.5	110	H
221	3.4	− 20.0	78	63.0	110	−12.0	−2.2	100	H
222	3.5	− 28.0	70	92.0	80	−13.0	0.6	90	H
239	5.2	− 64.0	70	186.0	70	−11.0	7.6	70	H
240	7.6	− 80.0	70	248.0	70	− 8.0	11.0	67	H
241	9.5	−103.0	73	345.0	68	− 7.0	13.0	63	H
242	12.0	−116.0	58	390.0	63	− 6.0	16.1	58	H

of the center of gravity of the head taken relative to a fixed system on the torso.

TOLERANCE LEVELS FOR THE NECK IN FLEXION - Various tolerance levels for neck flexion are indicated in Fig. 26. The volunteer LMP withstood a static moment about his occipital condyles of 26 ft lb. This reaction was generated by the volunteer's neck structure without any contribution from the chin contacting the chest. No tests were conducted to establish a static moment tolerance level with the chin in contact with the chest.

A dynamic tolerance level for the moemnt about the occipital condyles for the initiation of pain occurred during Run 78 in which the maximum equivalent resisting moment was 44 ft lb. There was contact between the chin and the chest during the run.

The maximum dynamic moment generated by the volunteer

LMP of 65 ft lb occurred during Run 79. This torque level produced sharp pain in the neck and upper back region with soreness persisting for several days. The 65 ft lb includes the moment produced by the chin reaction as well as the moments produced by the neck reactions. It is considered as noninjurious, but close to the injury threshold.

The maximum tolerable moment was not determined because the volunteer experienced no injuries of any consequence during any of his sled rides. The cadavers were exposed to a much more severe environment than the volunteer. Maximum moments of 85, 116, 130, and 140 ft lb were experienced by Cadavers 1548, 1530, 1538, and 1404, respectively. There was a progressive loosening of the neck structures of the cadavers with repeated exposures which is assumed to be due to the tearing of hardened connective tissue of the neck. Cadaver 1404 had the loosest neck structure and

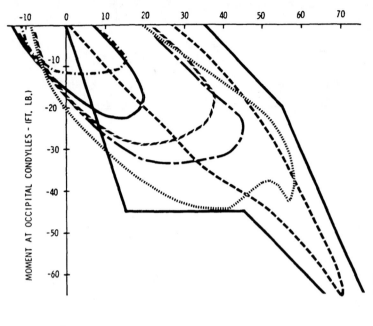

Fig. 21 - Comparison of the proposed flexion response envelope and the volunteer response curves

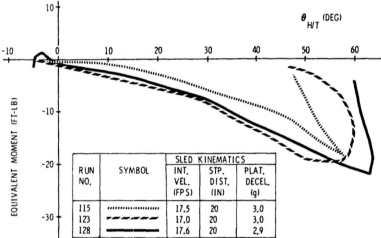

RUN NO.	SYMBOL	SLED KINEMATICS		
		INT. VEL. (FPS)	STP. DIST. (IN)	PLAT. DECEL. (g)
115	....	17.5	20	3.0
123	- - - - -	17.0	20	3.0
128	————	17.6	20	2.9

Fig. 22 - Comparison of neck response curves for identical 3 g plateau sled deceleration runs, Cadaver 1404, helmet only

showed little progressive loosening. None of these cadavers had any observable ligamentous, disc, or bone damage as noted from analysis of x-rays of their neck structures.

Based on this cadaver data, it would appear that the 50th percentile human could withstand equivalent moments of 140 ft lb without suffering ligamentous or bone damage. However, there is no guarantee that severe muscle injuries would not be produced at a lower value of equivalent moment. Consequently, the value of 140 ft lb should be used with discretion.

The static head rotational limit of 66 deg for the volunteer LMP compares favorably with his maximum dynamic head angle of 70 deg. The small difference between these two limits indicates that chin contact with the chest provides a stop for forward head rotation. Consequently, the position of the head relative to the torso is not a good physical measurement to be used in evaluating neck trauma. The equivalent moment about the occipital condyles is a better indicator.

The volunteer LMP withstood an A-P force of 190 lb with only his neck structure reacting. Dynamically, the volunteer LMP generated an A-P force of 177 lb, which includes the contribution of the chin contact force. None of these force levels produced any injury. The cadaver exposures can be used to obtain an approximation of a noninjurious level.

Cadavers 1548, 1404, 1530, and 1538 generated maximum A-P force levels of 320, 357, 390, and 473 lb, respectively. None of these force levels produced any observable neck damage. Consequently, an injury threshold of 450 lb is suggested for the A-P force acting on the head during hyperflexion when the chin is in contact with the chest.

DYNAMIC NECK RESPONSES AND TOLERANCE LEVELS FOR HYPEREXTENSION RESPONSE CHARACTERISTICS - Mertz and Patrick (1) presented data pertaining to the dynamic response and tolerance levels of the neck in hyperextension, including volunteer and cadaver data.

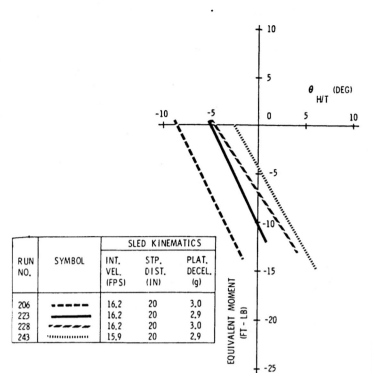

RUN NO.	SYMBOL	SLED KINEMATICS		
		INT. VEL. (FPS)	STP. DIST. (IN)	PLAT. DECEL. (g)
206	-----	16.2	20	3.0
223	———	16.2	20	2.9
228	─ ─ ─	16.2	20	3.0
243	··········	15.9	20	2.9

Fig. 23 - Comparison of neck response curves for identical 3 g plateau sled deceleration runs, Cadaver 1530, helmet only

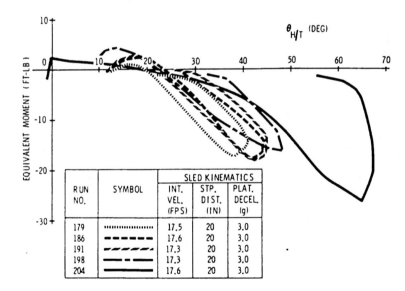

RUN NO.	SYMBOL	SLED KINEMATICS		
		INT. VEL. (FPS)	STP. DIST. (IN)	PLAT. DECEL. (g)
179	··········	17.5	20	3.0
186	-----	17.6	20	3.0
191	─ ─ ─	17.3	20	3.0
198	—·—·—	17.3	20	3.0
204	———	17.6	20	3.0

Fig. 24 - Comparison of neck response curves for identical 3 g plateau sled deceleration runs, Cadaver 1538, helmet only

The same volunteer was used in the earlier hyperextension experiments (1) and the hyperflexion experiments reported herein. The same data analysis techniques were also employed in the two programs. Therefore, the following development of hyperextension tolerance levels and a dynamic response envelope has a base consistent with the same development for hyperflexion.

The torque computed at the occipital condyles as a function of the position of the head relative to the torso is shown in Fig. 27 for the volunteer LMP and Cadavers 1035 and 1089. Only one curve is shown for the volunteer LMP because he did not wish to be exposed to higher severity sled runs for fear of neck injury. The response curves for the cadavers indicated a progressive loosening of their neck structures with increased severity of the run.

An approximation of a hyperextension response envelope shown in Fig. 27 was made based on this volunteer and cadaver data, the static neck strength data for hyperextension (Table 1) and the dynamic neck strength data given by Carroll, et al. (10).

The initial slope of the response envelope was taken as 1 ft lb/deg which corresponds to the initial slope of the volunteer curve. This line is extended to a level of 22.5 ft lb. This level was chosen by noting that Carroll, et al. (10) gives a ratio of dynamic to static response of the neck in extension of 1.28 and that the static strength of the neck in resisting hyperex-

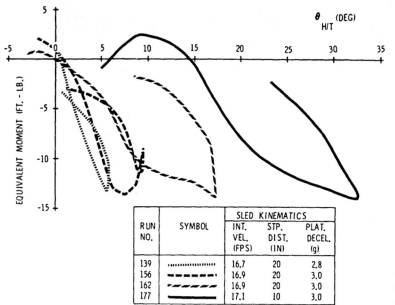

RUN NO.	SYMBOL	SLED KINEMATICS		
		INT. VEL. (FPS)	STP. DIST. (IN)	PLAT. DECEL. (g)
139	,,,,,,,,,,,,,,,,,,	16.7	20	2.8
156	- - - - - - -	16.9	20	3.0
162	·—·—·—·	16.9	20	3.0
177	————	17.1	10	3.0

Fig. 25 - Comparison of neck response curves for identical 3 g plateau sled deceleration runs, Cadaver 1548, helmet only

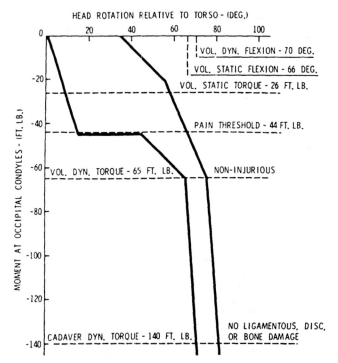

Fig. 26 - Head-neck response envelope for flexion and various tolerance levels

tension was 17.5 ft lb as indicated in Table 1. The product of these two factors gives a value of approximately 22.5 ft lb.

Similar to the flexion response curve, a constant torque level is assumed until the neck structure begins to "bottom out." For the volunteer LMP, this occurred at approximately 60 deg, as noted previously. The next portion of the response envelope was taken parallel to the maximum response curve for Cadaver 1089 and extended to the 35 ft lb level at approximately 80 deg. The torque level of 35 ft lb is twice the static strength level and is the noninjurious torque tolerance level recommended by Mertz and Patrick (1). The envelope is ex-

tended from this point by a line which is parallel to the maximum slope of the unloading portion of the response curve of Cadaver 1089. In this portion of the response envelope, the neck behaves as a linear spring and very little energy is dissipated. Consequently, the loading and unloading curves would be coincident.

The unloading portion of this envelope passes through a point 10 deg greater than the loading curve at the voluntary tolerance level of 35 ft lb and is taken parallel to the maximum unloading slope of the response curve for Cadaver 1089. Again, the 10 deg is arbitrary as it was in the flexion envelope.

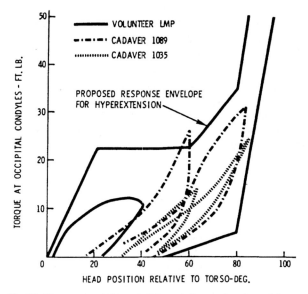

Fig. 27 - Torque about the occipital condyles as a function of the position of the head relative to the torso for the volunteer LMP and Cadavers 1035 and 1089

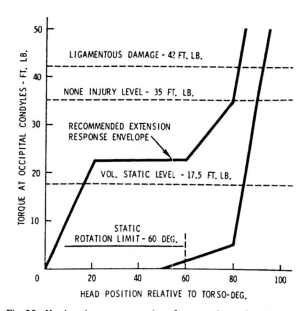

Fig. 28 - Head-neck response envelope for extension and various tolerance levels

This line is extended to the 5 ft lb and 80 deg point. At this point, the envelope is arbitrarily extended by a straight line to the abscissa at 50 deg. The envelope encloses the majority of the cadaver and volunteer response curves, except for the initial response curve of Cadaver 1089 in which the neck was quite stiff.

The logic employed in developing this response envelope is similar to that used to develop the response envelope for flexion. Until additional data is provided to indicate otherwise, this response envelope is recommended to describe the torque-angle response of the neck of a 50th percentile adult male in hyperextension.

The majority of the energy is dissipated by the neck muscles and occurs prior to bottoming out of the neck structure. There is no reason to believe that the ratio of the area between the loading curve and the abscissa will be any different for extension than flexion. Thus, the relationship developed for flexion (this ratio should not be less than 0.50) will be used for extension.

The response envelope shown in Fig. 27 places the following requirements on the response characteristics of any mechanical simulations of the human neck:

1. The evaluation of the neck structure must be made with an appropriate head structure.

2. The torque must be measured about the occipital condyles.

3. The torque-angle relationships must be determined dynamically and both the loading and unloading portions of the neck's response curve must lie within the response envelope of Fig. 27.

4. To insure adequate damping, the ratio of the area between the loading and unloading curves to the area between the loading curve and the abscissa should not be less than 0.5 for that portion of the response curve below the 22.5 ft lb moment level.

Again, these conditions apply to the extension response of the 50th percentile adult male and are necessary, but not sufficient conditions, to describe a unique head-neck response to extension. As mentioned in the discussion of the flexion response, the geometrical and inertial properties of the head and neck must be prescribed as well as the linear displacement range of the CG of the head measured relative to a prescribed system on the torso.

TOLERANCE LEVELS FOR EXTENSION - Some of the tolerance levels have been described in developing the response envelope for hyperextension. A static torque level of 17.5 ft lb was presented and a noninjurious dynamic torque level of 35 ft lb was recommended by Mertz and Patrick (1).

An injury tolerance level can also be obtained from the data presented by Mertz and Patrick. Cadaver 1035 suffered minor ligamentous damage between the third and fourth cervical vertebrae. The maximum neck torque experienced by Cadaver 1035 when this damage occurred was 24.6 ft lb. Since this cadaver was much smaller than the volunteer LMP, the torque value must be scaled in order to be applied to the volunteer LMP. Using the scaling technique presented by Mertz and Patrick (1), this torque would have been 42 ft lb for the volunteer LMP. Consequently, until data become available to indicate otherwise, an injury threshold for the torque developed at the occipital condyles of 45 ft lb is recommended for ligamentous damage of a 50th percentile adult male subjected to hyperextension.

The various injury thresholds for hyperextension are given for the response envelope shown in Fig. 28. It should be noted as it was for hyperflexion, that the measurement of the relative angle of the head to the torso is not an adequate physical measurement for describing hyperextension tolerance levels. The torque is a better indicator because, in the injury realm, a small error in angle measurement produces a large change in torque level.

CONCLUSIONS

The angle between the head and the torso is not a good physical measurement of trauma for neck extension or flexion because when the neck is hyperflexed or hyperextended, a small change in angle produces a large change in resisting torque. The best indicator for the degree of severity of neck flexion is the equivalent moment which consists of the moments of the neck and chin contact forces taken with respect to the occipital condyles. This is the same indicator used by Mertz and Patrick (1) for hyperextension. The resultant A-P and S-I forces acting on the head were well below tolerable levels and are not considered critical parameters.

Various tolerance levels proposed for neck flexion of a 50th percentile adult male are:

1. Equivalent moment about the occipital condyles of 44 ft lb for the initiation of pain.

2. Maximum voluntary equivalent moment about the occipital condyles of 65 ft lb based on volunteer tests.

3. Equivalent moment of 140 ft lb and A-P force level of 450 lb without producing ligamentous or bone damage based on cadaver responses with the chin in contact with chest. There is no guarantee that muscle injury will not occur at a lower level.

Various tolerance levels proposed for neck extension of a 50th percentile adult male are:

1. Noninjurious torque about the occipital condyles of 35 ft lb. Ligamentous damage at 42 ft lb.

2. Noninjurious shear force of 190 lb based on voluntary static strength tests. Noninjurious static strengths of the neck in tension and compression were 255 and 250 lb. These are considered lower bounds for the corresponding dynamic strengths.

There is no unique neck response curve for an individual because of the degrees of muscle tone that a person is capable of generating. The most repeatable neck response occurs when a person tenses his neck muscles. This condition produces the least amount of neck flexion for a given severity of exposure. Consequently, the muscle tensed response curves should be used for specifying performance of mechanical simulations of the human neck. This would optimize the protective systems for the head, neck, and torso.

Two necessary conditions required for the response of a mechanical simulation of the human neck of a 50th percentile adult male are:

1. The relationships between the equivalent moment and angular displacement of the head relative to the torso for loading and unloading must lie within the response envelopes shown in Figs. 26 and 28 for hyperflexion and extension. To evaluate the mechanical neck, it must be mounted between an appropriate dummy chest and head, and the testing must be done in a dynamic environment.

2. To insure adequate damping, the ratio of the area between the dynamic loading and unloading curves to the area between the loading curve and the abscissa must not be less than 0.5 for that portion of the particular response curve lying below constant plateaus (45 and 22.5 ft lb) of the flexion and extension response envelopes, respectively.

REFERENCES

1. H. J. Mertz and L. M. Patrick, "Investigation of the Kinematics and Kinetics of Whiplash." Proceedings of Eleventh Stapp Car Crash Conference, P-20, paper 670919. New York: Society of Automotive Engineers, Inc., 1967.

2. H. J. Mertz, "The Kinematics and Kinetics of Whiplash." Ph.D. Dissertation, Wayne State University, 1967.

3. C. L. Ewing, D. J. Thomas, G. W. Beeler, and L. M. Patrick, "Dynamic Response of the Head and Neck of the Living Human to $-G_x$ Impact Acceleration." Proceedings of Twelfth Stapp Car Crash Conference, P-26, paper 680792. New York: Society of Automotive Engineers, Inc., 1968.

4. C. L. Ewing, D. J. Thomas, L. M. Patrick, G. W. Beeler, and M. J. Smith, "Living Human Dynamic Response to $-G_x$ Impact Acceleration II—Accelerations Measured on the Head and Neck." Proceedings of Thirteenth Stapp Car Crash Conference, P-28, paper 690817. New York: Society of Automotive Engineers, Inc., 1969.

5. C. Tarriere and C. Sapin, "Biokinetic Study of the Head to Thorax Linkage." Proceedings of Thirteenth Stapp Car Crash Conference, P-28, paper 690815. New York: Society of Automotive Engineers, Inc., 1969.

6. C. A. Buck, F. B. Dameron, M. J. Dow, and H. V. Skowlund, "Study of Normal Range of Motion in the Neck Utilizing a Bubble Goniometer." Archives of Physical Medicine and Rehabilitation, Vol. 40 (September 1959).

7. A. D. Granville and G. Kreezer, "The Maximum Amplitude and Velocity of Joint Movements in Normal Male Human Adults." Human Biology, Vol. 9 (1937).

8. D. Ferlic, "The Range of Motion of the Normal Cervical Spine." Hopkins Hospital Bulletin 110, 1962.

9. L. E. Morehouse, "The Strength of a Man." Human Factors, Vol. 1 (April 1959).

10. D. F. Carroll, J. A. Collins, J. L. Haley, and J. W. Turnbow, "Crashworthiness Study for Passenger Seat Design—Analysis and Testing of Aircraft Seats." AVSER Report No. 67-4, May 1967.

973318

Injury Risk Curves for Children and Adults in Frontal and Rear Collisions

Harold J. Mertz
General Motors Corp.

Priya Prasad
Ford Motor Co.

Annette L. Irwin
General Motors Corp.

ABSTRACT

This paper describes the development of injury risk curves for measurements made with the CRABI and Hybrid III family of biofidelic child and adult dummies that are used to evaluate restraint systems in frontal and rear-end collision simulations. Injury tolerance data are normalized for size and strength considerations. These data are analyzed to give normalized injury risk curves for neck tension, neck extension moment, combined neck tension and extension moment, sternal compression, the rate of sternal compression, and the rate of abdominal compression for children and adults. Using these injury risk curves dummy response limits can be defined for prescribed injury risk levels. The injury risk levels associated with the various injury assessment reference values currently used with the CRABI and Hybrid III family of dummies are noted.

INTRODUCTION

A number of investigators have developed injury risk curves for various dummy response measurements for frontal impacts. Prasad and Mertz (1) and Mertz et al (2, 3) have published injury risk curves for skull fracture and for AIS ≥ 4 brain injury due to forehead impacts based on the 15 ms HIC criterion and for skull fracture based on peak head acceleration (Figures A1 - A3 of the Appendix). These curves represent the injury risks for the adult population since adult cadavers were used to obtain the biomechanical data which were not normalized for size and mass effects. Mertz et al (4) have developed an injury risk curve for AIS ≥ 3 thoracic injury based on the sternal deflection of the Hybrid III mid-size adult male dummy being restrained by an automotive 3-point belt system (Figure A4). Viano and Lau (5) have proposed an injury risk curve for AIS ≥ 4 thoracic organ injury based on the maximum value of the instantaneous product of the ratio of sternal compression normalized by the thoracic depth and the rate of sternal compression, the Viscous Criterion (Figure A5). This curve can be applied to adults and children because equal viscous criterion levels experienced by

both children and adults will produce equal thoracic organ stresses. Rouhana et al (18) have published injury risk curves for AIS ≥ 3 and 4 abdominal injuries which can be used to assess the severity of abdominal loading measured by the crushing of a special foam abdominal insert which is available with the Hybrid III small female and midsize male dummies. Mertz and Weber (6) have published injury risk curves for measurements made with the 3-year old "airbag" dummy (7). These curves were based on animal and child dummy data of Mertz et al (3) obtained from tests where the subjects were exposed to forces produced by inflating passenger airbags. These data will be combined with similar data obtained by Prasad and Daniel (9) to obtain normalized injury risk curves that can be used with the CRABI and Hybrid III family of child and adult dummies. In addition, the blunt thoracic cadaver impact data of Neathery et al (10) will be updated and normalized to give injury risk curves for AIS ≥ 3 and AIS ≥ 4 thoracic injury based on maximum sternal deflection of the various child and adult dummies.

NECK INJURY RISK CURVES

Mertz et al (8) and Prasad and Daniel (9) have conducted tests to assess the effects of deploying passenger airbag interactions with animals (10-week old pigs) that were chosen to represent the size, weight and state of tissue development of 3-year old children. In their studies, a series of matched tests was conducted where for every pig test a similar test was conducted using the 3-year old "airbag" dummy. This allowed the various injury severities experienced by the pig to be correlated with corresponding dummy response measurements. The neck injuries observed in both studies initiated by the tearing of small blood vessels of the membranes encasing the occipital condylar joint capsules and progressed to rupture of the alar ligament, damage to the spinal cord and brain stem, and finally to fatality as the impact severity increased. Blood in the synovial fluid of the occipital condylar joint capsules was rated as AIS = 3 and occurred in all the cervical neck injuries rated as AIS ≥ 3.

Based on the location and nature of the neck injuries, neck tension, neck extension moment, and a combination of tension and extension moment measured at the occipital condyles of the 3-year old air bag dummy were proposed as indicators of neck injury severity. Both studies showed that neck tension was the best indicator of the onset of AIS $\geq$ 3 neck injury with no AIS $\geq$ 3 neck injury occurring below a neck tension load of 1160 N. However, the severity of the neck injury that corresponded to the neck tension of 1160 N was fatality. As noted in the introduction, Mertz and Weber (7) analyzed the Mertz et al (8) data and provided an injury risk curve for AIS $\geq$ 3 neck injury based on neck tension measured with the 3-year old airbag dummy. This injury risk curve will be updated by analyzing the combined data sets of Mertz and Prasad. Injury risk curves for extension moment and the combination of tension and extension moment will be developed based on the combined data sets. Since both of these data sets are estimates of the tolerance of a 3-year old child, they will be normalized for size and strength considerations to give estimates of injury risk curves for any ages of children or sizes of adults.

The following is a discussion of the development of injury risk curves for AIS $\geq$ 3 neck injury for tension/extension loading of the neck based on peak neck tension, peak neck extension moment and the combination of tension and extension moment which can be used with the CRABI and Hybrid III family of child and adult dummies.

SIZE SCALE FACTORS - Neck circumference was used to characterize neck size. Table 1 gives the neck circumferences for 6-month, 12-month, 18-month, 3-year old, 6-year old, small adult female, mid-size adult male and large adult male which are based on various anthropometry studies (11-14). Since the reference tolerance data pertains to the 3-year old, the neck size scale factor, λ_c, will be defined as,

$$\lambda_c \qquad \frac{\text{Neck Circumference of Subject}}{\text{Neck Circumference of 3-year old}} \qquad (1)$$

The λ_cs for the various children and adults are given in Table 1.

STRENGTH SCALE FACTORS - A search was done of the biomechanical literature for dynamic failure data of ligamentous tissue as a function of age. While static data was found for the adult, no child data and no dynamic data were found. In the absence of such data, maximum tissue stress was chosen as the failure criterion and the failure stress level was assumed to be independent of age. These assumptions allow the elastic modulus of the ligament to vary with age and consequently the strain at failure to vary with age. If, at a later date, child and adult data are obtained for ligamentous failure at appropriate loading rates, then the analyses given in this paper can be updated.

However, it should be noted that the three neck criteria, maximum neck tension, maximum neck extension moment and the maximum value of the combination of neck tension and extension moment are measures of "macro" structural load carrying capacities of the neck structure for tension - extension loading of the neck. As such their failure values will be dependent not only on the ligament load, but also on the corresponding load being transmitted by the muscle groups. This latter load will be dependent on the degree of muscle tension

which is assigned to the occupant and its variation could mask the variation in ligament strength.

NECK TENSION - The relationship between the ratios of neck tension forces, λ_F, the sizes of the necks λ_A, and the average tensile stresses, λ_σ, can be expressed as,

$$\lambda_F = \lambda_\sigma \ \lambda_A \qquad (2)$$

For children and adults, the average stresses based on the cross-sectional areas of their necks will be taken as equal for equal injury severity, ie, $\lambda_\sigma = 1$. The ratio of their cross-sectional areas, λ_A, will be taken as the ratio of the square of their neck circumferences, λ_c^2. From Equation 2, the ratio of the tensile forces that corresponding to equal injury severity is,

$$\lambda_F = \lambda_c^2 \qquad (3)$$

Since data for neck tension forces and corresponding neck injury severities for the 3-year old child exist, the neck tension forces causing the same injury severity in other size occupants can be determined by simply multiplying the 3-year old forces by the square of the ratio of their circumferences.

The neck tension force and corresponding neck injury severity data of Mertz et al (8) and Prasad and Daniel (9) are given in Table A1 of the Appendix. These data were analyzed using the Mertz/Weber Method (6) to obtain the injury risk curve shown in Figure A6 for AIS $\geq$ 3 neck injury based on neck tension forces experienced by a 3-year old child. For convenience, the neck forces are normalized by the tension force, F_1, that corresponds to a 1 percent risk of AIS $\geq$ 3 neck injury. For the 3-year old, this force is 1070 N. For any size person, the corresponding value of F_1 can be determined from Equation 3, or

$$F_1 = \lambda_c^2 \ 1070 \text{ N} \qquad (4)$$

where λ_c is defined by Equation 1.

The normalized injury risk curve for the 3-year old child is identical to the normalized risk curve for, any size occupant provided the normalized force is computed by the relationship given by Equation 4. Figure 1 gives the injury risk curve for AIS $\geq$ 3 neck injury based on normalized neck tension for any size person. Normalizing values for various child and adult dummies are given in the legend. These values were calculated using Equation 4 and the values of λ_c given in Table 1. Note that the normalized values are the neck tensions that produce a 1 percent risk of AIS $\geq$ 3 neck injury for the corresponding dummy. Further, this curve gives an estimate of the injury risk when the neck is being loaded in tension and extension which is the loading mode experienced by the pigs and child dummy in the biomechanical tests.

NECK EXTENSION MOMENT - For structures whose cross-sectional area can be characterized by a single length scale factor, λ_L, Mertz et al (13) have shown that the relationship between the ratios of the internal bending moments, λ_M, and the internal bending stresses, λ_σ, can be expressed as,

$$\lambda_M = \lambda_\sigma \lambda_L^3 \qquad (5)$$

Again we specify $\lambda_\sigma = 1$ for equal injury severity and $\lambda_L = \lambda_c$. From Equation 5, the ratio of neck extension moments associated with neck injury is,

$$\lambda_M = \lambda_c^3 \qquad (6)$$

Since data for neck extension moments and corresponding neck injury severities for the 3-year old child exist, the neck extension moments causing the same injury severity in other size occupants can be estimated from Equation 6.

The neck extension moment and corresponding neck injury severity data of Mertz et al (8) and Prasad and Daniel (9) are given in Table A1 of the Appendix. These data were analyzed using the Mertz/Weber Method (6) to obtain the injury risk curve shown in Figure A7 for AIS $\geq$ 3 neck injury based on the neck extension moments experienced by a 3-year old child. The neck extension moments are normalized by the moment, M_1, that corresponds to a 1 percent risk of AIS $\geq$ 3 neck injury. For the 3-year old, this moment is 13 Nm. For any size person, the corresponding value of M_1 can be determined from Equation 6, or

$$M_1 = \lambda_c^3 \ 13.0 \ Nm \qquad (7)$$

where λ_c is defined by Equation 1.

The normalized injury risk curve for the 3-year old child is identical to the normalized risk curve for any size occupant provided the normalized moment is computed by the relationship given by Equation 7. Figure 2 gives the injury risk curve for AIS $\geq$ 3 neck injury based on normalized neck extension moment for any size person. Normalizing values for various child and adult dummies are given in the legend. These values were calculated using Equation 7 and the values of λ_c given in Table 1. Note that the normalized values are the neck extension moments that produce a 1 percent risk of AIS $\geq$ 3 neck injury for the corresponding dummy. Further, this curve gives an estimate of the injury risk when the neck is being loaded in tension and extension, which is the loading mode experienced by the pigs and child dummy in the biomechanical tests.

COMBINED TENSION AND EXTENSION MOMENT - The following is an approach to combining the tension and extension moment loadings. Let A be the cross-sectional area of the membrane and D be the distance from the anterior surface of the atlas to its posterior surface. Assume that one half the measured tensile force is carried by the membrane and that the membrane tensile force produced by the extension moment is equal to the measured extension moment divided by D. With these assumptions, the total force in the anterior membrane is,

$$P = M_E / D + F_T / 2 \qquad (8)$$

and the stress is,

$$\sigma = P/A = (AD)^{-1} [M_E + DF_T/2] \qquad (9)$$

In terms of neck scale factor, λ_c, defined by Equation 1, Equation 9 can be written as,

$$\sigma = (\lambda_c^3 \ A_3 \ D_3)^{-1} [M_E + \lambda_c \ D_3 \ F_T / 2] \qquad (10)$$

where A_3 and D_3 are the 3-year old dimensions.

Now a kernel, K, of M_E and F_T can be defined as

$$K = [M_E + \lambda_c \ D_3 \ F_T / 2] = \sigma \ \lambda_c^3 \ A_3 \ D_3 \qquad (11)$$

and represents lines of constant stress with a slope of $-\lambda_c \ D_3 / 2$ on a plot of M_E versus F_T. Table A2 gives values of K calculated from the data of Mertz et al (8) and Prasad and Daniel (9) along with the corresponding neck injury severity ratings. Since these data are representative of a 3-year old child, $\lambda_c = 1$ and $D_3 = 25.2$ mm were used in the calculations. These data were analyzed by the Mertz / Weber Method (6) to give an injury risk curve shown in Figure A8 for AIS $\geq$ 3 neck injury based on K values for a 3-year old child. For a 1 percent risk of AIS $\geq$ 3 neck injury, K = 20.0. The corresponding stress, σ_1, can be calculated from Equation 11 and is,

$$\sigma_1 = 20.0 / A_3 \ D_3 \qquad (12)$$

The stress level given by Equation 10 can be normalized by σ_1, the stress level corresponding to a 1 percent risk of AIS $\geq$ 3 neck injury or,

$$N_{TE} = \sigma / \sigma_1 = [M_E + \lambda_c \ D_3 \ F_T /2] / (20.0 \ \lambda_c^3) \qquad (13)$$

The normalized stress can be expressed in terms of the ordinate value, M_C, and the abscissa value, F_C, of the constant stress line corresponding to 1 percent risk of AIS $\geq$ 3 neck injury for any size occupant, or,

$$N_{TE} = M_E / M_C + F_T / F_C \qquad (14)$$

where from inspection of Equation 13,

$$M_C = 20.0 \ \lambda_c^3 \ Nm \qquad (15)$$

$$F_C = 1590 \ \lambda_c^2 \ N \qquad (16)$$

The injury risk curve for AIS $\geq$ 3 neck injury for combined normalized neck tension and extension moment for the 3-year old child can be obtained by dividing the tension and extension moments given in Table A2 by the corresponding values of M_C and F_C given by Equations 15 and 16 noting that $\lambda_c = 1$ for the 3-year old child. The resulting curve is identical to the injury risk curve for any size occupant provided M_C and F_C are calculated using Equations 15 and 16.

Figure 3 gives the injury risk curve for AIS $\geq$ 3 neck injury based on the normalized stress, N_{TE}, produced by combined neck tension and extension moment for any size person. Values of M_C and F_C for various sizes of child and adult dummies are given in the legend. These values were calculated using Equations 15 and 16 and the values of λ_c given in Table 1.

THORACIC INJURY RISK CURVES

RATE OF STERNAL COMPRESSION - Table A3 in the Appendix gives data of Mertz et al (8) and Prasad and Daniel (9) for the injury severities experienced by the heart and lungs of their animals and the corresponding maximum rates of sternal deflection measured with the 3-year old child dummy. These data were analyzed with the Mertz/Weber Method (6) to give the injury risk curve for AIS ≥ 3 heart/lung injury as a function of the maximum rate of sternal deflection for a 3-year old child shown in Figure 4. Since no age dependent, dynamic loading failure stress data were found in the literature, the failure stresses of the heart and lung tissues were assumed to be independent of age. With this assumption, the risk curve of Figure 4 can be used for all size occupants since equal rates of sternal deflection produce equal stress levels in the heart and lungs (13, 14).

An estimate of the risk curve for AIS ≥ 4 heart/lung injury can be obtained from the data given in Table A3. Because of the limited amount of AIS ≥ 4 data, the Mertz/Weber Method is used only to estimate the rate of sternal deflection corresponding to a 50 percent risk of AIS ≥ 4 injury. This value is 10.2 m/s. The corresponding standard deviation is assumed equal to that of the AIS ≥ 3 risk curve. The resulting AIS ≥ 4 risk curve is shown on Figure 4.

STERNAL COMPRESSION - Neathery et al (10) have summarized the thoracic impact data of various investigators who have subjected cadavers to distributed chest impacts. The data were presented in terms of ratios of peak chest compression divided by chest depth and the corresponding thoracic injury severities. These data were reviewed and the AIS ratings were updated. In addition, two cadavers which were impacted twice were deleted from the data set. The revised data are given in Table A4 of the Appendix.

The chest deflections of the mid-size adult male were obtained by multiplying the P/D values by 229 mm which is the chest depth of the mid-size adult male. To account for the effect of compression of the flesh covering the sternum, the chest compressions were reduced by 13 mm to get estimates of the sternal deflections which are also given in Table A4. These values of sternal deflections and the associated injury severities were analyzed by the Mertz/Weber Method to get injury risk curves shown on Figure A9 for AIS ≥ 3 and 4 thoracic injury for the mid-size adult male. The following is a discussion of how these risk curves were extended to other size adults and children.

AIS ≥ 3 INJURY RISK CURVE - Rib fractures are the predominant injury in the AIS ≥ 3 data set. Since the bending modulus of bone varies with age, it will affect the amount of sternal deflection required to produce rib fracture. Again, because of a lack of age-dependent, dynamic loading failure data, an equal failure stress level was assumed, or

$$\lambda_\sigma = \lambda_E \, \lambda_\varepsilon = 1 \tag{17}$$

where λ_σ is the ratio of failure stresses, λ_E is the ratio of bending moduli, and λ_ε is the ratio of bone strains. Now bone strain, ε, can be defined in terms of the sternal deflection, δ, and the thoracic depth, D, or,

$$\varepsilon = \delta / D \tag{18}$$

Using Equations 17 and 18, the relationship between sternal deflections that produce the same rib stress is,

$$\delta_i = \lambda_E^{-1} \, \lambda_x \, \delta_M \tag{19}$$

where

$$\lambda_x = D_i / D_M \tag{20}$$

$$\lambda_E = E_i / E_M \tag{21}$$

Equations 19, 20 and 21 provide the necessary relationships to calculate injury risk curves for AIS ≥ 3 rib fractures for any size occupant using mid-size adult male data given in Table A4 and knowledge of the elastic bending moduli of the ribs.

For rib fractures, we have chosen to normalize the sternal deflection by the sternal deflection, δ_c, corresponding to a 5 percent risk of AIS ≥ 3 injury. From Figure A9, $\delta_c = 47.7$ mm for the mid-size adult male. The δ_c for any size occupant can be calculated from Equation 19, or,

$$\delta_c = \lambda_E^{-1} \, \lambda_x \, 47.7 \text{ mm} \tag{22}$$

where λ_E and λ_x are given in Table A5 for various sizes and ages of occupants.

Figure 5 gives the injury risk curve for AIS ≥ 3 thoracic injury based on the normalized sternal deflection for any size occupant. This curve was developed by dividing the sternal deflections for the mid-size adult male given in Table A4 by 47.7 mm and then analyzing the normalized values and corresponding injury severity values using the Mertz/Weber Method. The legend gives the sternal deflections for 5 percent risk of AIS ≥ 3 thoracic injury for the various size dummies. Note that no values are listed for the CRABI and Hybrid III child dummies because rib fracture, which is the predominant AIS ≥ 3 injury in the data set, is unlikely to occur with children of these ages due to the low elastic bending moduli of their ribs.

AIS ≥ 4 INJURY RISK CURVE - Heart and/or aortic rupture are the predominant AIS ≥ 4 injury. These types of injury are dependent on the ratio of sternal deflection to chest depth being equal between different size subjects. The relationship between sternal deflection that produce equal stress in the heart is,

$$\delta_i = \lambda_x \, \delta_M \tag{23}$$

where λ_x is the ratio of chest depth of the two subjects. Again, the sternal deflection will be normalized by the deflection corresponding to a 5 percent risk of AIS ≥ 4 heart injury. From Figure A9, this value is 64.3 mm for the mid-

injury. From Figure A9, this value is 64.3 mm for the mid-size adult male. For any size occupant, the corresponding sternal deflection, δ_c, can be computed from Equation 23, or.

$$\delta_c = \lambda_x \ 64.3 \ mm \qquad (24)$$

Figure 5 gives the injury risk curve for AIS ≥ 4 heart injury based on normalized sternal deflection for any size person. This curve was obtained by dividing the sternal deflections of the mid-size adult male given in Table A4 by 64.3 mm. These normalized values and the corresponding AIS ≥4 values were analyzed by the Mertz/Weber Method to obtain the curve of Figure 5. The legend gives the sternal deflections for 5 percent risk of AIS ≥4 thoracic injury for various dummy sizes. Note that even the 6-month, 12-month and 18-month old children have a risk of experiencing an AIS ≥4 heart rupture due to crushing of the chest.

ABDOMINAL INJURY RISK CURVES

RATE OF ABDOMINAL COMPRESSION - Table A3 gives the data of Mertz et al (8) and Prasad and Daniel (9) for the injury severities experienced by the abdominal organs of the animals and the corresponding maximum rates of abdominal compression measured with the 3-year old child dummy. These data were analyzed with the Mertz/Weber Method to give injury risk curves for AIS ≥3 and AIS ≥ 4 abdominal injury as function of the maximum rate of abdominal compression (Figure 6). For the AIS ≥ 4 injury. the 50 percent value was obtained by the Mertz/Weber Method, but the standard deviation of the AIS ≥3 curve was used. Note that there is very little difference between the risk curves. Again, these risk curves can be used for all size occupants since equal rates of abdominal compression will produce equal stresses in the abdominal organs.

EFFICACY OF INJURY RISK CURVES

The efficacies of the various injury risk curves are the best when used to assess risks for subjects of the same age and size as that of the original test subjects. The normalization for size and material strength consideration is based on the laws of classical structural mechanics. The size scale factors have excellent efficacy since they are based on average dimensions taken from anthropometry studies. The strength scale factors lack rigorous supporting data since tissue failure stress data based on variations of age and dynamic loading rate were not found in the technical literature. In lieu of such data, it was assumed that equal stress would produce equal injury severity, independent of age. This assumption allows for variation of the elastic modulus with age and consequently variations in the strain level at failure as a function of age. For materials with time-dependent properties, this is a necessary requirement.

For the neck, the load it can carry prior to injury will be dependent not only on the failure stress level of the ligament, but more importantly, the degree of muscle tensing that has occurred prior to and during loading. It is the neck muscles that protect the neck ligaments from being overloaded. To assess the potential for neck injury based on

measured internal reactions between the head and the neck, the degree of muscle tone must be specified in order to determine how the internal neck load is distributed among the muscle groups and ligaments. The animals whose data were used to develop the neck injury risk curves were anesthetized. They had some passive muscle reaction which was well below maximum active muscle tension. Upper bounds of neck injury risk curves corresponding to maximum active muscle tension can be obtained by adding such levels to the critical values listed in the legends of each risk curve. Based on analysis of human volunteer tests, Mertz et al (19, 20) noted that statically the average size man could resist 1100 N (255 lb) in pure tension and 23.7 Nm (17.5 ft-lb) moment when resisting neck extension. Using Equations 3 and 6 and the neck circumference data given in Table 1, corresponding static neck muscle strength values were calculated for the other sizes of people and are given in Table 2. These static strength values were added to the critical values for neck injury risk curves (Figs. 1 - 3) to give critical values for maximum muscle tensing which are given in Table 2. Thus. neck injury risks can be calculated with minimal or maximal neck tension by using the critical values shown in the legends of Figures 1 - 3, or by using instead the values given in Table 2. respectively.

To assess the efficacy of the neck injury risk curves for peak tension and for peak extension moment, injury risks associated with the various published Injury Assessment Reference Values (IARV) for the neck (15-17) for minimal and maximal muscle tensing were obtained from the graphs of Figures 1 and 2 and are given in Table 3. Note that if an IARV is not exceeded, then the risk of significant neck injury is judged to be unlikely; i.e., ≤ 1 percent risk for children and ≤ 5 percent risk for adults. Based on the comparisons given in Table 3, there was excellent agreement between the calculated risks, both with and without muscle tone, and the IARVs for neck extension moment for adult and children, and for neck tension of children. For neck tension IARVs for adults, the risks are quite high if muscle tension is minimal, but quite low if muscle tension is maximal. This clearly points out the need to have a consensus on the degree of muscle tone to be assumed when using the risk curves. Perhaps the minimal muscle tension risk curves could be used for children, while 80 percent of the maximum static muscle tension levels could be used for the adult.

In general, it would seem appropriate to prescribe the desired protection level for a given simulation condition and then use the injury risk curves to set the performance limits for various dummy measurements.

SUMMARY

Biomechanical tolerance data for the neck, thorax and abdomen have been normalized for body size and tissue failure properties to give normalized injury risk curves for neck tension, neck extension moment, combined neck tension and extension moment, sternal compression, rate of sternal compression and the rate of abdominal compression for children and adults. The efficacies of the risk curves are best when used to assess risks of dummy data where the dummy represents the size and age of the original test subjects. Of the three neck criteria, neck tension is the preferred criterion since the correlation with the animal injury was the best for this measure. Once the degree of muscle tone and desired level of protection are prescribed for a given collision simulation condition, the injury risk curves can be used to prescribed design limits for the corresponding dummy measurements.

REFERENCES

1. Prasad, P. and Mertz, H. J., "The Position of the United States Delegates to the ISO Working Group 6 on the Use of HIC in the Automotive Environment", SAE 851246, 1985.
2. Mertz, H. J., Prasad. P. and Nusholtz, G., "Head Injury Risk Assessment for Forehead Impacts", SAE 960099, February, 1996.
3. Mertz, H. J., Prasad, P. and Nusholtz, G., "Head Injury Risk Assessments Based on 15 ms HIC and Peak Head Acceleration Criteria", Proceeding of AGARD Meeting on Impact Head Injury, November 7-9, 1996.
4. Mertz, H. J., Horsch, J. D., Horn, G., and Lowne, R. W., "Hybrid III Sternal Deflection Associated with Thoracic Injury Severities of Occupants Restrained with Force - Limiting Shoulder Belts", SAE 910812, February, 1991.
5. Viano, D. V. and Lau, I. V., "Thoracic Impact: A Viscous Tolerance Criterion", Proceeding of the Tenth Experimental Safety Vehicle Conference, July, 1985.
6. Mertz, H. J. and Weber, D. A., "Interpretations of the Impact Responses of a 3-Year Old Child Dummy Relative to Child Injury Potential", Proceedings of the Ninth International Technical Conference on Experimental Safety Vehicles, Kyoto, Japan, November 1-4, 1982. (Also published in SAE 826048, SP-736 Automatic Occupant Protection Systems, February, 1988).
7. Wolanin, M. J., Mertz, H. J., Nyznyk, R. S., and Vincent, J. H., "Description and Basis of a Three-Year-Old Child Dummy for Evaluation Passenger Inflatable Restraint Concepts", Proceedings of the Ninth International Technical Conference on Experimental Safety Vehicles, Kyoto, Japan, November . 1-4, 1982 (Also published in SAE 826040, SP-736, February, 1988).
8. Mertz, H. J., Driscoll G. D., Lenox, J. B., Nyquist, G. W., and Weber. D. A., "Responses of Animals Exposed to Deployment of Various Passenger Inflatable Restraint System Concepts for a Variety of Collision Severities and Animal Positions", Proceedings of the Ninth International Technical Conference on Experimental Safety Vehicles, Kyoto, Japan, November 1-4, 1982. (Also published in SAE 826047, PT31).
9. Prasad, P. and Daniel, R. P., "A Biomechanical Analysis of Head, Neck and Torso Injuries to Child Surrogates Due to Sudden Torso Acceleration", Twenty-Eighth Stapp Car Crash Conference, SAE 841656, November, 1984.
10. Neathery, R. F., Kroell, C. K. and Mertz, H. J., "Prediction of Thoracic Injury from Dummy Responses", Nineteenth Stapp Car Crash Conference, SAE 751151, November, 1975.
11. Weber, K. and Lehman, R. J., "Child Anthropometry for Restraint System Design", UMTRI-85-23, June, 1985.
12. Schneider, L. W., Robbins, D. H., Pflüg, M. A., and Snyder, R. G., "Development of Anthropometrically Based Design Specification for an Advanced Adult Anthropomorphic Dummy Family", Volume 1, UMTRI-83-53-1, December, 1983.
13. Mertz, H. J., Irwin, A. L., Melvin, J. W., Stalnaker, R. L., and Beebe, M. S., "Size, Weight and Biomechanical Impact Response Requirements for Adult Size Small Female and Large Male Dummies", SAE 890756, March, 1989.
14. Mertz, H. J., and Irwin, A. L., "Biomechanical Basis for the CRABI and Hybrid III Child Dummies", Forty-First Stapp Car Crash Conference November, 1997.
15. Mertz, H. J., "Anthropomorphic Test Devices", Accident Injury - Biomechanics and Prevention, Springer-Verlag, N. Y., 1993.
16. "Anthropomorphic Dummies for Crash and Escape System Testing", AGARD-AR-330, July, 1996.
17. Melvin, J. W., "Injury Assessment Reference Values for the CRABI 6-Month Infant Dummy in a Rear-Facing Infant Restraint with Airbag Deployment", SAE 950872, February, 1995.
18. Rouhana, S. W., Jedrzejczak, E. A. and McCleary, J. P., "Assessing Submarining and Abdominal Injury Risk in the Hybrid III Family of Dummies: Part II - Development of the Small Female Frangible Abdomen", Thirty-Fourth Stapp Car Crash Conference, SAE 902317, November, 1990.
19. Mertz, H. J. and Patrick, L. M., "Investigation of the Kinematics and Kinetics of Whiplash", Eleventh Stapp Car Crash Conference, October, 1967
20. Mertz, H. J. and Patrick, L. M., "Strength and Response of the Human Neck", SAE 710855, Fifteenth Stapp Car Crash Conference, November 1971.

Table 1 - Neck Circumferences for Various Ages of Children and Sizes of Adults.

Dummy	Circum. (mm)	λ_C	Ref.
CRABI 6	221	0.906	11
CRABI 12	226	0.918	11
CRABI 18	226	0.926	11
H III - 3 Yr.	244	1.000	11
H III - 6 Yr.	264	1.082	11
H III - Sm. Fem.	304	1.246	12
H III - Mid-Male	383	1.570	12
H III - Lg. Male	421	1.725	12

Table 2 - Critical Values for Neck Injury Risk Curves Based on Maximum Static Muscle Strengths.

| Dummy | Maximum Static Muscle Strengths | | Critical Values for Risk Curves Based on Maximum Static Muscle Strengths | | | |
	Tension (N)	Ext. Mom. (Nm)	F_1 (N)	M_1 (Nm)	F_C (N)	M_C (Nm)
CRABI 6	366	4.6	1246	14.3	1676	19.5
CRABI 12	383	4.9	1303	15.2	1723	20.4
CRABI 18	383	4.9	1303	15.2	1743	20.8
H III - 3 Yr.	446	6.1	1516	19.1	2036	26.1
H III - 6 Yr.	522	7.8	1772	25.6	2382	33.1
H III - Sm. Fem.	693	11.9	2353	37.0	3163	50.6
H III - Mid-Male	1100	23.7	3740	74.0	5020	101.1
H III - Lg. Male	1330	31.5	4520	98.3	6060	134.2

Table 3 - Risks of AIS ≥ 3 Neck Injury Associated with Injury Assessment Reference Values (IARV)

| Dummy | Neck Tension | | | Neck Ext. Mom. | | |
| | IARV (N) | Risk AIS ≥ 3 (%) | | IARV (Nm) | Risk AIS ≥ 3 (%) | |
		Min. Muscle	Max. Muscle		Min. Muscle	Max. Muscle
CRABI 6	500	< 0.1	< 0.1	5	0.2	< 0.1
CRABI 12	920	1.0	< 0.1	7	0.4	< 0.1
CRABI 18	920	1.0	< 0.1	7	0.4	< 0.1
H III - 3 Yr.	1000	0.2	< 0.1	10	0.5	0.2
H III - 6 Yr.	1300	2.0	< 0.1	13	0.4	0.1
H III - Sm. Fem.	2200	50	0.2	31	2.0	0.5
H III - Mid-Male	3300	30	< 0.1	57	1.4	0.4
H III - Lg. Male	4050	35	< 0.1	78	1.6	0.4

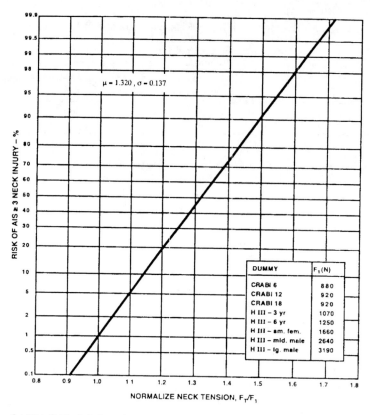

Figure 1 - Risk of AIS ≥ 3 Neck Injury for CRABI and Hybrid III Dummy Families as a Function of Normalized Neck Tension for Tension - Extension Loading of the Neck.

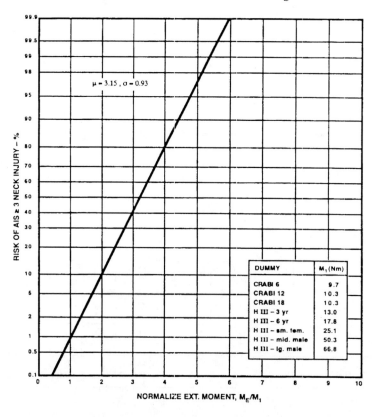

Figure 2 - Risk of AIS ≥ 3 Neck Injury for CRABI and Hybrid III Dummy Families as a Function of Normalized Neck Extension Moment for Tension - Extension Loading of the Neck.

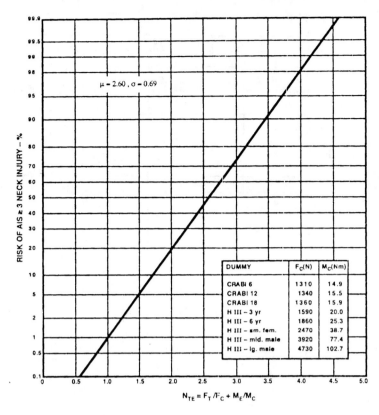

Figure 3 - Risk of AIS ≥ 3 Neck Injury for CRABI and Hybrid III Dummy Families as a Function of Combined Normalized Neck Extension Moment and Tension.

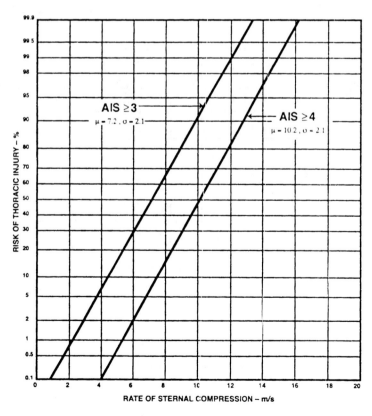

Figure 4 - Risk of AIS ≥ 3 and AIS ≥ 4 Heart/Lung Injury as a Function of Rate of Sternal Compression.

221

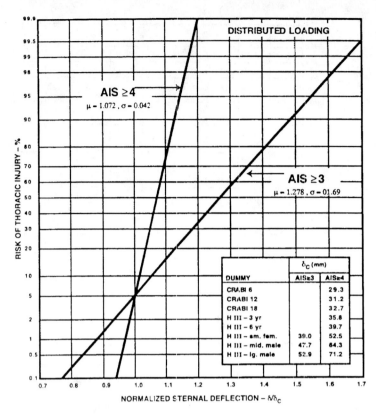

Figure 5 - Risk of AIS ≥ 3 and AIS ≥ 4 Thoracic Injuries for Distributed Chest Impacts as a Function of Normalized Sternal Deflection.

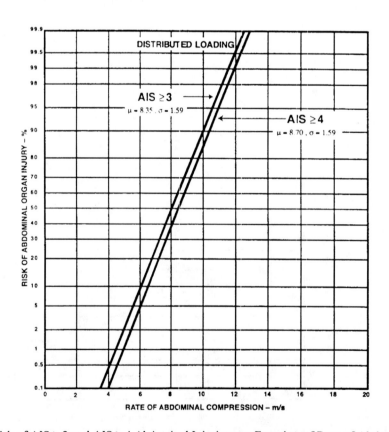

Figure 6 - Risk of AIS ≥ 3 and AIS ≥ 4 Abdominal Injuries as a Function of Rate of Abdominal Compression.

Table A1 - Mertz et al (8) and Prasad & Daniel (9) Data for Peak Neck Tension and
Peak Neck Extension Moment with Corresponding Neck Injury Severities.

Pk. Neck Tension				Pk. Neck Extension Moment			
Tension (N)	AIS	Tension (N)	AIS	Ext. Mom. (Nm)	AIS	Ext. Mom. (Nm)	AIS
400	0	1410	0	11.3	0	26.0	0
525	0	1430	6	14.0	0	26.0	0
525	0	1445	3	15.0	4	26.0	0
525	0	1460	0	15.8	0	29.4	6
560	0	1480	4	17.0	0	30.0	0
574	0	1490	4	18.0	1	33.9	6
588	0	1500	3	18.0	3	37.3	4
625	0	1500	4	18.0	0	37.3	0
635	0	1530	3	20.0	4	42.4	4
680	0	1570	0	20.0	3	46.0	5
805	1	1920	4	20.0	4	46.3	4
813	0	1925	4	20.0	0	46.3	6
813	0	1925	6	20.0	0	47.5	6
855	0	2270	4	20.0	0	63.0	3
938	0	2270	6	23.0	0	64.0	4
943	2	2270	6	23.0	0	64.0	6
960	0	2680	3	23.2	0	64.0	6
1050	0	2820	6	24.0	5	66.0	4
1150	0	2960	4	25.0	0	66.0	3
1160	6	3040	5	25.0	0	67.0	2
1250	4	4100	5	25.4	0	80.0	5
1260	5			25.4	0		

Table A2 - Mertz et al (8) and Prasad & Daniel (9) Data for Instantaneous Peak Kernel (K) of Neck Extension Moment and Neck Tension with Corresponding Neck Injury Severity.

Ext. Mom (Nm)	Tension (N)	K (Nm)	AIS	N_{TE}	Ext. Mom. (Nm)	Tension (N)	K (Nm)	AIS	N_{TE}
5.6	1030	18.6	0	0.93	18.0	1530	37.3	3	1.87
14.0	680	22.6	0	1.13	20.0	1445	38.2	3	1.91
0	1925	24.3	4	1.22	20.0	1490	38.8	4	1.94
18.0	574	25.2	0	1.26	15.0	1920	39.2	4	1.96
7.9	1460	26.3	0	1.32	30.0	938	41.8	0	2.09
20.0	588	27.4	0	1.37	22.6	1660	43.5	0	2.18
15.8	960	27.9	0	1.40	46.3	0	46.3	6	2.32
20.0	635	28.0	0	1.40	37.3	1164	52.0	4	2.60
18.0	805	28.1	1	1.41	37.8	1254	53.6	0	2.68
23.0	625	30.9	0	1.55	47.5	760	57.1	6	2.86
26.0	525	32.6	0	1.63	46.0	1260	61.9	5	3.10
26.0	525	32.6	0	1.63	24.0	3040	62.3	5	3.12
26.0	525	32.6	0	1.63	67.0	943	78.9	2	3.95
29.4	313	33.3	6	1.67	42.4	2960	79.7	4	3.99
26.4	560	33.5	0	1.68	66.0	1500	84.9	3	4.25
23.0	855	33.8	0	1.69	66.0	1500	84.9	4	4.25
33.9	0	33.9	6	1.70	64.0	2270	92.6	4	4.63
20.0	1150	34.5	0	1.73	64.0	2270	92.6	6	4.63
25.0	813	35.2	0	1.76	64.0	2270	92.6	6	4.63
25.0	813	35.2	0	1.76	63.0	2680	96.8	3	4.84
20.0	1250	35.8	4	1.79	80.0	4100	131.7	5	6.59

Note: $K = M_E + 0.0126 F_T$ (See Equation 11)

$N_{TE} = K/20.0 = M_E / M_C + F_T / F_C$ (See Equations 14, 15 & 16)

Table A3 - Rates of Sternal and Abdominal Deflections and Corresponding Thoracic and Abdominal Injury Severities of Mertz et al (8) and Prasad & Daniel (9).

Sternal Deflection Rate (m/s)				Abdominal Deflection Rate (m/s)			
$\dot{\delta}$	AIS	$\dot{\delta}$	AIS	$\dot{\delta}$	AIS	$\dot{\delta}$	AIS
0.6	2	5.6	1	0.6	2	4.0	2
1.1	2	5.6	2	0.8	2	4.6	0
1.4	1	6.4	2	1.1	2	4.7	2
1.4	2	6.4	3	1.1	0	4.9	2
1.7	2	6.7	1	1.1	0	5.0	1
1.9	2	7.8	4	1.2	2	5.6	3
2.2	1	8.3	4	1.4	1	5.6	2
2.2	1	8.5	3	1.4	2	5.8	0
2.5	2	8.6	3	1.5	0	5.8	0
3.0	3	8.6	4	2.2	0	5.8	4
3.3	3	8.6	5	2.2	0	6.1	3
3.6	1	9.7	2	2.5	0	6.1	5
3.6	1	9.7	2	2.5	2	6.1	6
3.9	1	10.7	3	2.5	0	6.7	4
3.9	2	10.7	2	2.8	0	7.6	5
4.7	2	11.3	3	3.1	1	7.8	4
4.7	1	11.3	3	3.3	0	8.2	3
4.9	2	11.3	2	3.3	0	8.5	3
5.0	3	11.6	4	3.3	2	8.5	0
5.3	1	11.9	3	3.4	0	9.8	3
5.3	1	12.5	3	3.4	0	10.7	3
5.3	0	12.8	4	3.6	0	11.6	2
				3.9	2		

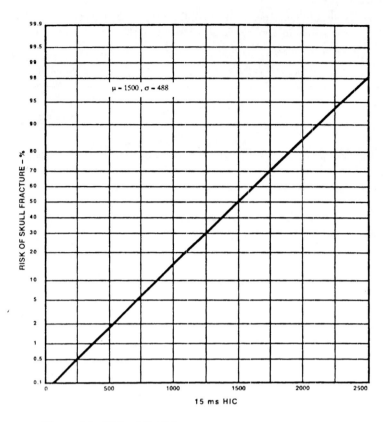

Figure A1 - Risk of Skull Fracture as a Function of 15ms HIC.

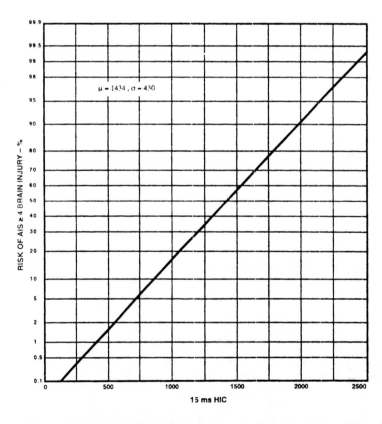

Figure A2 - Risk of AIS ≥ 4 Brain Injury as a Function of 15ms HIC.

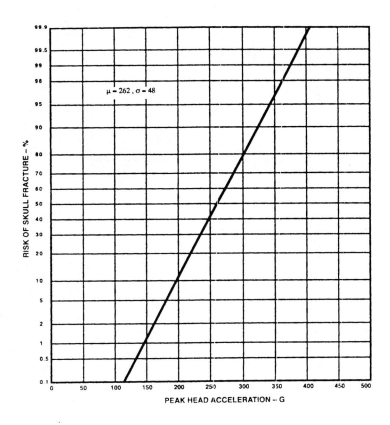

Figure A3 - Risk of Skull Fracture as a Function of Peak Resultant Acceleration of the
Center of Gravity of the Head.

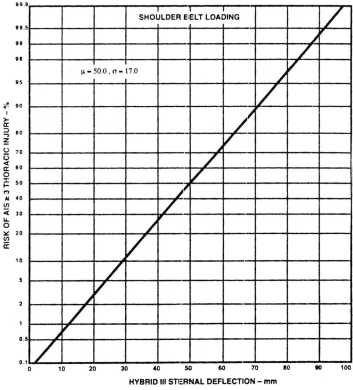

Figure A4 - Risk of AIS ≥ 3 Thoracic Injury Due to Shoulder Belt Loading as a Function
of Hybrid III Sternal Deflection.

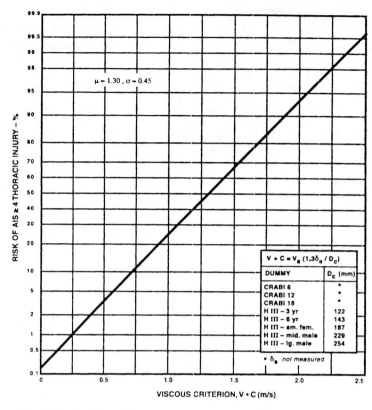

Figure A5 - Risk of AIS ≥ 4 Thoracic Injury as a Function of the Viscous Criterion, V*C.

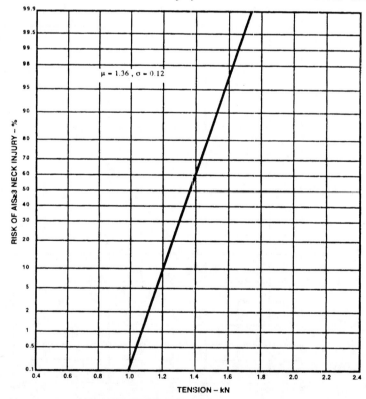

Figure A6 - Risk of AIS ≥ 3 Neck Injury for 3 Year Old Child as a Function of Neck Tension for Tension - Extension Loading of the Neck.

228

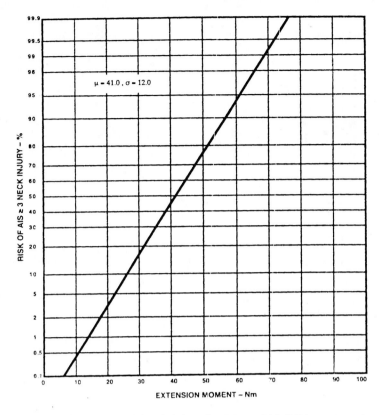

Figure A7 - Risk of AIS ≥ 3 Neck Injury for 3 Year Old Child as a Function of Neck Extension Moment for Extension - Tension Loading of the Neck.

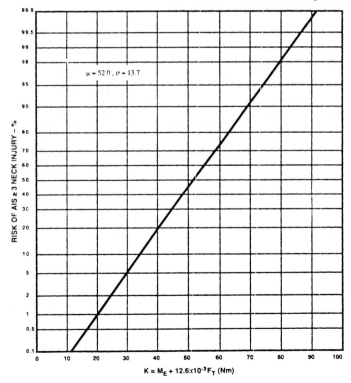

Figure A8 - Risk of AIS ≥ 3 Neck Injury for 3 Year Old Child as a Function of Combined Neck Extension Moment and Neck Tension for Extension - Tension Neck Loading.

Table A4 - Normalized Chest Compression (P/D), Thoracic Injury Severity Ratings, and Sternal Deflection (δ_M) of Mid-Size Adult Male Based on Cadaver Data Summarized by Neathery et al (10).

P/D	AIS	δ_M (mm)	P/D	AIS	δ_M (mm)
0.185	0	29	0.371	3	72
0.194	0	31	0.375	2	73
0.257	2	46	0.393	4	77
0.269	3	49	0.395	4	77
0.310	1	58	0.418	4	83
0.310	2	58	0.425	5	84
0.315	3	59	0.428	5	85
0.321	1	61	0.435	5	87
0.346	4	66	0.444	5	89
0.350	1	67	0.447	4	89
0.363	3	70	0.459	5	92

Note: $\delta_M = 229 \, (P/D) - 13$

Table A5 - Chest Depth and Rib Bending Modulus Scale Factors for Various Sizes and Ages of Occupants Based on the Data of Mertz and Irwin (14).

Subject	λ_X	λ_E
6-Month Old	0.455	0.282
12-Month Old	0.485	0.322
18-Month Old	0.508	0.362
3-Year Old	0.556	0.473
6-Year Old	0.618	0.667
Sm. Adult Fem.	0.817	1.000
Mid-Size Male	1.000	1.000
Lg. Adult Male	1.108	1.00

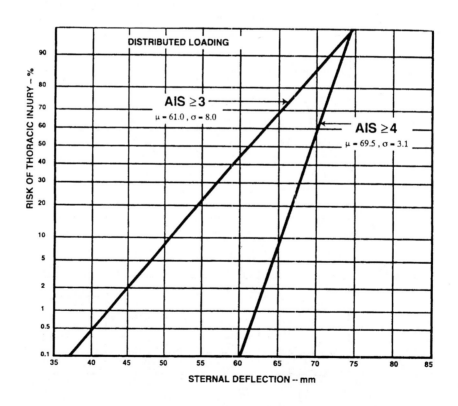

Figure A9 - Risk of AIS ≥ 3 and AIS ≥ 4 Thoracic Injuries for Distributed Chest Loading as a Function of Sternal Deflection of a Mid-Size Adult Male.

High Speed X-Ray Assessment of the Bony Kinematics of the Cervical Spine During Frontal Impacts

S. A. Millington, E. Tomasch, E. Mayrhofer, H. Hoschopf, M. Hofinger and H. Steffan
Frank Stronach, Vehicle Safety Institute, Technical University of Graz

E. P. Leinzinger and M. Darok
The legal Institute for Medicine, Medical University, Graz

ABSTRACT

The aim of this study was to assess the feasibility of using a high speed x-ray system (capable of 1000 frames/sec) to evaluate the bony kinematics of post mortem human surrogate (PMHS) cervical spines at real world speeds during frontal impact.

Whole body frontal impact sled tests were performed on two fresh PMHS specimens. Screws were inserted into the tips of the spinal processes to optimize contrast on the high speed cine x-ray. Head, T1 and sternum accelerations, as well as shoulder, and lap belt forces were recorded.

Vertebral motion was captured using a modified mobile c arm x-ray system, and an image intensifier linked to a high speed camera (Kodak motion corder analyzer, model SR 1000C, Kodak, San Diego, CA, USA)

The variable parameters for the tests were camera frame rate and sled velocity. Tests were performed with delta-V's (Δv) of approximately 15 kph (8G) and 21 kph (10G). Cine x-rays of the tests were recorded at 250, 500 and 1000 frames /sec. Motion tracking analysis of the bony elements of the cervical spine was performed.

Only tests recorded a 250 frames/sec produced high enough quality images to perform motion analysis.

The results show high speed cine x-ray can be used to directly visualize the bony skeleton, for the assessment of skeletal kinematics during impact. Analysis of bony kinematics using high speed cine x-ray reduces errors caused by extrapolating data from external surface makers. Additionally, the system also offers the potential for detecting the site at which fractures originate. It also allows visualization of the timing of fractures, and the order they occur in.

Further research and testing is required to optimize the use of the system, to increase the field of view and improve image quality at the highest frame rates (> 500 frames / sec).

INTRODUCTION

Accident conditions and injury mechanisms for cervical spine disorders following whiplash have been extensively studied. However, we still lack a true understanding of the injury mechanism. The Quebec Task Force on Whiplash-Associated Disorders, (Spitzer et al 1995) reported that over 3000 papers have been published in relation to understanding whiplash injuries and that limited diagnostic techniques were a considerable limitation to our understanding of the injury.

The majority of whiplash associated disorder (WAD) research has focused on rear impact mechanisms. There has been considerably less investigation of injury mechanisms in frontal and lateral impacts. However, several studies have shown that a significant proportion of whiplash cases occur in frontal collisions. Galasko et al (1993), found that 27% of claimed whiplash cases occurred in frontal collisions. Langwieder and Hell (1996) studied a database of 15,000 cases and found 11% of WAD cases were as a result of a frontal collision. An even higher percentage of neck complaints cases (53%) resulted from frontal collisions in a study of severe collisions by Temming and Zobel (1998).

A review of vertebral column injuries, in lap shoulder belt restrained vehicle occupants, in frontal collisions, by Huelke et al (1992, 1995) and Larder et al (1995), concluded that cervical spine injuries were the result of forward flexion with or without distraction, and with or without some rotation. They also concluded that these injuries may result purely from the inertial forces of the mass of the head acting on the neck. In frontal collisions with belted occupants, both Ewing et al (1975) and Walz and Moser (1995) concluded that in the first phase of impact a purely translational head motion occurs followed by flexion of the cervical spine during the next phase of the collision. Kullgren et al (2000), studied the relationship between the crash pulse in frontal collisions and the likelihood of sustaining a long term (> 6 months) soft tissue neck injury. They concluded that when a large decrease in the delta V occurred between the middle third and last third of the crash pulse the probability of a long term injury increased. Furthermore, the timing and nature of contact of the torso with the belt was felt to play a significant role in determining the long term outcome.

A variety of test methods have been used to study different aspects of cervical spine kinematics and cervical spine soft tissue injuries (table 2). Most testing has concentrated on rear impact studies. Panjabi and co-workers (Panjabi et al 1998 a, b, c, d and Cholewicki et al 1998), tested whole cervical spine specimens to assess the flexibility of the cervical spine. They utilized a bench top test system, and used a steel head to simulate loading during a rear end impact. Whilst studies of this nature provide invaluable information, and are relatively inexpensive, they are limited as the effect of interaction with the torso is eliminated.

Authors	Year	Method	Strengths	Limitation
Panjabi et al & Cholewiciki et al	1998	Bench top tests	Inexpensive Detailed spinal instrumentation possible	Effect of torso eliminated
Matsushita et al	1994	Volunteer sled tests with high speed x-ray. Multiple impact directions	Volunteers with active muscle tone were used	Tests only in non-injurious range. No detailed analysis
Kaneoka et al	1999	Volunteer rear impact sled tests with high speed x-ray.	Detailed kinematic analysis of each vertebra	Inclined sled track. Tests only in non-injurious range. Rear impacts only
Luan et al	2000	PMHS rear impact sled tests with high speed x-ray	Higher frame rate (250 fr/sec) used	Only performed Low speed rear impacts tests.

Table 1 summaries the major strengths and the limitations of studies reported in the literature, which used similar methods for investigating cervical spine kinematics.

Cineradiographic studies of the cervical spine have been previously reported literature, but there are no reported investigations performed at real world speeds. Matsushita et al, were the first to use cineradiography (90 frames per second) to assess the bony kinematics of the neck. The vast majority of their tests focused on rear impacts However, four frontal impact tests were performed in volunteers who were asked to tense their neck muscles. Matsushita provided tracings of the vertebral motion for belted and unbelted subjects, but no detailed analysis was reported. Kaneoka et al (1999) analyzed the motion of each vertebrae of volunteers during a low speed rear end impact. In Kaneoka et al's study, a short (4 m) track inclined at 10° was used to produce the acceleration to the desired speed. The incline results in both horizontal and vertical acceleration of the subject before impact; therefore, the data obtained may not reflect real life data as collisions do not necessarily occur on an incline. Both, Matsushita et al and Kaneoka et al showed a significant effect of interaction between the torso and the cervical spine. However, as low speed, non-injurious volunteer tests

were performed, an extrapolation must be made to obtain data in the injurious range. Luan et al (2000), used a higher speed cineradiographic system (250 frames per second), to qualitatively study cervical spine kinematics in a PMHS during a low speed rear end impact. This study also demonstrated the influence of the torso on cervical spine kinematics. Luan et al, reported three distinct phases of motion: 1) Initial flexion with loss of the cervical lordosis; 2) S-shaped curve of the cervical spine with the upper vertebrae gradually moving into extension and 3) the entire neck in extension with tensile and shear forces acting at all levels.

The aim of this study was, therefore, to assess the feasibility of using a high speed x-ray system for qualitative analysis of the bony kinematics of the cervical spine, during a simulated frontal impact at real world speed. It is anticipated this study will provide an insight into the mechanisms of whiplash injuries occurring in frontal impacts.

METHOD

Two fresh post mortem human surrogates (PMHS) were obtained through the Legal Institute for Medicine, Medical University, Graz, Austria, in accordance with National laws. Ethical approval was provided by the Medical University, Graz, research ethics committee. There was one male, and one female subject. Table 2 provides relevant anthropometric data for the test subjects.

Subject	Sex	Age (years)	Weight (Kg)	Height (m)	Neck circ (m)	Head circ Vertical (m)	Head circ horizontal (m)
1	Male	60	85	1.77	0.56	0.65	0.63
2	Female	50	69	1.61	0.35	0.62	0.54

Table 2. Anthropometric data of the PMHS's tested.

Rigor mortise was broken by moving the neck and major joints through a range of normal movement. The PMHS was then placed in a prone position, and a longitudinal midline incision was made in the neck from the base of the skull to T1. The incision was made directly down onto the tips of the cervical spinous processes. A small area of soft tissue was stripped from the spinous processes, and markers (small steel screws, 2.5 mm x 10 mm) were screwed into the tips of the spinous processes. The markers optimized contrast on the high speed cineradiographs facilitating motion tracking. During preparation of each PMHS care was taken to ensure minimal disruption of the neck musculature occurred, and that the spinal ligaments were maintained intact. The incision was closed, and the PMHS was instrumented with accelerometers. The accelerometers were positioned at T1 and the mid-sternum; a tri-axial accelerometer was positioned on the left side of the skull at the centre of gravity in the sagittal plane.

The PMHS was positioned in a rigid, sled mounted seat, with a head rest, in a simulated driving position. A 3 point belt restraint system, without a pre-tensioning device, was used. Belt tension gauges were mounted on the shoulder and lap belts. As there is no active muscle control in PMHS's a piece of tape was applied to the head cover, and ran to the top of the head rest. The tape acted to counter balance the flexion moment due to the weight of the head, but the attachments were light enough not to interfere with head movement following impact.

The sled apparatus consisted of: a rigid seat and head rest, a 3 point belt restraint system, a sled accelerometer and an 'on board' data acquisition system. The sled is a pneumatically driven system that runs backwards on a horizontal track. The data acquisition system recorded data at 12.5 kHz from the sled, T1 and sternal accelerometers; the tri-axial head accelerometer and the belt tension gauges. We used an acceleration based trigger with a threshold of 1G to define T=0. In an active reverse sled system an acceleration based trigger is a more sensitive and rapid trigger mechanism than a contact strip mechanism. A threshold of 1G is used to avoid premature triggering caused by vibration and noise during the pressurization of the firing piston and brake release.

Each PMHS was tested at: 1) a peak velocity of 15 kph, and peak acceleration of 8G and 2) a peak velocity of 21 kph, and peak acceleration of 10G.

The high speed cineradiography system consists of a modified mobile surgical c arm x-ray system, (System BV 25 Family-N/HR, Phillips, The Netherlands), in combination with a 40 cm diameter image intensifier system, (SIRECON, Siemens, Erlargen Germany). The manual dose rate control was set to 100kV / 3.0 mA, and the system was activated using a foot pedal. To optimise the field of view the image intensifier was positioned as closely as possible to the specimen without interfering with the sled system. When the x-ray generator was activated the stream of x-rays produced a continuous image on the projection screen of the image intensifier. The real time x-ray image on the image intensifier screen was recorded using a high speed camera system (Kodak motion corder analyzer, model SR 1000C, Kodak, San Diego, CA, USA). The camera frame rate settings are adjustable, and initial tests were performed at 3 different frame rates 250, 500 and 1000 frames per second. The high speed camera was connected to a handheld digital video camera, so the images from the high speed camera could be played back easily. This assisted in fine positioning of the x-ray system to optimize visualization of the cervical spine. The hand held digital video camera also allowed quick, easy storage and transfer of the high speed camera data. The data acquisition system and the high speed camera unit were manually started before firing the sled. The x-ray system was activated by the test co-ordinator immediately prior to firing the sled, and it was stopped immediately following the test. As the high speed camera was started before firing the sled, a flash was triggered when the sled accelerometer measured 1G. The flash was used to indicate t=0 on the recorded images. This allowed the synchronization of the cineradiography data with the other sensors.

Motion analysis was performed using an automated software system, 'Target Tracking Ver 1.1', (Dr Steffan Datentechnik, DSD, Linz, Austria). The heads of the markers were manually defined at time zero. The system subsequently automatically detects the position of the markers from frame to frame. The program also offers the opportunity to edit the tracked points. The data output is in the form of pixel co-ordinates of the markers for each frame.

The aim of this study was to assess the feasibility of using high speed x-ray for qualitative analysis of cervical spine bony kinematics. Therefore, data from one test was analyzed to describe the kinematics of the cervical spine and the time sequence of events during a frontal impact. It was believed that to assess the data from different PMHS's, at different velocities would cause statistical variations too large to produce any meaningful data to improve the existing knowledge base. A meaningful statistical analysis may be performed once a sufficiently large data set has been acquired. This study reports the results of a representative test at 21kph using subject 1 (see table 2).

RESULTS

The sled reached a peak acceleration of 9.53 g at 43 ms after impact and simulated a Δv 21.1 km/h (5.86 m/s) frontal impact. Sled acceleration and velocity as a function of time are shown in figure 1. The Δv's seen in this test series were well above the injury threshold of, Δv = 16kph, suggested by Ferrari (1999)

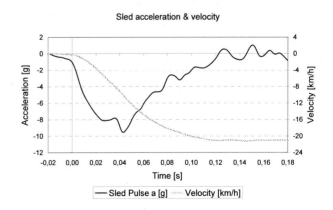

Sled acceleration & velocity

Figure 1 shows the time history of the sled acceleration and velocity. Acceleration and velocity are negative as the sled motion was rearward using a reverse sled system

The linear accelerations of the head centre of gravity in the anterior-posterior (x) and superior-inferior (z) directions, with respect to time are shown in figures 2a &

b, respectively. The greatest acceleration occurs in the vertical upward direction, (figure 2b), with the toe region of the curve beginning at approximately 65 ms, horizontal (*x*) head acceleration begins approximately 10 ms later.

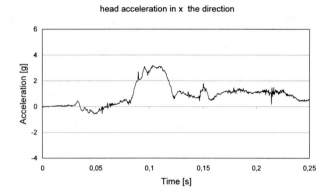

fig. 2a

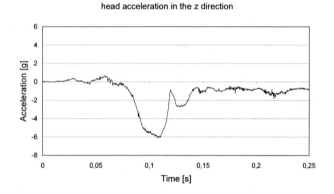

fig. 2b.

Figures 2 a & b show the time histories of the linear accelerations of the head in the *x* and *z* directions, respectively. Data was filtered using a CFC 1000 filter

The Linear acceleration of T1 and the chest in the *x* direction are shown in figures 3 and 4, respectively. The initiation of chest acceleration and a sharp increase in T1 acceleration can be seen at 40 ms, corresponding with the rise in shoulder belt forces (figure 5) as the restrain system comes into effect. The initiation of T1 acceleration in the *x* direction precedes the initiation of head acceleration in the *x* direction by 58 ms. This delay in initiation of the head x acceleration causes a posterior translation of the lower cervical vertebrae relative to the upper cervical vertebrae and head.

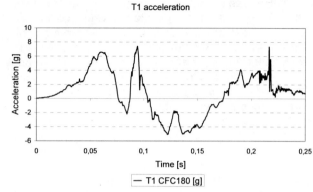

Figure 3 shows the linear acceleration of T1, with respect to time. The data was filtered with a CFC 180 filter

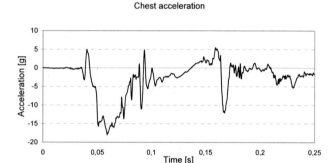

Figure 4 shows the acceleration time history of the chest. The data was filtered using a CFC 180 filter

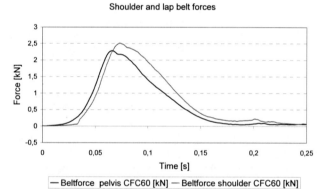

Figure 5 shows the force time histories recorded by the belt tension gauges on the lap and shoulder belts.

The cineradiography system was unable to generate a sufficiently bright image, for the high speed camera to record images with sufficient definition for motion analysis when the frame rate was set at 500 or 1000 frames per second. At 250 frames per second the image definition was suitable for motion analysis; therefore, all further analysis refers to radiographs recorded at 250 frames per second. Cineradiographic images, of the main phases of motion recorded at 250 frames/second, are shown in figure 6.

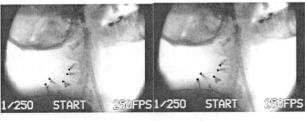

a) Time = 0 ms   b) Time = 52 ms

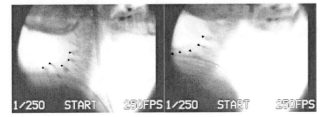

c) Time = 72 ms   d) Time = 88 ms

Figures 6 a, b, c and d show typical radiographic images, of the cervical spine, taken using the high speed cineradiography system. The frame rate was set at 250 frames/sec. Arrows highlight the heads of the markers.

Cineradiography clearly showed that the first phase of motion of the cervical spine involves an upward movement of the C7 vertebrae caused by the upward movement of the torso. As a result of the upward movement of C7, compression is generated in the cervical spine causing an initial increase in the cervical lordosis(figure 6b). The absolute motion, and the relative motion between vertebrae determined by motion tracking of the markers confirms that the first movements occur in the inferior cervical vertebrae, and upward movement precedes horizontal motion (figure 7 a, b, c and d). The spatial trajectories of the vertebral markers, in both the x and z directions, are shown in figure 8. The upward movement of C7 is transmitted through the cervical spine to the head producing upward acceleration of the head starting at 65 ms (figure 2b). The upward motion of the torso, combined with the delay between initiation of acceleration at T1, and the initiation of head acceleration in the x direction leads to combined compression and shear loading of the soft tissues of the neck, such as, the ligaments and intervertebral discs. This motion results in a straightening of the cervical lordosis following the small initial increase in the cervical lordosis (figure 6c).

The posterior translation of a cervical vertebra relative to its adjacent superior vertebra generates a shear force, causing the soft tissues of the neck to become taut. Once taut, the soft tissues cause the translation of the adjacent superior vertebrae. In this manner, the translation proceeds up the cervical spine until it reaches the head (figure 6d). As time is required for the soft tissues to become taut, before they cause a translation of the adjacent superior vertebrae, there is a delay before the head begins to move in the horizontal (x) direction. This offers a possible explanation of 'head lag' phenomenon. The displacement of each vertebra relative to C3 (the most superior marker) during the

frontal impact is shown in figures 7c and d; the results illustrate that the largest and fastest displacements, with respect to C3, occur at the lower vertebrae.

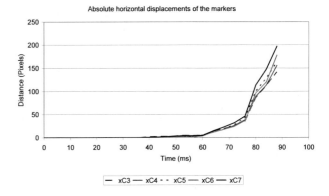

Figure 7 a shows the absolute horizontal (x) displacement time histories of the vertebral markers

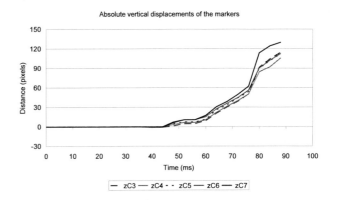

Figure 7 b shows the absolute vertical (z) displacement time histories of the vertebral markers

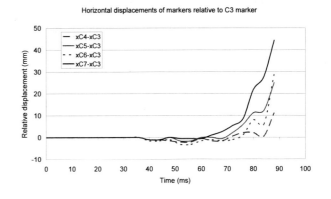

Figure 7 c shows the relative horizontal (x) displacement of each vertebral marker with respect to C3. A negative value indicates a vertebra was closer to C3 than at t=0.

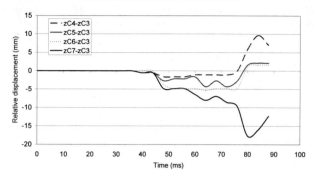

Figure 7 d shows the relative vertical (z) displacement of each vertebral marker with respect to C3. A negative value indicates a vertebra was closer to C3 than at t=0.

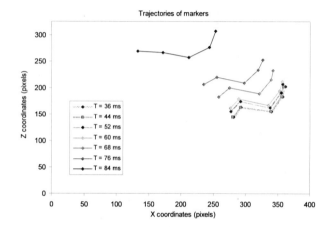

Figure 8 shows the spatial trajectory of the markers at multiple time steps between the start of motion at 36 ms until the markers moved out of the x-ray field of view

DISCUSSION

In situ PMHS sled tests simulating a frontal impact were performed, and the cervical vertebral motion was recorded using a high speed cineradiography system. Data from one representative test; a male PMHS, at 21.1 km/h (5.86 m/s) 9.53g, was used for detailed qualitative kinematic analysis. The results indicate that the kinematics of the cervical spine are complex during frontal impacts. The spine experienced compression, tension, shear, flexion and extension in different regions at different times.

Compression of the neck and hence, the vertical acceleration of the head is as a result of upward motion of the torso, with respect to the relatively static head, in the initial stages of the impact. The z acceleration time history of the head (figure 2), does not show the neck going into tension. However, a true assessment is difficult to make as we measured linear accelerations, but these were superimposed on rotation of the head. Tension in the soft tissues connecting the spinous processes can be determined though as the relative

displacement between C3 and C7 increased over time, in both the x and z directions.

Shear forces generated as a result of the relative movement between the torso, restrained by the belt system, and the head generates tension in soft tissues i.e. ligaments and facet joint capsules. Tension in the ligaments and joint capsules transmits forces upwards through the neck one level at a time causing the cervical lordosis to straighten.

Due to insufficient brightness of the x-ray source, image quality was insufficient for motion analysis at 500 and 1000 frames per second. At 250 frames per second, it was possible to perform motion analysis. However, as the Δv used in this study far exceeded those used in earlier cineradiographic studies, the image quality was suboptimal. In future, this problem may be addressed through the use of more advanced high speed cameras, which capture high quality images in conditions with as little as one quarter of the light intensity, need by the camera used in this study. Additionally, despite using a larger image intensifier than used in standard surgical c-arm systems, motion analysis was restricted to the early phase of the cervical spine motion. This is because: the high sled velocity meant cervical vertebral motion could only be completely tracked for 90 ms. However, sled acceleration, belt forces, T1 and chest acceleration all reached there peak values before this point, and the peak head accelerations typically occurred at approximately 90 ms. The results of the study indicate that the important features of the cervical spine bony kinematics can be captured in this time frame. This is in keeping with the findings of Luan et al (2000), who believed that the early phase of motion was the most critical as peak values for soft tissue stretching occurred during this time in rear end impacts.

Motion tracking of the markers provides useful quantitative data to characterize the absolute and relative displacements in the x and z directions. However, it must be assumed that no out of plane occurs as there was only one marker on each vertebra, which prevented assessment of out of plane rotation. Motion analysis results indicate initial vertebral motion is a vertical translation with a slight increase in the cervical lordosis. This differs from the conclusions of earlier studies, which reported that the initial motion is a pure horizontal translation, (Ewing et al 1975, Walz and Muser 1995). Subjective analysis of the radiographs, on a frame by frame basis, also suggests that flexion occurs at C7 at a very early stage, and that flexion in the superior vertebrae occurs in sequential order. However, it was not possible to quantify the degree of flexion or its exact timing as only one marker per vertebra was used. Later phases of the cervical vertebral motion involve a posterior translation of each vertebra, with respect to its adjacent superior vertebra, resulting in a straightening of the cervical lordosis before the head rotation causes the cervical spine to flex.

Motion analysis could be improved to include a quantitative rotational analysis by using additional markers on each vertebral body. Unfortunately, this would require more tissue stripping and disruption to vital structures if a posterior approach was used. If an anterior incision was used, extensive dissection of the neck would be required. As a result, either of these approaches may alter the behaviour of the cervical spine.

Previously, researchers have been critical of PMHS testing in impact biomechanics; believing, realistic results can only be obtained by using volunteers as muscle activity is absent in PMHS tests, (Kaneoka *et al* 1999). Other researchers performing electromyographic (EMG) studies concluded that muscle activity is not activated until 200 ms, during low speed rear impacts, (Ono *et al* 1997). In frontal impacts it may be argued vehicle occupants are more likely to tense their neck muscles, in anticipation of an impact, as they can often see what they are going to impact. However, in many collisions the occupant is unsighted in relation to what they strike. The most important aspect of performing PMHS impact tests is the ability to acquire data at realistic velocities, above injury thresholds, which is not possible in volunteer tests.

As a small sample size (2 PMHS's) was used in this study, meaningful quantitative kinematics analysis is not possible, and the effects of sled velocity and camera frame rate cannot be quantified. A small sample size was used because the aim of the study was to assess the feasibility of using a high speed x-ray system to evaluate cervical spine bony kinematics at speeds above suggested injury thresholds. A greater number of sled tests must be performed in order to produce quantitative results.

This study represents the first qualitative analysis of the cervical spine kinematics in a frontal impact at real world speeds using high speed cineradiography. The results highlight the need for further research and testing to optimise testing methods. By using more advanced cameras, additional vertebral markers and including fixed calibration markers in the x-ray field of view it will be possible to perform rotational analysis, assess out of plane motion, improve the accuracy of motion tracking, and include an assessment of errors in motion tracking measurements.

CONCLUSIONS

Sled test data was analyzed to assess the feasibility of using a high speed cineradiographic system, for qualitative kinematic analysis of the human cervical spine during a high speed frontal impact. Using the current experimental setup, analysis at 500 frames per second or above is not possible.

High speed cine x-ray offers an advantage over current methods used in high speed kinematic studies which require extrapolation of data from motion tracking of external makers to determine bony kinematics. The system also offers the potential for detecting the origin of fractures, as well as, allowing visualization of the timing of fractures and the order they occur in, in multi-fracture injuries.

The study findings indicate that further research and testing is required to optimize the use of the system, to increase the field of view and improve image quality at the highest frame rates (> 500 frames / sec).

ACKNOWLEDGMENTS

This project was funded by the European Community under the 'competitive and sustainable growth programme'. Acronym: Whiplash II; contract number G3RD-CT-2000-00278; project number G3RD-2000-00278.

REFERENCES

Cholewicki J, Panjabi MM, Nibu K, Babat LB, Grauer JN, Dvorak J. Head kinematics during in vitro whiplash simulation. Accident Analysis and Prevention. 30(4):469±79. (1998)

Ewing CL, Thomas DJ, Lustick L, Becker E, Willems G, Muzzs III WH: The effect of the initial position of the head and neck on the dynamic response of the human head and neck to Gx – impact acceleration. Proceedings of the 19th Stapp car crash conference, San Diego, 487-512 (1975)

Ferrari R. In :The whiplash encyclopaedia. ISBN 0834216612, Jones and Bartlett Publishers (1999)..

Galasko CSB, Murray PM, Pitcher M, Chamber H, Mansfield M, Madden M, Jordan C, Kinsella A, Hodson M. Neck sprains after road traffic accidents: a modern epidemic. Injury 24: 155-157. (1993)

Huelke DF, Mackay GM, Morris A, Bradford M. Non-head impact cervical spine injuries in frontal car crashes to lap shoulder-belt occupants. International congress and exposition, Detroit, Michigan. SAE paper number 920560. (1992)

Huelke DF, Mackay GM, Morris A. Vertebral column injuries and lap-shoulder belts. Journal of trauma, injury, infection and critical care. 38(4): 547-556. (1995)

Kaneoka K, Ono K, Inami S, Hayashi K. Motion analysis of cervical vertebrae during whiplash loading. Spine 24(8): 763-770. (1999)

Kullgren A, Krafft M, Nygren A, Tingvall C. Neck injuries in frontal impact: influence of crash plse characteristics on injury risk. Accident analysis and prevention 32: 197-205. (2000)

Langwieder K and Hell W. Neck injuries in car accidents, causes, problems and solutions. Proceedings of the ATA conference, Capri. P131-146 (1996).

Larder DR, Twiss MK, Mackay GM. Neck injury to car occupants using seat belts. Proceedings of the 29th

Association for the Advancement of Automotive Medicine conference. (1995)

Luan F, Yang KH, Deng B, Begeman PC, Tashman S, King AI. Qualitative analysis of neck kinematics during low speed rear-end impact. Clinical Biomechanics. 15: 649-657. (2000)

Matsushita T, Sato TB, Hirabayashi K, Fujimura S, Asazuma T, Takatori T. X-ray study of the human neck motion due to head inertia loading. Proceedings of the 38th Stapp Car Crash Conference, Fort Lauderdale, Florida. p55–64. (1994)

Ono K, Kaneoka K, Wittek A, Kajzer J. Cervical injury mechanism based on analysis of human cervical vertebral motion and head-neck-torso kinematics during low speed rear impacts. In: Proceedings of the 41st Stapp Car Crash Conference, SAE paper number 973340. (1997).

Panjabi MM, Cholewicki J, Nibu K, Grauer J, Babat LB, Dvorak J. Critical load of the human cervical spine: an in vitro experimental study. Clinical biomechanics 13:239±49. (1998a)

Panjabi MM, Cholewicki J, Nibu K, Babat LB, Dvorak J. Simulation of whiplash trauma using whole cervical spine specimens. Spine 23(1):17±24. (1998b)

Panjabi MM, Cholewicki J, Nibu K, Grauer J, Vahldiek M. Capsular ligament stretches during in vitro whiplash simulations. Journal of Spinal Disorders 11(3):227±32. (1998c)

Panjabi MM, Nibu K, Cholewicki J. Whiplash injuries and the potential for mechanical instability. European Spine Journal. 7(6):484±92. (1998d)

Spitzer WO, Skovron ML, Salmi LR, Cassidy JD, Duranceau J, Suiss S, Ziess E. Scientific monograph of the Quebec task force on whiplash associated disorders: redefining whiplash and its management. Spine 20 (Apr Suppl) S1-73. (1995)

Temming J and Zobel R. frequency and risk of cervical spine distortion injuries in passenger car accidents: significance of human factors data. Proceedings of the International Reasearch Council On biomechanics of Impact (IRCOBI) p.219-233 (1998)

Walz FH and Muser MH. Biomechanical aspects of cervical spine injuries. SAE international congress and exhibition, Detroit Michigan. SAE paper 950658. (1995)

CONTACT

Univ. Prof Dr H Steffan PhD, Frank Stonach Institute, Technical University of Graz, Austria. Infeldgasse 21/B III. h.steffan@tugraz.at

2009-01-0395

Inertially-Induced Cervical Spine Injuries in the Pediatric Population

Tara Moore, Michael Prange and Catherine Corrigan

Exponent, Inc.

ABSTRACT

This study integrates data from multiple sources to obtain a more complete understanding of inertially-induced pediatric cervical spine injury risk and the role of impact severity and restraint type. Data from previously conducted frontal crash and sled tests using a variety of anthropomorphic test devices (ATDs) in various restraint configurations were compiled and compared to injury assessment reference values (IARVs). The data show that neck loads in frontal collisions increase with increasing delta-V. At high delta-Vs, the neck loads correspond to a relatively high risk of neck injury regardless of restraint configuration. Pediatric inertial cervical spine injury risk in frontal collisions is governed primarily by the energy involved in the collision.

INTRODUCTION

Motor vehicle crashes are a common cause of cervical spine injuries in young children [1, 2]. While pediatric cervical spine injuries are rare, they are associated with a high mortality rate [1, 3]. Investigations using field accident data have found that the change of velocity (delta-V) of the vehicle during a crash is a strong predictor of injury risk for children [4, 5].

Current guidelines regarding child restraint usage from the American Academy of Pediatrics and the National Highway Traffic Safety Administration (NHTSA) specify that a child aged from 1 to about 4 years old and weighing 20 to 40 pounds should be placed in a forward-facing child restraint system (CRS) in the rear seat when riding in a motor vehicle. A child up to 80 pounds or more, between approximately four and eight years old, and shorter than four feet nine inches tall should be seated in a belt-positioning booster (BPB) in the rear seat of the vehicle. Children over 80 pounds and four feet nine inches are restrained using the vehicle seatbelts.

The majority of children involved in motor vehicle crashes are restrained by some means [6, 7]. However, rates of child restraint misuse and inappropriately restrained children are significant [4, 6-10], and improperly restrained children have an increased risk of injury [4]. Proper use of child restraints has resulted in a significant reduction of the risk of head injuries; however, spinal injury risk was not significantly reduced [11].

Serious pediatric cervical spine injuries are rare in real-world crashes, and data are primarily available as anecdotal case studies. The biomechanical literature contains descriptions of real-world crashes involving pediatric cervical spine injuries, including cases where properly and improperly restrained children in a variety of restraint systems have sustained cervical spine injuries (fractures, dislocations and/or spinal cord damage) without evidence of head contact [12-19]. Many of these cases involved young children involved in severe frontal crashes (delta-V $\geq$ 45 kph). Inertial loading of the cervical spine has been postulated as a mechanism for these injuries [13, 15]. In addition, the authors have personally investigated several frontal crashes where forward-facing restrained children (restrained in vehicle restraints only, forward-facing CRS, and booster seats) have sustained cervical spine injuries without evidence of head contact. Other groups have stated that in their series, cervical spine injuries have not occurred to properly restrained children in the absence of head impact [20].

In a frontal crash, the body of a restrained occupant is slowed by the restraint system while the head initially continues to move at its original speed. This forward

SAE Customer Service: Tel: 877-606-7323 (inside USA and Canada)
                          Tel: 724-776-4970 (outside USA)
                          Fax: 724-776-0790
                          Email: CustomerService@sae.org
SAE Web Address: http://www.sae.org

Printed in USA

motion of the head relative to the body causes inertial loading of the neck in tension and, as the head rotates, in flexion. Inertially-induced cervical spine injuries associated with these mechanisms, particularly in the pediatric population, have been a subject of debate in the biomechanical literature.

Differences in anatomy and anthropometry of the young pediatric population, compared to adults, make them particularly vulnerable to inertial neck injuries [13, 17]. As observed in a number of previous studies, children are not scaled-down adults. Young children have a relatively large head and smaller neck relative to adults [21, 22]. Also, the pediatric spinal ligaments are more lax, the neck musculature is less developed, and the vertebral bodies are not completely ossified with relatively horizontal facet joints. These anatomic characteristics allow for increased cervical spine motion and more compliance of the pediatric cervical spine relative to the adult spine [17]. This decreased neck strength and increased relative head size make young children more susceptible to inertial neck injuries.

Currently, Federal Motor Vehicle Safety Standard (FMVSS) 213, which pertains to pediatric injury risk, includes head and chest injury criteria, but does not include neck injury criteria. This standard presently requires child restraint systems (CRS) to be tested using peak sled accelerations of 19-25 g with a pulse duration of 75-90 ms. In 2002, the NHTSA published a notice of proposed rule making (NPRM) that suggested amending FMVSS 213 to use the FMVSS 208 Nij criterion, without the peak tension and compression limits, as a neck injury criterion [23]. However, no neck injury criterion was incorporated into the final rule. Injury Assessment Reference Values (IARVs) have been developed for the neck and scaled values are available for infant and child anthropomorphic test devices (ATDs) [24].

Several research groups have performed frontal crash and sled testing using child ATDs to evaluate pediatric injury risk, including measurements of neck load, in a wide variety of restraint configurations, including misuse configurations. Some tests were conducted to replicate conditions in a real-world crash; others were performed as part of a series of tests to systematically evaluate the effect of various parameters (e.g. restraint type, crash severity, CRS attachment to vehicle seat) on injury measures.

The purpose of this study is to review pediatric cervical spine loads in frontal collisions without head contact at different crash severities (using change in velocity (delta-V) as a measure of severity) and using different restraint systems, including restraint misuse. Data from multiple sources, including testing performed by the authors and that available in the biomechanical literature, were compiled to evaluate inertially-induced neck injury risk for forward-facing child occupants in different restraint configurations as a function of crash severity.

METHODS

A thorough search of the biomechanical literature was conducted to find published studies reporting neck loading of child ATDs in frontal crashes. Papers that reported neck loads from inertial motion of the head without head contact using upper or lower neck load cells were identified. These data were synthesized together and classified by the various ATD types (ages), crash severity, and the type of restraint used.

A series of sled tests were previously conducted. The details of the testing have been reported elsewhere [25, 26] and are summarized here. Nine sled tests were conducted using a deceleration sled. A minivan second row bench seat and the outboard restraint systems were rigidly secured to the sled. During each test, a 3-year-old Hybrid III ATD was placed on the right outboard position, and a 6-year-old Hybrid III ATD was placed in the left outboard position.

Tests were conducted using three changes of velocity (32, 48, and 64 kilometers per hour) and three restraint configurations for each ATD (Figure 1). The 3-year-old ATD was restrained in a properly installed forward-facing CRS, in the same model CRS with an improper installation (without applying tension to webbing during installation), and in the vehicle seatbelt without a CRS. The 6-year-old ATD was restrained in a high-back belt-positioning booster seat in the vehicle seatbelt with a booster seat, and in the vehicle seatbelt with the shoulder portion of the webbing placed behind the ATD.

Head accelerations and upper and lower neck loads were measured for each test. Peak resultant acceleration, HIC15, HIC36, and neck injury criteria (Nij) were calculated from the acquired data.

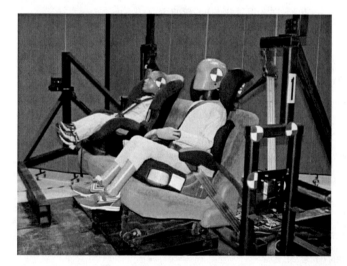

Figure 1: Photo showing test setup

In addition, a number of frontal crash and sled tests using pediatric ATDs were identified [15, 19, 25-51]. Our results and the available published results were compiled to obtain an overview of pediatric occupant response. The upper and lower neck forces and moments were analyzed by crash severity (delta-V), restraint type, and ATD size. Not all data were available for each test. In addition, the specifics of the crash pulses were examined to determine if changes in crash pulse affect injury risk measures.

Peak neck tensions as a percentage of the in-position IARV for the corresponding age were calculated [24]. Statistical analyses were performed to assess the effects of delta-V, restraint type, and ATD size. Upper neck tension and %IARV were compared across these factors using a 3 way analysis of variance (ANOVA). Tukey multiple comparison tests were used to compare groups within each factor. Restraint types that had less than 10 data points available were not included in this statistical analysis (Table 1).

RESULTS

Results from 399 frontal sled and crash tests using forward-facing ATDs conducted in 29 research studies with known ATD sizes and restraint types and reported neck loads were compiled [15, 19, 25-51]. Tests were performed using ATD sizes ranging from the 6-month-old to the 6-year-old. Different models of ATDs were used in different tests. In addition, the model and modifications of the neck load cells in some tests varied. The ATDs were restrained using a wide variety of restraint configurations (Table 1).

Restraint Type	Number of Tests
Forward-facing CRS (FF CRS)	222
High back booster (HBB)	64
Lap and shoulder belt	34
No back booster (NBB)	34
Shield booster	18
Lap belt	11
Lap and harness	4
3-pt belt, torso belt behind back	3
Wheelchair, 3-pt belt	3
4-pt belt	2
Booster (type not provided)	2
Lap and harness or Booster	2

Table 1. Restraint types

Some restraint types were tested by multiple research groups using a variety of ATD sizes and crash severities; others were tested in a single study. Within each restraint type, the specifics of the restraint (CRS model, location of seat belt anchor points, etc.) varied. The test delta-Vs ranged from 24 kph (15 mph) to 74 kph (46 mph). However, most of the tests had delta-Vs in the 37- to 56-kph range. Many tests that were not based on a real-world crash used either a standardized crash pulse (e.g. NCAP, FMVSS 213, ECE-R 44) or a scaled version of a standardized pulse. Not all neck parameters were measured in all tests.

Peak upper neck load cell tension increased with increasing crash severity (p<0.01) (Figure 2). Considerable scatter was observed in the data. The peak neck tensions in the 6-year-old ATD tended to be larger than those measured in smaller ATDs. There was considerable overlap between peak neck tensions for the smaller ATDs and the differences in peak neck tension were not significant. Normalized neck tension (peak neck tension divided by the appropriate IARV) decreased with age (p<0.01). The lap belted, lap and shoulder belted, and high back booster (HBB) tests resulted in the highest normalized neck tension (p<0.01). The peak neck tensions normalized by IARV from the forward facing child restraint systems (FF CRS), no back booster (NBB), and shield booster were not significantly different from one another.

Seventeen frontal sled tests were performed using forward-facing 6-month old ATDs, most at delta-Vs close to 50 kph (Figure 3) and all using FF CRS with and without tethers. In the majority of these tests, only resultant neck forces and moments were reported. In the three tests where neck tension and shear were reported [19, 46, 47], the peak neck tension was 93% to 99% of the peak resultant force.

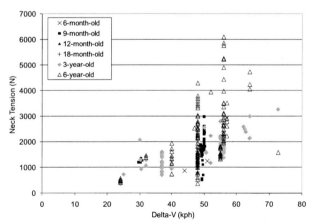

Figure 2. Peak upper neck tension as a function of delta-V. All tests for a particular ATD size were classified as a single category, regardless of restraint type.

Most of the sled tests involving 9-month-old ATDs (47 of 51 tests) were also performed at 50 kph, with four tests performed at 30 kph (Figure 4). Neck forces were not available for three of the 50-kph tests. Four tests were performed using a ballasted ATD (Figure 4, grey symbols) where the mass of the head and upper torso were both increased by 0.45 kg [28]; the results of those tests fell within the range of the 50-kph neck tensions for an unballasted ATD and were similar to the unballasted neck tensions at 30 kph.

All available tests with neck load data using forward-facing 12-month-old ATDs were conducted using tethered and untethered CRS. The tests used crash pulses with a delta-V of 48 kph and a peak acceleration of 20 g [30] (Table 2). The majority of the tests using the 18-month-old ATD used tethered and untethered CRS at the same pulse [30]; however, four additional tests were conducted using different restraint systems at the same delta-V and a higher peak acceleration of 27 g [27]. The tests using lap belts and lap/shoulder belts, which also had higher peak accelerations, resulted in higher peak neck tensions. The mean peak tensions at this severity exceeded the 12- and 18-month-old ATD in-position IARVs of 990 and 1080 N respectively [24].

Age	Restraint	n	Mean (kN)
12	CRS w/ tether	24*	1.28 (0.19)
12	CRS w/o tether	5**	1.54 (0.23)
18	CRS w/ tether	26	1.13 (0.20)
18	CRS w/o tether	4	1.38 (0.18)
18	Lap belt	2	2.1, 2.5
18	Lap/shoulder belt	2	2.5, 2.6

*Two tests did not report neck tension data
**One test did not report neck tension data

Table 2. Mean peak upper neck tension data for testing using 12- and 18-month ATDs. Numbers in parentheses represent standard deviations. No means are reported in the last two rows due to the small sample size.

Considerably more data (144 tests) were available for the 3-year-old ATD for various restraint types and crash severities (Figure 5). The peak upper neck tensions increased with increasing severity. For a given crash severity, peak neck tensions were typically within a 1-2 kN (225-450 lb) range, regardless of restraint type. For clarity, all CRS systems were classified in a single category, regardless of tether usage and changes in CRS or vehicle belt tension. The effects of changes in tethering and webbing tension will be reported separately below.

There were also extensive data (121 tests) for the 6-year-old ATD (Figure 6). The peak neck tensions were larger for this ATD than for the smaller ATDs discussed previously. The largest measured neck tensions for this ATD were nearly twice as large as those measured for any other ATD in this study. The range of peak neck tensions for a given crash severity was broader for this ATD than for other, smaller ATDs (Figure 6).

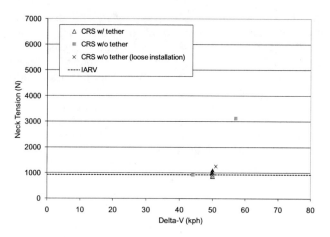

Figure 3. Peak resultant upper neck force as a function of delta-V for 6-month-old ATDs, by restraint type. The dashed line indicates the in-position tensile IARV.

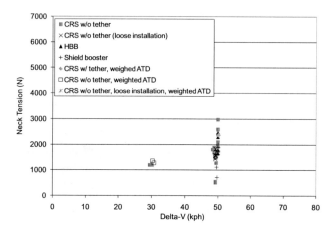

Figure 4. Peak upper neck tension as a function of delta-V for 9-month-old ATDs, by restraint type.

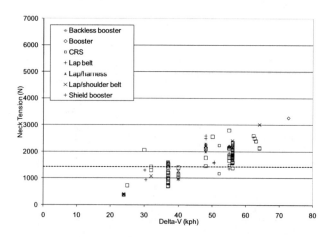

Figure 5. Peak upper neck tension as a function of delta-V for 3-year-old ATDs, by restraint type. Forward-facing CRS were classified as a single group, regardless of tethering or changes in webbing tension. In one test involving a booster-seated occupant, the type of booster was not identified. The dashed line indicates the in-position tensile IARV.

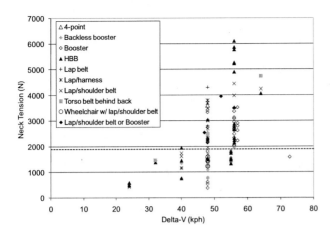

Figure 6. Peak upper neck tension as a function of delta-V for 6-year-old ATDs, by restraint type. The dashed line indicates the in-position tensile IARV.

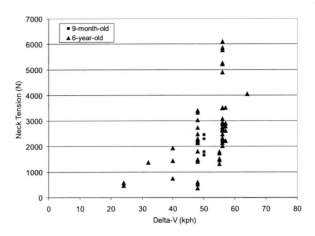

Figure 8. Peak upper neck tension in occupants seated in high-back boosters.

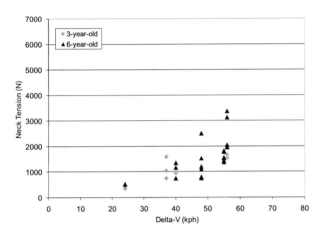

Figure 7. Peak upper neck tension in occupants seated in backless boosters.

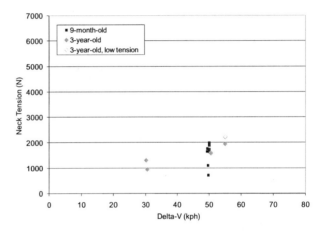

Figure 9. Peak upper neck tension in occupants seated in shield boosters.

For occupants seated in backless booster seats, there was considerable overlap in peak neck tensions measured in the 3- and 6-year-old ATDs (Figure 7). The peak neck tensions increased with increasing delta-V; however, the values at the highest delta-V using this restraint configuration fell in the lower part of the range of neck tensions measured for all restraint types at this crash severity.

Most tests using high back boosters were performed using the 6-year-old ATD; 6 tests used a 9-month-old ATD. Peak neck tensions increased with increasing crash severity (Figure 8). The values for the 9-month-old ATD, which were all performed at the same crash severity, were in the same range as those measured for the 6-year-old ATD at comparable crash severities.

Sled tests were performed using 9-month-old and 3-year-old ATDs seated in shield booster seats (Figure 9). Peak neck tension increased with increasing crash severity. The peak neck tension measured for all tests was 2.2 kN. All of the tests using the 9-month-old ATD were performed at similar crash severities; the peak neck tensions were comparable to those measured for the 3-year-old ATD at the same severity. A single test performed using a shield booster with low tension in the webbing attaching the booster to the vehicle seat had a higher peak neck tension than a test at the same severity without loose webbing.

Data were available for 18-month-old, 3-year-old, and 6-year-old ATDs restrained by lap and shoulder belts (Figure 10). Peak neck tension increased with increasing crash severity, with larger amounts of scatter in the data for higher severities. Peak tensions for the 6-year-old were generally higher than those measured for the other ATD sizes.

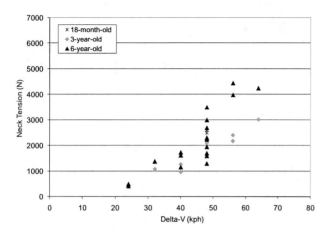

Figure 10. Peak upper neck tension in occupants restrained by lap and shoulder belts.

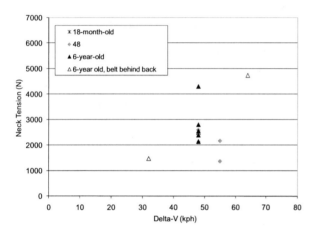

Figure 11. Peak upper neck tension in occupants restrained by lap belts.

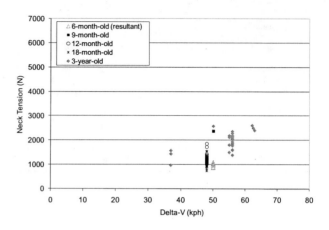

Figure 12. Peak upper neck tension in occupants restrained in FF CRS with tethers.

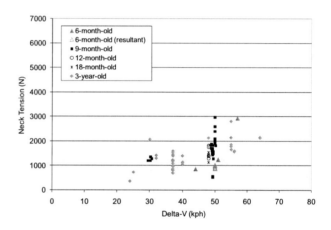

Figure 13. Peak upper neck tension in occupants restrained in FF CRS without tethers.

18-month-old, 3-year-old, and 6-year-old ATDs restrained by lap belts were tested in frontal crashes. In addition, three tests were performed with a 6-year-old ATD restrained by a three-point belt, with the torso portion of the restraint placed behind the back [26]. The limited data available at multiple crash severities for this restraint configuration prevent evaluation of the relationship between neck tension and crash severity (Figure 11).

Testing using forward-facing CRS were performed using all ATD sizes except the 6-year-old. Peak neck tensions increased with increasing delta-V for both tethered and untethered CRS (Figures 12, 13). For a given crash severity, the scatter in peak tension was approximately 1500 N or less for tethered CRS (Figure 12) and slightly larger for untethered seats (Figure 13). There was overlap between peak neck tensions measured for ATDs of different sizes. The two lowest tensions measured for the 9-month-old ATD at 49 kph in untethered CRS involved an ATD with a rigidly fixed neck [29].

For purposes of this paper, CRS configurations with slack or minimal tension in the CRS internal harness, in the vehicle belt, or in the LATCH attachment, were classified as a loose installation. Changes in tether webbing tension were not assessed. Increases in peak neck tension with increasing crash severity were observed for ATDs restrained in FF CRS without (Figure 14) and with (Figure 15) loose installation. At higher crash severities, tests with loose installations tended to have slightly lower peak tensions. Peak tensions and peak tensions normalized by IARV were not significantly different for CRS configurations with and without tethers and with and without loose installations.

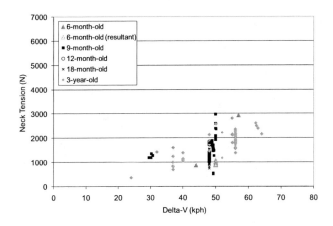

Figure 14. Peak upper neck tension in occupants restrained in FF CRS without loose installation.

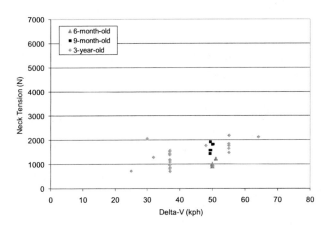

Figure 15. Peak upper neck tension in occupants restrained in FF CRS with loose installations.

Peak upper neck shear values were reported for fewer tests than peak neck tensions. Most of the available data were for the 12-month-old, 18-month-old, and 3-year-old ATDs. Peak neck shear tended to increase with increasing crash severity (Figure 16). There was considerable scatter in the data. The IARVs for upper neck shear range from 690 N in the 6-month-old ATD to 1410 N in the 6-year-old ATD [24]. For some tests, particularly at delta-Vs over 50 kph, the upper neck shear IARV was exceeded.

Neck moments tended to increase with increasing crash severity (Figures 17, 18); however, large amounts of scatter were observed. Many tests showed high extension moments (Figure 17), which typically occurred relatively early in the test before substantial head motion had occurred [37]. The IARVs (ranging from 13 to 30 N-m) were exceeded in some tests with a delta-V over 50 kph. A small number of tests exceeded the flexion IARVs (25 to 60 N-m, depending on ATD size) (Figure 18).

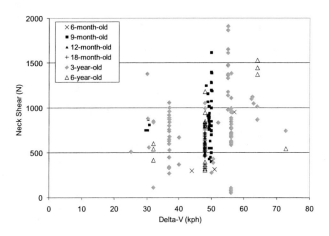

Figure 16. Peak upper neck shear v. delta-V.

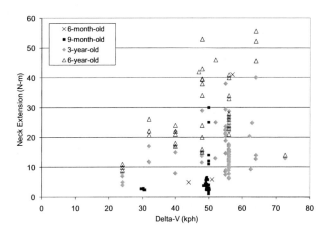

Figure 17. Peak upper neck extension v. delta-V.

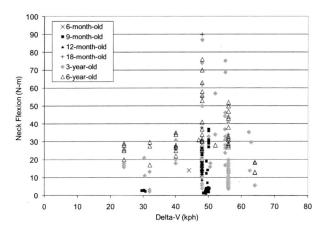

Figure 18. Peak upper neck flexion v. delta-V.

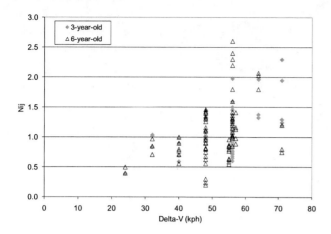

Figure 19. Nij v. Delta-v.

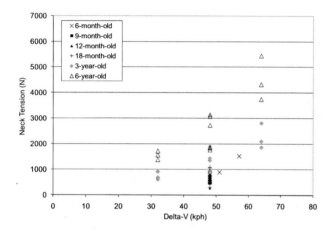

Figure 20. Peak lower neck tension v. delta-V.

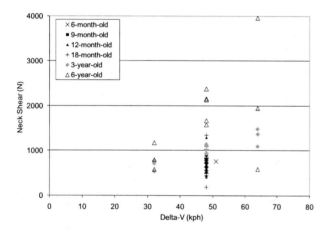

Figure 21. Peak lower neck shear v. delta-V.

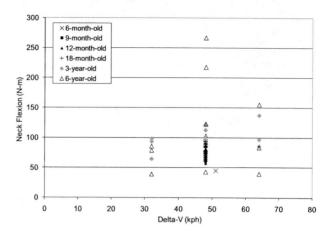

Figure 22. Peak lower neck flexion v. delta-V.

Neck injury criterion (Nij) data were available for some tests involving the 3-year-old and 6-year-old ATDs (Figure 19). Nij values tended to increase with increasing delta-V. A substantial number of tests at approximately 48 and 56 kph exceeded the IARV of 1.0. All of the 64-kph tests exceeded the IARV.

Fewer data were available for lower neck loads. The available data suggest that peak lower neck tension and shear forces increase with increasing crash severity (Figures 20, 21). The largest lower neck flexion moments occurred at the highest delta-V (64 kph) for the 3-year-old ATD but at a lower delta-V (48 kph) for the 6-year-old ATD (Figure 22). Lower neck extension moments, when reported, were typically small.

For a given crash severity, the peak sled acceleration varied, depending on the crash pulse used in the test. Tests performed with delta-Vs from 48-51 kph included 178 tests with a range of peak accelerations. Other delta-V ranges (37 kph, 40 kph, 55-57 kph) had relatively large numbers of tests, but had few reported peak accelerations or a limited range of accelerations. When sufficient data were available to evaluate the relationship, the correlations between peak sled acceleration and peak neck loads were low.

DISCUSSION

This study used data from frontal crash and sled testing to evaluate pediatric neck injury risk as a function of crash severity and restraint configuration. The peak neck loads increased with increasing crash severity, regardless of restraint configuration. All neck load components increased with increasing crash severity; however, no correlations between individual neck force and moment components were observed.

During a frontal collision, occupants initially continue to move at their original speed as the vehicle decelerates beneath them. This causes the occupants to move forward relative to the vehicle until their motion is slowed and eventually stopped by contact with vehicle structures in front of them, including the restraint system.

In all the restraint configurations discussed in this study, the lower body was restrained. All restraint configurations except the lap belt also provided some restraint of the torso and one or both shoulders. In the absence of head contact during a frontal collision, the head continues to move forward as the body is restrained. The neck will be inertially loaded as the body slows relative to the still-moving head.

In some tests, large upper neck extension moments were observed. A review of moment-time data [25, 26, 37] indicated that upper neck extension preceded upper neck flexion. This may reflect complex bending modes in the ATD and the human neck. As head motion continued, the upper neck experienced tension and flexion. Lower neck moments were primarily flexion in all cases.

We expect that neck loads will increase with increasing ATD size, due to increasing head mass. Based on the scaling factors reported by Mertz et al. [24] and Irwin and Mertz [52], the largest differences should be observed between the 18-month-old ATD and the 3-year-old ATD and between the 3-year-old ATD and the 6-year-old ATD, with the 6-month-old, 12-month-old, and 18-month-old ATDs having smaller differences in scaling factors. While no scaling factors or IARVs are reported for the 9-month-old ATD, we expect that these values will fall between the 6- and 12-month-old values. However, the observed peak neck tensions did not show a significant increase with increasing ATD size across all crash severities and restraint types. When neck tension was normalized by the IARV, the normalized value decreased with age.

For all neck measures, there was considerable scatter in the data, particularly for the more severe tests. There was also considerable overlap between different ATD sizes and restraint configurations. The peak neck tensions were high for the 6-year-old ATD (Figures 2, 20), consistent with expectations based on geometric scaling factors, but were also high for the ballasted and unballasted 9-month-old ATD. The five tests involving a ballasted 9-month-old ATD [28] had higher loads than the unballasted ATD. Because the 6-year-old ATD was restrained in different configurations (booster seats, vehicle restraints) than those commonly used for the smaller ATDs (FF CRS), the contribution of restraint configuration to the increases in neck tension could not be evaluated.

In the lower neck, the peak neck tension was also higher for the 3-year-old ATD than in the smaller ATDs. However, the peak upper neck tensions measured in the 3-year-old ATD overlapped those measured in the smaller ATDs, despite expected differences due to large differences in scaling factors. All tests involving the 12- and 18-month-old ATDs were conducted at the same crash severity [30]. The peak upper neck tensions for 12-month-old ATD were slightly higher than those measured for the 18-month-old ATD (Table 2); but other neck loads were higher in the 18-month-old ATD. The

authors of that study reported the results for both ATDs and stated that the head masses of the two CRABI ATDs used were identical [30].

The relative magnitudes of the tensile and shear loads depend in part on the ATD orientation. In a frontal crash, an ATD with some torso flexion will have a large neck tensile component because the neck is more aligned with the crash pulse. The 6-year-old ATD had the highest peak lower neck shear, but did not have high upper neck shear. This may be due to a number of factors, including increased torso flexion in the 6-year-old ATD's restraint configurations and the biofidelity of this ATD, which has been discussed by other authors [37, 40].

There was also considerable scatter in the neck loads measured with ATDs restrained in different configurations. Occupants seated in backless boosters (Figure 7) and shield boosters (Figure 9) had peak neck tensions at the lower end of the range of values for a given ATD size and crash severity, but those values overlapped with neck loads measured in other restraint configurations. Peak neck tensions measured for ATDs seated in tethered and untethered CRSs (Figures 12, 13) were similar, as were those measured for ATDs restrained with and without loose installation (Figures 14, 15). However, the amount of tension and/or slack varied between tests, ranging from no webbing tension to several inches of slack, and in some cases, the amount of slack was not quantified, making comparisons between "loose" and "tight" installation conditions difficult.

Most of the testing at delta-Vs of 50 kph and above exceeded the peak in-position IARVs reported by Mertz et al. [24], as did some tests at lower severities. While the IARVs referenced earlier refer to the peak tension, neck injury risk depends on both the magnitude of the tensile load and its duration. A review of force-time data from Sherwood et al. [37] and from testing performed by the authors [25, 26] indicate that some tests that do not exceed the peak tensile IARV may exceed the time-dependent criteria.

The majority of the data for the 6-month-old ATD reported resultant forces and moments without reporting individual components [19, 46, 47]. Review of force-time data [47] showed that at the time of peak upper neck resultant force, the primary component was tension. In the three tests with force component data available, the peak upper neck tensions were 93% - 99% of the peak resultant. As such, we used the resultant force as an over-predictive estimate of peak upper neck tension for the 6-month-old ATD.

This study encompassed a wide variety of restraint configurations, which were broadly classified into 11 categories (Table 1). Five of these categories were used in very few tests. Within the other six groups, there is additional variation, including make and model and internal adjustments (recline angle, webbing adjustment, location of buckle or clip relative to ATD, etc.) of any child restraint, the method of CRS attachment to the vehicle seat, vehicle seat geometry, vehicle restraint type, geometry, and mechanical properties, vehicle and restraint webbing tension, the presence of and tension in any tethers used, and other factors. Many of these factors are not reported in the available literature or are not reported in detail. In addition, some of child restraints and vehicles used in prior testing are no longer available. Changes in these factors and interactions between factors can affect restraint performance, including neck loads. This variation in restraints likely contributes to some of the scatter seen within the broad categories defined in Table 1. For example, Legault et al. [30] reported increasing neck loads with increasing tether slack for forward-facing CRSs and a further increase for untethered CRSs tested at the same crash pulse. Bohman et al. [33, 53] found reductions in peak neck tension and Nij in HBB, backless boosters and integrated boosters used in conjunction with pretensioners and with pretensioners and load limiters compared with the same booster/vehicle combinations used without pretensioners or load limiters when tested at the same crash pulse.

In addition to variation among restraint types and configurations, there was also variation in ATD types. The tests reviewed here used different ATD models, including ATDs used in American and European compliance testing. Different types of neck load cells and different load cell adjustments were used as well, which may contribute to some variation in measured loads. For example, Janssen et al. [29] performed four tests using a 9-month-old ATD with a rigidly fixed neck. This non-biofidelic configuration resulted in low peak upper neck tensions and shears and high upper neck moments. In addition, in some tests, the ATD chin contacted the chest. When chin-to-chest contact was identified, the maximum loads occurred at this time [37]. This contact is discussed in detail in some studies [37, 40, 41]; in other studies, there is no indication whether this contact occurred. When available, pre-contact values are reported. However, because chin-to-chest contact, or a lack of contact, was not identified in many studies, it was not possible to determine the peak loads for the pre-contact inertial phase in all tests where contact occurred. The inclusion of chin-to-chest contact loads tends to overestimate the peak inertial loads. This is a limitation of this study.

Crash pulses used in the testing presented here were obtained from multiple sources, including American and European compliance testing and reconstructions of real-world crashes. The shape of the crash pulse, including pulse width and peak acceleration, varied for tests conducted at a given change in velocity. Many studies did not report specifics of the crash pulse shape. In general, we expected that crashes with higher accelerations would have higher neck loads than crashes with the same delta-V but lower accelerations. In some studies [33], more severe accelerations did result in higher neck loads; however, there were no overall correlations observed between peak acceleration and neck loads in the currently available data.

There was wide variation in ATD sizes, vehicle and child restraint configurations, and crash pulses in the tests reported here. Real-world frontal crashes involve children of all ages, in many different combinations of vehicles, CRS types, and crash configurations, which cannot be completely represented in a limited number of crash and sled tests. The large number of independent variables, including the testing of different restraint types with ATDs of different sizes, and relatively small sample sizes makes it difficult to perform rigorous statistical comparisons on the available data from the testing reviewed in this study. However, the variation in testing parameters is more representative of real-world crashes than test series with fewer variables. The data presented here can be used as an estimate of the expected range of neck loads in the crash severities studied. To date, there are limited neck load data available from tests involving restraint misrouting misuses that have been identified in real-world child occupants (e.g. belt under arm, behind back). Further testing of these scenarios is warranted.

This study specifically evaluated pediatric neck injury measures in the absence of head contact. It did not examine injuries occurring as a result of head impacts to vehicle structures. Changes in restraint configuration, including low webbing tension and webbing slack, affect head excursion, and increases in head excursion can increase the risk of a head impact with forward vehicle structures during a real-world crash. Additionally, this work did not evaluate injury measures to parts of the body other than the neck, which are affected by restraint configuration. For example, studies have shown a reduction in injury risk for children restrained in child restraints and booster seats compared to those restrained in only the vehicle three-point seat belt [54-56]. Other studies have identified non-neck injury mechanisms to child occupants in frontal crashes that are not addressed in this work [57-62]. The current study provides a detailed evaluation of neck injury measures for pediatric occupants but must be integrated with information regarding other injuries, as well as vehicle- and crash-specific factors, to form a "whole-body" approach to understanding the effect of restraint type on pediatric injury in frontal crashes.

CONCLUSION

Increased crash severity increases the risk of inertially-induced cervical spine injury in children. This occurs for all restrained children, regardless of the type of restraint.

REFERENCES

1. Givens, T., Polley, K., Smith, G., Hardin, W. (1996). "Pediatric cervical spine injury: a three-year experience." J Trauma. 41: 310-314.
2. Zuckerbraun, B., Morrison, K., Gaines, B., Ford, H., Hackam, D. (2004). "Effect of age on cervical spine injuries in children after motor vehicle collisions: effectiveness of restraint devices." Journal of Pediatric Surgery. 39: 483-486.
3. Patel, J., Tepas, J., Mollitt, D., Pieper, P. (2001). "Pediatric cervical spine injuries: defining the disease." Journal of Pediatric Surgery. 36: 373-376.
4. Brown, J., McCaskill, M., Henderson, M., Bilston, L. (2006). "Serious injury is associated with suboptimal restraint use in child motor vehicle occupants." J Paediatr Child Health. 42: 345-349.
5. Nance, M., Elliott, M., Arbogast, K., Winston, F., Durbin, D. (2006). "Delta V as a predictor of significant injury for children involved in frontal motor vehicle crashes." Ann Surg. 243: 121-125.
6. Arbogast, K., Durbin, D., Cornejo, R., Kallan, M., Winston, F. (2004). "An evaluation of the effectiveness of forward facing child restraint systems." Accid Anal Prev. 36: 585-589.
7. Durbin, D., Chen, I., Smith, R., Elliott, M., Winston, F. (2005). "Effects of seating position and appropriate restraint use on the risk of injury to children in motor vehicle crashes." Pediatrics. 115: e305-309.
8. Bull, M., Stroup, K., Gerhart, S. (1988). "Misuse of car safety seats." Pediatrics. 81: 98-101.
9. Winston, F., Chen, I., Elliott, M., Arbogast, K., Durbin, D. (2004). "Recent trends in child restraint practices in the United States." Pediatrics. 113: e458-464.
10. Hummel, T., Lanwieder, K., Finkbeiner, F., Hell, W. (1997). "Injury risks, misuse rates and the effect of misuse depending on the kind of child restraint system." in Second Child Occupant Protection Symposium: Society of Automotive Engineering.
11. Valent, F., McGwin, G., Hardin, W., Johnston, C., Rue, L. (2002). "Restraint use and injury patterns among children involved in motor vehicle collisions." J Trauma. 52: 745-751.
12. Huelke, D., Mackay, G., Morris, A., Bradford, M. (1992). "Non-head impact cervical spine injuries in frontal car crashes to lap-shoulder belted occupants." SAE 920560.
13. Huelke, D., Mackay, G., Morris, A., Bradford, M. (1992). "Car crashes and non-head impact cervical spine injuries in infants and children." SAE Yes.
14. Huelke, D., Mackay, G., Morris, A., Bradford, M. (1993). "A review of cervical fractures and fracture-dislocations without head impacts sustained by restrained occupants." Accid Anal Prev. 25: 731-743.
15. Newman, J., Dalmotas, D. (1993). "Atlanto-occipital fracture dislocation in lap-belt restrained children." SAE 933099.
16. Howard, A., McKeag, A., Rothman, L., Mills, D., Blazeski, S., Chapman, M., Hale, I. (2005). "Cervical spine injuryes in children restrained in forward-facing child restraints: a report of two cases." Journal of Trauma. 59: 1504-1506.
17. Fuchs, S., Barthel, M., Flannery, A., Christoffel, K. (1989). "Cervical spine fractures sustained by young children in forward-facing car seats." Pediatrics. 84: 348-354.
18. Stalnaker, R. (1993). "Spinal cord injuries to children in real world accidents." SAE Skimmed.
19. Trosseille, X., Tarriere, C. (1993). "Neck injury criteria for children from real crash reconstructions." SAE 933103.
20. Henderson, M., Brown, J., Paine, M. (1994). "Injuries to restrained children." in 38th Annual Proceedings of the Association for the Advancement of Automotive Medicine. Lyon, France
21. Burdi, A., Huelke, D., Snyder, R., Lowrey, G. (1969). "Infants and children in the adult world of automobile safety design: pediatric and anatomical considerations for design of child restraints." J Biomech. 2: 267-280.
22. Huelke, D. (1998). "An overview of anatomical considerations of infants and children in the adult world of automobile safety design." in Association for the Advancement of Automotive Medicine, 42nd Annual Conference. Charlottesville, Virginia: AAAM.
23. Van Arsdell, W. (2005). "The evolution of FMVSS 213: Child Restraint Systems." Society of Automotive Engineers. SAE 2005-01-1840.
24. Mertz, H., Irwin, A., Prasad, P. (2003). "Biomechanical and scaling bases for frontal and side impact Injury Assessment Reference Values." Stapp Car Crash Journal. 47: 155-188.
25. Prange, M., Newberry, W., Moore, T., Peterson, D., Smyth, B., Corrigan, C. (2007). "Inertial neck injuries in children involved in frontal collisions." SAE 2007-01-1170.
26. Moore, T., Prange, M., Newberry, W., Peterson, D., Smyth, B., Corrigan, C. (2007). "Inertial neck injuries in children involved in frontal collisions: sled testing using the 6-year-old ATD." in Proceedings of the ASME 2007 Summer Bioengineering Conference Keystone, Colorado SBC2007-176671.
27. Henderson, M., Brown, J., Griffiths, M. (1997). "Children in adult seat belts and child harnesses: crash sled comparisons of dummy responses." SAE 973308.
28. Janssen, E., Huijskens, C., Verschut, R., Twisk, D. (1993). "Cervical spine loads induced in restrained child dummies II." SAE 933102.
29. Janssen, E., Nieboer, J., Verschut, R., Huijskens, C. (1991). "Cervical spine loads induced in restrained child dummies." SAE 912919.

30. Legault, F., Gardner, B., Vincent, A. (1997). "The effect of top tether strap configurations on child restraint performance." SAE 973304.

31. Planeth, I., Rygaard, C., Nilsson, S. (1992). "Synthesis of data towards neck protection criteria for children." in International Conference on the Biomechanics of Impact. Verona, Italy

32. Czernakowski, W., Otte, D. (1993). "The effect of pre-impact braking on the performance of child restraint systems in real life accidents and under varying test conditions." SAE 933097.

33. Bohman, K., Bostrom, O., Olsson, J., Haland, Y. (2006). "The effect of a pretensioner and a load limiter on a HIII 6 Y, seated on four different types of booster cushions in frontal impacts." in IRCOBI. Madrid, Spain.

34. Ha, D., Bertocci, G. (2007). "Injury risk of a 6-year-old wheelchair-seated occupant in a frontal motor vehicle impact--'ANSI/RESNA WC-19' sled testing analysis." Med Eng Phys. 29: 729-738.

35. Kapoor, T., Altenhof, W., Wang, Q., Howard, A. (2006). "Injury potential of a three-year-old Hybrid III dummy in forward and rearward facing positions under CMVSS 208 testing conditions." Accident Analysis and Prevention. 38: 786-800.

36. Menon, R., Ghati, Y. (2007). "Misuse study of latch attachment: a series of frontal sled tests." Annu Proc Assoc Adv Automot Med. 51: 129-154.

37. Sherwood, C., Shaw, C., Van Rooij, L., Kent, R., Crandall, J., Orzechowski, K., Eichelberger, M., Kallieris, D. (2003). "Prediction of cervical spine injury risk for the 6-year-old child in frontal crashes." Traffic Inj Prev. 4: 206-213.

38. Glaeser, K. (1992). "New test conditions for child restraint systems." SAE 922516.

39. Malott, A., Parenteau, C., Arbogast, K., Mari-Gowda, S. (2004). "Sled test results using the Hybrid III 6 year old: an evaluation of various restraints and crash configurations." SAE 2004-01-0316.

40. Menon, R., Ghati, Y., Ridella, S., Roberts, D., Winston, F. (2004). "Evaluation of restraint type and performance tested with 3- and 6-year-od Hybrid III dummies at a range of speeds." SAE 2004-01-0319.

41. Menon, R., Ghati, Y., Roberts, D. (2005). "Performance evaluation of various high-back booster seats tested at 56 kph using a 6-year-old Hybrid III dummy." in ESV 05-0366.

42. National Highway Traffic Safety Administration. "Tables for CRS RFC." (cited September 4, 2008). Available from: http://www.nhtsa.gov/Cars/rules/rulings/CRS-Rate/Tables3.htm.

43. Whitman, G., Yannaccone, J., Bandak, F., Sicher, L., D'Aulerio, L., Shanahan, D., Cantor, A., Roberts, D., Moss, S. (1999). "A method for the assessment of tethered and untethered child restraint systems using the Hybrid III three year old dummy." in 27th International Workshop on Injury Biomechanics Research. San Diego, California.

44. Belcher, T., Newland, C. (2007). "Investigation of lower anchorage systems for child restraints in Australia." in ESV.

45. Brun Cassan, F., Page, M., Pincemaille, Y., Kallieris, D., Tarriere, C. (2993). "Comparative study of restrained child dummies and cadavers in experimental crashes." SAE 933105.

46. Weber, K., Dalmotas, D., Hendrick, B. (1993). "Investigation of dummy response and restraint configuration factors associated with upper spinal cord injury in a forward-facing child restraint." SAE 933101.

47. Tanner, C., Wiechel, J., Morr, D., Guenther, D. (2002). "Response of the 6-month-old CRABI in forward facing and rear facing child restraints to a simulated real world impact." SAE 2002-01-0026.

48. Oster, J., Trommier, B. (1996). "Comparison of the six-year-old Hybrid III, Part 572 and TNO P6 child dummies." SAE 962437.

49. Kuppa, S., Saunders, J., Fessahaie, O. (2005). "Rear seat occupant protection in frontal crashes." in 19th International Technical Conference on the Enhanced Safety of Vehicles. Washington, D.C. 05-0212.

50. Charlton, J. L., Fildes, B., Laemmle, R., Koppel, S., Fechner, L., Moore, K., Smith, S., Douglas, F., Doktor, I. (2005). "Crash performance evaluation of booster seats for an Australian car." in 19th International Technical Conference on the Enhanced Safety of Vehicles. Washington, D.C. 05-0425.

51. Saul, R. A., Pritz, H. B., McFadden, J., Backaitis, S. H., Hallenbeck, H., Rhule, D. (1998). "Description and performance of the Hybrid III three year old, six-year-old and small female test dummies in restraint system and out-of-position air bag environments." in 16th International Technical Conference on the Enhanced Safety of Vehicles. Windsor, Ontario, Canada 98-S7-O-01.

52. Irwin, A., Mertz, H. (1997). "Biomechanical basis for the CRABI and Hybrid III child dummies." SAE 973317.

53. Bohman, K., Bostrom, O., Osvalder, A., Eriksson, M. (2007). "Rear seat frontal impact protection for children seated on booster cushions - an attitude, handling and safety approach." in ESV.

54. Durbin, D., Elliott, M., Winston, F. (2003). "Belt-positioning booster seats and reduction in risk of injury among children in vehicle crashes." JAMA. 289: 2835-2840.

55. Elliott, M., Kallan, M., Durbin, D., Winston, F. (2006). "Effectiveness of child safety seats vs seat belts in reducing risk for death in children in passenger vehicle crashes." Arch Pediatr Adolesc Med. 160: 617-621.

56. Miller, T., Zaloshnja, E., Hendrie, D. (2006). "Cost-outcome analysis of booster seats for auto occupants aged 4 to 7 years." Pediatrics. 118: 1994-1998.

57. Arbogast, K., Kent, R., Menon, R., Ghati, Y., Durbin, D., Rouhana, S. (2007). "Mechanisms of abdominal organ injury in seat belt-restrained children." J Trauma. 62: 1473-1480.

58. Shepherd, M., Hamill, J., Segedin, E. (2006). "Paediatric lap-belt injury: a 7 year experience." Emerg Med Australas. 18: 57-63.

59. Durbin, D., Arbogast, K., Moll, E. (2001). "Seat belt syndrome in children: a case report and review of the literature." Pediatr Emerg Care. 17: 474-477.

60. Edgerton, E., Orzechowski, K., Eichelberger, M. (2004). "Not all child safety seats are created equal: the potential dangers of shield booster seats." Pediatrics. 113: e153-158.

61. Lutz, N., Arbogast, K., Cornejo, R., Winston, F., Durbin, D., Nance, M. (2003). "Suboptimal restraint affects the pattern of abdominal injuries in children involved in motor vehicle crashes." J Pediatr Surg. 38: 919-923.

62. Nance, M., Lutz, N., Arbogast, K., Cornejo, R., Kallan, M., Winston, F., Durbin, D. (2004). "Optimal restraint reduces the risk of abdominal injury in children involved in motor vehicle crashes." Ann Surg. 239: 127-131.

CONTACT

Tara Moore is a Managing Engineer at Exponent, Inc. She can be reached at tmoore@exponent.com

962433

Neck Injuries in the UK Co-operative Crash Injury Study

Andrew P. Morris and Pete Thomas
ICE Ergonomics

Abstract

This study examines some of the factors associated with soft tissue neck injuries in the UK. The data were drawn from a retrospective study of vehicle crash injuries in which the overall soft tissue neck injury rate was 16%. This study shows how although it is commonly assumed that such injuries are a rear impact phenomenon, over 50% of the injuries occur in frontal crashes. In front and rear impacts, these injuries are undoubtedly associated with seat-belt use. The incidence of neck injury has been shown to double over the ten-year period of the study with the effect more prominent in females. Such injuries are also more likely to be self-reported than clinically diagnosed. Head restraints have not been found to mitigate neck injuries in either front or rear impacts at a statistically significant level. A slight but non-significant trend towards reduced neck injury rates is observed in cases of seat back yielding in a rear impact.

Introduction

Whilst Spitzer et al (1995) define 'whiplash' as 'an acceleration-deceleration mechanism of energy transfer to the neck', a universally agreed definition of soft tissue neck injury is less easy to come by. The term in this study relates to a range of symptoms that arise following a motor vehicle crash as a result of neck hyper-extension, neck hyper-flexion or a combination of both of these mechanisms. Such symptoms may include complaint of neck pain, stiffness and/or tenderness together with more vague clinical manifestations which are thought to be associated with damage to the soft tissue constituents of the head/neck/shoulder complex. As Pearce (1989) notes, few topics provoke so much controversy based on so little fact as this type of injury. It is an injury shrouded in mystery and 'creates clinical insecurity in those who attempt to explain its mechanism, prognosis and treatment'. Medical impairment associated with soft tissue neck injury is apparently an increasing universal problem which has some association with the global progression towards mandatory restraint use (Cameron, 1981; Thomas, 1990). Whilst the biomechanics of neck injuries have been previously described in-depth elsewhere, (Walz and Muser, 1995; Bogduk, 1986; Melvin, 1979) it has become clear in recent times that there is a need for a greater understanding of the vehicle, collision and occupant parameters that are prevalent in neck injuries (and particularly soft tissue neck injuries) before preventative measures can be adopted.

Neck Injury and Seat Belt Usage

Rutherford et al (1985) indicated a relative increase of 18% in neck 'sprains' co-incident with the UK seat-belt use increasing from 26% to 93% after the introduction of the UK seat belt law and this data was supported in a study by Galasko et al (1993) who found a corresponding increase in incidence of soft tissue neck injuries. They studied occupants in car crashes between the years 1982 and 1991 and found that in 1982, 12 months before the introduction of the compulsory seat belt legislation, the incidence of soft tissue neck injury was 8%. Immediately following the introduction of the law, the incidence of soft tissue neck injury rose to 21% and thereafter it rose steadily each year. By 1991, the incidence had risen to approximately 46% .

Maag et al (1990) found that neck sprains were relatively more numerous among belted occupants compared to unbelted occupants by a factor of 1.68 while Larder et al (1985) found that 93% of drivers and 96% of front passengers who sustained a neck injury wore their seat-belts in a population of predominantly unrestrained occupants.

Neck Injury and Head Restraint

Nygren et al (1985) examined the performance of different types of head restraints in rear-end collisions and suggested that they do have a certain influence on the incidence of neck injury in this type of collision but they observed that further studies were necessary since the knowledge of function is generally low. Olsson et al (1990) observed that 'most modern cars are equipped with head restraints designed to prevent whiplash injuries but neck injuries are still common in rear impacts indicating that restraints are not functioning satisfactorily'. They concluded that 'a distance of more than 10cm between the head and the head restraint correlates with an increase risk of neck injury'. This work was supported by Svensson et al (1993) who found that the head-to-head-restraint gap was the largest influence on head-neck motion in a rear impact and Parkin et al (1993), in an observational study of driver head-restraint position found that 50% of the driving population had the restraint positioned greater than 15cm from the back of the head horizontally. Only 5% of drivers had the restraint correctly positioned vertically while 50% of the population positioned 10cm or more below the centre of the head. Similarly Viano and Gargan (1995) found in their observational study that only 10% of drivers had head-restraints in the most favourable position to prevent neck extension in a rear-end crash. In the same study, they conducted a series of simulated rear-end collisions and found that the lowest response of neck extension occurred with a small gap between head and head-restraint and also a high head-rest.

Neck Injury and Seat Yield

Some authors have postulated that a controlled yield of the seat-back during an impact has a beneficial effect in terms of neck injury outcome since such a design reduces the recoil of the body during the latter stage of the impact thereby diminishing the risk of hyperflexion injury to the cervical spine. Kihlberg (1969) reviewed rear impact data from the US Automotive Crash Injury Research (ACIR) study and found that the frequency of flexion/torsion neck injuries was substantially less than for those cases in which the seat remained intact while Parkin et al (1995) found that with the presence of yield, the occupant was less likely to suffer an AIS 1 neck injury whilst a seat which exhibited no residual damage was more likely to result in AIS 1 neck injury. Overall, AIS 1 neck injuries were approximately twice as frequent in an undamaged seat than in a yielding seat. Von Koch (1995) observed a considerable difference in relative rebound velocity between seats of different stiffness coefficients.

Impact Classification

Larder et al (1985) found that in frontal impacts, 17% of occupants sustained a neck injury whilst for rear impacts, the rate was 31% and Lovsund et al (1988) found that more than 10% of car occupants involved in a rear-end collision sustained a neck injury. Galasko et al (1993) found that for drivers, 52% were injured in a rear impact, 27% in a front impact and 16% in a side impact. Comparable rates were found for front seat passengers.

Methodology

The data used in this study are from a study of vehicle crash performance and occupant injury (the Co-operative Crash Injury Study) which commenced in the UK in 1983 and ended in May 1992. In all, the CCIS database holds information on some 6,973 vehicles involved in crashes containing 11,866 occupant who sustained between them 42,876 injuries.

Each vehicle in the study was inspected within a few days of the collision. The general sampling criteria of the CCIS study are;

(i)   that the vehicle involved was towed away from the scene of the accident to a garage or recovery yard.

(ii)   that the vehicle was less than six years old at the time of the collision

(iii) that there was an injury in the vehicle according to the UK Police system of injury classification.

About 80% of serious and fatal accidents in each study area were investigated along with 10-15% of slight accidents according to the UK Police system of injury classification. The resulting sample represents all levels of injury outcome while being biased towards more serious injuries. Consequently, there may be a degree of under-reporting of the phenomenon of soft tissue neck injury due to the masking effect of serious injury.

Medical data concerning each occupant was obtained from hospitals and each occupant was also requested to complete a questionnaire which provided additional data several days after the crash. Injuries were coded according to the Abbreviated Injury Scale, 1985 revision (American Association for Automotive Medicine; 1985).

A more comprehensive overview of the Co-operative Crash Injury Study can be attained in Mackay et al (1985).

Results

Soft tissue neck injury in this study is defined according to the Abbreviated Injury Scale 1985 revision. The strict definition is 'Strain, Acute (no fracture or dislocation)'. The AIS code number is 70101.1.

1. Database Characteristics

1(a): Impact Classification

The impact classifications where these are known for all occupants including those with and without soft tissue neck injury are as follows;

Table 1: Impact Classifications for Occupants with and Without Soft Tissue Neck Injury.

Impact Classification	Soft Tissue Neck Injury		No Soft Tissue Neck Injury		Total
	N	%	N	%	
Rear	237	38	393	62	630
Front	1,045	15	5,869	85	6,914
Left-side	188	12	1,351	88	1,539
Right-side	280	15	1,608	85	1,888
Top	97	15	545	85	642
Unclassified	40	16	213	84	253
Totals	1,887	16	9,979	84	11,866

As can be seen from table 1, the soft tissue neck injury rate according to this definition is **16%** (i.e. 1887 occupants out of 11,866) but for rear impacts, the rate is **38%**; more than twice the rate than for any other impact class. However although the risk is much higher in rear impacts, it should be noted that over 50% of soft tissue neck injuries do occur in frontal impacts.

Of occupants who sustained a soft tissue neck injury, 51% were male and 49% were female while the overall gender breakdown in the CCIS database is 61% male and 37% female (2% of occupants not known).

Of the occupants in the database, there were 6,403 front seat occupants who were seated in the driver or front seat passenger positions in vehicles which sustained only one impact.

1(b) Collision Severities for Soft Tissue Neck Injury

Figures 1 & 2 show the speed distributions for soft tissue neck injury/no soft tissue neck injury in both frontal and rear impacts. MAIS 1 injuries only are included. As can be seen from both figures, at the MAIS 1 level of injury severity, soft tissue neck injuries occur at relatively lower mean and median collision severities compared with all other injuries.

Figure 1

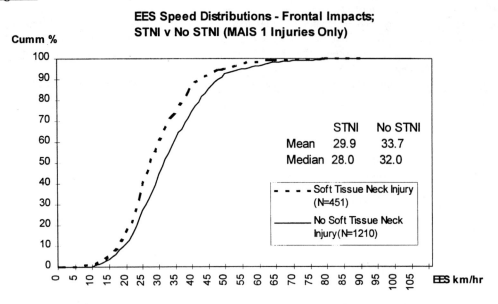

**EES Speed Distributions - Frontal Impacts;
STNI v No STNI (MAIS 1 Injuries Only)**

	STNI	No STNI
Mean	29.9	33.7
Median	28.0	32.0

Soft Tissue Neck Injury (N=451)

No Soft Tissue Neck Injury(N=1210)

Figure 2

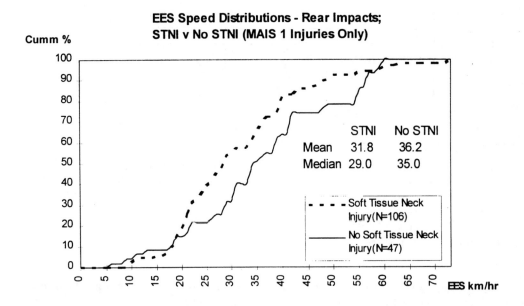

**EES Speed Distributions - Rear Impacts;
STNI v No STNI (MAIS 1 Injuries Only)**

	STNI	No STNI
Mean	31.8	36.2
Median	29.0	35.0

Soft Tissue Neck Injury(N=106)

No Soft Tissue Neck Injury(N=47)

1(c); Maximum Abbreviated Injury Score (MAIS)

Injury risks were evaluated for both soft tissue neck injured occupants and for all occupants. These are shown in table 2. As can be seen from this table, for 76% of occupants, MAIS 1 is the highest severity of overall injury level assessed by threat to life. This would suggest that the severity in terms of impairment is somewhat under-rated.

Table 2: Maximum Abbreviated Injury Scores for Occupants With and Without Soft Tissue Neck Injury - CCIS Study

MAIS	Occupants with Soft Tissue Neck Injury		All Occupants	
	N	%	N	%
1	1,437	76	4,923	51
2	361	19	2,193	23
3	64	3.5	815	8
4	6	0.4	291	3
5	2	0.1	331	3
6	0	0	262	3
9 (not known)	17	1	911	9
Totals	1,887	100	9,726*	100

* (2,140 Occupants sustained no injuries).

1(d): Restraint Use

The issue of restraint use for soft tissue neck injuries was then analysed across the whole database. This is shown in table 3.

Table 3: Restraint Use by Soft Tissue Neck Injury (All Impact Types)

	Soft Tissue Neck Injury		No Soft Tissue Neck Injury		Total
Restraint Use	N	%	N	%	
Used	1,530	20	6,264	80	7,794
Not Used	208	8	2,268	92	2,476
Not Known	149	9	1,447	91	1,596
Total	1,887	16	9,979	84	11,866

The figures are for all occupants in all vehicles and the high numbers of unrestrained occupants reflect the predominantly unrestrained population of rear seat occupants. The effect of seat-belt wearing on neck injury is explored later in the results section but it is interesting to note that 20% of restrained occupants sustained a soft tissue neck injury compared to 8% of unrestrained occupants.

1(e): Gender

The issue of gender was also addressed. Again occupants who sustained soft tissue neck injury were compared to all occupants in the database. The results are shown in table 4. As can be seen from this table, females are at greater risk of sustaining soft tissue neck injury than males. The reasons for this are by no means clear. Physiological and anatomical differences may in part explain these differences.

Table 4: Gender of Occupants - Soft Tissue Neck Injury v All Occupants (all Impact Directions)

	Soft Tissue Neck Injury		No Soft Tissue Neck Injury		Totals
Gender	N	%	N	%	
Male	955	13	6,273	87	7,228
Female	928	21	3,453	79	4,381
Not Known	4	2	253	98	257
Totals	1,887		9,979		

In this section, the incidence of soft tissue neck injuries according to differing sources of medical data was examined. The results are shown in table 5. Each percentage figure represents the number of soft tissue neck injuries expressed as a percentage of the total number of injuries that occurred in that classification. When casualties sustained only minor injuries of maximum AIS 1, the rates of soft tissue neck injury were higher than when more severe injuries were sustained. This could be explained by the presence of more severe injuries masking the presence of a soft tissue neck injury.

Of note is the fact that the rate of neck injury (7%) was higher if the occupant self-reported compared to the rate for clinical diagnosis at the hospital (3%). This is also true when only MAIS 1 injuries are investigated (12% compared to 8%). This could imply that the injury may not present itself until some time after the accident has occurred in which case the questionnaire is the most accurate means of reporting the injury. The issue of fraudulent claims should not however be overlooked.

Table 5; Data Sources for Soft Tissue Neck Injuries

	All Injuries	MAIS 1	MAIS 2	MAIS 3
Data Source	%	%	%	%
All data sources	5	9	3	1.5
Hospital Records Only	3	8	3	1
Questionnaire Only	7	12	3	1
Hospital Records & Questionnaire	6	9	4	2
Police Data	8	11	3	3

2. Soft Tissue Neck Injury as a Function of Time

The first analysis in this category examines soft tissue neck injury as a function of time.

Figure 3 shows the changing rate of the incidence of soft tissue neck injury. As can be seen from the graph, neck injury rates increased almost linearly over the study-period. The rates for years 1983 and 1992 have not been included as data were collected for only part of each of these years.

Figure 3

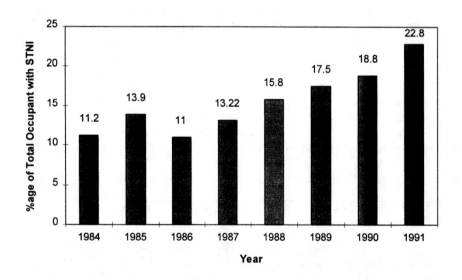

Increase in STNI Rates Over Time

This relationship has also been examined for differing impact classifications. Of interest is the fact that soft tissue neck injuries have increased for all impact classifications since the beginning of the study although the picture is somewhat confusing in the case of rear impacts. However, a clear relationship exists particularly for frontal impacts. It is also worth observing that the data were collected over a period of time just before the general introduction of improved restraint technology (seat-belt pretensioners, seat-belt grabbers etc.) in the UK. This relationship is shown in figure 4.

Figure 4

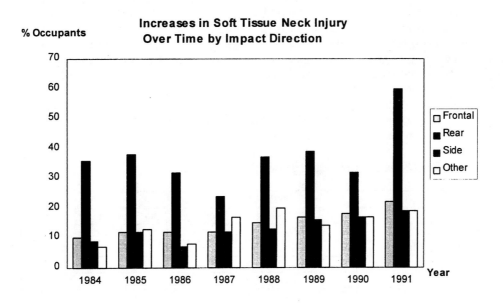

Gender differences in the increase of rate of STNI over time have also been observed. Figure 5 shows this relationship. All occupants in all impact types are included in this analysis. In can be clearly seen that the STNI rates for females have increased at a greater rate than that for males. The rate for females was 14% in 1984 and had risen to 31% by 1991 while the rate for males was 10% in 1984 and had risen to 18% by 1991.

Figure 5

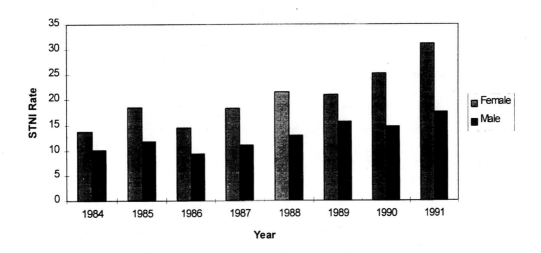

3. Frontal Crash Occupant, Vehicle and Collision Characteristics

This section explores differing occupant and collision characteristics associated with soft tissue neck injury in <u>frontal</u> impacts. All analyses were performed using front seat occupants only. Vehicles were selected for analysis on the basis that they sustained only <u>one</u> impact so that the occupant kinematics were as predictable as possible.

3(a); Seat Belt Effectiveness.

The effect of belt use on soft tissue neck injury in frontal impacts was studied for front seat occupants only. All occupants in the study were restrained by 3-point seat-belts. Virtually all seat-belts had upper outboard and lower outboard anchorage points located on the B-pillar and inboard anchorages on both the seat and vehicle body-shell. There were no vehicles which contained integrated seat and seat-belt systems. However, a comprehensive examination of seat-belt geometry was not possible in this study since both fore/aft adjustment and seat-belt positions were not investigated in each case. Furthermore, rear seat occupants were excluded from this analysis since belt usage rates over the period of the study was very low primarily because the UK mandatory rear seat belt wearing law was not introduced until very late in the study period (July 1991). In each case, the occupants experienced one impact only. The results are shown in table 6

Table 6; Seat Belt Use in Frontal Impacts; STNI-Vs-No STNI

MALES							
	Used		Not Used		Not Known		Total
Injury Condition	N	%	N	%	N	%	
Soft Tissue Neck Injury	344	85	38	9	22	6	404
No Soft Tissue Neck Injury	1,692	70	460	19	274	11	2,426
FEMALES							
	Used		Not Used		Not Known		Total
Injury Condition	N	%	N	%	N	%	
Soft Tissue Neck Injury	340	87	38	10	15	4	393
No Soft Tissue Neck Injury	968	69	310	22	130	9	1,408

Males were significantly more likely to sustain a neck injury if they were restrained than if they were unrestrained ($\chi2 = 26.2$, df = 1, p<0.000). The same relationship was observed for women ($\chi2 = 35.34$, df = 1, p<0.000) so overall, neck injuries are significantly more likely in restrained occupants in frontal crashes compared to unrestrained occupants.

3(b). Head Restraint Type and Effectiveness

Table 7 shows the effectiveness of head restraints in frontal impacts. As can seen from this table, slight but non-statistically significant trends emerge which suggest that head-restraints are slightly detrimental in terms of neck injury outcome in frontal crashes. The data also suggest a slight trend towards adjustable head restraints being slightly less detrimental in this type of impact but again this is not statistically significant.

Table 7 Head Restraint Type and Effectiveness

Frontal Impacts					
	No Neck Injury		Neck Injury		χ^2	p
Head Restraint	N	%	N	%		
None	442	83	90	17	2.28	0.20
Fixed	328	79	86	21		
None	442	83	90	17	1.081	0.30
Adjustable	2068	81	481	19		
None	442	83	90	17	1.449	0.20
Fixed & Adjust.	2396	81	567	19		

4. Rear-end Crash Occupant, Vehicle and Collision Characteristics

4(a) Seat Belt Effectiveness

The effect of belt use on soft tissue neck injury in rear impacts was studied for front seat occupants only. Rear seat occupants were excluded from this analysis since belt usage rates over the period of the study was very low primarily because the UK mandatory rear seat belt wearing law was not introduced until very late in the study period (July 1991). In each case, the occupants experienced one impact only. The results are shown in table 10.

Table 8 ; Seat Belt Use in Rear Impacts; STNI-Vs-All Injuries

MALES							
	Used		Not Used		Not Known		Total
Injury Condition	N	%	N	%	N	%	
Soft Tissue Neck Injury	42	78	3	6	9	16	54
No Soft Tissue Neck Injury	43	46	14	15	36	39	93
FEMALES							
	Used		Not Used		Not Known		Total
Injury Condition	N	%	N	%	N	%	
Soft Tissue Neck Injury	48	58	10	12	25	30	82
No Soft Tissue Neck Injury	33	59	13	23	10	18	56

Males were significantly more likely to experience neck injury if the seat belt was worn in a rear impact ($\chi 2 = 5.8$, df=1, p<0.02). However, no statistically significant association was found for females ($\chi 2 = 1.77$, df=1, p=n.s.). In both cases, the overall risks could be confounded by high numbers of rear crashes where the seat belt usage was unknown.

4(b) Head Restraint Type and Effectiveness

A comparison was made of head restraint performance on injury outcomes. In each case, the occupants sustained only 1 impact and only drivers and front seat passenger were included. Males and females were studied collectively. Table 9 shows the data for head restraint effectiveness in rear impacts

Table 9; Head Restraint Type and Effectiveness

Rear Impacts						
	No Neck Injury		Neck Injury		χ^2	p
Head Restraint	N	%		%		
None	18	58	13	42	0.126	0.75
Fixed	12	63	7	37		
None	18	58	13	42	0.433	0.70.
Adjustable	69	51	65	49		
None	18	58	13	42	0.283	0.70
Fixed & Adjust.	81	53	72	47		

Table 9 shows no evidence that fixed or adjustable head restraints had any statistically significant effect on soft tissue neck injury. The probability that the results observed were derived by chance was at least 70%. We were aware that occupant anthropometry (e.g. height, weight etc.) in relation to the seat and head-restraint may have affected the overall results in rear-end crashes. However, it was not possible to quantify this effect since detailed information regarding the occupant was only available in approximately 50% of cases.

5. Seat Back Yield and Soft Tissue Neck Injury

The next analysis examines seat back damage in relation to neck injury in a rear impact where there was only one impact to the vehicle. Instances of seat back damage which were generated by loading from the seat occupant were compared to examples of undamaged seats for neck-injury occupants and no neck-injury occupants. The results are shown in table 10.

Table 10: Seat Back Damage and Neck Injury Outcome.

	Seat Back Damaged		No Damage		Total
	N	%	N	%	
Neck Injury	43	39	67	61	110
No Neck Injury	61	46	72	54	133
Total	104	43	139	57	243

$\chi^2 = 1.1403$, d.f. = 1, $p < 0.20$

As can be seen from the table there is a statistically non-significant trend towards reduced neck injury outcomes in instances of seat back damage when there is only one rear impact to the vehicle.

Discussion

Soft tissue neck injuries are still a major concern but it is a common misconception that they occur more or less exclusively in rear impacts. In this study, it has been shown that although the risk of sustaining soft tissue neck injury is greatly increased in a rear impact, over 50% of these injuries are in fact occurring in frontal impacts with a further 25% occurring in side impacts. It is perhaps therefore necessary to establish whether the more impairing neck injuries occur more in one impact type compared to another, although the data in this study do not provide scope for doing this.

What is certain is that soft tissue neck injuries can occur at comparatively low-speeds and are associated with seat-belt use. They also occur more frequently to females compared to males and are an injury that has a greater tendency to be self-reported than clinically diagnosed. The reasons for gender differences are also unclear. It is suggested that anatomical and physiological differences may imply that the biomechanical tolerance of the female neck is compromised more readily than with males. However this is confounded by the fact that there is at present only a vague notion of the precise mechanism of injury and also what exactly the injury is (indeed if the injury exists at all). Certainly increasing incidences of soft tissue neck injury would suggest that there is a need for a more reliable diagnostic procedure as there remains the possibility that the crash population is becoming increasingly sophisticated and what in fact is emerging is a trend toward increasing numbers of fraudulent claims of soft tissue neck injury. This may be represented by the fact that self-reporting does produce higher incidence of neck injury but it should be remembered that the symptoms of the injury frequently do not manifest themselves until some 12-24 hours after the crash in which case self-reporting is in fact the most accurate means of diagnosis.

The explanation of increased potential for soft tissue neck injury with seat belt usage requires some clarification. It needs to be established whether the injury mechanism is the same in the case of a restrained occupant in both a front and rear impact. If so, then this would suggest that specifically in the case of rear impacts, the injury is a phenomenon of rebound from an unyielding seat creating hyperflexion during interaction with the belt as happens in a frontal impact and this phenomenon has been described in an earlier study by Krafft et al (1996). The data in this study, while not conclusive, show a trend toward reduced levels of soft tissue neck injury when the seat yields and this finding agrees with data from many other studies. What is necessary in this respect now is to establish precisely how a yielding seat affords protection against neck injury. If clear evidence emerges, then it is obvious that there is very real potential for reducing the incidence of soft tissue neck injury by introducing seating which deforms rearwards in a controlled manner in rear impacts.

The introduction of frontal crash airbags to modern vehicles was greeted with optimism since it was anticipated that head ride-down on the airbag would reduce the force and extent of hyperflexion in restrained occupants. However, a preliminary study of airbag deployments in the UK by Morris et al (1996) has shown that neck injuries occur at the rate of 20% for restrained drivers in frontal crashes in airbag-deployed vehicles. This compares to the overall soft tissue neck injury rate in this present study of 16% so it would appear that airbags have only limited effect. It will be interesting to establish the soft tissue neck injury rate when more data becomes available.

While it is clear that head restraints, not unexpectedly, have no overall effect in frontal crashes in this study, it is surprising to find no evidence of benefit in rear crashes. Nevertheless, although there is no evidence in this historical data-set that either fixed or adjustable head restraints reduce STNI, it cannot be inferred that alternative designs will not be effective. The explanation for limited effectiveness of adjustable head restraints must include poor positioning of the occupant in relation to the head restraint (i.e. horizontal clearance from head restraint to head) and also poor vertical positioning of the restraint itself. This generally supports the evidence provided by Parkin et al (1993) and Viano and Gargan (1995) who found that only 5% and 10% respectively of occupants had correctly adjusted restraints. However a confounding factor here is that in this study, the mean height of female occupants with soft tissue neck injury is approximately 12cm less than their male counterparts. It therefore follows that the female occupants are in a 'better' situation in terms of head restraint height/seat back height combination yet females are still sustaining higher rates of neck injury. We are aware that only a cursory examination of occupant anthropometry was possible in this study because occupant height and weight was available in only about 50% of cases. However, we intend to explore this issue more thoroughly in a future study.

Overall, the data show a clear value in conducting future work in engineering circles with regard to vehicle design. A useful follow-up would be a clear evaluation of the current design of head restraints to assess how they compare to older designs. A more rigorous comparison of fixed and adjustable head restraints is also necessary and there is a definite requirement for a closer examination of the relationship between occupant height, head restraint height and injury. The concept of yielding seats should also be explored further. If clear evidence emerges to support the case for seat yielding, then this may have implications for alternative seat design. Consideration also needs to be made of the influence of improved restraint technology on soft tissue neck injury rates as more data become available. Furthermore, as this study has examined predominantly frontal and rear impacts, some benefit would be attained by an in-depth examination of soft tissue neck injuries in side impact and rollover accidents.

Finally, we recognise that in a retrospective study such as this, only the residual effects of the crash can be investigated and we have attempted to limit our study to examining these effects. However we are aware that there are also important events that are associated with the dynamics of the crash and unfortunately we were unable to investigate these factors in this study.

Acknowledgements

The Co-operative Crash Injury Study is managed by the Transport Research Laboratory (TRL) on behalf of the Department of Transport (Vehicle Standards and Engineering Division) who fund the project with Ford Motor Company Limited, the Rover Group Limited and Toyota Motor Europe. The data were collected by teams at the Vehicle Safety Research Centre at Loughborough University, the Accident Research Centre at The University of Birmingham and from the Vehicle Inspectorate.

References

(1) Spitzer, W O; Skovron, M L; Salmi, L R; Cassidy, J D; Duranceau, J; Suissa, S and Zeiss, E. **'Scientific Monograph of the Quebec Task Force on Whiplash-Associated Disorders: Redefining 'Whiplash' and its Management'**. In Spine Vol. 20 No. 8S, April 1995.

(2) Pearce, J M. **'Whiplash Injury: A Reappraisal'**. J. Neurol Neurosurg Psychiatry No 52 pp 1329-31, 1989.

(3) Cameron, M H. **'The Effect of Seat Belts on Minor and Severe Injuries Measured on the Abbreviated Injury Scale'**. In Accident Analysis and Prevention Vol 13 No 1 pp17-28, 1981.

(4) Walz, F H and Muser, M H. **'Biomechanical Aspects of Cervical Spine Injuries'**. SAE Paper Number 950658, Society of Automotive Engineers, Warrendale, PA, 1995.

(5) Bogduk, N. **'The Anatomy and Pathophysiology of Whiplash'**. In Clinical Biomechanics Vol 1 pp 92-101, 1986.

(6) Melvin, J W. **'Human Neck Injury Tolerance'**. In The Human Neck - Anatomy, Injury Mechanisms and Biomechanics, Congress and Exposition, pp 45-46, 1979.

(7) Rutherford, W H; Greenfield, A A and Hayes, H R M. **'The Medical Effects of Seat Belt Legislation the UK'**. Department of Health and Social Security Research Report 13, London, HMSO, 1985.

(8) Galasko, C S B; Murray, P M; Pitcher, M; Chambers, H; Mansfield, M; Madden, M; Jordan, C; Kinsella, A and Hodson, M. **'Neck Sprains After Road Traffic Accidents: A Modern Epidemic'**. In Injury Vol 24 no 3 pp155-157, 1993.

(9) Maag, U; Desjardins, D; Bourbeau, R and Laberge-Nadeau, C. **'Seat Belts and Neck Injuries'**. In Proceedings of the IRCOBI Conference, Sept12-14, Lyon , France, 1990.

(10) Larder, D R; Twiss, M K and Mackay, G M. **'Neck Injury to Car Occupants Using Seat Belts'**. In Proceedings of the AAAM Conference, Des Plaines, Illinois, pp153-165, 1985.

(11) Nygren, A; Gustafsson, H; and Tingvall, C. **'Effects of Different Types of Head Restraints in Rear-End Collisions'**. In Proceedings of the 10th International Conference on Experimental Safety Vehicles, Oxford, England, pp85-90, 1985.

(12) Olsson, I; Norin, H and Ysander, L. **'An In-depth Study of Neck Injuries in Rear-end Collisions'**. In Proceedings of the IRCOBI Conference, Sept12-14, Lyon , France, 1990.

(13) Svensson, M; Lovsund, P; Haland, Y and Larsson, S. **'The Influence of Seat-Back and Head Restraint Properties on the Head-Neck Motion During rear Impact'**. In Proceedings of the IRCOBI Conference, Sept 8-10, Eindhoven , Holland, 1993.

(14) Parkin, S; Mackay, G M and Cooper, A. **'How Drivers Sit in Cars'**. In Proceedings of the AAAM Conference, San Antonio, Texas, pp375-388, 1993.

(15) Viano, D C and Gargan, M F. **'Headrest Position During Normal Driving: Implications to Neck Injury Risks in Rear Crashes'**. In Proceedings of the AAAM Conference, Chicago, Illinois, pp215-229, 1995.

(16) Kihlberg, J K. **'Flexion-Torsion Neck injury in Rear Impacts'**. In Proceedings of the AAAM Conference, Minneapolis, Minnesota, , 1969.

(17) Parkin, S; Mackay, G M; Hassan, A M and Graham, R. **'Rear-End Collisions and Seat Performance - To Yield or Not To Yield'**. In Proceedings of the AAAM Conference, Chicago, Illinois, pp231-244, 1995.

(18) V Koch, M; Kullgren, A; Lie, A; Nygren, A and Tingvall, C. **'Soft Tissue Injury of the Cervical Spine in Rear-End and Frontal Collisions'**. In Proceedings of the IRCOBI Conference, Sept 13-15, Brunnen, Switzerland, pp273-283, 1995.

(19) Lovsund, P; Nygren, A; Salen, B and Tingvall, C. **'Neck Injuries in Rear-End Collisions Among Front and Rear Seat Occupants'**. In Proceedings of the IRCOBI Conference, Sept 14-16, Bergisch-Gladbach, Germany, pp319-325, 1988.

(20) Association for the Advancement of Automotive Medicine. **'The Abbreviated Injury Scale, 1985 Revision'**. AAAM, Arlington Heights, Illinois, 1985.

(21) Mackay, G M; Galer, M D; Ashton, S J and Thomas, P. **'The Methodology of In-depth Studies of Car Crashes in Britain'**. SAE Paper No. 850556, Society of Automotive Engineers, Warrendale, PA, 1985.

(22) Morris, A P; Thomas, P; Foret-Bruno, J-Y; Thomas, C; Otte, D; Ono, K and Brett, M. **"A Review of Driver Airbag Deployments in Europe and Japan To Date".** Paper No. 96-S1-O-03 In Proceedings of the Enhanced Safety in Vehicles Conference, Melbourne, Australia, 1996.

(23) Krafft, M; Kullgren, A; Nygren, A; Lie, A and Tingvall, C. **"Whiplash Associated Disorder-Factors Influencing the Incidence in Rear-end Collisions".** Paper No. 96-S9-O-09 In Proceedings of the Enhanced Safety in Vehicles Conference, Melbourne, Australia, 1996.

A STUDY OF CURRENT NECK INJURY CRITERIA USED FOR WHIPLASH ANALYSIS. PROPOSAL OF A NEW CRITERION INVOLVING UPPER AND LOWER NECK LOAD CELLS.

D. Muñoz[1], A. Mansilla[1,2], F. López-Valdés[1,2]. R. Martín[1]
[1] Fundación CIDAUT
[2] Department of Mechanical Engineering and Engineering of Materials. University of Valladolid
Spain
Paper Number 05-0313

ABSTRACT

Nowadays several injury criteria are being used in the analysis and evaluation of whiplash risk in automotive rear impacts (NIC, Nkm, LNL, etc.). This study presents a review of the most accepted injury mechanisms and evaluates the advantages and inconveniences of the commonest criteria at present. Taking into account the conclusions arrived at during this comparison, a new criterion is proposed using the signals registered in the upper and lower neck load cells of a crash test dummy, trying to minimize the disadvantages previously found in the other criteria. In order to validate this study a series of sled tests with a BioRID-II dummy have been performed and its results analyzed, confirming the assumptions made during the review of the present criteria and showing a very promising response to the new one. In conclusion, the use of injury criteria involving the load cells situated in both ends of the neck at the same time is recommended as the best way to deal with the dynamics produced during the whiplash movement in a rear impact.

INTRODUCTION

In a rear-end car crash, even at low speed, the head of the occupants of the struck vehicle normally suffers a motion related to the torso that produces sudden distortions of the neck. Although in the most severe cases this movement can produce the fracture of cervical vertebrae, the commonest related lesions are only classified as minor injuries (AIS 1) [1]. Nevertheless, these lesions, known as whiplash-associated disorders (WAD) or simply whiplash, produce painful and often long-term or even chronic symptoms, causing huge economic costs to the society at the same time.

During the last few years a certain number of experimental procedures have appeared trying to evaluate the capacity of the automotive seats to protect the occupants in a rear-end crash. Currently the most accepted of these procedures (IIWPG [2][3], Folksam [4], ADAC, etc.) are using dynamic sled tests and the crash test dummy BioRID-II [5][6][7]. One of the main problems in the development of this kind of procedures has been related to the lack of a full understanding of whiplash injury mechanisms, even though several theories have been proposed trying to give an explanation to the observed symptoms. At the same time, a certain number of injury criteria have been developed looking for a correlation with the different proposed mechanisms. At present there is still a debate about which of these criteria should be taken into account to describe the ability of a seat to protect the neck of the occupants in a rear-end impact properly. In this situation, the groups that are developing new test procedures are adopting either several criteria simultaneously ([4]) or none of them, basing their assessment on the direct comparison of loads and accelerations ([3]). At this point, the lack of a criterion unifying the different injury mechanisms that can be used easily on a test protocol is clear.

The main objective of the presented work was to make a critical review of the commonest injury criteria used at present, trying to analyze the advantages and disadvantages of each one of them. The results would provide a better understanding about the different criteria themselves and, if possible, give guidelines for the definition of a new criterion solving the possible problems found.

METHODS

Keeping this objective in mind, the first question is: how do we evaluate a whiplash injury criterion? or even better, what do we expect from it?. The points found by the authors to answer this question are the following:

1. The criterion must be representative of one or more injury mechanisms, indicating and quantifying the probability of injury. It must be sensitive to the factors related to these injuries and able to give an assessment about different impact conditions. It must be able, for instance, to determine which seat is safer for an occupant with regard to the considered mechanisms when using a particular acceleration pulse.

2. At the same time it should be repeatable and stable. Values measured in similar situations should not be too different.

3. It should not be sensitive to other processes

different to the mechanism analyzed. Variables not related with the injury mechanism should not have a great influence on its value.

4. When possible, for practical reasons, the criterion should be easily and quickly calculated. It should use values directly measured during the test and avoid non automatic operations.

These points evidence that in order to proceed to the evaluation of the different criteria it is convenient to get the best possible understanding about what happens where and when in a typical rear-end impact. The dynamics of the neck and head have been studied both in the literature and with results of tests using the BioRID-II dummy. In addition, a review of the most accepted injury mechanisms has been done.

After these reviews, the most common injury criteria have been analyzed trying to understand their weak and strong points. A series of four sled tests with seat, dummy and seat belt have been done in order to validate the obtained conclusions. All the tests have been carried out at CIDAUT, using a MTS inverse catapult and a BioRID-II Rev.f fully instrumented dummy. The forces at the seat belt were measured using a Messring belt load cell, in order to get extra information about the rebound phase. The seating procedure was based on [2]. The position of several characteristic points of the dummy was registered with a FaroArm portable 3D measurement system, in order to guarantee its reproduction when using similar seats. The sled was accelerated using the IIWPG 16 Km/h pulse [2] (Figure 1 shows the acceleration measured in the different tests). Four Redlake high-speed digital cameras were used during the tests in both on-board and off-board positions, taking images at 1000 fps. When necessary, image analysis was done using the software Falcon eXtra. All the signs and axis mentioned on the present paper are according SAE J1733 standard ([8]).

Two models of seats have been chosen for the tests. As none of them has been specifically designed to prevent whiplash, we will refer to them as Seat "A" and Seat "B". Seat "A" is a common car driver seat, while Seat "B" is a minibus rear seat with an integrated 3-point seat belt. This forces its structure to be very rigid and, therefore, is expected to give worst results with regard to whiplash protection. Three tests were done with "A" type new seats (numbers 001, 002 and 003), and a fourth one was done with a seat "B", also new (number 004). In this way we could analyze the repeatability and sensitivity of the different criteria. Figure 2 shows the rotation of the backrest in the tests, measured from the high speed images. The difference of stiffness between both models of

seats appears clearly here (the rotation on the fourth test has been quite lower than on the other ones). The variability of the behaviour of the "A" seats can also be observed, even when using similar acceleration pulses. This can be used as a reference when studying the repeatability of the criteria.

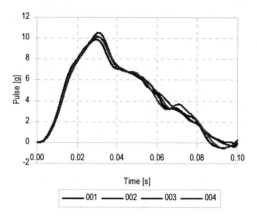

Figure 1. Acceleration pulses of the tests.

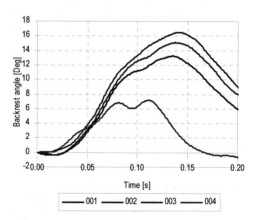

Figure 2. Rotation of the backrest during the tests.

In order to have numeric values to compare the sensitivity and the repeatability of the different criteria, a method has been defined using the Russell criterion for comparison of curves [9]. This criterion is normally used to compare two different series of data $f_1(i)$ and $f_2(i)$ defined by N points each, giving a numeric value ε_c closer to 0 when the curves are similar and greater when the curves are different. The expressions used are the following:

$$A = \sum_{i=1}^{N} f_1(i)^2$$

$$B = \sum_{i=1}^{N} f_2(i)^2$$

(1).

$$C = \sum_{i=1}^{N} f_1(i) f_2(i)$$

$$m = \frac{A - B}{\sqrt{AB}}$$

$$p = \frac{C}{\sqrt{AB}}$$

$$\varepsilon_p = \frac{\cos^{-1}(p)}{\pi}$$

$$\varepsilon_m = sign(m) \cdot \log^{10}(1 + |m|)$$

$$\varepsilon_c = \sqrt{x\left(\varepsilon_m^2 + \varepsilon_p^2\right)}$$

The values ε_m and ε_p represent respectively the errors associated to differences in magnitude and phase, and x is a reference constant that, in this case, has been defined as $\pi/4$.

To get an indicator of the repeatability of the injury criteria the first three tests have been compared to each other (001 to 002, 001 to 003 and 002 to 003), obtaining three ε_c values as results. The average of these values has been considered to be representative of the repeatability. The indicator for sensitivity has been calculated in a similar way, comparing the three first tests with the fourth one and calculating the average of the three obtained ε_c. As defined, the repeatability is assumed to be better when its indicator is closer to zero, and the sensitivity is better when its indicator is higher. To be used as a reference, the indicators of repeatability of the acceleration pulses (high repeatability and low sensitivity) and the rotations of the backrest (relative low repeatability and high sensitivity) were 0.028 and 0.108 respectively, while its sensitivities were 0.022 and 0.482.

HEAD-NECK MOVEMENTS DURING A REAR-END IMPACT

In order to be able to analyze the results of the tests and to try to identify the time when the possible injury mechanisms happen, it is indispensable to understand the kinematics of the neck and the head during a typical low speed rear-end impact. This movement is well documented and has been described by several authors using different techniques ([5], [10], [11], [12] and [13] among others). The main phases of the motion are shown in Figure 3.

In the initial state the subject is seated on the seat in normal position. When the vehicle is struck, the acceleration of the structure is transmitted to the seat through its anchorages, producing a movement forward with regard to the occupant. The first zone of the subject in receiving the pressure of the seat is normally the pelvis and the lumbar zone, followed by the thorax. When the spine, originally curved according to its physiological shape, is pushed forward, it tends to straighten, moving the base of the neck (vertebra T1) upwards and producing some compression on it. This phenomenon can be amplified by the movement upwards of the whole thorax due to the angle of the seat and the acceleration of the base. This is commonly called "ramping up". Although the thorax begins to move, the head at this point remains in its original position. The T1 vertebra, which was originally situated behind the centre of gravity of the head, passes to be in front of it, and the previous compression of the neck becomes traction, with the thorax pulling on the head. The movement of T1 makes the cervical vertebrae work as a chain, transmitting the motion from the lower end upwards, while at the upper end the inertia of the head produces resistance to the movement. The combination of these effects produces a transitory biphasic state known as "s-shape" in which the lower part of the neck (vertebrae C5-C7) presents a very pronounced extension, while the upper part is in flexion. The rearwards movement of the head referred to T1 is called retraction

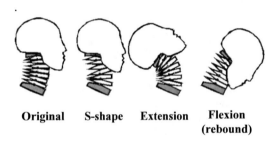

Original   S-shape   Extension   Flexion (rebound)

Figure 3. Different phases of the motion of the head in a typical rear-end impact.

When finally the head begins to rotate, the whole neck arrives in a state of extension with the head being pulled on by the thorax. When the acceleration of the base drops, the elastic energy stored on the seat and the occupant begins to be released, producing a rebound movement with a rotation forward of all the torso of the subject around the pelvis. The seat belt begins to tense over the pelvis and the thorax approximately when the body returns to its original position, producing a violent flexion of the neck. Finally, due to the tension of the belt, the body is stopped, and returns to the backrest.

THEORIES ABOUT WHIPLASH INJURY MECHANISMS

Up to the present a wide number of research works have been done trying to identify the origin of the symptoms related to whiplash associated disorders. As a result of these studies several injury mechanisms have been proposed, the coexistence of some of them being the most accepted hypothesis. If we want to analyze the different injury criteria it is necessary to understand the origin of the lesions as well as possible, in order to be able of relate them with the magnitudes measured in the lab. A review of the most accepted mechanism has been done keeping this idea in mind. Some of the main ones appear below:

Hyperextension

The hyperextension of the neck was the first hypothesis trying to explain the whiplash phenomenon. It was proposed in the sixties by Macnab [14], and suggested the movement of extension of the neck to be the cause of the whiplash injuries, producing lesions on the lower cervical spine. In 1969 the incorporation of head restrains in the new cars sold in USA was made compulsory, trying to limit this movement. However, this fact did not reduce the number of reported whiplash cases in the expected proportion, making evident the necessity for further research. Although hyperextension is still a possible cause of injuries, today the extended use of head restraints has limited it to particular cases, such as misuse or failure of the headrest.

Cervical flexion during the rebound phase

Opposite to the previous mechanism, Macnab also proposed the flexion of the neck due to the movement produced by the head when the seat belt acts on the rebound phase as a probable origin of injuries [15]. This was suggested after the observation of a higher frequency of cervical injuries on people using seat belt, and later confirmed by other authors ([16], [17] and [18] among others).

Pressure gradients on the spinal canal

In 1986 Aldman [19] predicted that volume changes produced inside the spinal canal during sudden movements of the cervical spine on the sagittal plane could be the origin of injuries in the intervertebral tissues. In 1993, Svensson et al. [20] confirmed this hypothesis, measuring the pressure changes on the spinal canal of anesthetized pigs and reporting damage to the spinal ganglia that could explain many of the typical symptoms of whiplash. In these experiments the highest pressure oscillations were related to the phase shift from the s-shape to the extension, and the highest pressures were registered at the level of the C4 vertebra during the s-shape.

Localized cervical compression and tension during the s-shape

Nowadays the most accepted cause of whiplash injuries is probably the one related to the hyperextension observed in the lower part of the neck during the formation of the s-shape (vertebrae C5, C6 and C7). In 1998 Panjabi et al. [21] reported that the intervertebral movements observed at these levels during in vitro tests exceeded their physiological limits, being the cause of lesions in the capsular ligaments and facet joints at the C5-C6 level. Similar findings have been done later by other authors ([22], [23] and [24] among others).

COMPARISON OF THE MOST USED CRITERIA

Figure 4 shows the sensors that at present are being included in a BioRID-II dummy as normal instrumentation. The signals of these sensors and the measurements done by image analysis on the sequences registered with high speed cameras are the current available tools to quantify the ability of a seat to protect the neck of an occupant during a low speed rear impact. Several criteria have been developed in order to quantify the risk of having whiplash related disorders, based either on accelerations, displacements or loads. The most accepted among these criteria have been evaluated critically by the authors trying to understand their virtues and defects. Below the results of the evaluation and its application to the tests are presented:

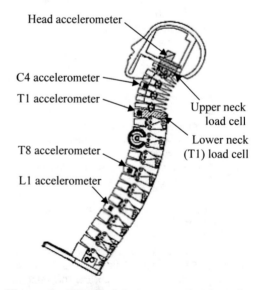

Figure 4. Standard instrumentation in spine and head of BioRID-II (Adapted from R. A. Denton drawing 5834 www.dentonatd.com).

NIC

NIC (Neck Injury Criterion) was proposed by Boström et al. in 1996 [25], as a value to correlate the movement of the head related to the base of the neck (T1 vertebra) with the damage found in the cervical spinal ganglia produced by transient pressure changes in the spinal canal. It uses the difference of accelerations in the longitudinal direction (x axis) between the centre of gravity of the head and the T1 vertebra, being therefore representative of the movement of the neck during the retraction phase. NIC is calculated as follows:

$$NIC = a_{rel} \cdot 0.2 + v_{rel}^2$$
$$a_{rel} = a_x^{T1} - a_x^{Head} \qquad (2).$$
$$v_{rel} = \int a_{rel} dt$$

The maximum reached by this expression during the first 150 milliseconds of the test is called NIC_{max}, and for years has been considered as one of the main indicators of whiplash.

Figure 5 and Table 1 show the NIC values achieved during the tests. The repeatability of the results of the first three tests is very good (with an indicator of 0.084), and even impressive looking at the maximum values. It is necessary to mention here that such a high repeatability of the maximum values is not that common in practice. On the other hand, the different behavior of the seats A and B has been well characterized, having a value of 0.407 on the sensitivity indicator.

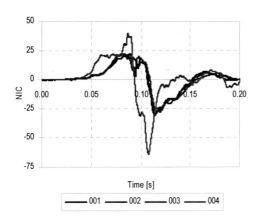

Figure 5. NIC.

Table 1. Maximum NIC values.

Test	001	002	003	004
NIC_{max}	21.94	21.94	21.93	40.08
Time (ms)	88.6	79.9	81.8	86.7

When analyzing the causes that can produce different accelerations in the longitudinal axis between the head and the T1 vertebra and, therefore, cause a modification on the value of the NIC, we observe that this difference can not only be produced by distortions in the neck, but also by any rotation of the head and T1 around the transversal axis (Y) as a rigid body. This movement does not cause any deformation in the neck, and, apart from extreme cases, should not be a direct cause of injury. We can see a scheme of this in Figure 6.

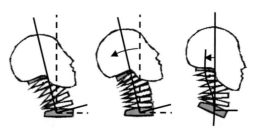

**Full motion = Rigid Body + Deformation
Rotation**

Figure 6. Decomposition of movements producing NIC values.

The influence of this effect can be estimated dividing the relative acceleration used in the NIC definition in two terms:

$$a_{rel} = a_{rotation} + a_{deformation} \qquad (3).$$

If we refer to the angular acceleration of T1 as α and the distance between the centre of gravity of the head and the accelerometer at T1 as d, we can then calculate the acceleration term corresponding to the deformation:

$$a_{deformation} = a_{rel} - \alpha \cdot d \qquad (4).$$

Although d is not fixed for all the configurations of the neck (it is deformable) we can consider 0.2 metres as an average, and we can estimate α from the double derivation of the angle of the T1 vertebra measured on the images (Figure 7).

If we use $a_{deformation}$ instead a_{rel} in expression (2) we get the curves shown in Figure 8. We will refer to these values as NIC*, calculated only with the term related to deformation. Table 2 shows that the maximum values obtained in this way can differ up to 30% from the original NIC values. This variation is produced by factors not directly related to the distortion of the neck.

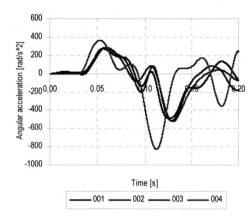

Figure 7. T1 angular acceleration.

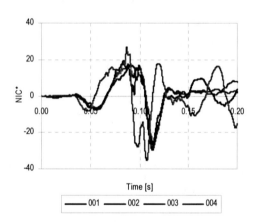

Figure 8. NIC* (without effect of T1 rotation).

Table 2. Maximum NIC* values and deviation with regard to the original NIC.

Test	001	002	003	004
NIC*$_{max}$	17.30	17.72	17.20	26.95
Deviation	21.2%	19.2%	21.6%	32.8%

This fact supports the observations made by Suffel during the fourth BioRID User Meeting [26], who reported the carrying out of some tests blocking the movement of the neck relative to the T1 vertebra, but obtaining NIC values around 8 m^2/s^2.

In short, NIC has shown to have a good repeatability and distinguishes well between the two different seats. It also takes into account the kinematics of the head with regard to the thorax trying to describe the retraction movement, but on the other hand, it is sensitive to effects not related to the distortion of the neck, due to the use of accelerations for its calculation (for instance, the rotation of the seatback produces the effect previously described).

Nkm

In 2001 Schmitt et al. [27] proposed the N$_{km}$ criterion, based on the linear combination of shear forces (F_x) and sagittal bending moments corrected to the occipital condyle ($M_{y\ OC}$), measured with the upper neck load cell. This criterion distinguishes among four possible situations depending on the sign of My and F_x (see Table 3)

Table 3. Cases of Nkm.

Case	M_y	F_x
N$_{fa}$ (Flexion Anterior)	> 0	> 0
N$_{fp}$ (Flexion Posterior)	> 0	< 0
N$_{ea}$ (Extension Anterior)	< 0	> 0
N$_{ep}$ (Extension Posterior)	< 0	< 0

The criterion is calculated as follows:

$$N_{km} = \frac{|F_x|}{F_{int}} + \frac{|M_{yOC}|}{M_{int}}$$
$$F_{int} = 845N \tag{5}$$
$$M_{yOC} > 0 \Rightarrow M_{int} = 88.1N \cdot m$$
$$M_{yOC} < 0 \Rightarrow M_{int} = 47.5N \cdot m$$

Figure 9 shows two possible representations of the results of N$_{km}$ applied to the tests, and Table 4 the maximum values achieved. After these results we can see that the criterion distinguishes both models of seats very well. With regard to the repeatability, it seems to be lower than that observed on the NIC. The maximum on the test 002 is reached during the phase of extension anterior (N$_{ea}$), instead of during the phase of flexion anterior (N$_{fa}$), as happens in the tests 001 and 003. This makes the time of the maximum differ between them. The indicators of repeatability and sensitivity have worse values than the ones obtained for the NIC, being 0.137 and 0.307 respectively.

Table 4. Maximum N$_{km}$.

Test	001	002	003	004
N$_{km}$ max.	0.33	0.20	0.27	0.62
Time (ms)	128.6	112.6	128.1	108.7
Case	FA	EA	FA	FA

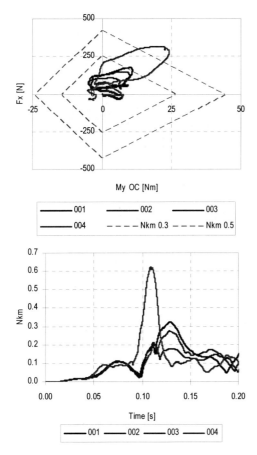

Figure 9. Two representations of N_{km} criterion.

The main advantage found for this criterion is the use of forces and moments, directly related to the loads of the neck, not being affected by other effects such as rotations. Another positive point is its definition in cases, depending on the sign of F_x and $M_{y\ OC}$. This allows the criterion to consider different values and limits depending on the load case. On the other hand, a possible disadvantage is related to the use of the signals measured only on the upper neck load cell, located at the occipital condyle, while the most common injuries have been described between the vertebrae C5 and C7, nearer to the base of the neck. Despite this, the combination of F_x and $M_{y\ OC}$ seems to correlate well with the time in which the s-shape is produced, at least in the studied cases.

Additionally, although the observed influence is not high, it was noticed that the mathematical definition of the criterion as a lineal combination depending on the load case can produce local minimums, oscillations or variations on the tendencies (discontinuities on the derivatives of the curves) at the points of change of case. This can be understood more easily by looking at the first representation of the criterion in Figure 9. The rhomboidal lines represent the points with 0.3 and 0.5 constant values of Nkm. If we intended to have a continuous value of the criteria on the zones of the corners (change of case) we should follow a line with this shape, producing a change in the tendencies of some of the magnitudes (force or moment, depending on the corner) when changing the case, and therefore a discontinuity on its derivative. In practise, the change of the definition of the N_{km} results in discontinuities on its derivative and possible local minimums related only to its mathematical formulation, although, as mentioned above, the influence of this effect has not been decisive in any of the studied cases.

LNL

In 2002 (one year after the proposal of the N_{km} criterion) the prototype of a new load cell placed on the T1 vertebra of the BioRID-II dummy was presented, designed to give information about the loads on the lower end of the neck, next to the vertebrae that had been more often related to injury mechanisms (C5-C7). In March 2003 the version "f" of the dummy was released, already equipped with this load cell. Taking advantage of this new instrument the LNL criterion (Lower Neck Load) was proposed, defined as follows:

$$LNL = \frac{\left| M_{y_{lw}} \right|}{C_{moment}} + \frac{\left| F_{x_{lw}} \right|}{C_{shear}} + \frac{\left| F_{z_{lw}} \right|}{C_{tension}} \quad (6).$$

In this expression $M_{y\ lw}$, $F_{x\ lw}$ and $F_{z\ lw}$ are the moment and forces measured with the T1 load cell, and C_{moment}, C_{shear} and $C_{tension}$ reference values (15 N·m, 250 N and 900 N respectively). The value to be used for the evaluation of the seats is the maximum of this curve.

The curves obtained when applying this expression to the data of the tests are shown in Figure 10, and the maximums in Table 5. Looking at these results, we can see that the repeatability for the first three tests is excellent throughout the curves (with an indicator value of 0.044), including the maximums, but the criterion has not been able to differentiate well between seats A and B, at least in the maximum values. The indicator for sensitivity has a value of 0.250.

Table 5. Maximum LNL.

Test	001	002	003	004
LNL max.	3.98	4.09	4.01	3.88
Time (ms)	119.3	123.3	120.7	107.3

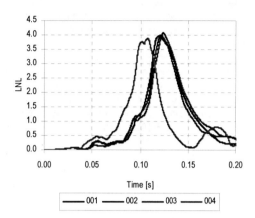

Figure 10. LNL.

The advantages found for this criterion are very similar to the ones found for the N_{km}. It is a criterion based directly on loads, and therefore easy to implement, and does not have the influence of other effects. It is also defined by segments (because of the modulus in the mathematical expression), although it only changes the sign of the reference values for positive and negative data. Besides, the load cell used is the nearest one to the vertebrae where the incidence of injuries is supposed to be higher, and the repeatability shown is very good. On the other hand, the definition by segments presents the same problem already mentioned for the N_{km}, and it has not been able to differentiate between two seats supposed to be very different in terms of whiplash protection.

Neck displacement based criteria (ND)

Viano and Davidsson have proposed a criterion based on the displacements and rotations of the occipital condyle with regard to the T1 vertebra [13]. This criterion, called Neck Displacement Criterion (NDC), was developed from the analysis of the kinematics of volunteers, and is based on two graphs, with the vertical displacement and rotation of the occipital condyle in abscissa and the rearwards horizontal movement of the occipital condyle in ordinate, all of them referred to the T1 vertebra (Z_{OC-T1}, θ_{OC-T1} and X_{OC-T1} respectively). According to the zones occupied by the curves the behaviour is classified as excellent, good, acceptable or poor. This classification was done considering the natural range of motion of both dummies and volunteers.

In order to get numeric values to compare with other criteria, Tencer, Mirza and Huber [28] have defined $Nd_{distraction}$, $Nd_{extension}$ and Nd_{shear} as the quotient between the data used by the NDC criterion and reference values, as described in (8):

$$Nd_{distraction} = \frac{Z_{OC-T1}}{-15mm}$$

$$Nd_{extension} = \frac{\theta_{OC-T1}}{25°}$$

$$Nd_{shear} = \frac{X_{OC-T1}}{35mm} \qquad (7).$$

Using experimental results with volunteers and in vitro tests, and comparing several criteria, they arrived at the conclusion that the best predictors of injury are Nd_{shear}, $Nd_{extension}$ and $Nd_{distraction}$, following this order, instead of other criteria such as N_{km} or NIC, and therefore they recommended the use of criteria based on displacements.

Figure 11 shows the Nd_{shear} calculated for the tests. We can see that the curves of the tests 001, 002 and 003 have a repeatability worse than the previous criteria (0.163), and seat B has been well differentiated (sensitivity of 0.343). Table 6 shows the relative maximums achieved during the formation of the s-shape (100-150 ms).

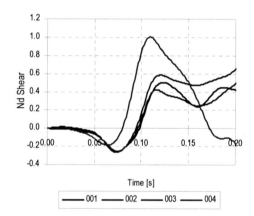

Figure 11. Ndshear.

Table 6. Maximum Ndshear.

Test	001	002	003	004
Ndshear **max.**	0.50	0.42	0.59	1.00
Time (ms)	124	116	121	109

The main advantage of these criteria is that they represent the real kinematics of the neck, taking into account the whole movement of the head with regard to T1. On the other hand, the main disadvantage seems to be the necessity of displacements measurement using motion analysis software, which, although available, represents additional operations, time of analysis and cost in practise.

Rebound

Several authors ([15], [16], [17] and [18] among others) have reported the risk of injury during the rebound phase when the seat is not able to absorb energy during the impact. This phase can be divided into two different stages. In the first one the dummy receives the released elastic energy from the seat, moving forward freely. The second phase starts when the seat belt begins to act on the dummy, stopping the pelvis and the thorax, and producing a sudden flexion of the neck. Figure 12 shows the data measured with the seat belt load cell during the tests and the rotation of the occipital condyle referred to the T1 vertebra measured by image analysis. We can observe how a violent flexion of the neck is produced when the forces in the seat belt grow. This is reflected also in the loads of the neck, as can be seen in the N_{km} values on this phase (Figure 13). It can be observed also that the maximum values in some of the cases (tests 001, 002 and 003) are considerably higher than the ones registered when observing only the first stages of the movement.

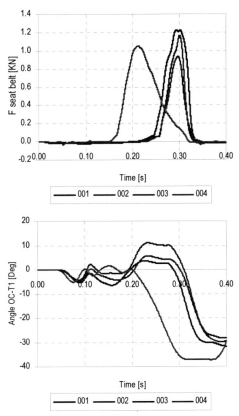

Figure 12. Forces measured at the seat belt and angle of the occipital condyle referred to T1.

At present the capacity of the seats to prevent injuries during the rebound phase is evaluated mainly by measuring the speed of the centre of gravity of the head when it comes back to the position that it occupied at the beginning of the movement (the results of this operation for the fulfilled tests are shown in Table 7). This is supposed to happen just before the seat belt begins to work, so the behaviour of the seat belt is not taken into account. Normally this approximation should be enough, when using seat belts with similar mechanical characteristics on the strap and spool out (the loads are too low so as to be affected by load limiters working at common levels), but this can change in special cases, such as when using pretensioner systems or, as in the case of the seat "B" (test 004), when the points of fixation of the seat belt are fixed to parts of the seat that displace during the impact. Having a look at Table 7 and Figure 13 we can see that, while the rebound speeds are similar for all the tests, the loads on the neck at the rebound are somewhat higher on the fourth test. This fact points to the convenience of reproducing the seat belt configuration in the injury assessment in this phase, at least in the mentioned particular cases.

Table 7. Rebound velocity and time of measure.

Test	001	002	003	004
Rebound velocity (m/s)	3.98	3.96	3.75	4.04
Time (ms)	242	260	259	184

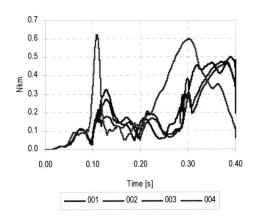

Figure 13. N_{km} extended to the rebound phase.

In conclusion for the rebound phase, a criterion based on loads seems to give more information for injury assessment than the calculation of the speed at a particular point. Considering that the possible injuries of the neck in this phase are better understood (the movements are similar to the ones produced in frontal crashes), a general criterion could be used, such as maximum loads at occipital condyle or N_{km}. Besides, the current method to calculate the rebound velocity supposes normally the use of image analysis, with the practical disadvantages already commented on in the case of displacement based criteria.

RESULTS

The study of these criteria has evidenced weak and strong points in all the cases. The advantages more esteemed by the authors on the underlying concepts of the different criteria have been the following:

1. Capacity to describe the dynamics of the whole neck, taking into account the upper and the lower parts (NIC and ND). Conceptually this should provide a better description of multiphasic states of the neck, particularly the s-shape.

2. Avoidance of distortions due to facts not directly related to the studied injury mechanism (NDC and Rebound speed), such as the angular accelerations found in the NIC (produced by the use of accelerometers at different points) or the mathematical definition in the change of case for the N_{km} and LNL.

3. Facility of calculation (NIC, LNL and N_{km}), avoiding the use of image analysis or complicated algorithms. For practical reasons, the results of the criteria should be available to be analyzed immediately after the test without extra operations.

Other considerations, such as the repeatability or the capability to distinguish different seats are not chosen in the design of the criterion, but are a consequence of the selection of the magnitudes or expressions used in the calculation.

Considering all this, we can draw some guidelines to be applied in the definition of a whiplash injury criterion, focusing on the advantages and avoiding the disadvantages of the studied ones:

1. It should be representative of the dynamics of the whole neck. Taking into account the importance given to the s-shape by the currently accepted injury mechanisms, it should work with values at both ends of the neck in order to be able to detect and quantify this biphasic state.

2. It should avoid the use of accelerations in more than one point, in order to eliminate the sensitivity to the rotations of the seatback.

3. For practical reasons, it should also avoid the use of displacements or velocities measured by image analysis.

4. It would be desirable that its mathematical expression was simple, avoiding the definition in segments.

Taking into account these guidelines and the current instrumentation of the BioRID-II dummy, the simplest solution seems to be the use of the two load cells that the dummy has in the upper and lower ends of neck within only one simple mathematical expression.

PROPOSAL OF A NEW WHIPLASH INJURY CRITERION (WIC)

Having described the previous guidelines, the next step was to determine whether the complex movement of the neck during a rear-end impact could be described by only one mathematical expression using just load magnitudes. As most authors coincide in pointing to the s-shape of the neck as the most probable cause of whiplash injuries, it was decided to look for a function that had a maximum when it happened. As we have seen, the s-shape is a biphasic stage in which the upper end of the neck suffers a flexion at the same time as a hyperextension occurs at the lower end. When using the sign convention stated by the SAE J1733 recommended practice [8], the extension movement is characterized by positive moments in the sagittal axis (Y) of both neck load cells, while the flexion moment is defined by negative moments. Therefore, during the s-shape of the neck, there must be a positive Y moment on the upper end of the neck and a negative Y moment on the lower end (see Figure 14). Taking this into account, the function WIC (Whiplash Injury Criterion) was defined as the most evident solution to the problem:

$$WIC = M_{y_{OC}} - M_{y_{lw}} \qquad (8).$$

In this expression $M_{y\ OC}$ represents the Y moment around the Occipital Condyle (at the upper end of the neck), and $M_{y\ lw}$ represents the Y moment measured at the T1 load cell.

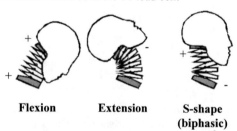

| Flexion | Extension | S-shape (biphasic) |

Figure 14. SAE J1733 sign convention for neck moments in "Y" axis.

Figure 15 shows the result of the application of this function to the data obtained in the tests. The maximum values registered were 25.10 Nm, 19.34 Nm and 22.32 Nm respectively for the three first tests (seat "A") and 38.67 Nm for the fourth test (seat "B").

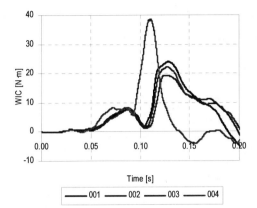

Figure 15. WIC.

After the evaluation of the results we can make the following observations:

1. The curves are very clear and easy to understand. There is a first peak corresponding to the time of the ramping up and spine straightening effects (50-100 milliseconds), coincident with the compression force measured on the lower neck load cell, and a second one, much more marked, during the time when the s-shape is more accentuated (100 to 150 milliseconds).

2. The repeatability in the curves for similar seats (tests 1, 2 and 3) is quite good, having an indicator value of 0.097 (Table 8 shows a comparison of the different values achieved by the indicators of repeatability and sensitivity by the different criteria). We can see also in Table 9 that maximum values for these first three tests happen at very similar times, within a range smaller than that observed by any other criterion.

3. There is a clear differentiation between the curves of the two different seats (sensitivity value of 0.359). The criterion has proved to be sensitive to the seat used and has indicated correctly the inferior seat with regard to neck protection.

4. Looking at the biomechanical aspects, the criterion was designed seeking a function to describe the s-shape, based on the studies that pointed to it as the origin of the more common

whiplash injuries. Figure 16 shows a detail of the neck and head of the dummy at the times when the s-shape seemed to be more pronounced visually. We can appreciate that, as expected, the s-shape was significantly more accentuated in the fourth test (the seat was much more rigid than in the other tests, so the thorax accelerated before and the retraction movement happened more violently).

5. Table 9 presents the times in which the different criteria had a maximum, compared to the times when the most accentuated s-shape in the videos were observed (Figure 16). We can see how the proposed criterion was in general, next to N_{km}, the nearest one to the observed times. Besides, it quantified the magnitude of the loads, indicating clearly which seat produced a more pronounced s-shape.

6. Finally, it is easily implemented, neither image analysis being necessary, nor additional instrumentation or complicated algorithms. It can also be easily applied to previously done tests using the version "f" of the dummy (the first one implementing the lower neck load cell) or later.

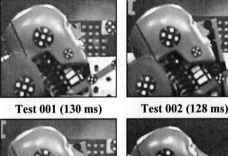

Test 001 (130 ms)   **Test 002 (128 ms)**

Test 003 (129 ms)   **Test 004 (112 ms)**

Figure 16. Detail of the head and neck at the time of the most accentuated observed s-shape during the tests.

Table 8. Indicators for repeatability and sensitivity.

	Repeatability (lower better)	Sensitivity (higher better)
WIC	0.097	0.359
NIC	0.084	0.407
N_{km}	0.137	0.307
LNL	0.044	0.250
Nd_{Shear}	0.163	0.343

Table 9. Times of maximum values (ms).

Test	001	002	003	004
Observed S-shape	130	128	129	112
WIC	<u>128.6</u>	<u>126.3</u>	127.7	<u>110.2</u>
NIC	88.6	79.9	81.8	86.7
N_{km}	<u>128.6</u>	112.6	<u>128.1</u>	108.7
F_x (upper)	128.4	121.2	121.2	108.1
F_z (upper)	115.3	123.1	119.9	98.9
LNL	119.3	123.3	120.7	107.3
Nd_{Shear}	124	116	121	109

This study could not have been finished without a critical review of the new criterion. The observed points were the following:

1. This criterion only takes into account the injury mechanisms associated with the formation of the s-shape in the neck. It does not reveal other possible mechanisms such as, for instance, damages produced during the rebound phase or simple hyperextension. To consider them it should be complemented with another criterion for general use (for instance, maximum loads on the occipital condyle or N_{km})

2. The dynamics of the whole neck has only been represented by the two sagittal moments. Of course, this is a simplification, and a more complex criterion could be defined using additional parameters, such as forces, one acceleration or derivative terms. On the other hand, the criterion has shown to be able to detect and quantify the formation of the s-shape, which was its main objective. This could be enough to evaluate the protection for the most accepted whiplash injury mechanisms. Further studies are suggested in order to analyze this point and the convenience of developing a more complete criterion using this one as a base.

CONCLUSIONS

The original aim of this study was the critical review of the commonest injury criteria used to evaluate whiplash protection, analyzing the advantages and disadvantages of each one of them in order to get a more thorough knowledge of their use. A review of the current theories about the motion of the head and the injury mechanisms was done in order to provide a better understanding of the whiplash phenomenon. Four tests with a BioRID-II dummy were fulfilled to provide data to be used in the comparison of the criteria. As a result, some guidelines to define a new criterion were drawn up focusing on the advantages and avoiding the disadvantages of those previously studied. To resume, it should be based on measurements done at both ends of the neck, in order to be able to describe accurately the biphasic state of the s-shape, and, at the same time, it should avoid the use of several accelerometers or image analysis. Therefore, the clearest solution was to use the upper and lower neck load cells at the same time.

Following these directives a new criterion called "WIC" (Whiplash Injury Criterion) was proposed and evaluated under the same conditions that had been used for the study of the other criteria. The results have been very promising, having shown a good repeatability, sensitivity to the seat and capacity to predict and quantify the s-shape of the neck.

In conclusion, some ideas are suggested for future studies:

1. Further evaluation of the new criterion with previously done tests, in order to confirm the first results.

2. Definition of limit values for evaluation of seats, based either on biomechanical studies, on statistical results (taking into account the values given by different types of seats, as done by IIWPG to define their current limits [3]), or using either tests or simulations of real-world accidents with known injury outcomes and recorded crash pulses, as done by Eriksson and Kullgren [29] or Linder et al. [30].

3. More in depth biomechanical analysis, researching into the convenience or not of defining a more complex criterion based on the same guidelines.

REFERENCES

[1] (1998) "The Abbreviated Injury Scale (AIS) 1990 - Update 98". Association for the Advancement of Automotive Medicine (AAAM). Des Plaines, USA.

[2] (2004) "IIWPG Protocol for the dynamic testing of motor vehicle seats for neck injury prevention DRAFT V 1.4". International Insurance Whiplash Prevention Group (IIWPG). iiwpg.iihs.org.

[3] (2004) "Rationale for IIWPG ratings of seats and head restraints for neck injury prevention". International Insurance Whiplash Prevention Group (IIWPG). iiwpg.iihs.org.

[4] (2004) "Standard test method for rear end

impact crash tests". Folksam, Swedish national road administration. Sweden. www.vv.se.

[5] Davidsson J. (2000) "Development of a mechanical model for rear impacts: Evaluation of volunteer response and validation of the model", Chalmers University of Technology, Goteborg, Sweden.

[6] (2002) "BioRID II User's guide". Denton ATD, Inc. USA.

[7] Svensson M. Y., Boström O., Davidsson J., Hansson H.A., Håland Y., Lövsund P., Suneson A. and Säljö A. (1999) "Neck injuries in car collisions – A review covering a possible injury mechanism and the development of a new rear-impact dummy". Whiplash Associated Disorders World Congress, Vancouver, Australia.

[8] (1994) "SAE J1733. Sign Convention for vehicle crash testing. Issued DEC94". Society of Automotive Engineers (SAE) USA.

[9] Russell D. M. (1997) "Error measures for comparing transient data – Part I: Development of a comprehensive error measure". 68th Shock and Vibration Symposium. Hunt valley, Meryland. USA.

[10] McConnell W. E., Howard R. P., Poppel J. V., Krause R., Guzman H. M., Bomar J. B., Raddin J. H., Benedict J. V. and Hatsell C. P. (1995) "Human head and neck kinematics after low velocity rear-end impacts – Understanding whiplash". Society of Automotive Engineers (SAE) Paper Number 982724. USA.

[11] Grauer J. N., Panjabi M. M., Cholewicki J., Nibu K., Dvorak J. (1997) "Whiplash produces an S-Shaped curvature of the neck with hyperextension at lower levels". Spine 22:2489-2494.

[12] Ono K., Kaneoka K., Wittek Q., Kajzer J. "Cervical injury mechanism based on the analysis of human cervical vertebrae motion and head-neck-torso kinematics during low speed rear impacts" 41st STAPP Car Crash Conference. Society of Automotive Engineers (SAE) Paper Number 933340. USA.

[13] Viano D. C., Davidsson J. (2002). "Neck displacements of volunteers, BioRID P3, and Hybrid III in rear impacts: Implications to whiplash assessment by a neck displacement criterion (NDC)". Traffic Injury Prevention Vol. 3, Number 2, 105-116.

[14] Macnab I. (1966). "Whiplash injuries of the neck". Manitoba Med. Rev. 46. 172-174.

[15] Macnab I. (1971) "The whiplash syndrome". Orthop. Clin. North. Am. 2: 389.

[16] Von Koch M., Nygren Å. and Tingvall C. (1994) "Impairment pattern in passenger car crashes, a follow-up of injuries resulting in long term consequences". 14th ESV conference. Munich. Germany.

[17] Muser M. H., Walz F. H. And Zellmer H. (2000) "Biomechanical significance of the rebound phase in low speed rear end impacts". International IRCOBI conference on the Biomechanics of Impacts, Montpellier, France.

[18] Hell W., Langwieder K., Walz F., Muser M. Kramer M. and Hartwig E. (1999) "Consequences for seat design due to rear end accident analysis, sled tests and possible test criteria for reducing cervical spine injuries after rear-end collision" International IRCOBI conference on the Biomechanics of Impacts, Sitges, Spain.

[19] Aldman B. (1986). "An analytical approach to the impact biomechanics of head and neck". Proceedings of the 30th Annual AAAM Conferences.

[20] Svensson M. Y., Aldman B., Lövsund P., Hansson H. A., Seeman T., Suneson A. and Örtengren T. (1993). "Pressure effects in the spinal canal during whiplash extension motion – a possible cause of injury to the cervical spinal ganglia". International IRCOBI conference on the Biomechanics of Impacts, Eindhoven, The Netherlands.

[21] Panjabi M. M., Cholewicki J., Nibu K., Grauer J. N., Babat L. and Dvorak J. (1998). "Mechanism of whiplash injury". Clinical Biomechanics 13 239-249.

[22] Kaneoka K., Ono K., Inami S., Ochiai N. and Hayashi K. (2002) "The human cervical spine motion during rear-impact collisions: A proposed cervical facet injury mechanism during whiplash trauma". Journal of Whiplash & Related Disorders, Vol 1(1), 85-97.

[23] Cusick J., Pintar F. A., and Yoganandan N. (2001) "Whiplash syndrome, Kinematic factors influencing pain patterns". Spine 26, 1252-1258.

[24] Siegmund G. P., Myers B. S., Davis M. B., Bohnet H. F., Winkelstein B. A. (2000). "Human cervical motion segment flexibility and facet capsular ligament strain under combined posterior shear, extension and axial compression". STAPP

[25] Boström O., Svensson M. Y., Aldman B., Hansson H., Håland Y., Lövsund P., Seeman T., Suneson A., Säljö A. and Örtengren T, (1996) "A new neck injury criterion candidate based on injury findings in the cervical spine ganglia after experimental neck extension trauma". International IRCOBI conference on the Biomechanics of Impacts, Dublin, Ireland.

[26] (2003) "Minutes of 4th European BioRID user meeting (BUM)". BioRID Users Meeting, Stadthagen, Germany. www.rcar.org.

[27] Schmitt K. U., Muser M. H. and Niederer P. (2001) "A new injury criterion candidate for rear-end collisions taking into account shear forces and bending moments". 17th ESV Conference. Amsterdam, Netherlands.

[28] Tencer A. F., Mirza S. and Huber P. (2003) "A comparison of injury criteria used in evaluating seats for whiplash protection". 18th ESV Conference. Tokyo. Japan.

[29] Eriksson L., Kullgren A. (2003) "Influence of seat geometry and seating position on NIC_{MAX} and N_{km} AIS1 neck injury predictability". International IRCOBI conference on the Biomechanics of Impacts, Lisbon, Portugal.

[30] Linder A., Avery M., Kullgren A., Krafft M. (2004) "Rear-world rear impacts reconstructed in sled tests". International IRCOBI conference on the Biomechanics of Impacts, Graz, Austria.

On the Structural and Material Properties of Mammalian Skeletal Muscle and Its Relevance to Human Cervical Impact Dynamics

Barry S. Myers, Chris A. Van Ee, Daniel L. A. Camacho, C. Todd Woolley, and Tom M. Best

Duke Univ.

ABSTRACT

The absence of constitutive data on muscle has limited the development of models of cervical spinal dynamics and our understanding of the forces developed in the cervical spine during impact injury. Therefore, the purpose of this study is to characterize the structural and material properties of skeletal muscle. The structural responses of the tibialis anterior of the rabbit were characterized in the passive state using the quasi-linear theory of viscoelasticity ($r = 0.931 \pm 0.032$). In passive muscle, the average modulus at 20% strain was 1.75 ± 1.18, 2.45 ± 0.80, and 2.79 ± 0.67 MPa at test rates of 4, 40, and 100 cm·s^{-1}, respectively. In stimulated muscle, the mean initial stress was 0.44 ± 0.15 MPa and the average modulus was 0.97 ± 0.34 MPa. These data define a corridor of responses of skeletal muscle during injury, and are in a form suitable for incorporation into computational models of cervical spinal dynamics.

INTRODUCTION

An absence of data on the material properties of skeletal muscle has resulted in limitations in the utility of human surrogates in evaluating injury potential during head and neck impact, as discussed below. Specifically, both cadaveric studies and computational models are limited by a lack of constitutive data on skeletal muscle. While constitutive properties of most tissues of the body have been well characterized, difficulty in measuring local strain, the need to use a large strain formulation, variation in cross-sectional area with length, and the need to preserve the neurologic and vascular supplies to the tissue have precluded these measurements in skeletal muscle. Further, while these responses have been recognized as nonlinear and viscoelastic, the importance of these effects on human cervical impact dynamics has not been investigated. Therefore, the purpose of this paper is two fold. First, a quasi-linear viscoelastic model is developed to describe and predict the passive structural response of the New Zealand White rabbit tibialis anterior muscle. Second, the consti-

tutive properties of skeletal muscle and the effect of elongation rate on these properties are determined for both the passive and stimulated states of the muscle. A review of the literature, and recommendations for modeling the effects of skeletal muscle on cervical spine impact dynamics are discussed.

CADAVERIC STUDIES – In order to quantify head and neck response to impact, investigators have relied on experimental data derived from studies of volunteers and human surrogates. Volunteer studies have provided useful data on the reactions within the physiological range of motion and low AIS injury; however, these data are limited in their ability to predict motion beyond this range and the evaluation of high AIS injuries [1,2]. To better understand catastrophic cervical injury, the human cadaver has frequently been utilized as a human surrogate [1,2,3,4,5,6,7,8,9]. Given the anthropometric similarity of the cadaver and human, the utility of the cadaver response in representing the human response is governed by the degree to which the constitutive properties of the cadaver tissues match those of human tissues at the material level.

A number of studies have compared the response of cadaver to human in an effort to validate the cadaver model. Fitzgerald [10,11] investigated the postmortem changes in skeletal muscle, cancellous bone, and intervertebral disk. He found significant decreases in compliance of muscle greater than 90%, while changes in bone and intervertebral disk were less than 10% and 25%, respectively. A sharp transition region for these properties occurred at 5 to 6 hours postmortem at a point he terms "the life-to-death transition period." Effects of freezing and long term storage have also been investigated and have been shown to not adversely affect mechanical properties of spinal segments [7], cortical bone [12], ligaments [13,14] and intervertebral disks [15,16,17].

In contrast to the properties of ligament, bone, and tendon, skeletal muscle experiences large changes in mechanical properties postmortem. A recent study by Gottsauner-Wolf [18] compared the mechanical properties of bone-muscle-bone units after being frozen to those of fresh specimens. Peak load at failure decreased 59% and stiffness decreased 47%. Comparisons of the responses of volunteers to those of whole cadavers during sled decelerations support these observations [1,2,19,20]. Specifically, larger head excursions and neck bending moments have been consistently observed in the cadaver response and have been attributed to a lack of muscular tone. Nusholtz [5] compared the behavior of live and postmortem primates and noted differences in both head impact response and epidural pressure. These differences were also attributed to postmortem changes to the muscle and other soft tissue mechanical properties. The microstructural changes responsible for these alterations of muscle material properties have been the subject of a great deal of investigation by the meat industry [21,22,23,24]. Changes in ionic concentration and enzymatic activity taking place postmortem cleave skeletal muscle proteins located at the z-line (titin, desmin, nebulin, α-actinin), resulting in a lack of structural support within the individual muscle fibers. While variations in temperature, including freezing, alter the time course of degradation, the final result is the same.

With other tissues in the cadaver maintaining their mechanical properties, the large changes in skeletal muscle properties represent the single largest limitation of the cadaver model. As a result, experimental investigations of axial skeletal injury dynamics are forced to neglect the effects of skeletal muscles, and only report the behavior of the ligamentous spine [4,6,9,25,26]. In an attempt to improve the biofidelity of the cadaver response, Crandall et al. [27] and Wismans et al. [20] treated whole cadavers with fixative agents. While these agents are able to cross-link collagen, their direct effects on muscle constitutive properties are not well characterized. Further, while several studies have characterized the structural responses of skeletal muscle, few studies have measured the constitutive properties of live skeletal muscle. As a result, a basis for comparison of the postmortem changes in skeletal muscle does not exist.

COMPUTATIONAL STUDIES – Due to the limitations of cadaver testing, which include tissue availability, time intensive preparation, and advanced experimental facilities, investigators are developing computational models to describe human dynamics. While computational models of the neck have the potential to both explain normal neck behavior and offer insight into cervical spine tolerance during injury-producing situations, the reliability of model predictions critically hinges on an accurate mechanical representation of the component tissues. Given the lack of more complete constitutive data on skeletal muscle, the ability to incorporate the complex behavior of living muscle in computational models of the neck has been limited.

Belytschko et al. [28] developed a three-dimensional model of the spine to study the pilot ejection problem. The cervical vertebrae were modeled as rigid bodies, the intervertebral disks as linearly elastic beam elements, the ligaments as linear spring elements, and the facets as hy-

drodynamic elements. The neck musculature was not represented. Reber and Goldsmith [29] presented a two-dimensional lumped parameter head-neck system for whiplash simulation. The head and vertebrae were modeled as rigid bodies, and the ligaments, disks, facets, and muscles were modeled by combinations of linear and nonlinear springs and dashpots. The passive muscles were modeled as Kelvin-Voigt solids with nonlinear springs and discrete origins and insertions. The effects of muscle stimulation were not examined. Modeling constants for the passive responses were derived from studies of cadaveric muscles [30,31]. The authors acknowledged the lack of accurate tissue parameters and the need for extended testing of material properties.

Deng and Goldsmith's [32] three-dimensional lumped parameter neck and similar models [33,34,35] have been used to simulate head and neck responses to impact. The head and vertebrae were treated as rigid bodies. The actions of the disk, ligaments, and facets between a pair of vertebrae were collectively represented by a single stiffness and damping matrix. Fifteen sagittally symmetric pairs of passive neck muscles were modeled as three-point spring elements with nonlinear constitutive relationships. The intermediate node in each element allowed muscles to curve around other structures in a more anatomically representative manner. Muscle model constants were also derived from cadaveric sternocleidomastoid muscle data [30].

Williams and Belytschko [36] incorporated one of the most sophisticated representations of muscle into their three-dimensional model of the cervical spine. The head and vertebrae were modeled as rigid bodies, while the disks, ligaments, and facets were modeled by combinations of linear and nonlinear deformable elements. Twenty-two major neck muscle groups were represented by three-parameter viscoelastic solids with discrete origins and insertions. Intermediate sliding nodes were included to allow the muscles to realistically curve around bones. The axial force of the spring elements could be activated independently of elongation to mimic contraction of the muscle. The actual state of contraction of the muscle was dynamically scaled by a factor which depends on the influx of a molecule due to depolarization following muscle stimulation. Muscle model constants were obtained by fitting the results of Inman and Ralston [37], who conducted in vivo studies on human amputees. Actual forces in each muscle were derived by multiplying the calculated stress by the estimated muscle cross-sectional area. The head-neck model was exercised in frontal and lateral impact situations with and without a stretch reflex response in the muscles. The results suggest that the response of a living subject, with the stretch reflex response, is significantly different from the response of a cadaver subject.

Recently, the finite element method has been applied to the study of cervical spine mechanics. While more realistic geometric representations of the structures are possible with this technique, accurate material definitions of the tissues, muscle in particular, remain a primary limitation. Liu [38], Saito [39], Kleinberger [40], and Yoganandan [41] have all developed detailed finite element models of the ligamentous cervical spine, but none have incorporated the effects of the neck musculature, citing an absence of these constitutive properties in the reported literature. Dauvilliers *et al.* [42] accounted for the passive action of muscles in their finite element head-neck model by increasing the moduli of the linearly elastic elements representing the posterior and lateral ligaments. Clearly, without more definitive mechanical characterization of the active and passive behavior of living muscle, realistic representation of dynamic muscle behavior in these models is not possible. Despite its importance, we are unaware of a previous publication which has provided the constitutive responses of passive and stimulated skeletal muscle to elongation based on local measures of stress and strain suitable for incorporation into these models.

MECHANICAL PROPERTIES OF SKELETAL MUSCLE – The structural and material properties of skeletal muscle have been investigated by researchers for many years. One of the first recorded skeletal muscle models was postulated in 200 B.C. by Erasistratus [43]. Despite improvements in experimental techniques, the characterization of skeletal muscle remains a challenge. The activity is complicated by the need to reproduce the *in vivo* conditions of the tissue in the laboratory. Muscle properties are sensitive to hydration, temperature, and enzymatic break down in the perimortem period. Muscle also exhibits nonlinear, rate sensitive behavior, is capable of large strains during physiologic loading, and has a nonuniform cross-sectional area. Analysis of the mechanical properties of skeletal muscle is further complicated by the contributions of the contractile component.

Previous studies of skeletal muscle modeling have met with varying degrees of success. Huxley-type molecular models are based on the chemical kinetics of muscular contraction [44] and are not integrated into lumped parameter or finite element models of musculo-skeletal dynamics. The phenomenological models used are usually of the three parameter type Hill posed in 1938 [45]. These contain a contractile element, a series element, and a parallel elastic element. A fundamental assumption of this model is that the instantaneous force and the contractile element velocity are related at a given muscle length and level of activation. This, however, is generally not true [46,47]. Modifications of this model are abundant but in general cannot predict the strain stiffening response, large relaxation, and a rate insensitivity to hysteresis energy characteristic of skeletal muscle.

The quasi-linear viscoelastic theory (QLV) has been useful in describing tissues demonstrating a nonlinear stiffening response, large relaxation, and rate insensitivity to hysteresis energy. QLV, which uses a reduced relaxation function to describe the time dependence and superimposes a nonlinear elastic function to describe the relationship between load and deflection, has been proposed to model similar responses in a variety of soft tissues including smooth muscle [48,49,50,51,52,53,54]. In this theory, the passive load response, $F(t)$, of a tissue to an applied deformation history, $\delta(t)$, is expressed as the convolution integral of a reduced relaxation function, $G_r(t)$, and the elastic function $F^e(\delta)$,

$$F(t) = \int_o^t G_r(t-\tau)\frac{\partial F^e(\delta)}{\partial \delta}\frac{\partial \delta}{\partial \tau}d\tau \qquad (1)$$

with $F(t)=0$ and $\delta(t)=0$ for $t < 0$, and $G_r(t)$ given by:

$$G_r(t) = \frac{1 + C\left[E_1(t/\tau_2) - E_1(t/\tau_1)\right]}{1 + C\ln(\tau_2/\tau_1)} \qquad (2)$$

where E_1 is the exponential integral function.

The elastic function and the reduced relaxation modeling constants – C, τ_1, and τ_2 – may be derived from relaxation testing [55]. The inability to perform an infinite rate, infinite duration ramp-and-hold displacement in experimental systems results in errors in the generation of these modeling constants and limitations in model performance. Efforts to quantify and reduce these errors have been developed [3,53,54,56].

Structural responses of muscles have been predicted using estimates of the maximum tetanic stress and physiologic cross-sectional area (PCSA). This technique has the advantage of not requiring mechanical testing of the muscle of interest. This method calculates a theoretical (physiologic) cross-sectional area based on muscle volume, fiber length and fiber pennation angle. Assuming the entire muscle acts along a single line of action, PCSA is multiplied by a maximum tetanic stress to provide an estimate of the maximum force carried by the tendon of a tetanized muscle. This method has been used successfully in the appendicular skeleton to describe muscles of different architectures and size. This method, however, is less amenable for use in the spinal musculature because of the diffuse origins and insertions and variations in the pennation angle of those muscles [38,57]. In addition, this technique does not provide a location for the center of pressure resulting from a muscle with a broad origin or insertion. Further, this method does not provide data on the relationship between stress and strain within the muscle during elongation, nor does it provide data on the stress-strain relationships of the passive structure.

In contrast, a solid mechanics approach, in which the true muscle geometry is combined with constitutive properties, could more effectively predict the responses of muscles with broad origins and insertions. In addition, it can account for the effects of elongation, stimulation, and the effects of elongation rate. Further, this approach is more readily incorporated into computational models of axial skeletal dynamics. Previous investigations have successfully obtained true geometry of muscle tissue with the use of CT, MRI, and ultrasound techniques [58,59,60]. However, we are unaware of a publication in which the engineering stress-large strain constitutive properties of skeletal muscle during elongation are measured.

EXPERIMENTAL METHODS

Institutional Animal Use Review Panel approval was obtained prior to all experiments. New Zealand White rabbits, mass 2.50 ± 0.03 kg, were obtained and anesthetized with an intramuscular preparation of Ketamine (100 mg·kg^{-1}), Xylazine (12.5 mg·kg^{-1}) and Acepromazine (3 mg·kg^{-1}). Both hind limbs were prepped and shaved and an anterior midline incision was made from proximal to the knee to distal to the ankle. The tibialis anterior (TA) muscle was exposed, and care was taken to preserve the neurovascular supply. Supporting connective tissue and fascia were removed to expose the muscle. An initial length was established by flexing the ankle to 90 degrees and measuring the distance between the femoral condyles to a mark placed on the tendon at the second cruciate ligament of the ankle [61]. This initial length corresponds to a sarcomere length of 2.46 μm [62]. The distal tendon was then cut at its bony insertion 1.5 cm distal to the mark and placed in a clamp to reestablish the initial length. Proximal bone fixation was achieved by placing one drill bit transversely through the midshaft of the femur and another through the femoral condyles. The muscle and tendon were kept hydrated with normal saline throughout testing. Additional anesthesia was administered during testing as required. Physiologic temperature of the muscle was maintained by covering the muscle with saline soaked gauze and maintaining a warm environment using a heat lamp. Muscle stimulation was achieved by direct stimulation of the deep peroneal nerve. The nerve was exposed below a muscle flap lateral to the thigh just proximal to the knee. It was dissected free of connective tissue, rinsed with normal saline and stored below the muscle flap until the time of stimulation. The nerve was stimulated with a Grass Model SD9 stimulator and a 2 mm gap tip probe using a 1 Volt 40 Hz square wave. Following the completion of testing, the animal was sacrificed with an intravenous injection of Lethalis (100 mg·kg^{-1}).

Mechanical testing was performed with a servo-controlled MTS load frame (MTS Systems Corp.; Minneapolis, MN). Force was measured at the distal tendon using one of two load cells (Figure 1). In low velocity tests, force was measured using a strain gauge load cell and signal conditioner ($\pm$ 0.03% FS, Omega Engineering Inc.; Stamford, CT). To reduce the magnitude of inertial effects at high loading rates, a low mass piezoelectric load cell and electrostatic charge amplifier were used (11 mV·N^{-1}, $\pm$ 0.009 N resolution, PCB Piezotronics; Depew, NY). Axial length changes were measured using a linearly variable differential transformer (MTS Systems Corp.; Minneapolis, MN). Data were collected on a PC based digital data acquisition system with a 1 MHz aggregate sampling rate (RC-Electronics; Santa Barbara, CA). Force data were inertially compensated. Data analysis was performed using a Sun Microsystem computer (Sun Microsystems Inc.;

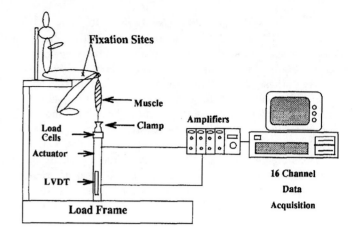

Figure 1. Experimental Apparatus

Mountain View, CA).

The muscle cross-section was imaged with a 7.5 MHz linear phased array ultrasound transducer operating on the Siemens Quatum QAD1 ultrasound machine (Sieman Quantum, Issaquah, WA). Muscle cross-sections were traced on a real time muscle image and areas were calculated by the QAD1. The transducer was fitted with a plastic cylindrical water filled stand off which was sealed with a thin polyvinyl sheet that conformed to the muscle surface without distorting or stretching the muscle. The standoff maintained the muscle at the transducer focal length. To optimize coupling and enhance the surface image, the muscle was coated with a slurry containing ultrasound transmission gel (E-Z-GEL, E-Z-EM; Westbury, NY) and glass microballons (Grade IG-25, Emerson and Cuming; Canton, MA). Following areal measurements, the TA was rinsed with normal saline to remove the gel. Accuracy was determined by measuring the area of acoustical phantoms and precision was determined by repeated measurements of the same image. Errors in conduction velocity were evaluated by measuring linear distances between echogenic markers with and without interposed muscle.

Surface displacements were quantified by the movement of approximately 40 one mm diameter black marks placed along the long axis of the muscle using a 27 gauge needle (Figure 2). The muscle was filmed during the test by a Kodak EktaPro EM-2 1000 Hz digital image processor with a 238x192 pixel CCD camera (Eastman Kodak Company; San Diego, CA). The accuracy of the displacement data was obtained by measuring the distance between known marks, and the repeatabliity was measured by repeated digitizations of the same image. Camera data, in binary TIF format, were downloaded and digitized on a Silicon Graphics Personal Iris workstation (Silicon Graphics Inc.; Mountain View, CA).

VISCOELASTIC STRUCTURAL PROPERTIES – A viscoelastic test protocol consisting of a relaxation test and six constant velocity tests was then initiated on a total of seven animals to determine the passive viscoelastic

Figure 2. High speed digital image of the TA during passive elongation. Deflections in both the lateral and axial directions were quantified by digitizing the black surface marks.

structural response of the muscle. The muscle was preconditioned using a 1 Hz haversine for 50 cycles at a displacement of 0.6 cm. This displacement was chosen because it produced approximately 5% of the expected failure load (55 ± 3 N) as determined by preliminary tests. A 0.15 second ramp of 1.1 cm displacement and 300 second hold was used to approximate the step function of a relaxation test. The pseudoelastic function (the force–displacement history determined from the ramp portion of the relaxation test) was fitted with a fourth order polynomial and used to approximate the infinite rate elastic function. Reduced relaxation constants were derived from the hold portion of the relaxation test. C was -2.303 times the slope of the central portion of the relaxation in log time fitted using a least squares method. τ_1 was defined by a 5% deviation of the measured relaxation from the line C extrapolated to zero time. τ_2 was defined as the duration of the test, 300 s. Using the applied deflection-time history as an input file and the relaxation test-derived modeling constants, the quasi-linear model was used to predict the force-time history at six loading rates (2 Hz, 1 Hz, 0.5 Hz, 0.2 Hz, 0.1 Hz, and 0.01 Hz). Individual tests were separated by 3 minutes to allow for the relaxation of load from the previous loading cycle. A similar prediction of hysteresis energy was also performed at each frequency.

MATERIAL AND STRUCTURAL FAILURE PROPERTIES – To determine the effect of rate of elongation on material properties, 42 TA from 21 animals were studied. Each muscle was preconditioned using a 1 Hz haversine for 20 cycles at 5% of the expected force at failure. Each animal was then randomly assigned to one of three constant velocity elongation to failure test rates: 4 cm·s^{-1}, 40 cm·s^{-1}, or 100 cm·s^{-1}. From each animal, one randomly selected muscle was stimulated 0.5 s prior to elongation to

failure, and the other muscle was elongated in the passive state. The camera data were digitized to give the displacement history along the length of the muscle, $u(x, y, t)$, for each mark on the muscle. Because of the large deformations, the Lagrangian finite strain relation, $E_{xx}(x, t)$, was used to compute axial strain:

$$E_{xx}(x,t) = \frac{\partial u}{\partial x} + \frac{1}{2}\left[\left(\frac{\partial u}{\partial x}\right)^2 + \left(\frac{\partial v}{\partial x}\right)^2 + \left(\frac{\partial w}{\partial x}\right)^2\right] \quad (3)$$

Preliminary tests showed the derivatives of the lateral deformations (v, w) were small compared to the derivatives of the axial deformations and could be neglected. The $u(x, t)$ data were fitted with a fourth order polynomial and $E_{xx}(x, t)$ was calculated.

Force data were filtered with a finite impulse response filter at 700 Hz. Using the ultrasound cross-sectional area from the midbelly of the muscle (corresponding to 42 mm distal to the muscle origin) an engineering stress-time history was calculated. Using the stress-time history and the associated large strain-time history, the stress-strain response was calculated for the midbelly. Engineering stress-large strain responses for stimulated and passive muscle were compared using a one way ANOVA to determine the significance of strain rate at each of the elongation rates studied. Structural properties at failure – including force, displacement, and energy absorbed – were also compared using a one way ANOVA. All data presented are of the form of mean $\pm$ standard deviation.

RESULTS

VISCOELASTIC STRUCTURAL RESPONSE – The fourth order polynomial fit of the response to the ramp input was used to represent the pseudoelastic function with high correlation ($r > 0.99$). The pseudoelastic function of the specimens showed a typical low load region followed by a stiffening response characteristic of biological tissue (Figure 3).

The reduced relaxation response showed initial rapid relaxation with decreasing rates of relaxation with increasing time (Figure 4). Viewed in log time, the reduced relation function showed a central linear region typical of a spectral relaxation formulation (Figure 5). Accordingly, the relaxation response was fit with the exponential integral form of the relaxation function resulting in average modeling constants of: $C_{mean} = 0.0618 \pm 0.0076$, $\tau_{1\,mean} = 0.0124 \pm 0.0047$ s, and $\tau_2 = 300$ s.

Using these modeling parameters to predict the response of the tissue to the constant velocity tests, a mean correlation coefficient of 0.926 ± 0.037 was attained (Figure 6). The model predicted that compressive forces were required to return the muscle to its original length during the unloading phase of the constant velocity tests. As

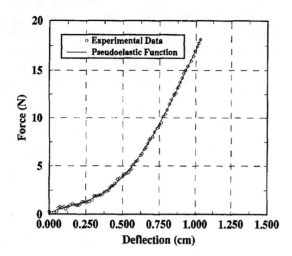

Figure 3. The response of the TA to a 0.15 s ramp of 1.1 cm displacement. A fourth order fit was used to approximate the response (r = 0.99).

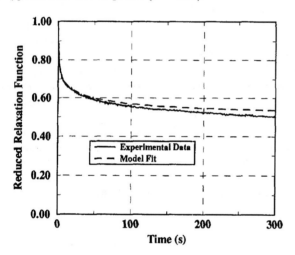

Figure 4. The reduced relaxation response with the spectral damping model in linear time demonstrating the large initial relaxation with decreased relaxation with increasing time.

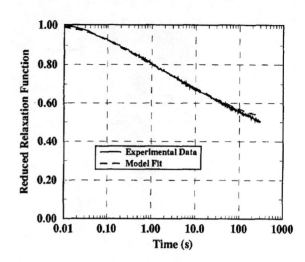

Figure 5. The reduced relaxation response with the spectral damping model in log time demonstrating the central linear region indicative of a spectral relaxation model.

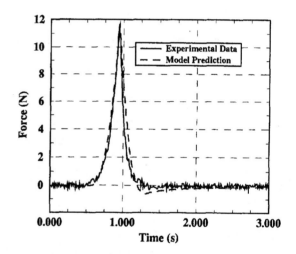

Figure 6. The experimental force-time response of the muscle and the quasi-linear model prediction for a 1 Hz constant velocity test.

Table 1. Hysteresis Energy

Frequency (Hz)	Experemental	Theory
2	35.3 ± 5.3%	28.1 ± 3.2%
1	36.8 ± 4.9%	27.3 ± 2.2%
0.5	38.1 ± 4.9%	32.4 ± 3.7%
0.2	39.8 ± 4.9%	31.7 ± 4.3%
0.1	42.0 ± 3.6%	31.0 ± 3.5%
0.01	43.8 ± 3.7%	31.0 ± 4.2%

the experimental apparatus could not apply compressive forces, the model was constrained to only predict tensile forces. This constraint improved the correlation coefficient to 0.931 ± 0.032. The muscle exhibited rate dependence with peak force varying by as much as 60% between the low and high velocity tests. The experimental mean hysteresis energy, expressed as a percent of maximum stored strain energy, was 39.3 ± 5.4% and varied by less than 25% over the range of frequencies tested (Table 1). The model predicted the relative rate insensitivity of hysteresis energy, but underestimated hysteresis in all tests with a mean predicted hysteresis energy of 30.2 ± 4.0%.

MATERIAL AND STRUCTURAL FAILURE PROPERTIES – Muscle cross-sectional area at midbelly was 43.9 ± 5.2 mm². Areal measurements were made with ± 0.5 mm² accuracy and ± 1.0 mm² repeatabliity. No errors due to conduction velocity were found when measuring linear distances between echogenic markers with and without in-

terposed muscle. At the given focal length, this optical system has a linear measurement accuracy of ± 0.125 mm, and a precision of ± 0.1 mm. Strain was found to vary significantly along the muscle length with an average variation of 19.0% ± 8.8% over a 5 mm region surrounding the muscle midbelly (Figure 7).

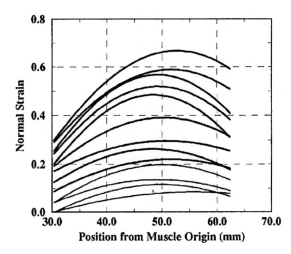

Figure 7. Strain distibution from near the muscle origin ($x = 30.0$) to the distal myotendinous junction ($x = 60.0$) at sequential 4 ms intervals during constant velocity elongation at 100 cm·s^{-1}. Strain increases monotonically with time; however, the strain varies considerably with postion along the muscle creating the need for a continuous full field measurement technique.

Table 2. Failure Properties of the Tibialis Anterior

Rate (cm/s)	State	Force (N)	Deformation (cm)	Energy (N–cm)
4	Passive	54.5 ± 4.7	2.89 ± 0.27	73 ± 18
	Stimulated	55.0 ± 2.9	3.05 ± 0.25	122 ± 13
40	Passive	68.3 ± 4.5	2.94 ± 0.24	98 ± 25
	Stimulated	70.6 ± 3.8	2.99 ± 0.42	145 ± 20
100	Passive	71.5 ± 7.3	2.93 ± 0.41	87 ± 21
	Stimulated	68.0 ± 11.0	2.68 ± 0.61	133 ± 38

Structural failure properties of the muscle were sensitive to elongation rate. Force to failure was found to be rate sensitive in both the passive and stimulated states of the muscle ($F = 10.84$, $p < 0.01$, and $F = 9.51$ $p < 0.01$, respectively, Table 2). Significant differences in the force to failure were found between the 100 and 4 cm·s^{-1} and the 40 and 4 cm·s^{-1} ($p < 0.05$), but the 100 and 40 cm·s^{-1} tests were not significantly different. Energy to failure and displacement to failure did not vary significantly with loading rate. The site of mechanical failure was found to vary with the rate of elongation. At the rates of 4 and 40 cm·s^{-1}, the muscle failed at the myotendinous junction, while at the 100 cm·s^{-1} rate it failed in the distal muscle belly, approximately 0.5 cm proximal to the myotendinous junction.

The passive stress-strain response showed a typical low-load toe region followed by a stiffening response with increasing strain (Figure 8). The average modulus at a strain of 20% was 1.75 ± 1.18, 2.45 ± 0.80, and 2.79 ± 0.67 MPa at the test rates of 4, 40, 100 cm·s^{-1} respectively. In the stimulated test, because the muscle was stimulated prior to elongation, an initial isometric stress of 0.44 ± 0.15 MPa was observed (Figure 9). Further elongation of the stimulated muscle generated a linear response with

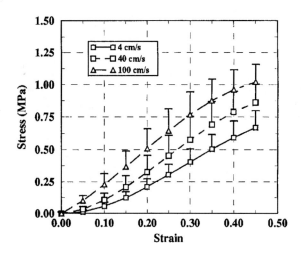

Figure 8. Engineering stress–large strain response of passive muscle for a range of elongation rates showing the significant effect of elongation rate on the responses over the entire range of strain.

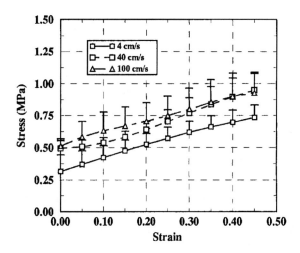

Figure 9. Engineering stress–large strain response of stimulated muscle for a range of elongation rates. The effect of elongation rate are significant, though less so than for passive muscle.

an average modulus of 0.97 ± 0.34 MPa (average correlation coefficient of 0.977 ± 0.037). Load sharing between the active and passive elements was present in the stimulated muscle. At small strains a majority of the stress was carried in the active element, while at larger strains the passive element carried a majority of the stress (Figure 10).

Stress–strain responses were found to be rate sensitive in both the passive and stimulated tests (Figures 8 and 9). ANOVA revealed a significant difference between the 4 and the 100 cm·s^{-1} tests ($p < 0.01$). In the stimulated tests, for strains under 25%, the F_{ave} was 7.32 ± 2.13 (range = 4.48 to 10.31, $p < 0.01$). For strains larger than 25%, this difference was significant, though not as pronounced with $F_{ave} = 3.80 \pm 0.41$ (range = 3.40 to 4.48, $p < 0.05$). In contrast, the effects of elongation rate were more

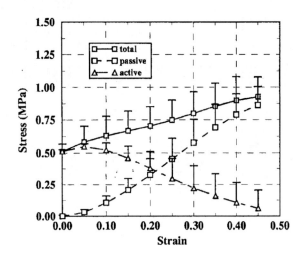

Figure 10. Engineering stress–large strain response of passive and stimulated skeletal muscle (100 cm·s^{-1}). The active response, the difference between the passive and stimulated responses, carries a majority of the load at small strain while the passive element dominates at larger strain.

pronounced in the passive response, ($F_{ave} = 8.32 \pm 1.91$ (range $= 6.01$ to 11.3, $p < 0.01$)). Multiple comparison testing demonstrated no significant differences comparing the stress strain responses between the 4 and 40 cm·s^{-1} and the 40 and 100 cm·s^{-1} tests.

DISCUSSION

Characterizing structural and material properties of skeletal muscle has presented a challenge to investigators. Skeletal muscle exhibits nonlinear, rate sensitive behavior and is capable of undergoing large strains. In addition, skeletal muscle properties are sensitive to the physiological and biochemical environment. These variables impose both time and testing constraints on the study of muscle. Several investigators have demonstrated that muscle significantly alters the dynamic responses of the spine to impact, particularly in bending [2,36]. Currently, the lack of constitutive data on the mechanical properties of skeletal muscle has limited investigators' ability to incorporate skeletal muscle into models of axial musculo-skeletal dynamics [35,40,42,57,63,64]. Accordingly, the purpose of these experiments is to provide more complete constitutive data on muscle to allow for a better characterization of spinal injury. By measuring the passive and stimulated responses, these data define a corridor of possible muscle states during injury dynamics. These data are also presented to serve as a basis for comparison of the changes which occur in muscle in the perimortem and postmortem periods.

VISCOELASTIC STRUCTURAL RESPONSE – The structural properties of skeletal muscle were modeled using QLV with strong correlation. Many investigators have used QLV to predict the nonlinear viscoelastic responses of soft tissue [48,49,50,51,52,53,54]. Results from the current

study suggest that passive skeletal muscle structural properties can be predicted by QLV. Relaxation testing of skeletal muscle has shown the tissue to exhibit both significant initial relaxation and long time relaxation until testing has ceased. The tissue also exhibits a relative rate insensitivity to hysteresis energy. These properties cannot be modeled accurately with discrete time constants typical of lumped parameter models. For this reason, a continuous spectral relaxation function was used to model the tissue. The nonlinear elastic response of the tissue could also be included in this model which is a limitation of linear spring models. QLV has the additional advantage of allowing nonlinearity without the need for several multi-step relaxation tests as are required with other nonlinear viscoelastic formulations [65].

The model correctly predicted that compressive loads would need to be applied to return the tissue to its original length at the rates tested. Our testing apparatus, and the *in vivo* environment do not apply compressive loads to the tissue. Modification of the model to predict only tensile forces increased the model's predictive ability. In contrast, the model consistently underestimated the hysteresis energy of the tissue. This is likely the result of the inability of the finite rate relaxation test to accurately measure the contribution of relaxation time constants which are of shorter duration than the ramp duration and the unmeasured relaxation which occurs during the finite rate ramp. This phenomenon has been noted and methods have been developed to reduce these effects [3,56]. Another limitation of the current model is its inability to accommodate the stimulated response of muscle. The model uses an infinite rate elastic response function $F^e(\delta)$ which is not a function of stimulation or velocity. In contrast, Leiber [66] noted that compared to a constant length condition, stimulated muscle produces between 0 and 1.8 times the force depending on the velocity. Theoretically, the elastic function can be modified to account for these effects, and the reduced relaxation function altered to account for muscle fatigue; however, this hypothesis has yet to be developed mathematically or evaluated experimentally.

MATERIAL AND STRUCTURAL FAILURE PROPERTIES -- In an effort to understand the constitutive behavior of skeletal muscle, the stress-strain relationships were determined for the muscle at the muscle midbelly. In contrast to averaging strain across a region surrounding the midbelly, a method was used which measured the nonuniform continuous normal strain field along the long axis of the muscle [67]. By fitting the measured displacement field, and differentiating the fitted function to determine the strain field, individual discretization errors are minimized. Further, errors associated with spatial averaging of a nonuniform strain field are removed. This technique also allows for the accurate determination of large strain. Under *in vivo* conditions, Zajac [68] reported that the rabbit TA can change in length from 15% to 20%. This corresponds to a large strain of approximately 25%. Goslow *et al.* [69] reported changes in muscle length from 10% to 50% in cat gait. Merrill *et al.* [63], using a computational model, suggested that strains in the cervical musculature may be as large as 40%. Clearly, a systematic underestimation of strains will occur if the small strain relations are used.

Engineering stress was computed using true cross-sectional area measured with phased array ultrasound. This method allowed a complete description of the muscle geometry prior to testing. Previous investigations have used gravimetric, or mechanical measures based on an assumed geometry and direct contact measurements. While these may be suitable for some biologic structures, these methods are prone to error in structures with complex cross-sections and low stiffness [70]. Optical methods have also been described. These methods typically require a large arc of visibility which is unavailable in the *in situ* skeletal muscle, and they are not able to account for the concave surfaces of the deep side of the muscle. In contrast to these techniques, phased array ultrasound allows access to the tissue area with minimal contact pressure and a small exposed arc of visibility. The instrument is also capable of measuring area along concave surfaces.

Limitations of the technique are the high cost of the instrumentation and the time and computation intensive data reduction. It is also assumed that stress is constant across the cross-section of the muscle during elongation. Further, engineering stress was reported because the frequency response the ultrasound phased array (30 Hz) and the movement of the midbelly during failure testing precluded real time measurement of cross-sectional area. However, true stress can be derived from the engineering stress-strain data presented by assuming constant volume. An additional potential limitation is the proposed use of rabbit muscle constitutive data to describe the behavior of human muscle. However, studies using rat, rabbit, cat, monkey, and human skeletal muscle have demonstrated stuctural and mechanical similarity amongst mammalian species supporting the use of animal muscle constitutive data to describe human muscle. Specifically, the skeletal muscle microstructure across these species was found to be similar with only slight variations in thin filament length [71,72]. Further, sarcomere length-isometric force relationships are also similar across mammalian species. Finally, the anesthesia used provided a neural muscular block, thereby removing spinal mediated muscle reflex responses. However, as cervical muscle reflex response times are typically in excess of 50 to 70 ms [73], spinal mediated reflexes are also absent during spinal impact injuries.

Engineering stress-large strain behavior for skeletal muscle has not been reported elsewhere; however, modulus and isometric stress values are reported in the literature. Tidball [74] reports a complex modulus of 3.29 MPa for single muscle fibers tested in rigor at a strain rate of 0.0628 to 0.169 s^{-1}. In our experiments, using a muscle fiber length of 4.1 cm as a gauge length, average strain rate was considerably higher (1.0, 10, and 25 s^{-1}) and the muscle was not tested in rigor. Modulus varied from 0.276 MPa to 2.04 MPa for passive and from 0.252 to 1.42 MPa for active tests. Tetanic stress values based on PCSA, as reported by Roy *et al.* [75] and Spector *et al.* [76] for the soleus and gastrocnemius, are 0.225 MPa. Winter [77] summarizes the isometric stress values to be between 0.20 to 1.00 MPa. In these experiments the mean tetanized stress was 0.44 MPa based on a true cross-sectional area which compares favorably with these reported values.

Both the structural QLV model and the constitutive relations presented in this paper demonstrate that the material properties of skeletal muscle are rate sensitive. However, the impact injury environment imposes a relatively small range of high rate, large–strain elongations to the muscles of the spine. In the absence of a significant rate effect, muscle response could be modeled with a nonlinear elastic formulation using a pseudoelastic function generated from a single rate test at an appropriate loading rate [78]. ANOVA revealed that both the passive and stimulated responses of muscle were sensitive to average strain rate. This indicates that significant differences were observed between the slow 4 cm·s^{-1} tests and the highest rate tests, 100 cm·s^{-1}. Smaller increases in average strain rate, from 40 to 100 cm·s^{-1} or from 4 to 40 cm·s^{-1}, did not result in statistically significant changes in the response. This relative rate insensitivity is consistent with the spectral nature of soft tissue viscoelasticity [3,79,80] and suggests that within small ranges of average elongation rate, a pseudoelastic function derived from an experiment of near equal average elongation rate may be an appropriate choice for representation of muscle constitutive behavior. This is particularly true in the stimulated responses in which the sensitivity to elongation rate was smaller than in passive responses.

CONCLUSIONS

1. Passive responses of mammalian skeletal muscle can be modeled with the quasi-linear theory of viscoelasticity.

2. The passive and stimulated stress-strain responses of skeletal muscle to elongation have been quantified in a form suitable for incorporation into computational modeling of cervical impact dynamics. These responses define an upper and lower bound to the possible responses of muscle during injury.

3. While the stress-strain responses of muscle are sensitive to rate of elongation over a wide range of loading rates, smaller changes in loading rate do not result in significant changes in the stress-strain responses. These data suggest that for the purpose of cervical impact injury modeling, muscle may be modeled using a pseudoelastic formulation derived from an appropriate rate test. This is particularly true for stimulated muscle in which the responses are less sensitive to elongation rate.

ACKNOWLEDGMENTS

This study was supported by the Virginia Flowers Baker Chair and the Department of Health and Human Services, Centers for Disease Control Grant **R49/CCR402396-08**.

REFERENCES

1. Mertz, H.J. Jr., Patrick, L.M., "Investigation of the Kinematics and Kinetics of Whiplash," *Biomechanics of Impact Injury and Injury Tolerances of the Head-Neck Complex*, SAE Paper no. 670919, 1967.

2. Mertz, H.J. Jr., Patrick, L.M., "Strength and Response of the Human Neck," *Biomechanics of Impact Injury and Injury Tolerances of the Head-Neck Complex*, SAE Paper no. 710855, 1971.

3. Myers, B.S., McElhaney, J.H., and Doherty, B.J., "The Viscoelastic Responses of the Human Cervical Spine in Torsion: Experimental Limitations of Quasi-linear Theory, and a Method for Reducing These Effects," *J. Biomech*, **24**(9):811-817, 1991.

4. Myers B.S., McElhaney, J.H., Richardson, W.J., Nightingale, R.W., Doherty, B.J., "The Influence of End Condition on Human Cervical Spine Injury Mechanisms," *Proceedings of the 35th STAPP Car Crash Conference*, SAE Paper no. 912915, 1991.

5. Nusholtz, G.S., Melvin, J.W., Alem, N.M., "Head Impact Response Comparisons of Human Surrogates," *Biomechanics of Impact Injury and Injury Tolerances of the Head-Neck Complex*, SAE Paper no. 791020, 1979.

6. Panjabi, M.M., White A.A., Johnson, R.M., "Cervical Spine Mechanics as a Function of Transection of Components," *J Biomech*, **8**:327-336, 1975.

7. Panjabi, M.M., Krag, M., Summers, D., Videman, T., "Biomechanical Time Tolerance of Fresh Cadaveric Human Spine Specimens," *J. Orthop. Res.*, **3**:292-300, 1985.

8. Pintar F., Sances A. Jr., Yoganandan N., Reinartz, J., Maiman D.J., Suh J.K., Unger, G., Cusick J.F., and Larson S.J. "Biodynamics of the Total Human Cervical Spine," *Proceedings of the 34th Stapp Car Crash Conference*, SAE Paper no. 902309, 1990.

9. Yoganandan, N., Sances, A., Pintar, F., Maiman, D.J., Reinartz, J., Cusick, J.F., Larson, S.J., "Injury Biomechanics of the Human Cervical Column," *Spine*, **15**:10, 1990.

10. Fitzgerald, E.R., "Dynamic Mechanical Measurements During the Life to Death Transition in Animal Tissues," *Biorheology*, **12**, 1975.

11. Fitzgerald, E.R., "Postmortem Transition in the Dynamic Mechanical Properties of Bone," *Med. Phys.*, **4**, 1977.

12. Sedlin, E.D., "A Rheologic Model for Cortical Bone. A Study of the Physical Properties of Human Femoral Samples," *Acta Orthop. Scand.*, **36**(Suppl. 83):1-77, 1965.

13. Viidik, A., Sandqvist, L., Magi, M., "Influence of Postmortal Storage on Tensile Strength Characteristics and Histology of Rabbit Ligaments," *Acta Orthop. Scand.*, **36**(Suppl. 79):1-38, 1965.

14. Woo, L.-Y., Orlando, C.A., Camp, J.F., Akeson, W.H., "Effects of Postmortem Storage by Freezing on Ligament Tensile Behavior," *J. Biomech* **19**:399-404, 1986.

15. Nachemson, A., "Lumbar Interdiscal Pressure," *Acta Orthop. Scand.*, **31**(Suppl. 43):1-104, 1960.

16. Favenesi, J.A., Gardeniers, J.W.M., Huiskies, R., Slooff, T.J.J., "Mechanical Properties of Normal and Avascular Cancellous Bone," *Trans Orthop. Res. Soc.*, **9**:198, 1984.

17. Gardeneirs, J.W.M., "Behavior of Normal, Avascular and Revascularizing Cancellous Bone," Thesis, Catholic University of Nijmegen, Nijmegen, 1988.

18. Gottsauner-Wolf, F., Grabowski, J.J., Chao, E.Y.S., An, K.-N., "Effects of Freeze/Thaw Conditioning on the Tensile Properties and Failure Mode of Bone-Muscle-Bone Units: A Biomechanical and Histology Study in Dogs," *J. Orthop. Res.*, **13**(1), 1995.

19. Bendjellal, F., Terriere, C., Gillet, D., Mack P., Guillon, F., "Head and Neck Responses under High G-level Lateral Deceleration," *Biomechanics of Impact Injury and Injury Tolerances of the Head-Neck Complex*, SAE Paper no. 872196, 1987.

20. Wismans, J., Phiippens, M., van Oorswchot, E., Kalliers, D., Mattern, R., "Comparison of Human Volunteer and Cadaver Head-Neck Response in Frontal Flexion," *Biomechanics of Impact Injury and Injury Tolerances of the Head-Neck Complex*, SAE Paper no. 872194, 1987.

21. Takahashi, K., "Non-enzymatic Weakening of Myofibrillar Structures During Conditioning of Meat: Calcium Ions at 0.1 mM and Their Effect on Meat Tenderization," *Biochemie*, 74, 1992.

22. Koohmaraie, M., "The Role of Ca-dependent Proteases (Calpains) in Post Mortem Proteolysis and Meat Tenderness," *Biochemie*, 74, 1992.

23. Ouali, A., "Proteolytic and Physicochemical Mechanisms Involved in Meat Texture Development," *Biochemie*, 74, 1992.

24. Croall, D., Demartino, G.N., "Calcium-Activated Neutral Protease (Calpain) System: Structure, Function, and Regulation," *Physiological Reviews*, **71**(3), 1991.

25. Nightingale, R.W., McElhaney, J.H., Richardson, W.J., Myers B.S., "Dynamic Response of the Head and Cervical Spine to Axial Impact Loading," *J. Biomech.* In Press.

26. Pelker, R.R., Duranceau, J.S., Panjabi, M.M., "Cervical Spine Stabilization, A Three-Dimensional, Biome-

chanical Evaluation of Rotational Stability, Strength, and Failure Mechanisms," *Spine*, **16**:2, 1991.

27. Crandall, J.R., Pilkey, W.D., "The Preservation of Human Surrogates for Impact Studies," *Proceedings of the Thirteenth Southern Biomedical Engineering Conference*, 1994.

28. Belytschko, T., Schwer, L., and Privitzer, E., "Theory and Application of a Three-Dimensional Model of the Human Spine," *Aviat. Space Environ. Med.*, **49**(1):158-165, 1978.

29. Reber, J.G., Goldsmith, W., "Analysis of Large Head-Neck Motions," *J. Biomech.*, **12**:211-222, 1979.

30. Yamada, H., edited by Evans, F.C., *Strength of Biological Materials*, Williams and Wilkins, Baltimore, 1970.

31. Ohnishi, T., "Rheology of Glycerinated Muscle Fibers.," *Biorheology*, **1**:83-90, 1963.

32. Deng, Y.C., Goldsmith, W., "Response of a Human Head/Neck/Upper-Torso Replica to Dynamic loading–II. Analytical/Numerical Model," *J. Biomech.*, **20**(5): 487-497, 1987.

33. Luo, Z., Goldsmith, W., "Reaction of a Human Head/Neck/Torso System to Shock," *J. Biomech.*, **24**(7):499-510, 1991.

34. Yang, K.H., Latouf, B.K., King, A.I., "Computer Simulation of Occupant Neck Response to Air Bag Deployment in Frontal Impacts," *J. Biomech. Engng.*, **114**:327-331, 1992.

35. de Jager, M., "Mathematical Modeling of the Human Cervical Spine: A Survey of the Literature," *International Conference on the Biomechanics of Impacts*, 213-217, 1993.

36. Williams J.L., Belytschko, T.B., "A Three Dimensional Model of the Human Cervical Spine for Impact Simulation," *J. Biomech. Engng.*, **105**:321-331, 1983.

37. Inman, V.T., Ralston, H.T., edited by Klopsteg, P.E., Wilson, P.D., *Human Limbs and Their Substitutes*, McGraw-Hill, New York, 1964.

38. Liu, Y.K., Goel, V.K., Clark, C.R., "Origins and Insertions of Cervical Muscles," *Advances in Bioengineering*, 40-41, 1986.

39. Saito, T., Yamamuro, T., Shikata, J., Oka, M., Tsutsumi, S., "Analysis and Prevention of Spinal Column Deformity Following Cervical Laminectomy I," *Spine*, **16**(5):494-502, 1989.

40. Kleinberger, M., "Application of Finite Element Techniques to the Study of the Cervical Spine," *Proceedings of the 37th Stapp Car Crash Conference*, SAE Paper no. 933131, 1993.

41. Yoganandan, N., Voo, L., Pintar, F.A., Kumaresan, S.,

Cusick, J., Sances, A. Jr., "Finite Element Analysis of the Cervical Spine," *Proc. Fifth Annual Injury Prevention Through Biomechanics Symposium*, 149-155, 1995.

42. Dauvilliers, F., Bendjella, F., Weiss, M., Lavaste, F., "Development of a Finite Element Model of the Neck," *Proceedings of the 38th Stapp Car Crash Conference*, SAE Paper no. 942210, 1994.

43. Thunnissen, J., *Muscle Force Prediction During Human Gait*, CIP-GEGEVENS Koninklijke Bibliotheek, Den Haag, Netherlands, 1993.

44. Apter, J.T., and Graessley, W.W., "A Physical Model for Muscular Behavior," *Biophys J*, **10**:539-555, 1970.

45. Hill, A.V., "The Heat of Shortening and the Dynamic Constants of Muscle," *Proc Roy Soc (London), Ser B*, **126**:136-195, 1938.

46. Katz, B., "The Relation Between Force and Speed in Muscular Contraction," *J Physiol*, **96**:45-64, 1939.

47. van Ingen Schenau, G.J., Bobbert, M.F., Ettema, G.J., de Graaf, J.B., and Huijing, P.A., "A Simulation of Rat EDL Force Output Based on Intrinsic Muscle Properties," *J Biomech*, **21**:815-824, 1988.

48. Fung, Y.C.B., "Elasticity of Soft Tissues in Simple Elongation," *Am J Physiol*, **213**(6):1532-1544, 1967.

49. Beskos, D.E., and Jenkins, J.T., "A Mechanical Model for Mammalian Tendon," *J. Appl. Mech.*, Paper no. 75-WA/APM-20, 1975.

50. Lanir, Y., "A Microstructural Model for the Rheology of Mammalian Tendon," *J. Biomech. Engng.*, **102**:332-339, 1980.

51. Mow, V.C., Kuei, S.C., Lai, W.M., and Armstrong, C.G., "Biphasic Creep and Stress Relaxation of Articular Cartilage in Compression: Theory and Experiments," *J. Biomech. Engng.*, **102**: 73-84, 1980.

52. Pinto, J.G. and Patitucci, P.J., "Visco-elasticity of Passive Cardiac Muscle," *J. Biomech. Engng.*, **102**:57-61, 1980.

53. Lin, H-C., Kwan, M.K-W., and Woo, S.L-Y., "On the Stress Relaxation of the Anterior Cruciate Ligament," *Advances in Bioengineering.*, **5/6**, 1987.

54. Spirt, A.A., Mak, A.F., Wassell, R.P., "Nonlinear Viscoelastic Properties of Articular Cartilage in Shear," *J. Orthop. Res.*, **7**:43-49, 1989.

55. Dortmans, L.J.M.G., Sauren, A.A.H.J., and Rousseau, E.P. M., "Parameter Estimation Using the Quasi-Linear Viscoelastic Model Proposed by Fung," *J. Biomech. Engng.*, **106**: 198-203, 1984.

56. Nigul, I., and Nigul, U., "On Algorithms of Evaluation of Fung's Relaxation Function Parameters," *J. Biomech.*, **20**(4):343-352, 1987.

57. Rab, G.T., Chao, E.Y.S., Stauffer, R.N., "Muscle Force Analysis of the Lumbar Spine," *Orthop. Clin. No. Am.*, **8**:193-199, 1977.

58. McGill, S.M., Patt, N., Norman, R.W., "Measurement of the Trunk Musculature of Active Males Using CT Scan Radiography: Implications for Force and Moment Generating Capacity about the L4/L5 Joint," *J. Biomech*, **21**(4):329-341, 1988.

59. Goel, V.K., Liu, Y.K., Clark, C.R., "Quantitative Geometry of the Muscular Origins and Insertions of the Human Head and Neck" in *Mechanisms of Head and Spine Trauma*, Aloray, Goshen, New York, 1986.

60. Breau, C., Shirazi-Adl, A., de Guise, J., "Reconstruction of a Human Ligamentous Lumbar Spine Using CT Images – A Three Dimensional Finite Element Mesh Generation," *Ann. Biomed. Engng.*, **19**:291-302, 1991.

61. Garrett, W.E., Jr., Safran, M.R., Seaber, A.V., Glisson, R.R., Ribbeck, B.M., "Biomechanical Comparison of Stimulated and Nonstimulated Skeletal Muscle Pulled to Failure," *Am. J. Sports. Med.*, **15**(5):448-454, 1988.

62. Chow, G.H., LeCroy C.M., Seaber, A.V., Ribbeck, B.M., Garrett, W.E., "Sarcomere Length and Maximal Contractile Force in Rabbit Skeletal Muscle," *Orthop. Res. Soc.*, 547, 1990.

63. Merrill, T., Goldsmith, W., Deng, Y.C., "Three-dimensional Response of a Lumped Parameter Head-Neck Model Due to Impact and Impulsive Loading," *J. Biomech.*, **17**:81-95, 1984.

64. Goel, V.K., Kong, W.Z., Han, J.S., Weinstein, J.N., Gilbertson, L., "A Combined Finite Element and Optimization Investigation of Lumbar Spine Mechanics With and Without Muscles," *Spine*, **18**:1531-1541, 1993.

65. Lockett, F.J., *Nonlinear Viscoelastic Solids*, Academic Press, London · New York, 1972.

66. Lieber, R.L., *Skeletal Muscle Structure and Function, Implications for Rehabilitation and Sports Medicine*, Williams & Wilkins, Baltimore, 1992.

67. Fischer, G., "Versuche uber die Wirkung von Kerben an elastische beanspruchten Biegestaben," In Durelli, A.J., Phillips, E.A., Tsao, C.H. *Introduction to the Theoretical and Experimental Analysis of Stress and Strain*, McGraw-Hill Book Company, New York, 1958.

68. Zajac, F.E., "Muscle and Tendon: Properties, Models, Scaling and Application to Biomechanics and Motor Control," *Critical Reviews in Biomedical Engineering*, **17**:359-411, 1989. Bourne, JR editor.

69. Goslow, G.E. Jr., Reinking, R.M., Stuart, D.G., "The Cat Step Cycle: Hind Limb Joint Angles and Muscle Lengths During Unrestrained Locomotion," *J. Morph.*, **141**:1-42, 1973.

70. Woo, S.L.-Y., Danto, M.I., Ohland, K.J., Lee, T.Q., Newton, P.O., "The Use of a Laser Micrometer System to Determine the Cross-sectional Shape and Area of Ligaments: A Comparative Study with Two Existing Methods," *J. Biomech. Engng.*, **112**:426-431, 1990.

71. Herzog, W., Kamal, S., Clarke, H.D., "Myofilament Lengths of Cat Skeletal Muscle: Theoretical Considerations and Functional Implications," *J. Biomechanics*, **25**:945-948, 1992.

72. Gordon A.M., Huxley A.F., Julian F.J., The Variation in Isometric Tension with Sarcomere Length in Vertebrate Muscle Fibers, *J. Physiol.* **184**:170-192, 1966.

73. Foust, D.R., Chaffin, D.B., Snyder, R.G., Baum, J.K., "Cervical Range of Motion and Dynamic Response and Strength of Cervical Muscles," *Proceedings of the 17th Stapp Car Crash Conference*, SAE Paper no. 730975, 1973.

74. Tidball, J.G., "Energy Stored and Dissipated in Skeletal Muscle Basement Membranes During Sinusoidal Oscillations," *Biophys. J.*, **50**:1127-1138, 1986.

75. Roy, R.R., Meadows, I.D., Baldwin, K.M., Edgerton, V.R., "Functional significance of compensatory overloaded rat fast muscle," *J. App. Phys.*, *Respiratory, Environment, Exercise Physiology*, **52**: 473-78, 1982.

76. Spector, S.A., Gardiner, P.F, Sernicke, R.F., Roy, R.R., Edgerton, V.R. "Muscle architecture and force–velocity characteristics of cat soleus and medial gastrocnemius: implications for motor control," *J. Neurophys.*, **44**, 951-60, 1980.

77. Winter, D.A., *Biomechanics and Motor Control of Human Movement*, A. Wiley & and Sons, Inc., New York, 1991.

78. Fung, Y.C., Fronek, K., Patitucci, P., "Pseudoelasticity of Arteries and the Choice of Its Mathematical Expression," *Am. J. Physiol.*, **237**: H620-H631, 1979.

79. Fung, Y.C.B., "Mathematical Representation of the Mechanical Properties of the Heart Muscle," *J. Biomech.*, **3**:381-404, 1970.

80. Kwan, M.K., Lin, T.H., Woo, S.L., "On the Viscoelastic Properties of the Anteromedial Bundle of the Anterior Cruciate Ligament," *J. Biomech.*, **26**:447-452, 1993.

973344

The Dynamic Responses of the Cervical Spine: Buckling, End Conditions, and Tolerance in Compressive Impacts

Roger W. Nightingale, James H. McElhaney, and Daniel L. Camacho
Duke University

Michael Kleinberger
National Highway Traffic Safety Administration

Beth A. Winkelstein and Barry S. Myers
Duke University

ABSTRACT

This study explores the dynamics of head and cervical spine impact with the specific goals of determining the effects of head inertia and impact surface on injury risk. Head impact experiments were performed using unembalmed head and neck specimens from 22 cadavers. These included impacts onto compliant and a rigid surfaces with the surface oriented to produce both flexion and extension attitudes. Tests were conducted using a drop track system to produce impact velocities on the order of 3.2 m/s. Multiaxis transduction recorded the head impact forces, head accelerations, and the reactions at T1. The tests were also imaged at 1000 frames/sec.

Injuries occurred 2 to 30 msec following head impact and prior to significant head motion. Head motions were not found to correlate with injury classification. Decoupling was observed between the head and T1, resulting in a lag in the force histories. Cervical spine loading due to head rebound constituted up to 54 ± 16 percent of the total axial neck load for padded impacts and up to 38 ± 30 percent for the rigid impacts. Dynamic buckling was also observed; including first order modes and transient higher order modes which shifted the structure from a primarily compressive mode of deformation to various bending modes. The average load at failure was 2243 ± 572 N for males and 1061 ± 273 N for females. The difference between male and female tolerance was significant $p = 0.0015$). Impacts onto the padded surfaces produced significantly larger neck impulses ($p = 0.00023$) and a significantly greater frequency of cervical spine injuries than rigid impacts ($p = 0.043$). The impact angle was also correlated with injury risk ($p < 0.00001$).

These experiments demonstrate that in the absence of head pocketing, the head mass can provide sufficient constraint to cause cervical spine injury. The buckling modes illustrate the kinematic complexity of cervical spine dynamics and may explain why injuries can occur at different vertebral levels and with widely varying mechanism in compressive head impacts. These experiments also suggest that highly deformable padded contact surfaces should be employed carefully in environments where there is the risk for cervical spine injury; however, the orientation of the head, neck, and torso relative to the impact surface is of greater significance.

INTRODUCTION

Injuries to the cervical column and to the spinal cord continue to be a major social, economic, and clinical problem. More than 10,000 new spinal cord injuries are reported each year in the United States [1], and the total annual cost of these injuries has been estimated to be as high as 7.6 billion dollars [2]. A number of investigations have found motor vehicle accidents to be the cause of between 52 and 68 percent of cervical spine injuries [3, 4, 5, 6, 7]. The incidence of cervical spine injury among motor vehicle fatalities has been reported to be between 21 and 24 percent [8, 9, 5, 10]. Hadley *et al.* suggested that 25-40 percent of patients with upper cervical spine injuries actually die prior to admission to clinical facilities and are, therefore, often omitted from epidemiological studies. A comprehensive review of cervical spine injury mechanisms and their epidemiology has been published by Myers and Winkelstein [11].

The dynamics of the cervical spine during compressive impact loading are complex. Previous studies have identified a number of factors which influence cervical spine behavior. These include end conditions, inertia, buckling, and the initial orientations of the head, neck, and torso relative to the impact surface [12, 13, 14, 15, 16, 17, 18]. Although previous studies have examined these variables and provided invaluable insights into the dynamic behavior of the cervical spine, few had the statistical power or sample size to make statistically confident conclusions.

Failure loads for the cervical spine are reported to range from 0.3 kN to 14.7 kN, prompting some investigators to suggest that a compressive tolerance does not exist [19]. However, with a sufficient understanding of the factors which influence dynamic compressive behavior, it may be possible to reduce the scatter in the failure data and formulate a tolerance for predominantly compressive loading.

The purpose of this study was to use a highly instrumented cadaver model to test specific hypotheses related to compressive cervical spine injuries. These include the following: 1) the inertia of the head and neck influences cervical spine dynamics, 2) buckling influences the cervical spine kinematics and injury mechanism, 3) constraints on head motion influence dynamics and tolerance, and 4) the initial orientations of the head, neck, and torso relative to the impacted surface influence dynamics and tolerance.

MATERIALS AND METHODS

EXPERIMENTAL APPARATUS - An experimental apparatus was designed to model cervical spine injury resulting from vertical head impact with a following torso (Figure 1). A steel carriage was mounted to a drop track using two linear bearing sliders and was weighted to simulate an effective torso mass of 16 kg. The value for the torso mass is based on the GEBOD output for the 50th percentile male upper torso and is an estimate of the portion of the total body mass which acts on the neck during dynamic loading. GEBOD is a public domain program developed by the Armstrong Aerospace Medical Research Laboratory which calculates body segment inertial properties (AL/CF-TR-1994-0051). The specimen preparations were mounted to the carriage in an inverted position.

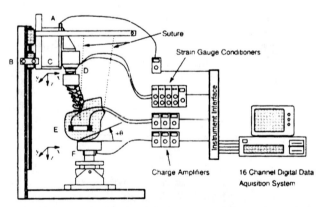

FIGURE 1 A diagram of the test apparatus showing the accelerometer on the torso mass (A), the optical velocity sensor (B), the carriage and torso mass (C), the six-axis load cell at T1 (D), the head accelerometers (E), and the anvil and three-axis load cell (F).

The impact surface was a 4 cm thick steel anvil with a diameter of 15.25 cm. Variation in impact angle about the y-axis (that axis normal to the sagittal plane) was achieved by mounting the anvil on a locking clevis. The impact angle was varied between -15° (posterior head impact) and +30° (anterior head impact), according to the

sign convention shown in Figure 1. The anvil was covered with 3 mm of Teflon sheet to simulate impacts onto a rigid, frictionless surface. Impacts onto a padded surface were simulated by attaching 3.8 cm thick foams to the anvil with duct tape. The Teflon covered steel surface simulated the unconstrained head end condition in 10 tests. A more constrained head end condition was simulated in the remaining tests (n=12) using either an expanded polystyrene foam (EPS) (E =2096.1 kPa, σ_y =206.2 kPa, ρ =0.0284 g/cm^3) or a less stiff, open cell polyurethane foam (OPU) (E =158.6 kPa, σ_y=7.0 kPa, ρ =0.0277 g/cm^3).

Multiaxis transduction was used to fully quantify the forces and moments acting on the head and neck during the impact event. Head impact forces were quantified using a Kistler 9067 three-axis piezoelectric load cell mounted under the impact surface. A GSE Model 6607-00 six-axis load cell mounted to the specimen was used to measure forces and moments at T1. A PCB 302A02 uniaxial accelerometer measured torso deceleration. Sagittal plane kinematics were quantified using two PCB 306A06 accelerometers which were mounted to the head of the specimen. Impact velocity was recorded using an MTS optical sensor. The sixteen channels of transducer data were sampled at 62.5 kHz using a PC-based acquisition system. Each test was also imaged using a Kodak Ektapro EM-2 digital camera at 1000 frames per second.

SPECIMEN PREPARATION Unembalmed human heads with intact spines were obtained shortly after death. All specimen handling was performed in compliance with both NHTSA 700-4 guidelines and CDC guidelines [20]. The specimens were sprayed with calcium buffered isotonic saline, sealed in plastic bags, frozen and stored at -20°. All donor medical records were examined to ensure that there were no preexisting pathologies, such as degenerative diseases or spinal pathologies, which could affect the structural responses of the specimens. Donor age ranged from 35 to 80 years.

The specimens were prepared for testing in a 100 percent relative humidity chamber. The muscular tissues were removed while keeping all the ligamentous structures intact (with the exception of the ligamentum nuchae). The specimens were transected at T3-T4 and the bottom two vertebrae were cleaned, defatted, and cast into aluminum cups with reinforced polyester resin. Care was taken that the first uncast motion segment (C7-T1) was free of resin and was allowed full range of motion. The C7-T1 intervertebral disc was oriented at 25 degrees to horizontal to preserve the resting lordosis of the cervical spine [21]. Following casting, a triaxial accelerometer array was attached to exposed parietal bone using dental acrylic and bone screws. A jig was used to ensure that the array was positioned parallel to the sagittal plane. The position of the array relative to the Frankfort anatomical plane was determined radiographically using the auditory meati and the inferior margins of the orbits. Finally, photographic target pins (4.0 mm diameter) were inserted in the anterior vertebral bodies, the spinous processes, and lateral masses of C2-C7. The pins were used for photogrammetric analysis of the vertebral motions.

EXPERIMENTAL PROTOCOL - Cadaveric specimens were inverted and mounted to the carriage of the drop track system in the anatomically neutral position.

Break-away sutures were passed through the ear lobules and nasal septum and were tied to the suspension frame to support the mass of the head and to maintain the neutral orientation of the cervical spine. Each specimen was raised to the desired height and the cervical spine was mechanically stabilized by manual exercise through a flexion-extension range of 60° for fifty cycles [22]. The specimens were dropped from a height of 0.53 m, which were less than that required to cause a skull fracture, yet sufficient to produce cervical spine injury [23]. Following impact testing, anteroposterior and lateral radiographs were obtained and the specimen was disarticulated at O-C1 and the head was weighed. To document the injuries, dissection was performed on both the heads and cervical spines.

DATA ANALYSIS All transducer data were uploaded to a Sun workstation for analysis. Digital filtration was performed in accordance with the Society of Automotive Engineers standard for head impacts (SAE J211b Class 1000). In order to determine inertial head loading and evaluate the risk of head injury, both linear and angular accelerations of the center of gravity of the head were determined. This was done using the data from the two triaxial accelerometers which were mounted to the head.

The traditional definition of failure was used to identify the occurrence of injury - a decrease in axial load with continued axial compression. However, in impact testing, events other than failure can be associated with a drop in the axial force. These include buckling, slip in the end conditions, and unloading of the neck due to rebound of the torso. Before a decrease in axial force was attributed to injury, the high speed video images and the load cell data were examined to ensure that the so-called injury was not due to buckling. In addition, head acceleration and video data were studied to determine if there was slip of the head on the impact surface. Finally, the video images and load cell data were compared to validate the correlation in time between the increases in axial deformation and the decreases in axial load.

The clinical stability of a cervical spine injury is a measure of its severity. Stable injuries generally do not result in a neurological deficit and are treated conservatively. Unstable injuries are much more likely to have neurological involvement and are treated by open or closed reduction and fixation. The stability of all the injuries produced in the drop tests was assessed using two methods. First, the injured motion segments were manipulated and obvious gross motions were defined as unstable. Second, the injuries were evaluated using the "one column plus one element" stability criteria outlined by Panjabi et al. [24, 25]. If all the elements of the anterior column (anterior longitudinal ligament, disc, and posterior longitudinal ligament) and any element of the posterior column (facets, ligamentum flavum, interspinous ligament, and supraspinous ligament) are disrupted, then the injury is considered to be potentially unstable. Similarly, the injury is classified as unstable if the posterior column and any element of the anterior column are disrupted.

The impulse of the compressive component of force at T1 was calculated for all the impacts by integrating the axial force history. Differences in axial impulse between the padded and rigid tests were evaluated using two-way ANOVA. The effect of padding on the frequency of injury was examined using a Fisher's exact test. The effects of padding on the peak head and neck forces were examined using t-tests.

RESULTS

A summary of the subject data and the drop test results is presented in Table 1. Results for 11 of the 22 tests have been reported previously [18, 26]. The average impact velocity was 3.14 ± 0.19 m·s^{-1}. All but one of the padded impacts, and five of the ten rigid impacts produced cervical spine injury (Table 2). In the padded tests, injuries occurred 18.0 ± 4.5 milliseconds after impact at a resultant force of 1905 ± 860 N (mean $\pm$ standard deviation). In the rigid tests, injury occurred 5.5 ± 2.4 milliseconds after impact at a resultant force of 2038 ± 374 N. Since the cervical spine can regain its ability to react a compressive load after injury, the values of neck force at the time of injury are not necessarily the peak forces recorded during the test. In addition, in some tests cervical spine buckling occurred at a higher peak force than the failure (Figure 2). The rigid impacts had larger peak head forces and shorter impulse durations than the padded impacts (Figure 2). The peak head forces in the padded impacts (4024 ± 1335) were significantly lower ($p < 0.0001$) than those in the rigid impacts (7977 ± 1795); however, the resultant neck forces at the time of neck injury were not significantly different ($p = 0.75$).

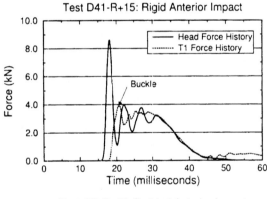

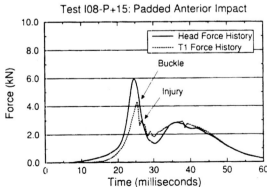

FIGURE 2 Plots showing head and neck force histories for a rigid and a padded impact. Peak head forces were greater in the rigid impacts and impulse durations were greater in the padded impacts.

TABLE 1: Subject Data and Drop Test Results

TEST[†]	Age, Sex	Vel. (m/s)	Angle (deg.)	Res.[a] Head Force (N)	Res.[a] Neck Force (N)	Axial[b] Neck Force (N)	Res.[c] Neck Force (N)	Time[d] (msec)	Impulse (N · s)	Lag (msec)	HIC
Rigid surface											
N05-R+30	36,M	3.23	Anterior (+30)	8790	1971	1552	1593	8.3	35.9	1.9	497
N18-R+15	–,M	3.26	Anterior (+15)	7498	2553	1863	1895	6.4	62.6	1.4	1935
D41-R+15	69,M	3.11	Anterior (+15)	8604	3885	N I	N I	N I	56.6	1.9	–
I32-R+15	78,M	3.18	Anterior (+15)	8234	2921	2416	2612	3.9	38.9	1.2	1361
N26-R+0	65,M	2.43	Vertex (0)	7638	4189	N I	N I	N I	47.7	1.4	–
N24-R+0	62,M	3.20	Vertex (0)	8566	2643	1839	1973	2.2	40.7	1.6	–
N22-R+0	71,M	3.26	Vertex (0)	8111	3010	1955	2120	6.5	46.9	1.9	490
N11-R-15	55,M	3.14	Posterior (-15)	11621	2891	N I	N I	N I	24.1	2.2	543
N13-R-15	35,F	3.28	Posterior (-15)	5615	2079	N I	N I	N I	20.6	1.2	704
UK3-R-15	62,M	3.13	Posterior (-15)	5093	4084	N I	N I	N I	30.7	1.4	1783
Padded surface											
N21-P+30	61,M	3.13	Anterior (+30)	1760	1762	1632	1662	14.8	42.7	5.6	50
N23A-P+30	46,M	3.03	Anterior (+30)	3608	2128	N I	N I	N I	39.7	5.4	77
N23B-P+30*	46,M	3.51	Anterior (+30)	3857	2595	2240	1698	18.7	31.7	6.0	197
I08-P+15	80,M	3.15	Anterior (+15)	5946	4309	2915	2918	30.5	78.6	5.4	110
I11-P+15	63,F	3.20	Anterior (+15)	3115	2096	967	972	14.0	71.9	5.5	118
I04-P+15	63,M	3.19	Anterior (+15)	3383	2091	1675	1698	18.0	74.1	5.8	
N03-P+0	75,M	3.08	Vertex (0)	5664	3701	3172	3509	18.2	62.7	3.8	84
N02-P+0	75,F	3.14	Vertex (0)	3452	2016	715	793	14.7	76.0	4.9	122
D40-P+0	53,F	3.16	Vertex (0)	4187	2491	1438	1440	16.7	81.1	2.3	
N19-P-15	42,F	3.07	Posterior (-15)	2604	1289	1011	1037	18.8	22.6	6.8	175
NA2-P-15	61,M	3.16	Posterior (-15)	4749	2091	1968	2091	15.6	35.1	3.8	270
I25-P-15	59,M	3.07	Posterior (-15)	5963	3448	2558	2574	18.4	39.7	3.5	384

N I The specimen had No Injury.

[†] The first three symbols are the specimen ID, the fourth is the impact surface (P=padded, R=rigid), and the last two are the impact angle.

[a] Peak resultant force.

[b] The magnitude of axial neck force at injury.

[c] The magnitude of the resultant neck force at injury.

[d] The elapsed time between impact and injury.

* Specimen N23 was dropped twice due to an error in the initial drop height.

TABLE 2: Injury Results

TEST	Pathology[†]	Class (Allen, 1982)	Stability	Head Motion
Rigid surface				
N05-R+30	C3 burst fx.,	VC	unstable	extension
	C3–C4 disc and ALL,	DE	stable	
	C4–C5 ALL	DE	stable	
N18-R+15	C1 lateral mass fracture,	VC	stable	extension
	C2 hangman's, C2-3 disc and ALL,	DE	unstable	
	C6-7 bilat. facet dislocation	DF	unstable	
D41-R+15	None	–	–	extension
I32-R+15	C5-6 disc, L capsular lig., ALL	DE	stable	extension
N26-R+0	None	–	–	extension
N24-R+0	C1 2 part posterior arch fracture,	CE	stable	flexion
	C2 hangman's fracture	DE	stable	
N22-R+0	C1 3 part comminuted fracture	VC	unstable	extension
N11-R-15	None	–	–	flexion
N13-R-15	None	–	–	flexion
UK3-R-15	None	–	–	flexion
Padded surface				
N21-P+30	C1 anterior ring fx.,	CF	stable	extension
	C4 spinous process fx.,	CE	stable	
	C5 spinous process fx.,	CE	stable	
	C5–C6 disc, left capsular lig. and ALL	DE	stable	
N23A-P+30	None	–	–	extension
N23B-P+30	C1 2 part right aspect fx.,	VC	stable	extension
	C3–C4 disc and ALL,	DE	stable	
	C4–C5 disc and ALL	DE	stable	
I08-P+15	C2 hangman's fracture and burst	DE	unstable	none
I11-P+15	C2 type III dens + comminution,	–	unstable	none
	C4 body, R lamina	VC	stable	
I04-P+15	C1 2 part posterior arch fracture,	CE	stable	none
	C2 hangman's, C2-3 disc, ALL,	DE	unstable	
	C7-T1 posterior ligs.	DF	stable	
N03-P+0	C4-5 capsular lig.,	–	stable	flexion
	C5-6 disc,	DE	stable	
	C6-7 bilat. facet dislocation	DF	unstable	
N02-P+0	C1 ant. ring,	CF	stable	none
	C2 hangman's + type III dens,	DE	unstable	
	C6 R lamina and pedicle,	VC	stable	
	C7 burst	VC	unstable	
D40-P+0	C1 3 part comminuted fracture,	VC	unstable	none
	C3-4 disc, ALL, spinous proc.,	DE	stable	
	C5 (burst), C5-6 PLL,	VC	unstable	
	C6 R lamina and pedicle	VC	stable	
N19-P-15	C2-3 disc, ALL, C2 ant. avul. fracture,	DE	stable	flexion
	C3-4 disc, ALL C3 ant. avul. fracture	DE	stable	
NA2-P-15	C3-4 disc, ALL, L capsular lig.,	DE	stable	flexion
	C5-6 disc, ALL	DE	stable	
I25-P-15	C1-2 L capsular lig.,	–	stable	flexion
	C3-4 disc, ALL, PLL, L capsular lig.	DE	unstable	

* Upper cervical spine injuries were classified using a system similar to that of Allen *et al.* [18].

† PLL = posterior longitudinal ligament   ALL = anterior longitudinal ligament

There was a statistically significant difference between the failure loads of the male and female specimens ($p = 0.0015$). The female tolerance in this study was 1061 ± 273 N for injured specimens with mean age of 58.3 ± 14.1 years. For the injured males, the tolerance was 2243 ± 572 N and the mean age was 61.8 ± 11.9 years.

In all the tests there was a delay in the onset of measured neck load with respect to the head load. This lag in response at T1 was 1.6 ± 0.3 msec for the rigid impacts and 4.9 ± 1.3 msec for the padded impacts (Figures 2 and Table 1). The lag is evidence that the head and cervical spine are not rigidly coupled despite a compressive preload.

The head impact response for almost all of the tests exhibited the bimodal behavior which have been seen in previous studies [27, 14, 28, 26] (Figure 2 and 9-16). The first peak in the head force is the head loading phase, and subsequent peaks are due to loading of the head and neck by the torso mass. For the rigid impacts, the first peak is attributed almost entirely to stopping the head and was not reflected in the neck force history. During this first mode, the head impact force reached a maximum with no concomitant neck force. For the padded impacts, the head contact times were significantly increased (Figure 2). Therefore, the first mode reflects both the head inertia and loading by the torso. For both impact surfaces, the second mode represents head and cervical spine loading by the inertia of the torso mass.

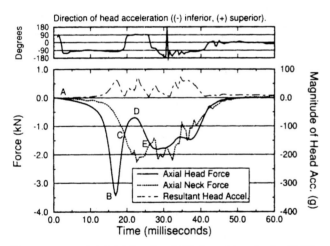

FIGURE 3 Axial (z) head and T1 force histories for N23A-P+30 illustrating rebound loading after the first mode in the head force. At point C, the forces measured by the load cells at the head and at T1 are equal and the vertical acceleration of the head at point C must be 0 to satisfy equilibrium. At C, the head center of gravity is subjected to a downward (180°) acceleration as the cervical spine becomes loaded by the inertial force of the rebounding head. Between points C and E, the difference between the head force and T1 force is the neck force due to head rebound.

Head rebound in the vertical (-z) direction accounted for a significant portion of the compressive neck loading during the injury interval and was observed in almost all of the impacts (Figure 3). Rebound accounted for as much as 54 ± 16 percent of the total neck load in the padded impacts and 38 ± 30 percent of the total neck load in the rigid impacts. The difference between these

percentages for the rigid and padded impacts was not significant using a two-sample t-test ($p > 0.3$).

INJURIES – These experiments produced injuries of widely varying mechanisms and at all vertebral levels (Table 2). These included burst fractures, bilateral facet dislocations, and hangman's fractures. Almost all of the disc injuries in these test were transverse tears through the anterior half of the disc due to distractive extension and were associated with anterior longitudinal ligament disruption. These were all stable and would most likely not be diagnosed on a plain radiographic evaluation.

The injuries were classified using a scheme similar to one developed by Allen (1982). The classifications were as follows: VC = vertical compression, DE = distractive extension, DF = distractive flexion, CE = compressive extension, and CF = compressive flexion. The configuration of the cervical spine during the all the injuries was the post buckled mode discussed in the subsequent section and illustrated in Figure 6. In almost every case, the injury classification listed in Table 2 is consistent with the post buckled deformation.

HEAD MOTION AND NECK INJURY – Cervical spine injuries were produced prior to the occurrence of large head motions (translations and/or rotations). The time between impact and cervical injury was from 2 to 9 msec for the rigid impacts, and from 14 to 30 msec for the padded impacts. In contrast, the time required to produce large head motions (greater than 20 degrees of rotation or 2.5 cm of translation) was between 20 and 100 msec after impact. Flexion or extension head rotations greater than 90 degrees occurred at least 90 msec following impact. For all the impacts with an incident angle of 30°, the motion of the head consisted of posterior translation and extension rotation. However, for the vertex and posterior impact surface orientations, the head motion was that of flexion rotation with anterior translation.

Head motion did not explain the local mechanism of injury. For example, head extension was associated with compression-extension and also flexion type injuries (Table 2). Further, specimens suffered multiple noncontiguous cervical injuries, with differing mechanisms. Head flexion was produced in eight tests and head extension was produced in nine tests. Injuries produced with head flexion included a vertical compression injury, eight extension injuries, and one flexion injury. Moreover, for the cases with head extension, injuries included those classified as flexion, vertical compression, and extension.

BUCKLING BEHAVIOR – Dynamic buckling of the cervical spine was observed regardless of surface type or orientation. Buckling, as evidenced by an abrupt decrease in measured compressive load with increasing deflection and moment (Figure 4) or a "snap-through" which is characterized by a visible and rapid transition from one equilibrium configuration to another (Figure 5), was observed in all the specimens tested. The deformation pattern of the post-buckled mode is shown schematically (Figure 6) together with the injuries produced, and can also be seen in the high speed image data (Figure 6). It was characterized by extension with extension type injuries in the C3-C5 motion segments (C2-C3-C3-C5 and the connecting ligaments), and flexion with flexion type injuries in the C7 and C8 motion segments (C6-C7-T1) and the connecting ligaments. The buckle occurred 3 to 8 msec after impact and the resulting mode of deformation provides the mechanism of the injury in the C2-C7 vertebrae; including those tests which produced multiple, noncontiguous injuries of different mechanisms. For example, some tests resulted in concomitant flexion and extension injury mechanisms consistent with the post-buckled deformation pattern. Test N18-R+15 produced a C2 Hangman's bipedicular fracture (due to local extension) and a C6-C7 bilateral facet dislocation (resulting from local flexion). Test N21-P+30 produced a C1 anterior ring fracture due to local flexion and C5-C6 concomitant fractures of the C4 and C5 spinous processes due to extension.

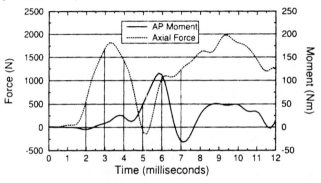

FIGURE 4 Axial neck force and anteroposterior moment histories for N13-R-15 showing cervical spine buckling between 3 and 5 msec. Each vertical line corresponds to a video frame in Figure 8. Compressive force has a positive sign in this figure so that it may be more easily compared with the moment. During buckling, the cervical spine loses axial stiffness and the compressive load decreases. This results in a drop in the measured axial force (4 msec). Increasing eccentricity of the axial load, increasing shear, and cervical spine inertia give rise to flexion bending moments. Once the new equilibrium position is reached (5.5 msec), the axial load rapidly increases and reaches its peak value for the test. Note the tensile force at 5+ msec due to the inertia of the cervical spine. Test N13-R-15 had no cervical spine injury.

In four of the rigid impacts (N11-R-15, N13-R-15, N18-R+15, and N05-R+30), a transient, higher order buckling mode was observed prior to the first order mode described above (Figure 5, 3 msec and Figure 7). Two of these specimens escaped injury, and the other two sustained injuries later in the impact. The unstable higher order mode was observed during the ascending portion of the neck loading curve, and lasted 2 to 8 milliseconds.

THE EFFECTS OF PADDING – In order to examine the effects of padding on injury risk, a subset of the 22 tests was considered. The impacts that were 30 degrees anterior to the vertex of the head were not included in the analysis because there was an insufficient number of tests to perform a 2-way ANOVA. Only the +15, -15, and 0 degree impacts were considered. Of these, all of the padded impacts and four of the nine rigid impacts produced cervical spine injury (Table 2). A total of 38 vertebral and vertebral motion segment injuries were produced in these 18 experiments. The injuries included bilateral facet dislocations, burst fractures, atlas fractures, and traumatic spondylolisthesis of the axis. Most of the injured specimens (10 of 13) had injuries to more than one vertebral level and three specimens (D40-P+0, N18-R+15 and N02-P+0) had unstable injuries at more than one level. A total of

eight injuries were produced in the rigid impacts, of which three were classified as unstable. In contrast, a total of 25 injuries were produced in the padded impacts, of which nine were classified as unstable. Of those nine, four were at C2. The frequency of injury in the padded impacts was significantly greater than the frequency of injury in the rigid impacts ($p = 0.043$).

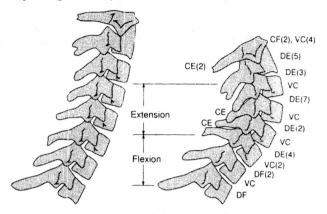

FIGURE 6 Illustration of the initial posture and buckling deformation of the cervical spine. Extension was observed at the third through sixth cervical motion segments and flexion at C7-C8. The injuries produced are shown by level and by classification. (CF = compressive flexion, CE = compressive extension, VC = vertical compression, DE = distractive extension, DF = distractive flexion).

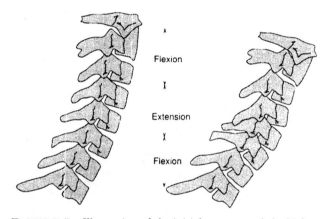

FIGURE 7 Illustration of the initial posture and the higher-order buckling mode of the cervical spine.

For the rigid impacts, the axial impulses were 41.0 ± 14.2 N-s. The momentum of the carriage was approximately 51 N-s; therefore, this result is consistent with our observation that the carriage was not completely arrested in some of the rigid impacts. For the padded impacts, the axial impulses were 60.2 ± 21.9 N-s. The axial impulses were grouped based on impact angle and impact surface in order to perform a two-way analysis of variance. The impulses in the padded impacts were significantly larger than the impulses in the rigid impacts ($p = 0.00023$). Two-way ANOVA also found significant differences between axial impulses as a function of impact angle ($p < 0.00001$). The posterior impacts produced the smallest impulses and the anterior and vertex impacts produced larger impulses (Table 1).

Anterior Impacts – All three of the padded anterior (+15 degrees) impacts produced unstable injuries. The rigid anterior impacts produced injuries in two of three

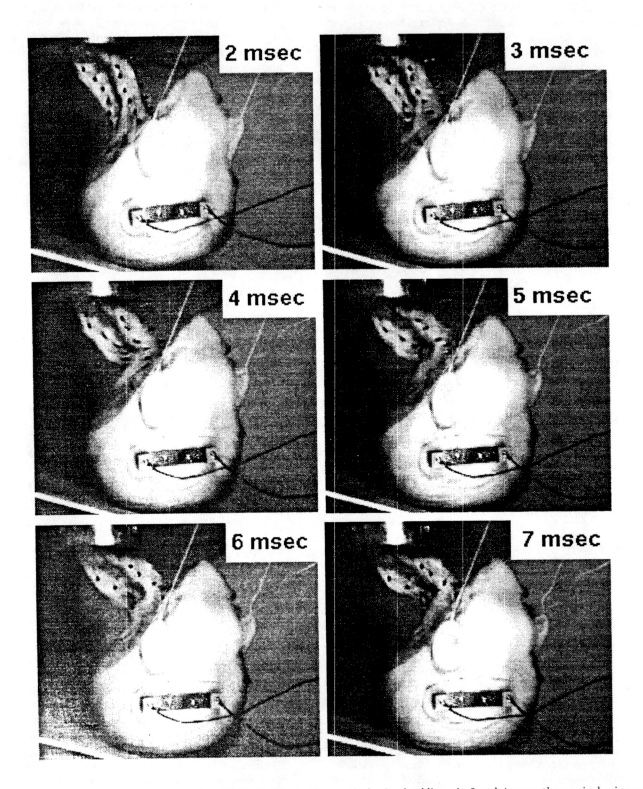

FIGURE 5 Six frames of video data (Test N13-R-15) illustrating cervical spine buckling. At 3 and 4 msec, the cervical spine snaps through to a new equilibrium position (6 msec) which is locally flexed at C6 through T1 and extended at C1 through C5. The force and moment histories for this test are shown in Figure 4. This test did not produce a cervical spine injury, illustrating that the buckle was stable.

specimens, one of which was unstable. In the padded impacts the head struck the contact surface and became pocketed in the foam. As a result, there was little or no head motion after the impact and the neck was essentially trapped between the head and the following torso (Figure 8). In the two rigid impacts which produced injury, the head began to translate posteriorly, however, large neck loads and injuries developed before appreciable head motion could occur. In the uninjured case (D41-R+15), the rigid contact surface and the neck produced sufficiently large posteriorly directed forces to push the head into extension and out of the path of the torso without injuring the cervical spine.

Vertex Impacts – All three of the padded vertex impacts (0 degrees) produced unstable cervical spine injuries. The rigid vertex impacts produced injuries in two of three specimens, however, it should be noted that the uninjured specimen (N26-R-0) had a lower impact velocity. The motions of the head and neck in the padded vertex impacts were very similar to those in the padded anterior impacts: the head hit the foam, stopped, and the neck was compressed between the head and the following torso. All the rigid vertex impacts were characterized by 5 to 10 degrees of head extension after contact, followed by anterior translation of the head and, finally, head and neck flexion. After contact, the thrust from the neck pushed the head anteriorly; however, all the injuries occurred before there was time for appreciable head motion.

Posterior Impacts – All three of the padded posterior (−15 degrees) impacts produced injuries, with one being classified as unstable. In contrast, none of the rigid posterior impacts produced injuries. The head motions in both the padded and rigid posterior impacts were characterized by a small amount of extension rotation followed by anterior translation and flexion of the head and neck. In these posterior impacts, the head and neck had a component of velocity in the anterior direction relative to the impact surface. This, combined with the anteriorly directed neck force, accelerated the head anteriorly. In the padded impacts; however, the foam exerted posteriorly directed forces which opposed head motion. Consequently, the head was constrained for a sufficient amount of time to subject the neck to larger impulses and allow the development of cervical spine injuries.

PADDED IMPACT (I08-P+15)

RIGID IMPACT (D41-R+15)

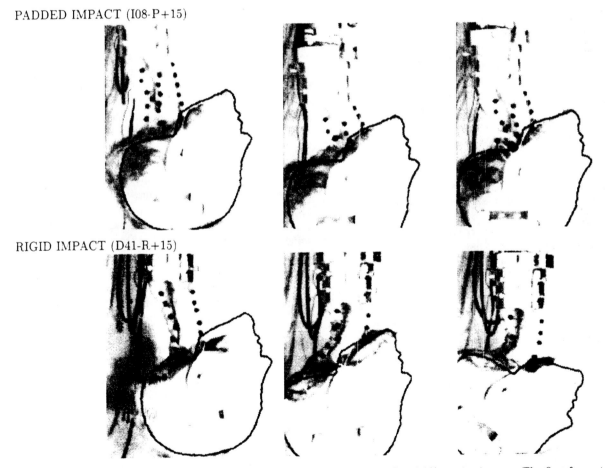

FIGURE 8   Three frames of image data for a padded anterior impact and a rigid anterior impact. The first frame in each sequence shows the position of the head and spine at impact, and the next two frames are taken at 35 msec and 70 msec respectively. The images are overexposed to give good pin contrast and a head tracing is provided to orient the reader. In the padded impact, there is very little head motion after impact and a great deal of cervical spine compression. In the rigid impact, the head begins to translate posteriorly because the forces applied by the extending neck and the impact surface are not opposed by surface padding. As a result, the head and neck moved out of the path of the following torso and the specimen escaped uninjured.

DISCUSSION

This study explores the dynamics of the cervical spine during compressive impact with the specific goals of determining the effects of head and neck inertia and the effects of impact surface properties on injury risk and injury mechanisms. Experiments were conducted on 22 cadaveric specimens and were designed to control as many variables as possible and to maximize repeatibility. It was not our intention to simulate any particular accident situation, but rather to examine the worst case scenario where the neck is forced to manage the energy of the following torso.

LIMITATIONS - The primary limitation of cadaver based biomechanical studies is the absence of the forces applied by the musculature. Forces due to passive muscle tone undoubtedly play a role in cervical spine stabilization and energy absorption during traumatic loading; however, these effects are minimized in impacts which produce primarily compressive loading. This was the rationale for limiting this series of experiments to predominantly compressive impacts with the head, neck, and torso aligned in the anatomically neutral position. The potential significance of active, reflex mediated muscle forces is mitigated by the rapidity with which the injuries occur. Reflex mediated electrical activity in the neck muscles has been shown to occur from 50 to 65 msec following head loading in studies of volunteers [29, 30]. In contrast, 5.5 ± 2.4 msec were required to produce injury in the rigid impacts, and 18.0 ± 4.5 msec were required to produce injury in the padded impacts.

HEAD INERTIA - These experiments demonstrate that the head inertia can impose forces which oppose cervical spine motion and produce injury in the absence of a pocketing impact surface. This was evidenced by the production of cervical spine injuries in the experiments where the head impacted a rigid, lubricated Teflon surface (an unconstrained end condition). This was particularly true for vertex (0°) impacts, and anterior (+15°) head impacts in which the neck loading vector had only a small component in a direction necessary to accelerate the head and neck out of the path of the following torso. Thus, while head pocketing may increase cervical spine injury risk, it is not requisite for injury during impact.

As a consequence of head inertia, the injuries occurred prior to any significant head motion. The more dramatic head motions, which have often erroneously been associated with the injury mechanism [31, 32, 25], did not occur until after injury. The type of injury produced was dependent on the cervical spine deformations which occurred shortly after head impact. These deformations determine the local injury mechanism and are consistent with published mechanistic classifications of cervical spine injury [33, 25]. In contrast, head motion following injury depended primarily on the type of surface, the surface orientation, and the direction of thrust from the injured, post-buckled spine. Our finding, that observed head motion cannot be used to predict or classify impact injury, is supported in other studies [27, 34].

CERVICAL SPINE BUCKLING - It has been suggested by a number of authors that buckling may play a role in spine mechanics [35, 17, 36]. Buckling has been demonstrated during quasi-static testing in the human cadaver cervical spine at low compressive loads [17]; however, this is the first experimental study to demonstrate buckling in the dynamically loaded cervical spine. In this series of experiments, buckling was observed in all the impacts, and sometimes without injury. The first order buckling mode (Figure 6) indicates that the neutrally positioned cervical spine assumes a deformation which predisposes the C3 through C6 motion segments to extension type injuries and the C7 and C8 motion segments to flexion injuries. Clearly, flexion injuries at C3 through C6 are observed clinically. Previous studies [22, 37. 28] have suggested that preflexion may be one mechanism for producing compressive flexion injuries in this region; however, the results of this study suggest that these mid-cervical injuries may be produced in a lordotic spine at higher impact velocities. The higher order buckling modes observed in some of these impacts produced mid-cervical flexion (Figures 7); however, the cervical spine made a transition to the first order mode before injurious levels of force had time to develop. It is possible that at higher impact velocities, injurious force levels may develop before the cervical spine can move from the higher order mode to the characteristic flexion extension shape which produced the injuries observed in these experiments. Thus, while buckling does not necessarily result in fracture or dislocation, it plays a central role in the pre-injury kinematics and may, in part, be the basis for the poor relationship between head motion and local cervical spine injury. It should be recognized that passive muscle tone would increase the buckling load and possibly alter the modes of deformation due to the stiffening effects of the muscle itself, and the nonlinear stiffening effect of muscle loads on ligamentous spine.

EFFECTS OF PADDING AND ORIENTATION - There were significant differences in axial neck impulses among the three impact angles ($p < 0.00001$) with the posterior impacts having the lowest impulses (Table 1). Since the neck impulse is a measure of the amount of torso momentum managed by the cervical spine, these results indicate that the posterior impacts are the least severe. This is consistent with the injury results. None of the posterior rigid impacts produced injury and the posterior padded impacts produced ligamentous injuries, with only one being classified as unstable (Table 2). In contrast, the frequency and severity of injury were greater for the vertex and anterior impacts. Accordingly, the results suggest that the orientation of the head and neck relative to the surface is a very important factor in determining the risk for neck injury. Orientations in which the cervical spine is nearly perpendicular to the impact surface (e.g. anterior +15 deg. impacts) place the neck at greater risk than those where the cervical spine is more oblique (posterior -15 deg. impacts).

This study was designed to simulate extremes in impact surface materials to determine if a highly deformable, pocketing material could increase neck injury risk in head impacts with an unrestrained, following torso. The materials used in this experiment do not reflect the effects of paddings, liners, and covers that are currently in use by the automotive and other safety industries.

Analysis of the injury data shows that there was a significantly greater frequency of injury in the padded impacts than in the rigid impacts ($p = 0.043$). This supports the hypothesis that padded surfaces may increase the risk of cervical spine injury. The effect of padding on injury risk was particularly apparent in the posterior impacts: none of the rigid impacts produced injury, and all of the padded impacts produced injury. In the posterior impacts, the head and neck have an initial component of velocity in the anterior direction relative to the impact surface. The forces applied to the head by the impact surface and by the neck are directed such that they increase this anterior escape velocity. In the rigid posterior impacts, the resulting anterior component of neck force and the initial anterior component of the velocity of the head were sufficient to move the head out of the path of the following torso before injurious loads could develop in the neck. However, in the experiments where a padded surface was used, the deformed pad applied posteriorly directed forces which decreased the anterior head escape velocity. As a result, the neck was subjected to significantly more of the torso momentum ($p = 0.00023$); that is, the neck was forced to stop the moving torso, and the neck sustained more injuries. While less pronounced in the other surface orientations, the overall frequency and severity of injury was greater in the padded impacts than in the rigid impacts. Despite increasing the risk of neck injury, the surface padding used in this study was able to significantly reduce the magnitude of the measured head impact force and therefore the risk for head injury. HIC values were reduced from an average of 1045 to an average of 159.

These experiments suggest that highly deformable padded contact surfaces should be employed carefully in environments where there is the risk for cervical spine injury. In addition, the results suggest that the orientation of the head, neck, and torso relative to the impact surface is of equal if not greater importance in neck injury risk. Additional research and a large number of specimens is necessary to identify which head and neck positions and impact angles are most likely to produce injury, and which padding materials and material thickness provide the optimal balance between head protection and neck protection. Given the prohibitive cost of cadaver studies, perhaps these questions will be best answered using computational models.

TOLERANCE The scatter in the existing compressive tolerance data for the cervical spine is primarily due to differences in experimental technique. Since neck responses are very sensitive to end conditions, inertia, and orientations, it is important to use data from studies which employ similar experimental setups. Motion segment studies are limited because the specimens are too short to exhibit buckling behavior. Studies of isolated spines are also limited because they do not include the important dynamic effects of the intact head. In the studies which use whole heads and cervical spines, the most confounding problem leading to the scatter in tolerance data is the use of force data from the head or impactor. The force measured at the point of head contact is considerably larger than the force transmitted to the neck because the neck force is reduced by the product of head mass and the head acceleration.

Accordingly, data must be interpreted with the realization that neck loading will be greatly reduced and will depend on the inertial characteristics and accelerations of the head and impactor. This effect is evident in the studies which have simultaneously measured the head impact load and the neck load. Yoganandan et al. [16] reported peak head loads of 5.9 ± 3.0 kN and peak neck loads of 1.7 ± 0.57 kN. Pintar et al. reported similar data, with mean peak impact loads of 11.78 ± 5.7 kN, while the corresponding neck loads were 3.5 ± 1.95 kN [28]. In our current study, the mean head impact load is 5.8 ± 2.5 kN and the resultant neck failure load was 1.9 ± 0.7 kN. If analysis is limited to the studies which; a) are dynamic, b) include the head mass, c) are unconstrained, and d) include instrumentation to measure the loads in the cervical spine, the scatter in the data is sufficiently reduced to allow the formulation of a tolerance for predominantly compressive loading. The studies which meet these criteria are this one, Pintar et al. (1990), Yoganandan et al. (1990), Yoganandan et al. (1991), and Pintar et al. (1995).

Because males and females constitute different populations, they should not be grouped in studies of cervical spine tolerance. This study shows a significant difference between the failure loads of the male and female spine ($p = 0.0015$). This is consistent with the experience of Pintar et al. [38], who report mean failure loads of 2.30 ± 1.10 kN for eight females with a mean age of 65 and 3.81 ± 0.97 kN for 11 males with a mean age of 62. The mean failure loads reported by Pintar et al. [38] for both the males and females are higher than the values reported in this study. However, the Pintar et al. data were obtained using preflexed spines while the data in this study was obtained using curved, neutrally positioned spines. As a result, our tolerance data reflects the effects of stresses due to bending while these stresses are minimized in the Pintar et al. study. Accordingly, it is not surprising that the Pintar et al. observed larger mean forces to generate the same material stresses and failures. This indicates that the cervical spine tolerance is also a function of the degree preflexion (straightening of the cervical spines secondary to flexion). Since the initial curvature of the cervical spine in the injury environment is unknown, it seems reasonable to combine the results of the two studies to formulate a tolerance for males and females. Averaging the means or the two studies results in the following tolerance values: 1.68 kN for females, and 3.03 kN for males. These values will be lower for a neutrally positioned spine and higher for a straightened spine; however, they are a representation of the mean resultant force required for a catastrophic cervical spine injury in any given impact situation.

Finally, it is important to recognize that the age of the cadaveric specimen contributes to the determined failure loads. The mean age of donors of the specimens in the tests described above are typically in the fifth decade, though specimens from donors above age 90 and below age 30 have been reported. McElhaney et al. [39] suggested that vertebral cancellous bone from 20 year old donors is 45 percent stronger then that from 50 year old donors. Keaveny and Hayes [40] reported that cortical bone is approximately 10 percent stronger when comparing these same age groups. While the contributions of cortical and cancellous bone to whole cervical vertebral strength is unknown, it would seem appropriate to use a scaling factor of 1.2 to 1.3 when formulating a tolerance for the younger age group. Therefore, we suggest that the cervical spine tolerance for the young human male is in the range of 3.64 to 3.94 kN.

CONCLUSION

1. The inertia of the head can oppose cervical spine motion and produce cervical spine injuries in the absence of a pocketing impact surface.

2. Head motions are not predictive of the local injury mechanism in the cervical spine in impact.

3. Dynamic stable, asymmetric buckling of the cervical spine was observed. Buckling modes may be transient and are not, in and of themselves, injury. Buckling alters the kinematics and position of the cervical spine prior to injury and, therefore, plays a role in determining which injuries are produced.

4. Constraint of head motion due to pocketing in a padded surface may increase the risk of cervical spine injury; however, the orientations of the head, neck, and torso relative to the impact surface are of equal if not greater importance. It should be noted that padding significantly reduces peak head forces.

5. Loading of the cervical spine due to head rebound can contribute to cervical spine injury.

6. There is a significant difference between the failure loads of the male and female cervical spines which indicates that male and female specimens should be treated as separate populations in biomechanical testing of human necks. According to this study, the female failure load is 1.03 kN and the male failure load is 2.15 kN.

7. Cervical spine compressive failure is a function of a number of different variables, and several methods exist to combine data into an average human tolerance. Once such method results in a tolerance range of 3.58 to 3.73 kN for the young human male.

ACKNOWLEDGMENTS

We would like to acknowledge Dr. Frank Pintar and Dr. Narayan Yoganandan for providing gender information not published in [38]. This study was supported by the Department of Transportation NHTSA Grant **DTNH22-94-Y-07133**, the Department of Health and Human Services CDC Grant **R49/CCR402396-10**. The views and opinions expressed in this paper are not necessarily those of the funding organizations.

REFERENCES

[1] J. H. McElhaney and B. S. Myers. Biomechanical aspects of cervical trauma. In A. M. Naham and J. W. Melvin, editors, *Accidental Injury*, pages 311–361. Springer-Verlag, New York, 1993.

[2] T. R. Miller, J. B. Douglass, M. S. Galbraith, D. C. Lestina, and N. M. Pindus. Costs of head and neck injury and a benefit-cost analysis of bicycle helmets. In R. S. Levine, editor, *Head and Neck Injury*, pages 211–240. Society of Automotive Engineering, New York. 1994.

[3] D.F. Huelke, E.A. Moffat, R.A. Mendelsohn, and J.W. Melvin. Cervical fractures and fracture dislocations - an overview. *Proceedings of the 23rd Stapp Car Crash Conference*, (790131):462–468, 1979.

[4] D. C. Burke, H. T. Burley, and G. H. Ungar. Data on spinal injuries–part ii. outcome of the treatment of 352 consecutive admissions. *Aust N Z J Surg*, 55(4):377–82, Aug 1985.

[5] M. N. Hadley, V. K. Sonntag, T. W. Grahm, R. Masferrer, and C. Browner. Axis fractures resulting from motor vehicle accidents. the need for occupant restraints. *Spine*, 11(9):861–4, Nov 1986.

[6] A. M. Levine and C. C. Edwards. Fractures of the atlas. *Journal of Bone and Joint Surgery*, 73(5):680–691, 1991.

[7] S. A. Shapiro. Management of unilateral locked facet of the cervical spine. *Neurosurgery*, 33(5):832–7; discussion 837, Nov 1993.

[8] G. J. Alker, Y. S. Oh, E. V. Leslie, J. Lehtotay, V. A. Panaro. and E.G. E. G. Eschnev. Post-mortem radiology of head and neck injuries in fatal traffic accidents. *Radiology*, 114:611–617, 1975.

[9] R. W. Bucholz, W. Z. Burkhead, W. Graham, and C. Petty. Occult cervical spine injuries in fatal traffic accidents. *Journal of Trauma*, 19(10):768–771, 1979.

[10] N. Yoganandan, M. Haffner, D. J. Maiman, H. Nichols, Frank A. Pintar, J. Jentzen, S. S. Weinshel, S. J. Larson, and Anthony Sances. Epidemiology and injury biomechanics of motor vehicle related trauma to the human spine. *Proc. 33rd Stapp Car Crash Conference*, 1989.

[11] B. S. Myers and B. A. Winkelstein. ' Epidemiology, classification, mechanism, and tolerance of human cervical spine injuries. *Crit. Rev. Biomed. Eng.*, 23(5&6):307–409, 1995.

[12] R.J. Bauze and G. M. Ardran. Experimental production of forward dislocation of the human cervical spine. *Journal of Bone and Joint Surgery*, 60-B:239–245, 1978.

[13] V. R. Hodgson and L. M. Thomas. Mechanisms of cervical spine injury during impact to the protected head. *Proceedings of the 24th Stapp Car Crash Conference*, pages 17–42, 1980.

[14] G. S. Nusholtz, D.E. Huelke, P. Lux, N. M. Alem, and F. Montalvo. Cervical spine injury mechanisms. *Proc. 27th Stapp Car Crash Conference*, pages 179–197, 1983.

[15] N. M. Alem, G. S. Nusholtz, and J. W. Melvin. Head and neck response to axial impacts. *Proceedings of the*

28th Stapp Car Crash Conference. (841667):275–288, 1984.

[16] N. Yoganandan, A. Sances Jr., D. J. Maiman, J. B. Myklebust, P. Pech, and S. J. Larson. Experimental spinal injuries with vertical impact. *Spine,* 11(9):855–860, 1986.

[17] B. S. Myers, J. H. McElhaney, W. J. Richardson, R. W. Nightingale, and B. J. Doherty. The influence of end condition on human cevical spine injury mechanisms. *Proc. 35th Stapp Car Crash Conference.* (912915):391–399, 1991.

[18] R. W. Nightingale, J. H. McElhaney, W. J. Richardson, T. M. Best, and B. S. Myers. Experimental cervical spine injury: relating head motion, injury classification, and injury mechanism. *Journal of Bone Joint Surgery,* 78-A(3):412–421, 1996.

[19] N.M. Alem, G.S. Nusholtz, and J.W. Melvin. Superior-inferior head impact tolerance levels. Technical Report UMTRI-82-42, UMTRI, 1982.

[20] J. M. Cavanaugh and A. I. King. Control of transmission of HIV and other bloodborne pathogens in biomechanical cadaveric testing. *Journal of Orthopaedic Research,* 8:159–166, 1990.

[21] T. Matsushita, T. B. Sato, K. Hirabayashi, S. Fujimura, T. Asazuma, and T. Takatori. X-ray study of the human neck motion due to head inertia loading. *Proceedings of the '38th Stapp Car Crash Conference,* (942208):55–64, 1994.

[22] J. H. McElhaney, J. G. Paver, H. J. McCrackin, and G. M. Maxwell. Cervical spine compression responses. *Proc. 27th Stapp Car Crash Conference,* (831615):163–177, 1983.

[23] J. H. McElhaney, John D. States R. G. Snyder, and M. A. Gabrielsen. Biomechanical analysis of swimming pool injuries. *Society of Automotive Engineers,* 790137:47–53, 1979.

[24] M. M. Panjabi, A. A. White 3d, and R. M. Johnson. Cervical spine mechanics as a function of transection of components. *J Biomech,* 8(5):327–36, 1975.

[25] A. A. White III and M. M. Panjabi. *Clinical Biomechanics of the Spine.* JB Lippincott Company, Philadelphia, 1990.

[26] R. W. Nightingale, J. H. McElhaney, W. J. Richardson, and B. S. Myers. Dynamic responses of the head and cervical spine to axial impact loading. *Journal of Biomechanics,* 29(3):307–318, 1996.

[27] G. S. Nusholtz, J. W. Melvin, D. F. Huelke, N. M. Alem, and J. G. Blank. Response of the cervical spine to superior-inferior head impact. *Proc. 25th Stapp Car Crash Conference,* pages 197–237, 1981.

[28] F. A. Pintar, A. Sances, N. Yoganandan, J. Reinartz, D. J. Maiman, G. Unger, J. F. Cusick, and S. J. Larson. Biodynamics of the total human cadaveric cervical spine. *Proc. 34th Stapp Car Crash Conference.* (902309):55–72, 1990.

[29] D. R. Foust, D. B. Chaffin, R. G. Snyder RG, and J. K. Baum. Cervical range of motion and dynamic response and strength of cervical muscles. *Proceedings of the 17th Stapp Car Crash Conference,* pages 285–308, 1971.

[30] L. W. Schneider, D. R. Foust, B. M. Bowman, R. G. Snyder, D. B. Chaffin, T. A. Abdelnovr, and L. K. Baum. Biomechanical properties of the human neck in lateral flexion. *Proceedings of the 19th Stapp Car Crash Conference,* page 3212, 1975.

[31] J. L. Babcock. Cervical spine injuries, diagnosis and classification. *Arch Surg,* 111(6):646–51, June 1976.

[32] R. Braakman and L. Penning. Causes of spinal lesions In *Injuries of the cervical spine.* Excerpta Medica, Amsterdam, 1971.

[33] B. L. Allen, R. L. Ferguson, T. R. Lehmann, and R. P. O'Brien. A mechanistic classification of closed indirect fractures and dislocations of the lower cervical spine. *Spine,* 7(1):1–27, 1982.

[34] D. F. Huelke and G. S. Nusholtz. Cervical spine biomechanics: a review of the literature. *J Orthop Res,* 4(2):232–45, 1986.

[35] J. J. Crisco and M. M. Panjabi. Euler stability of the ligamentous human lumbar spine. *Transactions of the 36th Annual Meeting of the ORS,* 15:607, 1990.

[36] J. S. Torg and H. Pavlov Axial load teardrop fracture. In *Athletic Injuries to the Head, Neck and Face.* Lea & Febiger, Philadelphia, 1991.

[37] J. S. Torg, J. J. Vegso O., M. J. Neill, and B. Sennett. The epidemiologic, pathologic, biomechanical, and cinematographic analysis of football-induced cervical spine trauma. *Am J Sports Med,* 18(1):50–57, Jan 1990.

[38] F. A. Pintar, N. Yoganandan, M. Pesigan, L. Voo, J.F. Cusick, D.J. Maiman, and A. Sances Jr. Dynamic characteristics of the human cervical spine. In *Proc. 39th Stapp Car Crash Conference,* number 95722, pages 195–202, 1995.

[39] J. H. McElhaney, J. L. Fogle, J. W. Melvin, R. R. Haynes, V. L. Roberts, and N. M. Alem. Mechanical properties on cranial bone. *J Biomech,* 3(5):495–511, Oct 1970.

[40] T. M. Keaveny and W. C. Hayes. A 20-year perspective on the mechanical properties of trabecular bone. *J Biomech Eng,* 115(4B):534–42, Nov 1993.

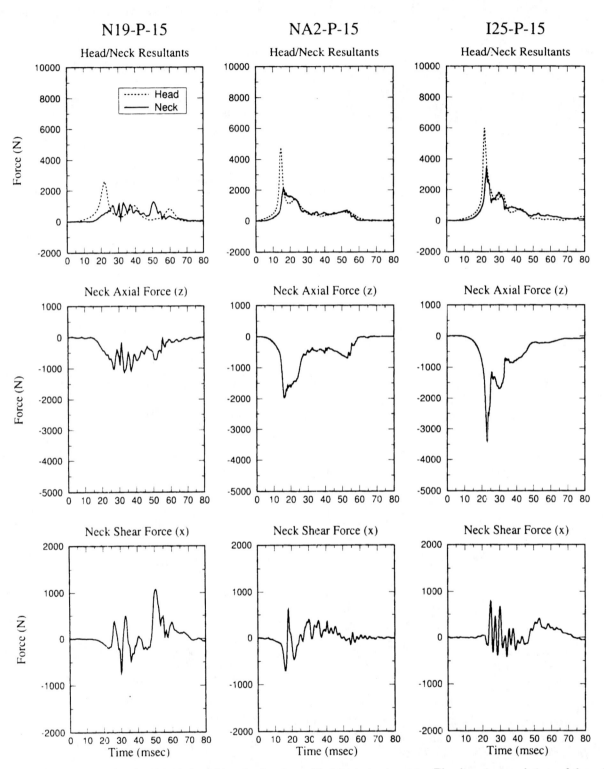

FIGURE 9 Plots showing head and neck force histories for the padded posterior impacts. The directions and signs of the axial and shear components of force coincide with the load cell at T1 (Figure 1).

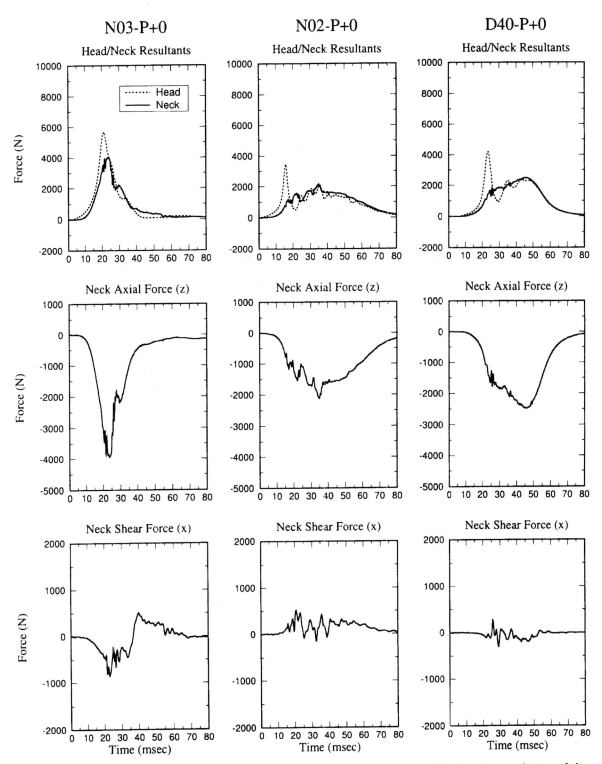

FIGURE 10 Plots showing head and neck force histories for the padded vertex impacts. The directions and signs of the axial and shear components of force coincide with the load cell at T1 (Figure 1).

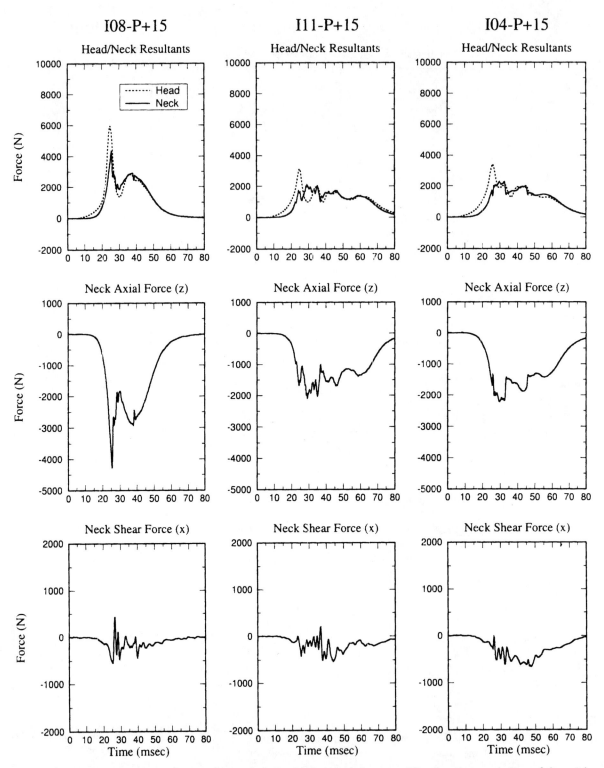

FIGURE 11 Plots showing head and neck force histories for padded anterior impacts. The directions and signs of the axial and shear components of force coincide with the load cell at T1 (Figure 1).

308

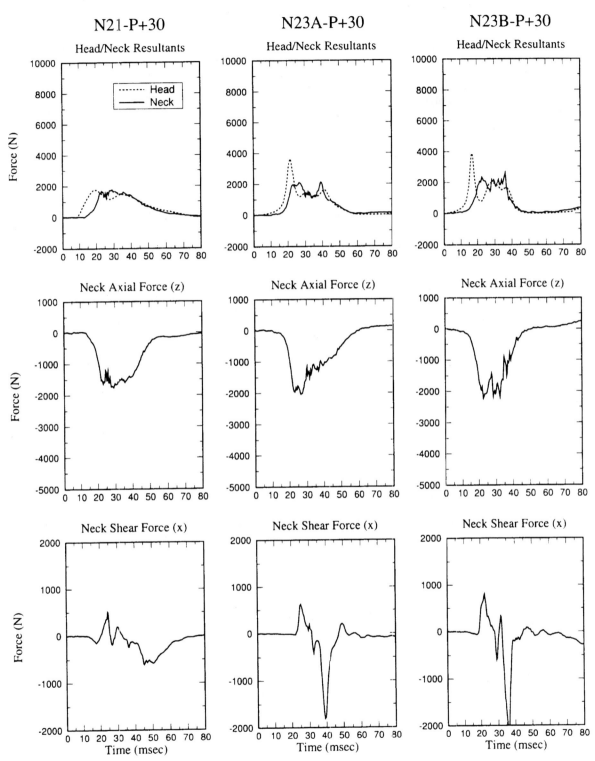

FIGURE 12 Plots showing head and neck force histories for padded anterior impacts. The directions and signs of the axial and shear components of force coincide with the load cell at T1 (Figure 1).

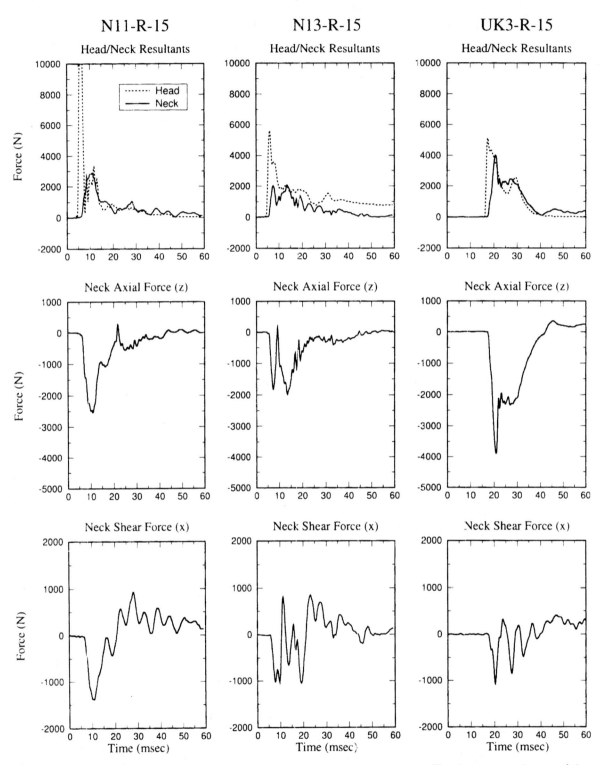

FIGURE 13 Plots showing head and neck force histories for the rigid posterior impacts. The directions and signs of the axial and shear components of force coincide with the load cell at T1 (Figure 1).

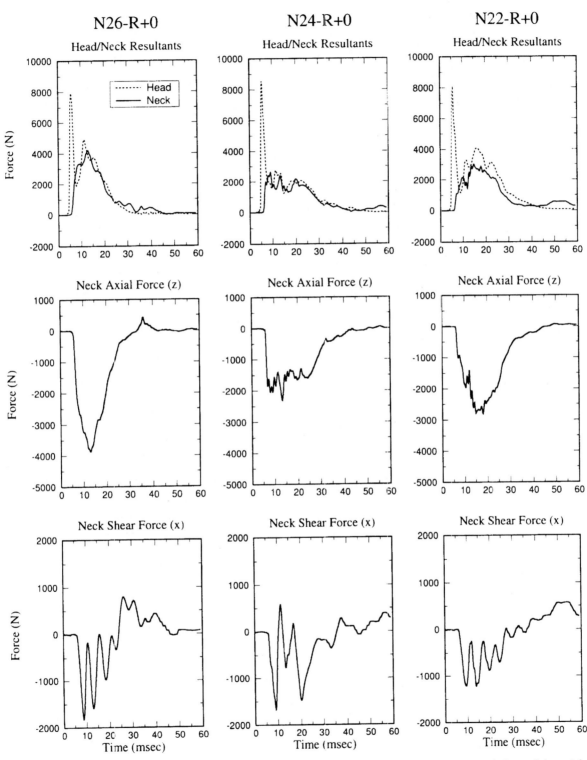

FIGURE 14 Plots showing head and neck force histories for the rigid vertex impacts. The directions and signs of the axial and shear components of force coincide with the load cell at T1 (Figure 1).

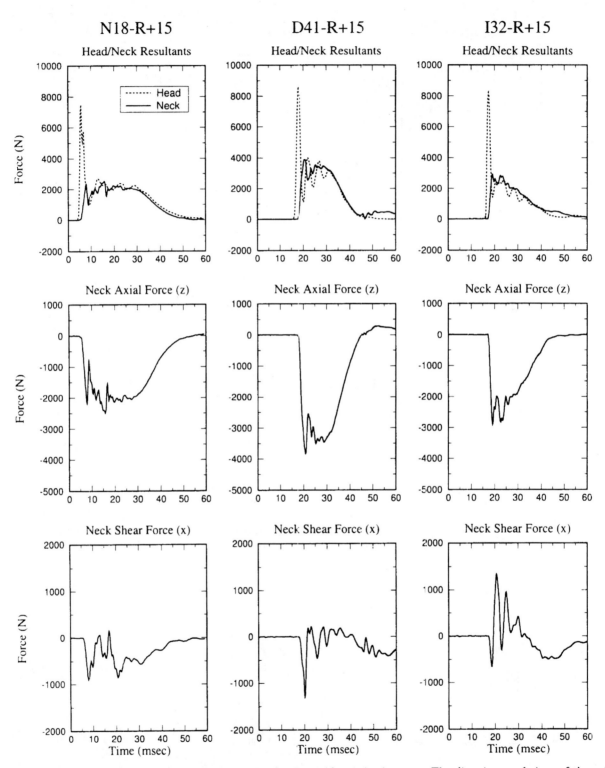

FIGURE 15 Plots showing head and neck force histories for the rigid anterior impacts. The directions and signs of the axial and shear components of force coincide with the load cell at T1 (Figure 1).

N05-R+30

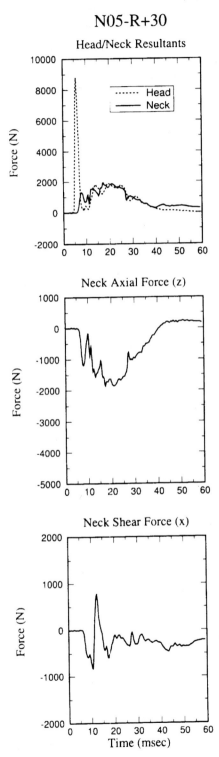

FIGURE 16 Plots showing head and neck force histories for the rigid anterior impacts. The directions and signs of the axial and shear components of force coincide with the load cell at T1 (Figure 1).

Biomechanical Response of Head/ Neck/ Torso and Cervical Vertebral Motion to Lateral Impact Loading on the Shoulders of Volunteers

Koshiro Ono
Susumu Ejima
Japan Automobile Research Institute
Koji Kaneoka
Makoto Fukushima
Department of Orthopaedic Surgery, University of Tsukuba
Shintaro Yamada
Sadayuki Ujihashi
Tokyo Institute of Technology
Japan
Paper Number 07-0294

ABSTRACT

To understand the response of the head, neck and torso during a lateral collision, and to investigate the relation between cervical vertebral motion and the occurrence of neck injuries, lateral impact experiments were conducted on the shoulder areas of human volunteers. Test subjects consisted of 8 volunteers (5 males and 3 females). For the analysis of cervical vertebral motions of each subject, a cineradiography system was used. A VICON motion photographic device was also used for the three-dimensional analysis of head/neck/torso motions. In the experiment, 3 levels of impact force (400N, 500N, and 600N) were applied considering both the presence and absence of muscle tension.

Cervical vertebral rotations all started at 35 ms, but the time required to reach the peak rotation increased toward the upper vertebrae, with C7 and T1 peaking at 120 ms and the final peak in the head at 120 ms. At around 35-80 ms, the rotation angle of C5 surpassed those of the head and C4 showing that the cervical spine was bending into an S-curve. This phenomenon shows the same type of cervical vertebral motions causing whiplash during a rear-end collision. Also, extreme compression was at work in the vertebral disc and/or the facet joint in C6/C7 and C7/T1, suggesting a high probability of injury occurring in the neck.

INTRODUCTION

Vehicle occupants involved in automobile accidents but saved from fatality with injury severity level reduced to serious - minor are increasing, owing probably to the implementation of automobile safety measures and advances made in emergency medical treatments. It can be deduced that the increase in number of those with severe - minor injuries is attributable to the abovementioned developments.

In order to keep pace with this development, active studies are being made for further enhancement of automobile safety, particularly against vehicle frontal collisions. Despite such efforts, the number of those injured by rear-end collisions is increasing significantly (Kraft et al., 2002), which is considered by some researchers as a "trade-off" between the number of fatalities and the number of "severe - minor injuries", with the priority set on the reduction of the fatalities. Regarding neck injuries, such increase were found not only in rear-end collisions but also in lateral-collisions (Hell et al., 2003). The same as in the case of rear-end collisions, the neck injury mechanism in lateral-collisions has not been clearly determined, with many questions still remaining unsolved (Kumar et al., 2005, Ito et al., 2004, Yoganandan et al., 2001). One of the reasons is the scarcity of biomechanical studies conducted on human head/neck/torso impact responses in lateral-collisions. In this regard, a new test equipment called "head/neck inertia impactor" was used in this study in order to analyze the "human head/neck junction" while applying a lateral impact to the shoulder. To be more specific, volunteers were impacted on their shoulders to simulate automobile lateral-collisions, and study human head/neck/torso impact responses as well as cervical vertebral motions. Differences in neck muscle responses between the male and female volunteers were also investigated.

EXPERIMENTAL METHODS

Lateral Inertia Impactor

An inertia impactor (Figure 1) specially designed for this study was used in order to investigate head/neck/torso responses and cervical vertebral motions of subjects submitted to a lateral inertia

Fig.1 Lateral inertia impactor

impact. The test equipment consists of a compressed air storage/coil spring unit to eject the impactor, the impactor height adjuster, and the test subject sitting position adjuster (forward/ backward & up/down). The front plate, pushed against the impactor front was fixed to the piston through the piston rod. The compressed air is stored in the cylinder with the piston fixed to the air chuck located at the rear end. The impactor mass is 8.5 kg. The impactor is ejected by opening the air chuck, and impact is applied to the back of test subject. A coil spring is provided to control the impactor stroke and the rise of impact load. The stroke setting and the rise of impact load can be varied per test.

Head/Neck/Torso Visual Motions

In order to record the kinematics of the head/neck/torso of each subject during impact, a high-speed video camera with a photographic capability of taking 500 frames/s was used. The head rotation angle and the displacement relative to the torso (the first thoracic vertebra: T1) were calculated by tracing the motion of each marker adhered to the subject according to the photographic images. A VICON motion photographic device (125 frames/s) was also used for the three-dimensional analysis of head/neck/torso motions.

Cervical Vertebral Motions Using Cineradiography System

For the analysis of cervical vertebral motions of each subject during impact, a cineradiography system (Philips: BH500) was used. The system is capable of taking cervical vertebral images at the rate of 60 frames per second with 16.67 ms intervals.

Experimental Conditions

Using five healthy male and three healthy female adults as human volunteers, experiments on the head/neck/torso impact responses and the cervical vertebral motions upon lateral inertia impact was conducted. Table 1 shows anthropometric data on human volunteers. The impact loading direction was set vertical (0 deg inclination) against the shoulder on one side (Figure 2). To be more specific, each test subject sat on one side of the impactor, with the back set practically straight against the stiff seat, so that the impact direction become parallel to the line connecting the acromion and the lower part of the cervical vertebrae. In order to analyze the differences in impact loading directions, the impact was also applied from 15 deg forward and 15 deg backward directions (Figure 2), in addition to the 0 deg direction. The impactor surface is rectangular with an area of 100 mm x 150 mm. The impact loading location against the subject's shoulder was set so that the position of acromion would become the same as that of the impactor upper surface. In order to find the difference in effects of neck muscle response on the head/neck/torso motions, the states of muscle were set in tensed and relaxed conditions, respectively. The impact load was set at 3 different levels such as 400 N, 500 N and 600 N in order to find the differences in head/neck responses to the lateral impacts. For the direction with 0 deg inclination, impact responses were compared between cadaver tests and those on the volunteers. Table 2 shows the different test conditions classified

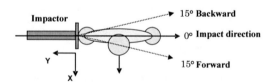

Fig.2 Impact directions

by differences in sex (male and female), impact loading levels, impact directions and states of muscle, with different combinations of test conditions.

Table 1 Anthropometric data of the subjects

	Age	Sex	Height (cm)	Weight (kg)	Sitting Height (cm)	Mass of head (estimate) (kg)	Inertia of head (estimate) (10^{-2}kgm^2)
1	25	M	172	67	97	4.28	2.21
2	23	M	170	63	94	4.14	2.14
3	22	F	162	46	83	3.63	1.85
4	23	F	166	51	88	3.77	1.93
5	24	F	161	58	86	3.98	2.04
6	23	M	180	85	91	4.97	2.59
7	24	M	174	61	90	4.07	2.10
8	24	M	181	77	96	4.64	2.42

Table 2 Test conditions

No. of Subject	Sex	Impact force (N)	Impact direction	Muscle condition
8	Male Female	400	15° forward	Relaxed Tensed
		500	0 degree	
		600	15° backward	

Informed Consent for Volunteers

The informed consent procedure in line with the Helsinki Declaration (WHO/CIOMS, 1988) was conducted in order for the volunteers to be fully informed of the purpose and method of experiments and also to ensure their full consent. The details/contents of the experiments were subjected to the approval of Special Committee of Ethics, Medical Department, Tsukuba University.

ANLYTICAL METHODS

Impact Force Applied to Head/Neck

Head acceleration was measured with the head 9 channel accelerometer, first thoracic vertebra (T1) acceleration was measured with 3-axis accelerometer, and electromyogram was analyzed. The measuring instruments were the head 9ch accelerometer (X, Y & Z), head angular velocity sensor (X, Y & Z), T1 accelerometer (X, Y & Z) and the pelvis accelerometer. The locations where the sensors were attached are shown in Figure 3. A mouth-piece suitable for the teeth profile (teeth impression) was prepared for each test subject. Assuming that the head is rigid, the head coordinate system was set in line with the location of anatomical center of gravity. The 9 channel acceleration measurement method (Ono et al., 1980) was applied according to the

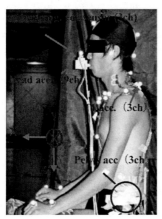

Fig.3   Mounting of accelerometers and rotational velocity sensors

coordinates of each accelerometer in this system, and the rotational and linear accelerations at the head CG were calculated.

Torso Acceleration (T1)

For the measurement of acceleration at T1, a three-axial accelerometer was attached onto the skin over a spinous process of T1.

Three-dimensional Motions of Head/Neck/Torso

The three-dimensional motions of head/neck/torso were measured by means of a VICON Motion Capture. Then the right-shoulder strain (displacement), left-shoulder strain (displacement), head rotation angles (X, Y & Z), T1 rotation angles (X, Y & Z) and the head rotation angles relative to T1 were analyzed.

RESULTS

Characteristic Aspect of Neck Impact Loading & Visual Motions

A 600 N impact loading experiment (in relaxed muscle condition) is shown in Figure 4, with the sequential photographs of the head/neck/torso motions during impact. X-ray of the neck motions under the same test conditions are shown in Figure 5. Figure 6 shows the corridors of the impact forces, the impact velocities, and the impact accelerations of impactor measured in 600 N impact loading experiment (in relaxed muscle condition). The linear and the angular accelerations at the head CG (X, Y & Z) calculated from the values measured with the head 9 channel accelerometer, the accelerations (X, Y & Z) at the T1 are also shown. Figure 7 shows the neck forces (Fx, Fy, Fz, Mx, My & Mz), and the visual head (head displacements and head rotational angles) motions in relation to the T1. Figure 8 shows the visual motions in relation to the shoulder strains (at the sternum upper end and the right or the left acromion) of the right shoulder (right acromion) and the left acromion).   On the other hand, the rear view and the lateral views of spine trajectories by the VICON are shown in Figure 9.

<u>**Phase 1 [0-50 ms]**</u> - The duration of impact for each one of 8 test subjects were 70 ms or so (Figure 6a). The impact load peak levels were fluctuating, as the impactor and the shoulder were not in complete contact in the initial stage of impact. This presumably resulted in the relatively low impact peak level in the initial stage and the relatively high peak level in the secondary stage. The T1 accelerations, on the other hand, showed that the maximum value was found around 50 ms (Figures 6j-6l), while that of the head

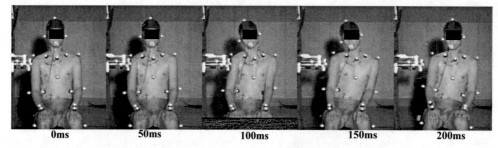

| 0ms | 50ms | 100ms | 150ms | 200ms |

Fig.4 Sequential motions of head/neck/torso (Impact forces: 600N, Relaxed condition)

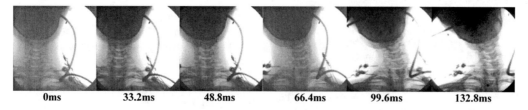

| 0ms | 33.2ms | 48.8ms | 66.4ms | 99.6ms | 132.8ms |

Fig.5 Sequential images of cervical vertebrae by cineradiography (Impact force: 600N, Relaxed condition)

was around 60 ms (Figures 6d-6f). The maximum values of T1 and the head in the Y-axial direction were 55 m/s^2 and 18 m/s^2, respectively. It is deduced that the axial forces between the T1 and the head were acting in opposite direction of compression, as the accelerations of T1 and the head in the Z-axial direction were reversed around 50 ms. The rotations of the head and T1 around the X-axis were reversed around 30 ms. The rotations around the Z-axis were also reversed. The neck shear force (in Y-axial direction) and the neck moments around X-axis and Z-axis did not show their maximum values around 50 ms (Figures 7m), 7q), 7r)), but the axial force of neck in Z-axis showed the maximum value at 50 ms or so. The right shoulder strain (on the impact side) showed the maximum value around 70 ms (Figure 8a). A slight torsion of upper cervical vertebrae was found around the Z-axis (Figure 5).

Phase 2 [50-100 ms] - The impact was continually set up to 70 ms or so (Figure 6a), and the shoulder was separated from the impactor due to the torso inertia. Hence, the acceleration at each portion of the head drops thereafter (Figures 6d-6f). However, the head rotates laterally against the torso, and the acceleration in the X-axial direction starts to increase around 90 ms, as the head is subjected to a restriction by the lateral bending at the same time. The head rotation angles found from the three-dimensional motion analysis by means of VICON Motion Capture showed the maximum values around 100 ms in both X and Z axial directions (Figures 9a-9b). The timing was roughly the same as the timing when the head rotational angle relative to T1 was highest. The maximum value around the X-axis was 32 deg, and 25 deg around the Z-axis. Similar to this trend of head acceleration, the neck shear force decreases

around 90 ms, but increases again as the head acceleration was restricted by the lateral bending. The displacements of right and left shoulders and the strains start resuming at the initial states around 80 ms, while the upper cervical vertebral torsion and the lateral extension which occurs mainly at the lower cervical vertebra also started (Figure 5).

Phase 3 [100-300 ms] - The impact loading already stopped, but the entire body keeps rotating clockwise due to inertia. The T1 acceleration in Y-axial direction converged around 150 ms, whereas the head acceleration remains up to 200 ms or so (Figures 6d-6f). The T1 rotation angle around the Y-axis showed gradual changes after 100 ms, while the head keeps on rotating. The lateral extension of cervical vertebrae starts to end, resuming the initial states while maintaining the torsion in the Z-axial direction. It was found from the three-dimensional motion data obtained with VICON that the torsion angle around the Z-axis resumed the initial state at 300 ms or so (Figure 9b). The lateral extension of cervical vertebrae started to resume in the initial state while maintaining the torsion in the Z-axial direction (Figure 5).

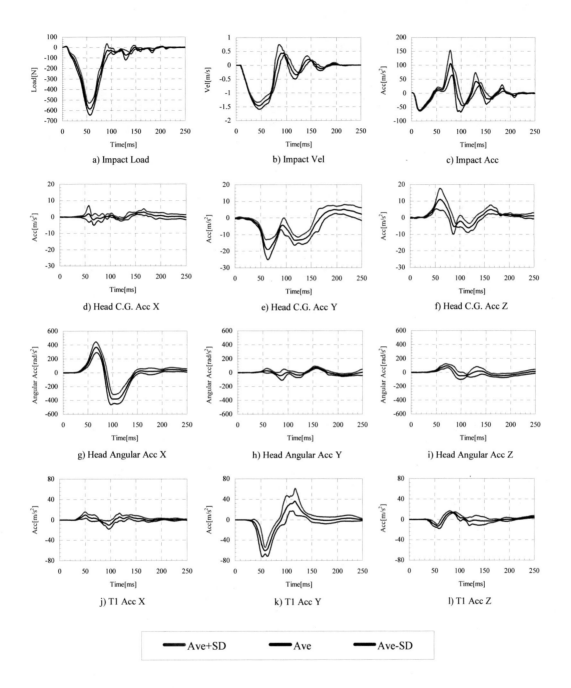

Fig. 6 Impact load, Head C.G. Acc., Head angular Acc., and T1 Acc., (Relax, 600N)

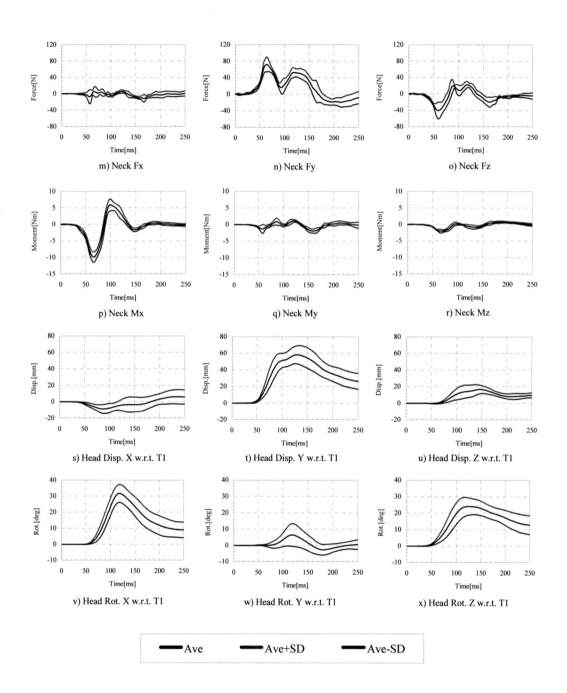

Fig. 7   Neck force, Neck moment, Head Disp. and Rot Ang. w.r.t. T1 (Relax, 600N)

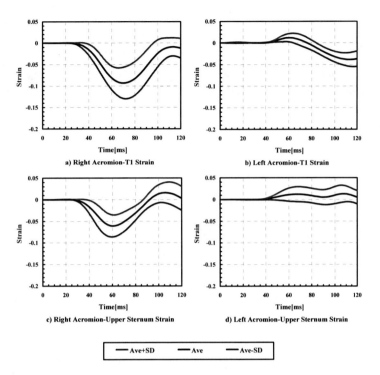

a) Right Acromion-T1 Strain

b) Left Acromion-T1 Strain

c) Right Acromion-Upper Sternum Strain

d) Left Acromion-Upper Sternum Strain

Ave+SD — Ave — Ave-SD

Fig. 8 Shoulder strain at the sternum upper-end and the right or left acromion

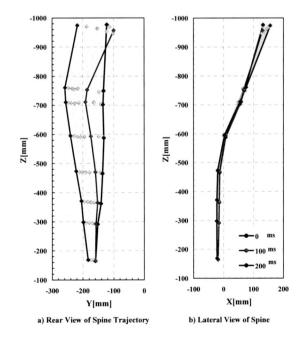

a) Rear View of Spine Trajectory

b) Lateral View of Spine

Fig. 9 Views of spine trajectories by the VICON

DISCUSSION

Effect of Differences in Muscle Functions of Head/Neck/Torso Impact Responses

The average value of T1 acceleration for tensed/relaxed muscle conditions with an impact load of 600N is shown in Figure 10. The maximum of T1 acceleration becomes 60m/s^2 in the case of the relaxed muscle condition. On the other hand, the maximum of T1 acceleration becomes 50m/s^2 in the case of tensed muscle condition. Suppression of T1 acceleration under the different muscle conditions was observed. Generally in the case of tensed muscle condition, impact force is transmitted easily to the T1 region when stiffness of the shoulder structure increase. The T1 acceleration rapidly increases according to this phenomenon, and its value becomes greater. Furthermore, effective mass of the shoulder region which was impacted showed higher stiffness. As a result, T1 acceleration decreased and there was

an increase in muscle tone, thus, impact force acting on the upper neck is reduced (Fig.11) at an average of 15%. Furthermore, in the case of tensed muscle condition, the motion of head rotation is suppressed so that the stiffness of neck structure itself is increasing (Fig.12 and Fig.13). According to this result, it can be said that the impact motion responses of head/neck/torso easily change based on the different state of muscle conditions.

Effects of Differences between Male & Female on Head/Neck Impact Responses

The maximum of T1 acceleration and head C.G. acceleration under the relaxed muscle condition with impact force of 600N (three males and two females) are shown in Fig.14 and Fig.15. As for the head C.G. acceleration, female subjects showed greater value than male subjects. For the T1 acceleration, no difference was seen between male and female. As a result, even if the force level in lateral impact is

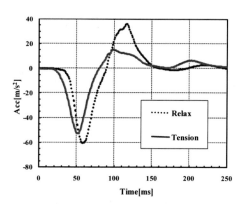

Fig. 10 Comparison of T1 acceleration between relaxed and tensed muscle conditions

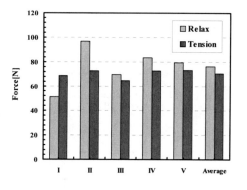

Fig. 11 Comparison of neck shear force (Fy) between relaxed and tensed muscle conditions

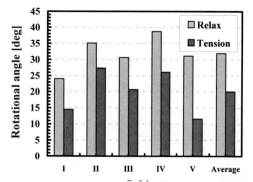

Fig. 12 Comparison of head rot. ang (Y) w.r.t. T1 between relaxed and tensed muscle conditions

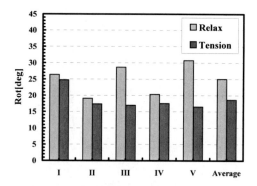

Fig. 13 Comparison of head rot. ang (Z) w.r.t. T1 between relaxed and tensed muscle conditions

almost same, difference of the head/neck motion is observed between male and female. This could probably be due to the smaller head mass of females compared to males. Furthermore, it is thought that the structure size of cervical vertebrae of a female being small might be the cause. The maximum head displacement relative to T1 in the Y-axis and the maximum head rotational angle relative to T1 in the X-axis under the relaxed muscle condition with impact force of 600N are shown in Fig.16 and Fig.17, respectively. The displacement and rotation of head/neck for both male and female were suppressed by doing muscle tone. However, the displacement and the rotation of the head/neck for two female subjects were greater than those of male values under the tensed condition, whereas no difference was observed between male and female under the relaxed condition. According to this situation, it is suggested that under tensed muscle condition, stronger muscular strength of males in general can greatly depress the head/neck/torso motions. On the other hand, females who have weak muscular strength, has difficulty in suppressing the global motion. According to the difference in responses of head/neck/torso between males and females, it is supposed that there will be a higher risk of neck injury for females.

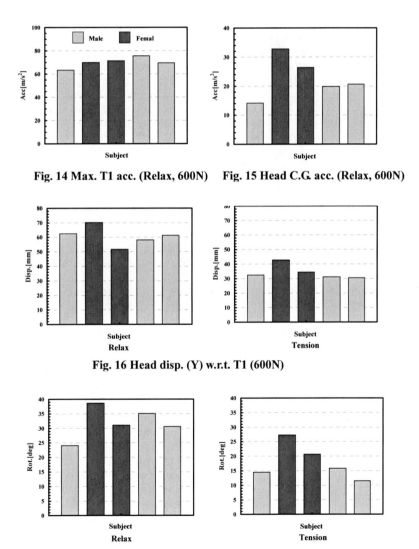

Fig. 14 Max. T1 acc. (Relax, 600N)   Fig. 15 Head C.G. acc. (Relax, 600N)

Fig. 16 Head disp. (Y) w.r.t. T1 (600N)

Fig. 17 Head rot. ang. (X) w.r.t. T1 (600N)

Effect of Shoulder Structural Deformation on Head/Neck Impact Responses

Sabine et al (2002, 2003, and 2004) reported that a difference of motion such as the clavicle and the shoulder blade etc. was clarified in the experimental studies on the PMHS lateral shoulder impacts. In the lateral collision, the impact which went from the shoulder takes the influence of the shoulder structure greatly before reaching neck region when an impact acts on the occupant's shoulder region through the vehicle inside structure such as a door panel. And, a change in the impact energy dispersion of the shoulder region, the impact transmission direction of the torso and so on occurs at the same time. The shoulder structure which influences the motion responses of head/neck in the lateral impact was examined here.

When a lateral impact is imposed to the shoulder region, it is transmitted to the clavicle and thorax, the sternum through the shoulder blade, and it influences the neck region consequently through T1 region (Fig.9). The compression strain between the right acromion and the T1 was greater (Fig.8a), Fig.8c)). This corresponds to the result of the PMHS experiment by Sabine et al (2002, 2003, and 2004). It is not compressed comparatively because the clavicle exists between the acromion and the sternum and it is fixed firmly when an impact is imposed from the lateral direction to the shoulder region. The shoulder blade may slide behind the aperture thoracic superior by the impact, and greatly compress in the acromion and the T1. In other words, the acromion and the aperture thoracic superior though an impact is transmitted directly, and the transmission of the impact is delayed in the acromion and the T1. The rising time of the lateral displacement of shoulder markers were shown in Table 3. Displacement between the sternum top-end and the left acromion almost started at the same time, and the motion of T1 was delayed. This shows a difference in the impact transmission mechanism that the neighborhood of the bone structure on the torso front side such as the sternum and clavicle followed by the movement of the neighborhood of the bone structure on the torso rear side such as T1.

Table 3 Rising time of the lateral displacement of the shoulder markers (Relax, 600N)

Subject	Rising Time of Displacement (ms)			
	Right Acromion	Upper Sternum	T1	Left Acromion
I	8	16	26	14
II	8	28	30	22
III	8	18	22	16
IV	10	14	28	22
V	8	20	28	20
Average	8.4	19.2	26.8	18.8

It is understood that the different motion response was due to the structural difference of the rear and front torso as described above. An impact was introduced to the left acromion directly without deformation between the left acromion and the sternum top-end though the impact transmitted to the top-end of the sternum was transmitted to the left acromion through the clavicle on the opposite impact side. In other words, the left acromion was imposed an impact through the top-end of the sternum, and the left acromion was displaced backward. It can be considered that the strain of the left acromion and T1 showed slight tension at first, and as a result showed compression.

Characteristics of Cervical Vertebral Motions during Lateral Impacts

The head rotation was delayed for about 30ms to the neck, after which, head rotation begins. The rotation of C4 was lower than that of C5 in 35-80ms (Fig.18). It can be considered that the torso moves first, and then the left lateral moment acts to the upper neck as shown in Fig.19. Furthermore, C4/C5 which is the relative rotational motion of cervical vertebrae as shown in the Fig.20 showed a negative value in the early stage of impact. This indicated that the tension of the left cervical vertebral joint in C4/C5 and the compression of the right cervical vertebral joint in C4/C5 occurred. It was estimated that the rotational angle of C1~C3 which can not be analyzed in this experiment will be delayed from that of the lower cervical vertebra, and the rotational angle of the upper cervical vertebra will exceed that of the lower cervical vertebra. The rotation angle of C5 suppressed those of the head and C4, showing that the cervical spine has a bi-phases curvature form such as an S-curve. An S-shape form with relative left extension of upper cervical vertebra and relative right flexion of lower cervical vertebra was presented concretely, and it can be considered that the right bending moment was acting on upper cervical vertebra and the left bending moment was also acting on the lower cervical vertebra. This phenomenon shows the same type of cervical vertebral motions causing the whiplash during a rear-end collision.

Moreover, tension on the left side of the cervical vertebra always shows an increase tendency as shown in Fig.21. On the other hand, compression on the right side of the cervical vertebra (C4/C5~C7/T1 in 90-120ms) shows a constant value (Fig.22). The rotation angle of the cervical vertebra was depressed by restricted motion of the facet joint on the right of cervical vertebra, and it can be considered that larger compression acts on this area at the latter half of impact. The compression of the intervertebral disk decreased with the elasticity of the neck itself due to a decrease in compression and the axial force applied on the upper neck shifted to tension force after 130ms (Fig.23).

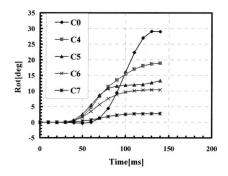

Fig. 18 Vertebral angle w.r.t. T1

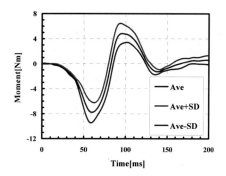

Fig. 19　Neck Moment

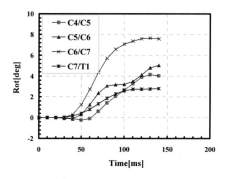

Fig. 20 Vertebral angle w.r.t. lower vertebra

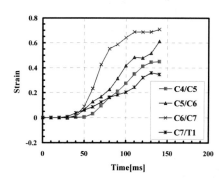

Fig. 21 Left side strain of intervertebral disc

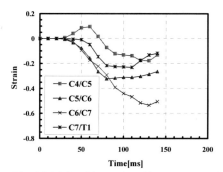

Fig. 22 Right side strain of intervertebral disc

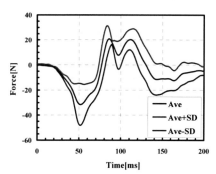

Fig. 23 Neck force Z

325

CONCLUSIONS

Using five healthy male and three healthy female adults as human volunteers, experiments on the head/neck/torso impact responses and the cervical vertebral motions upon lateral inertia impact have been conducted, with the impact forces set at 400 N, 500 N and 600 N, respectively. The findings obtained from the above are as follows:

Effect of Differences in Muscle Functions of Head/Neck/Torso Impact Responses

The suppression of head/neck/torso motions was greater in tensed muscle than in relaxed condition. The T1 displacement (18%) and the head displacement (48%) relative to T1 were more suppressed in the tensed condition than in relaxed condition.

Effects of Differences between Male & Female on Head/Neck Impact Responses

Regardless of the state of muscle tension, the displacement of acromion with respect to the first thoracic vertebra (T1) tends to be greater for male than for female subjects. As female shoulders tend to have less flexibility against impact than male, the female cervical vertebral motions are likely to show longer lateral extensions than male. It is suggested that the differences in muscle responses should be taken into account, in addition to the differences in shoulder anatomical structures, as marked differences between male and female.

Effect of Shoulder Structural Deformation on Head/Neck Impact Responses

When an impact is applied to a shoulder, the head/neck/neck impact responses become different even if the magnitude of impact on the torso is the same. Thus, it is suggested that the differences in head/neck/torso motions are caused by the differences in shoulder anatomical shape and/or front-rear structural differences. A shoulder has high three-dimensional flexibility and a wide range of movability, owing to the gleno-humeral and sternoclavicular joints, which facilitate vertical and lateral motions against lateral impacts. However, the shoulder movability would be restricted, if the direction of the lateral impact roughly aligns with the line connecting the acromio-clavicular joint and the sternoclavicular joint - i.e., the longitudinal direction of the clavicle.

Characteristics of Cervical Vertebral Motions during Lateral Impacts

Cervical vertebral rotations all started at 35 ms, but the time required to reach the peak rotation increased toward the upper vertebrae, with C7 and T1 peaking at 120 ms and the final peak in the head at 120 ms. At around 35-80 ms, the rotation angle of C5 surpassed those of the head and C4 showing that the cervical spine was bending into an S-curve. This phenomenon shows the same type of cervical vertebral motions causing whiplash during a rear-end collisions. Also, extreme compression was at work in the vertebral disc and/or the facet joint in C6/C7 and C7/T1, suggesting a high probability of injury occurring in the neck.

AKNOWLEGMENTS

The authors would like to give special thanks to the volunteers who understood the aim of the research and its contribution to current social needs.

REFERENCES

Compigne S., Caire Y., and Verriest J., Three Dimensional Dynamic Response of the Human Shoulder submitted to Lateral and Oblique Glenohumeral Joint Impacts, Proceedings of IRCOBI Conference, Sept. 2002, pp.377-379

Compigne S., Caire Y., Quesnel T., and Verriest J., Lateral and Oblique Impact Loading of the Human Shoulder 3D Acceleration and Force-Deflection Data, IRCOBI Conference, Lisbon (Portugal), September, (2003), pp. 265-279

Compigne S., Caire Y., Quesnel T., and Verriest J., Non-Injurious and Injurious Impact Response of the Human Shoulder Three-Dimensional Analysis of Kinematics and Determination of Injury Threshold, Stapp Car Crash Journal, Vol.48 (November 2004), pp.89-123

Hell W., Hopfl, F., Langweider K., and Lang D., Cervical Spine Distortion Injuries in Various Car Collisions and Injury Incidence of Different Car Types in Rear-end Collisions, Proceedings of IRCOBI Conference, 2003, pp.193-206

Kraft M., Kullgren A., Ydenius A., Tingval C., Influence of Crash Pulse Characteristics on Whiplash Associated Disorders in Rear Impact-Crash Recording in Real-Life Crashes. Journal of Crash Prevention and Injury Control, 2002

Narayan Yoganandan, Srirangam Kumaresan, Frank Pinter, Review Paper, Biomechanics of the cervical spine Part2. Cervical spine soft tissue responses and biomechanical modeling, Clinical Biomechanics 16, 2001, pp.1-27

Ono K., Kikuchi A., Nakamura M., Kobayashi H., and Nakamura N., Human Head Tolerance to Sagittal Impact - Reliable Estimation Deduced from Experimental Head Injury Using Subhuman Primates and Human Cadaver Skulls, 1980 Transactions of the Society of Automotive Engineers, SAE Paper No. 801303, 1980, pp.

3837-3866

Shigeki Ito, Paul C. Ivancic, Manohar M. Panjabi, and Bryan W. Cunningham,, Soft Tissue Injury Threshold During Simulated Whiplash A Biomechanical Investigation, SPINE Volume29, Number9 , 2004, pp979–987

Shrawan Kumar, Robert Ferrari, and Yogesh Narayan, Looking Away From Whiplash: Effect of Head Rotation in Rear Impacts, SPINE Volume30, Number7, 2005, pp760–768

WHO/CIOMS proposed guidelines for medical research involving human subjects, and the guidelines on the practice of ethics committees published by the Royal College of Physicians, The Lancet, November 12, 1988, pp.1128-1131

973340

Cervical Injury Mechanism Based on the Analysis of Human Cervical Vertebral Motion and Head-Neck-Torso Kinematics During Low Speed Rear Impacts

Koshiro Ono
Japan Automobile Research Institute

Koji Kaneoka
University of Tsukuba

Adam Wittek and Janusz Kajzer
Nagoya University

ABSTRACT

Twelve volunteers participated in the experiment under the supervision of Tsukuba University Ethics Committee. The subjects sat on a seat mounted on a newly developed sled that simulated actual car impact acceleration. We selected impact speeds (4, 6 and 8 km/h), seat stiffness, neck muscle tension, and alignment of the cervical spine for the parameter study of the head-neck-torso kinematics and cervical spine responses. The effects of those parameters were studies without headrest. The muscle activity was measured with surface electromyography. The cervical vertebrae motion was recorded by cineradiography (90 frames/s X-ray) and analyzed to quantify the rotation and translation of cervical vertebrae at impact. Furthermore, the motion patterns of cervical vertebrae in the crash motion and in the normal motion were compared.

Subject's muscles in the relaxed state did not affect the head-neck-torso kinematics upon rear-end impact. The ramping-up motion of the subject's torso was observed due to the inclination of seatback. An axial compression force occurred when this motion was applied to the cervical spine, which in turn developed the initial flexion, with the lower cervical vertebral segments extended and rotated prior to the motions of the upper segments. Those motions were beyond the normal physiological cervical motion, which should be attributed to the facet joint injury mechanism. Furthermore, the more rigid the seat cushion, the greater was the axial compression force applied to the cervical spine. On the other hand, the torso rebounding caused by the softer seat cushion tended to intensify the shearing force applied to the upper vertebrae. It was also deduced that the difference in alignment of the cervical spine affected the impact responses of head and neck markedly. Also the kyphosis of the cervical spine caused the upper travel of rotation center of the lower cervical vertebral segments and its rotational motion resulting in a higher neck injury incidence. Based on the differences in the alignment of the cervical spine between male and female occupants, it is also pointed out that the female neck injury incidence tends to become higher than that of male, as the female cervical spines take the kyphosis position more often than the male cervical spines.

INTRODUCTION

It is believed that the extension of cervical spine would not exceed the normal physiological motion range, and the hyperextension of cervical spine would not occur as long as the occupant is using a headrest upon rear impact. It is reported, however, that neck injuries caused by vehicle rear-end collisions are still occurring with high incidence rates. Neck injuries have been on the upward trend in Japan, and automobile occupants who sustained neck injuries by rear-end impacts increased from 42.7% in 1991 to 44.0% in 1994 (Ono and Kanno, 1997). The effectiveness of headrest is reported to be 18% or so in the USA (O'Neill et al., 1972), roughly 14% for the adjustable type headrest and 24% or so for the fixed type in Sweden (Nygren et al., 1984, 1985). It is reported in a recent investigation conducted by Morris & Thomas in 1996 that the effectiveness is not statistically clear. It is also reported that the incidence of neck injuries is higher for female occupants than for male occupants (Carlson et al., 1985; Lovsund et al., 1988; Muser et al. 1994). According to such incidence rates of neck injuries, it is pointed out that headrests including current seat systems do not have sufficient functions and structures for the suppression of neck injuries (LTB van Kampen, 1993). The improper adjustment of relative positions of occupant's head and headrest, difference in seating position are also pointed out as other factors of neck injuries (Foret-Bruno et al.,1991, Olsson et al. 1990).

The head and torso of an occupant are subjected to various kinds of outer forces upon rear impact, which are related with the incidence of neck injuries. A number of experimental studies on the factors that may affect the neck injuries using anthropomorphic dummies have been conducted so far. According to those studies, Melvin et al. (1972), classified the factors of neck injuries into four main potential kinematics of the car occupant's body, head and neck; namely, 1) head displacement, rotation and acceleration, 2) difference motions of the head and torso into the deflected seatback, 3) occupant ramping up the seatback, and 4) occupant rebound. However, most of experimental studies on neck injury mechanism conducted in the past were limited to cadaveric experiments done at high impact speeds or those on frontal collisions, whereas studies on rear-end collisions were quite few. It is said that the clarification of correlations among the neck muscular response, motions of cervical vertebrae, intervertebral disc and intervertebral articular injury is necessary for further pursuit of the injury factors including those on severer disabilities from neck injuries as well as those of minor neck injuries. A number of experimental studies using volunteers (Ono and Kanno 1993; McConnell et al. 1993; Szabo et al. 1994; McConnell et al. 1995; Geigel et al. 1994, 1995; Szabo and Welcher 1996) or anthropomorphic dummies (Foret-Bruno et al. 1991; Viano 1992) have been also carried out so far, including some reports that analyzed the cervical vertebral motions (Matushita et al. 1994; Yoganandan and Pinter 1997), but none of them has compared such motions with normal physiological motions for the clarification of neck injury mechanism on rear-end impacts.

In this regard, we decided to conduct experiments using volunteers to simulate actual car rear-end impacts at low speeds. X-ray cineradiography was used in order to compare and analyze the motions of cervical vertebrae upon impact and during physiological conditions, in an attempt to clarify the neck injury mechanism according to the characteristic motions of cervical vertebrae during impact. Furthermore, the effect of volunteers' muscular tension on the head and neck motions; effects of seat cushion characteristics on the head, neck and torso; the differences in head-neck impact responses due to the difference in alignment of the cervical spine in terms of seating position, and how the differences in the alignment of the cervical spine for male and female would influence the incidence of neck injuries will also be discussed in this paper.

METHODS OF EXPERIMENTS

Volunteers and Informed Consent - Twelve healthy male volunteers without history of cervical spine injury participated in this study. Their average age was 24 years, and it was confirmed through X-rays that they had no degenerative cervical spine. The study protocol was reviewed and approved by the Tsukuba University Ethics Committee (Notice No.28), and all volunteers submitted their informed consent in writing according to the Helsinki Declaration (WHO/CIOMS Guidelines 1988).

Sled Apparatus for Simulation of Rear End Impact - The newly developed sled was designed to overcome such restrictions as limited space for the installation of a sled system, limited field of vision of cineradiography, and the necessity to transfer the sled system quickly for emergency clinical use of cineradiography. This was installed at the Tsukuba University Hospital. The system was capable of sliding the sled freely on a 4-meter long rail angled at 10 degrees and colliding the sled against a damper with a maximum speed of 9 km/h. The sled was designed according to the actual car rear-end impact experiments conducted in the past, and an oil shock damper was installed to simulate accelerations applied to cars colliding with other cars. A standard seat sold in the market or a rigid seat made of wood was mounted on the sled. A grip handle was installed in front of such a seat to reproduce the extent of each subject's muscular tension. A schematic diagram and the specifications of the sled system are shown in Figure 1.

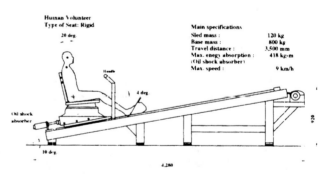

Figure 1. Sled test apparatus and main specifications

Cineradiography - The column of radiation probe of the cineradiographic system (cine-system; Angiorex made by Toshiba Medical Inc.; cine-camera: Arritechno 35, NAC Inc.) can be rotated 180 degrees on a horizontal plane, and the probe itself can be also rotated by ± 180 degrees. The cineradiographic range is 30 cm x 30 cm, and the probe position can be adjusted vertically in the range of 105 to 130 cm as shown in Figure 2. A position adjuster capable of

Figure 2. Cineradiography system and test set-up

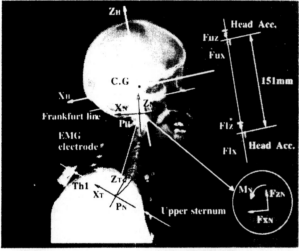

X_H : X axis of Head	C.G.: Center of Gravity of Head
Z_H : Z axis of Head	P H : Intersection point on head-neck joint
X_N : X axis of Neck	P N : Intersection point on neck-torso joint
Z_N : Z axis of Neck	U.S.: Upper sternum point
X_T : X axis of Torso	Fux: Frontal upper x-axis accelerometer
Z_T : Z axis of Torso	Fuz: Frontal upper z-axis accelerometer
M_N : Neck Moment	Flx: Frontal lower x-axis accelerometer
F_{XN} : Neck Shear Force	Flz: Frontal Lower z-axis acceleometer
F_{ZN} : Neck Axial Force	

Figure 3. Coordinate system and lateral view of the head/neck/torso with mounted accelerometers, EMG electrodes, VTR targets and the marked points for X-ray picture

positioning the volunteer's neck within those ranges on impact was installed on the sled apparatus. The cervical spine motion was recorded by cineradiography at 90 frames per second and the dose of exposure was 0.073 mG per frame. Approximately 25 frames were recorded for one crash motion.

Sled Acceleration Measurement - The sled acceleration was measured with three-axial accelerometers installed on the sled floor, while the sled impact speed was measured with phototubes. The data processing and analysis of the acceleration measurement as described herein were done in accordance with SAE J211.

Head Acceleration Measurement - The subject's motion in this experiments can be assumed as the two-dimensional motion of the X-Z plane, and four-channel accelerometers were used for the measurement of head acceleration, since the six-degree of freedom component measuring method (Ono et al. 1980; Ono and Kanno 1993) was applied. The shear and axial forces and the bending moment acting on the neck upper region (C1) were measured with this method. The fixture shown in Figures 2 and 3 was fabricated for the installation of accelerometers to the subject's head. A tooth form made of a dental resin molded specifically for each subject was set at the lower portion of the fixture while a fastening tape was attached at the upper portion to fix the subject's head at both the upper and lower portions. Head dimensions and locations of accelerometers installed on each subject were determined by means of a dimensional measurement using X-ray films as shown in Figure 3.

Location of Head C.G. - The location of anatomic center of gravity of the head was assumed by the determination methods which are reported by Walker et al. (1973) and Beier et al. (1980). The position of the head C.G. was located 5 mm in front of the external auditory meatus and 20 mm above the Frankfurt line which connects the lower orbital margin and the center of auditory meatus.

Chest Acceleration Measurement - Tri-axial accelerometers were installed on the front chest around the sternum region. The accelerometers were adhered onto an aluminum plate attached to the front chest with rubber bands.

Electromyography - In contrast to experiments using cadavers, in experiments using volunteers, the muscular responses tend to vary due to the mental states of the subjects. As this effect cannot be ignored, we have conducted experiments under two conditions of muscle tension - when they were relaxed, and when they intentionally became tense - in order to find out how the muscle tension affects their head and neck motions. Muscle activities was measured by means of surface electromyogram compared with the cineradiography. EMG electrodes were attached onto the skin over sternoclaidomastoid muscles, paravertebral muscles and trapezius muscles on the subject's left side, and sternoclaidomastoid muscles on the right side. The electrodes with diameter of 5 mm were arranged as bi-polar electrodes with a distance of roughly 2 cm between the electrode centers. The EMG signal was sampled at 5000 Hz and then recorded on OVDAS (On Vehicle Data Acquisition System) data

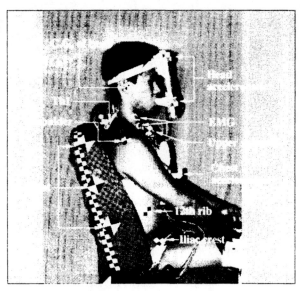

Figure 4. Lateral view of the head/neck/torso with mounted accelerometers, EMG electrodes, targets for high-speed video

acquisition system. Processing of the EMG signal was done with MATLAB 5 program (Math Works, 1996). In order to remove artifacts, the EMG signal was high-pass and low-pass filtered at frequencies of 8 Hz and 450 Hz, respectively. These frequencies were selected from values recommended by Basmajian and De Luca, (1985).

Visual Motion Analysis of Head-Neck-Torso Kinematics and Cervical Vertebrae - For the analysis of subject's motions upon impact, target marks were adhered over the head accelerometers, Th1 accelerometer and chest accelerometer, near the position of 5 mm below the auditory meatus around the center of gravity of the subject's head, over the surfaces of each shoulder (acromion of scapula), the 12th rib (on the lateral side), the iliac crest and the upper sternum (Figure 4). For the analysis of relative motions of the head, neck and torso, small lead balls of about 3 mm diameter were used as the reference points for the determination of the head-neck joint and the neck-torso joint in X-ray photographs (Figure 3). They were adhered between individual target marks and the surfaces of auditory meatus, Th1 and the upper sternum. Target marks were also adhered to the upper and lower portions of seatback side, and photographs were taken with high-speed video cameras (made by MEMORECAMNAC Inc.) at the speed of 500 frames per second. The photographed images were incorporated in ImageExpress (NAC Inc.) and analyzed.

Definitions of Head, Neck and Torso Angles, Head-Neck and Neck-Torso Joints - The head-neck joint is defined as the intersection point of the extended line from the posterior

tangent of cervical vertebra C2 and the skull base tangent parallel to the Frankfurt line on the head. The neck-torso joint is defined as the intersection point of the extended line of the curve formed by the posterior of vertebra and the line connecting the small lead balls on the surfaces of Th1 and the upper sternum. This point is located around the upper end of Th1. The head-neck link is defined as the line connecting the head-neck joint and the neck-torso joint, while the inclination of the line is defined as the neck rotational angle (Figures 3 and 4). The rotational angle of upper torso portion is defined as the torso rotational angle formed with the inclination of the line connecting Th1 and the upper sternum. The head rotational angle is defined as the inclination of the line passing through the location of head C.G. parallel to the Frankfurt line.

Cervical Vertebral Motion Analysis - The cervical vertebral images taken by cine-cameras were digitized and analyzed. Although it is desirable to analyze the cervical vertebral motions over the entire range of C1 to Th1 in reference to the first thoracic spine, the analysis was done mainly in the range of C2 to C6 due to the limited cineradiographic field of vision. Templates suitable for the shapes of individual cervical vertebrae were produced as shown in Figure 5. This was done to fit them precisely over the individual cervical vertebrae and the spinous process which should move sequentially with time. Based on these, the system of coordinates of the inferior anterior and posterior points were determined accordingly. From those coordinate values, the angles from the horizontal plane and the vertical translations of individual vertebral segmental bodies were calculated. It was decided to represent the vertical translation by taking the midpoint between the inferior anterior and posterior points of each vertebral body. The motions of the entire cervical vertebrae were represented by the changes in the relative rotational angle and translation of each cervical vertebra from the sixth cervical vertebra. The direction of cervical spine extension was decided as the positive rotational direction, and the upward motion was designated as the positive vertical distance.

There were some cases in which analyzable images could not be obtained from the impact experiments due to the limited cineradiographic field of vision. In such cases, the obtainable images were deemed as the initial values.

Experiment Matrix - Four series of experiments were conducted on 12 volunteers under different conditions without headrest in every series. Due to the overlapping of the same subjects used in different experimental series, the total number of subjects shown in the tables is not 12. The series-1 experiments used ten subjects to find the variation of motions among individual subjects under the same impact conditions

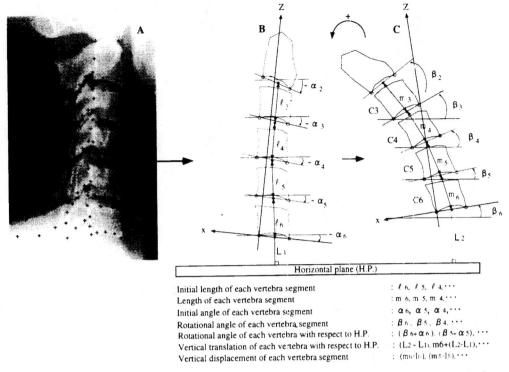

Initial length of each vertebra segment   : $\ell_6, \ell_5, \ell_4, \cdots$
Length of each vertebra segment   : $m_6, m_5, m_4, \cdots$
Initial angle of each vertebra segment   : $\alpha_6, \alpha_5, \alpha_4, \cdots$
Rotational angle of each vertebra segment   : $\beta_6, \beta_5, \beta_4, \cdots$
Rotational angle of each vertebra with respect to H.P.   : $(\beta_6 + \alpha_6), (\beta_5 + \alpha_5), \cdots$
Vertical translation of each vertebra with respect to H.P.   : $(L_2 - L_1), m_6 + (L_2 - L_1), \cdots$
Vertical displacement of each vertebra segment   : $(m_6 - \ell_6), (m_5 - \ell_5), \cdots$

Figure 5. Template method and measurement items for vertebral motion analysis

Table 1. Series 1 - Primary Experiments

	Impact Velocity	Sitting Position	Type of Seat	Headrest
10 Adult Male	4 km/h	Standard	Standard	Without

Standard sitting position : seatback angle 20 degrees from the vertical line

Table 2. Series 2 - Muscle Tension Experiments

	Impact Velocity	Sitting Position	Type of Seat	Tension	Headrest
3 Adult Male	6 km/h	Standard	Rigid	Without	Without
				With	

Standard sitting position : seatback angle 20 degrees from the vertical line

Table 3. Series 3 - Neck Alignment Experiments

	Impact Velocity	Sitting Position	Type of Seat	Neck Alignment	Headrest
1 Adult Male	6 km/h	Standard	Rigid	Flexion	Without
				Neutral	
				Extension	

Standard sitting position : seatback angle 20 degrees from the vertical line

Table 4. Series 4 - Seat Stiffness Experiments

	Impact Velocity	Sitting Position	Type of Seat	Headrest
3 Adult Male	4 6 km/h 8	Standard	Rigid	Without
			Standard	

Standard sitting position : seatback angle 20 degrees from the vertical line

(4 km/h; with the standard seat - Table 1). Three subjects were used in the series-2 experiments (Table 2) to clarify the differences in level of muscular tension on the subjects' motions at an impact speed of 6 km/h . One subject was used in the series-3 experiments (Table 3) to verify the differences in initial relative positions of head and neck alignment on the subject's motions at the impact speed of 6 km/h. Three subjects were used in the series-4 experiments (Table 4) to find the differences in seatback characteristics (standard or rigid) on the subjects' motions. Visual motions by high-speed video of each subject could not be recorded in experiments using cineradiography. Therefore, experiments were repeated under the same impact conditions. Namely, the first experiment was conducted to record the cervical vertebral motions by means of cineradiography, while the second experiment was done to record high-speed video visual motions.

RESULTS

Subject's Motions and Responses of Head, Neck and Torso - The impact phenomenon as the typical case at speed 8 km/h, with rigid seat, and without head restraint can be described by the following: 1) the motions observed by high speed video, 2) acceleration at each region of subject, moment, forces on the neck and motions of the head, neck and torso, 3) cervical vertebral motions observed by cineradiography, and

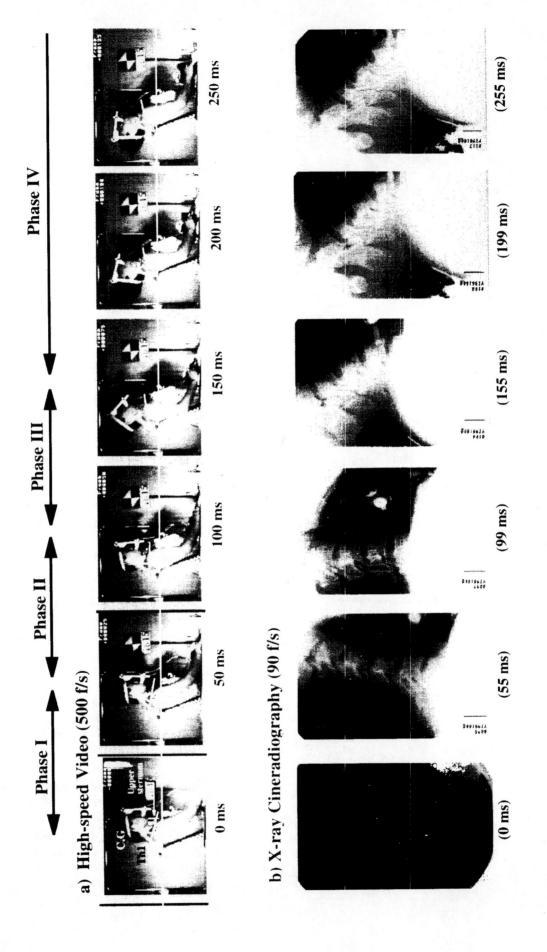

Figure 6. Head-neck-torso motion by high-speed video and cervical vertebrae motion by X-ray cineradiography

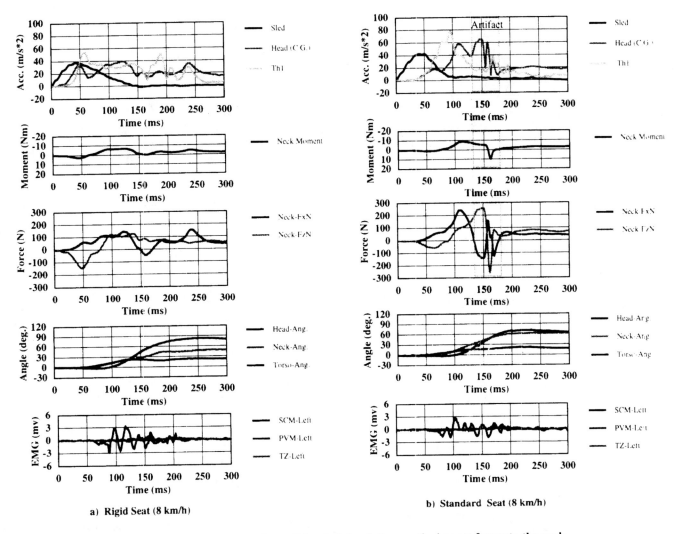

Figure 7. Time-histories of accelerations of the sled, head, thorax, the impact forces to the neck, and the examples of EMG (SCM, PVM, TP - Left) on the same subject

4) the electromyographic response. Subject's motions and head-neck-torso responses are divided here in four phases as present in Figure 6. Figure 6 a) shows the sequential visual motions of the head, neck and upper torso of the subject photographed with high speed video. Figure 6 b) shows sequential images of cervical vertebral motions taken by cineradiography under the same impact conditions.

Figures 7 a) and 7 b) show the time histories of resultant accelerations of the sled, head and Th1, the moment and the forces acting on the neck (around C1), and rotational angles of the head, neck and torso taken by X-ray photography and VTR under the conditions of rigid and standard seats. The compressive axial force on the neck is considered negative in Figures 7 a) and 7 b). Figures 8 and 9 show the trajectories of the neck-torso joint analyzed by high-speed video based on the reference points on the X-ray photography. The reading

error of cervical vertebral images taken by cineradiography is shown with the standard deviation of 10 measurements taken on the same cervical vertebra of the subject. The mean value of deviation was 0.24 mm.

Phase I (0-50 ms, Initial Response Phase) - 1) The subject's back starts to be pushed against the seatback. 2) About 25 ms after impact, the head and Th1 start to accelerate around the same time with the head slightly faster than Th1 probably by initial flexion of the cervical vertebrae due to the upward-backward rotaion motion of the torso. The sled acceleration reaches maximum around 45 ms. 3) No significant motion of head or neck is found in this phase. However, the straightening of the spine is observed and the low level of the neck bending moment is also observed in flexion phase. 4) Neck muscular response is not found.

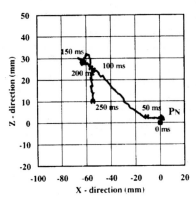

Figure 8. Motion of neck-torso joint (Rigid Seat)

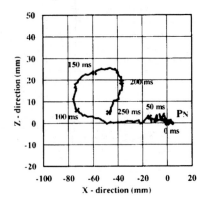

Figure 9. Motion of neck-torso joint (Standard Seat)

Phase II (50-100 ms, Principal Neck Axial Force Motion, Flexion Phase) - 1) The subject's upper torso is pushed against the seatback and the extension of spine reaches maximum. The head is moving backward in parallel to the torso due to its inertia. The spine extension results in the arching of the torso. The neck which is the link between the head and the torso starts showing an S-shape formation. 2) The ramping-up motion and the backward rotation of torso continue. The neck rotation as the head-neck link becomes faster and greater than the head rotation. The axial compression force applied to the cervical spine shows the maximum value around 50 ms due to the torso ramping-up motion, and the shear force starts increasing gradually. The Th1 acceleration shows the maximum value around 60 ms, while the neck moment in extension shows the maximum value around 100 ms and starts showing a plateau. 3) A lower cervical vertebra (C6) reaches the phase of initial flexion. The rotation of upper cervical vertebrae (C3, C4 and C5) starts later on. 4) In relation with these neck link motions, the discharge of sternoclaidomastoid muscles starts around 70 ms.

Phase III (100-150 ms, Principal Neck Shear Force Motion, Flexion-Extension Phase) - 1) The torso is going to be pushed against the seatback, and the arched rotation of the upper torso shows the maximum value around 130 ms, and the principal extension rotation of the head starts. The ramping-up of torso becomes maximum around 150 ms and it can be observed to coincide with a significant extension of the spine. 2) Acceleration of the head becomes maximum around 130 ms. The neck-torso joint motion is characterized by the ramping-up-and-down motion of torso around that time. The head rotational angle starts becoming larger than that of the neck, and the shear force of neck becomes maximum. 3) About 100 ms after impact, the sixth cervical vertebral extension angle becomes maximum, and the upward translation also becomes maximum. The upper cervical vertebrae follow the extension of C6, and the extension of

aligned cervical vertebrae starts. 4) The discharge of sternoclaidomastoid muscles, paravertebral muscles and trapezius muscles continues.

Phase IV (150 ms, Final Response Phase: Maximum Extension Phase) - 1) The rotational extension angles of the head and neck become maximum around 250 ms, then they start resuming the original positions thereafter. 2) Head acceleration remains roughly constant after 150 ms, but the torso ramping-up motion continues. This ramping-up motion influences the Th1 acceleration, neck bending moment, and shear and axial compression forces, but this effect gradually becomes smaller. 3) Individual vertebra rotates while maintaining roughly the same extension alignment. 4) The muscular discharge of the neck disappears around 250 ms.

Experiments with Standard Seat - Characteristic phenomena of the experiments with the standard seat differ from the phenomena which was described above. The accelerations, head, neck and torso angles and neck forces are shown in Figure 7 b). However, the subject's teeth force gripping the fixture of accelerometers slackened in the time between 130 ms and 180 ms, which was found as an artifact. Compared with the experiments using the rigid seat, the torso motions in general and the rebound motion in particular are greater. The head rotational angle is smaller, and the axial compression force is also smaller but the timing for the axial compression force to reach the maximum value is later in the experiments using the rigid seat. The timing to reach the maximum value of shear force is also later than in the case with the rigid seat, but the maximum value itself tends to become larger.

Motions and Muscular Responses of Head and Neck - For experiments with neck muscles (Table 2) relaxed before impact, the maximum head extension angle was in a range of 40-56 degrees (Figure 10a. and 10b.). The average value of

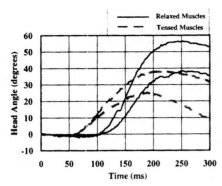

Figure 10a. Head extension angle-time histories

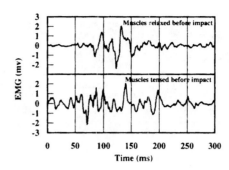

Figure 10b. Typical example of Sternoclaidmastoid
Muscles EMG-time histories

the maximum head angle was about 50 degrees, and standard deviation was 7 degrees. When the subjects tensed their muscles before impact, the maximum angle ranged from 25 to 40 degrees . Thus, pre-impact tension of the neck flexors decreased the maximum angle of the head extension by about 30-40%. Furthermore. when the subjects activated their muscles before impact, the head and the neck began to rotate sooner than for the relaxed muscles. Two explanations for this phenomenon are possible. The first is that the pre-impact muscle tension increased stiffness of joints connecting the torso and the head-neck complex. Accordingly, "coupling" between the head and the torso motion increased. The second explanation is that pre-tension of the extensor muscles could tend to pull the head back sooner. Electromyographic activity of the paravertebral neck extensor and trapezius muscles was low even when the subjects were asked to do tensed runs. For this reason electromyographic responses are analysed only for the sternocleidomastoid muscle (Figure10b and Figure 11). The start time of the sternoclaidomastoid discharge was in a range of 76-93 ms. The average start time was 79 ms, and standard deviation was 9 ms. Thus, muscle activity started not sooner than about 100-180 ms before the head extension angle reached its maximum (Figure 11).

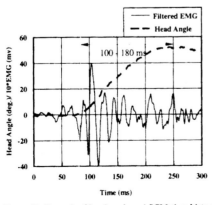

Figure 11. Example of head angle and SCM-time histories
from neck muscles relaxed before impact

Effect of Difference in Alignment of the Cervical Spine - Differences in seating posture, seatback inclination and positions of head-neck are factors that affect the load applied to the cervical spine. The seatback inclination was set at the standard sitting position in this study, while the neck was set at the lordosis, neutral and kyphosis positions, respectively to find out how the difference in alignment of the cervical spine would affect the loads applied to the head-neck kinematics (Table 3). As for the cervical spine alignment, the position of the neck where the head becomes roughly vertical as shown in Figure 12 was deemed as the neutral position, the position by which the head inclines forward by 30 degrees as the flexion position, and the position by which the head inclines backward by 40 degrees as the extension position. It is found that the cervical alignment becomes kyphosis at the flexion position, roughly linear at the neutral position, and lordosis at the extension position.

Figure 13 shows the head, neck and torso rotational angles caused by the different cervical alignment. These angles are derived from the individual anatomical reference frames. The initial position is adjusted to be 0 degree based on the individual anatomical reference frames. Figure 14 shows relative angles of head-neck rotations, while Figure 15 shows the time-histories of rotational angles of individual cervical vertebral segments at different positions of cervical alignment. The rotational angles of the head and C2 are similar implying that there is no relative rotational motion between the head and C2. Figure 16 shows the time-histories of bending moment, axial compression force and shear force applied to the upper neck in different states of cervical alignment. It should be noted that the computed neck loads shown in Figure 16 include the effect of unknown muscles forces and therefore not net loads. It is found from Figure 14 that the head rotational angle becomes approximately 80 degrees at the flexion position, while it becomes approximately 38 degrees at the extension position. indicating that the head rotational angle becomes larger as the head-neck forward inclination angle becomes larger.

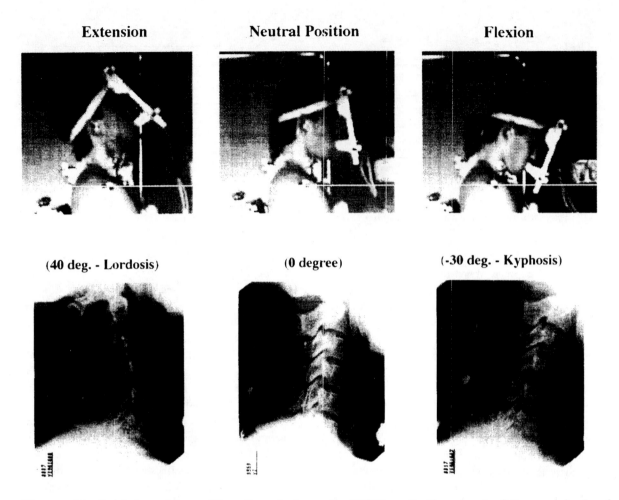

Figure 12. Initial position of head-neck-torso by VTR and alignment of the cervical spine by X-ray views

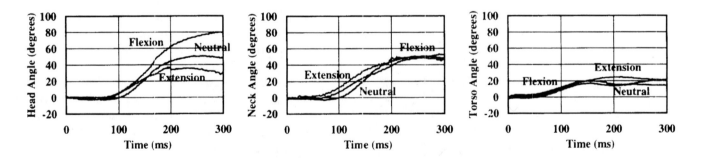

Figure 13. Comparison of head, neck and torso angles for the different alignments of the cervical spine (6 km/h)

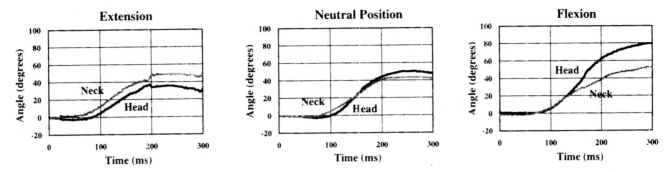

Figure 14. Relative angles of head-neck for the different alignments of the cervical spine

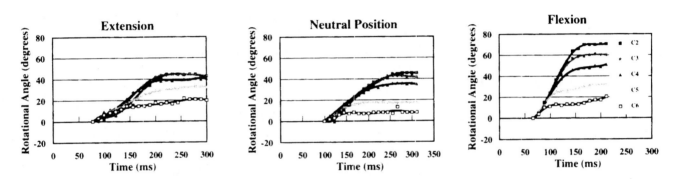

Figure 15. Different patterns of rotational angle of each vertebra (from the horizontal plane) for the different alignments of the cervical spine

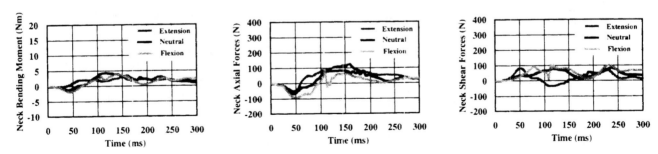

Figure 16. Comparison of neck forces and moments for the different alignments of the cervical spine

(Note: The neck reaction forces contained the influence of unknown neck muscles.)

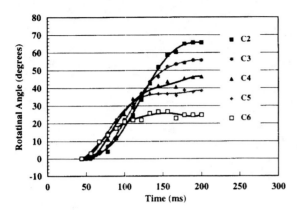

Figure 17. Crash extension motion - Pattern of rotational angle of each vertebra (from the horizontal plane)

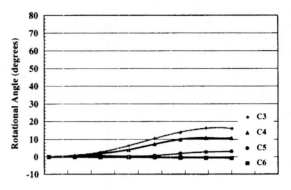

Figure 18. Normal extension motion - Pattern of rotational angle of each vertebra (from the horizontal plane)

On the other hand, no remarkable differences are found in rotational angles of neck and torso by the difference in cervical alignment. In terms of change in head-neck relative angle, however, the variation of relative angle becomes smaller as the initial position changes from flexion to neutral then to extension. In other words, non-alignment of relative motions of head and neck (S-shape formation) becomes greater in the state of flexion, whereas the head rotates in link with the rotation of cervical spine in the state of extension, resulting in smaller loads to be applied to the intervertebral joints and intervertebral discs. Such phenomena can be explained well by the change in rotational angle of each cervical vertebral segment shown by cineradiography in Figure 15. The relative displacement between the head and neck is shown by the variation in the extent of difference in intervertebral rotational angle. By observing such phenomena in terms of variations in bending moment, axial compression and shear forces applied to the cervical spine, the findings may be summarized as follows:

1) Bending moment : In the state of flexion or neutral, a negative moment acts on the cervical spine around 50 ms, but it changes into a positive moment around 80 ms or so. In the state of extension, on the other hand, the moment is in positive direction from beginning. It indicates that the cervical vertebral motion changes from the state of flexion to extension for both lordosis and neutral alignments, while the cervical vertebral motion stays in the state of extension from the beginning in case of kyphosis alignment.

2) Axial compression force: The force reaches maximum roughly at the same time for flexion, extension and neutral. The value itself is the greatest in the state of flexion, and it becomes smaller as the state changes into neutral then to extension in that order. This indicates that the torso ramping-up motion causes more stress in the state of flexion

as in the typical S-shape formation, resulting in the greatest axial compression force.

3) Shear force: The shear force onset time is the earliest for extension, followed by neutral and flexion in that order. Needless to say, it reflects the difference in the timing of torso pushed against the seatback.

Symptoms on Subjects After the Experiments - A clinical doctor had personal interviews with the subjects at the time of MRI prior to the experiment, on the experiment day, one day, one week, and one month after experiment, and handed out questionnaires to the subjects. Presence/absence of any subjective symptoms and details of such symptoms in daily life, if any, were recorded accordingly. Although one subject out of 12 recognized neck discomfort next day of experiment, the discomfort disappeared within a few days. No other symptom was recognized thereafter.

DISCUSSION

Muscle Effects - The current results indicate that only muscles activated before impacts can influence kinematics of the head-neck complex in rear-end impacts at low speeds. In the present study pre-impact activity of the neck muscles reduced the maximum angle of the head extension by about 30-40%. Significant decrease of the head angle due to muscle tension in rear-end impacts has been also reported by Mertz and Patrick (1968; 1971). However, they did not monitor muscle activity. Thus, exact information about the muscle effect cannot be obtained from their study.

The average start time of the neck flexors discharge was measured here to be 79 ms. Since there is about 70-100 ms delay between the EMG onset and the time when muscle force can reach its maximum (Tennyson and King, 1976;

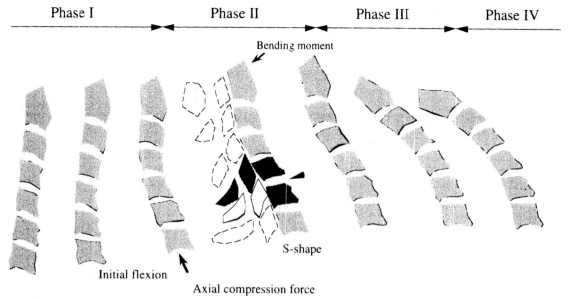

| Phase I | Phase II | Phase III | Phase IV |

Bending moment

S-shape

Initial flexion

Axial compression force

The cervical spine shows flexion in the Phase 1 and S-shape formation in the Phase II. At this phase, the motion segment at the apex of convex curvature (C5-C6) suffers from the stress of intervertebral articulations and stretch of the anterior longitudinal ligament and forced to collide with each articular facet (arrow heads).

Figure 19. Pattern of cervical injury mechanism based on each vertebra motion during rear-end impacts

Herzog, 1994), and the head angle reached its maximum at 200-250 ms after the start of an impact, we conclude that muscle effect on kinematics of the head-neck complex was insignificant when the neck muscles were relaxed before impact.

The start time of muscle discharge measured in the present study agrees well with values reported by Foust et al., (1973) who found that the average start time for neck flexors is 56-92 ms. On the other hand, the current values for the start time of the muscle discharge are about 20-40% lower than the results of Szabo and Welcher (1996). They reported values of 100-130 ms for discharge of sternoclaidomastoid. One explanation of this difference can be that in our study a steady training was performed in order to teach the subjects to keep their muscles in relaxed state prior to impacts. Since this state changed individual reaction patterns of the subjects, it could also affect the muscle discharge start time.

In the current study the muscle effect on the head angle only was analyzed. However, for understanding of a possible muscle effect on the injury mechanisms in rear-end impacts, in-depth knowledge about muscle influence on neck kinematics is needed. Therefore we propose to investigate the effects of muscle tension on movements of cervical vertebrae during impacts for further study.

Cervical Injury Mechanism Deduced From the Cervical Vertebral Motions - A subject's torso shows the ramping-up motion by the inclined seatback during rear-end impact. As the head remains in its original position due to inertia in the initial phase of impact, an axial compression force is apt to be applied to the cervical spine which in turn moves upward and the flexion occurs at about same time. The lower vertebral segments (C6, C5 and C4) are extended and rotated earlier than the upper vertebral segments. Those motions are beyond the normal physiological range of motion. It is found by comparing the motions during crash (Figure 17) with the normal extension motions (Figure 18) of the same subject that the rotational angle pattern in crash is reversed by the pattern in the normal state around 100 ms. The lower the vertebral segment, the larger the rotational angle becomes. That is, the rotational angle between the fifth and sixth vertebral segments is the largest of all. This is an non-physiological motion of the vertebral segments. On the other hand, Bogduk and Marsland (1988), April and Bogduk (1992), Lord et al. (1993) reported that the cervical facet joints are the most common sources of chronic cervical pain which is related to the lower cervical vertebrae. Kaneoka and Ono (1997) hypothesized that such non-physiological motions in rear impact were attributed to the mechanism of facet joint injury (zygapophyseal joint injury). A pattern of the mechanism is shown in Figure 19, based on the cervical

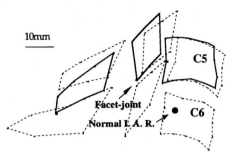

The C5 inferior articular facet surface rotates smoothly around the normal I. A. R.

Figure 20. Normal extension of C5/6.

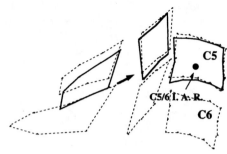

The posterior edge of C5 inferior articular facet shows downward movement toward the C6 superior articular facet surface.

Figure 21. Crash extension of C5/6.

vertebral motion in rear impact. This pattern of the cervical spine motions represents the so-called "whiplash motions of cervical spine" with the S-shape formation of cervical vertebrae.

It can be deduced easily that such whiplash motions are caused by an improper setting of the clearance between the headrest and the head. This bi-phasic curvature is also found in the simulated whiplash experiments using cadavers (Panjabi, 1996; Yoganandan et al., 1996). According to a recent study done by Yang et al. (1997), the stiffness of intervertebral articulations is reduced to half when a compression force is applied. As for the intervertebral articular mobility, it can be pointed out that the rotational angle becomes larger for lower cervical vertebrae than for upper ones, which makes the lower cervical vertebrae easier to move than others. Hence, it can be said that the motions of vertebral segments in impact are beyond the normal physiological movable range, and the cervical vertebrae can move more easily when a bending moment and a compression force are applied between C5/C6 segments and the I.A.R. (instantaneous axis of rotation) of the segment C5 is moved upward (Figures

20 and 21). Such loads at low speeds in particular would tend to concentrate on the lower vertebral segments.

It can be hypothesized that such non-physiological motions of vertebral segments increase stress of the intervertebral articulations and stretch of the anterior longitudinal ligaments. Based on this hypothesis, it can be deduced that facet joint injury or intervertebral disc injury may be caused by the motions if the impact speed is increased enough or the subject is disadvantaged due to degenerative changes in the spine.

Effect of Seat Stiffness on Head-Neck-Torso Kinematics and Cervical Vertebral Motions - Injuries caused by the hyperextension or hyperflexion of cervical spine in particular have been evaluated by two parameters - neck bending moment and head rotational angle. It is found, however, that the axial compression of cervical spine plays an important role, as well as the extension of cervical spine, in the determination of the amount of impact applied to the neck as already described.

The trajectories in Figures 22 and 23 show comparisons

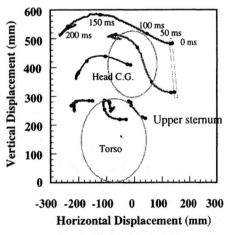

Figure 22. Head/neck/torso trajectories for Rigid Seat

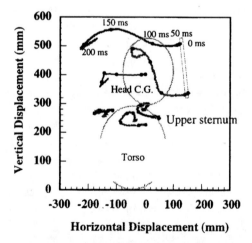

Figure 23. Head/neck/torso trajectories for Standard Seat

of subjects' motions at the impact speed of 8 km/h for the standard and rigid seats. Great differences in subjects' motions are found. Namely, the comparison of torso ramping-up motion by the trajectories of the target marks adhered to the upper sternum indicates that a sharp ramping-up motion starts with the rigid seat in the early phase of impact, while a sharp ramping-up motion and rebound are found around 100 ms with the standard seat, though no significant difference is found in the upward displacement (it is approximately 50 mm for both of them). Looking back at the trajectories for the neck-torso joint over time (Figures 8 and 9), it is found that the ramping-up motion is sharper with the rigid seat, which agrees with the phenomenon mentioned above. The rebound of standard seat caused the sharp rotation of head around 150 ms (Figure 23).

The interpretation of these variations in terms of neck moment, shear and axial compression forces reveals that the axial compression force applied to the cervical spine is approximately 150 N with the rigid seat around 100 ms in the early phase of impact, which is about twice greater than that of the standard seat (Figure 7 a)). The shear force, on the other hand, is 241 N with the standard seat around 110 ms when the rebound of torso has occurred, which is roughly 1.6 times greater than the value of 152 N with the rigid seat (Figure 7 b)).

Based on the first stage of experiments done by Kaneoka and Ono (1997), the effect of the compression force on the cervical spine in the early phase of impact was compared in terms of the rotation of C3 against C6 at 100 ms after impact for speeds of 4 and 6 km/h, respectively, and the variations of motions at anterior and posterior points. It can be pointed out that the motions at both anterior and posterior points were greater with the rigid seat than the standard seat, and the compression of cervical spine was also greater with the rigid seat.

The stiffness of seat is an important parameter for the optimum design of occupant restraint system aiming at the enhancement of performance. Parkin et al. (1995) pointed out that the incidence of minor neck injuries would tend to increase if the seat stiffness was increased. Svennson et al. (1993) pointed out that the elastic rebound caused by the seatback would facilitate the incidence of whiplash injury.

From the current study, it may be said that the difference in seat stiffness affects the torso motions upon impact, which in turn markedly affects the load to be applied to each cervical vertebral segment. If the stiffness is increased, the torso ramping-up motion in the initial phase of impact in particular will become sharper, and the axial compression force on the cervical spine tends to become greater. Even if the stiffness is low, however, the head-neck link will be displaced if the rebound is great particularly in the latter half of impact. It is presumed that this phenomenon causes an

intense shear force against the upper cervical spine.

Different Alignments of the Cervical Spine and Incidence of Neck Injuries - As has been described in the foregoing, the difference in initial head-neck position - i.e., difference in position of alignment of cervical spine as the difference in seating position - greatly affects the motions of cervical vertebrae and loads applied to them. In this study, a rigid seat has been used, but the torso ramping-up motion and rebounding are apt to change if a different type of seat is used. It is also presumed that the effect of alignment of the cervical spine is equivalent to or higher than the effect of seat characteristics in some cases. In this regard, more attention should be paid to the cervical spine alignment than any other parameter affecting the occupant's seating position such as seat stiffness and seatback inclination angle, when considering parameters for the evaluation of neck injuries.

It has been also reported that the incidence of neck injuries was higher for female occupants than for male occupants, and the difference in seatback characteristics was pointed out as the probable cause of the difference in neck injury incidence (Carlson et al., 1985; Lovsund et al., 1988; Muser et al. 1994).

However, Matsumoto et al. (1997) in a recent study conducted on the relationship between cervical curvature and dis degenerations using 495 subjects reported that lordosis position accounts for 35% or so as the cause of such injuries among female occupants younger than 40, while kyphosis including linear position accounts for 65% or so. He also showed that the percentage of kyphosis position is much higher for female than male. In the case of male occupants the lordosis position accounts for 75% or so. Based on our experimental study, it can be pointed out that the rotational angle of the cervical vertebrae becomes obviously larger at the kyphosis position. This may explain the higher minor neck injury incidence for occupants with the kyphosis position.

Under such circumstances, it will be difficult to evaluate the safety and effectiveness of headrest or seatback unless investigations and analysis are conducted in proper manner so that not only the relative positions of headrest and head but also the seating position - alignment of the cervical spine in particular - can be estimated, as suggested in this study.

CONCLUSIONS

A new low speed impact test setup was developed, and the compact, light sled was fabricated, and experiments were conducted by means of X-ray cineradiography and EMG measurements, using twelve volunteers. As a result, the effect of tension/relaxation of volunteers' muscles on their head and neck motions was verified. Motions of cervical vertebrae during impact and those in physiological conditions were also compared, and the facet joint injury mechanism was deduced from the characteristic motions of cervical vertebrae, as the mechanism of occurrence of neck injuries. Based on this mechanism, the effect of seat cushion characteristics on the impact responses of head, neck and torso, the relationship between the alignment of the cervical spine with the seating position and neck injuries, and the effect of correlation between the alignment of the cervical spine and male/female difference on the incidence of neck injuries have been clarified as described below:

1) It was found from the current study that the effect of the difference in muscle tension did not affect the head-neck-torso kinematics, since volunteers were relaxed as they became familiar with the atmosphere of the experiments. However, it was also found that the head backward rotational angle could be reduced by 30 to 40 % when the subject made the cervical muscles tensed. It will be, therefore, necessary to determine how the muscle tension/ relaxation affects the neck injury mechanism.

2) Subject's torso exhibits a ramping-up motion due to the inclination of seatback. Hence an axial compression force is applied to the cervical spine, resulting in the initial flexion. The lower cervical vertebral segments (C6, C5 and C4) are extended and rotated prior to the upper cervical vertebral segments. These are non-physiological motions of individual cervical vertebral segments, which are beyond the physiological motion range. When a bending moment is applied to C5/C6, it becomes easier for C5/C6 to rotate with the rotational center of C5 moved upward. This motion was hypothesized as "Facet Joint Injury Mechanism". Owing to this hypothesis, the tendency of concentration of loads on the cervical spine in low speed rear-end impact can be explained easily. It is deduced that this is a cause of intervertebral articular injury.

3) Difference in seat cushion stiffness affects the torso motion upon impact, which influences loads applied to the cervical vertebrae. The higher the seat cushion stiffness, the more significant the torso ramping-up motion becomes and the greater the axial compression force applied to the cervical spine becomes. On the other hand, the lower the seat cushion stiffness, the greater the torso rebound becomes, resulting in non-alignment of head and neck in the latter half of impact which would cause a great shear force against the upper cervical spine.

4) It is found that the initial state of neck position (difference in cervical alignment position) markedly affects the impact responses of head and neck. Namely, in the state of flexion (kyphosis), the lower cervical vertebral segments are affected by the effect of torso ramping-up motion, resulting in a greater difference in relative motions (S-shape formation). Considering this phenomenon and the facet

joint injury mechanism, it can be deduced that the upward travel of the rotation center and the amount of rotation of C5 vertebral segment increased in the state of flexion . This facilitates the incidence of neck injury.

5) It is hypothesized that facet joint contact occurs easily in the state of flexion for the kyphosis position. This may expalin the higher incidence rate of neck injuries for occupants with the kyphosis position.

6) It is suggested that not only the relative positions of the headrest and head but also the alignment of the cervical spine (seating posture) can be estimated for the evaluation of the effectiveness of headrest or seatback.

7) It is also concluded that carefully reviews will be needed for the knowledge and understanding of the-state-of-art on the incidence of minor neck injuries based on accident investigation and analysis.

ACKNOWLEDGMENTS

The authors thank specially the volunteers who understood the aim of the research as contribution for current social needs. The authors also wish to acknowledge Dr. Satoshi Inami and Dr. Yokoi Naoyuki of Department of Orthopaedic Surgery, Institute of Clinical Medicine, University of Tsukuba, and Mr. Tsuguhiro Fukuda, Mr. Kyouichi Miyazaki, Mr. Masahiro Ito, and Mr. Hiroyuki Mitsuishi of Third Division, Japan Automobile Research Institute for their technical support and the data processing.

REFERENCES

April C. and Bogduk N., The Prevalence of Cervical Zygapophyseal Joint Pain: A First Approximation. Spine 17: 744-747,199229) Basmajian, J. V. and De Luca, C. J., Muscles Alive: Their Functions Revealed By Electromyography, The William's & Wilkins. 1985.

Beier G., Schuler E., Schuck M., Ewing C.L., Bécker E.D. and Thomas D.J., Center of Gravity and Moments of Inertia of Human Heads, Proceedings of International IRCOBI Conference Biomechanics of Impacts, 218-228, 1980

Bogduk N. and Marsland A., The Cervical Zygapophyseal Joints as a Source of Neck Pain. Spin 13 (6); 610-617,1988

Carlsson G., Nilsson S., Nilsson-Ehle A., Norin H., Ysander L., Ortengren R.; Neck Injuries in Rear End Car Collisions. Biomechanical Considerations to Improve Head Restraints. Proceedings International 1985 IRCOBI/AAAM Conference

Foret-Bruno J.Y. et. al., Influence of the seat and head rest stiffness on the risk of cervical injuries in rear impact, 13th ESV Conference, 1991

Foust, D. R., Chaffin, D. B., Snyder, R. G. and Baum, J. K.,

Cervical Range of Motion and Dynamic Response and Strength of Cervical Muscles. Proc. of the Seventeenth Stapp Crash Conference, Oklahoma City. Society of Automotive Engineers, 285-308. 1973.

Geigel B.C., Steffan H., Leinzinger P., Muhlbauer M., Bauer G., The Movement of Head and Cervical Spine During Rearend Impact, Proceedings of the International IRCOBI Conference on the Biomechanics of Impact. Lyon, 1994:127-137

Geigel B.C., Dippel Ch., Muser M.H., Walz F., and Svensson M.Y., Comparison of Head-Neck Kinematics During Rear End Impact Between Standard Hybrid III, Rid Neck, Volunteers and PMTO's, Proceedings of the International IRCOBI Conference on the Biomechanics of Impact. Brunnen, 1995:127-137

Herzog, W. , Muscle. Biomechanics of the Musculo-Skeletal System, Edited by B. M. Nigg and W. Herzog, John Wiley & Sons, 154-187. 1994.

Kaneoka K. and Ono K., Human Volunteer Studies on Whiplash Injury Mechanisms, "Frontiers in Head and Neck Trauma: Clinical and Biomechanical" , Publisher :IOS Press, Harvard, MA in press, 1997

Kaneoka K. and Ono K., Motion Analysis of Cervical Vertebrae in Low Impact Speed Rear-end Collisions, The Society of Japan Clinical Biomechanics, in press, 1997

Lord S. Barnsley L., Bogduk N., Cervical Zygapophyseal Joint Pain in Whiplash, In Spine: Cervical Flexion-Extension/Whiplash Injuries 7(3):355-372, 1993

Lovsund P., et.al., Neck Injuries in Rear End Collisions among Front and Rear Seat Occupants. Proceedings of International IRCOBI Conference Biomechanics of Impacts, 319-325, 1988

L.T.B. van Kampen, Availability and (Proper) Adjustment of Head Restraints in the Netherlands, 1993 IRCOBI Conference, pp 367-377, Eindhoven, The Netherlands

Math Works, MATLAB: Signal Processing User's Guide, The Math Works. Math Works 1996.

Matsumoto M, Fujimura Y., Suzuki N., Ono T., Ishikawa M., and Yabe Y., Relationship Between Cervical Curvature and Disc Degeneration in Asymptomatic Subjects, Journal of the Eastern Japan Association of Orthopaedics and Traumatology in Japanese Journal, Volume 9:1-4, 1977

Matsushita T, Sato TB, Hirabayashi K, Fujimura S, Asazuma T, Takatori T. X-ray Study of the Human Neck Motion due to Head Inertia Loading. Proceedings of the 38th Stapp Car Crash Conference. Fout Lauderdale: Society of Automotive Engineers, Inc., 1994:55-64.

McConnel W. E., Howard R.P., Guzman H.M., Bomar J.B., Raddian J.H., Benedict V., Smith H.L, and Hatsell C.P., Analysis of Human Test Subject Kinematic Responses to Low Velocity Rearend Impacts. SAE Paper No. 930889, 1993

McConnel W. E., Howard R.P., Poppel J.V., Krause R., Guzman H.M., Bomar J.B., Raddian J.H., Benedict V., and Hatsell C.P., Human Head and Neck Kinematic After Low Velocity Rearend Impacts - Understanding "Whiplash". SAE Paper No. 952724, 1995

Mertz, H. J. and Patrick, L. M.. Investigation of the kinematics and kinetics of whiplash. Proc. of the 11th Stapp Car Crash Conference, Anaheim, USA. Society of Automotive Engineers, 269-317. 1967.

Melvin, J. W. and McElhaney, J. H., "Occupant Protection in Rear-End Collisions." SAE Technical Paper 720033, SAE Warrendale, PA, 1972

Mertz, H. J. and Patrick, L. M. . Strength and response of the human neck. Proc. of the Fifteenth Stapp Car Crash Conference, Coronado, California. Society of Automotive Engineers, 207-255. 1971.

Morris, A.P. & Thomas, P., A Study of Soft Tissue Neck Injuries in the U.K., Proceedings of Annual Conference on Enhanced Safety Vehicles, 1996

Muser M. H., et. al., Neck Injury Prevention by Automatically Positioned Head Restraint. Proceedings 1994 AAAM/IRCOBI Conference Joint Session, Lyon, France.

Nygren A., Injuries to Car Occupants. Some Aspects of Interior Safety of Cars - A Study of 5 years Material from an Insurance Company.. Acta Otolaryngol Suppl (Stockholm) 1984 : (Suppl 395)

Nygren A. et. al., Effects of Different Types of Headrests in Rear-End Collisions. 10th International Conference on Experimental Safety Vehicles, NHTSA, USA. 1985, pp 85-90

Olsson, I., Bunketorp, O., Carlsson, G.. Gustafsson, C., Planath, I., Norin, H. & Ysander, L., An in-Depth Study of Neck Injuries in Rear End Collisions. Proceedings of Annual IRCOBI Conference, Bron, France pp. 269-278, 1990

O'Neill, B, Haddon, W., Kelley, A.B. & Sorenson, W.W., Automobile Head Restraints :Frequency of Neck Injury Insurance Claims in Relation to the Presence of Head Restraints. American Journal of Public Health,1972

Ono K., Kikuchi A., Nakamura M., Kobayashi H., and Nakamura N., Human Head Tolerance to Sagittal Impact - Reliable Estimation Deduced From Experimental Head Injury Using Subhuman Primates and Human Cadaver Skulls - Proceedings of 24th

Stapp Car Crash Conference, SAE Paper 801303, 101-160

Ono K, Kanno M. Influences of the Physical Parameters on the Risk to Neck Injuries in Low Impact Speed Rear-end Collisions. Proceedings of the International IRCOBI Conference on the Biomechanics of Impact. Eindhoven, 1993:201-212.

Ono K. and Kanno M., Human Neck Response in Low Impact Speed Rear-end Collisions, "Frontiers in Head and Neck Trauma: Clinical and Biomechanical" , Publisher :IOS Press. Harvard, MA in press, 1997

Parkin S., Mackey G.M., and Cooper A., Rear End Collisions and Seat Performance - To Yield or Not To Yield. Proceedings of the 39th Annual AAAM Conference, Chicago, Illinois., pp231-244, 1995

Panjabi M., Whiplash Trauma Injury Mechanisms, A Biomechanical Viewpoint, Prc. International Symposium Whiplash '96, Belgium, 196 pp.19

Svensson, M. Y., Lovsund, P., Haland, Y. & Larsson, S., Rear-End Collisions - A Study of the Influence of Backrest Properties on Head-Neck Motion Using a New Dummy Neck. SAE 930343 pp 129-138, 1993

Szabo T. J., Welcher J. B., Anderson R. D., Rice M. M., Ward J. A., Paulo L. R. and Crapenter N. J, Human Occupant Kinematic Response to Low Speed Rear-End Impacts, Proceedings of the 38th Stapp Car Crash Conference. Fout Lauderdale: Society of Automotive Engineers, Inc., 1994:23-35.

Szabo TJ, Welcher JB. Human Subject Kinematics and Electromyographic Activity During Low Speed Rear End Impacts. Proceedings of the 40th Stapp Car Crash Conference. Albuquerque: Society of Automotive Engineers, Inc., 1996:295-315.

Tennyson, S. A. and King, A. I.. A Biodynamic Model of the Human Spinal Column. Proc. of the SAE Mathematical Modeling Biodynamic Response to Impact, Dearborn, Michigan, USA. Society of Automotive Engineers, 31-44. 1976.

Viano D. C., Restraint of a Belted or Unbelted Occupant by the Seat in Rear-End Impacts, SAE Paper No. 922522, 36th STAPP Conference, pp. 157-164

Viano D. C., Influence of Seatback Angle on Occupant Dynamics in Simulated Rear-End Impacts, SAE Paper No. 922521, 36th STAPP Conference, pp. 157-164

Walker, L. M. et. al., Mass, Volume, Center of Mass, and Mass Moment of Inertia of Head and Neck of Human Body, SAE Paper 730985

WHO/CIOMS proposed guidelines for medical research involving human subjects, and the guidelines on the practice of ethics committees published by the Royal College of Physicians , The Lancet, November 12, 1988,1128-1131

Yang K. H., Begman P. C., Muser M., Niederer P. and Walz F., On the Role of Cervical Facet Joints in Rear End Impact Neck Injury Mechanisms, SAE SP-1226, Motor Vehicle Safety Design Innovations, SAE Paper 970497, 127-129

Yoganandan N., Pinter F. A., and Cusick J. K, Cervical Spine Kinematics under Inertial Flexion Extension, Proc. North American Spine Society. Canada, 1996. pp 265-266

Yoganandan N. and Pinter F. A., Internal Loading of the Human Cervical Spine, Journal of Biomechanics Engineering. May 1997

983164

Upper Neck Response of the Belt and Air Bag Restrained 50th Percentile Hybrid III Dummy in the USA's New Car Assessment Program

Brian T. Park, Richard M. Morgan, James R. Hackney, Susan Partyka, Michael Kleinberger and Emily Sun
National Highway Traffic Safety Administration

Heather E. Smith and Johanna C. Lowrie
Conrad Technologies, Inc.

ABSTRACT

Since 1994, the New Car Assessment Program (NCAP) of the National Highway Traffic Safety Administration (NHTSA) has compiled upper neck loads for the belt and air bag restrained 50th percentile male Hybrid III dummy. Over five years from 1994 to 1998, in frontal crash tests, NCAP collected upper neck data for 118 passenger cars and seventy-eight light trucks and vans. This paper examines these data and attempts to assess the potential for neck injury based on injury criteria included in FMVSS No. 208 (for the optional sled test).

The paper examines the extent of serious neck injury in real world crashes as reported in the National Automotive Sampling System (NASS). The results suggest that serious neck injuries do occur at higher speeds for crashes involving occupants restrained by belts in passenger cars. Results of this paper also suggest that neck tension and neck extension can reach levels in the NCAP frontal crash tests that are higher than those allowed in FMVSS No. 208 for the sled test. Furthermore, neck tension and neck extension are generally higher in light trucks and vans than in passenger cars. In addition, when the neck responses are examined as a function of the Head Injury Criterion (HIC) for driver and passenger dummies, no correlation is found.

INTRODUCTION

Apart from the safety performance standards, one of the successful safety programs that NHTSA has developed is called the New Car Assessment Program (NCAP). In 1979, NHTSA began assessing the occupant-protection capabilities of new cars by conducting high speed, frontal barrier crash tests. NCAP has two primary goals. The first is to provide consumers with a measure of the rela-

tive safety of automobiles. The second is to establish market forces that encourage vehicle manufacturers to design higher levels of safety into their vehicles. The USA NCAP represents the first program ever initiated to provide relative crashworthiness information to consumers on the safety performance of passenger vehicles [1, 2, 3].

The frontal USA NCAP test conditions represent a severe frontal crash condition that can result in serious or fatal injury to occupants. In these controlled crash tests, measurements assess the levels of potential injury. These measurements are taken from two instrumented anthropomorphic test devices (dummies) that simulate the 50th percentile adult male. These dummies are positioned in the driver and front-right passenger seats and are restrained by the vehicle's safety belts and air bags, if available [1].

During the crash test, measurements are taken from each dummy's head, chest, and upper legs. A composite of acceleration values measures the injury potential to the head known as the Head Injury Criterion, or HIC. The injury potential to the chest is measured by the resultant chest deceleration. For the upper legs, the injury potential is measured by compressive axial forces on each of the femur bones [1].

Besides from recording responses of the head, chest, and upper legs, NCAP has been collecting upper neck responses since 1994. The purpose of collecting neck data is to understand how the neck responds in higher speed crashes. Over the past five years, neck data for 118 passenger cars and seventy-eight light trucks and vans were compiled. Of the neck data, most of the readings, about 98 percent for driver and 83 percent for passenger, are from the belt and air bag restrained dummies, as shown in Appendix A.

The objective of this paper is to examine the upper neck responses of belt and air bag restrained 50th percentile male Hybrid III dummies used in 56 kmph frontal barrier tests from 1994 through 1998. Neck injuries also occur in real-world crashes, as is shown by the estimates from investigated crashes included in this paper. After presenting the data, a closer look was given to the relationship between neck tension and extension.

MEASUREMENT OF UPPER NECK RESPONSES

The dummy used in the USA NCAP frontal crash tests is the 50th percentile male Hybrid III. This neck is a single structure with four asymmetric rubber members connected to aluminum plates and to plates on the ends. A cable passes through the center of the neck and attaches to end plates. The uppermost plate is connected to the head at a single pivot joint [4]. The location of the upper neck load cell is shown in Figure 1 [5]. A closer view of the neck and attached load cell design is shown in Figure 2 [6].

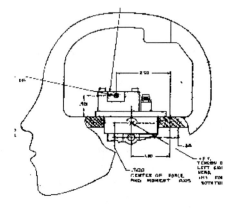

Figure 1.   A head and neck load cell configuration of the 50th percentile Hybrid III dummy.

Figure 2.   A neck and attached load cell configuration of the 50th percentile Hybrid III dummy.

In the NCAP tests, the upper neck load cells measure three forces and three moments about x, y, and z axes. These measurements which evaluate frontal restraint effectiveness are fore-and-aft shear force, compression, tension, flexion moment, and extension moment. Shear is

defined when the head moves forward or rearward with respect to the torso. Tension is defined as the upward movement of the head with respect to the torso. Compression is defined as the opposite force, or when the head is pushed downward. Neck flexion is defined when the neck bends forward and the chin moves toward the sternum (relative to the torso). Extension is defined when the neck bends backward and the chin moves away from the sternum. In other words, neck extension occurs when the head is rotated in the posterior direction in the mid-sagittal plane.

Neck injury criteria were included as part of the temporary amendment to FMVSS No. 208 (for the optional sled test) [7]. The reference values for the neck injury criteria are listed in Table 1. These criteria are similar to Injury Assessment Reference Values (IARV) [8, 10].

In References 8 and 10, the neck fore-and-aft shear, compression, and tension are proposed for assessment by time independent criteria. The analysis in this paper follows the practice of using the maximum neck reading [9]. All results presented in this paper have been filtered according to the filter class specified in Table 1. The neck data from the NCAP tests will be assessed by using the reference values in Table 1. (New neck injury assessment values are being contemplated in an advanced air bag notice of proposed rulemaking and will be considered in future studies.) In addition, the reported neck readings of flexion and extension moments that are in this report have been corrected about the occipital condyle [9]. Shear was filtered at SAE Filter Class 600 for these bending moment corrections.

Real-world crash data were used to provide perspective for the neck loads measured by crash test dummies. This paper uses National Automotive Sampling System (NASS) data as evidence that serious neck injuries do occur in higher speed crashes in the real world. This paper then uses NCAP frontal crash data to analyze the frequencies of these neck injuries under the loading of shear and axial forces and of bending moments.

NATIONAL AUTOMOTIVE SAMPLING SYSTEM STUDY

NASS collects detailed data on a statistical sample of light passenger vehicles (passenger cars, pickup trucks, vans, and sport utility vehicles) towed because of damage received in a police-reported traffic crash [13]. During the nine years from 1988 through 1996, NASS researchers selected a sub-sample of people who met specific criteria. First, they were front-outboard occupants (that is, drivers and right-front passengers). Second, they were at least 15 years old (and so "adult-sized"). Third, they were using a lap-and-shoulder belt. Fourth, they were not ejected from the vehicle. Fifth, the vehicle in which they were riding was a light passenger vehicle that was towed from the scene because of damage received in a nonrollover frontal crash (which was identified in terms of the general area of damage). The sample size is a total of 10,026 people and an estimate of the ΔV (the change in vehicle velocity during the impact) was available for the vehicle.

NASS collects detailed injury data for each occupant, including the location and Abbreviated Injury Scale (AIS) value for each injury. As used here, the term "serious neck-type injury" includes injuries with an AIS of at least 3 to the neck and to the lower part of the head. Additional information on the identification of these injuries is included in Appendix B.

The available data — both the number of investigated cases and the annualized national estimates—are shown in Table 2 for ranges of the total ΔV (of the vehicle in which the occupant was seated). The annualized estimates, i.e., estimates that reflect the average over the 5 years of data, have not been adjusted for missing data (including the large number of investigated cases for which there was no vehicle inspection or for which ΔV could not be estimated).

NASS data indicate that 0.08 percent of belted occupants in frontal crashes who met the criteria received at least one serious neck-type injury. The percentage with a neck-type injury increased rapidly with increasing ΔV: from 0.001 percent for ΔV values that did not exceed 16 kmph to 11 percent for ΔV values of at least 65.6 kmph. In comparison, the ΔV in a frontal NCAP test is about 62 kmph.

Most (an estimated 79 percent) NASS light vehicle occupants were occupants of passenger cars, and most (an estimated 87 percent) passenger car occupants did not have an air bag available at that seat location. The data are shown in Appendix B. These data are insufficient for comparisons between vehicle types or between those with versus without an available air bag. A future year of NASS data collection will provide more light truck and more air bag cases.

Table 1. Neck Injury Criteria for the 50th percentile Hybrid III dummy used in the temporary amendment to FMVSS No. 208 (for the optional sled test).
(** Time independent criteria)

Loading Mechanism	Reference Values	SAE Filter Classes
Flexion bending moment	190 Nm	600
Extension Bending	57 Nm	600
Axial Tension**	3300 N	1000
Axial Compression**	4000 N	1000
Fore-and-Aft** Shear**	3100 N	1000

Table 2. Serious "Neck-Type" Injuries among Nonejected, Lap-and-Shoulder-Belted, Front-Outboard Occupants at Least 15 Years Old in Towed Light Vehicles in Front Nonrollover Crashes.
(1988-1996 NASS Investigated Cases and Annualized Estimates)

Total Delta V (kmph)	Investigated Cases		Annualized Estimates		Percent with Neck-Type Injury
	Occupants Selected for Study	With Serious Neck-Type Injury	Occupants Selected for Study	With Serious Neck-Type Injury	
1.0 - 16.0	1,991	1	154,979	1	0.001
17.6 - 32.0	5,834	8	269,233	86	0.032
33.6 - 48.0	1,687	15	36,890	101	0.274
49.6 - 64.0	401	25	3,941	97	2.461
65.6 +	113	15	762	86	11.286
Total	10,026	64	465,805	372	0.080

OVERALL PATTERN OF NECK LOADING

To examine the overall profile for the five neck criteria, Figures 3 through 7 are plotted for all neck readings from the belt and air bag restrained occupants—193 drivers and 163 passengers. Of the driver, 118 are from passenger cars and 78 are from light trucks and vans (LTVs). Of the right front passenger, 111 are from the passenger cars and 52 are from the LTVs. Many earlier LTV models, 1994 and 1995, did not have passenger air bags. The individual neck readings (for the driver and right front passenger from the model years 1994 to1998) are tabulated in Appendix C. These figures are plotted with the driver readings on the vertical axis and the passenger readings on the horizontal axis, for shear, compression, flexion, tension, and extension. The readings are normalized, i.e., each force and moment is divided by the appropriate reference value given in Table 1.

The absolute value of the maximum shear loadings recorded for the driver and the right front passenger in the NCAP tests are shown in Figure 3. These data in Figure 3 have been normalized, i.e., each shear reading has been divided by the 3100 N reference value of Table 1. The data show that all of the neck readings remained well below the established reference values.

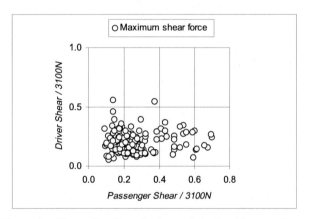

Figure 3. Normalized neck shear data for driver and passenger from MY1994 to MY1998.

Figure 4 represents neck compression data recorded from model year 1994 through model year 1998. As with the shear data in Figure 3, the compression data have been normalized by the reference value presented in Table 1. The recordings here show that most individual neck readings remained well below the established reference values. Here, only one neck compression force of the driver was greater than the neck injury criterion.

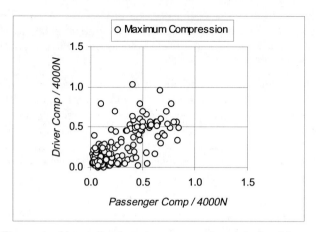

Figure 4. Normalized neck compression data for driver and passenger from MY1994 to MY1998.

Figure 5 shows all data recorded from model year 1994 to model year 1998 for neck flexion. The data represent normalized readings from the frontal test crashes. As seen previously in fore-and-aft shear and axial compression, the data show values well within the limits of the established reference value. In Figure 5, there are no tests where the individual reading exceeded this criterion.

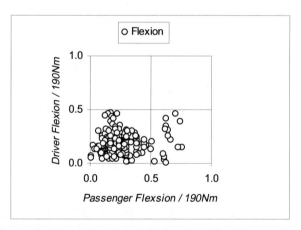

Figure 5. Normalized neck flexion data for driver and passenger from MY1994 to MY1998.

Based on these high speed frontal crashes of the NCAP vehicles, almost all of the neck readings — for fore-and-aft shear, compression, and flexion moment criteria — are below the reference value of Table 1. In contrast, neck readings for tension and extension moments show a different pattern.

Figure 6 shows data for normalized neck tension readings in the NCAP test crashes examined in this study. Approximately 3 percent of the Hybrid III neck readings are equal to or exceed the tension force reference value of 3300 N — eight drivers and two passengers.

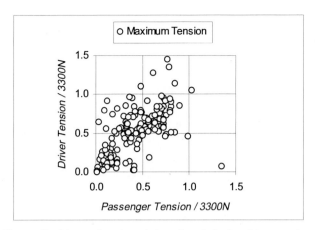

Figure 6. Normalized neck tension data for driver and passenger from MY1994 to MY1998.

Figure 7 is a plot of normalized moments due to neck extension for all the high-speed NCAP frontal crash test data. Similar to the preceding plot of neck tension for the Hybrid III dummies, approximately 12 percent of the Hybrid III neck readings are equal to or exceed the extension moment reference value of 57 Nm — 20 drivers and 21 passengers. Neck extension trauma has been shown to occur at about 85 degrees of rotation, based on a response envelope developed from results of an experiment conducted on a human volunteer and a cadaver [17]. For those cases where the neck extension moment was high, the NCAP test film was viewed to establish that the neck had undergone an excessive amount of extension relative to the torso. In addition, based on high speed film analyses, no evidence of an air bag being caught between the chin and the neck column was found during the crash tests.

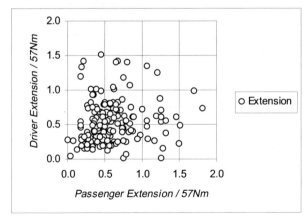

Figure 7. Normalized neck extension data for driver and passenger from MY1994 to MY1998.

In this section of the paper, the overall pattern of loading to the neck of the Hybrid III dummy has been analyzed for the NCAP high-speed frontal crashes. Based on the biomechanically-based understanding of neck injury summarized in Table 1, the data suggest that in NCAP tests, fore-and-aft shear, axial compression, and bending in flexion are associated with a low risk of neck trauma. On the other hand, these same frontal crash tests suggest that axial tension and bending in extension are high relative to the criteria [8, 9, 10]. Therefore, the analysis in the rest of the paper will focus only on the neck axial tension and extension bending.

NECK EXTENSION MOMENT AND TENSION

In order to examine the relationship between extension moment and neck tension, the neck tension readings are plotted as a function of neck extension moments. These plots are shown for the driver and passenger for model years 1994 - 1998, in Figures 8 through 12, respectively.

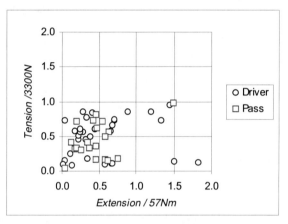

Figure 8. Neck tension versus extension for model year 1994.

Figures 8 through 12 show three types of neck readings that exceed the criteria in Table 1. The first type of significant neck loading is that in which tension loading goes above the reference value of Table 1, while the extension moment remains relatively small, i.e., below the given criteria. That is, the head is being pulled away from the torso in the superior direction. For example, in Figure 9, a data point for a driver is below the reference value for extension, while the tension is a multiple of 1.3 times the limit of 3300 N. This type of loading occurred in five of the 193 driver dummy exposures — no similar pattern was found for the passenger.

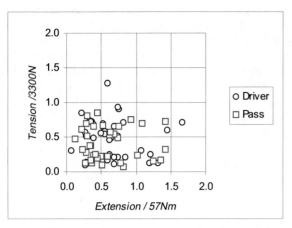

Figure 9. Neck tension versus extension for model year 1995.

In the second type, the neck extension exceeds its recommended limit, while neck tension stays below the given reference value. In this situation, the head is rotated in the posterior direction in the midsagittal plane. In reviewing the high speed films, the extension appeared to be due to the air bag contacting the face and rotating it away from the sternum. For example, Figure 10 shows a data point for the passenger dummy that has low tension but an extension moment one and one-half times the recommended limit. Extension above the reference value and tension below the value occurred in 38 of the 356 dummy exposures — 17 drivers and 21 passengers. This type of neck loading generally occurred when an air bag was deploying from in front of the dummy's head, but not from below the head.

A third type of significant neck loading occurred when both extension moment and tension force exceed the established reference values. In this case, the driver's face hits the air bag, and extends during the rebound. The extension is first, followed by tension. High extension moment and high tension force occurred in three of the 193 dummies — no similar pattern was found for the passenger.

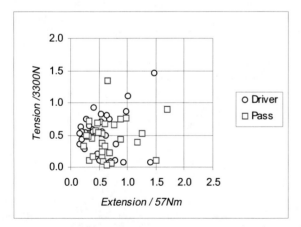

Figure 10. Neck tension versus extension for model year 1996.

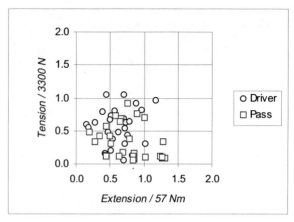

Figure 11. Neck tension versus extension for model year 1997.

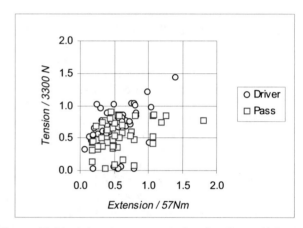

Figure 12. Neck tension versus extension for model year 1998.

Figures 13 and 14 show the calculated average neck extension moment for driver and passenger for all model years. Note that the calculated average is made based on the data from air bag equipped vehicles. In general, the average extension moment for the LTV group, for both driver and passenger, is greater than for the car group.

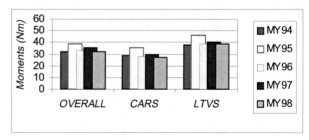

Figure 13. Average neck extension moments of the driver dummy for the model years 1994 - 1998.

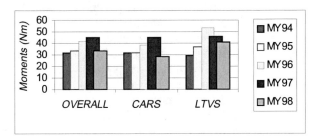

Figure 14. Average neck extension moments of the passenger dummy for the model years 1994 - 1998.

Figures 15 and 16 exhibit the average tension for both driver and passenger over five model years. Again, as seen in the neck extension results, the average neck tension for LTVs is higher than for passenger cars.

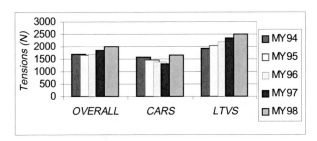

Figure 15. Average neck tension of the driver dummy for the model years 1994 - 1998.

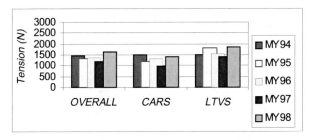

Figure 16. Average neck tension of the passenger dummy for the model years 1994 - 1998.

The preceding four figures suggest that there are different trends between passenger cars and LTVs. The tension and extension moment are higher in the LTV class on average. Moreover, it is interesting to note that of the model year 1998 vehicles, all but one of the twelve neck readings that exceeded the criteria, either in extension or tension, are from the LTVs — only one passenger car exceeds the criteria. Note that all model year 1998 tested under the NCAP had redesigned air bags [18].

NECK EXTENSION MOMENTS, AXIAL TENSION, AND HIC

In Figure 17, the normalized peak neck extension readings are plotted as a time peak for all model years. Here, time zero is at the initiation of the crash event. As shown, most of the high neck extensions occur early, between

50 - 100 msec for the driver and between 50 - 120 msec for the passenger. Based on the NCAP test film analysis, most of the high neck extension readings occurred during the interaction with air bags.

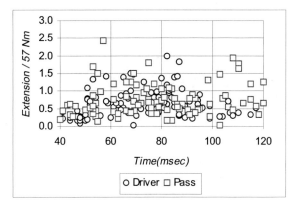

Figure 17. Time and magnitude of peak neck extension moment for driver and passenger.

In Figure 18, the normalized neck tension readings are plotted as a function of time. As mentioned before, there are few readings that exceeded the values in Table 1. Again, as shown, most or all of the high neck tension readings occurred relatively early in time.

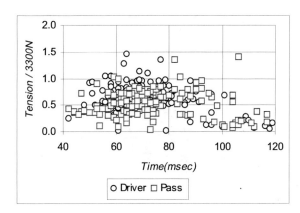

Figure 18. Time and magnitude of peak neck tension for driver and passenger.

Since most of the high neck extension readings are occurring at earlier times, when the head and neck are interacting with air bags, it is of interest to find out if there is any relationship between Head Injury Criterion (HIC) values and the neck extension. In Figure 19, the neck extensions for drivers and passengers are plotted as a function of the HIC. It shows that high neck extensions occur within a wide range of HIC values. In fact, some of the highest neck readings for right front passengers occur when the corresponding HIC values are below 500. Also, Figure 19 shows that low neck extensions occur at HIC values above 1000. This suggests that even with good head protection, there exists the possibility of the dummy obtaining high neck extension moments. Furthermore, statistical analysis was performed to find any correlation using a linear regression routine. Results gave R-squared values of 0.0594 and 0.0380 for driver and passenger, respectively. Thus, again, this suggests that there is no correlation between HIC and neck extension moments.

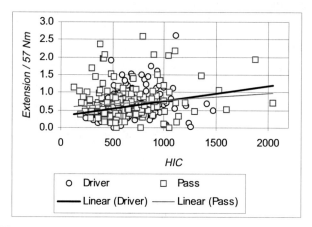

Figure 19. Neck extension moment and HIC readings.

In Figure 20, the neck tensions are plotted as a function of the HIC for the driver and passenger. The data show that low neck tension occurs at the HIC values above 1000. Statistical analysis was performed to find any correlation using a linear regression routine. Results gave low R-squared values of 0.0289 and 0.0024 for driver and passenger, respectively. Again, this suggests no correlation between HIC and neck tension forces.

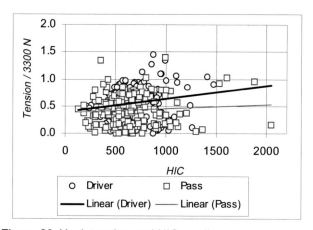

Figure 20. Neck tension and HIC readings.

CONCLUSIONS

Since 1994, the New Car Assessment Program of the National Highway Traffic Safety Administration has compiled upper neck loads for the belt and air bag restrained 50th percentile Hybrid III dummy. Over the past five years, frontal NCAP collected the upper neck data for 118 passenger cars and seventy-eight light trucks and vans.

Regarding real world crash statistics, the National Automotive Sampling System indicates that there is little evidence of serious neck trauma for lower speed vehicle crashes. The percentage of real world crash occupants with a neck injury increases rapidly with increasing V. When the V of the vehicle is 65.6 kmph or over, the NASS data produce an estimate that 11 percent of the belted occupants have a serious neck injury. In this paper, because most of the NASS occupants are restrained by safety belts without air bags, a correlation cannot be made between the NASS and NCAP data at this time.

This paper analyzed the forces and moments applied to the neck of the Hybrid III dummy in high speed frontal crashes. The Hybrid III dummies were seated in the driver position or the right front passenger position of the vehicle. Five types of neck loading were analyzed: fore-and-aft shear, compression force, tension force, extension moment and flexion moment. These five types of loadings to the neck were compared with existing biomechanically based values. The values are referred to as "criteria" in a frontal crash standard and as "injury assessment reference values" in industry publications. In the NCAP crash tests, shear, compression force, and moment of flexion were generally below the limits.

For neck extension moments and tension forces, when the neck interacts with an air bag during the crash, there are three types of readings that exceed the established reference values. High extension loading, with low tension, occurred in 11% of the occupants. High tension force, with a low moment of extension, occurred in 3% of the dummy occupants. About 1% of the dummy occupants experienced both a high moment of extension and a high tension force.

Average neck moment of extensions and average tension forces were calculated for each of the five years of vehicle testing. The averages were presented by the driver and passenger position and by passenger car class and LTV class. The results show that the average extension moment and average tension force are generally higher for the LTV group when compared with the passenger car group.

Neck extensions and tensions were further evaluated in terms of the Head Injury Criterion. The analysis was done for both the driver and passenger dummies. These data suggest that there is no correlation between the Head Injury Criterion and either neck extension moment or tension force.

REFERENCES

1. Hackney, J. R., *The Effects of FMVSS No. 208 and NCAP on Safety as Determined from Crash Test Results*, Proceedings of the Thirteenth International Conference on Experimental Safety Vehicles, Paris, France, November 1991.

2. Kahane, C. J., Hackney, J. R., and Berkowitz, A. M., *Correlation of NCAP Performance with Fatality Risk in Actual Head-On Collisions*, Report No. DOT HS 808 061, National Highway Traffic Safety Administration, Washington DC, 1994.

3. Hackney, J. R., Kahane, C. J., and Chan, R., *Activities of the New Car Assessment Program in the United States*, Proceedings of the Fifteenth International Conference on Experimental Safety Vehicles, Melbourne, Australia, May 1996.

4. Foster, J. K., Kortge, J. O., and Wolanin, M. J., "Hybrid III - A Biomechanically Based Crash Test Dummy," Twenty-First Stapp Car Crash Conference, SAE 770938, October 1977.

5. Robert A. Denton, Inc., Drawing C-1709, "Six Axis Neck Transducer".

6. "Hybrid III: The First Human-Like Crash Test Dummy," Society of Automotive Engineers, Inc., 1994.

7. Safety Automotive Engineering, Report J1733.,"Sign Convention for Vehicle Testing"., March 1994.

8. "Anthropomorphic Dummies for Crash and Escape System Testing," North Atlantic Treaty Organization, AGARD, Advisory Report 330, July 1996.

9. National Highway Traffic Safety Administration, U.S. Department of Transportation, Motor Vehicle Safety Standard No. 208, Docket No. 74-14 as amended by Notice 114 (Federal Register, Volume 62, No. 53, pg. 12960, 19 March 1997).

10. Mertz, H. J., "Anthropomorphic Test Devices," *Accidental Injury: Biomechanics and Prevention*, Nahum, A. M., and Melvin, J. M., eds., Springer-Verlag, New York, 1993.

11. Evelyn J. Benton and Linda K. Brandon (editors), *1988 NASS Injury Coding Manual*, NHTSA, September 1987.

12. NPRM for neck extension.

13. *National Accident Sampling System, 1988 Crashworthiness Data System, Data Collection, Coding, and Editing Manual*, HS 807 196, NHTSA, January 1988.

14. *The Abbreviated Injury Scale, 1980 Revision*, Association for the Advancement of Automotive Medicine, Des Plaines, IL, 1980.

15. *The Abbreviated Injury Scale, 1990 Revision*, Association for the Advancement of Automotive Medicine, Des Plaines, IL, 1990.

16. *National Accident Sampling System, 1993 Crashworthiness Data System, Injury Coding Manual*, DOT HS 807 969, NHTSA, January 1993.

17. Mertz, H. J. and Patrick, L. M., "Strength and Response of the Human Neck", Fifteenth Stapp Car Crash Conference, SAE 710855, November 1971.

18. Park, B. T., Morgan, R. M., Hackney, J. R., and Lowrie, J. C. *The Effect of Redesigned Air Bags on Frontal USA NCAP*, Proceedings of the Sixteenth International Conference on Enhanced Safety of Vehicles, Windsor, Ontario, Canada, June 1998.

APPENDIX A.

The purpose of Appendix A is to supply the reader with information on the percentages of vehicles that had air bags for model years 1994 - 1998.

PERCENTAGES OF VEHICLES WITH AIR BAGS DURING MY 1994-1998

PASSENGER CARS

MY	Total	# with air bags		% with air bags	
		driver	passenger	driver	passenger
1994	24	24	19	100.0	79.0
1995	29	29	27	100.0	93.0
1996	21	21	21	100.0	100.0
1997	18	18	18	100.0	100.0
1998	26	26	26	100.0	100.0
1994-1998	118	118	111	100.0	94.1

LTVs

MY	Total	# with air bags		% with air bags	
		driver	passenger	driver	passenger
1994	12	9	2	75.0	16.7
1995	11	11	5	100.0	45.5
1996	13	13	7	100.0	53.8
1997	20	20	16	100.0	80.0
1998	22	22	22	100.0	100.0
1994-1998	78	75	52	96.2	66.7

APPENDIX B.

There are two purposes of Appendix B. One is is to list specifics about the case selection criteria used in the National Automotive Sampling System Study section of this paper. The second is to provide statistical data on LTVs and passenger cars, and passenger cars with either an air bag/belt or belt restraint system alone.

Until 1992, information for each injury was reported in terms of the Occupant Injury Classification (OIC) [11] and the Abbreviated Injury Scale (AIS) [14] severity rating. Beginning in 1993, injury data are reported in terms of the 1990 revision of the AIS [15] (AIS90), as modified for use by NASS [16]. The 10,026 investigated occupants that are relevant to this study include 64 that received at least one serious (that is, with an AIS rating of three or higher) neck-type injury. These "neck-type" injuries were defined for this study to include all injuries to the neck (except those that involved only the skin) plus the following injuries to the lower part of the head:

Table B-1. (See References 11 and 16 for definitions)

1988-1992: OIC =	HIFS, HICB, HILB; and
1993-1996: AIS90 =	121099.3, 121002.5, 122899.3, 122802.5, 140202.5, 140204.5, 140210.5, 140212.6, 150200.3, 150202.3, 150204.3, 150206.4.

The data for light trucks (including pickup trucks, vans, and sport utility vehicles) are shown in Table 2, and the data for passenger cars are shown in Table 3. The NASS data for belted occupants in frontal crashes indicate that an estimated 0.049 percent of those in light trucks and 0.086 percent of those in passenger cars received at least one serious neck-type injury. NASS estimates based on small numbers of investigated cases are subject to large sampling variability, so the difference between the estimates for light trucks and for passenger cars may not be statistically significant.

Table B-2. Serious "Neck-Type" Injuries among Nonejected, Lap-and-Shoulder-Belted, Front-Outboard Occupants at Least 15 Years Old in Towed Light Trucks in Frontal Nonrollover Crashes. (1988-1996 NASS Investigated Cases and Annualized Estimates)

	Investigated Cases		Annualized Estimates		
Total Delta V (kmph)	Occupants Selected for Study	With Serious Neck-Type Injury	Occupants Selected for Study	With Serious Neck-Type Injury	Percent with Neck-Type Injury
1.0 - 16.0	436	0	26,240	0	0.000
17.6 - 32.0	1,207	2	45,690	21	0.046
33.6 - 48.0	372	1	9,293	3	0.032
49.6 - 64.0	75	4	830	15	1.807
65.6 +	22	1	109	1	0.917
Total	2,112	8	82,162	40	0.049

Table B-3. Serious "Neck-Type" Injuries among Nonejected, Lap-and-Shoulder-Belted, Front-Outboard Occupants at Least 15 Years Old in Towed Passenger Cars in Frontal Nonrollover Crashes. (1988-1996 NASS Investigated Cases and Annualized Estimates)

	Investigated Cases		Annualized Estimates		
Total Delta V (kmph)	Occupants Selected for Study	With Serious Neck-Type Injury	Occupants Selected for Study	With Serious Neck-Type Injury	Percent with Neck-Type Injury
1.0 - 16.0	1,555	1	128,738	1	0.001
17.6 - 32.0	4,627	6	223,543	65	0.029
33.6 - 48.0	1,315	14	27,596	98	0.355
49.6 - 64.0	326	21	3,112	82	2.635
65.6 +	91	14	654	84	12.844
Total	7,914	56	383,643	331	0.086

The data for passenger car occupants with an airbag at that seat position are shown in Table 4, and the data for passenger car occupants without an airbag at that seat position are shown in Table 5. (There were also two investigated occupants for whom airbag availability could not be determined; neither of these had a serious neck-type injury.) The NASS data for belted passenger car occupants in frontal crashes indicate that an estimated 0.029 percent of those with an airbag available and 0.096 percent of those without an airbag available received at least one serious neck-type injury. Estimates based on small numbers of investigated cases are subject to large sampling variability, so the difference between the estimates for passenger car occupants with and without an available airbag may not be statistically significant.

Table B-4.   Serious "Neck-Type" Injuries among Nonejected, Lap-and-Shoulder-Belted, Front-Outboard Occupants at Least 15 Years Old with an Airbag Available at that Seat Position in Towed Cars in Front Nonrollover Crashes. (1988-1996 NASS Investigated Cases and Annualized Estimates)

Total Delta V (kmph)	Investigated Cases		Annualized Estimates		Percent with Neck-Type Injury
	Occupants Selected for Study	With Serious Neck-Type Injury	Occupants Selected for Study	With Serious Neck-Type Injury	
1.0 - 16.0	244	0	19,125	0	0.000
17.6 - 32.0	595	0	31,495	0	0.000
33.6 - 48.0	171	1	4,807	9	0.187
49.6 - 64.0	36	1	332	5	1.506
65.6 +	7	2	22	2	9.091
Total	1,053	4	55,780	16	0.029

Table B-5.   Serious "Neck-Type" Injuries among Nonejected, Lap-and-Shoulder-Belted, Front-Outboard Occupants at Least 15 Years Old without an Airbag Available at that Seat Position in Towed Cars in Frontal Nonrollover Crashes. (1988-1996 NASS Investigated Cases and Annualized Estimates)

Total Delta V (kmph)	Investigated Cases		Annualized Estimates		Percent with Neck-Type Injury
	Occupants Selected for Study	With Serious Neck-Type Injury	Occupants Selected for Study	With Serious Neck-Type Injury	
1.0 - 16.0	1,311	1	109,613	1	0.001
17.6 - 32.0	4,031	6	192,044	65	0.034
33.6 - 48.0	1,143	13	22,785	90	0.395
49.6 - 64.0	290	20	2,780	77	2.770
65.6 +	84	12	632	83	13.133
Total	6,859	52	327,855	316	0.096

APPENDIX C.

The purpose of Appendix C is to list the Hybrid III Dummy neck readings from the New Car Assessment Program for model year 1994 through 1998.

Table C-1. Hybrid III Dummy neck readings from frontal New Car Assessment Program for model year 1994. Blank spaces indicate no data. * Denotes vehicle without a passenger air bag. ** Denotes vehicle without a driver and passenger air bag. The neck readings from a vehicle without a passenger air bag were not included in the analysis.

| Make | Model | Driver | | | | | | Passenger | | | | | |
		Shear (N)	Comp (N)	Tension (N)	Exten (Nm)	Flex (Nm)	HIC	Shear (N)	Comp (N)	Tension (N)	Exten (Nm)	Flex (Nm)	HIC
Buick	Regal*	481	1972	360	38	9	503	1597	3369	529	33	41	2044
Cadillac	Deville	556	243	1869	16	37	753	426	262	2035	27	47	663
Chevrolet	Caprice	280	680	2842	16	15	712						1058
Chevrolet	Corsica	428	3181	519	2	33	587	1499	3100	128	2	45	1297
Chevrolet	S-10**	2005	661	3105	113	26	1116	2498	86	1845	36	96	1115
Chevrolet	Sport Van*	1062	3839	808	6	71	1027	1657	2652	381			1049
Chrysler	New Yorker	1025	420	2420	76	55	826	750	250	2384	11	70	568
Dodge	Dakota*						329	1715	49	2684	26	97	857
Dodge	Caravan	826	224	2829	68	13	514	708	157	1057	10	55	422
Dodge	Spirit*	845	1160	3154	82	4	778	1658	341	2392	24	101	807
Dodge	Ram 1500*	744	1238	2838	51	4	586	2168	6071	4624	110	59	
Ford	Escort Wagon*	919	114	2422	2	39	533						
Ford	F150 PU*	346	486	1925	10	26	446	525	606	2349	31	31	
Ford	Mustang	427	661	1520	13	33	473	1886	476	3247	85	39	419
Ford	Probe	484	566	1504	18	39	238	806	434	1071	11	97	538
Ford	Thunderbird	392	1697	358	39	8	475	425	2751	548	26	30	302
Ford	Bronco*	432	624	1923	37	41	299						
Honda	Civic	357	356	2552	19	25	867	705	585	1632	33	44	684
Honda	Accord	853	1948	443	86	19	618	763	2267	1200	26	32	935
Hyundai	Elantra*	762	164	1664	13	49	575	1373	287	3360	20	20	1602
Infiniti	J30	475	250	1630	22	37	530	718	256	1000	15	62	290
Jeep	Wrangler**	3108	5682	339	1	68	1260	1279	2577	610	43	47	726
Mazda	626	325			15	32	485	689	618	1093	21	49	300
Mercedes	C220	365	418	2787	23	22	647	696	150	1350	20	46	654
Mitsubishi	Galant	250	1600	251	8	19	553	648	150	1350	7	82	533
Nissan	Altima	604	2365	312	33	23	564	573	1645	494	35	14	906
Nissan	Quest*	312	488	2404	40	24	688						
Oldsmobile	Achieva*		1983	411	104	24	844	2198	3062	626	93	28	1103
Pontiac	Grand Prix	505	413	2015	25	36	426	783	312	1849	24	44	725
Pontiac	Trans Sport*	573	159	2180	39	5	454	1710	512	2625	21	67	1003
Toyota	Camry	240	217	1979	14	8	607	396	342	1904	45	19	881
Toyota	Corolla		448	1851	14	13	384	545	364	1124	18	64	433
Toyota	Previa	403	1553	600	20	20	402	633	2600	1900	56	38	532
Toyota	T100**	607	1026	2465	40	30	601						
Volvo	850		628	1692		46	434	763	661	1286	19	50	421
VW	Jetta	325	100	2000	14	33	725	1007	200	2312	36	88	637

Table C-2. Hybrid III Dummy neck readings from frontal New Car Assessment Program for model year 1995. Blank spaces indicate no data. * Denotes vehicle without a passenger air bag. ** Denotes vehicle without a driver and passenger air bag. The neck readings from a vehicle without a passenger air bag were not included in the analysis.

| Make | Model | Driver | | | | | | Passenger | | | | | |
		Shear (N)	Comp (N)	Tension (N)	Exten (Nm)	Flex (Nm)	HIC	Shear (N)	Comp (N)	Tension (N)	Exten (Nm)	Flex (Nm)	HIC
Acura	Integra	743	1064	2402	20	47	675	751	395	1237	21	66	664
Audi	A6	565	1835	539	22	13		0	1707	760	34	22	
BMW	325i	746	260	1900	16	34		644	250	900	15	49	
Chevrolet	C-1500*	1435	24	2261	43	90	487	433	299	2134	22	33	539
Chevrolet	Lumina	200	264	1700	7	27	394	434	255	2213	31	31	560
Chevrolet	Monte Carlo	473	2275	416	30	21	684	864	2522	764	42	58	783
Chevrolet	S-10 PU*	798	2275	844	58	42	1065	1503	3294	231	68	48	1359
Chevrolet	S10 Blazer*	876	1581	2336	99	56	932	1716	2836	3167	95	47	1874
Chevrolet	Cavalier	961	509	1924	47	42	814	500	97	1566	35	73	788
Dodge	Avenger	719	126	1586	13	88	408	1280	1082	1044	37	134	277
Dodge	Ram Van*	480	1	3000	25	6	1474	2091	518	2815	43	116	572
Dodge	Stratus	606	484	4200	12	49	858	475	185	2021	34	19	
Ford	Aspire	319	307	1622	34	29	487	913	665	741	22	145	720
Ford	Crown Victoria	390	751	2175	42	22	633	727	435	2213	35	31	276
Ford	Explorer	320	278	2294	30	15	525	656	281	1800	30	53	448
Ford	Ranger	728	868	2449	28	50	508						537
Ford	Contour	796	215	1500	39	49	471	340	133	1600	30	32	357
Ford	Escort	972	2213	147	33	25	707	1266	2600	725	19	56	622
Ford	Winstar*	394	1321	315	44	18	518	772	1900	419	15	45	231
Geo	Metro	629	90	1769	19	57	467	775	227	1244	31	66	186
Honda	Odyssey	425	2555	403	21	31	637	963	1613	433	75	30	644
Hyundai	Sonata	694	2414	675	61	15	793	937	1887	615	48	18	511
Isuzu	Trooper II	828	431	2325	22	49	853	712	309	2161	47	41	735
Jeep	Cherokee*	1696	542	2793	81	33	484	1162	1031	2402	12	46	682
Mazda	323 Protege	422	282	2595	25	16	846	943	357	2278	42	58	996
Mazda	Millenia	358	1619	688	49	14	433	585	1859	643	42	35	202
Mazda	MPV	1489	1345	3066	62	7	682						768
Misubishi	Eclipse	701	106	1853	24	80	552	1200	697	830	28	120	503
Mitsubishi	Montero	1733	189	2399	53	82	640	428	651	2476	20	36	549
Nissan	240 SX	300	540	1951	35	40	900	875	163	1761	83	56	404
Nissan	Maxima	464	2250	678	29	19	747	815	2541	332	39	17	783
Oldsmobile	Aurora*	1144	1829	385	81	74	687	1345	3953	1071	39	141	936
Plymouth	Neon	811	2401	998	77	53	1088	641	2692	539	61	24	924
Saab	900	718	723	1690	17	58	640	407	467	1704	16	13	578
Saturn	SL2	853	2013	813	71	29	633	589	1950	500	69	20	506
Subaru	Legacy	380	1816	449	20	14	482	499	1962	565	16	31	532
Suzuki	Sidekick	914	398	2800	17	42	1214	1856	1427	2649	12	126	886
Toyota	Tercel	790	324	2166	36	40	907	576	247	1884	39	40	561
Volkswagen	Passat	452	2099	780	16	40		655	1502	407	0	59	
Vw	Jetta	325	200	1950	15	33	725	1007	250	2312	4	88	637

Table C-3. Hybrid III Dummy neck readings from frontal New Car Assessment Program for model year 1996. Blank spaces indicate no data. * Denotes vehicle without a passenger air bag. The neck readings from a vehicle without a passenger air bag were not included in the analysis.

Make	Model	Driver						Passenger					
		Shear (N)	Comp (N)	Tension (N)	Exten (Nm)	Flex (Nm)	HIC	Shear (N)	Comp (N)	Tension (N)	Exten (Nm)	Flex (Nm)	HIC
Acura	TL 4-dr	519	79	1711	22	59	740	675	749	1223	32	57	628
Audi	A4	416	534	1807	32	30	665	449	265	1470	21	37	432
Chevrolet	Astro Van*	687	188	1684	22	53	613	1498	828	2777	138	33	412
Chevrolet	C-1500 PU*	297	260	2486	16	35	498	1481	885	2266	27	82	487
Chevrolet	Tahoe*	225	2164	368	29	18	683	1827	2930	287	101	58	899
Dodge	Caravan	280			28	12	879	862	68	349	86	116	403
Dodge	Neon	738	517	2060	10	75	610	376	173	2383	19	32	531
Dodge	Ram Van*	875	270	4803	84	66	874	1825	800	2509	56	120	764
Ford	Crown Victoria	391	1912	933	14	19	499	645	1559	749	39	19	218
Ford	Mustang	378	653	1637	35	37	216	926	2463	1444	50	65	128
Ford	Taurus	564	230	1733	18	24	541	684			25	57	438
Geo	Tracker	482	2623	300	40	45	930	849	2157	200	41	58	666
Honda	Civic 4-dr	425	2797	1175	9	35	375	1097	925	1086	13	86	531
Honda	Civic 2-dr	470	292	1771	15	45	480	867	610	1267	67	26	329
Hyundai	Accent	973	1365	246	80	12	928	279	3355	4450	37	22	348
Hyundai	Elantra	443	127	1920	22	42	528	417	362	1738	32	34	773
Isuzu	Rodeo	584	2270	352	45	43	528	417	2299	286	33	31	782
Isuzu	Tropper II	300	853	2724	31	45	668	400	598	2238	34	37	843
Jeep	Cherokee*	911	455	3634	58	1	952	1196	587	1591	18	119	554
Landrover	Discovery	822	400	2837	56	41	825	2153	390	2969	97	34	379
Lexus	ES300	426	199	1412	11	44	432	618	400	1095	34	19	902
Lincoln	Town Car	301	684	2311	31	13	600	776	531	1716	72	39	181
Mazda	Miata	365	2770	1202	46	16	710	574	2028	608	32	39	672
Mazda	MPV	165	343	2121	19	17	593	360	414	1851	25	51	409
Mitsubishi	Mirage	864	438	2471	39	17	516		1625	606		115	997
Nissan	Altima	541	2076	240	53	21	710	449	2087	347	19	16	777
Nissan	4X2 PU	544	342	3040	23	49	758	2031	1118	2457	50	122	653
Nissan	Sentra	489	1867	214	38	27	583	709	1937	80	36	20	599
Pontiac	Grand Am 4-dr	621	41	1800	11	59	535	775	306	1631	16	122	604
Subaru	Impreza	260	942	1798	22	25	491	315	924	1789	28	19	515
Toyota	4-runner	582	1082	2670	36	41	920	606	479	2152	44	48	601
Toyota	Avalon	323	1850	364	33	41	517	796	776	772	32	49	243
Toyota	Camry	335	2067	714	27	20	627	1153	2317	526	22	74	475
Toyota	Tacoma*	450	2788	1700	9	35	1240	1892	2910	674	84	41	1041

Table C-4. Hybrid III Dummy neck readings from frontal New Car Assessment Program for model year 1997. Blank spaces indicate no data. * Denotes vehicle without a passenger air bag. The neck readings from a vehicle without a passenger air bag were not included in the analysis.

Make	Model	Driver						Passenger					
		Shear (N)	Comp (N)	Tension (N)	Exten (Nm)	Flex (Nm)	HIC	Shear (N)	Comp (N)	Tension (N)	Exten (Nm)	Flex (Nm)	HIC
Buick	LeSabre	572	409	1813	30	29	566	612	1477	399	45	31	686
Cadillac	Deville	535	2263	507	24	22	656	553	2007	262	49	20	552
Chevrolet	Cavalier 2-Dr	1221	1480	415	47	32	646	922	1651	560	39	33	885
Chevrolet	Venture	339	710	3033	50	14	692	851	2162	408	71	33	704
Chevrolet	Malibu	1227	1945	1007	58	60	810	453	1733	395	36	28	546
Chevrolet	Tahoe	293	1002	2245	40	28	823	915	660	2265	38	58	545
Chevrolet	S-10 Ext PU*	875	4114	1270	31	61	955	620	1602	1238	45	45	1205
Chevrolet	Blazer*	1029	1189	3186	67	47	595	1605	2702	3011	43	73	1525
Chevrolet	CK Ext PU	650	1100	1426	43	45	468	754	1666	550	49	39	689
Chevrolet	CK PU	581	212	1977	9	47	314	830	2174	1105	73	46	381
Chrysler	Sebring	1041	1032	1585	35	70	654	582	522	2500	51	36	698
Dodge	Ram Ext PU*	666	843	2058	42	31	793	987	125	1092	16	15	1004
Dodge	Dakota	480	480	2630	32	24	669	800	1750	300	75	34	603
Dodge	Caravan	752	357	2400	29	61	773	837	755	1403	20	118	419
Ford	F150 PU	333	700	2096	45	24	548	930		2101	37	34	474
Ford	Escort		2102	372		17	959	934	2320	400	48	76	436
Ford	Ranger	253	293	2237	37	52	724	846	718	2393	33	60	711
Ford	Expedition	740	1999	685	29	32	693	617	1804	203	48	29	393
Ford	Club Wagon	602	703	1795	41	52	932	859	231	2109	37	37	565
Ford	Winstar	435	42	1352	7	59	363	668	2176	48	37	31	294
Honda	Accord 2-Dr	434			39	21	447	616	1767	430	25	33	716
Hyundai	Accent	499	1805	200	40	20	917	860	1372	305	72	35	252
Jeep	Cherokee	714	389	3089	35	19	692	501	531	1927	31	39	512
Jeep	Wrangler		130	2680	55	13	566		257	758		2	488
Kia	Sportage	825	496	3611	41	20		2248	828	2597	57	142	
Kia	Sephia	1035	427	2092	23	56	872	1504	306	2520	112	27	406
Mitsubishi	Galant	913	831	946	1	8	526	887	1499	340	72	19	487
Nissan	Pathfinder	835	428	3477	31	41	1107	913	351	3374	50	31	797
Nissan	200SX	565	155	1802	22	43	423	618	439	1927	33	51	543
Pontiac	Grand Am 4-Dr	747	174	1573	17	30	517	525	1059	1657	54	13	604
Pontiac	Grand Prix	369	228	2058	16	33	719	560	773	1582	11	4	528
Pontiac	Grand Am 2-Dr	924	1797	594	32	20	626	537	962	540	30	19	373
Toyota	Paseo		1995	1162		13	632	752	2009	329	58	31	655
Toyota	RAV4	474	785	2579	22	28	919	911	468	1402	28	74	747
Toyota	Tacoma Ext PU*	354	886	3453	40	11	1411						
Toyota	Tercel	249	446	2227	28	12	470	349	415	1861	25	27	
Toyota	Camry	643	91	1827	10	89	625	472	1822	367	25	28	501
Volvo	960	517	789	1582	27	42	511	292	1673	1005	29	25	699

Table C-5. Hybrid III Dummy neck readings from frontal New Car Assessment Program for model year 1998. Blank spaces indicate no data. All vehicles have driver and passenger an air bags. The neck readings from a vehicle without a passenger air bag were not included in the analysis.

| Make | Model | Driver | | | | | | Passenger | | | | | |
		Shear (N)	Comp (N)	Tension (N)	Exten (Nm)	Flex (Nm)	HIC	Shear (N)	Comp (N)	Tension (N)	Exten (Nm)	Flex (Nm)	HIC
Chevrolet	Cavalier 4 dr	840	406	2112	38	27	514	692	444	1054	27	32	751
Chevrolet	Cavalier 2 dr	1053	85	1671	18	46	643	414	181	1137	28	30	620
Chevrolet	CK-Ext PU	358	279	2190	42	30	726	1171	457	2378	103	45	693
Chevrolet	Camero	758	347	2372	38	27	469	772	436	1999	27	54	328
Chevrolet	Malibu	1011	849	1350	58	55	691	549	214	1516	22	35	473
Chevrolet	S-10 Ext PU	774	714	2686	36	48	634	999	501	2315	68	57	450
Chevrolet	Suburban	347	500	2601	32	35		984	389	2620	61	74	
Chevrolet	Venture	912	600	2982	17	10	538	1339	1212	2652	62	1	962
Chevrolet	Blazer	810	236	2769	47	64	668	786	442	2256	29	40	506
Chevrolet	Lumina	440	175	2362	26	16	679	670	221	982	10	48	495
Dodge	Caravan	352	535	2610	47	3	870	672	821	1584	18	57	788
Dodge	Dakota	550	963	3040	59	35	550	944	194	1814	35	45	570
Dodge	Durango	848	763	4448	80	16		1002	632	2595	46	74	
Dodge	Grand Caravan	416	2671	200	34	30	1026	536	2210	517	36	23	994
Dodge	Neon	978	404	1904	19	83	655	534	140	2405	22	34	533
Dodge	Ram	516	760	2528	35	8	691	1508	525	2608	72	36	295
Dodge	Stratus	1114	1378	1889	22	84	873	570	1182	2401	21	24	641
Ford	Escort	786	2241	81	31	25	681	667	3207	1315	10	7	532
Ford	Expedition	559	392	1843	24	61	544	318	430	1569	33	25	569
Ford	Explorer	924	488	3237	43	41	567	644	820	2314	19	47	558
Ford	F-150	438	1113	2770	23	40	497	480	1048	2156	35	31	615
Ford	Contour	338	178	89	11	49	514	666	65	1367	11	72	617
Ford	Crown Victoria	397	600	3174	14	43	602	902	376	1908	23	54	335
Ford	Mustang	389	513	1728	15	32	436	776	392	1855	43	33	364
Ford	Ranger	338	654	2045	32	30	442	617	778	1095	31	52	545
Ford	Taurus	415	370	3148	28	31	577			1307	19	45	486
Ford	Winstar	620	241	993	4	50	535	452	82	1454	45	33	471
Honda	Accord 4 dr	442	1819	106	32	12	631	552	1403	225	45	39	596
Honda	Civic	354	1157	2140	23	28	619	1138	524	92	21	85	531
Honda	CRV	680	376	2342	41	58		839	611	1321	23	59	
Honda	Accord 2 dr	547	244	1193	16	45	454	793	98	1454	27	43	642
Isuzu	Rodeo	609	962	2536	32	44	650	590	1452	1597	61	40	561
Lexus	ES300	697	85	1435	9	58	512	528	420	1028	15	30	478
Nissan	Frontier	559	692	3193	45	29		585	723	1188	28	138	
Nissan	Sentra	651	344	2399	18	40	898	817	315	2097	13	24	797
Nissan	Maxima	338	156	1502	18	25	565	682	459	2459	25	11	654
Nissan	Altima	616	2069	187	28	26	887	643	2559	398	10	38	1119
Oldsmobile	Intrigue	287	236	1588	8	43		847	189	1153	19	85	
Saturn	SL	721	262	1662	11	88	435	860	396	2643	32	42	585
Subaru	Legacy	672	206	1761	22	39	525	503	1332	1632	42	33	623
Toyota	Camry	608	1258	78	45	18	525	320	1182	256	29	17	480
Toyota	4-runner	752	3156	3749	56	61	760	542	408	2797	26	40	743
Toyota	Corolla	863	957	2349	35	79	722	433	478	1168	27	37	566
Toyota	Avalon	650	291	2034	12	36	504	427	242	1133	15	21	577
Toyota	RAV4	485	270	1564	31	41	434	704	361	1660	29	46	355
Toyota	Sienna	568		1447	27	54		770	89	1266	61	42	
Toyota	Tacoma	912	1268	3124	46	34	731	446	774	2610	33	55	683
Volvo	S-70	530	86	1679	11	35	259	394	305	1161	27	32	294

952722

Dynamic Characteristics of the Human Cervical Spine

Frank A. Pintar, Narayan Yoganandan, Liming Voo, Joseph F. Cusick, Dennis J. Maiman, and Anthony Sances, Jr.
Medical College of Wisconsin and the Department of Veterans Affairs Medical Center

ABSTRACT

This paper presents the experimental dynamic tolerance and the force-deformation response corridor of the human cervical spine under compression loading. Twenty human cadaver head-neck complexes were tested using a crown impact to the head at speeds from 2.5 m/s to 8 m/s. The cervical spine was evaluated for pre-alignment by using the concept of the stiffest axis. Mid cervical column (C3 to C5) vertebral body wedge, burst, and vertical fractures were produced in compression. Posterior ligament tears in the lower column occurred under flexion. Anterior longitudinal ligament tears and spinous process fractures occurred under extension. Mean values were: force at failure, 3326 N; deformation at failure, 18 mm; stiffness, 555 N/mm. The deformation at failure parameter was associated with the least variance and should describe the most accurate tolerance measure for the population as a whole.

INTRODUCTION

Cervical spine injuries occur as a result of motor vehicle crashes, falls, and athletic related incidents. These injuries can be severe and costly both to the individual and to society as a whole. Our understanding of the mechanism and biomechanics associated with these injuries comes from an analysis of the epidemiological literature as well as a limited number of experimental studies.

Despite improvements in pre-hospital stabilization and transport as well as the addition of comprehensive centers for definitive treatment, spinal injury continues to represent a health care problem of major significance. Spinal cord injuries carry very high risks of disability and fatality. From epidemiological studies we understand that cervical spinal injuries can result from a number of different motor vehicle crashes including frontal, side, rear and rollover. Survivors of spinal cord injury require extensive medical treatment and long term care. Previous epidemiological studies have indicated that the majority of survivors of cervical spinal cord injury have mid cervical spinal column fracture/dislocations (C4-C6) resulting from compression or compression-flexion related trauma [5, 18].

The importance of compressive related cervical spine injury has also been reflected in the experimental literature as a number of previous studies experimentally evaluated compressive related trauma to the head-neck complex of cadavers [1, 4, 7, 8, 10-13, 15-17, 19, 21, 22]. Many of these previous studies noted that the pre-alignment condition of the cervical column with respect to the head was an important factor in determining the resulting spinal column injury [10-13]. From our previous experimental work [13, 14, 20] as well as the theoretical work by Liu and Dai [6], it can be surmised that the "stiffest axis" of the cervical spine results when the cervical lordosis is removed and the spinal vertebrae are aligned vertically. Cervical alignment along this axis prior to compressive loading has been noted to produce compression related injuries to the mid column (C4-5-6) vertebral bodies [3, 13, 20].

For the application of automotive related trauma, a necessary focus of cervical spine injury research should be the development of anthropomorphic test devices (ATD). A first step in the process of creating design parameters to develop biofidelic ATD's is to obtain dynamic tolerance data and force-deformation corridors that describe the behavior of the cervical spine. The purpose of this paper therefore, is to present the dynamic force-deformation corridors under compressive loading for the human cadaver head-neck complex, to describe the injury tolerance of the human cadaver cervical column under compression related forces, and to describe the affect of pre-alignment condition of the cervical spinal column on the resulting injury mechanism and biomechanical parameters.

MATERIALS AND METHODS

SPECIMEN PREPARATION AND MOUNTING- Dynamic compression tests were performed on a total of 20 human cadaver head-neck complexes according to a previously established protocol [13]. Some results from the first six preparations have been reported previously [13] and were re-analyzed for the purpose of the present study. The age ranged from 29 to 95 years (mean = 62 ± 15); there were nine females and 11 males and all were cardiopulmonary related deaths. Briefly, human cadaver head-neck complexes were mounted in fixative at T1-T2 inferiorly; the cranium was left intact superiorly. The inferior of the preparation was mounted to a six-axis load cell and firmly fixed to the platform of an electrohydraulic piston apparatus (MTS Systems Corp., Minneapolis, MN). The cervical spine was pre-aligned to remove the lordosis of the column by including 15 to 30 degrees of head flexion. The head was held in place using pulleys and deadweights. A schematic of the test setup is shown (Figure 1). A flat plate with 20 mm thick ensolite padding was fixed to the piston of the testing device and served as the impact surface.

Recorded transducer measurements included an in-series load cell and displacement gauge in the piston and an accelerometer on the impactor. A six-axis load cell (Denton, Inc., Rochester Hills, MI) recorded the generalized force histories at the inferior of the preparation. Retroreflective pin targets were placed in the vertebrae anteriorly in the vertebral body, in the lateral mass, and in the spinous process. A 16 mm high-speed camera was used to document the kinematics of the event in the

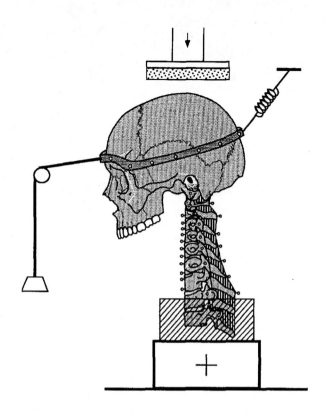

Figure 1: Schematic diagram of the human cadaver ligamentous head-neck preparation. The set-up included a pulley-weight system anteriorly, and a spring posteriorly to pre-align the specimen. A six-axis load cell was mounted inferiorly and the electrohydraulic piston applied loads to the crown of the head. Retroreflective pin targets were used to document the kinematics.

sagittal plane. Preparations were dynamically impacted at speeds from 2.5 m/s to 8 m/s.

All specimens were tested to failure with piston contact displacements from 25 to 40 mm. Transducer data were recorded using a digital data acquisition system at sampling rates from 8 to 12.5 kHz according to SAE J211b specifications, and post-processed with a SAE class 1000 digital filter.

ANALYSIS - Preparations were examined for pre-alignment condition by evaluating pretest x-rays taken prior to impact. The zero position for the "stiffest axis" was defined as the occipital condyles aligned vertically over the middle of the T1 vertebral body (Figure 2). Preparations with occipital condyles anterior to the T1 vertebral body were defined as positive pre-alignment and preparations with occipital condyles aligned posteriorly to the T1 vertebral body were assigned negative pre-alignment (Figure 2).

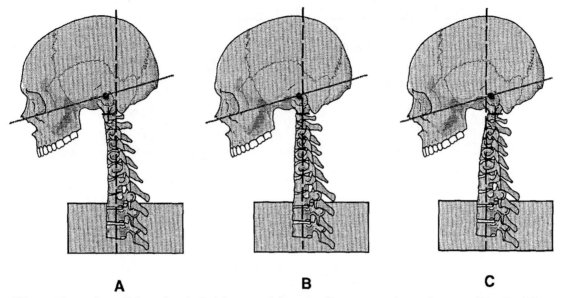

Figure 2: Illustrations describing the definition used in pre-alignment of specimens. The "stiffest axis" was defined as in B where the occipital condyles are aligned over the center of the T1 vertebral body. Specimens aligned with occipital condyles anterior to the stiffest axis (A) were assigned positive values, and specimens aligned posterior (C) were assigned negative values. All preparations had 15°-30° of head flexion.

The mechanism of injury was defined from viewing the 16 mm film and movements of the retroreflective pin targets. At the location of the primary injury, as defined by clinical evaluation of post-test x-rays, computed tomography (CT), and cryosections, the localized movements of the spinal column were defined as either extension, compression, or flexion. For example, specimen N07 demonstrated an anterior vertebral body fracture of C4 from examination of the CT's and x-rays; the kinematic analysis of the targets indicated downward movements of each component (vertebral body, lateral mass, spinous process) and was thus classified as a compression injury.

Each primary injury was also defined as either minor, moderate, or severe as follows:

Minor Vertebral and/or soft tissue trauma not requiring appreciable clinical intervention, i.e., neither internal or external intervention.

Moderate Vertebral and/or soft tissue trauma requiring moderate clinical intervention with external and possibly internal (surgical) intervention.

Severe Vertebral and/or soft tissue trauma requiring appreciable clinical intervention including both internal (surgical) and external (orthosis) intervention.

Each test was analyzed to obtain the dynamic force-deformation characteristics. The superior inferior force from the inferiorly mounted six-axis load cell was plotted against the deformation of the piston testing device. Since displacement of the piston occurred prior to contact with the cranium, the in-line force recorded on the piston was used to define the cranium contact. At the point of initial rise of the superior piston force, time zero was noted. The initiation of neck loading was determined using the kinematic film data by tracking the target on the head at the level of the occipital condyles. The first detection of movement was taken as time zero for neck loading and the resulting dynamic force-deformation responses began at this time. Dynamic stiffness was obtained by taking the slope of the force-deformation response in the linear phase just before failure.

RESULTS

A summary table of the various derived and recorded parameters is provided (Table 1). Out of the 16 preparations that were pre-aligned close to the stiffest axis (± 0.5 cm), 11 produced directcompression injuries. Upon analysis of the films from the remaining five preparations, the

Table 1: Summary of Biomechanical Data

I.D.	Pre-Align (cm)†	Mech#	Stiffness (N/mm)	Failure load (N)*	Neck disp @ failure (mm)	Injury Severity‡	Injury
N01	-0.5	C	235	1183	16.4	minor	vertical fracture of C3 vertebral body
N02	0.0	C	957	3678	12.6	severe	compressive burst fracture of C5 vertebral body
N03	0.0	C	168	744	15.0	severe	wedge fracture of C4, compressive fractures of C2, C3
N04	0.0	C	619	5005	18.8	moderate	wedge fracture of C6 vertebral body
N05	0.5	E	1375	6431	19.6	moderate	C3-C4 anterior longitudinal ligament tear with avulsion fracture of C3 vertebral body
N06	0.5	F	357	3445	19.7	severe	C2-C3 dislocation with ligamentous rupture
N07	0.5	C	370	4580	17.7	moderate	anterior vertebral body fracture of C4 vertebral body
N08	0.0	C	897	3906	22.3	moderate	vertical fracture of C3 vertebral body with lamina fracture
N09	0.0	C	1222	5179	17.5	minor	anterior-superior chip fracture of C3 vertebral body
N10	0.0	C	735	3744	16.5	moderate	compression fractures of C4, C7 vertebral bodies
N11	-0.5	E	652	4799	19.0	moderate	C3-C4 anterior longitudinal ligament tear; C4, C5 spinous process fractures
N13	2.5	F	512	2884	13.9	minor	interspinous ligament tear at C6-C7
N14	0.0	C	515	3086	17.4	severe	compressive burst fracture of C5 vertebral body
N15	-0.5	E	125	1341	19.6	moderate	C6-C7 anterior longitudinal ligament tear with lamina fractures of C6, C7
N16	3.0	F	282	2545	20.9	moderate	entire posterior ligaments torn at C6-C7
N17	0.0	C	453	2732	14.5	moderate	compression fracture of C5 vertebral body
N18	0.5	F	516	2691	21.5	minor	mild compression of C7 vertebral body
N19	2.0	F	273	2707	14.8	moderate	entire posterior ligaments torn at C7-T1
N20	1.0	C	251	2410	24.8	severe	compression burst fracture of C5 with posterior ligament rupture at C5-C6
N21	0.5	C	596	3699	16.5	severe	compression burst fracture of C4 with posterior ligament rupture at C3-C4

NOTES: * Superior-inferior load from inferior load cell
Mechanism of Failure as detected from high speed film; C=compression, F=flexion, E=extension
† Pre-alignment condition defined as horizontal offset of occipital condyles with respect to center of T1 vertebral body
‡ Injury Severity as defined from a clinical perspective, see text

---

failure mechanism was dependent upon the initial slight curvature of the spine. Extension failures occurred in preparations with slight initial lordosis of the cervical spine; flexion injuries occurred in preparations with slight kyphosis of the column. The three preparations that were pre-aligned two or more centimeters anterior to the "stiffest axis" all produced similar lower column flexion injuries with no bony injury.

Injuries occurring under compression were consistently mid-column fractures of the vertebral bodies, with burst, wedge, and vertical fractures (Figure 3). Injuries occurring in the flexion mode were more ligamentous having a component of posterior ligament involvement. The three spines that failed under an extension mechanism all had anterior ligament involvement (Figure 4). The load

duration to failure for all specimens was less than ten msec.

Generally, the spines that demonstrated compression failure mechanisms had greater values of failure force (mean = 3334 N) and stiffness (mean = 585 N/mm) than spines failing in flexion or extension (exception: specimen N05, 38 yr. old, extension). Neck displacement at failure was generally consistent, with no particular patterns with regard to mechanism of injury, injury severity, or age (overall mean = 18 ± 3 mm).

When classifying the injuries in terms of severity, there was no one biomechanical parameter that described a trend. The mean stiffness values decreased slightly but not significantly from minor (621 N/mm) to moderate (578 N/mm) to severe (474 N/mm). The dynamic force-deformation

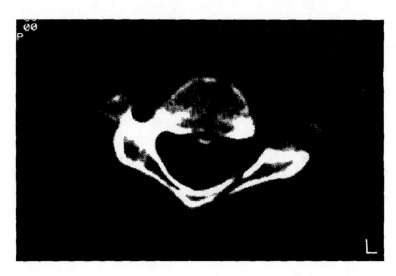

Figure 3: Axial CT through C3 vertebra of specimen N08 post test. Note the fracture line through the vertebral body and the lamina.

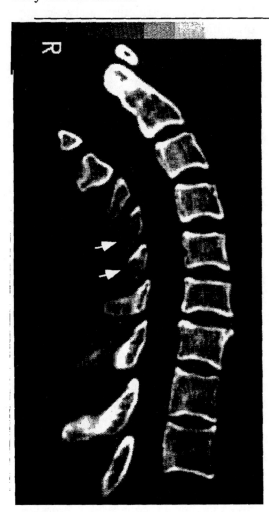

Figure 4: Sagittal CT of specimen N11 post test. Arrows indicate spinous process fractures of C4 and C5. There was also an anterior longitudinal ligament tear at C3-C4.

responses for all samples are plotted in figure 5. The resulting suggested corridor and mean response curve are depicted in figure 6. The mean response curve was derived from the mean force at failure (3326 N), mean deformation at failure (18 mm), and mean stiffness (555 N/mm). This stiffness depicts the upper end of the curve; the average stiffness (3326 N divided by 18 mm) is 185 N/mm.

DISCUSSION

The human cervical spinal column has to resist physiologic as well as traumatic forces. Under these conditions, it is important to maintain the normal functional relationships between the bony elements of the spine and the cervical spinal cord. Clinical studies indicate that under traumatic situations such as motor vehicle crashes, surviving victims with cervical injuries often have compression related trauma to the mid column [2, 18]. These injuries have significant societal costs. To reduce the risk of injury to the occupant of a motor vehicle, safer vehicles must be designed. A major tool in the development of safer vehicles is the anthropomorphic test devices. To increase the biofidelity of the ATD, human volunteer and cadaver experimentation must be done. This paper describes the compressive tolerance of the human cadaver head-neck under axial loading to the crown of the head. The dynamic force-deformation corridor has been determined as a guide for further improvement to the ATD neck.

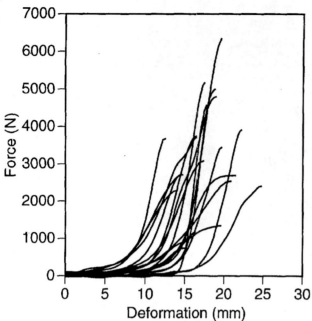

Figure 5: Dynamic human cervical spine force-deformation response curves for 20 cadaver specimens loaded under axial compression with impact to the crown of the head.

In the present series of tests, several different biomechanical parameters were recorded. Previous investigators examining cervical spine injuries noted that the pre-alignment condition of the cervical spine is a major contributor to the resulting injury pattern [3, 11]. The previous studies from our laboratory indicated that axial compressive burst fractures were experimentally reproduced only under one particular pre-alignment condition [13, 20, 22]. To provide a definition for pre-alignment the term "stiffest axis" was adopted as theoretically described by Liu and Dai [6]. The stiffest axis can be defined experimentally as the pre-alignment condition that results when a given amount of load produces the least amount of deformation [14]. This was determined for the cervical spine through previous experimental studies as the occipital condyles vertically aligned over the center of the T1 vertebral body accompanied by 15 to 30 degrees of head flexion. A similar "center of minimum stiffness" for the cervical spine in torsion has been reported [9].

In the present series of tests, the pre-alignment was measured with respect to the stiffest axis definition. Of the 16 preparations that were aligned on or close to the stiffest axis (± 0.5 cm), 11 produced axial compression injuries of the mid-

column. For the three preparations that were aligned greater than 1.5 cm anterior to the stiffest axis, all resulted in flexion injuries to the lower column. It appears that there is a "window" of alignment that is necessary to produce mid-column compression injuries, and outside of that window, distinctively different injuries occur. This may be the "cone" of the second stiffest axis suggested previously [6]. From the present series of tests, mid column bony compression injuries occurred only when pre-alignment was within ± 1 cm of the stiffest axis.

Although the resulting injury patterns were affected by pre-alignment, the biomechanical parameters within this sample population were not significantly different. Statistical analyses did not reveal any significant correlations between any of the measured biomechanical parameters and pre-alignment. The dynamic force-deformation responses all fit into a fairly narrow corridor. Perhaps when pre-alignment becomes drastically different than what was used in the present series of tests will the biomechanical parameters demonstrate significant variations. It can be stated with confidence however, that any alignment outside the window of the stiffest axis, will never produce mid-column compressive bony injuries of the cervical spine. Thus, if it is the intent to address these kinds of clinically relevant injuries that are routinely demonstrated in surviving victims of motor vehicle crashes, the present resulting corridors are valid.

To determine the mechanisms of injury, the resulting pathology as determined by gross examination, x-ray, CT, and cryomicrotomy, was evaluated with respect to the high-speed film of each test. The retroreflective targets inserted into bony landmarks of the vertebrae were examined according to previous methods [13, 20]. Kinematic analysis of adjacent targets documented when the injury occurred, and the mode (flexion, extension, or compression) of the affected area of the column was determined to be the injury mechanism. The rest of the column may have been under a different mode. For example, specimen N09 produced a fracture of the C3 vertebral body under local compression, however the lower aspect of the column went into extension without injury. It can also be noted that for essentially the same loading vector, (crown impact to the head) a variety of different injury mechanisms can be produced, depending more on the configuration of the spine prior to loading. This implies that without sufficient external data (e.g., scalp laceration) an assumed mechanism of injury from retrospective evaluation of radiographs may not be correct.

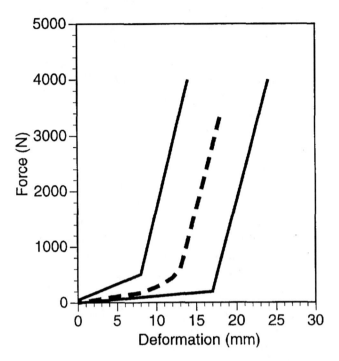

Figure 6: Derived human neck dynamic force-deformation corridor. The dashed line indicates the mean response curve derived from the mean force at failure (3326 N), mean deformation at failure (18 mm), and mean stiffness (555 N/mm). The mean values of force and deformation at failure describe the compression tolerance of the spine.

The dynamic force-deformation response curves indicated a region of softer response initially, followed by a significantly stiffer region before failure. This type of curve is typical of biological tissues. The initial less-stiff region represents contributions from the soft tissues (e.g., intervertebral discs) and the more-stiff region from the bony components. On a microscopic level, soft tissue structures respond by initial straightening of the collagen fibers as they are stretched in response to an applied load. In the present study, kinematic analysis revealed the initial compression of the intervertebral discs at this time followed by relevant bony compressions during the stiffer phase. Deformation of each disc results in the 5-10 mm of soft response. Others have also shown this initial soft response with cadaver cervical spines tested in compression [10].

The biomechanical parameters were analyzed with respect to injury severity and mechanism of injury. There was a wide variation of values for force at failure and stiffness. These did not seem to be dependent upon any singular factor although mean stiffness values decreased slightly, but not significantly, from minor to moderate to severe injuries. This may confirm the intuitive notion that the stiffer column is more resistant to severe injuries. Failure loads ranged from 744 N to 6431 N, perhaps indicating the varying nature of the population in terms of age (range 29 to 95) and gender. A second order polynomial regression analysis between failure force and age increase demonstrated decreasing tendencies, however with weak correlation ($R^2 = 0.33$). The mean failure force of 3326 N can be described as the compressive neck loading tolerance of this population (mean age: 62 yrs). A more consistent measure of tolerance may be the value of neck deformation to failure however, which had much less variance (mean: 18 mm) with a standard deviation of three mm. The suggested compressive loading corridor given in figure 6 also includes the mean response curve from all 20 specimens using the mean from failure force, failure deformation, and stiffness variables. This corridor describes the structural characteristics of the human cadaver neck in direct compression. Perhaps a preliminary design of an ATD neck useful for compressive loading could follow this loading corridor and include a single measurement of axial deflection to describe the tolerance criteria.

ACKNOWLEDGEMENT

We gratefully acknowledge the technical assistance of Michael Schlick and Wendy Pietz. This study was supported in part by PHS CDC-R49CCR-507370, DOT NHTSA DTNH22-93-Y-17028, and the Department of Veterans Affairs Medical Research. The views expressed herein do not necessarily reflect the views of the funding agencies.

REFERENCES

1. Alem NM, Nusholtz GS, Melvin JW: "Head and neck response to axial impacts." *28th Stapp Car Crash Conf.* SAE paper #841667:275-278, 1984.

2. Bohlman HH: "Acute fractures and dislocations of the cervical spine." *Journal of Bone and Joint Surgery*. 61A(8):1119-1140, 1979.

3. Burstein AH, Otis JC, Torg JS. "Mechanisms and pathomechanics of Athletic Injuries to the cervical spine." In: <u>Athletic Injuries to the Head, Neck and Face</u>, J.S. Torg, ed. Philadelphia: Lea and Febiger, pp 139-154, 1982.

4. Ewing CL, Thomas DJ, Sances A Jr, Larson SJ, eds. <u>Impact Injury of the Head and Spine</u>. Springfield, IL: Charles C. Thomas, 1983.

5. Kraus JF. "Epidemiological aspects of acute spinal cord injury: A review of incidence, prevalence, causes, and outcome." In: <u>Central Nervous System Trauma Status Report - 1985</u>, D. Becker and J. Povlishock, eds.. Washington, D.C.: National Insitute Neurol and Communicative Disorders and Stroke, National Institutes of Health, pp 313-322, 1985.

6. Liu YK, Dai QG: "The second stiffest axis of a beam-column: Implications for cervical spine trauma." *Journal of Biomechanical Engineering*. 111(2):122-127, 1989.

7. Maiman DJ, Sances A Jr, Jr, Myklebust JB, et al.: "Compression injuries of the cervical spine: A biomechanical analysis." *Neurosurgery*. 13(3):254-260, 1983.

8. McElhaney JH, Paver JG, McCrackin JH, Maxwell GM: "Cervical spine compression responses." *27th Stapp Car Crash Conf*. SAE paper #831615:163-178, 1983.

9. Myers BS, McElhaney JH, Dohery BJ, Paver JG, Nightingale RW, Ladd TP: "Responses of the human cervical spine to torsion." *33rd Stapp Car Crash Conf*. SAE paper #892437:215-222, 1989.

10. Myers BS, McElhaney JH, Richardson WJ, Nightingale RW, Doherty BJ: "The influence of end condition on human cervical spine injury mechanism." *35th Stapp Car Crash Conf*. SAE paper #912915:391-399, 1991.

11. Nusholtz GS, Huelke DF, Lux P, Alem NM, Montalvo F: "Cervical spine injury mechanisms." *27th Stapp Car Crash Conf*. SAE paper #831616:179-198, 1983.

12. Nusholtz GS, Melvin JW, Huelke DF, Alem NM, Blank JG: "Response of the cervical spine to superior-inferior head impact." *25th Stapp Car Crash Conf*. SAE paper #811005:197-237, 1981.

13. Pintar FA, Sances A Jr, Yoganandan N, et al.: "Biodynamics of the total human cadaver cervical spine." *34th Stapp Car Crash Conf*. SAE paper #902309:55-72, 1990.

14. Pintar FA, Yoganandan N, Pesigan M, Reinartz JM, Sances A Jr, Cusick JF: "Cervical vertebral strain measurements under axial and eccentric loading." *Journal of Biomechanical Engineering*. (In Press) 1995.

15. Pintar FA, Yoganandan N, Sances A Jr, Reinartz J, Harris GF, Larson SJ: "Kinematic and anatomical analysis of the human cervical spinal column under axial loading." *33rd Stapp Car Crash Conf*. SAE paper #892436:191-214, 1989.

16. Sances A Jr, Myklebust JB, Maiman DJ, Larson SJ, Cusick JF, Jodat R: "The biomechanics of spinal injuries." *CRC Crit Rev Bioeng*. 11:1-76, 1984.

17. Sances A Jr, Thomas DJ, Ewing CL, Larson SJ, Unterharnscheidt F, eds. <u>Mechanisms of Head and Spine Trauma</u>. Goshen, NY: Aloray, 1986.

18. Yoganandan N, Haffner M, Maiman DJ, et al.: "Epidemiology and injury biomechanics of motor vehicle related trauma to the human spine." *SAE Transactions*. 98(6):1790-1807, 1990.

19. Yoganandan N, Pintar FA, Sances A Jr, Maiman DJ: "Strength and motion analysis of the human head-neck complex." *J Spinal Disorders*. 4(1):73-85, 1991.

20. Yoganandan N, Pintar FA, Sances A Jr, Reinartz JM, Larson SJ: "Strength and kinematic response of dynamic cervical spine injuries." *Spine*. 16(10S):511-517, 1991.

21. Yoganandan N, Sances A Jr, Pintar FA: "Biomechanical evaluation of the axial compressive responses of the human cadaveric and manikin necks." *Journal of Biomechanical Engineering*. 111(3):250-255, 1989.

22. Yoganandan N, Sances A Jr, Pintar FA, et al.: "Injury biomechanics of the human cervical column." *Spine*. 15(10):1031-1039, 1990.

Head/Neck/Torso Behavior and Cervical Vertebral Motion of Human Volunteers During Low Speed Rear Impact: Mini-sled Tests with Mass Production Car Seat

Jonas A. Pramudita[1], Koshiro Ono[2], Susumu Ejima[2],
Koji Kaneoka[3], Itsuo Shiina[3], and Sadayuki Ujihashi[1]
1) Tokyo Institute of Technology, Japan
2) Japan Automobile Research Institute
3) University of Tsukuba, Japan

ABSTRACT

The purpose of this study is to clarify the neck injury mechanism during low speed rear impact. Low speed rear impact tests on human volunteers were conducted using a mini-sled apparatus with a mass production car seat. Head/neck/torso behavior and cervical vertebral motion were analyzed and strains on the cervical facet joint were determined. The effect of differences on gender, seat back angle, sled acceleration and muscle condition to the head/neck/torso behavior and the cervical vertebral motion were also discussed. Moreover, the risk of neck injury in a mass production car seat environment was evaluated by comparing the strain values to the results from a previously reported test that used a rigid seat environment.

Keywords: Neck, Sled Tests, Volunteers, Mass Production Car Seat, Cervical Vertebral Motion

NECK INJURIES have a higher incidence during rear impact collisions compared to other collision patterns (Ono et al., 1996; Eis et al., 2005). Therefore, clarification of the neck injury mechanism during rear impact calls for urgent attention.

Studies simulating rear impact using human volunteers have been conducted to analyze the occupant's behavior during rear impact collisions (van den Kroonenberg et al., 1998; Brault et al., 1998; Siegmund et al., 2000; Tencer et al., 2001; Hernandez et al., 2005; Dehner et al., 2007). However, in almost all of the studies, human kinematics were only considered and subsequently used for predicting the occurrence of neck injuries. In a previously reported study, low speed rear impact tests on human volunteers using a mini-sled apparatus with a rigid seat and without a head restraint (Ono et al., 2006) were conducted. The cervical vertebral motion of the human volunteer was recorded using a cineradiography system and the strains on intervertebral disc and facet joint were also calculated. However, in a real car condition, a mass production car seat is used instead of a rigid seat. The structure of the mass production car seat, the shape of the head restraint, and the mechanical properties of the seat have a significant effect on the head/neck/torso behavior and the cervical vertebral motion during low speed rear impact.

Welcher et al. (2001) conducted tests on human volunteers and determined the effects of seats with different properties on the human subject's kinematics. Kleinberger et al. (2003) conducted tests using dummies and evaluated the influence of the seat back and the head restraint properties on the occupant's dynamics using an existing neck injury criterion. However, there has been almost no reported research on the effect of a mass production car seat on cervical vertebra motion during rear impact.

Mertz et al., (1971) reported that the tension on the neck muscles reduced the head flexion during rear impacts. Authors also reported that the head flexion was decreased by 30% to 40% in the tension condition (Ono et al., 1999). Moreover, from the tests with a rigid seat, it was found that the cervical vertebral motion could not be restrained during the tension condition, although the head behavior was restrained significantly (Ono et al., 2006). However, this observation requires to be clarified in mass production seat environments.

In this study, low speed rear impact tests on human volunteers in a mass production seat environments were conducted to determine the influence of the differences in gender, the seat back angle, the sled acceleration and the muscle condition to head/neck/torso behavior and the cervical vertebral motion. The results were used to evaluate the risk possibility of neck injuries during rear impacts in mass production car seat environments.

EXPERIMENTAL METHODS

MINI-SLED APPARATUS

A mini-sled apparatus was designed by considering the maximum acceleration and duration of impact, based on the previous low speed rear impact testing (Ono et al., 1999) to ensure the volunteers' safety. Fig.1 shows a schematic view of the mini-sled apparatus. A carriage was set on horizontal rails and a mass production car seat was mounted on the carriage. The seat has a Whiplash Injury Lessening (WIL) technology (Sawada et al., 2005) developed by Toyota Motor Corporation, that allowed the occupant's torso to sink into the seatback during rear impact, controlling the relative motion between the head and torso. The rear impact was simulated by utilizing a force, from a motor placed on the anterior part of the apparatus, to pull the carriage frontward. A damper was used on anterior part of the rails to decelerate the carriage.

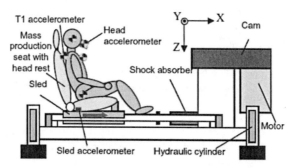

Fig.1 Schematic view of the mini-sled apparatus

PHOTOGRAPHY OF THE HEAD/NECK/TORSO BEHAVIOR USING A HIGH SPEED VIDEO CAMERA

In order to obtain the head/neck/torso behavior of the volunteers during the rear impact, a high speed video camera (Redlake MotionXtra HG-100K) with a photographic capability of 500fps was used. Markers that were attached to the volunteers' bodies were able to be tracked and analyzed.

PHOTOGRAPHY OF THE CERVICAL VERTEBRAL MOTION USING A CINERADIOGRAPHY SYSTEM

In order to obtain the cervical vertebral motion, a cineradiography system (Philips BH-5000) from the University of Tsukuba Hospital, with a photographic capability of 60fps was used. Using this system, 1024 x 1280 pixels of sequential images was obtained. Based on the sequential images, the cervical vertebrae were digitized and then analyzed.

MEASUREMENT METHOD

The sled acceleration was measured by mounting an accelerometer on the carriage. In order to calculate forces acting on the head/neck/torso during the rear impact, a 3-axis accelerometer and a 3-axis angular velocimeter were placed in the volunteer's mouth and on the body surface above the T1 spinous process, this enabled measurement of translational acceleration and angular velocity of head and T1. Furthermore, strain gauges were attached to the head restraint pole to determine the reaction force of the head restraint. The relationship between the strain measured and reaction force generated was determined through a calibration test. In order to obtain the muscle activities of the volunteers, electrodes were attached to the body surface above the representative muscles including the neck muscles. Electromyography (EMG) of the muscles was measured during the tests. In addition, the contact condition between volunteer's body and the seat was measured by attaching touch sensors to the surfaces of the occipital region of the head, back, head restraint and seat back. This sensor was capable of monitoring the start time and finish time of the contact between the volunteer's body and the seat.

TEST CONDITION

The sled accelerations used for the experiments were 28m/s² (females only), 33m/s² (both) and 40m/s² (males only). Fig.2 shows the time history of the sled accelerations and velocities. The seat back angle was set to 20° and 25°. Subjects were asked to perform the tests in both relaxed and tensed muscle states. The average and standard deviation of the backsets measured during the initial condition are shown in Fig.3.

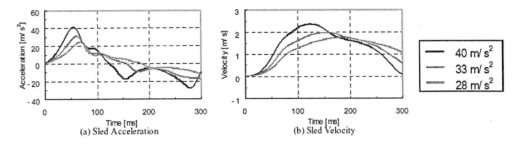

(a) Sled Acceleration   (b) Sled Velocity

Fig.2 Time history of the sled accelerations and velocities

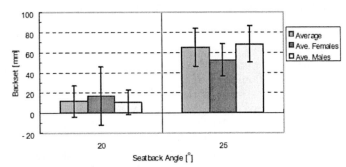

Fig.3 Average and standard deviation of the backsets measured during initial condition

VOLUNTEERS

6 adult males and 3 adult females were chosen as volunteers for the tests. Table 1 shows the physical information of volunteers. Tests with cineradiography measurements were conducted only using the male volunteers. The consent of the volunteers was obtained before testing commenced. The test protocol was subjected to the approval of the Special Committee of Ethics, Medical Department, University of Tsukuba.

Table 1. Details of the volunteers

Volunteer ID	Age [Year]	Sex	Height [cm]	Weight [kg]	Sitting height [cm]
I	23	F	164	50	85
II	22	F	160	45	86
III	24	F	162	46	85
IV	22	M	172	69	86
V	25	M	173	70	90
VI	24	M	176	65	94
VII	26	M	166	61	89
VIII	24	M	172	64	90
IX	25	M	172	67	92

ANALYTICAL METHODS

ANALYTICAL METHOD OF THE HEAD/NECK/TORSO BEHAVIOR

Coordinate system : An absolute coordinate system, as shown in Fig.1, was used for analyzing the head/neck/torso behavior. Positive values of the X-axis and the Z-axis indicate the forward direction and downward direction respectively. In addition, counter clockwise rotation around Y-axis was set as positive.

Physical quantities representing the head/neck/torso behavior : The head displacement and rotation angle were calculated as physical quantities representing the head behavior. The head behavior relative to upper torso was obtained by calculating the head displacement and rotation relative to the T1. The neck rotation relative to the T1 and head rotation relative to the neck were calculated as physical quantities representing the relative neck behavior. The T1 displacement and rotation were calculated as physical quantities of the upper torso behavior. Moreover, In order to determine the torso extension during the impact, the change in distance between the T1 and the hip point was calculated.

ANALYTICAL METHOD OF CERVICAL VERTEBRAL MOTION

Template of the cervical vertebra and coordinate systems : A template of the cervical vertebra for each volunteer was created from an x-ray image of each volunteer's cervical vertebra (Fig.4). The cervical vertebral motion was determined by superimposing the template over the sequential images from the cineradiography. Fig.5 shows the outer edge of the facet joint and the local coordinate system utilised.

Physical quantities representing the vertebral body motion : The vertebral rotation relative to the C7 and the lower vertebra were calculated as physical quantities representing the relative rotation between the vertebrae. The vertebral displacement relative to the C7 and the lower vertebra were also calculated to represent the shear between the vertebrae. The vertebral rotation and the vertebral displacement relative to the C7 were defined respectively as the rotation and the displacement of the lower edge midpoint of the local coordinate system of each vertebra relative to the local coordinate system of the C7. The vertebral rotation and displacement relative to the lower vertebra were defined respectively as the rotation and displacement of the lower edge midpoint of the local coordinate system of each vertebra relative to the local coordinate system of vertebra located directly below. Also the vertebral displacement relative to the lower vertebra was normalized by the linear distance of the lower edge anterior point and the posterior point of the lower vertebra. A positive value represented flexion, forward displacement and downward displacement.

Physical quantities representing the facet joint behavior : Because it was impossible to measure the facet joint strains directly, the front/rear edge strain and the shear strain of the facet joint, as shown in Fig.6, were calculated as physical quantities representing the facet joint behavior. These strains were assumed to be equivalent with the strains caused by the facet joint capsule deformation. In addition, positive values represented tensile strain and rearward shear strain.

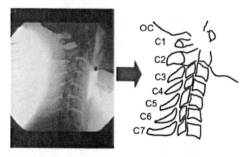

Fig.4 Template of the cervical vertebrae

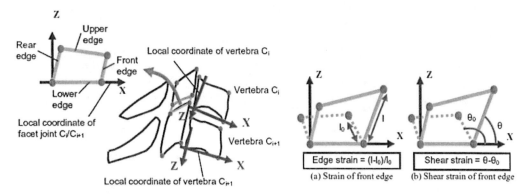

Fig.5 Local coordinate systems of the vertebra and facet joint

Fig.6 Strains representing the facet joint behavior

ANALYTICAL METHOD OF IMPACT FORCES ACTING ON HEAD/NECK

The head motion was assumed to be the 2-dimensional motion of a rigid body. The origin of head coordinate system was defined as the anatomical center of gravity of the head (Beier et al., 1980); located at a point which was 5mm in front of the ear hole on the Frankfurt line and 20mm perpendicular to this line. The X-axis was parallel to the Frankfurt line and Z-axis was perpendicular to X-axis (Fig.7). The acceleration responses of the head, and axial forces, shear forces, bending moments acting on the upper neck were determined from the values taken by the accelerometer and velocimeter placed in the subject's mouth and also by the values measured by the strain gauges attached on the head restraint pole.

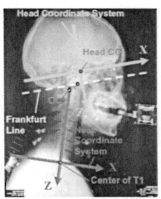

Fig.7 Definition of the Head and T1 Coordinate System

RESULTS

THE HEAD/NECK/TORSO BEHAVIOR AND THE CERVICAL VERTEBRAL MOTION IN A MASS PRODUCTION CAR SEAT ENVIRONMENT

The head/neck/torso behavior (Fig.8) and the cervical vertebral motion (Fig.9) during the impact were divided into the first phase; when the volunteer's body was sinking into seat, and the second phase; when the volunteer's body was rebounding forward. In this paper, the average results of the male volunteers for tests with a seat back angle of 25°, a sled acceleration 40m/s² and a relaxed muscle condition are presented.

<u>First phase of the impact (0ms to 130ms)</u> :

The upper torso contacted the seat back after 85ms and the head contacted the head restraint approximately 10ms later (Fig.10(a)). The head and the upper torso moved rearward as the head restraint and the seat back cushioned the impact (Fig.10(a)). Therefore, it was found that the head flexed relative to the upper torso (Fig.10(c)). Extension of the head began at 110ms (Fig.10(c)). The rearward displacement of the head and the T1 reached a maximum value at 130ms (Fig.10(a)). In addition, the extension of the upper torso also reached a maximum at this time (Fig.10(c)).

From the average results of each of the vertebral bodies' behavior relative to the C7, it was found that the C0 to the C6 began to flex due to the contact between the upper torso and the seatback (Fig.11(c)). The flexion angle was found to increase in the upper vertebrae relative to the C7. The C0 to the C6 also sheared forward relatively to the C7 (Fig.11(a)). In addition, the C0 and the C1 moved relatively upward, but the other vertebral body moved relatively downward (Fig.11(b)).

From the average results of the strains of the facet joint, forward shear strains occurred on the C5/C6 and backward shear strains occurred on the C2/C3 to the C5/C6 at 100ms (Fig.12(c)). Moreover, the tensile strain also occurred on the C2/C3 to the C5/C6 and compression strain was found to occur on the C6/C7 at the same time (Fig.12(a) and Fig.12(b)).

Second phase of the impact (130ms to 200ms) :
The head and the upper torso were found to move forward after 130ms due to the rebound of the head and upper torso from the seat (Fig.10(a)). The head extended to a maximum at 150ms, then began to flex (Fig.10(c)). The upper torso left contact from seatback at 165ms and head left the head restraint 15ms later (Fig.10(a)). This phenomenon called "differential rebound" was also confirmed in low speed rear-end impact tests on human cadavers (Sundararajan et al., 2004).

From the average results of each vertebra's behavior relative to the C7, it was found that the inclination of the flexion angle of the C0 to the C4 increased with the rebound of the head from head restraint (Fig.11(c)). At 150ms, the C6 angle changed from flexion to extension and the inclination of the flexion angle of the C0 to the C3 increased (Fig.11(c)). At 165ms, the C4 and C5 angle changed from flexion to extension and the inclination of the flexion angle of the C0 to the C3 increased again (Fig.11(c)). Moreover, the C0 to the C6 also changed from forward shear to rearward shear (Fig.11(a)), the C0 and the C1 also changed from upward displacement to rearward displacement at this time (Fig.11(b)). The flexion angle of the C0 to the C3 reached a maximum at 180ms (Fig.11(c)).

From the average results of strains on the facet joint, the strain on the C5/C6 changed from tension to compression at 150ms and tensile strain on the C2/C3 to the C4/C5 began to decrease at 180ms (Fig.12(a) and Fig.12(b)). The shear strain on the C4/C5 to the C6/C7 changed from rearward to forward at 165ms (Fig.12(c)).

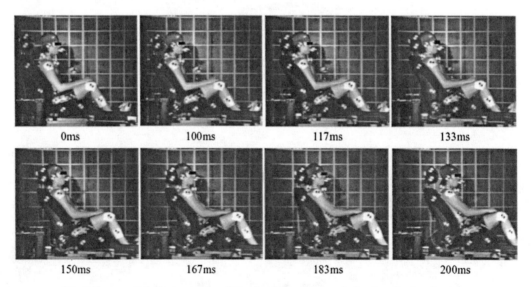

| 0ms | 100ms | 117ms | 133ms |

| 150ms | 167ms | 183ms | 200ms |

Fig.8 Sequential photographs of the head/neck/torso behavior during impact

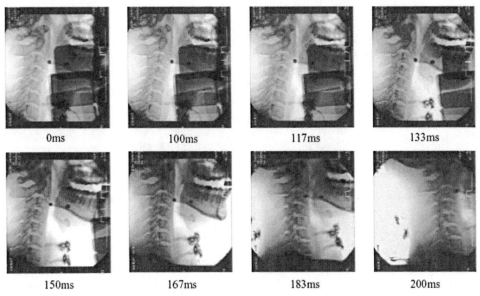

Fig.9 Sequential cineradiography images of the vertebral motion during impact

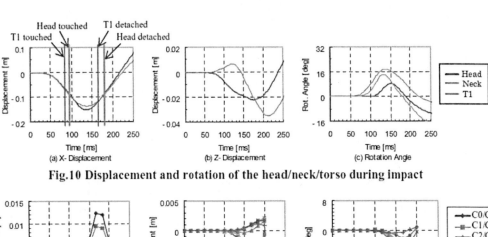

Fig.10 Displacement and rotation of the head/neck/torso during impact

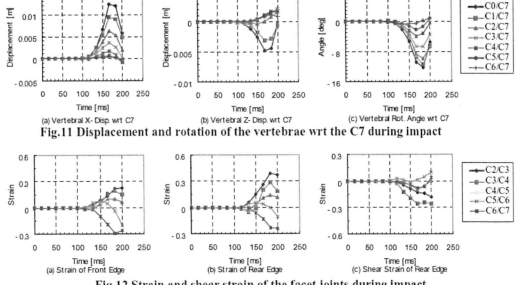

Fig.11 Displacement and rotation of the vertebrae wrt the C7 during impact

Fig.12 Strain and shear strain of the facet joints during impact

In addition, the head displacement and rotation angle relative to the T1, the head rotation angle relative to the neck, the neck rotation angle relative to the T1, and the change of the distance between the T1 and the hip point are shown in Fig.13. Vertebral displacement and the rotation angle relative to the lower vertebra are shown in Fig.14.

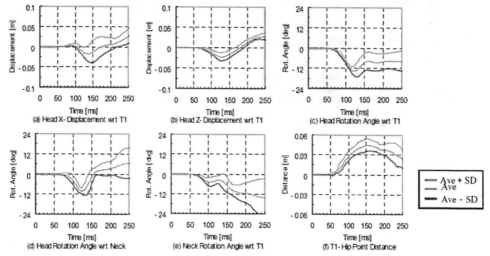

Fig.13 Relative displacement and rotation of the head/neck/torso during impact

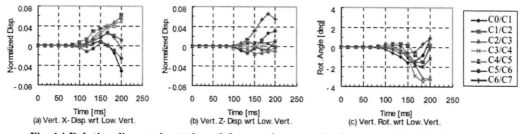

Fig. 14 Relative disp. and rotation of the vertebrae wrt the lower vertebra during impact

THE EFFECT OF GENDER DIFFERENCE ON THE HEAD/NECK/TORSO BEHAVIOR

As a result of the comparison of the average head/neck/torso behavior between males and females under same impact conditions, it was found that females have a larger head rearward displacement relative to the T1 compared to males as shown in Fig.15(a) ($p<0.1$). Furthermore, females were also found to have a greater head flexion angle relative to the neck compared with males as shown in Fig.15(b) ($p<0.05$). Meanwhile, the error bars in the figure show the standard deviation (SD) of each parameter.

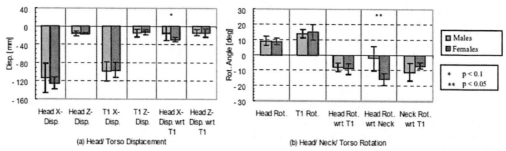

Fig.15 Comparison of the max. values of the parameters describing the average head/neck/torso behavior between male and female volunteers

EFFECT OF SEAT BACK ANGLE DIFFERENCE

The maximum value of the average head/neck/torso behavior and the average cervical vertebral motion between seat back angles of 20° and 25° under same sled acceleration and muscle condition were compared and the following results were obtained.

Head/neck/torso behavior :

The head and the T1 rearward displacement was smaller when the seat back angle was 20° compared to 25°, as shown in Fig.16(a) (p<0.05). Moreover, the head and the T1 extension angle, and the head flexion angle relative to the neck were restrained significantly when the seat angle was 20°, as shown in Fig.16(b) (p<0.05).

Cervical vertebral motion :

Since the C2 and the C3 flexion angles relative to lower vertebra were smaller when seat back angle was 20° (Fig.17(a)), the tensile strain on the rear edge of the C2/C3 and the C3/C4 facet joint also decreased (Fig.17(a)). Furthermore, because the C5 and the C6 extension angle relative to the lower vertebra were small (Fig.17(b)) and their rearward displacements are also small (Fig.17(c)), compression strain on the rear edge of the C5/C6 and the C6/C7 facet joint was better restrained when the seatback angle was 20°, as shown in Fig.18(b) (C5/C6: p<0.1). Despite forward shear strain on the upper vertebrae facet joint was increasing, (Fig.18(c)) due to an increase in the upper vertebrae forward shear relative to lower vertebra, as shown in Fig.17(c) (C3/C4: p<0.1), the rearward shear strain on the C5/C6 and the C6/C7 facet joint were restrained (Fig.18(d)) due to the decrease of the C5 and the C6 rearward shear relative to the lower vertebra as shown in Fig.17(d) (C5/C6: p<0.1).

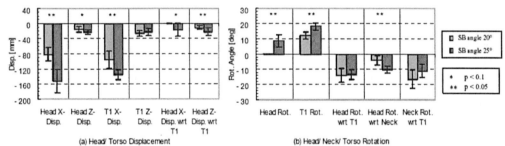

(a) Head/ Torso Displacement

(b) Head/ Neck/ Torso Rotation

Fig.16 Comparison of the max. values of the parameters describing the average head/neck/torso behavior between seatback angles of 20° and 25°

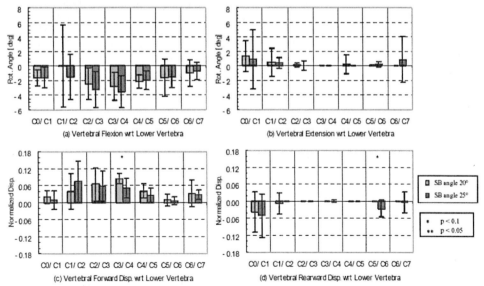

(a) Vertebral Flexion wrt Lower Vertebra

(b) Vertebral Extension wrt Lower Vertebra

(c) Vertebral Forward Disp. wrt Lower Vertebra

(d) Vertebral Rearward Disp. wrt Lower Vertebra

Fig.17 Comparison of the max. values of the parameters describing the average vertebral motion between seatback angles of 20° and 25°

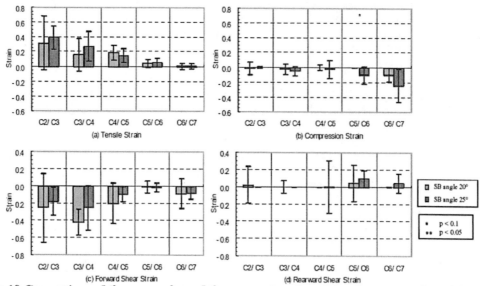

Fig.18 Comparison of the max. values of the parameters describing the average facet joint strains between seatback angles of 20° and 25°

EFFECT OF SLED ACCELERATION MAGNITUDE

The maximum value of the average head/neck/torso behavior and the average cervical vertebral motion between a sled acceleration of 33m/s² and 40m/s², under same seat angle and muscle conditions, were compared and the following results were obtained.

Head/neck/torso behavior :

The head (p<0.1) and the T1 (p<0.05) rearward displacements when the sled acceleration was 40m/s² were found to be greater than when the sled acceleration was 33m/s² (Fig.19(a)). Moreover, the T1 extension angle (p<0.05), the head flexion angle relative to the T1 (p<0.05) and the head flexion angle relative to the neck (p<0.1) indicated a similar tendency (Fig.19(b)).

Cervical vertebral motion :

As the sled acceleration increased, the flexion angle of the C0, and the C2 to C4 relative to the lower vertebra increased as shown in Fig.20(a) (C0/C1: p<0.1). The C5 rearward displacement relative to the C6 also increased, as shown in Fig.20(d). Therefore, the tensile strain on the rear edge of the C2/C3 to the C4/C5 facet joint and the compression strain on the rear edge of the C5/C6 (p<0.1) increased (Fig.21(a) and Fig.21(b)). However, the difference between the compression strain on the C6/C7 facet joints was not significant (Fig.21(b)). Furthermore, when the sled acceleration was 40m/s² the rearward shear strain on the C5/C6 facet joint increased (Fig.21(d)) due to an increase in the C5 rearward displacement relative to the C6 (Fig.20(d)).

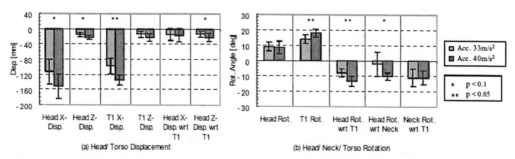

Fig.19 Comparison of the max. values of the parameters describing the average head/neck/torso behavior between sled acc. of 33m/s² and 40m/s²

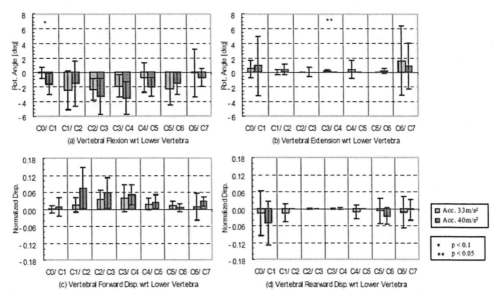

Fig.20 Comparison of the max. values of the parameters describing the average vertebral motion between sled acc. of 33m/s² and 40m/s²

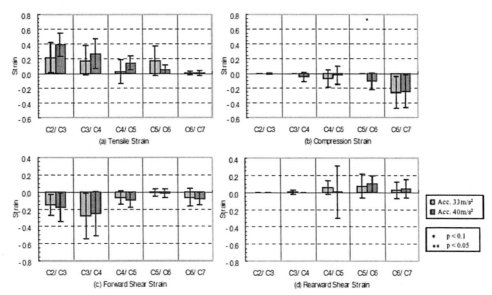

Fig.21 Comparison of the max. values of the parameters describing the average facet joint strains between sled acc. of 33m/s² and 40m/s²

EFFECT OF THE MUSCLE CONDITION

The maximum value of the average head/neck/torso behavior and the average cervical vertebral motion between relaxed and tensed muscle conditions, under same seat angle and sled acceleration, were compared and the following results were obtained.

Head/neck/torso behavior :

Comparing the results between relaxed and tensed muscle conditions, it was found that there were no significant differences on the rearward displacement and the upward displacement of the head and the T1 (Fig.22(a)). However, the head extension angle increased ($p<0.1$) in the tensed muscle condition, although it was also found that the head flexion angle relative to the T1, the head flexion angle relative to the neck and the neck flexion angle relative to the T1 were significantly restrained (p<0.05) (Fig.22(b)).

Cervical vertebral motion :

In the tensed muscle condition, the tensile strain on the rear edge of the C2/C3 to the C5/C6 facet joints were significantly restrained as shown in Fig.24(a) (C2/C3 and C4/C5: p<0.05, C5/C6: p<0.1) due to the decrease in flexion angle of the C2 to the C5 relative to the lower vertebra as shown in Fig.23(a) (C2/C3 and C4/C5: p<0.05, C3/C4: p<0.1). However, because the extension angle (C4/C5: p<0.05) and the rearward displacement of the C4 to the C6 relative to lower vertebra increased (Fig.23(b) and Fig.23(d)), there was nearly no restrictive effect in the compression angle on the rear edge of the C4/C5 to the C6/C7 facet joints. Conversely compression strain on the rear edge of the C4/C5 and the C5/C6 facet joints showed a tendency to increase (Fig.24(b)). Furthermore, in the tensed muscle condition, the forward shear strain on the C2/C3 to the C6/C7 facet joints decreased (Fig.24(c)) along with a decrease of the C1 to the C6 forward displacement relative to lower vertebra as shown in Fig.23(c) (C2/C3: p<0.1, C6/C7: p<0.05), though rearward shear strain on the C4/C5 to the C6/C7 increased (Fig.24(d)) due to the increase of the C4 to the C6 rearward shear relative to lower vertebra (Fig.23(d)).

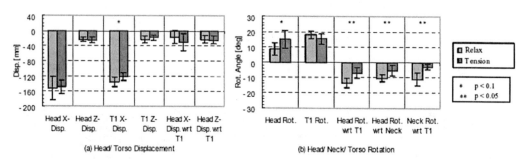

(a) Head/ Torso Displacement

(b) Head/ Neck/ Torso Rotation

Fig.22 Comparison of the max. values of the parameters describing the average head/neck/torso behavior between the relaxed and tensed conditions

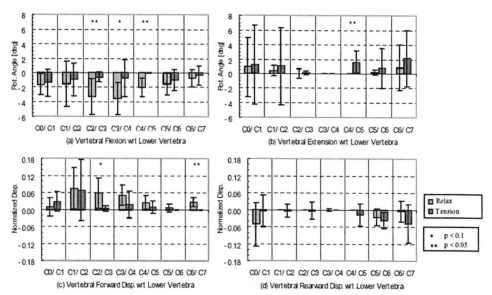

(a) Vertebral Flexion wrt Lower Vertebra

(b) Vertebral Extension wrt Lower Vertebra

(c) Vertebral Forward Disp. wrt Lower Vertebra

(d) Vertebral Rearward Disp. wrt Lower Vertebra

Fig.23 Comparison of the max. values of the parameters describing the average vertebral motion between the relaxed and tensed conditions

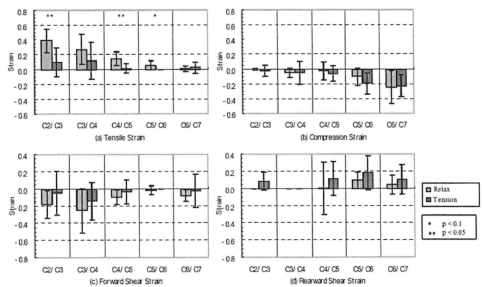

Fig.24 Comparison of the max. values of the parameters describing the average facet joint strains between the relaxed and tensed conditions

DISCUSSION

THE HEAD/NECK/TORSO BEHAVIOR AND THE CERVICAL VERTEBRAL MOTION CHARACTERISTIC

In the tests using a rigid seat without a head restraint, the imposition on torso by the rigid seat back and the inertial force acting on head caused flexion in the upper vertebrae and extension in the lower vertebrae, resulting in an S-shape curve of cervical vertebrae (Ono et al., 1997). In this study, since the upper torso sunk into the seat back and the head motion was restrained by the head restraint, a relative flexion occurred in almost all vertebrae. Moreover, ramping up by seat back caused a compression force on the intervertebral joint. It was assumed that the stiffness of intervertebral joint tend to decrease due to the compression force (Yang et al., 1997) and it might have caused the shear on the C2/C3 and the C5/C6.

In the second phase, the spine straightened and the head moved upward, so the acting point of the head restraint reaction force became lower than the OC (Occipital Condyles). Therefore, despite the head extension, the upper vertebrae showed relative flexion and forward shear behavior. Furthermore, the forward displacement made progress from the lower vertebrae, causing a rearward shear in the C4/C5 and the C5/C6, and a large extension in the C6 relative to the C7. Consequently, the tensile strain occurred on facet joints of the upper vertebrae and the compression strain occurred on the C6/C7 facet joint.

DIFFERENCES ON THE HEAD/NECK/TORSO BEHAVIOR DUE TO GENDER DIFFERENCE

Males have bigger cervical skeletal geometry and also the head moment of inertia is greater compared to females. This caused a flexion moment increase during the first phase of the impact, so the neck extension angles of the male volunteers were smaller when compared to the female volunteers. Moreover, another possible reason is that males are sinking into the seat back more than females due to being heavier. Van de Kroonenberg et al. (1998) reported that sinking more into seat back restrained the T1 acceleration, causing a decrease in head acceleration.

DIFFERENCES ON THE HEAD/NECK/TORSO BEHAVIOR AND THE CERVICAL VERTEBRAL MOTION DUE TO THE SEAT BACK ANGLE DIFFERENCE

As also reported by Latchford et al. (2005), the linear distance between the head and the head restraint (backset), and between the upper torso and the seat back in the initial condition tended to increase along with an increase of the seat back angle. As a result, if seat back angle is increased, it could be said that the rearward displacement of the head and the T1 would also increase. Furthermore, during the test condition using a small seatback angle, before the head extension occurred, the head collided against the head restraint first and then rebounded. Therefore, the extension behavior of the head was restrained significantly.

During the test condition when the seat back angle was 20°, the compression strain on the rear edge of the C5/C6 and the C6/C7 facet joints also decreased. This was caused by a decrease in the rearward shear and the extension of the lower vertebrae along with a decrease of the head rearward displacement. The decrease of the lower vertebrae extension angle due to a small backset corresponded well with the result of a simulation conducted, using the head/neck model of MADYMO by Stemper et al. (2006).

DIFFERENCES ON THE HEAD/NECK/TORSO BEHAVIOR AND THE CERVICAL VERTEBRAL MOTION DUE TO AN IMPACT LEVEL DIFFERENCE

As the sled acceleration increased, it was found that the impact forces acting on the volunteers also increased, causing an increase in almost all parameters representing the head/neck/torso behavior. However, the head extension angles of both conditions were nearly the same due to the limitation of the head restraint.

The flexion moment on the upper neck caused by the reaction force from the head restraint increased along with an increase of the sled acceleration. Therefore, the flexion angle of the upper cervical vertebrae increased, causing a subsequent increase in the tensile strain on facet joint. Furthermore, due to an increase in the impact force spreading from the T1 to a higher cervical vertebra, the rearward shear and extension of the C5 relative to the C6 also increased. As a result, the compression strain on the rear edge of the C5/C6 facet joint increased, following the compression strain on the C6/C7 facet joint.

DIFFERENCES ON THE HEAD/NECK/TORSO BEHAVIOR AND CERVICAL VERTEBRAL MOTION DUE TO THE MUSCLE CONDITION

The rotational motion of the neck and the T1 were significantly restrained in the tensed muscle condition. It was observed that the muscle forces limited the neck and the T1 motion. Moreover, the head restraint limitation caused nearly no differences on the rearward displacement of the head.

In the tensed muscle condition, the neck could be assumed to be one rigid link, therefore the relative motion between the vertebrae decreased significantly. As a result, the flexion angle and the forward displacement of each vertebra relative to the lower vertebra decreased, restraining the tensile strain and the forward shear strain of the middle vertebra facet joint. However, the boundary of the stiffness change on the connection region (joint) between neck and the torso caused the C5/C6 and the C6/C7 to become relatively susceptible to loading. Therefore, the restrictive effect of the muscle forces on the compression strain on the rear edge of this vertebrae facet joint was not significant.

RISK POSSIBILITY OF NECK INJURIES IN MASS PRODUCTION CAR SEATS

In previous research, the data obtained from tests with human volunteers using rigid seats were analyzed to determine the cervical vertebral motion using the same method as in this study. We established injury risk thresholds using the strains on the facet joints by weighing the sense of discomfort after tests and strains on facet joints during the impact (Ono et al., 2006). Here, we evaluated strains on the facet joints of each volunteer during tests with mass production car seats by the thresholds.

Fig.25 shows the strain and the shear strain on the front edge and on the rear edge of the C2/C3 to the C6/C7 facet joints of each of the volunteers, plotted at 16.7ms intervals. Dashed lines inside the figure indicate the thresholds of each facet joint strain.

All of the strains exceeding the thresholds were shear strain. It could be said that the compression strain and the tensile strain on the facet joints could be reduced using mass production car seats. However, there are some cases when the shear strain came close or exceeded the thresholds. In other words, although the average backset was short, or during the tensed condition when the neck motion was restrained, the risk possibility of a neck injury could be higher if the interaction condition between the occupant's head/neck/torso and the seat were not controlled appropriately.

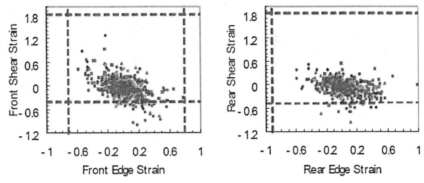

Fig.25 Comparison between the facet joint strains and the injury threshold

In this study, it was found that the visual head/neck/torso motion was suppressed by the motion limitation of the head restraint and seatback. However, high magnitudes of strains on certain facet joints were observed, showing that the relative motion between cervical vertebrae was not sufficiently depressed. In other words, it was difficult to predict or assess the occurrence of neck injuries based only on the visual head/neck/torso behavior. Development of a neck injury criteria and assessment based on the cervical vertebral motion is essential.

It can also be assumed that the differences on the interaction effect between the head/neck/torso and the seat influenced the risk of neck injury. Therefore, diversity in seating posture, occupant's body size, gender, etc should be taken into consideration during the design and development or the safety assessment of vehicle seats.

LIMITATIONS OF THE STUDY

In this study, the number of volunteers was small (6 males and 3 females). Thus, there was a lack of test data for universalizing the result of this study. Furthermore, since it was impossible to conduct tests with the impact level that would probably trigger a neck injury, it is difficult to predict the cervical spine motion which would cause a neck injury based on the results of this study.

CONCLUSION

By conducting low speed rear impact tests on human volunteers using a horizontal mini-sled apparatus with a mass production car seat, and by analyzing the head/neck/torso and the cervical vertebral motion during the impact, the following conclusions were made.

- In the first phase of the impact, since the upper torso sunk into the seat back and the head motion was restrained by the head restraint, the relative flexion occurred in almost all vertebrae. In the second phase of the impact, due to the reaction force from the head restraint and the forward motion of the upper torso, high magnitudes of tensile strain and compression strain occurred on facet joints of the upper vertebrae and the C6/C7 respectively.
- With a small seat back angle, the rearward displacement and the extension angle of the head were restrained; causing a decrease in the compression strain on facet joints of lower vertebrae. However, the tensile strain on facet joints of the upper vertebrae showed a tendency to increase.
- Although the impact level increased, the head flexion angle did not increase due to the head restraint limitation. However, the strain on the facet joint increased along with an increase in the impact level.
- In tensed muscle condition, the head displacement, the T1 displacement and the neck rotation decreased, restraining the motion of the middle vertebrae. However, the compression strain on the facet joint of the C6/C7 did not decrease significantly. Conversely the compression angle on the C5/C6 facet joint tend to increase.

- The neck extension of female subjects tends to increase in comparison to the males under the same test conditions.
- By comparing with the previous experimental results using a rigid seat, the compression strain and the tensile strain of facet joints were reduced significantly. But, in certain damping conditions between the occupant's head/torso and the seat, the shear strain on facet joints of upper vertebrae increased, indicating the possibility of increasing the risk of neck injuries developing.

ACKNOWLEDGEMENT

This study was supported by Toyota Motor Corporation. We want to use this space to show our appreciation.

REFERENCES

Eis V., Sferco R., and Fay P., A Detailed Analysis of The Characteristics of European Rear Impacts, Proceeding of the 19th ESV Conference, Washington D.C., Paper No. 05-0385, 2005

Beier G., Schuller E., Schuck M., Ewing C.L., Becker E.D., and Thomas D.J., Center of Gravity Moments of Inertia of Human Heads, Proceedings of International IRCOBI Conference Biomechanics of Impacts, 1980, 218-228

Brault J.R., Wheeler J.B., Siegmund G.P., and Brault E.J., Clinical Response of Human Subjects to Rear-End Automobile Collisions, Arch Phys Med Rehabil 79, 1998, 72-80

Dehner C., Elbel M., Schick S., Walz F., Hell W., and Kramer M., Risk of injury of the cervical spine in sled tests in female volunteers, Clinical Biomechanics 22, 2007, 615-622

Hernandez I.A., Fyfe K.R., Heo G., and Major P.W., Kinematics of Head Movement in Simulated Low Velocity Rear-End Impacts, Clinical Biomechanics 20, 2005, 1011-1018

Kleinberger M., Voo L., Merkle A., Bevan M., Chang S., and McKoy F., The Role of Seatback and Head Restraint Design Parameters on Rear Impact Occupant Dynamics, Proceedings of the 18th International Technical Conference on the Enhanced Safety of Vehicles, Paper No. 229, 2003

Latchford J., Chirwa E.C., Chen T., and Mao M., The Relationship of Seat Backrest Angle and Neck Injury in Low-Velocity Rear Impacts, Proc. IMechE Vol.219 Part D: J. Automobile Engineering, 2005, 1293-1302

Mertz H.J., and Patrick L.M., Strength and Response of Human Neck, Proceedings of the 15th Stapp Car Crash Conference, SAE Paper 710855, 1971

Ono K, Ejima S., Suzuki Y., Kaneoka K., Fukushima M., and Ujihashi S., Prediction of Neck Injury Risk Based on the Analysis of Localized Cervical Vertebral Motion of Human Volunteers during Low-Speed Rear Impacts, Proceedings of International IRCOBI Conference Biomechanics of Impacts, 2006, 103-114

Ono K., and Kaneoka K., Motion Analysis of Human Cervical Vertebrae During Low-Speed Rear Impacts by The Simulated Sled, Crash Prevention and Injury Control, Vol.1, Issue 2, November 1999

Ono K., Kaneoka K., Wittek A., and Kajzer J., Cervical Injury Mechanism Based on the Analysis of Human Cervical Vertebral Motion and Head-Neck-Torso Kinematics During Low Speed Rear Impacts, Proceedings of the 41st Stapp Car Crash Conference Proceedings, SAE Paper 973340, 1997

Ono K. and Kanno M., Influences of The Physical Parameters on The Risk to Neck Injuries in Low Impact Speed Rear-End Collisions, Accident Analysis and Prevention 28(4), 1996, 493-499

Sawada M. and Hasegawa J., Development of A New Whiplash Prevention Seat, Proceeding of the 19th ESV Conference, Washington D.C., Paper No. 05-0288, 2005

Siegmund G.P., Brault J.R., and Wheeler J.B., The Relationship Between Clinical and Kinematic Responses from Human Subject Testing in Rear-End Automobile Collisions, Accident Analysis and Prevention 32, 2000, 207-217

Stemper B.D., Yoganandan N., and Pintar F.A., Effects of Head Restraint Backset on Head-Neck Kinematics in Whiplash, Accident Analysis and Prevention Vol.38, 2006, 317-323

Sundararajan S., Prasad P., Demetropoulos C.K., Tashman S., Begeman P.C., Yang K.H., and King A.I., Effect of Head-Neck Position on Cervical Facet Stretch of Post Mortem Human Subjects During Low Speed Rear End Impacts, Stapp Car Crash Journal, Vol.48, Paper No. 2004-22-0015, 2004

Tencer A.F., Mirza S.K., and Bensel K., The Response of Human Volunteers to Rear-End Impacts: The Effect of Head Restraint Properties, Spine 22, 2001, 2432-2442

van den Kroonenberg A., Philippens M., Cappon H., and Wismans J., Human Head-Neck Response During Low-Speed Rear End Impacts, Proceedings of the 42nd Stapp Car Crash Conference, SAE Paper 983158, 1998

Welcher J.B., Szabo T.J., Relationships Between Seat Properties and Human Subject Kinematics in Rear Impact Tests, Accident Analysis and Prevention Vol.33, 2001, 289-304

Yang K.H., Begman P.C., Muser M., Niederer P., and Waltz F., On The Role of Cervical Facet Joints in Rear End Impact Neck Injury Mechanisms. SAE SP-1226, Motor Vehicle Safety Design Innovations, SAE Paper 970497, 1997, 127-129.

973342

Biofidelity of Anthropomorphic Test Devices for Rear Impact

P. Prasad, A. Kim, and D.P.V. Weerappuli
Ford Motor Company

ABSTRACT

This study examines the biofidelity, repeatability, and reproducibility of various anthropomorphic devices in rear impacts. The Hybrid III, the Hybrid III with the RID neck, and the TAD-50 were tested in a rigid bench condition in rear impacts with ΔVs of 16 and 24 kph. The results of the tests were then compared to the data of Mertz and Patrick[1]. At a AV of 16 kph, all three anthropomorphic devices showed general agreement with Mertz and Patrick's data [1]. At a AV of 24 kph, the RID neck tended to exhibit larger discrepancies than the other two anthropomorphic devices. Also, two different RID necks produced significantly different moments at the occipital condyles under similar test conditions.

The Hybrid III and the Hybrid III with the RID neck were also tested on standard production seats in rear impacts for a AV of 8 kph. Both the kinematics and the occupant responses of the Hybrid III and the Hybrid III with the RID neck differed from each other. Comparison testing of the Hybrid III and TRID necks were conducted on production seats with different NIF scores [2] at ΔVs of 8 and 16 kph. The two necks responded similarly and no significant differences were observed. The corrected lower neck moment of both necks were predictive of the seats' NIF scores.

The Hybrid III and the Hybrid III with the pedestrian pelvis were also tested with standard production seats to determine if hip joint stiffness affected the kinematics of the ATD in rear impacts. The standard Hybrid III is a molded pelvis while the pedestrian pelvis is a cut pelvis. The kinematics and occupant responses of the two Hybrid III configurations were similar to each other. From this study, it can be concluded that the standard Hybrid III dummy is suitable for rear impact testing.

INTRODUCTION

There are approximately 1.75 million rear end crashes per year in the United States, accounting for approximately 605,000 injuries. Whereas rear end crashes account for nearly 28.8% of all motor vehicle crash involved injuries, they account for only 4.9% of all motor vehicle involved fatalities [3]. The majority of injuries in rear crashes are minor (AIS 1) in severity. For belted occupants in rear crashes, Digges et al [4] have estimated that 25% of the total harm is due to non-contact injuries and another 17% is due to contact with the head restraint. 82% of the non-contact injuries are to the neck. The AIS 1 neck injuries are generally soft tissue sprains which are considered by many to be due to hyperextension of the neck. However, Prasad et al [5] have shown that the hyperextension of the neck, as measured by the change in the relative angle between the head and the first thoracic vertebra, may not accurately reflect the forces and moments developed at the head/neck and the neck/thorax junctions. Utilizing a two-dimensional model of the human spine subjected to rear impact in a rigid seat back and a plastic seat back, the authors showed that the loads and moments at various levels of the cervical spine were substantially reduced in the plastic seat back case in spite of a $3°$ increase in the head/T1 angle compared to those with the rigid seat back. Prasad et al[5] have also shown that various factors, such as seat back and seat back cushion characteristics (height and stiffness) and head restraint placement, affect the loads developed in the cervical spine in a rear impact. As a result, the experimental evaluation of various seat designs in rear impact should be conducted with Anthropomorphic Test Devices (ATDs) with adequate biofidelity and load measuring capability in the neck.

The primary objective of this paper is to report the results of the authors' investigations of the biofidelity and load measuring capability of a currently available ATD, the Hybrid III, and some experimental, prototype ATDs. The secondary objective of this paper is to report experimental results of rear impact sled tests with current

391

production seats at low-to-medium impact velocities in which AIS 1 type neck injuries are most likely to occur.

The test plan reported in this paper is shown below.

TEST PLAN

a. <u>Neck Design</u>   <u>Seat</u>
 - Hybrid III Dummy   Rigid Bench and
                                             Production Seats

 - RID land RID 2 on Hybrid III   Rigid Bench and
                                             Production Seats

 - TRID on Hybrid III   Production Seats

b. <u>Thorax Design</u>   <u>Seat</u>
 - Hybrid III   Rigid Bench Seat
 - Hybrid III neck on TAD-50   Rigid Bench Seat
   Articulated spine

c. <u>Pelvis Design</u>   <u>Seat</u>
 - Hybrid III molded pelvis flesh   Production Seats
 - Hybrid III Pedestrian pelvis,   Production Seats
   3-piece pelvis flesh

The biofidelity tests were conducted with rigid bench seats similar to those used by Mertz and Patrick [1]. These tests were conducted at 16 kph and 24 kph ΔVs to replicate the severity level of cadaveric and volunteer tests conducted by Mertz and Patrick [1].

RIGID BENCH TESTING

TEST METHOD - To determine their biofidelity in rear impacts, three ATD configurations were tested using an experimental protocol similar to that of Mertz and Patrick [1]. They consisted of the Part 572(b) Hybrid III dummy, a Hybrid III dummy in which the original neck was replaced by the Rear Impact Dummy (RID) neck developed by Chalmers University [6], and a prototype of an advanced ATD, the 50th% Thoracic Assessment Device (TAD-50) [7]. Two RID necks were tested in identical test conditions to determine the reproducibility of this device. The TAD-50 utilizes the existing Part 572(b) Hybrid III neck, but the chest construction of the dummy is substantially different from that of the Hybrid III dummy. Whereas the TAD-50 thoracic spine has an articulation representing the flexibility of the human thoracic spine, the Hybrid III dummy has a rigid thoracic spine. It was assumed that comparison between the TAD-50 and Hybrid III dummy responses would reveal the effect of thoracic spine flexibility on head/neck responses.

All test devices had instrumentation to measure accelerations, (head, chest, and pelvis), and neck loads at the occipital condyles and at the neck bracket. The neck responses were filtered with a 600 Hz Chebyshev filter. In some of the tests, the ATDs were also instrumented with a lumbar spine load cell and the responses were filtered with a 1000 Hz Chebyshev filter.

The sign convention for all instrumentation followed that in SAE J211.

The ATDs were consistently placed in normal seating positions on a rigid bench seat constructed of plywood and metal bracing, and were restrained with a 3-point belt system as shown in Figure 1. The seat back, which ended at shoulder level, had a 20° angle with respect to the vertical and the seat cushion plane had a 13.5" angle with respect to the horizontal. There was no head restraint. On two of the 24 kph AV runs, the seat back was cushioned with a thin foam mat to eliminate lumbar spine ringing caused by initial placement of the load cells against a rigid surface. This seat was installed in a rigidized car buck at approximate mid-seat track position.

The simulated rear impact collisions were run on a Hyge sled. As shown in Figure 2, the acceleration pulses used in these tests differ from that of Mertz and Patrick's [1], but comparable ΔVs of 16 and 24 kph were obtained. Throughout this paper, the referenced volunteer and cadaver curves are from "Investigation of the Kinematics and Kinetics of Whiplash" [1] and are compared with a typical response for each ATD configuration. The Hybrid III was tested four times at the lower speed and six times at the higher speed. The Hybrid III with a RID neck was tested twice at the lower speed and four times at the higher speed. The TAD-50 was tested twice at each speed.

TEST RESULTS - The test results of primary interest are the head rotation relative to T-1, neck extension moments at the occipital condyles, linear acceleration (Anterior-Posterior, (A-P) and Superior-Inferior, (S-I)) of the center of gravity of the head, and the angular acceleration of the head. These are the kinematic and kinetic responses of cadavers and volunteers reported by Mertz and Patrick [1]. It should be noted that for the ATDs, the neck torque at the occipital condyles was calculated by the following formula:

$$M_{Y@Occ. Cons.} = M_Y + (F_x \cdot D_z) + (F_z \cdot D_x)$$

where F, F, and M_Y are the loads and torques measured by the upper neck load cell, and D_x and D_z are the distances between the load cell and the head/neck joint. The distances from the head/upper neck joint to the upper neck load cell and from the lower neck/torso joint to the lower neck load cell for the three neck configurations are given in Table 1.

Table 1. - Correction distances for the Hybrid III, RID, and TAD-50.

	D_x (mm)	D_z (mm)
Hybrid III, Upper Neck	0.0	17.8
Hybrid III, Lower Neck	50.8	28.6
RID, Upper Neck	0.0	17.8
RID, Lower Neck	50.8	28.6
TAD-50, Upper Neck	0.0	17.8
TAD-50, Lower Neck	0.0	25.4

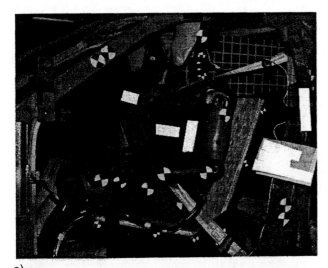

a)

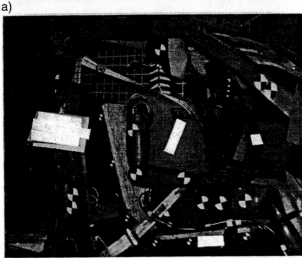

b)

c)

Figure 1. - Rigid bench set-up: a) Hybrid III, b) RID, c) TAD-50.

HYBRID III DUMMY RESPONSE - The Hybrid III dummy head/neck responses in the 16 kph AV tests are shown in Fig. 3 in which cadaver and volunteer responses reported by Mertz and Patrick are also shown. The head extension angle of the dummy was determined

from film analysis in a sled fixed coordinate system and assumed to represent extension of the head relative to the thoracic spine since the film analysis showed negligible rotation of the upper torso of the dummy. This definition of the head extension angle is similar to that used by Mertz and Patrick [1] in the cadaver and volunteer tests. To remove any biasing, the initial angle of the head relative to the sled fixed axis system was subtracted from the dynamic response in the current series of tests as well as in those reported by Mertz and Patrick [1] for the cadavers and volunteers.

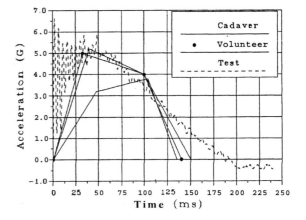

a)

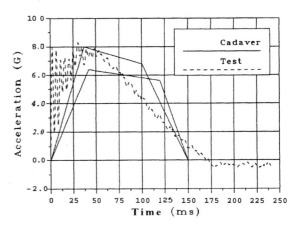

b)

Figure 2. - Sled acceleration pulses: a) AV = 16 kph, b) AV = 24 kph.

As seen in Fig. 3, the Hybrid III head extension angle is within measured cadaver responses and is slightly higher than that experienced by the volunteer. The difference between cadaver and volunteer responses can be explained by the lack of muscle tone in the cadavers and potential pre-impact tensing by the volunteer. The Hybrid III dummy peak extension is approximately 20 ms earlier than that seen with the cadaver responses and may be explained by the earlier onset rate in the current sled test series compared to that in Mertz and Patrick's series. Further kinematic

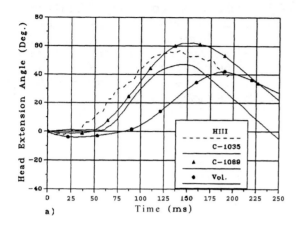

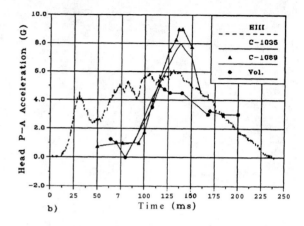

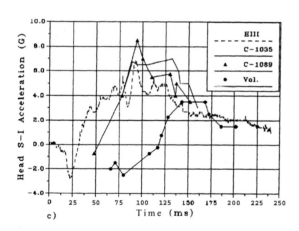

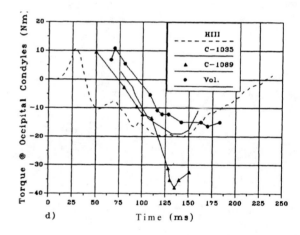

Figure 3. - Hybrid III vs. Mertz and Patrick's data, AV = 16 kph: a) head extension angle (Deg.), b) head P-A accel. (G), c) head S-I accel. (G), d) occipital condyle torque (Nm).

comparisons between the dummy and cadaveric/volunteer responses for head S-I and P-A accelerations are also shown in Fig. 3. The head S-I accelerations of the Hybrid III dummy compare well with those measured on the cadavers and are higher than those of the volunteer. The P-A acceleration responses of the Hybrid III dummy is of longer duration than those of the volunteer and the cadavers and has peak magnitudes between those of the cadavers and the volunteer. The peak moments at the occipital condyles of the dummy are within those experienced by the two cadavers and slightly higher than that of the volunteer.

Based on the above comparison between the Hybrid III dummy head/neck responses and those of the cadavers/volunteer, it can be concluded that the Hybrid III neck is somewhat softer than that of the tensed volunteer and closer in stiffness to that of the two cadavers tested by Mertz and Patrick at 16 kph AV.

Figure 4 shows head/neck responses in the 24 kph AV test. The peak head extension of the Hybrid III dummy is approximately 10° higher than those of the cadavers. The calculated angular acceleration of the Hybrid III head is in close agreement with the cadaver data in the 75-150 ms time range. The dummy data show a flexion spike around 45 ms in angular acceleration as well as in occipital condyle torque. Although Mertz and Patrick's data for the cadavers show such tendencies, the earlier part of these responses were not reported by Mertz and Patrick. At this severity of testing, the Hybrid III dummy responses like head accelerations in the S-I and P-A directions are very close to those of the cadavers. The occipital condyle torques in the Hybrid III dummy are also very close to those measured in the cadavers.

Based on the tests at 16 and 24 kph ΔVs with the Part 572(b) Hybrid III dummy, it can be concluded that the Hybrid III dummy neck has adequate biofidelity for determining head/neck kinematics in rear impacts of severity tested.

RID NECK RESPONSE - Two RID necks, RID 1 and RID 2, were tested. Only RID 1, the RID neck system which was borrowed courtesy of Chalmers University, was tested at the 16 kph AV. Figure 5 shows the responses of RID 1 as well as the cadaver and volunteer responses from Mertz and Patrick's study. The peak head extension of RID 1 is slightly higher and like the Hybrid III's response, earlier than those of the cadavers. At this test speed, RID 1's head P-A acceleration responses peak magnitudes, (which were higher than the volunteer's), and curve shapes correspond very well to the cadavers but are shifted about 25 ms earlier in the time frame. The RID 1's S-I head accelerations also compare well with the cadavers and are higher than the volunteer's. The peak torques at the occipital condyle of RID 1 fall within the corridors defined by the cadavers' responses but are higher than the volunteer's peak torques.

From these responses, it is judged that the RID neck is sufficiently biofidelic to predict head/neck kinematics in rear impacts at a AV of 16 kph. Like the Hybrid III's standard neck, the stiffness of the RID neck is more like those of the cadavers and is softer than the volunteer's.

Figure 6 shows the head and upper neck responses of RID 1 and a second RID neck, RID 2, at a 24 kph AV. The magnitude of the peak head extension of both RID 1 and RID 2 compare very well with those seen with the cadavers but their peak durations are about 50 ms shorter. The angular accelerations of the head were only calculated for RID 2 and are much higher and oscillate more than the calculated cadaver responses. Along the P-A axis, the head accelerations of both RID 1 and RID 2 are higher than those of the cadavers but in the S-I direction, the measured head accelerations of both RID 1 and RID 2 are very close to the cadavers. The peak occipital condyle moments measured by RID 1 are about 75% higher than the highest cadaver torque and even exceeds the injury assessment reference value of -57 Nm. In contrast, the torques measured by RID 2 at its occipital condyle are approximately a third of the magnitude of the lowest cadaver torque. There were no test conditions that could account for this large variation between the measured moments of RID 1 and RID 2 other than variations in the hardware of the necks themselves.

At the more severe sled test of 24 kph AV, the majority of the head/neck responses of RID 1 and RID 2 do not compare as well with the cadaver responses as they did at the 16 kph AV indicating poorer biofidelity in rear impacts at higher speeds. From the large difference in occipital condyle moments measured by the two different RID necks, it is evident that there is a significant reproducibility problem. This lack of reproducibility shows that the RID neck is not a reliable instrument for testing and measuring occupant responses in rear impacts at its current level of production.

TAD-50 RESPONSE - The head/upper neck responses of the TAD-50 at the 16 kph AV are shown in Figure 7. The peak head extension angles of the TAD-50 are noticeably larger than either the volunteer's or even the highest peak of the two cadavers. The TAD-50's head P-A accelerations have longer peak durations than either those of the volunteer or the cadavers with peak magnitudes which fall between the two. Along the S-I axis, the TAD-50 and the cadavers experience very similar head accelerations which occur sooner and are larger than the volunteer's. The maximum torques measured at the TAD-50's occipital condyle fall within the corridors defined by the cadavers and also have initial slopes closer to their responses than the Hybrid III.

The head/neck kinematics predicted by the TAD-50 at this lower speed of 16 kph AV compare reasonably well with the responses obtained in Mertz and Patrick's study. As expected, the TAD-50's neck, which is the standard Hybrid III neck, gives responses more like those seen with the cadavers than with the volunteer.

At the 24 kph AV the extension curves of the TAD-50's head, seen in Figure 8, match well with those of the cadavers. The head acceleration traces of the TAD-50 along the P-A axis have shapes similar to the cadavers' curves but with lower peak magnitudes. The

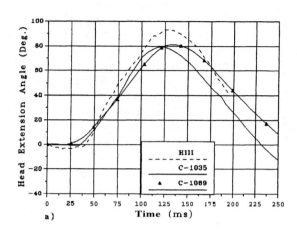

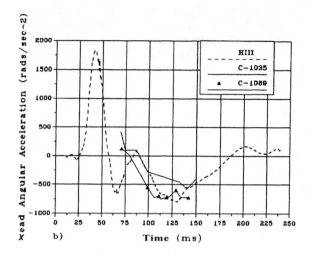

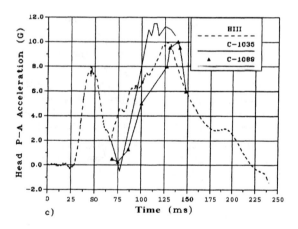

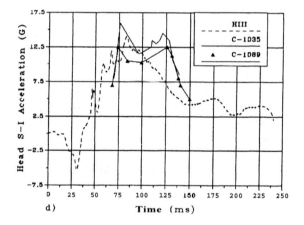

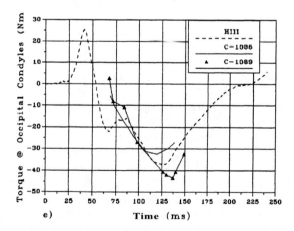

Figure 4. - Hybrid III vs. Mertz and Patrick's data, AV = **24** kph: a) head extension angle (Deg.), b) head angular accel. (rads/sec^2), c) head P-A accel. (G), d) head S-I accel. (G), e) occipital condyle torque (Nm).

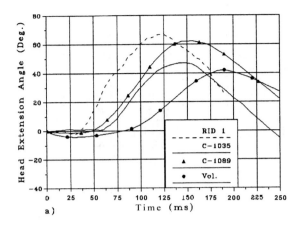

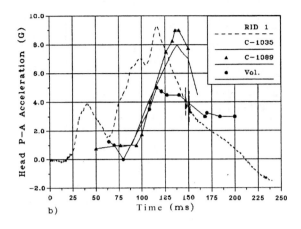

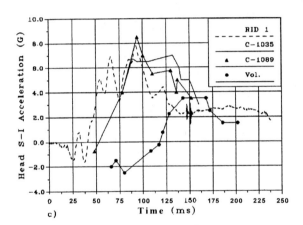

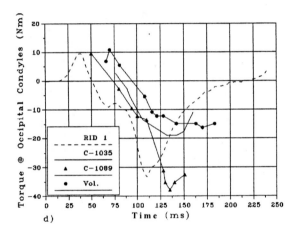

Figure 5. - RID vs. Mertz and Patrick's data, AV = 16 kph: a) head extension angle (Deg.), b) head P-A accel. (G), c) head S-I accel. (G), d) occipital condyle torque (Nm).

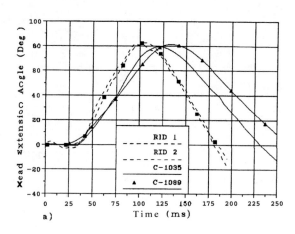

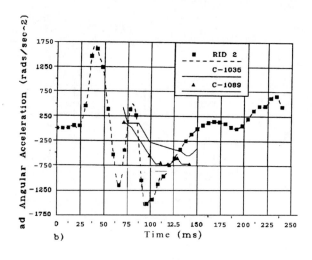

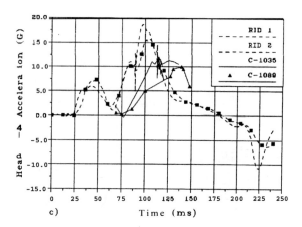

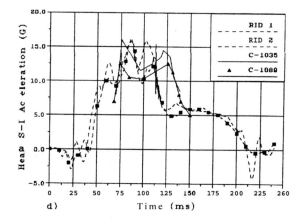

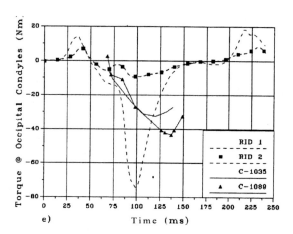

Figure 6. - RID vs. Mertz and Patrick's data, AV = **24** kph: a) head extension angle (Deg.), b) head angular accel. (rads/sec^2), c) head P-A accel. (G), d) head S-I accel. (G), e) occipital condyle torque (Nm).

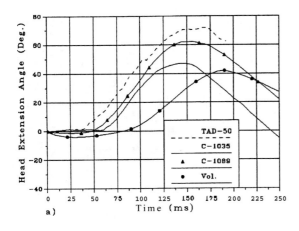

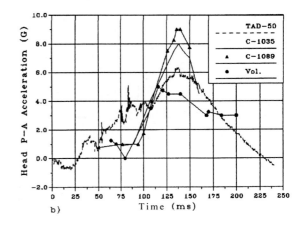

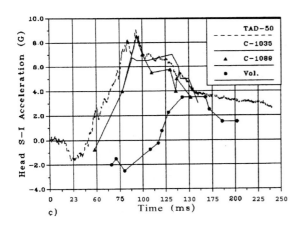

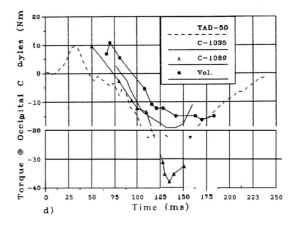

Figure 7. - TAD-50 vs. Mertz and Patrick's data, AV = 16 kph: a) head extension angle (Deg.), b) head P-A accel. (G), c) head S-I accel. (G), d) occipital condyle torque (Nm).

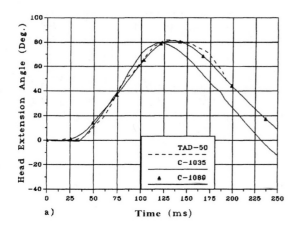

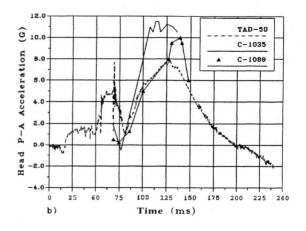

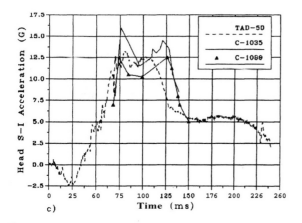

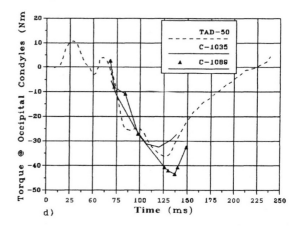

Figure 8. - TAD-50 vs. Mertz and Patrick's data, AV = **24** kph: a) head extension angle (Deg.), b) head P-A accel. (G), c) head S-I accel. (G), d) occipital condyle torque (Nm).

S-I head accelerations of the TAD-50 and the cadavers compare very well to each other as do the occipital condyle torques.

At the higher **AV** of 24 kph, the responses of the TAD-50 are slightly closer to those of the cadavers than the Hybrid III's. The ability to predict head/neck kinematics in rear impacts with the standard Hybrid III neck, on either the Hybrid III or TAD-50 body, is well proven from these comparisons. The TAD-50, with the additional articulation of the thoracic spine, seems to be slightly more biofidelic at higher impact speeds, but the TAD-50 is not yet in mass production due to design and durability issues.

ADDITIONAL RESPONSES - Additional dummy loads and torques for which there is no volunteer or cadaver data were also measured. The RID neck, due to its reproducibility problems, is unsuitable as a reliable test device and therefore, the additional test results of this dummy will not be discussed. The additional upper and lower neck responses of the Hybrid III and TAD-50 at a AV of 16 kph are shown in Figure 10. The upper neck shear and tensile/compressive forces are similar for both ATDs while the lower neck responses are more disparate. At the lower neck, the Hybrid III shear force curves have much steeper initial slopes than the TAD-50's and higher peak magnitudes. Along the vertical axis, the Hybrid III lower neck load cells measure a largely compressive force while the TAD-50's responses have a significant tensile portion. The lower neck moments, measured at the load cells and those corrected to give the moments at the base of the neck, are similar in magnitude for both crash dummies but the TAD-50's responses peak about 35-40 ms later than those of the Hybrid III.

At the 24 kph **AV**, the upper and lower neck responses of the two ATDs, shown in Figure 11, follow the same trends seen at the lower test speed. The lag between the initial slopes of the TAD-50's upper and lower neck responses and those of the Hybrid III is attributable to the extra flexibility of the TAD-50's articulated thoracic spine joint. This added flexibility also influences the neck and upper torso kinematics.

The differences between the lower neck axial responses are due to the added flexibility of the articulated thoracic joint. When there is a large rotation of the head/neck system, a downward shear force, F_s, is exerted on the neck at the occipital condyles. This force is resisted by a compressive force at the lower neck joint. For the Hybrid III, the rotation of the lower neck load cell is minimal, and therefore the measured lower neck axial force, F, would be equal to this compressive force, (Fig. 9). The added articulation of the TAD-50's spine allows the lower neck load cell to rotate more. This added rotation, Θ, changes the orientation of the load cell, and the measured axial force, F_z, now consists of both a compressive component, $(F_z \cdot \cos\Theta)$, and a tensile component, $(F_z \cdot \sin\Theta)$. This tensile component mitigates the compressive nature of the axial force, F_z measured by the TAD-50's lower neck load cell.

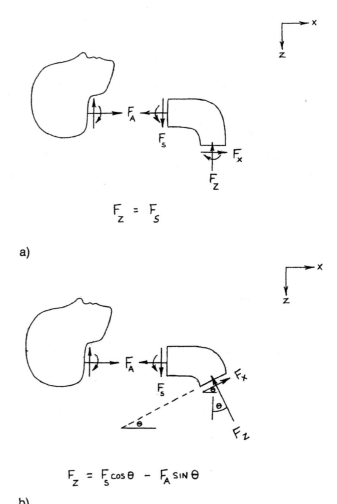

$$F_z = F_s$$

a)

$$F_z = F_s \cos\theta - F_A \sin\theta$$

b)

Figure 9. - Free body diagrams: a) Hybrid III, b) TAD-50.

The lumbar spine responses of the Hybrid III and the TAD-50 at the 16 kph and 24 kph **AV** test speeds are shown in Figure 12. The TAD-50's shear and tensile/compressive forces are much lower than those of the Hybrid III but its flexion moments are much higher. The TAD-50's additional thoracic articulation seems to mitigate the forces developed at the lumbar spine joint while slightly amplifying the moment as compared to those of the Hybrid III.

REAR IMPACT MODEL

To verify the existence of compressive forces at the Hybrid III neck load cell location, a mathematical model was exercised.

MODEL DESCRIPTION - A mathematical model of the rigid bench tests with the Hybrid III was made using the MADYMO program (Fig. 13). The geometry of the rigid bench seat, the buck's floor, roof, windshield, seat belt system, and ATD at T_0 were incorporated into the model's set up. The Hybrid III model was based on the standard 50th% Hybrid III database provided by TNO in the 5.1 version of MADYMO [8] with one major modification to the neck. In the TNO database, the entire

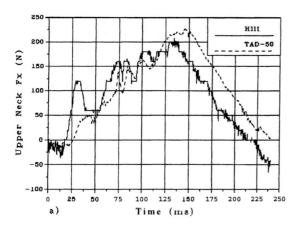

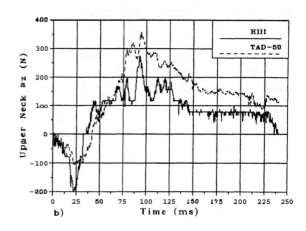

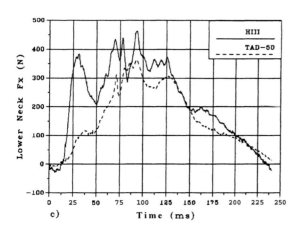

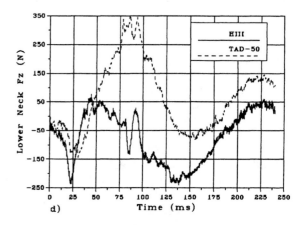

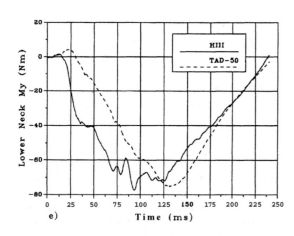

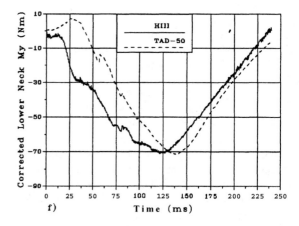

Figure 10. - Hybrid III and TAD-50, AV = 16 kph: upper neck F_x (N), b) upper neck F_z (N), c) lower neck F_x (N), d) lower neck F_z (N), e) lower neck M_y (Nm), f) corrected lower neck M_y (Nm).

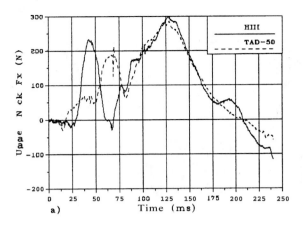

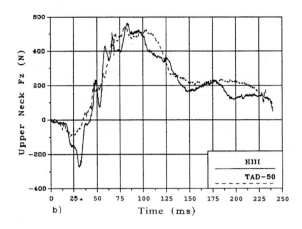

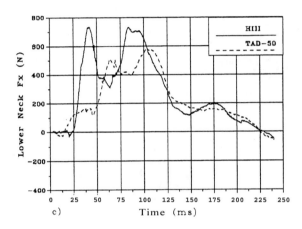

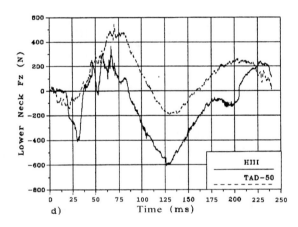

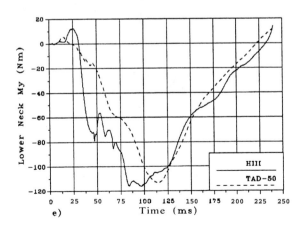

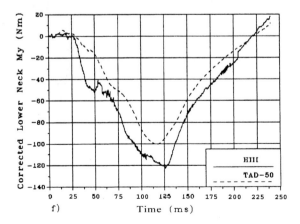

Figure 11. - Hybrid III and TAD-50, AV = 24 kph: a) upper neck F_x (N), b) upper neck F_z (N), c) lower neck F_x (N), d) lower neck F_z (N), e) lower neck M_y (Nm), f) corrected lower neck M_y (Nm).

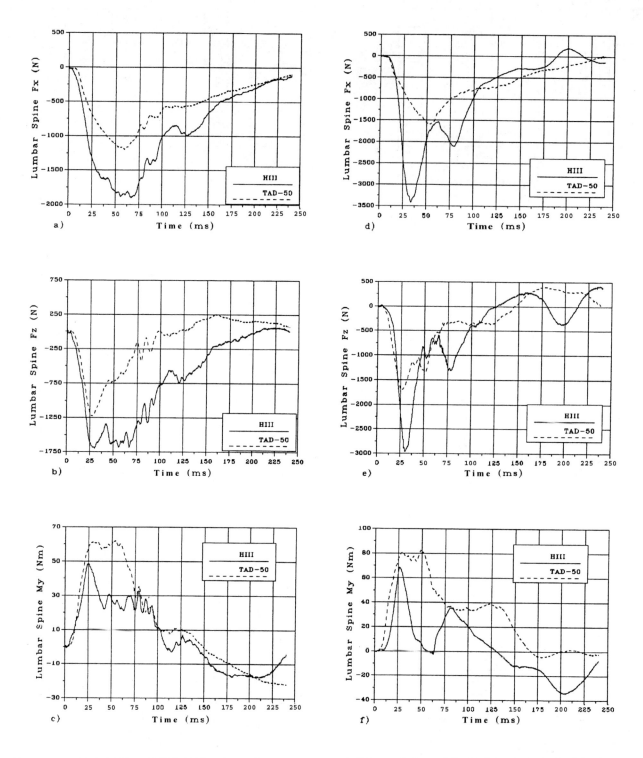

Figure 12. - Hybrid III and TAD-50 lumbar spine responses: a) AV = **16** kph, F$_x$ (N), b) AV = 16 kph, F, (N), c) AV = **16** kph, M$_y$ (Nm), d) ΔV = 24 kph, F$_x$ (N), e) ΔV = 24 kph, F$_z$ (N), f) AV = 24 kph, M$_y$ (Nm).

neck is represented by one rigid body but in this model, the neck consists of four separate bodies which allow movement between them, more accurately representing the flexibility of the actual Hybrid III dummy neck. The seat belt system was modeled with the Ford MADYMO Advanced Belt Model (ABM) which is capable of capturing the tensile behavior of belts. The ABM is based on the Articulated Total Body (ATB) harness model and can represent belts, retractors, pretensioners, buckles, and slip rings. The Hyge sled acceleration pulses were used as input.

Figure 13. ⁃ MADYMO3D rear impact model.

MODEL RESULTS - The MADYMO rear impact model simulations correlated reasonably well to their respective Hyge sled tests in regards to both occupant kinematics and responses. The movement of the Hybrid III dummy in the model animations was very similar to that seen in the test films. This is confirmed by the close agreement of the theoretical and experimental head extension curves (Fig. 15). The tensile/compressive responses of the model, for both the upper and lower neck, reflect the trends seen in the tests, including the compressive nature of the response at the lower neck.

PRODUCTION SEAT TESTING

HYBRID III NECK vs. RID NECK ⁃ In addition to the rigid bench seat testing, low speed rear impact Hyge sled tests were conducted using production bucket front seats. These seats were installed in the appropriate rigidized sled buck and set in mid-seat track position with a seat back angle of 21° and with the adjustable head restraints in the full up position, see Fig. 14. For each run, a Hybrid III and a Hybrid III with a RID neck, (RID 2), were placed in the front seats in normal 50th% sitting position. Both ATDs were restrained by 3-point belt systems.

a)

b)

Figure 14. ⁃ Production seat set-up: a) Hybrid III, b) RID.

At a 8 kph AV, the seat backs deflected dynamically but no permanent deformation or damage occurred to either seat. During the impact, the rotation of

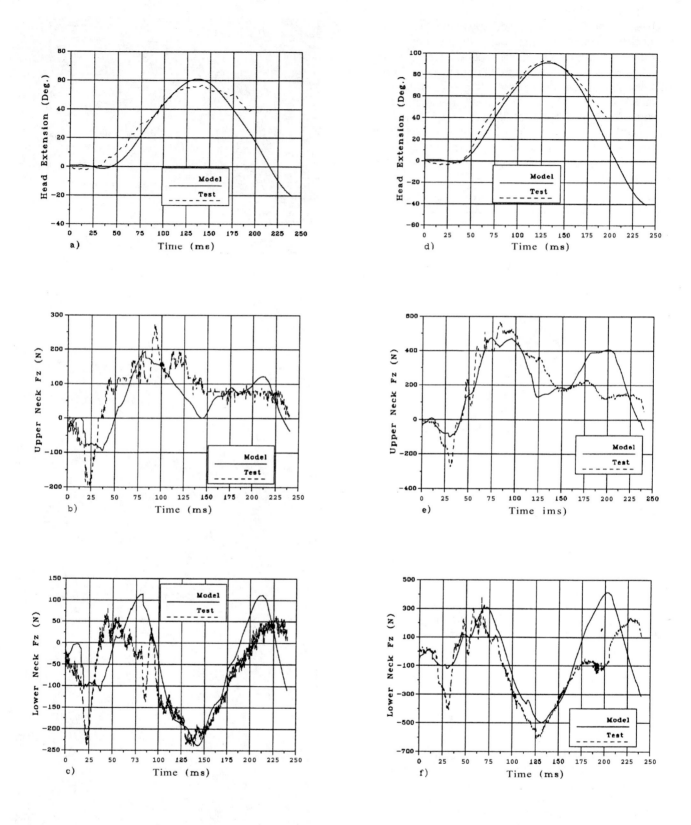

Figure 15. - Model vs. Test: a) A V = 16 kph, head extension angle (Deg.), b) A V = 16 kph, upper neck F_z (N), c) A V = 16 kph, lower neck F, (N), d) A V = 24 kph, head extension angle (Deg.), e) A V = 24 kph, upper neck F_z (N), f) A V = 24 kph, lower neck F_z (N).

the Hybrid III's head/neck system was distributed fairly evenly among the neck segments, including the upper and lower neck joints, while the RID 2 rotated more about the upper neck joint. This motion caused the RID 2 to exert a larger downward force on the head restraint which slipped during the crash pulse.

For this test condition, the responses of the two ATDs had similar shapes although there were evident differences. The RID 2 upper neck shear response was of shorter duration than that of the Hybrid III's and had slightly higher peak magnitudes (Fig. 17). The RID 2 peak upper neck tensile force was approximately 60% higher than the Hybrid III's but conversely, the peak occipital condyle moment of the Hybrid III was almost three times that of the RID 2. These differences between the upper neck responses are probably due to the reproducibility problem with the RID neck instead of any differences due to neck design. As seen in the rigid bench testing (Fig. 6), RID 2 significantly under-predicted the occipital condyle moment even when compared to another RID neck, RID 1.

At the lower neck joint, the peak magnitudes of the shear and tensile/compressive forces of the RID 2 were again higher than those of the Hybrid III. The lower neck moments of the two ATDs were similar but the RID 2 peaked slightly earlier and higher than the Hybrid III.

HYBRID III NECK vs. TRID NECK - The TRID, (TNO Rear Impact Dummy), neck was developed by TNO specifically for low severity rear impacts of speeds up to 25 kph [9]. It was designed to directly replace the Hybrid III neck with no alterations to the rest of the Hybrid III or its instrumentation. Comparison testing between ATDs with the two necks was conducted using three different types of production seats sold in the European market. The relative insurance claim frequency of these seats has been reported by Eichberger et al [2]. The authors have developed a scale called the NIF (Neck Injury Factor). An NIF of 1 would correspond to the average claim frequency; above 1 would correspond to higher and below 1 to lower than average claim frequency in the insurance files. The three seats used in this series of experiments were chosen to represent NIF = 0.7, 1.1 and 1.5. The aim of the experiments was to determine if the neck loads in tests with these seats would distinguish the NIF differences between them. The secondary aim was to determine the differences between the Hybrid III and the TRID necks in their sensitivity to differences in the three seat designs.

The ATDs were placed in normal seating position and restrained by 3-pt seat belts. The seats were mounted on a rigid platform buck and placed into mid-seat track position with a seat back angle of 21°, see Fig. 18. If the head restraint was adjustable, it was placed in the full up position. Each individual seat underwent two simulated Hyge sled rear impacts; one at a AV of 8 kph and one at a AV of 16 kph. Only one seat and one ATD was tested at a time. The sled acceleration pulses, seen in Figure 16, were derived from a 80.5 kph car-to-car rear impact acceleration trace.

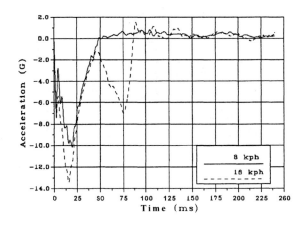

Figure 16. - Production seat Hyge sled acceleration pulses: ΔVs = 8 and 16 kph.

Figures 19 and 20 show the peak magnitudes of the upper and lower neck forces and moments of both ATDs for the respective ΔVs of 8 and 16 kph. They are also listed in Appendix A, Tables 1 and 2. In general, the Hybrid III and TRID necks responded similarly across all the tests. The only noticeable and consistent difference was that the Hybrid III measured lower peak occipital condyle moments in extension than the TRID, although in one case they were nearly equal.

In terms of the performance of the seats, none of the upper neck responses of either ATD, including the occipital condyle moment, were predictive of the trend in the NIF scores for both test speeds. For the Hybrid III, both the measured and corrected lower neck moment in extension showed the trend in the NIF scores (Fig. 21). The Hybrid III's positive lower neck shear force peak magnitudes also showed the trend in NIFs but are not as sensitive as the bending moment. With the TRID, only the corrected lower neck moment in extension is predictive of the NIF scores, although, at the higher 16 kph AV, the TRID's uncorrected lower neck moment also predicted the trend.

Even though the performance of the two different ATD necks were not identical, no major differences or contradictions were evident from their responses. Regardless of the neck used, the corrected lower neck moment seemed to be the best sorter of the NIFs.

EFFECT OF PELVIS DESIGN - The flesh of the standard Hybrid III pelvis is one foam piece which is molded into the seated position. Because of its construction, the standard pelvis hip joint has been postulated as being too stiff to allow the legs to rotate away from the torso as easily as a human's in rear impacts. As a result, the head/neck and upper torso kinematics in rear impacts could be different from those of human occupants. To see if there was any basis to this theory and how this would affect the neck responses, comparison testing with the pedestrian pelvis was performed. The flesh of the pedestrian pelvis consists of three separate, cut foam pieces which allow the Hybrid III

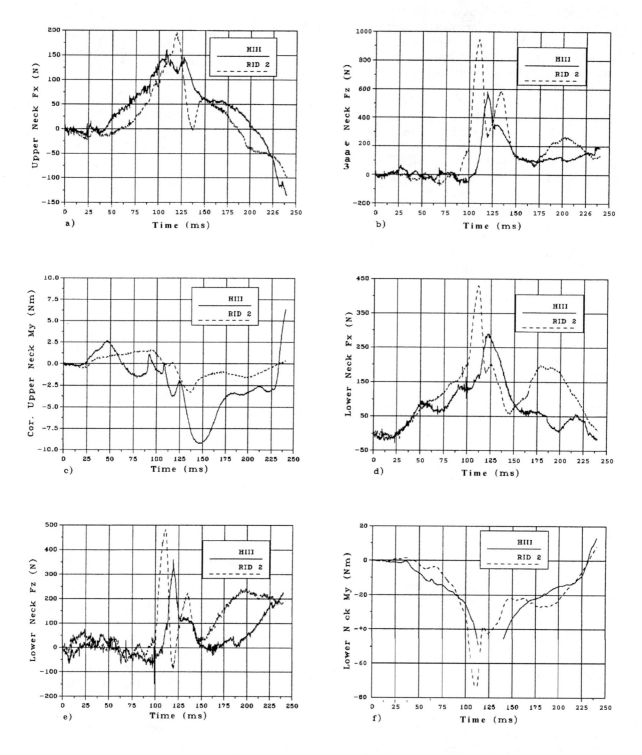

Figure 17. - Hybrid III vs. RID, production seat, AV = 8 kph: a) upper neck F_x (N), b) upper neck F_z (N), c) corrected upper neck M_y (Nm), d) lower neck F_x (N), e) lower neck F_z (N), f) lower neck M_y (Nm).

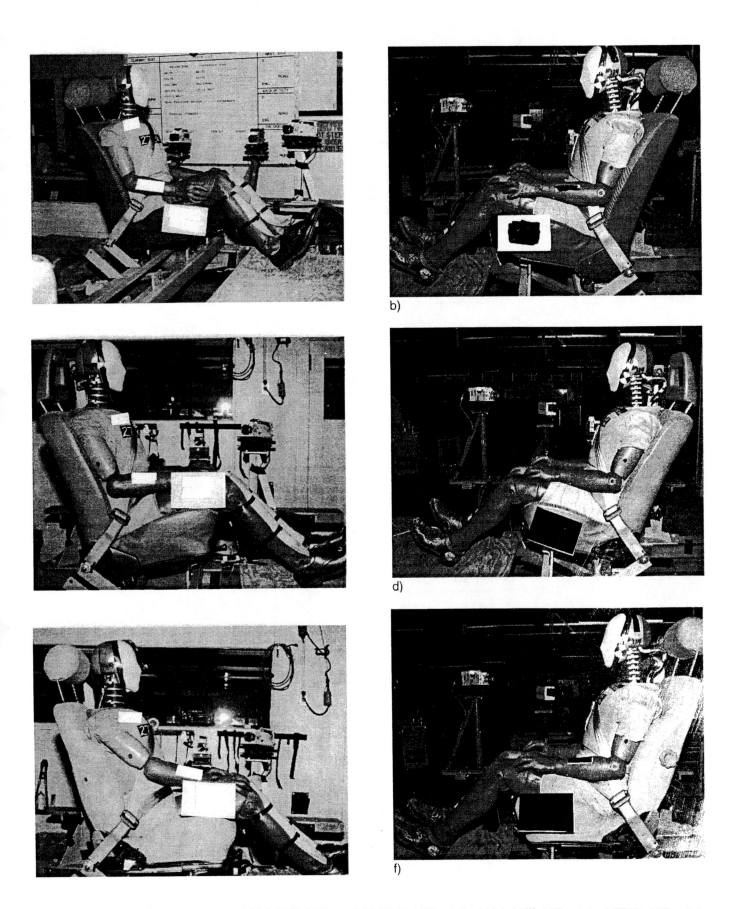

Figure 18. - Production seat set-up: a) Hybrid III, NIF = 1.5, b) TRID, NIF = 1.5, c) Hybrid III, NIF = 1.1, d) TRID, NIF = 1.1, e) Hybrid III, NIF = 0.7, f) TRID, NIF = 0.7.

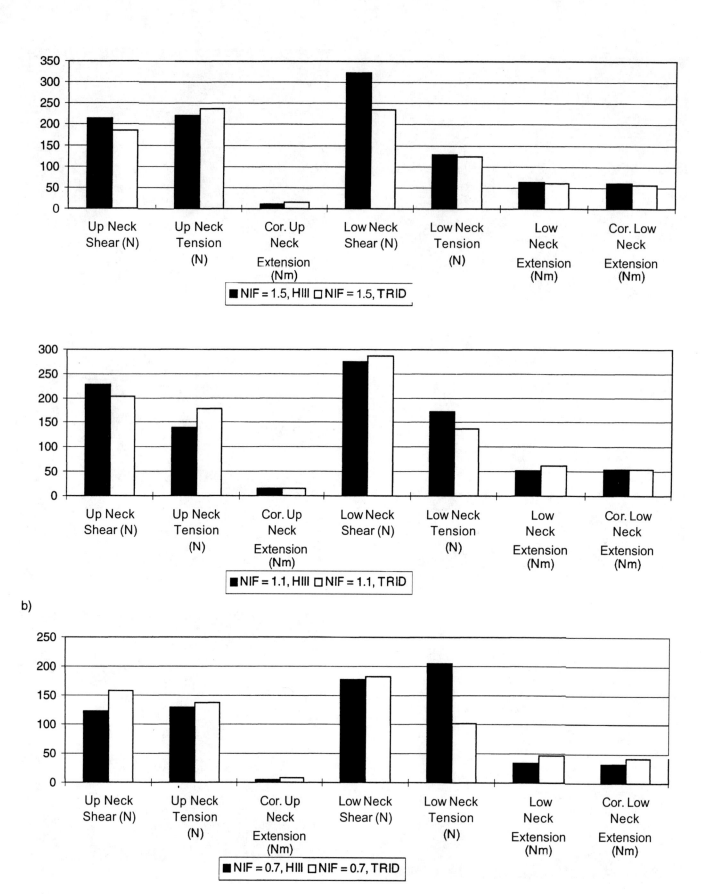

Figure 19. - Hybrid III vs. TRID neck responses, AV = 8 kph: a) NIF = 1.5 seats, b) NIF = 1.1 seats, c) NIF = 0.7 seats.

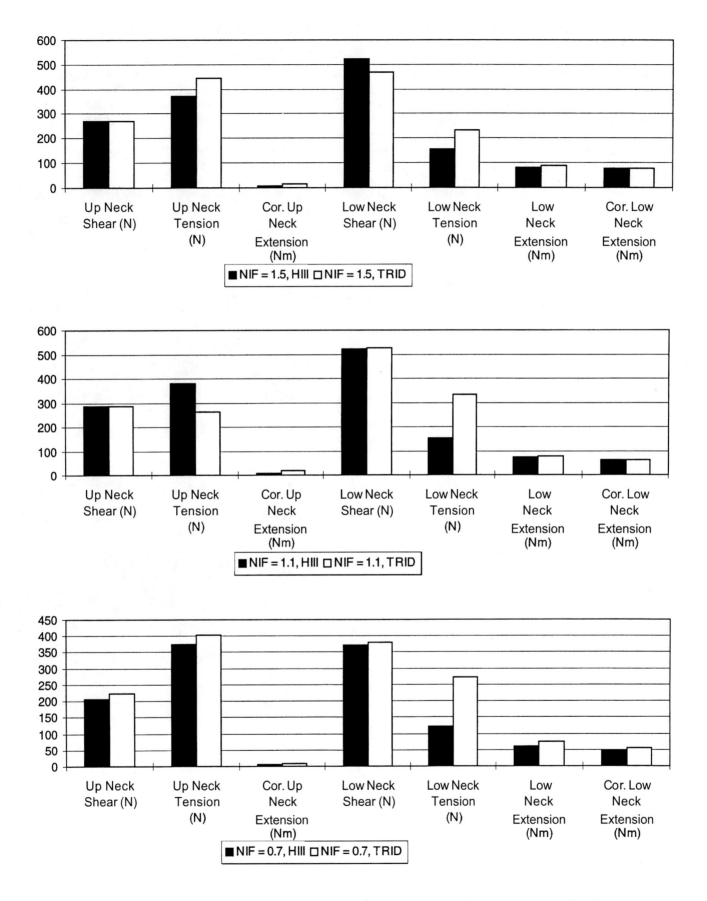

Figure 20. - Hybrid III vs. TRID neck responses, AV = 16 kph: a) NIF = 1.5 seats, b) NIF = 1.1 seats, c) NIF = 0.7 seats.

to be placed either in an erect standing position or a seated position.

Hyge sled tests with ΔVs of 8, 16, and 24 kph were conducted with two different Hybrid III dummy configurations: a Hybrid III with the standard pelvis and a Hybrid III with the pedestrian pelvis. The ATDs were placed in normal seating position in production bucket front seats, see Fig. 22. The seats were set in mid-seat track position with a seat back angle of 21° and with the head restraint in the full-up position. The Hybrid III with the standard pelvis flesh was restrained with a passive shoulder belt/active lap belt system, while the Hybrid III with the pedestrian pelvis was restrained by a 3-pt seat belt system.

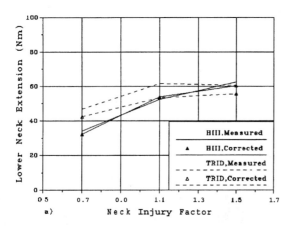

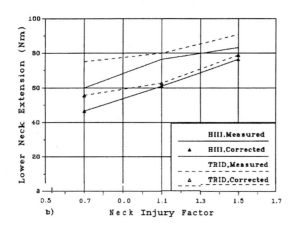

Figure 21. - Hybrid III and TRID lower neck extension moments (Nm) vs. NIF scores: a) AV = 8 kph, b) AV = 16 kph.

As seen in Figure 23, the kinematics of the Hybrid III, regardless of pelvis type, are similar at each test speed. At the 8 kph AV, both ATDs primarily moved rearwards with very little rotation. At the higher ΔVs of 16 and 24 kph, the Hybrid III with the standard pelvis does rotate more than the Hybrid III with the pedestrian pelvis, and conversely, the Hybrid III with the pedestrian pelvis ramps slightly higher up the seat back than the standard Hybrid III. These differences, however, may be due to

the different restraint systems rather than the pelvis flesh. The separate lap belt may have restricted the upward displacement of the standard Hybrid III more than the 3-pt belt system, thereby, resulting in more rotation and less ramping. It should be noted that the femurs of both Hybrid III configurations rotated upward and around the lap belt, restraining the lower torso. Also, the primary cause of this seems to be the rearward movement of the lower legs which forces the femurs to move upwards, rather than hip joint stiffness.

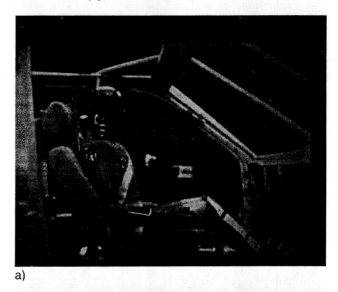

a)

b)

Figure 22. - Production seat set-up: a) Hybrid III pelvis, b) pedestrian pelvis

The neck responses of the two different Hybrid III configurations were similar for all the test speeds but some slight trends were apparent. Figures 24-26 show the peak magnitudes and the standard deviation of the samples for the different test speeds, (also listed in App. A, Tables 3-5). The peak positive upper neck shear forces of the Hybrid III with the pedestrian pelvis are slightly higher than those of the Hybrid III with the standard pelvis, but the converse is true of the upper neck tensile forces. At the lower neck, the Hybrid III with

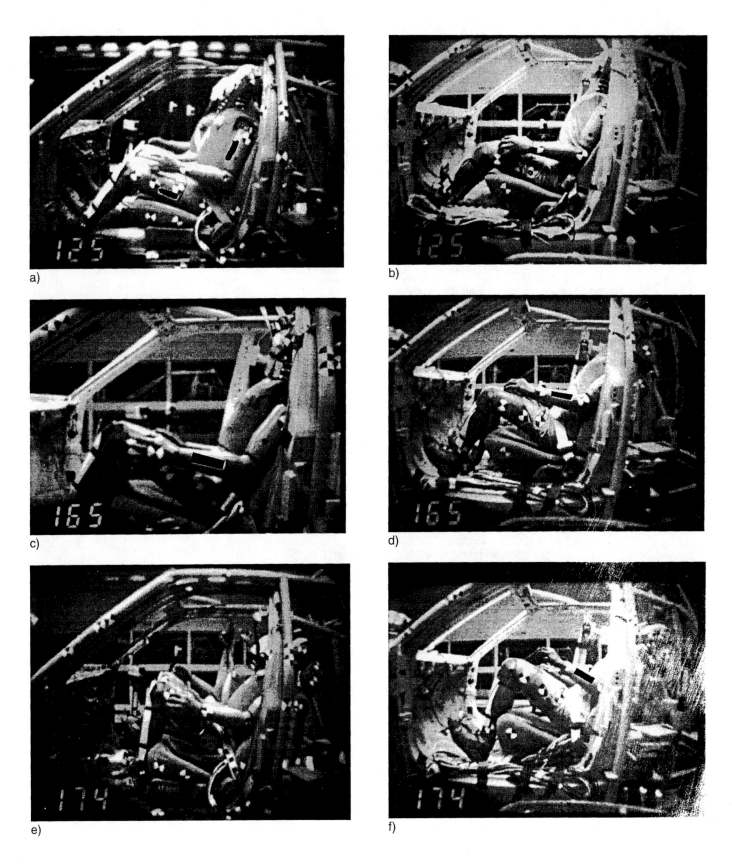

Figure 23. - Hybrid III pelvis vs. pedestrian pelvis kinematics: a) Hybrid III, ΔV = 8 kph, b) pedestrian, AV = 8 kph. c) Hybrid III, AV = 16 kph, d) pedestrian, AV = 16 kph, e) Hybrid III, AV = 24 kph, f) pedestrian, AV = 24 kph

413

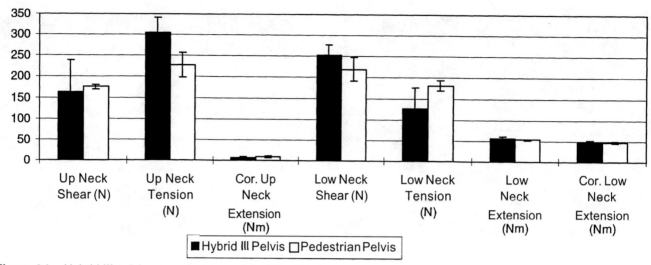

Figure 24. - Hybrid III pelvis vs. pedestrian pelvis, **AV** = 8 kph.

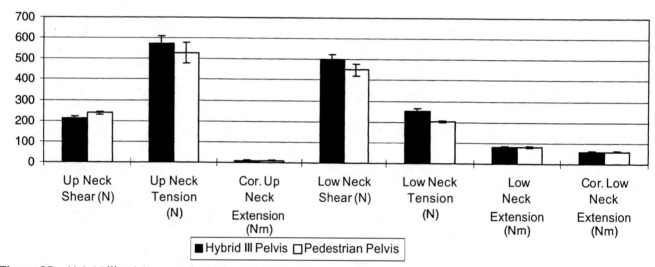

Figure 25. - Hybrid III pelvis vs. pedestrian pelvis, **AV** = 16 kph.

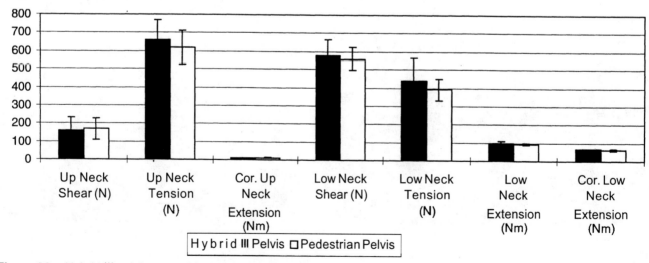

Figure 26. - Hybrid III pelvis vs. pedestrian pelvis, **AV** = 24 kph.

the pedestrian pelvis has lower shear force magnitudes and lower peak extension moments than the standard Hybrid III. Even though these differences were evident across the speed range, it should be noted that they were not significant across the range and usually fell within the standard deviations for each specific test.

The effect of pelvis type, standard Hybrid III or pedestrian, seems to have had a minimal effect on both the neck responses as well as the ATD's kinematics. In fact, the differences between the restraint system may have been an equal or larger contributor to the slightly increased occupant rotation seen with the Hybrid III with the standard pelvis than the pelvis itself. Whether or not the pelvis flesh was one piece of molded foam or three separate pieces did not seem to be significant.

DISCUSSION

Two different neck designs on the basic Hybrid III dummy have been tested in identical conditions. The head/neck responses of these necks were compared against known responses of cadavers at two test speeds (ΔVs of 16 and 24 kph) and a volunteer at one test speed (16 kph AV). Linear and angular accelerations of the Hybrid III dummy head compare well with those recorded on cadavers in similar test conditions. The moments developed at the head/neck junction of the Hybrid III dummy also compare well with those estimated at the head/neck junction of the cadavers at both speeds. The Hybrid III neck responses are closer to those of the cadavers than those of the volunteer. Based on the experimental data discussed earlier in this paper, the Hybrid III neck has adequate biofidelity for rear impact testing. The second neck tested, the RID, also has adequate biofidelity at the lower 16 kph AV, but lacks biofidelity at the higher, 24 kph AV. Additionally, the reproducibility of the RID neck was shown to be poor.

A third neck, the TRID, was tested with three different seats and its responses were compared to those of the Hybrid III neck. No significant differences in response trends were observed between these two dummy necks. Since the TRID neck was developed for low speed testing, the tests were conducted at ΔVs of 8 and 16 kph with production seats that were equipped with head restraints. Based on the comparison of the Hybrid III neck responses with those of the TRID neck, the Hybrid III is suitable for rear impact testing from ΔVs of 8 kph to higher speeds.

The above conclusion regarding the suitability of the Hybrid III dummy for rear impact testing is contrary to the opinions of Svensson and Lovsund [6] and of reference [8]. Svensson and Lovsund's [6] opinion is based on a previous paper by Seemann et al [12] in which it was claimed that the Hybrid III neck was too stiff in the sagittal plane. However, Seemann's study [12] compared the flexion responses of the dummy with those of trained, Navy volunteers in frontal impact sled simulations. No extension response studies were conducted by Seemann. Additionally, since the initial seated position of the human volunteers were erect and

the Hybrid III dummy neck is designed for seated posture in a passenger car, the conclusions reached by Seemann are not justified, even for flexion. Svensson and Lovsund [6] further cite a modeling study of Deng [13] that concluded that the torque response of the Hybrid III neck is similar to that of the human neck, but it has a higher shear response. In our testing with the TRID neck, which is a further refinement of the RID neck, no significant differences in shear response have been noted.

Two thoracic spine designs were tested and the test results have been described previously in this paper. Based on the available data, an articulated thoracic spine appears to somewhat improve correlation between dummy head/neck responses and cadaver responses. The peak bending moments and forces at the head/neck junction are nearly the same with both dummies. Differences in force and moment responses recorded at the lower neck load cells are different. However, these differences are due to the different locations of the load cells in the two dummy spines. It should be noted that the bending stiffness of the additional articulation in the thoracic spine of the TAD-50 was selected arbitrarily, and its biofidelity needs further investigation.

Two different pelvis designs were tested, a base Hybrid III and a pedestrian pelvis. The head/neck responses with the two pelvises are within test-to-test repeatability. The overall dummy kinematics are also the same. Both dummy pelvises show that the dummy femurs rotate up and load the lap belt.

LOWER NECK INJURY ASSESSMENT REFERENCE VALUES - The upper and lower neck responses are shown in Figures 27 and 28, (and listed in Appx. A, Tables 6 and 7), for the 16 and 24 kph AV rigid bench seat tests. Mertz and Patrick [1] have reported no discernible neck injuries in the 16 kph AV tests. In the 24 kph AV tests, a small sized cadaver showed minor ligamentous rupture at an estimated 34.4 Nm bending moment at the occipital condyle. The currently known Injury Assessment Reference Value, (IARV), for neck bending is based on this data [10]. An estimate of the bending moment at the C7-T1 junction was not made by Mertz. However, with the results of the current series of tests with the Hybrid III dummy, such an estimate can be made. The average bending moment at the occipital condyle of the Hybrid III dummy in 6 tests is 34.5 Nm. Assuming that the neck length and kinematics of the neck in the current series of tests are similar to those of the cadavers tested by Mertz and Patrick, the bending moments at the C7-TI junction in the cadaver test should be similar to those measured by the dummy. The average corrected lower neck extension moment was 112.1 Nm. This value can be tentatively associated with minor ligamentous damage for the population represented by the cadaver tested by Mertz and Patrick [1]. Since the upper neck extension moments in these tests averaged 34.5 Nm and the IARV for a 50th% adult male has been estimated by Mertz [10] to be 57 Nm, the IARV for the corrected lower neck extension moment should also be higher than the measured 112.1 Nm for a

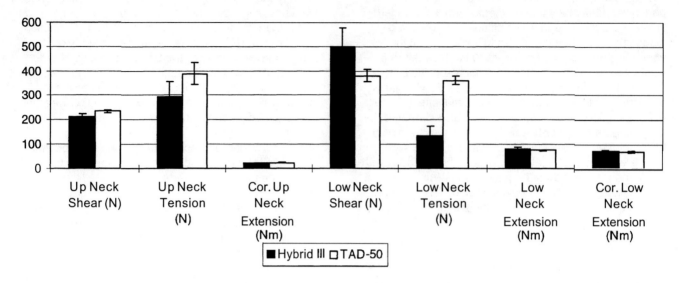

Figure 27. - Hybrid III and TAD-50 neck responses, rigid bench testing, **AV** = 16 kph.

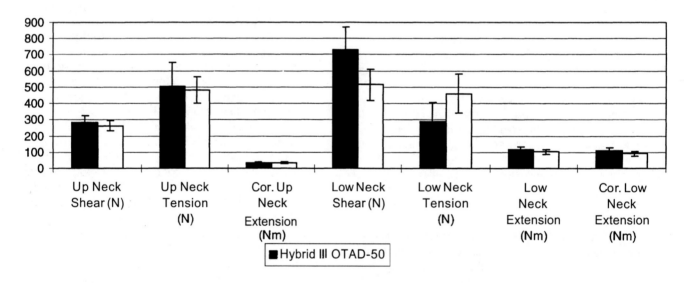

Figure 28. - Hybrid III and TAD-50 neck responses, rigid bench testing, **AV** = 24 kph.

50th% adult male. Comparing the corrected lower and upper neck moments for the Hybrid III dummy from Tables 6 and 7 in Appendix A, we find that the corrected lower neck moments are approximately 3.28 times the corrected upper neck moments. Therefore, if the upper neck extension IARV for the 50th% male adult is 57 Nm, the lower neck extension IARV could be 187 Nm. Since 187 Nm is associated with an injury, the threshold could be somewhat lower. Mertz and Patrick [11] have also estimated that 47 Nm is a non-injurious extension moment at the occipital condyles of a mid-size male. Using our ratio of 3.28 between the lower neck moment and occipital condyle moment, the non-injurious lower neck moment would be 154 Nm. Therefore, the threshold for ligamentous damage for a mid-size male would range between 154 Nm and 186 Nm. With similar reasoning, the lower neck extension IARV with the TAD-50, which has an approximate ratio of 2.79, should range between 131 Nm and 156 Nm. Further testing with smaller size dummies and cadavers are suggested to further verify the suggested IARV from these tests.

CONCLUSION

1. The head and neck responses of the Hybrid III dummy correlate very well with the cadaveric head/neck responses in the rigid bench seat tests with ΔVs of 16 and 24 kph reported by Mertz and Patrick [1].

2. The RID neck is biofidelic at the lower speed tested, (AV = 16 kph), with the rigid bench seat. It was bottoming out at the higher test speed, (AV = 24 kph), and had poor reproducibility.

3. Testing of production seats with the Hybrid III and the TRID necks showed no major differences in the responses of interests between the two necks.

4. The TAD-50 dummy, with an articulated thoracic spine, seems to be less biofidelic at the lower speed but slightly more biofidelic at the higher speed.

5. No significant differences in head/neck responses were found between the one-piece molded Hybrid III dummy pelvis and the three-piece pedestrian pelvis.

6. The corrected lower neck extension moment appeared to be the most sensitive response in the European production seat rear impact testing.

7. When utilizing the Hybrid III dummy, the threshold for ligamentous damage for a mid-size male ranges between 154 Nm and 186 Nm. With A TAD-50 dummy, the range is from 131 Nm to 156 Nm.

8. Based on the results of this study, the Hybrid III dummy is a suitable dummy for rear impact testing of automotive seat designs. Since all production seats have cushion in the seat back and are equipped with head restraints, the severity of impact up to which the Hybrid III is biofidelic for rear impacts is much higher than the 24 kph AV with the rigid seats. A companion paper [14] shows that even in 40 kph AV impacts with production seats, the load levels experienced by the dummy neck are below those developed in the rigid seat tests. As a result, the Hybrid III dummy is a suitable dummy for rear

impact testing at ΔVs exceeding 40 kph - a speed covering 94% of all NASS rear-end collisions [14].

APPENDIX A - DATA TABLES

Table 1. - Hybrid III vs. TRID neck responses, AV = 8 kph.

ΔV= 8 kph	NIF = 1.5		NIF = 1.1		NIF = 0.7	
(n = 1)	HIII	TRID	HIII	TRID	HIII	TRID
Up. Neck F$_x$ (N)	215.2	184.9	229.0	202.3	122.5	157.7
	-115.4	-91.0	-154.5	-85.6	-120.4	-108.4
Up. Neck F$_z$ (N)	221.6	237.1	139.9	179.0	129.7	137.0
	-17.0	-44.0	-116.0	-156.9	-45.6	-77.7
Cor. Up. Neck M$_y$ (Nm)	11.5	5.0	7.4	4.4	4.7	3.6
	-11.1	-14.8	-15.5	-15.5	-4.7	-8.5
Low. Neck F$_x$ (N)	324.0	233.8	275.8	286.2	177.2	182.4
	-128.3	-26.4	-128.4	-31.1	-94.3	-70.1
Low. Neck F$_z$ (N)	129.2	123.8	172.9	137.2	204.9	102.2
	-187.4	-106.3	-245.6	-128.3	-99.2	-56.5
Low. Neck M$_y$ (Nm)	30.7	13.8	25.3	8.7	13.1	14.9
	-62.6	-60.6	-52.4	-61.6	-34.1	-46.8
Cor. Low Neck M$_y$ (Nm)	30.6	17.3	30.0	15.9	21.8	18.5
	-60.3	-55.7	-53.9	-53.4	-32.3	-42.4

Table 2. - Hybrid III vs. TRID neck responses, AV = 16 kph.

ΔV=16 kph	NIF = 1.5		NIF = 1.1		NIF = 0.7	
(n = 1)	HIII	TRID	HIII	TRID	HIII	TRID
Up. Neck F$_x$ (N)	272.0	272.6	284.4	287.4	207.1	223.0
	-98.3	-103.7	-31.2	-81.0	-109.0	-104.6
Up. Neck F$_z$ (N)	371.5	446.6	380.0	261.5	374.3	402.8
	-81.3	-75.5	-79.1	-245.2	-117.1	-102.6
Cor. Up. Neck M$_y$ (Nm)	8.8	6.7	4.4	11.1	10.0	9.2
	-9.3	-15.9	-9.6	-19.5	-6.9	-9.2
Low. Neck F$_x$ (N)	520.9	469.3	520.0	526.9	371.1	381.7
	-79.0	-30.6	-55.4	-59.9	-167.9	-94.5
Low. Neck F$_z$ (N)	156.0	232.3	153.8	333.7	120.5	273.7
	-173.0	-56.9	-265.3	-158.0	-35.5	-72.15
Low. Neck M$_y$ (Nm)	17.0	11.2	0.1	5.6	21.4	14.3
	-83.2	-90.9	-76.4	-79.9	-60.0	-75.0
Cor. Low Neck M$_y$ (Nm)	20.2	20.8	0.9	10.0	18.5	16.7
	-76.3	-78.8	-61.0	-62.8	-46.6	-55.7

Table 3. - Hybrid III pelvis vs. pedestrian pelvis neck responses, AV = 8 kph.

ΔV = 8 kph	Hybrid III Pelvis (n =3)	Pedestrian Pelvis (n = 2)
Up Neck F_x (N)	151.9 +/- 10.5 -162.0 +/- 77.1	171.8 +/- 11.9 -175.2 +/- 6.3
Up Neck F_z (N)	305.2 +/- 36.8 -33.4 +/- 17.6	228.2 +/- 28.4 -36.6 +/- 17.9
Cor. Up Neck M_y (Nm)	9.5 +/- 4.5 -6.9 +/- 1.1	13.1 +/- 0.1 -8.4 +/- 1.9
Low Neck F_x (N)	252.9 +/- 24.4 -126.4 +/- 11.3	219.9 +/- 28.8 -102.6 +/- 33.2
Low Neck F_z (N)	126.3 +/- 49.4 -81.8 +/- 52.2	181.2 +/- 12.9 -136.9 +/- 52.9
Low Neck M_y (Nm)	35.6 +/- 16.4 -57.1 +/- 3.7	38.7 +/- 4.7 -54.0 +/- 1.2
Cor. Low Neck M_y (Nm)	35.1 +/- 18.2 -47.0 +/- 4.0	41.5 +/- 3.7 -47.6 +/- 2.0

Table 4. - Hybrid III pelvis vs. pedestrian pelvis neck responses, AV = 16 kph.

ΔV = 16 kph	Hybrid III Pelvis (n = 2)	Pedestrian Pelvis (n = 2)
Up Neck F_x (N)	211.9 +/- 10.0 -96.8 +/- 3.12	237.2 +/- 7.1 -98.4 +/- 6.5
Up Neck F_z (N)	572.2 +/- 35.6 -39.0 +/- 11.3	528.0 +/- 49.7 -24.4 +/- 10.9
Cor. Up Neck M_y (Nm)	7.1 +/- 0.7 -10.8 +/- 1.8	4.2 +/- 0.3 -9.9 +/- 1.7
Low Neck F_x (N)	499.1 +/- 24.6 -122.4 +/- 21.8	450.7 +/- 30.0 -71.4 +/- 19.6
Low Neck F_z (N)	251.9 +/- 14.7 -57.3 +/- 40.1	202.1 +/- 4.9 -86.1 +/- 12.7
Low Neck M_y (Nm)	19.4 +/- 1.8 -82.8 +/- 3.1	10.6 +/- 5.4 -81.7 +/- 2.9
Cor. Low Neck M_y (Nm)	20.7 +/- 5.2 -59.4 +/- 3.2	16.1 +/- 3.6 -60.2 +/- 2.5

Table 5. - Hybrid III pelvis vs. pedestrian pelvis neck responses, AV = 24 kph.

AV = 24 kph	Hybrid III Pelvis (n = 2)	Pedestrian Pelvis (n = 2)
Up Neck F_x (N)	161.1 +/- 73.4 -35.4 +/- 19.0	168.0 +/- 59.3 -30.3 +/- 6.8
Up Neck F_z (N)	662.5 +/- 108.6 -29.2 +/- 13.6	618.1 +/- 94.7 -50.8 +/- 27.2
Cor. Up Neck M_y (Nm)	4.8 +/- 0.02 -10.6 +/- 1.3	2.8 +/- 0.5 -12.5 +/- 0.8
Low Neck F_x (N)	580.5 +/- 85.0 -54.0 +/- 32.4	558.9 +/- 63.2 -27.2 +/- 11.5
Low Neck F_z (N)	436.8 +/- 133.4 -84.3 +/- 1.8	390.9 +/- 59.3 -42.9 +/- 46.0
Low Neck M_y (Nm)	2.1 +/- 1.3 -98.8 +/- 12.1	0.1 +/- 0.04 -93.9 +/- 6.5
Cor. Low Neck M_y (Nm)	4.7 +/- 1.5 -67.8 +/- 1.4	1.7 +/- 2.1 -62.2 +/- 3.4

Table 6. - Hybrid III vs. TAD-50 neck responses, rigid bench testing, AV = 16 kph.

AV = 16 kph	Hybrid III (n = 4)	TAD-50 (n = 2)
Up Neck F_x (N)	213.7 +/- 9.5 -94.1 +/- 43.7	234.43 +/- 7.1 -15.1 +/- 0.4
Up Neck F_z (N)	293.8 +/- 61.6 -159.1 +/- 29.3	388.6 +/- 43.8 -103.3 +/- 7.3
Cor. Up Neck M_y (Nm)	13.4 +/- 4.5 -21.7 +/- 2.0	9.0 +/- 0.3 -24.7 +/- 1.0
Low Neck F_x (N)	500.6 +/- 76.5 -47.5 +/- 16.2	379.6 +/- 25.1 -6.8 +/- 4.9
Low Neck F_z (N)	133.9 +/- 39.4 -253.3 +/- 23.5	361.0 +/- 17.8 -127.5 +/- 23.7
Low Neck M_y (Nm)	16.1 +/- 16.32 -80.4 +/- 8.8	5.3 +/- 1.5 -76.9 +/- 2.3
Cor. Low Neck M_y (Nm)	18.2 +/- 17.6 -72.0 +/- 5.4	6.2 +/- 1.5 -69.7 +/- 2.5

Table 7. - Hybrid III vs TAD-50 neck responses, rigid bench testing, AV = 24 kph.

AV = 24 kph	Hybrid III (n = 6)	TAD-50 (n = 2)
Up Neck F_x (N)	285.9 +/- 40.7 -120.0 +/- 42.2	262.4 +/- 32.3 -46.1 +/- 18.8
Up Neck F_z (N)	506.6 +/- 141.1 -242.11 +/- 78.3	482.1 +/- 81.3 -102.0 +/- 8.0
Cor. Up Neck M_y (Nm)	21.2 +/- 8.1 -34.5 +/- 4.8	9.6 +/- 1.2 -33.1 +/- 5.3
Low Neck F_x (N)	734.2 +/- 138.7 -71.9 +/- 29.9	514.2 +/- 95.6 -43.0 +/- 23.4
Low Neck F_z (N)	291.7 +/- 112.8 -496.5 +/- 126.8	459.8 +/- 119.0 -162.2 +/- 44.3
Low Neck M_y (Nm)	25.5 +/- 19.1 -113.5 +/- 17.7	9.1 +/- 6.2 -102.5 +/- 15.3
Cor. Low Neck M_y (Nm)	28.2 +/- 19.6 -112.1 +/- 16.8	8.0 +/- 5.7 -91.0 +/- 13.5

ACKNOWLEDGMENTS

The authors would like to thank the GTO/Hyge sled group for all the help they gave. We would also like to thank everybody who helped with the production of this paper, esp. T. Laituri for the drawings.

REFERENCES

1. Mertz, Jr. and L.M. Patrick, "Investigation of Kinematics and Kinetics of Whiplash," 11th Stapp Car Crash Conference, Society of Automotive Engineers, Inc., Warrendale, Pennsylvania, USA, SAE Trans., Vol. 76, 1967 670919.

2. Eichberger, A.; Geigl, B.C.; Moser, A.; Fachbach, B.; Steffan, H.; Hell, W.; Langwieder,K, "Comparison of Different Car Seats Regarding Head-neck Kinematics of Volunteers During Rear End Impact," Proceedings of 1996 International IRCOBI Conf. On the Biomechanics of Impacts, Dublin, Ireland, 1996.

3. "Traffic Safety Facts 1994: A Compilation of Motor Vehicle Crash Data from the Fatal Accident Reporting System and the General Estimates System," NHTSA, National Center for Statistics and Analysis, U.S. Dept. of Transportation, Washington, D.C., August 1995.

4. Digges, K.H.; Morris, J.H.; Malliaris, A.C., "Safety Performance of Motor Vehicle Seats," International Congress and Exposition; Society of Automotive Engineers, Inc., Warrendale, Pennsylvania, USA, SAE Trans., Vol. 94, Section 1, 1985-850090.

5. Prasad, P.; Mital, N.; King, A..I.; Patrick, L.M., "Dynamic Response of the Spine During +G_x Acceleration," Proceedings of the 19th STAPP Car Crash Conference, Society of Automotive Engineers, Inc., San Diego, CA, November 1975, 751172.

6. Svensson, M.Y.; Lovsund, P., "A Dummy for Rear-End Collisions - Development and Validation of a New Dummy-Neck," Proceedings of 1992 International IRCOBI Conf. On the Biomechanics of Impacts, Verona, Italy, 1992.

7. Schneider, L.W.; Ricci, L.L.; Salloum, M.J.; Beebe, M.S.; King, A.I.; Rouhana, S.W.; Neathery, R.F., "Design and Development of an Advanced ATD Thorax System for Frontal Crash Environments, Vol. 1: Primary Concept Development," U.S. Dept. Of Transportation, NHTSA, June 1992, DOT HS 808 138.

8. TNO Road-Vehicles Research Institute, "MADYMO Databases Manual Version 5.1," TNO Road-Vehicles Research Institute, Delft, Netherlands, 1994.

9. Thunnissen, J.G.M; van Ratingen, M.R.; Beusenberg, M.C.; Janssen, E.G., "A Dummy Neck For Low Severity Rear Impacts," Proceedings of 15th International Technical Conference on the Enhanced Safety of Vehicles, Vol. 2, Melbourne, Australia, 1996, 96-S10-O-12.

10. Mertz, H.J., "Anthropomorphic Test Devices," in "Accidental Injury, Biomechanics and Prevention," ed. by Nahum, A.M.; Melvin, J.W., Springer-Verlag, New York, 1993.

11. Mertz, H.J.; Patrick, L.M., "Strength and Response of the Human Neck," 15th Stapp Car Crash Conference; Society of Automotive Engineers, Inc., Warrendale, Pennsylvania, USA, SAE Trans., Vol. 80.1971 710855.

12. Seemann, M.R.; Muzzy, W.H.; Lustick, L.S., "Comparison of Human and Hybrid III Head and Neck Response," 30th Stapp Car Crash Conference; Society of Automotive Engineers, Inc., Warrendale, Pennsylvania, USA, ISBN 0-89883-451-1, paper 861892.

13. Deng, Y.C., "Anthropomorphic Dummy Neck Modelling and Injury Considerations," Accident Analysis & Prevention, Vol. 21, No 1, pp. 85-100, 1989.

14. Prasad, P.; Kim, A.; Weerappuli, D.P.V.; Roberts, V.; Schneider, D., "Relationships Between Passenger Car Seat Back Strength and Occupant Injury Severity in Rear End Collisions: Field and Laboratory Studies," Proceedings of the 41" STAPP Car Crash Conference, Society of Automotive Engineers, Inc., Orlando, Fl., November 1997.

BIOMECHANICAL INVESTIGATION OF INJURY MECHANISMS IN ROLLOVER CRASHES FROM THE CIREN DATABASE

Stephen A. Ridella, Ana Maria Eigen
National Highway Traffic Safety Administration
United States Department of Transportation

ABSTRACT

Previous research has described the types of injuries suffered by belted occupants of vehicles involved in rollover crashes. There has been much debate concerning these injuries, in particular the head, spine and thoracic injuries. Since rollovers may result in complex occupant kinematics, this paper analyzed extensive crash field and descriptive injury data from 55 belted occupants involved in 51 rollover crashes taken from the Crash Injury Research Engineering Network (CIREN) database. The paper discusses a methodology to deduce specific body region injury mechanisms of occupants in rollover crashes selected for crash as well as occupant characteristics.

Keywords: Rollover Crashes, Crash Investigations, Kinematics, Injury Severity, Mechanics

NUMEROUS PUBLICATIONS HAVE ATTEMPTED to understand the nature and mechanisms of occupant injuries as a result of rollover crashes. Early field analysis work by Huelke (1973, 1976) and Hight (1972) indicated a dominance of head and brain injuries in rollover crash occupants, however, most of these injuries occurred in unbelted, ejected occupants. When examining belted, unejected occupants, Huelke (1983), Digges (1993), Parenteau (2000) and Bedewi (2003) found the head and spine to be the most seriously injured body regions. The majority of these studies characterized the crash and occupant outcomes based on potential contact the belted occupant could have made with the vehicle interior. The head and neck regions are of most interest due to their typically severe outcomes and long recovery.

Digges and Eigen (2005) reported on injury outcomes for restrained and unrestrained occupants in various rollover crash conditions. They found that the severe injury (AIS => 3) rate for occupants in multiple-event rollover crashes (e.g., a planar crash followed by a roll) was double the rate for occupants in single vehicle rollover crashes. Additional analysis by Digges and Eigen (2006, 2007) examined injury severity in occupants involved in single vehicle or "pure" rollover crashes compared to these multi-event rollovers. They found that injury distributions were different for the two groups with the pure roll (single vehicle) cases having more head injury and the multi-event rollover crashes having more thoracic injury.

Others have attempted to identify injury causation with respect to rollover head and neck injuries. The National Highway Traffic Safety Administration (NHTSA) has published a significant body of rollover injury and fatality analysis using the National Automotive Sampling System (NASS) and Fatality Analysis Reporting System (FARS) databases. In their most recent work, Strashny (2007a) described rollover injuries in belted, unejected occupants relative to roof intrusion from single vehicle rollover crashes in the National Automotive Sampling System Crashworthiness Data System (NASS CDS) database. The analysis indicated that the maximum severity of head, neck (cervical spine and/or spinal cord) and face injury was significantly related to the amount of roof intrusion into the occupant compartment and the remaining post-crash headroom in the vehicle. Alternatively, Padmanaban's (2005) analysis of NASS CDS indicated no association of pre-crash effective headroom with serious head or neck (cervical spine and/or spinal cord) injury in belted occupants in single vehicle rollover crashes.

These unresolved associations have led some to test belted anthropometric test devices (ATDs) in rollover crashes to determine kinematics and potential for injury based on contact. Bahling et al.

(1990) found high axial neck loads in belted ATDs subjected to severe rollover crashes in both production and roof-strengthened vehicles as a result of the ATD's interaction with the roof during the event. Further analysis of this data by Friedman (2001) concluded that human cervical neck injury could occur from pure compressive forces on the top of the head aligned with the neck as a result of head contact to the roof during the roll event. Young et al (2007) reanalyzed the Bahling data to calculate a neck injury criterion, Nij (Eppinger, 1999), originally developed for assessment of neck injury potential in frontal impact. He suggested that this criterion could be used to predict the likelihood of neck injury between the production and roof strengthened vehicle in the rollover crash tests studied.

Brumbelow (2008) used police-reported, multi-state crash injury data to compare rollover crash injury severity to vehicle roof strength. While he concluded that injury risk was associated with roof strength, he was unable to derive injury mechanisms due to the type of data analyzed. Bedewi (2003) analyzed more detailed rollover crash case data from NASS CDS, but did not indicate specific mechanisms of injury to the belted occupants. Young (2007) pointed out that more research is required to understand the specific injury mechanisms as a result of occupant contact to interior structures during rollover.

The research described above has relied on large sample databases, severe crash tests or other sources to illustrate the rollover injury issue. This study intends to further the understanding of the underlying biomechanical processes that may describe the rollover crash injuries by examining specific cases from the CIREN database and applying a new tool within the CIREN program, the Biomechanics Table, or BioTab. The rich nature of the CIREN data, such as detailed medical injury and imaging information, coupled to extensive crash engineering data, can provide a unique perspective on the mechanisms for a particular injury. The BioTab tool provides an objective methodology to assess the injury and assign a specific injury mechanism based on factual evidence from the crash, clinical data, and prior biomechanical research.

METHODS
CIREN DESCRIPTION - This study is based on information drawn from the NHTSA-sponsored CIREN database. This database, now over 10 years old, has collected information from over 3500 crashes of late model year vehicles where at least one serious (AIS => 3) or two moderate (AIS = 2) injuries occurred to one of the vehicle occupants. Other selection criteria are involved, but the crashes are selected based upon injury severity of the occupant. A complete list of CIREN inclusion criteria may be found in the United States Federal Register (2004). The related vehicles are studied in a detailed crash investigation, forming the basis of the CIREN vehicle parameters, linked to the detailed medical records of the occupant's injuries. This includes radiological images and reports, clinical progress, notes during treatment, operating room reports, clinical photographs, occupant interviews, discharge reports, one year follow-up recovery assessment and other descriptions of the injuries. A thorough, multidisciplinary review of each case occurs to derive a biomechanical basis for the injuries based on physical evidence. A review team at each of the eight CIREN centers consists of an experienced NASS-trained crash investigator, a board-certified trauma physician, biomechanical engineer, data coordinator, and emergency response personnel. Other physicians (surgeon or radiologist) and engineering personnel may be consulted on individual cases. These reviews confirm the crash and injury assessment (including AIS coding) as well as complete the BioTab process (to be discussed below) for all AIS 3+ injuries suffered by the case occupant.

CASE SELECTION/DATA EXTRACTION - Data was extracted from the master CIREN data repository. Case selection was based on the following occupant and vehicle crash parameters: occupant was at least 16 years of age and older seated in the right/left front outboard seating position, confirmed wearing lap/shoulder belt and sustained serious and/or disabling injury, per CIREN requirement. The injury level provides injury source, confidence in the assignment of injury source, and AIS 90 injury coding (1998 update), with related disaggregation by injury severity, body region, and injury aspect. The vehicle rollover was characterized as either a pure rollover event or a multiple event rollover crash. The pure rollover crashes were characterized by a single-event non-planar crash of at least one quarter turn. The multiple event rollover crashes consisted of at least one planar event and one rollover event and generally experienced a planar crash event followed by a rollover event. Only rollovers up to eight quarter turns (two roof contacts) were selected to avoid uncertainty in injury mechanism analysis from extremely violent crashes. Generally, the multiple event rollover crashes sustained a planar event that precipitated the rollover crash. In these cases, the maximum injuries were disaggregated by event to determine when it was experienced. It was noted that cases did exist with a planar event following the rollover event. These were usually interrupted rollover crashes. The arrested rollover crash at the first event might be deemed as more injurious, from energy dissipation standpoint, than multiple quarter turn rollover events.

The data was received in Excel spreadsheets, as extractions from the master CIREN data repository. These data sets were then imported into SAS for analysis. This paper considers crash and vehicle parameters of a similar specificity to NASS CDS. The occupant and injury parameters, however, benefit from detailed injury reporting and review inherent in CIREN analysis. A synthesis of engineering and medical expertise comprises the conclusions reported in this data set and on which the conclusions of this paper are predicated. The BioTab, reports the disaggregation of injuries based upon the separate crash events. This concept is absent in the NASS CDS, however, with less detail this system provides the national estimates of tow-away crashes occurring on public roadways in the United States. While CIREN cannot be used to generate nationally-representative crash statistics, the use of CIREN and associated BioTab allows crash events to be analyzed individually as opposed to en masse where all injury are assigned to the most harmful event. In multi-event crashes, especially involving rollover, clear and distinct injury patterns can be differentiated with the appropriate data and expertise.

Upon dissagregation, the cases were subjected to a review of the crash summary, scene diagram, occupant contact points, collision deformation classification (CDC), and relevant injury radiology. This review is done initially by the case crash reconstruction and injury review team at the individual CIREN centers. From the case crash reconstruction and injury review done by the CIREN center for that case, consideration of the injury ranking was undertaken. The ranking was not necessarily the sole predictor of the maximum injury for the case. The authors gave consideration to whether the injury might be produced during the rollover event. Typically, this injury ranking resulted in the most severe injury, known as the Rank 1 injury, to be selected for further analysis. For this analysis in this paper, the Rank 1 injury was usually the highest AIS injury that occurred as a result of a rollover or a frontal or side collision. The data tables in the Appendices give the Rank 1 injury and associated crash information.

The data from the Excel extractions, as well as the full CIREN report were integrated to form a master workbook of crash data for each relevant occupant. The case was identified as a pure or multiple event rollover as described above. Basic demography relevant to occupant size and seating location, as well as vehicle roll direction and crash severity were included. The Rank 1 injury was assigned to the planar or rollover phase of the crash based upon the occupant kinematics, as described, scene diagram, photographs, and radiological evidence. This injury was then subjected to additional analysis using the BioTab which will be described below.

BIOTAB DESCRIPTION - The BioTab provides a means to completely and accurately analyze and document the physical causes of injury based on data obtained from detailed medical records and imaging, in-depth crash investigations, and findings from the medical and biomechanical literature. The BioTab was developed because the terminology and methods currently used to describe and document injury causation from crash investigations are vague and incomplete. For example, the terms direct and indirect loading are often used to describe how an injury occurred. However, there are situations where these terms are unclear, e.g., is a femoral shaft fracture from knee-to-knee bolster loading from direct loading of the knee or indirect loading of the femur through the knee. In addition, the term inertial loading is often used to describe how tensile neck injuries occur, however, using this terminology fails to document that neck tension would not have occurred unless the torso was restrained. The BioTab removes these ambiguities by providing a consistent and well-defined manner for coding injuries and recording the biomechanics of injury in crash injury databases.

Coding in the BioTab revolves around the definition of an Injury Causation Scenario (ICS), which is the set of crash, vehicle, occupant, and restraint conditions that were necessary for an AIS 3+ injury to have occurred as well as the factors that affected the likelihood and severity of the injury. The elements of an injury causation scenario will identify and describe the following:

1) Whether the injury was caused by another injury (e.g., a rib fracture causes a lung laceration),
2) The Source of Energy (SOE) that led to the occupant loading that caused the injury,
3) The Involved Physical Component (IPC) that caused injury by contacting the occupant and the body region contacted by the IPC,
4) The path by which force was transmitted from the body region contacted, through body components, to the site of injury, and
5) All other factors that contributed to the severity or likelihood of injury.

The BioTab documents the evidence supporting these elements and uses it to determine confidence levels for each IPC and ICS.

The BioTab also documents the specific "mechanisms" by which an injury is believed to have occurred. Importantly, the BioTab distinguishes between injury causation scenarios and injury mechanisms. While the latter are necessarily a function of a particular ICS, the mechanisms that produced an injury are specific descriptions of the physical response of an occupant to the applied loading. In the BioTab, mechanisms can be documented at the body-region and organ/component level and include physical events such as compression, torsion, acceleration, and bending. As with IPCs and ICSs, evidence for injury mechanisms is documented in the BioTab and may include specific injury data obtained from the crash investigation (e.g., an avulsion fracture of a long bone due to a tension mechanism), as well as information in the biomechanical and medical literature (e.g., ribs break in bending). As with ICSs and IPCs, confidence levels are assigned to injury mechanisms based on evidence. A flowchart of the process and further discussion on BioTab may be found in Scarboro et al. (2005).

The Biotab process described was applied to Rank 1 (or most serious) injury of the occupant from all the cases meeting the inclusion criterion for this study. In several cases, the BioTab injury mechanism analysis was already completed by the CIREN center team and these results were confirmed by the authors and support. In other cases, the BioTab process was completed by the authors and the CIREN dataset extended to reflect this analysis. The master workbook of all the vehicle and occupant data was summarized into a single-page dataset (including BioTab Involved Physical Component and injury mechanism) that could sustain integrated queries and forms the basis for the findings to follow. This data is found in Appendices 1 and 2 and forms the basis for the results to be presented.

RESULTS

DEMOGRAPHICS (Occupant) – Sixty four occupants in 59 vehicles involved in a rollover crash met the criteria for inclusion in this study. However, after review of each case, 3 occupants were eliminated from further analysis since the rollover was associated with a second, non-planar event that arrested the roll of the vehicle. These types of rollovers were not the intent of this study. An additional six occupants were eliminated from further analysis since their Rank 1 injury was an AIS 2 injury and the BioTab was designed especially for AIS 3+ injury mechanism analysis. Of the remaining 55 occupants in 51 vehicles, 19 were in a single event, or pure rollover crash and 36 were in a multiple event crash with rollover. Sample representative demographics for these 55 occupants are listed in Table 1 and additional details may be found in Appendices 1 and 2.

For the pure roll cases, 11 of the 19 occupants were male and 8 of 19 were female. Eight of the 11 males were considered overweight by Body Mass Index (BMI > 25) calculation and Centers for Disease Control tables while 5 of the 8 females were overweight. It should be noted that mean BMI for adult males and females, based on 1999-2002 NHANES data, was 27.9 and 28.1 respectively (Ogden et al., 2004). The males in the pure roll cases had the oldest average age for either gender by roll event categories. For the multi event roll case occupants, 19 were females and 17 were males and about half of each gender group was overweight.

Table 1. Case Occupant Demographics

Occupant	Pure Roll	MultiEvent
Males	Mean	Mean
Age (years)	49	35
BMI	28	25
Females	Mean	Mean
Age (years)	40	42
BMI	26	25
Drivers	N	N
Near Side	4	11
Far Side	10	18
Passengers	N	N
Near Side	2	2
Far Side	3	5

Drivers made up nearly 80% and right front seat passengers comprised about 20% of the occupants sampled. For both event types, more occupants experienced a far side rollover, that is, the roll was initiated on the opposite side of the occupant's seated position. Although not statistically significant, 64% of the multiple event rollover crash occupants experienced far side rollover orientation. Of the multiple event rollover crash occupants, 25% resulted in far side orientation with one quarter turn, zero vehicle inversions. Another 25% of the multiple event rollover crash occupants sustained one vehicle inversion, producing a far side orientation and 13% of the occupants experienced 2+ vehicle inversions. For the pure rollover occupants, approximately three-quarters of the occupants experienced far side rollover crashes. Nearly 60% of these crash occupants experienced one vehicle inversion while the rest experienced two inversions.

DEMOGRAPHICS (Vehicle) - Fifty-one rollover crashes were identified from the CIREN database. These selected crashes were of at least one quarter turn and did not exceed eight quarter turns. These crashes were disaggregated by single event, or pure, and multiple event rollover crashes as well as by vehicle type.

Nineteen (19) pure and 32 multi-event rollover crashes were identified. In terms of crash severity, none of the pure roll vehicles had a reported longitudinal delta-V due to the rollover damage indicated by the crash investigator. The average reported delta-V for the vehicles in the 32 multievent rollovers, inherent to the planar component of the crash and based upon damage measurements, was 37 kph, indicating a moderate to severe crash experience. Rollover severity, defined either as number of ¼ turns or roof inversions, was greater for the pure roll vehicles regardless of vehicle type. The pure roll cases averaged nearly 6 quarter turns (2 roof inversions) versus an average of 3 quarter turns (1 rood inversion) for the multi-event rollovers.

INJURIES – For the 55 occupants in this analysis, a total of 842 injuries at all severity levels (AIS 1 to 6) were reported. Twenty-nine percent of the 842 injuries were AIS 3+, however, the distributions of injuries were different between the pure roll and multiple event roll cases (Figure 1). Injuries to the spine, largely involving the cervical bony structure, accounted for 57% vs. 19% of the AIS 3+ injuries to pure roll event vs. multiple event roll occupants. Neck injuries in this analysis refer only to the soft tissue structures of the neck and accounted for less than 1% of the serious injuries for either group.

The percentage of spine injuries in the pure roll group is in contrast to the 11.5% figure for all AIS 3+ injuries in belted, single vehicle rollover occupants derived from figures in Digges and Eigen (2005) in their NASS CDS analysis. This is indicative of the CIREN sampling process for more severe injuries. Thoracic injuries were the dominant severely injured body region for the multi-event roll group, accounting for just below 30% of all the AIS 3+ injuries for that group. The same percentage of the AIS 3+ injuries to the pure roll occupants were thoracic injuries. Nearly equal percentages of serious head injuries were evident in both groups and lower extremity and abdominal injury were the remaining serious injury regions for the multi event rollover occupants. It is interesting to note that the pure roll serious injuries fall mainly in the head, spine and thorax regions while the multi-event occupant injuries are distributed to all body regions.

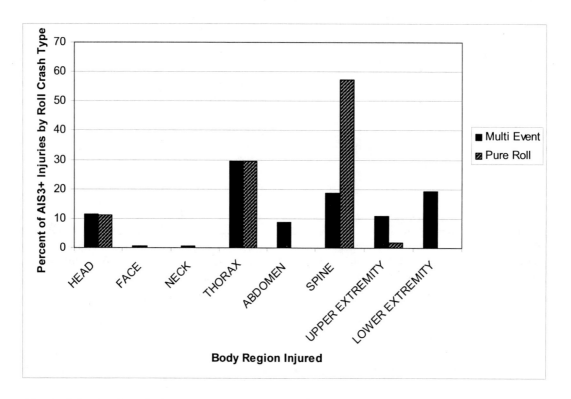

Fig. 1 – Distribution of AIS3+ Injuries by Body Region for Occupants by Rollover Crash Type

BIOTAB ANALYSIS - The Rank 1 or most severe AIS 3+ injury for each occupant was evaluated by the BioTab process. Appendices 1 and 2 give a complete listing of the relevant occupant and injury variables for the pure roll cases and the multiple event roll cases respectively, including the results of the BioTab analysis (Injury Event, i.e., the source of the energy for the injury, Involved Physical Component and Regional Injury Mechanism) for the Rank 1 injury in the case. Appendix 1 lists the occupants involved in a pure roll only and Appendix 2 lists the occupants in a multi-event rollover crash. It is apparent from Appendix 1 that the fracture of the cervical spine is the most frequent Rank 1 injury in this set accounting for eight of the nineteen Rank 1 injuries. Head injury accounted for six of the nineteen Rank 1 injuries with chest, thoracic, and lumbar spine injuries in the remaining five cases. The cervical injuries occur throughout the cervical spine structure from C2 to C7 and even involving T1. All of these eight cervical injury cases involved a complex kinematic and combined loading condition of the cervical spine during the injury event as the head interacted with either the roof or roof rail as the indicated IPC. The regional mechanism for all of the spinal injuries included compression combined with a flexion, extension, or lateral bending. The six head (brain) injuries involved a head to roof impact and a regional (head) compression mechanism of injury.

There was insufficient information to deduce the organ level (brain) injuries which could be a shearing or tension mechanism. The three chest injuries were from compression mechanism as a result of a direct thorax interaction with the roof or side interior.

For the multi-event rollover cases, BioTab analysis implicated twenty-nine of thirty-six (80%) Rank 1 injuries to the planar crash event (FI: frontal or SI: side impact) as the injury event or source of energy for the injury. The roll event was the injury producing event in eight cases and involved injuries from all the other body regions with no specific pattern. Chest and abdominal injuries accounted for 12 of the 36 Rank 1 injuries. This involved either rib fracture, lung contusion, or other organ injury. The head and entire spinal column each accounted for six Rank 1 injuries. There were four cases with severe femur fracture as the Rank 1 injury and the remaining Rank 1 injuries were distributed to the arm and lower leg. All of the chest, abdomen, and femur injuries involved a compression regional injury mechanism, usually due to interaction with the side door structure, belt, or air bag depending on the injury event. Of the six cases of spinal injury as the Rank 1 injury, only one involved the cervical spine bony structure. The spinal injuries were mainly lower thoracic and lumbar spine burst fractures caused by compression from loading the seat or seat belt during one of the planar events other than the rollover.

CASE DESCRIPTION – To better inform the reader of the analysis process involved for each case, the following example describes the crash, occupant kinematics and injuries, and BioTab analysis of an occupant from the pure roll event list in Appendix 1.

Case example (Appendix 1, Case 8): A 4-door sport utility vehicle (SUV) was traveling south in the left southbound lane of a rural, four-lane divided freeway (two lanes southbound, unprotected median, two lanes northbound). It was daylight, snowing and the road was slush covered. The driver of the SUV was overtaking a slower vehicle in the right lane when he lost control due to the slippery road surface. The SUV departed the left side of the road and entered the median. The driver attempted to regain control by steering to the right, but the vehicle began to rotate in a clockwise manner. The SUV tripped over its left wheels/tires and rolled six quarter turns (2 roof inversions) before coming to rest on its roof facing north-northwest. The impact was classified as severe but no WinSmash reconstruction was attempted due to the rollover nature of the crash. Figure 2 shows the post-crash condition of the vehicle and the interior space at the case occupant seating position.

The case occupant is the 155 cm (5' 1"), 86 kg (190 lb), 44-year-old female right-front passenger who was using the available three-point seat belt, but the dash-mounted air bag did not deploy. The case occupant was transported via ground ambulance to a local hospital and was later transferred to a regional level-one trauma center. The occupant kinematics were deduced such that she moved up and outboard, relative to the vehicle interior as the vehicle rolled. Evidence indicated her head interacted with the roof where she sustained a loss of consciousness, fractures to the left facets of C5 through C7, a left C6 lamina fracture, and a cervical spinal cord contusion resulting in quadriplegia. A computed tomography (CT) scan of her cervical spine (Figure 3) indicates the location and extent of the injuries. Note the asymmetry of the fracture pattern. Her other injuries included minor contusions to the neck and upper and lower extremities. For the BioTab process, the Injury Causation Scenario was defined and indicated the rollover as the source of energy. Based on the physical evidence of head contact to the roof, the roof was identified as the involved physical component (IPC). The path of the force was transmitted from the head to the lower cervical spine. Based on her kinematics and analysis of the fracture pattern of her cervical spine injuries from the CT scan (Figure 3), a compression/lateral bending mechanism to the region resulted in failure of the lower cervical spine structures. This pattern is consistent with mechanisms reported by Pintar (1989, 1995).

DISCUSSION

This paper discusses an approach using the results of queries from NASS CDS (others like the German In-Depth Accident Study (GIDAS) could also be used) to describe the vehicle crash modes and injuries of interest and then derives the comprehensive injury detail from CIREN cases to understand the body region injury mechanisms. In studies cited above by Digges and Eigen (2005,

Fig. 2 – Post-Crash Vehicle Photo of Pure Roll Case 8 (Appendix 1)

2006), the use of large sample databases such as NASS CDS provided perspectives on rollover crashes to begin the understanding of what types of injuries are associated with these crashes and what vehicle components may be involved in the injury process. However, the conclusions on specific body region injury mechanisms, as defined in this paper, can only be derived from a more detail-oriented (though more limited in cases) database such as CIREN. This paper clarifies the definition of injury mechanism based on biomechanical analysis of the entire crash event. The CIREN database offers a unique opportunity to expand upon the aggregate injury numbers and use evidence-based, objective methods to deduce injury mechanisms in all crash types, including rollover. CIREN data mimics other large sample data bases by indicating the severe injuries in rollover occur to the head, neck, thorax and extremities as shown in Figure 1. However, when further analyzed using the BioTab process, there are significant differences in what body regions are injured between pure roll and multiple roll events as well as the injury mechanisms indicated.

The majority of the injuries in the multi event rollovers was attributed to the planar crash and not the rollover and indicated a compression mechanism to the region of injury whether it was head, chest, abdomen or femur. Thoracic injuries were dominant and almost always indicated the planar event as the source of the energy. In contrast, complex injury mechanisms were noted in the injuries suffered by the occupants in a pure roll crash. Most often, these injuries were cervical spine injuries. The mechanisms identified for these cervical spine injuries involved compression combined with bending, extension, or flexion based on the physical vehicle evidence and the analysis of radiographic and other medical information on the occupant. Pre-crash maneuver, roll severity, and other occupant factors may play a role in the occupant's position and injury mechanism. Conroy et al (2006) compared occupant, crash and vehicle characteristics in rollover crashes that involved seriously injured occupants from the CIREN database to a control group of minor or uninjured occupant in rollover crashes from the NASS database. They found that the roof and side interior structures were a significant source of injury for seriously injured occupants, but did not employ the BioTab process to look at injury mechanisms.

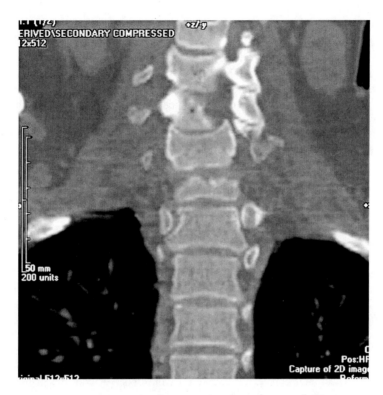

Fig. 3 – Radiology of Pure Roll Case Vehicle Occupant Showing Multiple Cervical Spine Injuries

Occupants involved in pure rollover crashes in this study had slightly more severe rollovers than multiple event rollover occupants based on the number of vehicle ¼ turns and inversions and this may have affected the injury distributions and patterns between the groups. These occupants were usually male and generally overweight, as defined by their BMI, however no meaningful statistics could be applied due to sample size. Strashny (2007b) calculated injury severity rates for occupants in single vehicle rollovers (relative to age, gender, and BMI). He found that restrained drivers, fatally injured in a rollover, had a higher BMI than unrestrained drivers. Since all occupants in this study were restrained, it was not possible to compare to unrestrained.

The dominance of cervical spine injuries among the pure roll occupants bears further inspection. The rollover crash severity, indicated by number of inversions, was higher for the crashes that involved cervical spine injury. This led to the roof or some roof structure as the involved physical component for these injuries. Biomechanics investigators have attempted to characterize the response of the human cervical spine under loading through the head. This paper is not an exhaustive review of this literature, but uses those conclusions to support the complex kinematic and biomechanical processes that contribute to the injuries observed in the occupants of this paper. Moffatt (1978) and Patrick (1987) have provided good overviews of automotive cervical trauma. Their observations of fracture patterns indicate the loading modes responsible for the injury. The work of Nusholtz (1981), Pintar (1989, 1995), and Nightingale (1997) derived the cervical spine failure modes based on initial orientation and loading of the head. A conclusion from this body of work can be that the complex loading of the neck through the head can induce buckling and injury in a variety of locations and mechanisms depending on initial occupant orientation and load path. The CIREN occupants in this study had failures along the entire cervical spine and even into the thoracic region and a variety of mechanisms were derived from the BioTab analysis based on the physical evidence. To derive surrogate predictors of injury that characterize these injuries would require additional analysis of existing biomechanical data and development of test surrogates that could mimic the occupant kinematics predicted in these crashes.

LIMITATIONS OF THE STUDY- The results of this study and conclusions drawn were from a highly selective subset of the CIREN database, which by design, is a censored sample of severely injured occupants admitted to one of eight participating Level-1 trauma centers in the USA. One cannot conclude the trends indicated are nationally representative based on the disparity of injury distributions between NASS and CIREN for the belted occupants in a pure rollover. In addition, this small sample of rollover cases limits some possible direct comparisons of non-injured to injured occupants since all the occupants had at least one severe injury.

CONCLUSIONS

Vehicular rollover is a result of a complex, chaotic crash event that results in a unique set of circumstances for each occupant injury. Conclusions on body region injury mechanisms, as defined in this paper, can not be drawn from large sample databases. This study provides a unique opportunity to utilize the CIREN database coupled with powerful analysis tools to deduce specific injury mechanisms. The data in this study reveals that rollovers need to be disaggregated based on number of crash events. The resulting dissimilarity in injury distribution helps to better understand how to describe the scenario that led to the injury. Thoracic, not just head and neck injury mechanisms need to be considered in these analyses. Also, this study indicates that cervical spine injury mechanisms are the result of complex loading combining compression with flexion, extension or lateral bending as a result of interaction with vehicle components, primarily during a pure rollover event. Thoracic injury mechanisms were compression based as a result of interaction with vehicle components during the planar (frontal or side impact) event of the multiple event roll crashes. This understanding could help describe a modeling or test environment for countermeasures, such as belt improvements described by Moffatt (1997, 2005) to mitigate such injuries. Further analysis of CIREN case data as well as research and development of appropriate surrogates to mimic human kinematics and injuries in rollover is required.

ACKNOWLEDGEMENTS

The authors gratefully acknowledge the invaluable support from Mr. Mark Scarboro (NHTSA) for accessing the CIREN cases and the additional BioTab analysis. The authors also thank Dr. Jonathan Rupp from the University of Michigan Transportation Research Institute for the BioTab process descriptions.

REFERENCES

Bahling, G.S., Bundorf R.T., Kaspzyk,G.S., Moffatt, E.A., Orlowski, K.F.,and Stocke, J.E. Rollover and Drop Tests : The influence of roof strength on injury mechanics using belted dummies. In: Proceedings of the 34th Stapp Car Crash Conference, 1990, pp. 101-112.

Bedewi, P.G., Godrick, D.A., Digges, K.H. and Bahouth, G.T. An Investigation of Occupant Injury in Rollover: NASS-CDS Analysis of Injury Severity and Source by Rollover Attributes. In: Proceedings of the 18th International Technical Conference on Experimental Safety Vehicles, Paper No. 18ESV-000419, 2003, pp. 1-15, Nagoya, Japan.

Brumbelow, M.L., Teoh, E.R., Zuby, D.S., McCartt, A.T. Roof Strength and Injury Risk in Rollover Crashes, Insurance Institute for Highway Safety, 2008, pp. 1-19.

Conroy, C., Hoyt, D.B., Eastman, A.B., Pacyna, S., Holbrook, T.L., Vaughan, T., Sise, M., Kennedy, F., and Velky, T. Rollover Crashes: Predicting Serious Injury Based on Occupant, Vehicle, and Crash Characteristics. Accident Analysis and Prevention, Vol. 38, 2006, pp. 835-842.

Digges, K.H., Malliaris, A.C. and DeBlois H.J. Rollover Injury Causation and Reduction. In: Proceedings of the 26th International Symposium on Automotive technology and Automation, 1993, pp. 681-685.

Digges, K., Eigen, A.M., Dahdah, S. Injury Patterns in Rollovers by Crash Severity. In: Proceedings of the 19th International Technical Conference on Experimental Safety Vehicles., Washington, D.C., 2005, Paper No. 05-0355, pp. 1-9.

Digges, K.E. and Eigen, A.M. The Injury Risk from Objects Impacted Before and During Rollovers. In: Proceedings of the International IRCOBI Conference on the Biomechanics of Impact, Madrid, Spain, 2006, pp. 385-388.

Digges, K. and Eigen, A.M. Injuries in Rollovers by Crash Severity. In: Proceedings of the 20th International Technical Conference on Experimental Safety Vehicles., Lyon, France, 2007, Paper No. 07-0236, pp 1-10.

Eppinger, R., Sun, E., Bandak, F., Haffner, M.,Khaewpong, N., Maltese, M., Kuppa, S., Takhounts, E., Nguyen, T., Zhang, A., Saul, R. Development of Improved Injury Criteria for the Assessment of Advanced Automotive Restraint Systems – II, National Highway Traffic Safety Administration, 1999.

Friedman, D. and Nash, C.E. Advanced Roof Design For Rollover Protection. In: Proceedings of the 17th International Technical Conference on Experimental Safety Vehicles. Amsterdam, The Netherlands, 2001, Paper No. 01-S12-W-94, pp. 1-10.

Hight, P.V, Siegel, A.W. and Nahum A.M. Injury Mechanisms in Rollover Collisions. In: Proceedings of the Sixteenth Stapp Car Crash Conference, 1972, SAE Paper No. 720966. Warrendale, PA., pp. 204-225.

Huelke, D.F, Marsh, J.C., Dimento, L, Sherman, H., Ballard, W.J. Injury Causation in Rollover Accidents. In: Proceedings of the 17th Conference of the American Association of Automotive Medicine, 1973, pp. 87-115.

Huelke, D.F., Lawson, T.E, and Marsh, J.C. Injuries and Vehicle Factors in Rollover Car Crashes. Accident Analysis and Prevention, Vol. 9, No. 2-B, 1976, pp. 93-107.

Huelke, D.F. and Compton, C.P. Injury Frequency and Severity in Rollover Car Crashes as related to Occupant Ejections, Contacts and Roof damage. Accident Analysis and Prevention Vol. 15, No. 5, 1983, pp. 398-401.

Moffatt. E.A., Siegel, A.W., Huelke, D.F., and Nahum, A.M. The Biomechanics of Automotive Cervical Fractures. In: 22nd Annual Proceedings of the American Association for Automotive Medicine, 1978, pp. 151-178.

Moffatt, E.A., Cooper, E.R., Croteau, J.J., Parenteau, C., and Toglia, A. Head Excursion of Seat Belted Cadaver, Volunteers and Hybrid III ATD in a Dynamic/Static Rollover Fixture. SAE Paper No. 973347, 1997, Warrendale, PA., pp. 509-525.

Moffatt, E.A. and James, M.B. Headroom, Roof Crush, and Belted Excursion in Rollovers. In: SAE 2005 Transactions Journal of Passenger Cars: Mechanical Systems, V114-6, 2005, 2005-01-0942, pp. 1037-1056.

Nightingale et al. The Dynamic Responses of the Cervical Spine: Buckling, End Conditions, and Tolerance in Compression Impact," Proceedings of the Forty-First Stapp Car Crash Conference, 1997, SAE Paper No. 973344, pp. 451-471.

Nusholtz, G.S., Melvin, J.W., Huelke, D.F., Blank, N.M. and Blank, J.G. Response of the Cervical Spine to Superior-Inferior Head Impact. In: Proceedings of the 25th Stapp Car Crash Conference, 1981, SAE Paper No. 811005, pp. 197-237.

Ogden, C.L., Fryar, C.D., Carroll, M.D. and Flega, K.M. Mean Body Weight, Height, and Body Mass Index, United States 1960–2002. Advanced Data (from Vital and Health Statistics), Vol. 347, 2004.

Padmanaban, J., Moffatt, E.A., Marth, D.R. Factors Influencing the Likelihood of Fatality and Serious/Fatal Injury in Single-Vehicle Rollover Crashes. In SAE 2005 Transactions Journal of Passenger Cars: Mechanical Systems, V114-6, 2005, SAE Paper No. 2005-01-0944., pp. 1072-1085.

Parenteau, C.S. and Shah, M. Driver Injuries in US Single-Event Rollovers. In: Side Impact Collision research, SP-1518, 2000, SAE Paper No. 2000-01-0633, pp 1-8, Warrendale, PA.

Patrick, L.M. Neck Injury Incidence, Mechanisms and Protection. In: 31st Annual Proceedings of the American Association for Automotive Medicine, 1987, pp. 409-31.

Pintar FA, Yoganandan N, Reinartz, et al. Kinetic and Anatomical Analysis of the Human Cervical Spinal Column under Axial Loading. Proceedings of the 33rd Stapp Car Crash Conference, 1989. SAE Paper 892436. Warrendale, PA: Society of Automotive Engineers, pp. 191-214.

Pintar et al. Dynamic Characteristics of the Human Cervical Spine. In: Proceedings of the Thirty-Ninth Stapp Car Crash Conference, Warrendale, PA, 1995, SAE Paper No. 952722, pp. 195-202.

Scarboro, M. Overview of the Enhancements and Changes in the CIREN Program. SAE Government/Industry Meeting, Washington, D.C., 2005.

Strashny, A. The Role of Vertical Roof Intrusion and Post-Crash Headroom in Predicting Roof Contact Injuries to the Head, Neck, or Face During FMVSS No. 216 Rollovers; An Updated Analysis. DOT HS 810 847, 2007a, pp. 1-18.

Strashny, A. An Analysis of Motor Vehicle Rollover Crashes and Injury Outcomes. DOT HS 810 741, 2007b, pp. 1-89.

United States Federal Register, Vol. 69, No. 235, 2004, pp. 71111-71112.

Young, D., Grzebieta, R., McIntosh, A., Bambach, M., and Frechede, B. Diving Versus Roof Intrusion: A Review of Rollover Injury Causation. IJCrash, Vol. 12, No. 6, 2007, pp. 609-628.

Appendix 1. Demographic Data, Injury, and BioTab Analysis for Pure Roll Event Occupants

Obs	Age	Gender	Height(cm)	Weight(kg)	BMI	Seat Pos	Near/Far	R/L	Inversions	AIS Rank1	Rank 1 Injury	Rank 1 Injury Description	Body Region	INJ EVENT	IPC	Regional Mechanism
1	64	Male	188	129	36	Row 1 Left	Near side	Roll left	1	3	4502143	Rib cage fracture 1 rib with hemo-/pneumothorax (OIS Grade I)	Chest	Roll	Roof	Compression
2	30	Female	165	69	25	Row 1 Left	Near side	Roll left	2	3	4414063	Lung contusion unilateral with or without hemo-/pneumothorax	Chest	Roll	Left side interior surface	Compression
3	19	Male	173	61	20	Row 1 Left	Far side	Roll right	2	4	4414104	Lung contusion bilateral with or without hemo-/pneumothorax	Chest	Roll	Roof	Compression
4	76	Male	180	82	25	Row 1 Right	Near side	Roll right	2	3	6502263	Cervical Spine fracture pedicle	C-spine	Roll	Roof	Compression/extension
5	53	Male	175	66	22	Row 1 Right	Far side	Roll left	1	3	6502283	Cervical Spine fracture odontoid (dens)	C-spine	Roll	Roof	Compression/extension
6	73	Male	182	130	39	Row 1 Left	Far side	Roll right	1	4	6402184	Cervical Spine Cord contusion incomplete cord syndrome with fracture and dislocation	C-spine	Roll	Roof	Compression/extension
7	33	Female	170	50	17	Row 1 Left	Far side	Roll right	2	3	6502223	Cervical Spine fracture facet	C-spine	Roll	Roof	Compression/lateral bend
8	44	Female	155	86	36	Row 1 Right	Far side	Roll left	2	5	6402285	Cervical Spine Cord contusion complete cord syndrome C-4 or below with frac. and dis.	C-spine	Roll	Roof	Compression/lateral bend
9	21	Male	183	79	24	Row 1 Right	Near side	Roll right	1	4	6402184	Cervical Spine Cord contusion incomplete cord syndrome with fracture and dislocation	C-spine	Roll	Roof	Compression/lateral bend
10	51	Male	191	95	26	Row 1 Left	Far side	Roll right	1	3	6502103	Cervical Spine dislocation facet unilateral	C-spine	Roll	Left roof rail	Compression/flexion
11	78	Female	157	86	35	Row 1 Left	Far side	Roll right	1	3	6402083	Cervical Spine Cord contusion with transient neurological signs with fx/dis	C-spine	Roll	Sunroof	Compression/flexion
12	41	Male	180	102	31	Row 1 Left	Far side	Roll right	2	4	1406404	Cerebrum hematoma/hemorrhage intracerebral small	Head	Roll	Left roof rail	Compression
13	60	Male	183	86	26	Row 1 Left	Far side	Roll right	1	5	1406545	Cerebrum hematoma/hemorrhage subdural small bilateral	Head	Roll	Ground	Compression
14	20	Female	163	61	23	Row 1 Left	Near side	Roll left	1	5	1406285	Cerebrum diffuse axonal injury (white matter shearing)	Head	Roll	Roof	Linear acceleration
15	38	Female	152	54	23	Row 1 Right	Near side	Roll left	1	4	1406524	Cerebrum hematoma/hemorrhage subdural small	Head	Roll	Roof	Compression
16	42	Female	165	68	25	Row 1 Right	Far side	Roll left	1	3	1406843	Cerebrum subarachnoid hemorrhage	Head	Roll	Roof	Compression
17	36	Male	175	84	27	Row 1 Left	Far side	Roll right	1	3	1908063	Scalp avulsion blood loss > 20% by volume	Head	Roll	Roof	Shear
18	47	Male	180	84	26	Row 1 Left	Far side	Roll right	2	3	6506343	Lumbar Spine fracture vertebral body major compression	L-spine	Roll	Seat	Compression/flexion
19	34	Female	157	68	28	Row 1 Left	Far side	Roll right	2	5	6404685	Thoracic Spine cord laceration complete cord syndrome with fracture and dislocation	T-spine	Roll	B-pillar	Compression/flexion

Appendix 2. Demographic Data, Injury, and BioTab Analysis for Multi-Event Roll Occupants

Obs	Age	Gender	Height(cm)	Weight(kg)	BMI	Seat Pos	Near/Far	R/L	Inversions	AIS Rank1	Rank 1 Injury	Rank 1 Injury Description	Body Region INJ	INJ EVENT	IPC	Regional Mechanism
1	29	Male	190	100	28	Row 1 Left	Far side	Roll right	1	4	5442264	Spleen laceration major (OIS Grade IV)	Abdomen	FI	Seat	Compression/flexion
2	32	Female	157	47	19	Row 1 Left	Near side	Roll left	0	4	5442264	Spleen laceration major (OIS Grade IV)	Abdomen	SI	Left side interior surface	Rate of compression
3	27	Male	183	91	27	Row 1 Left	Far side	Roll right	0	3	5442243	Spleen laceration moderate (OIS Grade III)	Abdomen	SI	Left armrest	Compression
4	66	Female	171	68	23	Row 1 Left	Near side	Roll left	0	4	5442264	Spleen laceration major (OIS Grade IV)	Abdomen	SI	Left armrest	Compression/Rate of compression
5	40	Female	163	59	22	Row 1 Right	Far side	Roll left	1	3	8534183	Tibia fracture posterior malleolus open/displaced/comminuted	Ankle	FI	Floor	Compression
6	40	Male	180	90	28	Row 1 Left	Far side	Roll right	1	3	7532043	Ulna fracture open/displaced/comminuted	Arm	FI	A-pillar	Compression
7	30	Female	163	64	24	Row 1 Left	Near side	Roll left	0	3	7528043	Radius fracture open/displaced/comminuted	Arm	FI	A-pillar	Compression
8	54	Female	168	98	35	Row 1 Right	Near side	Roll right	1	4	4502644	Rib cage flail chest with lung contusion (OIS Grade III or IV)	Chest	FI	Belt	Compression
9	21	Female	165	54	20	Row 1 Left	Far side	Roll right	2	3	4414063	Lung contusion unilateral with or without hemo-/pneumothorax	Chest	FI	Air bag	Rate of compression
10	39	Male	183	86	26	Row 1 Left	Near side	Roll left	0	3	4502503	Rib cage fracture open/displaced/comminuted (any or combination; >1 rib)	Chest	FI	Belt	Compression
11	41	Male	175	104	34	Row 1 Left	Far side	Roll right	0	3	4414063	Lung contusion unilateral with or without hemo-/pneumothorax	Chest	SI	Left side interior surface	Rate of compression
12	49	Female	168	79	28	Row 1 Left	Near side	Roll left	1	6	4410146	Heart (Myocardium) laceration perforation complex or ventricular rupture	Chest	FI	Air bag	Compression
13	42	Male	188	93	26	Row 1 Left	Far side	Roll right	0	6	4410166	Heart (Myocardium) multiple lacerations	Chest	FI	Air bag	Compression
14	62	Female	163	82	31	Row 1 Left	Near side	Roll left	1	3	4502143	Rib cage fracture 1 rib with hemo-/pneumothorax (OIS Grade I)	Chest	FI	Belt	Compression
15	57	Female	155	61	25	Row 1 Left	Far side	Roll right	0	4	4502324	Rib cage fracture >3 ribs on one side and <=3 ribs on other side, stable chest or NFS == with hemo-/pneumothorax	Chest	SI	Left side interior surface	Compression
16	51	Female	163	57	21	Row 1 Left	Far side	Roll right	2	3	6502263	Cervical Spine fracture pedicle	C-spine	Roll	Roof	Compression/lateral bend
17	25	Female	170	63	22	Row 1 Left	Near side	Roll left	0	3	7532043	Ulna fracture open/displaced/comminuted	Elbow	Roll	Ground	Compression
18	60	Male	183	64	19	Row 1 Left	Far side	Roll right	1	3	1502023	Base (basilar) skull fracture without CSF leak	Head	SI	Left sice interior surface	Compression
19	51	Female	160	68	27	Row 1 Left	Near side	Roll left	1	4	1406304	Cerebrum hematoma/hemorrhage epidural or extradural NFS	Head	Roll	Roof	Compression

Appendix 2. Demographic Data, Injury, and BioTab Analysis for Multi-Event Roll Occupants (con't)

Obs	Age	Gender	Height(cm)	Weight(kg)	BMI	Seat Pos	Near/Far	R/L	Inversions	AIS	Rank1 Rank 1 Injury	Rank 1 Injury Description	Body Region	INJ_EVENT	IPC	Regional Mechanism
20	59	Female	173	75	25	Row 1 Left	Far side	Roll right	1	3	1502503	Base (basilar) skull fracture with CSF leak	Head	Roll	Ground	Compression
21	17	Male	178	52	16	Row 1 Left	Far side	Roll right	0	5	1406285	Cerebrum diffuse axonal injury (white matter shearing)	Head	SI	Left side interior surface	linear acceleration
22	18	Male	180	79	24	Row 1 Right	Near side	Roll right	0	5	1406285	Cerebrum diffuse axonal injury (white matter shearing)	Head	SI	Other occupant	Compression/linear acceleration
23	41	Male	175	84	27	Row 1 Left	Far side	Roll right	2	4	1406524	Cerebrum hematoma/hemorrhage subdural small	Head	Roll	Roof	Angular acceleration
24	17	Male	191	69	19	Row 1 Left	Near side	Roll left	0	3	8534223	Tibia fracture shaft open/displaced/comminuted	Lower leg	Roll	Foot controls	Compression/bending
25	56	Male	183	87	26	Row 1 Left	Far side	Roll left	1	3	4502303	Rib cage fracture >3 ribs on one side and <=3 ribs on the other side, stable chest or NFS	L-spine	Roll	Seat	Compression
26	20	Female	152	43	19	Row 1 Left	Near side	Roll right	2	3	6506343	Lumbar Spine fracture vertebral body major compression	L-spine	FI	Seat	Compression
27	23	Female	163	50	19	Row 1 Right	Far side	Roll left	1	3	6506343	Lumbar Spine fracture vertebral body major compression	L-spine	FI	Belt	Compression/flexion
28	66	Male	183	91	27	Row 1 Left	Far side	Roll right	0	3	8518223	Femur fracture supracondylar	Thigh	FI	Knee bolster	Compression
29	33	Female	163	77	29	Row 1 Left	Near side	Roll left	1	3	8518143	Femur fracture shaft	Thigh	FI	Knee bolster	Compression
30	20	Male	180	68	21	Row 1 Left	Far side	Roll left	1	3	8518143	Femur fracture shaft	Thigh	FI	Knee bolster	Compression
31	39	Male	173	88	29	Row 1 Left	Far side	Roll right	1	3	8518143	Femur fracture shaft	Thigh	FI	Knee bolster	Compression
32	62	Female	165	78	29	Row 1 Right	Far side	Roll left	0	5	6404645	Thoracic Spine cord laceration complete cord syndrome with fracture	T-spine	FI	Belt	Compression/flexion
33	31	Female	150	57	25	Row 1 Right	Far side	Roll left	0	3	6504343	Thoracic Spine fracture vertebral body major compression	T-spine	FI	Belt	Compression/flexion
34	19	Male	193	76	20	Row 1 Right	Far side	Roll left	1	3	1602043	Length of Unconsciousness known to be <1 hr. with neurological deficit	Up arm	SI	Console	Compression
35	25	Male	188	86	24	Row 1 Left	Far side	Roll right	0	3	7526043	Humerus fracture open/displaced/comminuted	Up arm	SI	Left side interior surface	Compression
36	25	Female	163	73	27	Row 1 Left	Far side	Roll right	2	3	7526043	Humerus fracture open/displaced/comminuted	Up arm	FI	Air bag	Compression

Head/Neck Kinematic Response of Human Subjects in Low-Speed Rear-End Collisions

Gunter P. Siegmund, David J. King, and Jonathan M. Lawrence
MacInnis Engineering Associates

Jeffrey B. Wheeler, John R. Brault, and Terry A. Smith
Biomechanics Research & Consulting, Inc.

ABSTRACT

Limited data exist which quantify the kinematic response of the human head and cervical spine in low-speed rear-end automobile collisions. The objectives of this study were to quantify human head/neck kinematics and how they vary with vehicle speed change and gender during low-speed rear-end collisions. Forty-two human subjects (21 male, 21 female) were exposed to two rear-end vehicle-to-vehicle impacts (speed changes of 4 kmlh and 8 km/h). Accelerations and displacements of the head and torso were measured using 6 degree-of-freedom accelerometry and sagittal high speed video respectively. Velocity was calculated by integrating the accelerometer data. Kinematic data of the head and C7-T1 joint axis in the global reference frame, and head kinematic data relative to the C7-T1 joint axis are presented. A statistical comparison between peak amplitude and time-to-peak amplitude for thirty-one common peaks in the kinematic response was performed. Peak amplitudes and time-to-peak amplitude varied significantly with collision severity for most response peaks, and varied significantly with gender for about one quarter of the response peaks.

INTRODUCTION

Whiplash-Associated Disorders (WAD) comprised about 61 percent ($590 million) of all injury claims paid by the Insurance Corporation of British Columbia in 1995 [1,2]. In the United States, neck strains and sprains (assumed to be WAD) are the most serious injuries reported by 40 percent of claimants [3]. Despite the magnitude of this phenomenon, the injury mechanisms causing WAD remain unclear. The incomplete understanding of the injury mechanisms is partially the result of limited occupant kinematic data of the head, neck, and torso during low-speed rear-end collisions.

Head and torso kinematic data exist for human subjects exposed to frontal collisions [4], and has been used to construct and validate mathematical models [5]. For rear-end collisions, a common source of WAD [6], the data are less complete.

Previous experiments examining the head, neck, and torso kinematics in low-speed rear-end collisions have used cadavers, anthropomorphic test devices (ATD's), and human subjects. Cadavers and current ATD's lack biofidelity in low-speed collisions [7,8]. The complex intersegmental dynamics produced by a low-speed rear-end collision and the potential role of muscle force in whiplash kinematics currently make the use of human subjects the best method of evaluating occupant kinematics at lower collision severities.

Severy et al, exposed a male volunteer to two rear-end impacts in 1950's vintage vehicles not equipped with head restraints [9]. Horizontal head and shoulder accelerations, and head extension determined from film analysis were reported for both collisions and showed the differential kinematic response of the head and shoulders to a rear-end collision. Mertz and Patrick also exposed a single male volunteer to multiple rear-end impacts with the head both supported and unsupported by a head restraint [7]. The subject in these tests was seated in a rigid seat mounted to a laboratory test sled. Bi-axial head accelerations in the superior-inferior and anterior-posterior directions were recorded and resolved to the estimated center of mass of the head. This measurement technique allowed angular acceleration of the head about the medial-lateral axis to be quantified. These authors recognized that head kinematics relative to the torso were important and later proposed response envelopes for the head and neck based on torque at the occipital condyles as a function of head angle relative to the torso [10].

In two separate and more recent studies, McConnell et al, exposed eight male subjects to multiple impacts in 1980's vintage vehicles [11,12]. Based on their data, the authors developed a semi-quantitative description of the occupant kinematic response to a rear-end impact. Linear acceleration at a number of points on the head and angular acceleration and velocity about the medial-lateral axis of the head were reported. Transient symptoms of WAD were produced despite head extension remaining within the range of voluntary motion. Szabo et al, also in two separate studies, exposed seven male and three female subjects to rear-end impacts using 1970's and 1980's vehicles [13,14]. Resultant accelerations at the head's estimated center of mass were reported for both series of tests and the acceleration at the base of the cervical spine was reported for the first series of five subjects. Szabo et al, found that the peak head acceleration observed in their data was a result of head contact with the head restraint and was not necessarily an indicator of neck injury potential.

Matsushita et al, subjected 22 males and 4 females to sled tests simulating rear-end collisions [15]. Head acceleration in the posterior-anterior direction of the head and bi-axial chest accelerations were reported for numerous pre-impact postures. These authors also used cineradiography to quantify segmental motion of the cervical vertebrae and found that initial posture influenced the kinematic response.

Epidemiological studies have found that women suffer whiplash more frequently than men [16-21]. Most proposed explanations for gender differences are based on observations that females have a greater head mass for their neck area [22] or neck strength [23] than do males. Although WAD have been reported after various impact directions and severities, they occur most commonly after rear-end collisions [6,21,24], and overall injury risk in rear-end collisions generally increases with collision speed change [17].

Although these experimental studies have quantified some of the kinematic response, a detailed volunteer study with adequate sample size and instrumentation to compare and quantify the absolute and relative motion of the head and torso has yet to be performed. This paper presents detailed kinematic response data for 42 human subjects exposed to low-speed rear-end collisions at two severities, and specifically examines the effect of gender and collision severity on the peak kinematic response of the head, neck, and torso.

MATERIALS AND METHODS

SUBJECTS - Human subject protection policies and procedures were reviewed and approved by the Western Institutional Review Board of Olympia, Washington, USA. A more detailed description of the human subject handling procedures has been published elsewhere [25].

Subjects between 20 and 40 years old were recruited by newspaper and job-line advertisements. Potential subjects were screened by telephone for height

Table 1. Mean and standard deviation (SD) of subject age and selected anthropometry.

	Male	Female
N	21	21
Age - yr	26.4 (4.5)	27.1 (4.8)
Age range	21-37	20-40
Height - cm	175.4 (5.2)	164.3 (5.2)
Mass - kg	74.9 (9.7)	62.3 (8.8)
Head Circumference - cm	57.8 (2.2)	55.5 (1.7)

and weight (10 - 90th percentile) [26]. Subjects with a history of specific medical conditions or a prior or active injury claim were excluded. Potential subjects were then invited to the lab to undergo an initial screening process. Each subject was seated in an exemplar test seat to ensure that their head and not their neck contacted the head restraint. This criterion eliminated subjects with an erect seated height of 96 cm or greater (the median seated height for a 90th percentile male [27]), and excluded subjects with above-average seated heights who would otherwise have qualified based on standing height. After obtaining informed consent, subjects underwent a cervical magnetic resonance scan. Subjects with a disc bulge greater than 2 mm or degenerative findings deemed moderate or greater by a radiologist were excluded from the study.

Forty-two subjects (21 males and 21 females) successfully completed the interview, screening, and informed consent and participated in the test procedure. Table 1 contains subject age and selected anthropometry data.

INSTRUMENTATION - Head acceleration was measured using a nine accelerometer array (Kistler 8302B20S1; ±20g, Amherst, NY) arranged in a 3-2-2-2 configuration [28]. A uni-axial angular rate sensor (ATA Sensors ARS-04E; ±100 rad/s, Albuquerque, NM) was attached to the accelerometer unit and oriented along the medial-lateral axis. This redundant sensor was used as a check on the primary head acceleration sensors. The accelerometer arrangement was secured to the subject's head with two straps as shown in Figure 1. The mass of the complete head instrumentation package was 198 grams, including straps and 40 cm of cable. Torso acceleration was measured using a tri-axial accelerometer (Summit 34103A; ±7.5g, Akron, OH) and an angular rate sensor (ATA-Sensors DynaCube; ±100 rad/s). Both torso transducers were fastened to an aluminum plate, which was applied in the mid-sagittal plane to the chest immediately below the manubrium with adhesive. Straps over the shoulders and under the arms also secured the plate. The mass of the torso instrumentation package was 255 grams.

The location and orientation of the head and torso instrumentation was measured relative to anatomical landmarks using a three-dimensional digitizer (FaroArm B08-02, Lake Mary, FL) with single-point accuracy of k0.30 mm [29]. The accuracy of the

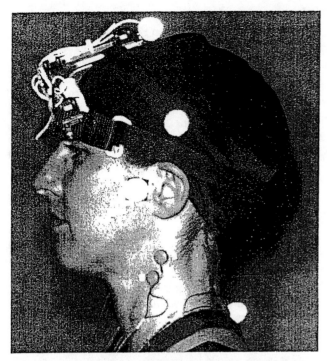

Figure 1. Head Instrumentation

Figure 2. Subject seated in the vehicle

FaroArm was certified according to the manufacturer's instruction before every test.

Vehicle Instrumentation - Vehicle speeds were measured with a 5th wheel (MEA 5th Wheel, Richmond, BC) attached to each vehicle. Uni-axial load cells (Sensotec Model 41; range ±45 000 N, Columbus, OH) inserted at both rear bumper mounts of the Honda recorded the longitudinal component of the impact force (lateral and vertical components were assumed to be negligible). Bumper contact onset and duration were detected with two ribbon switches (Nortel TapeSwitch 121BP; 2 N activation force, Scarborough, ON) connected in parallel and applied to the rear bumper of the target vehicle. Head restraint contact was detected by three force sensitive resistors (Interlink Electronics

Inc.; 0.2 N activation, Camarillo, CA) connected in parallel and applied to the front of the head restraint.

High-Speed Videography - Digital video of sagittal plane motion relative to earth was captured using an OmniSpeed HS motion capture system (Speed Vision Technologies, Solana Beach, CA) and high-speed camera (JCLabs 250; 512 x 216 lines resolution, Mountain View, CA). Video data were recorded at 250 frames per second (fps) using a shutter speed of 1/1000 s. Reflective targets were applied to the subject, seat, and vehicle (see Figure 2). Vehicle and seat targets were 25 mm in diameter and subject targets were 20 mm in diameter. Head targets were applied over the glabella, left temporomandibular joint, left lateral aspect of the cranium, and to the left side of the head accelerometer assembly; torso targets were applied in the mid-sagittal plane to the chest accelerometry and over the spinous process of the seventh cervical vertebrae (C7); seat targets were applied to the upper seat back and head restraint; and vehicle targets were applied to the interior surfaces of the right front door and upper door frame (roof rail).

Digital video data were digitized using OmniSpeed AutoTracker software with a combined experimental setup and video system accuracy of ± 2 mm at the vertical plane containing the seat centerline. Additionally, stationary video cameras (30 fps) were used to record front, overhead, and overall views of each test and an onboard video camera mounted to the driver's A-pillar captured an oblique view.

TEST PROCEDURE - With their head stabilized in an optometrist's forehead rest, anatomical landmarks (glabella, upper incisors, vertex, opisthocranion, occiput, external acoustic meati, and bilateral lower rims of the orbits) were measured in three-dimensions with the FaroArm for each subject. Head and torso accelerometry and video targets were applied and their locations measured relative to the previously-measured landmarks. These data were subsequently referenced to the Frankfort plane, defined by the digitized locations of the lower orbit rims and the external acoustic meati. The torso instrumentation and head video targets were also digitized relative to selected vehicle and seat locations with the subject seated in the vehicle. The torso accelerometry was referenced to the mid-sagittal plane, manubrium and C7 spinous process. The right acromioclavicular joint, greater trochanter, and lateral femoral epicondyle were also digitized to record the subject's seated posture.

The subjects were seated and restrained by a lap and shoulder seat belt in the front passenger seat of the test vehicle. Subjects were instructed to sit normally in the seat, face forward with their head level, place their hands on their lap, and to relax prior to impact.

Because of the potential effect of pre-impact neck muscle contraction on kinematics, special attention was devoted to depriving the subjects of visual and aural cues of the impending impact and to ensuring subjects were relaxed before the impact. A black felt curtain separated the target vehicle from the bullet vehicle and instrumentation equipment to eliminate visual cues. Foam ear plugs and music were used to defeat aural cues. No test personnel were visible to the subject in the minutes preceding the impact. The relaxed state of the occupant was confirmed visually with a live video feed from the A-pillar camera and by monitoring EMG signals from the sternocleidomastoid and cervical para-spinal muscles bilaterally for at least one-minute prior to impact.

An aligned collision between a rolling bullet vehicle and a stationary target vehicle was used for this study. Both vehicles were in neutral and their engines were not running. The bullet vehicle accelerated down a ramp and its front bumper squarely struck the rear bumper of the unbraked target vehicle. After impact, the target vehicle rolled into gravel located 3 meters ahead of the vehicle and was decelerated to rest at about 0.12 g.

Subjects were exposed to two impacts, one which produced a 4 km/h speed change on the target vehicle and another which produced an 8 km/h speed change. The order of impact-severity presentation was randomized. In all cases, the two impacts were separated by at least seven symptom-free days.

VEHICLE SPECIFICATIONS - The bullet vehicle was a 1981 Volvo 240DL station wagon (mass 1618 kg) and the target vehicle was a 1990 Honda Accord LX 4-door sedan (mass 1414 kg). The bullet vehicle was unaltered. The target vehicle's windshield, left doors, left B-pillar, driver's seat, and rear bench seat were removed. A hole was cut in the roof over the test subject to allow overhead filming. A custom B-pillar installed midway between the actual B- and C-pillar locations compensated for the reduced stiffness resulting from removal of the actual B-pillar. Mass was added to the vehicle to offset the removed parts. No damage (other than minor plastic straining of the Honda's rear bumper cover) was sustained by either vehicle over the 100-plus pre-study and study impacts.

Aside from minor modifications to accommodate seat instrumentation, the Honda's seat remained in its stock condition. The fore/aft seat adjustment was locked in the full rear position and the seat back angle was maintained at about 27 degrees from the vertical. The head restraint was adjusted and locked to the full-up position for all subjects. Detailed information regarding seat back modifications made to accommodate the head restraint instrumentation has been presented elsewhere [30].

DATA ACQUISITION - Accelerometer, angular rate sensor, load cell and contact switch data were acquired at 10 kHz and each data channel conformed to SAE J211, Channel Class 1000 [31]. Signal conditioners onboard the vehicle were tethered to four 16-channel, 12-bit, simultaneous-sample-and-hold Win30 DAQ cards (United Electronics Incorporated, Watertown, MA). Low-pass line-noise filters were inserted immediately before

the DAQ boards. All four DAQ boards were installed in the same computer and driven by a single external clock. Two seconds of data were acquired for each test, with a minimum of 0.4 s of pre-impact data. Fifth wheel data were acquired at 128 Hz and triggered by the bumper contact switch. Fifth wheel data were recorded simultaneously for both vehicles for 1 s before and 4 s after impact. A synchronization signal from the high-speed video camera was recorded by the DAQ system and LED's indicating bumper and head restraint contact were placed in each camera's field of view.

REFERENCE FRAMES - Kinematic parameters obtained from the occupant-mounted transducers were initially resolved to local head and torso reference frames. For analysis and presentation, these data were resolved to the global reference frame at the appropriate origin. The head and neck origins, and the direction of the global axes, were defined as follows (Figure 3):

The origin of the head was located at the estimated location of the head's center of mass, assumed to lie in the mid-sagittal plane. Its superior-inferior and anterior-posterior position was estimated for each subject based on regression equations published by Clauser et al [32]. The origin of the neck was located in the mid-sagittal plane at the C7-T1 joint axis (center of rotation of the base of the neck), estimated to be at the midpoint between the C7 spinous process and the manubrium [33].

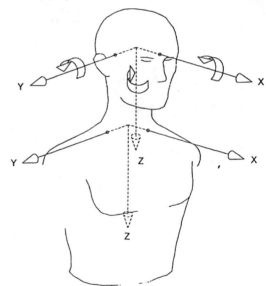

Figure 3. Reference frames for the head and C7-T1 joint axis. The broad arrows show the direction of positive rotation about each axis. (adapted from reference 34).

The z-axis of the global reference frame was defined parallel to the direction of the earth's gravity and positive down. The x-axis was defined such that the xz-plane was parallel to the longitudinal axis of the vehicle and was defined positive toward the front of the vehicle. The y-axis was positive to the right. The global origin was arbitrary, but for reporting purposes, it was assumed to be at the pre-impact origin of either the head or C7-T1 joint axis, whichever was appropriate. Flexion and extension of the head/neck and torso in the sagittal plane occurred about the y-axis, with extension defined positive and flexion negative.

DATA PROCESSING - The kinematic parameters extracted from the data were the linear and angular acceleration, velocity and position of the head center of mass and the C7-T1 joint axis. Since the instrumentation was mounted externally, the kinematic parameters were measured externally and then resolved to the estimated internal location of the head center of mass and the C7-T1 joint axis assuming rigid body kinematics. The treatment of each group of data is outlined below:

Head Kinematics - The head accelerometers were sensitive to DC and therefore the 1g field from the earth's gravity was subtracted from the data in order to yield the transient linear accelerations due to the impact. To determine the initial component of the 1g field on each channel, the mean signal of each accelerometer over the 100 ms preceding impact was assumed to be due to gravity. Because the 3-2-2-2 configuration yielded three independent measures of the three orthogonal components (in the 3-2-2-2 reference frame) making up the 1g field, the magnitude and direction of the 1g field were estimated from these redundant data by minimizing the sum of squares error. The resulting vector estimate of the 1g field defined the initial orientation of the head relative to the lab reference frame. An additional assumption that the x-axis of the head frame lay in the xz-plane of the global frame was required to obtain a unique rotation matrix between the two reference frames.

During and after the collision, the head frame translated and rotated relative to the lab frame. The time-varying three-dimensional orientation angle between the two frames was computed by first integrating the angular acceleration (a) to obtain the angular velocity (ω), and then using the orientation vector technique to update the transformation matrix between the body-fixed and inertial reference frames [35]. Because head and torso rotation were predominantly in the xz-plane, the orientation vector method produced a transformation matrix that was essentially identical to a direct double integration of the angular acceleration vector, or a method of Euler rates and angles [36].

The angular acceleration *(a)* required for the foregoing calculation was computed using Equation **1** [28]. The instantaneous angular acceleration thus obtained was independent of the instantaneous angular velocity, which minimized accumulated errors and yielded an estimate of the angular acceleration which remained reliable for a longer duration than other methods [28]. Angular velocity of the head about the medial-lateral axis was compared with the uni-axial angular rate sensor on the head, and sagittal head angle was compared with the sagittal high speed video for agreement.

$$\alpha_x = \frac{a_{z1} - a_{z0}}{2\rho_{y1}} - \frac{a_{y3} - a_{y0}}{2\rho_{z3}}$$

$$\alpha_y = \frac{a_{x3} - a_{x0}}{2\rho_{z3}} - \frac{a_{z2} - a_{z0}}{2\rho_{x2}} \qquad (1)$$

$$\alpha_z = \frac{a_{y2} - a_{y0}}{2\rho_{x2}} - \frac{a_{x1} - a_{x0}}{2\rho_{y1}}$$

where α_i = angular acceleration along axis i,
        a_i = acceleration at accelerometer i,
        ρ_i = distance between accelerometers.

The origin of the 3-2-2-2 assembly, could not be placed at the head center of mass of a human subject. Therefore, the head acceleration computed at the origin of the 3-2-2-2 assembly was resolved to the center of gravity of the head using Equation 2 [37].

$$a_B = a_A + \alpha \times r_{B/A} + \omega \times (\omega \times r_{B/A}) \qquad (2)$$

where a_B = acceleration at point B,
        a_A = acceleration at point A,
        α = angular acceleration
        ω = angular velocity, and
        $r_{B/A}$ = position vector between points A and B

Torso Kinematics - The tri-axial linear accelerometer used to measure chest acceleration was sensitive to DC and was also corrected for the uniform 1g field of the earth's gravity. Unlike the head assembly correction, redundant channels were not available. The mean of each axis of the linear accelerometer over the 100 ms preceding impact was assumed to define the initial direction of the 1g field. This assumption, combined with the assumption that the x-axis of the chest accelerometry assembly was contained in the xz-plane of the global reference frame, was used to determine the rotation matrix between the initial sensor reference frame and the global frame. Torso rotation angles were computed as described earlier and used to update the rotation matrix during impact-induced motion.

Unlike the head accelerometer assembly, the rotational kinematics of the torso were directly measured using a tri-axial angular rate sensor. The angular rate sensor data contained substantially more noise than the linear accelerometer data and were therefore optimally filtered [38,39] before computation. Digital low-pass filters based on the data of five subjects were used to filter the angular velocity data of all subjects. After filtering, the angular acceleration was computed from the angular velocity by calculating the slope between the instantaneous angular velocity 10 ms before and after the time of interest. Intervals shorter than 10 ms produced unrealistically short, high angular acceleration peaks. Because the contribution of the *(a × r)* term in

Equation 2 to the acceleration at the C7-T1 joint axis was small, the differentiation process was not refined.

The sensitive axes of the tri-axial linear accelerometer used for measuring the torso acceleration did not pass through a single point. Although torso transducer rotation was not large, the accelerations of the y- and z-axes were corrected to a single point on the sensitive x-axis using equation 2. This correction was typically less than 10 percent of the peak signal, but was performed because the internal offset of the sensitive axes was large (about 15 mm) relative to the distance from the sensor origin to the C7-T1 joint axis (typically about 90 mm). Linear accelerations at the C7-T1 joint axis were then calculated from the measured torso transducer signals in the same manner as for the head.

High Speed Video Data - Reflective target data extracted from the high-speed video were first corrected for camera lens barrel distortion using a 5th order polynomial (odd terms only) [40]. The polynomial coefficients were determined using a 10 cm by 10 cm grid covering the camera's field of view (about 0.90 m by 1.20 m at the centerline of the target vehicle's seat), and then minimizing the sum of squares error between the actual and digitized grid. Because the camera-to-subject distance was relatively short (5 m), the data for targets located off the mid-sagittal plane were also adjusted for parallax [40]. The camera axis was assumed to be perpendicular to the plane of vehicle and occupant motion. Since a target could not be positioned a priori over the origins of the head and torso reference frames, the path of these points was calculated assuming a fixed position relative to the actual markers.

Collision severity - Vehicle position data from the 5th wheels were differentiated across a 16 ms window, and speed change was then determined from scale plots of vehicle speed versus time [41]. The bumper load-cell sensor bias, estimated as the mean signal over the 100 ms preceding impact, was removed from the data before summing the left and right load cells to obtain total collision force. The sum was then integrated and divided by the vehicle mass to confirm vehicle speed change.

Some parameters were measured using multiple methods to confirm the response. For instance, head angle about the y-axis was computed from the 3-2-2-2 accelerometers and the uni-axial angular rate sensor, and then compared to the high-speed video data. The acceleration data presented in this paper were computed from the accelerometer and angular rate sensors, and the position and angle data were extracted from the high speed video. Velocity data were computed by integrating acceleration rather than differentiating displacement data because of the inherent random-noise-reducing effect of the integration process.

STATISTICAL ANALYSIS - Gender and speed change were the primary variables in this study. The null hypotheses were that neither gender nor speed change

affected the kinematic response of an occupant exposed to a low-speed, rear-end collision.

Peak amplitude and time-to-peak amplitude were extracted for peaks common to the absolute and relative linear and angular acceleration, velocity and position data of the head and C7-T1 of each subject. The effect of gender and speed change on the amplitude and time of each common peak in the kinematic response was tested using a single analysis of variance (ANOVA) for each of the two extracted measures. The covariance of peak amplitude and time-to-peak amplitude was examined to ensure that single tests were valid. A method of unweighted means was used to account for the unequal samples in each cell of the ANOVA [42].

To ensure the probability of a false positive was less than 0.05 across the 31 peaks examined, a Bonferroni adjustment was used [43]. Each of the 62 ANOVA tests (31 peak amplitudes and 31 time-to-peak amplitudes) was required to achieve a significance level of 0.0008 (0.05162) to be judged significantly related to speed change or gender.

RESULTS

Forty-two subjects were exposed to impacts. Three subjects withdrew between their 4 km/h and 8 km/h tests, and all subjects who underwent the 8 km/h test first completed the study. Impacts at the 4 and 8 km/h level were repeatably produced (Table 2).

Position and angle data obtained from high speed video were acquired for all subjects (Table 3a). Because of the nature of the accelerometer calculations, a failure of one transducer rendered the test data incomplete. Incomplete data were collected for six tests, which reduced the number of subjects for whom complete transducer data were recorded (Table 3b). The full data set was used to calculate the position-based

Table 2. Vehicle speed and collision properties at the 4 and 8 km/h level. All properties were significantly different at the two levels.

Property	4 km/h level	8 km/h level
Volvo impact speed (km/h)	4.86 (0.12)	10.02 (0.06)
Honda speed change (5th wheel)	3.95 (0.11)	8.10 (0.11)
(load cell)	4.04 (0.09)	8.07 (0.07)
Restitution	0.59 (0.01)	0.56 (0.01)
Collision duration (ms)	138 (4)	135 (2)
Time of peak force (ms)	42 (2)	35 (1)
Peak bumper force (kN)	27.0 (0.9)	48.5 (0.5)

Tables 3a and 3b. Number of tests used for the analysis of (a) position and angle from high speed video (left), and (b) acceleration and velocity from transducer data (right).

	Male	Female	Total		Male	Female	Total
4 km/h	21	21	42	4 km/h	19	20	39
8 km/h	20	19	39	8 km/h	19	17	36
Total	41	40	81	Total	38	37	**75**

results, whereas the reduced data set was used to calculate the acceleration and velocity results presented here.

Only sagittal plane motion was considered in this analysis, which yielded 27 kinematic response signals: nine signals (a, a, v_x, v, s_x, s, a, ω_y, θ_y) each for the absolute motion of the head, absolute motion of the C7-T1, and relative motion of the head with respect to C7-T1. Motion out of the sagittal plane, i.e., translational motion along the y-axis and rotational motion about the x and z axes, were small and varied considerably between subjects. The complete sagittal kinematic response data of all subjects is attached in Appendix A.

Good agreement between double-integrated accelerometer data and high-speed video position data was achieved for the kinematic response of the head and the translational components of the C7-T1 joint axis (Figure 4). Integrated angle data from the uni-axial angular rate sensor (ARS) on the head also compared well with the accelerometer and video data. Differences between accelerometer and high-speed-video data for the angle (θ) of the C7-T1 joint axis relative to the earth were likely caused by skin motion and/or filtering. Extraneous vertical motion of the torso transducers was visible in the video; however, manually digitized check measurements of the manubrium showed that the automatically-tracked video target data reliably measured upper torso angle. Skin motion was also present in the head data of some subjects, however to a much lesser degree than in the torso data.

In all tests, initial flexion between the head and the torso was observed. Although only slight flexion was present in some subjects, maximum flexion of 13 degrees from the initial orientation of the head relative to the C7-T1 joint axis was reached by some subjects.

Relative horizontal motion between the head and C7-T1 joint axis also occurred in all subjects. Forward acceleration of the C7-T1 origin typically began about 25 to 35 ms after impact, coincident with vertical acceleration of the head relative to the earth. Forward horizontal acceleration and angular acceleration of the head relative to the earth began about 10 to 30 ms after the onset of vertical head acceleration. Observable positive C7-T1 joint axis rotation relative to the earth began about 30 to 40 ms after impact and forward horizontal displacement began about 20 ms later. Positive head rotation relative to the earth was detected about 50 to 70 ms after impact, after 1 to 5 degrees of flexion between the head and C7-T1 joint axis had developed from torso motion.

Forward horizontal head displacement with respect to the earth was not observed until about 80 to 110 ms after impact. By this time, between 1 to 5 cm of rearward horizontal translation had developed between the head center of mass and the C7-T1 origin. Maximum rearward horizontal translation of the head's center of mass relative to the C7-T1 origin varied between 2.5 and 11 cm.

Head restraint contact was made in 80 of 81 tests. Excluding one male at the 4 kmlh level who did not contact the head restraint, the mean time from bumper contact to head restraint contact was 118 ± 18 ms at the 4 km/h level and 94 ± 13 ms at the 8 km/h level. The duration of head restraint contact was 95 ± 17 ms at the 4 kmlh level and 103 ± 13 ms at the 8 kmlh level.

At head restraint contact, the head was rotating rearward and just beginning to translate forward in the global reference frame. Relative to the C7-T1 joint axis, however, head restraint contact was made with the head in its initially flexed and retracted position. Rearward rotation (extension) of the head relative to the C7-T1 joint axis did not begin until after head restraint contact, and in about 20 percent of the tests, the head never extended rearward of its original orientation relative to the C7-T1 joint axis.

Peak horizontal head acceleration relative to the earth occurred during head restraint contact for all subjects who contacted the head restraint. Peak horizontal head speed relative to the earth averaged about 1.9 times the target vehicle speed change across all subjects and both speed changes. The peak horizontal speed of the C7-T1 joint axis relative to the earth averaged about 1.6 times the vehicle speed change.

Thirty-one response peaks common to all subjects were analyzed for the effect of speed change and gender on peak kinematic response (Figure 5). Some kinematic response signals contained multiple peaks common to all subjects, whereas other signals were excluded from the analysis because of dissimilar subject response. At the adjusted significance level of 0.0008, the amplitude of 28 peaks and the time of 27 peaks varied significantly with vehicle speed change (Tables 4a and 4b). The amplitude of seven peaks and the time to eight peaks varied significantly with gender. The interaction term was not significant for the amplitude or time of any of the selected peaks.

DISCUSSION

The number of subjects in this study allowed statistical comparisons to be made between the kinematic response of male and female subjects at the two selected speed changes in rear-end automobile collisions. The larger peak amplitudes demonstrated at the 8 km/h level compared to the 4 kmlh level for most of the kinematic parameters tested provides insight into the effect of collision severity on the kinematic response of human subjects in low-speed rear-end collisions. However, the importance of the absolute values of each peak and the relationship of each kinematic parameter to the injury mechanism causing WAD has not yet been established.

When addressing the potential for injury, the kinematics of the head relative to the C7-T1 joint axis may be more important than simple peak values of the head or C7-T1 joint axis relative to the earth. For

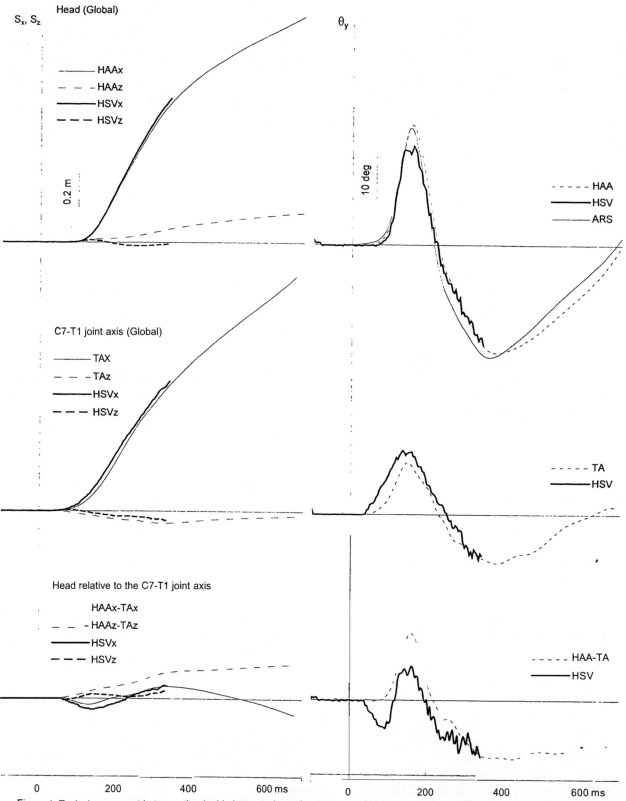

Figure 4 Typical agreement between the double-integrated accelerometery and high-speed video position (s_x,s_z) and angle (θ_y) data for the head (top row) and C7-T1 joint axis (middle row) in the global reference frame, and the head relative to the C7-T1 origin (bottom row). Abbreviations are as follows: HAA - Head accelerometer array. TA - torso accelerometers, HSV - high speed video.

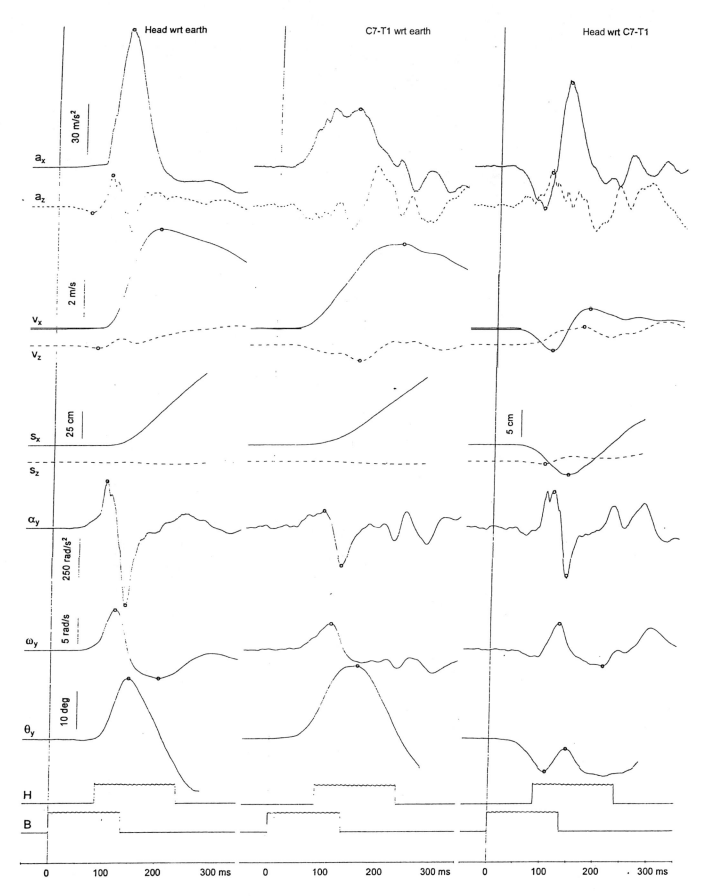

Figure 5. Exemplar kinematic response data for a female at the 8km/h level. The three graphs depict (left) the head response relative to earth, (center) the C7-T1 joint axis relative to earth, and (right) the head relative to the C7-T1 joint axis. H and B data depict head restraint contact and bumper contact respectively. Circles identify peaks used for analysis. Note scale change for s_x and s_y of the head relative to C7-Tl joint axis.

Table 4a. Mean, standard deviation (SD), and ANOVA results for amplitudes of selected peaks in the kinematic response data.

Kinematic Parameter[†]			Units	8 km/h Female	8 km/h Male	4 km/h Female	4 km/h Male	P-Values Gender	Severity	Interaction
Head	a_x	+	m/s^2	91.3(11.4)	84.0(12.5)	38.9(5.7)	29.3(6.2)	0.0002*	0.0000'	0.5955
	a_z	-		-3.8(2.1)	-3.3(2.6)	-2.9(1.1)	-2.3(1.0)	0.2140	0.0398	0.8988
		+		22.4(6.0)	16.9(5.2)	11.0(2.6)	8.5(2.3)	0.0001*	0.0000'	0.1438
	v_x	+	m/s	4.38(0.15)	4.27(0.18)	2.20(0.17)	1.96(0.22)	0.0002*	0.0000'	0.1251
	v_z	-		-0.13(0.07)	-0.11(0.10)	-0.10(0.05)	-0.08(0.05)	0.3092	0.1989	0.9435
	a_y	+	rad/s^2	293 (67)	319 (74)	152 (45)	140 (37)	0.5869	0.0000*	0.1721
				-673 (193)	-737 (201)	-319 (103)	-268 (80)	0.8546	0.0000*	0.1066
	ω_y	+	rad/s	9.26(2.66)	10.79(2.33)	5.17(1.42)	5.43(1.05)	0.0512	0.0000*	0.1638
				-7.11 (1.38)	-7.78(1.44)	-4.88(1.06)	-4.71(0.80)	0.3654	0.0000"	0.1349
	θ_y	+	deg	15.89(4.83)	23.81(5.02)	13.42(4.57)	17.26(3.73)	0.0000*	0.0000*	0.0505
C7-T1	a_x	+	m/s^2	42.8(5.8)	36.3(6.5)	20.8(3.0)	17.9(2.4)	0.0001*	0.0000*	0.1034
	v_x	+	m/s	3.79(0.16)	3.57(0.24)	1.76(0.28)	1.56(0.21)	0.0003*	0.0000'	0.8381
	v_z	-		-0.65(0.15)	-0.61(0.15)	-0.36(0.05)	-0.38(0.08)	0.5851	0.0000*	0.3275
	a_y	+	rad/s^2	113 (27)	102 (46)	56 (19)	50(15)	0.2158	0.0000*	0.6789
				-292 (87)	-206 (92)	-115 (46)	-88 (52)	0.0011	0.0000*	0.0804
	ω_y	+	rad/s	4.16(0.90)	3.80(0.99)	2.06(0.54)	2.02(0.76)	0.2799	0.0000*	0.4072
	θ_y	+	deg	16.83(2.58)	17.67(2.43)	10.28(1.56)	11.36(2.55)	0.0695	0.0000*	0.8209
Head w.r.t. C7-T1	a_x	-	m/s^2	-25.3(3.6)	-26.0(5.1)	-15.1(3.4)	-13.4(1.8)	0.5487	0.0000*	0.1664
		+		57.9(12.1)	59.9(15.2)	28.8(4.9)	26.4(6.2)	0.9466	0.0000*	0.3624
	a_x	+		28.1(6.1)	25.1(8.5)	13.9(5.0)	12.1(3.4)	0.0826	0.0000*	0.6731
	v_x	-	m/s	-0.94(0.21)	-1.14(0.18)	-0.67(0.13)	-0.71(0.09)	0.0026	0.0000*	0.0474
		+		0.84(0.14)	0.95(0.11)	0.61(0.14)	0.56(0.16)	0.3836	0.0000*	0.0169
	v_z	+		1.03(0.15)	0.90(0.21)	0.56(0.11)	0.51(0.09)	0.0127	0.0000*	0.2162
	s_x	-	m	-0.056(0.013)	-0.072(0.016)	-0.043(0.010)	-0.050(0.009)	0.0001*	0.0000*	0.1393
	s_z	-		-0.006 (0.002)	-0.005(0.001)	-0.004(0.002)	-0.003(0.001)	0.0378	0.0001*	0.9730
	a_y	+	rad/s^2	353 (168)	391 (133)	165 (75)	149 (64)	0.6834	0.0000*	0.3219
				-549 (250)	-678 (231)	-276 (131)	-243 (101)	0.2698	0.0000*	0.0653
	ω_y	+	rad/s	8.18 (3.90)	10.18(2.89)	4.41(1.84)	4.29(1.55)	0.1296	0.0000*	0.0885
				-5.39(1.44)	-7.40(1.70)	-4.40(1.66)	-4.73(1.47)	0.0021	0.0000*	0.0244
	θ_y	-	deg	-7.73(2.41)	-6.58(3.12)	-3.70(1.46)	-2.66(1.38)	0.0306	0.0000*	0.9114
		+		-0.41(3.86)	6.69(5.34)	3.56(4.56)	6.39(3.59)	0.0000*	0.0681	0.0345

† symbol "+" refers to a positive peak in the lab reference frame directions, "-" refers to a negative peak
* statistically significant

Table 4b. Mean, standard deviation (SD), and ANOVA results for time (milliseconds) to selected peaks in the kinematic response data.

Kinematic Parameter[†]			8 km/h Female	8 km/h Male	4 km/h Female	4 km/h Male	P-Values Gender	Severity	Interaction
Head	a_x	+	128 (7)	140 (6)	150 (10)	161 (9)	0.0000*	0.0000*	0.8575
	a_z	-	64 (5)	67 (7)	81 (4)	83 (6)	0.0855	0.0000*	0.7162
		+	108 (8)	112 (9)	133 (9)	137 (10)	0.1137	0.0000*	0.8362
	v_x	+	201 (13)	202 (10)	222 (18)	231 (25)	0.3156	0.0000'	0.3579
	v_z	-	80 (7)	80 (10)	99 (6)	101 (7)	0.6699	0.0000*	0.5680
	α_y		108 (7)	114 (10)	125 (9)	128 (13)	0.0633	0.0000*	0.5478
			144 (7)	156 (12)	168 (13)	178 (13)	0.0003*	0.0000'	0.7959
	ω_y	+	126 (8)	135 (8)	145 (10)	147 (10)	0.0096	0.0000*	0.1124
		-	184 (15)	206 (13)	218 (19)	226 (17)	0.0003*	0.0000*	0.0778
	θ_y	+	143 (8)	154 (9)	171 (13)	178 (13)	0.0008*	0.0000*	0.5615
C7-T1	a_x	+	128 (12)	119 (17)	110 (11)	117 (10)	0.7103	0.0015	0.0199
	v_x	+	228 (25)	226 (22)	233 (33)	251 (30)	0.2449	0.0277	0.1240
	v_z	-	138 (18)	135 (14)	139 (20)	139 (14)	0.7687	0.5357	0.7265
	α_y	+	86 (13)	88 (15)	109 (15)	106 (18)	0.8106	0.0000*	0.4880
		-	131 (9)	135 (11)	156 (18)	162 (23)	0.1915	0.0000*	0.7763
	ω_y	+	111 (8)	112 (12)	127 (15)	130 (21)	0.5513	0.0000*	0.7678
	θ_y	+	140 (10)	158 (17)	168 (12)	189 (22)	0.0000'	0.0000'	0.5621
Head w.r.t. C7-T1	a_x	-	86 (9)	93 (11)	103 (9)	105 (10)	0.0695	0.0000*	0.2239
		+	129 (9)	141 (6)	151 (12)	161 (11)	0.0000*	0.0000*	0.5187
	a_z	+	117 (12)	112 (11)	127 (16)	133 (11)	0.9288	0.0000*	0.0663
	v_x	-	105 (7)	114 (8)	124 (7)	130 (8)	0.0001*	0.0000*	0.4628
		+	178 (22)	190 (11)	219 (30)	221 (24)	0.1975	0.0000*	0.3496
	v_z	+.	139 (11)	135 (10)	154 (15)	152 (9)	0.2557	0.0000*	0.6639
	s_x	-	135 (11)	150 (9)	162 (15)	171 (15)	0.0002*	0.0000*	0.2984
	s_z	-	89 (12)	88 (12)	107 (11)	100 (18)	0.2398	0.0000*	0.3394
	α_y	+	117 (10)	118 (9)	126 (13)	132 (13)	0.2455	0.0001*	0.4268
		-	148 (7)	155 (9)	170 (13)	179 (17)	0.0086	0.0000*	0.6453
	ω_y	+	132 (8)	139 (9)	148 (14)	149 (10)	0.1215	0.0000*	0.2829
		-	191 (20)	205 (19)	215 (18)	222 (19)	0.0275	0.0000*	0.4441
	θ_y	-	102 (7)	102 (10)	116 (9)	106 (16)	0.0735	0.0013	0.0722
		+	148 (10)	158 (9)	175 (17)	181 (15)	0.0108	0.0000*	0.4704

† symbol "+" refers to a positive peak in the lab reference frame directions, "-" refers to a negative peak
* statistically significant

instance, female subjects had greater horizontal accelerations of the head and C7-T1 relative to the earth than male subjects, but this difference was not present for the head relative to the C7-T1 joint axis. This finding underscores the potential for misinterpretation of results referenced only to the global reference frame.

The greater and earlier peak horizontal acceleration of the head and C7-T1 joint axis (relative to the earth) of the female subjects and the larger and later peak head extension of the male subjects were consistent with the larger body mass and head size (and therefore head mass) of the male subjects and the correspondingly lower frequency response of the seat back/occupant system. Both seat back stiffness and occupant mass govern the frequency response of the seat back, and variations in seat back stiffness have been shown to amplify or attenuate differential motion between the head and neck [44]. Greater male peak head extension may also be related to a lower relative head restraint position for the male subjects than for the female subjects.

The reason for the gender differences found for other response peaks is less clear. A regression analysis which incorporates anthropometry, seated posture, and head restraint adjustment may yield insight into the parameters responsible for the gender differences observed in this study.

Differences in the horizontal distance between the back of the head and the front of the head restraint, known as "backset", have been shown to affect head neck motion using a Hybrid III dummy equipped with a RID neck [44]. Backsets greater than 10 cm have correlated with increased neck symptom duration [45], and lower vertical head restraint position has correlated with an increased incidence of neck injuries in rear-end collisions [46]. Because the adjusted seat back angle and head restraint position relative to the seat in the present study were fixed, inter-subject differences in anthropometry and posture resulted in variable horizontal and vertical head restraint positions. For all subjects, however, the top of the head restraint was above the ears and backset was less than 10 cm. Both head restraint backset and vertical position were potential confounding variables in this study and warrant additional investigation.

In an observational study of the motoring public, 15 percent of observed drivers had backsets less than 10 cm, and only 10 percent of drivers had a combination of backset less than 10 cm and top of the head restraint adjusted above the ears [47]. The subjects in the current study therefore represent the small segment of the motoring public with "optimal" head restraint protection.

The observed initial flexion between the head and torso was a result of torso rotation preceding head rotation: torso rotation began about 30 to 40 ms after impact, whereas head rotation was not observed until about 50 to 70 ms after impact. The position and angle of the head and the position of the upper torso initially remained stationary relative to the earth, while the pelvis and lower torso were accelerated forward by the seat

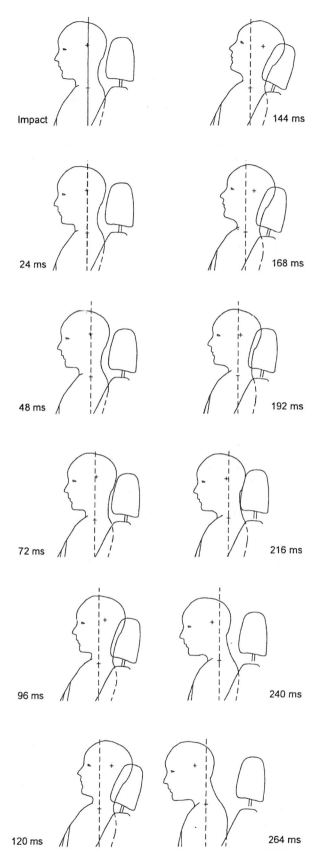

Figure 6. Exemplar kinematic response of the head and C7-T1 joint axis in 24 ms steps after initial contact for an 8 km/h speed change. Note the horizontal position of the head COG (+) in relation to the initial position (solid line) and instantaneous position (dashed line) of the C7-T1 joint axis.

Parameter	Units	Severy et al. [9][a]	Mertz et al. [7][b]	McConnell et al.[11,12]	Szabo et al.[13,14]	Matsushita et al.[15]	Siegmund et al	
Number of human subjects		1 M	1 M	8 M[c]	7 M, 3 F[c]	16 M, 3 F	21M, 21F	20M, 19F
Number of tests		2	2	24	17	19	42	39
Collision speed change	(km/h)	8.4, 9.5[d]	13.5, 14.3[e]	3.5 - 10.9	8 - 10	2.5 - 5.0	4	8
Test type		vehicle	sled	vehicle	vehicle	sled	vehicle	vehicle
Peak head acceleration	(g)	5.0. 2.9[f]	~ 7[g]	3 - 6[f]	6.6 - 16.6[g]	2.7 - 6.3[h]	1.6 - 5.0[f]	6.7 - 12.0[f]
Peak head angular acceleration (-y dir.)	(rad/s^2)		~ 200	400 - 600[i]			160 - 510	450 - 1260
Peak head angular velocity (-y direction)	(rad/s)			16 - 20[i]			2.4 - 7.3	5.4 - 17.7
Peak head extension from initial position	(deg)	34, 38	37	< ~ 60	7 - 30		4 - 27	9 - 33
Peak acceleration at top of torso	(g)				4.5 - 7.4[k]	1.6 - 2.9[k]	1.4 - 2.6[f]	2.7 - 5.9[f]

a. subject aware of second Impact
b. kinematic data for only one impact reported, subject tense and aware, no head restraint
c. data drawn from two test series
d. computed by integrating acceleration data
e. sled impact speed, rebound assumed to be zero

f. horizontal acceleration relative to the earth
g. resultant acceleration
h. acceleration in the anterior-posterior direction of the head
i. data from high-speed video
k. direction in global coordinates not known

back. This forward motion of the pelvis and lower torso set up a positive rotation of the torso about the y-axis , and resulted in flexion between the stationary head and rotating torso.

Rearward horizontal translation of the head relative to the C7-T1 joint axis was prominent in all subjects (Figure 6) and was greater in male subjects than female subjects. A comparison between the amount of rearward horizontal translation of the head center of mass relative to the C7-T1 joint axis and the active range of motion in retraction of the subjects in this study (3.1 ± 0.9 cm [25]) suggests that dynamic retraction may have approached or exceeded the subjects' active range of motion. Although some of the rearward horizontal translation between the head's center of mass and C7-T1 origin was the result of head extension relative to the C7-T1 joint axis, rearward horizontal translation occurred in all subjects whereas relative head extension did not. In subjects whose head remained flexed during the entire impact, horizontal translation between the top of the cervical spine (base of the skull) and the C7-T1 origin may have been larger than indicated by the horizontal translation between the head center of mass and the C7-T1 joint axis. It has been previously reported that small amounts of horizontal translation between the head and C7-T1 bring the craniovertebral junction into maximal flexion and that translation of the head relative to the torso may cause damage at the craniovertebral junction [48]. Additional work is needed to determine whether dynamic retraction of the top of the cervical column relative to the C7-T1 origin exceeded active range of motion, and whether this translation was related to symptom production.

Previous research has reported that the head of some test subjects never reached anatomical extension relative to the torso (extension beyond the head/torso orientation when standing upright) [13]. The current data showed that the head of some subjects never rotated rearward of its initial "anatomically flexed" position relative to the C7-T1 joint axis. Additional work is required to determine whether this pattern of motion was

associated with symptom production, and whether the flexed position of the head relative to the C7-T1 joint axis at head restraint contact contributed to a WAD injury mechanism.

Peak head extension relative to C7-T1 was less than 20 degrees from the initial flexed position of the head relative to the torso for all subjects. Active neck range of motion in extension for the subjects in this study was 70 ± 8 degrees from the anatomical neutral position [25]. None of the subjects exceeded their extension range of motion. This finding indicates that hyperextension of the neck was not the mechanism responsible for transient symptoms produced in this study [25], and is consistent with other studies producing transient symptoms without hyperextension [11,12,13].

The torso and head over-speed observed in the current data appear to be larger than that qualitatively reported in a previous human subject study [12]. Torso over-speed was previously described as slightly greater than the vehicle's post-impact speed and head over-speed was described as slightly greater than torso speed. Over-speed is likely a function of seat back elasticity and frequency response, and differences in seat back properties between studies may account for the greater over-speed found in the current data.

The statistical findings of this research cannot be compared directly to the results of previous kinematic investigations into low-speed rear-end collisions because no previous studies examined gender and collision severity effects, largely due to small test populations. A general comparison between peak kinematic parameters observed in this study and those reported in the literature was possible (Table 5). This comparison was restricted largely to the kinematic response of the head relative to the earth except for two studies [14,15] that reported acceleration measured at the approximate location of the C7 spinous process and on the frontal surface of the chest respectively.

Peak head acceleration varied widely between the studies in Table 5, perhaps due to differences between pre-impact muscle contraction, collision severity, seat backs and head restraints. Angular acceleration of the head relative to the earth was higher in the current study, however the data from other studies presented in Table 5 were calculated from high speed film rather than accelerometers. Higher peak head angular velocity and greater head extension in the data of McConnell et al, [12] may be related to a higher head position relative to the head restraint used in their tests. Head extension compared favorably with data reported by Szabo et al, [13,14] who used a head restraint height similar to that used in the present study. Peak acceleration at the base of the cervical spine also compared well with the limited previous data [14,15], although the comparison may not be valid because the data were not measured at or resolved to the same points.

This study examined the sagittal plane response of the head and torso of ideally seated occupants with a well-adjusted head restraint in an aligned vehicle collision within a single vehicle and seat position. Additional studies are needed to quantify the effect on the kinematic response of the many variables controlled in the present study.

SUMMARY

Head and torso kinematic response data for 21 male and 21 female subjects exposed to a controlled series of low-speed rear-end automobile collisions have been presented. Initial flexion between the head and torso was observed in all subjects. Retraction of the head center of mass relative to the C7-T1 joint axis was present in all subjects, whereas extension of the head relative to its initial position with respect to the C7-T1 joint axis was not present in all subjects.

Significant gender differences existed between the peak amplitude and time-to-peak amplitude for about one quarter of the thirty-one common peaks in the kinematic response data. Significant differences between the two collision severities were demonstrated for both the amplitude and time of most common peaks in the kinematic response data.

ACKNOWLEDGMENTS

This study is part of a larger study examining the kinematic, kinetic, and muscle responses, and clinical outcome of subjects exposed to low-speed rear-end automobile collisions. This study was partially funded by a grant awarded under the Technology BC program administered by the Science Council of British Columbia. Remaining funding was provided by MacInnis Engineering Associates and Biomechanics Research & Consulting, Inc.

The authors wish to thank Elaine Brault, Ian Brault, Jamie Catania, Brad Heinrichs, and Jeff Nickel for their valuable contributions, and the staff at MacInnis Engineering Associates and Biomechanics Research & Consulting, Inc for their assistance and patience.

REFERENCES

1 Quebec Task Force on Whiplash-Associated Disorders. Whiplash associated disorders (WAD), Redefining "whiplash" and its management. Quebec City, QC: Societe de l'assurance automobile du Quebec, January 1995.

2. Fockler SKF, Vavrik J & Kristiansen. Educating drivers to correctly adjust head restraints: Assessing the effectiveness of three different interventions. 39th Annual Proceedings of the Association for the Advancement of Automotive Medicine, pp. 95-109. Des Plaines, OH: Association for the Advancement of Automotive Medicine, 1996.

3. Insurance Research Council. Auto Injuries: Claiming behavior and its impact on insurance costs. Oak Brook, IL, 1994.

4. Thunnissen J, Wismans J, Ewing CL & Thomas DJ. Human volunteer head-neck response in frontal flexion: A new analysis (952721). 39th Stapp Car Crash Conference Proceedings (P-299), pp. 439-460. Warrendale, PA: Society of Automotive Engineers, 1995.

5 de Jager M, Sauren A, Thunnissen J & Wisman J. A global and a detailed mathematical model for head-neck dynamics (962430). 40th Stapp Car Crash Conference Proceedings (P-305), pp. 269-281. Warrendale, PA: Society of Automotive Engineers, 1996.

6 Sturzenegger M, DiStefano G, Radanov BP & Schnidrig A. Presenting symptoms and signs after whiplash injury: The influence of accident mechanisms. Neurology 1994; **44:** 688-693.

7 Mertz HJ & Patrick LM. Investigation of the kinematics and kinetics of whiplash (670919). Proceedings of the 11th Stapp Conference, pp. 175-206. Warrendale PA: Society of Automotive Engineers, 1967.

8. Scott MW, McConnell WE, Guzman HM, et al. Comparison of human and ATD head kinematics during low-speed rearend impacts (930094). Warrendale PA: Society of Automotive Engineers, 1993.

9. Severy DM, Mathewson JH & Bechtol CO. Controlled automobile rear-end collisions, an investigation of related engineering and medical phenomena. Canadian Services Medical Journal, pp. 727-759, November, 1955.

10. Mertz HJ & Patrick LM. Strength and response of the human neck (710855). 15th Stapp Car Crash Conference Proceedings, pp. 207-255. Warrendale, PA: Society of Automotive Engineers, 1971.

McConnell WE, Howard RP, Guzman HM, et al. Analysis of human test subject kinematic responses to low velocity rear end impacts (930889). Warrendale, PA: Society of Automotive Engineers, 1993.

12. McConnell WE, Howard RP, Van Poppel J, et al. Human head and neck kinematics after low velocity rear-end impacts - Understanding "whiplash" (952724). 39th Stapp Car Crash Conference Proceedings (P-299), pp. 215-238. Warrendale, PA: Society of Automotive Engineers, 1995.

13. Szabo TJ, Welcher JB, Anderson RD, et al. Human occupant kinematic response to low speed rear-end impacts (940532). In: Backaitis S, ed. Occupant Containment and Methods of Assessing Occupant Protection in the Crash Environment (SP-1045), pp. 23-36. Warrendale, PA: Society of Automotive Engineers, 1994.

14. Szabo TJ & Welcher JB. Human subject kinematics and electromyographic activity during low speed rear impacts (962432). 40th Stapp Car Crash Conference (P-305), pp. 295-315. Warrendale, PA: Society of Automotive Engineers, 1996.

15. Matsushita T, Sato TB, Hirabayashi K, et al. X-ray study of the human neck motion due to head inertia loading (942208). Warrendale, PA: Society of Automotive Engineers, 1994.

16. O'Neill B, Haddon W, Kelley AB & Sorenson WW. Automobile head restraints - Frequency of neck injury in relation to the presence of head restraints. American Journal of Public Health, March 1972, pp 399-406.

17. Kahane CJ. An evaluation of head restraints - Federal motor vehicle safety standard 202 (DOT HS-806 108). Washington DC: US Department of Transportation, National Highway Traffic Safety Administration, February 1982.

18. Lovsund P, Nygren A, Salen B & Tingvall C. Neck injuries in rear end collisions among front and rear seat occupants. Proceedings of 1988 International IRCOBI Conference on the Biomechanics of Impact, pp.319-325. Bron, France: IRCOBI Secretariat, 1988.

19. Kihlberg JK. Flexion-torsion neck injury in rear impacts. Proceedings of the Thirteenth Annual Conference of the American Association for Automotive Medicine, pp. 1-16. Des Plaines, OH: American Association for Automotive Medicine, 1969.

20. Balla JI. The late whiplash syndrome. Aust. N.Z. J. Surg. 1980; 50:6 p.610-614.

21. Otremski I, Marsh JL, Wilde BR, McLardy Smith PD & Newman RJ. Soft tissue cervical spinal injuries in motor vehicle accidents. Injury 1989; 20: 349-351.

22. States J, Balcerak J, Williams J, et al. Injury frequency and head restraint effectiveness in rear-end impacts (720967). Warrendale PA: Society of Automotive Engineers, 1972.

23. Snyder RG, Chaffin DB & Foust DR. Bioengineering study of basic physical measurements related to susceptibility to cervical hyperextension-hyperflexion injury (UM-HSRI-BI-75-6). Ann Arbor, MI: University of Michigan, Highway Safety Research Institute, September, 1975.

24. Olney DB & Marsden AK. The effect of head restraint and seat belts on the incidence of neck injury in car accidents. Injury 1986; 17: 365-367.

25. Brault JR, Wheeler JB, Siegmund GP, Brault EJ. Clinical response of human subjects to rear-end automobile collisions (in press). Archives of Physical Medicine and Rehabilitation.

26. Najjar MF & Rowland M. Anthropometric Reference Data and Prevalence of Overweight, United States, 1976-80, Data from the National Health Survey series 11, No. 238 (PHS) 87-1688. Hyattsville MD: National Center for Health Statistics, Department of Health and Human Services, October 1987.

27. Diffrient N, Tilley AR & Bardagjy JC. Humanscale Manual. The MIT Press,'Cambridge, MA, 1974.

28. Padgaokar AJ, Krieger KW & King AI. Measurement of angular acceleration of a rigid body using linear accelerometers (No. 75-APMB-3). Transactions of the American Society of Mechanical Engineers, pp 522-526, September 1975.

29. Faro Technologies Inc. FaroArm Bronze Series User's Manual. Lake Mary, FL, 1995.

30. Lawrence JM, Siegmund GP & Nickel JS. Measuring head restraint force and point of application during low-speed rear-end automobile collisions (970397). In: Irwin A & Backaitis S, ed. Occupant Protection and Injury Assessment in the Automotive Crash Environment (SP-1231), pp. 225-37. Warrendale, PA: Society of Automotive Engineers, 1997.

31. Society of Automotive Engineers. SAE recommended practice: Instrumentation for impact tests (SAE J211 Jun 88). 1989 SAE Handbook, Volume 4, On-highway vehicles and off-highway machinery, pp. 34.184 - 34.191. Warrendale, PA: Society of Automotive Engineers, 1989.

32. Clauser CE, McConville JT & Young JW. Weight, volume, and center of mass of segments of the human body (AMRL-TR-69-70). Yellow Springs, OH: Wright Patterson Air Force Base, Aerospace Medical Research Laboratory, August 1969.

Queisser F, Bluthner R & Seidel H. Control of positioning the cervical spine and its application to measuring extensor strength. Clinical Biomechanics 1994; **9**: 157-161, May.

34. Backaitis SH & Mertz HJ (Eds). Hybrid III: The first human-like crash test dummy (PT-44). Warrendale, PA: Society of Automotive Engineers, 1994.

35. Mital NK & King AI. Computation of rigid-body rotation in three-dimensional space from body fixed linear accelerometer measurements. Journal of Applied Mechanics 1979; **46**: 925-930, December.

36. Mital HK. Computation of rigid-body rotation in 3D space from body-fixed acceleration measurements. Ph.D. Dissertation, Detroit MI: Wayne State University, 1978.

37. Beer FP & Johnston ER. Vector mechanics for engineers, Statics and dynamics (3rd edition). McGraw Hill, 1978.

38. Rabiner LR & Gold B. Theory and application of digital signal processing. Prentice Hall, Englewood Cliffs NJ, 1975.

39. Press WH, Teukolsky SA, Vetterling WT & Flannery BP. Numerical recipes in C, The art of scientific computing (2nd Edition). Cambridge University Press, 1992.

40. Moffit FH & Mikhail EM. Photogrammetry, Third Edition. New York, NY: Harper & Row Publishers, 1980.

41. Siegmund GP, King DJ & Montgomery DT. Using barrier impact data to determine the speed change in aligned, low-speed vehicle-to-vehicle collisions (960887). Accident reconstruction: Technology and animation VI (SP-1150), pp. 147-167. Warrendale, PA: Society of Automotive Engineers, 1996.

42. Kleinbaum DG & Kupper LL. Applied regression analysis and other multivariable methods. Duxbury Press, Boston, MA, 1978.

43. Devore JL. Probability and Statistics for Engineering and the Sciences. Brooks/Cole Publishing Company, Monterey, CA, 1982.

44. Svensson MY, Lovsund P, Haland Y & Larsson S. The influence of seat-back and head-restraint properties on the head-neck motion during rear-impact. Accid. Anal. and Prev. 28: 221-227, 1996.

45. Olsson I, Bunketorp O, Carlsson G, Gustafsson C, Planath I, Norin H & Ysander L. An in-depth study of neck injuries in rear-end collisions. Proceedings of 1990 International IRCOBI Conference on the Biomechanics of Impact. Bron, France: IRCOBI Secretariat, 1990.

46. Nygren A, Gustafsson H & Tingvall C. Effects of different types of headrests in rear-end collisions. 10th Experimental Safety Vehicle Conference, p.

85-90. Washington, DC: National Highway Traffic Safety Administration, US DOT, 1985.

47. Viano DC & Gargan MF. Headrest position during normal driving: Implications to neck injury risks in rear crashes. 39th Annual Proceedings of the Association for the Advancement of Automotive Medicine, pp. 215-229. Des Plaines, OH: Association for the Advancement of Automotive Medicine, 1995.

48. Penning L. Acceleration injury of the cervical spine by hypertranslation of the head, Part I, Effect of normal translation of the head on cervical spine motion: A radiology study. European Spine Journal 1992, **1**: 7-12.

Appendix A - Summary of the absolute and relative kinematic response data for the head and C7-T1 joint axis.

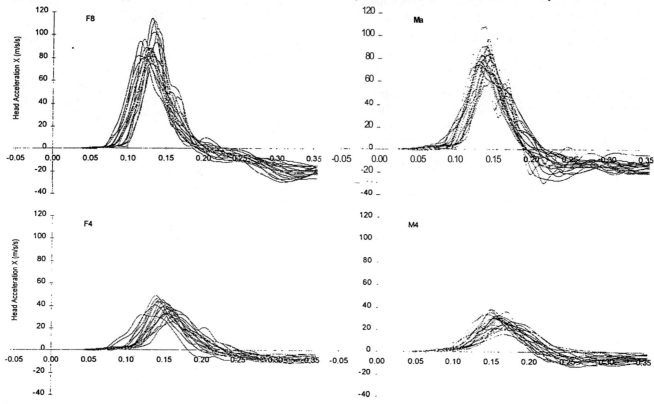

Figure A1. Head acceleration (m/s²) in the x-direction as a function of time (s). F = female. M = male. **4** = **4** krnlh level, 8 = 8 km/h level.

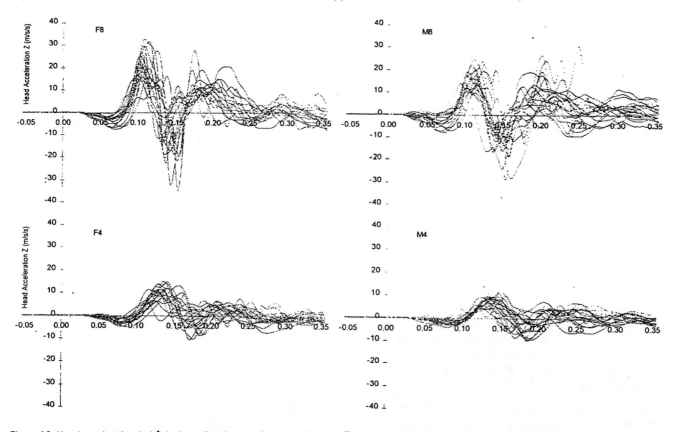

Figure A2. Head acceleration (m/s²) in the z-direction as a function of time (s). F = female, M = male, **4** = **4** km/h level, 8 = *8* km/h level.

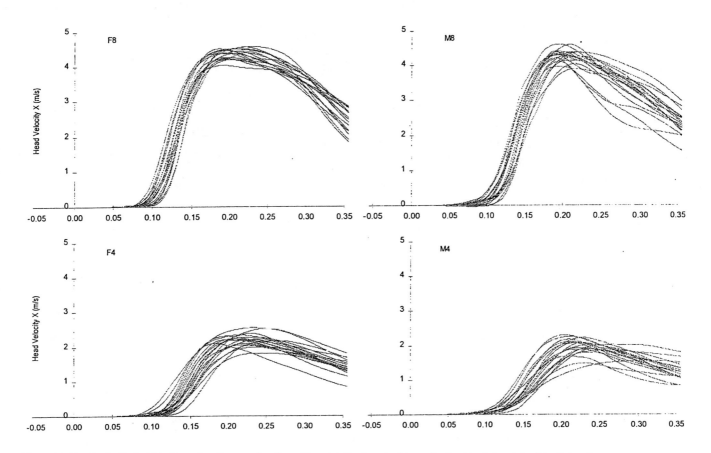

Figure A3. Head velocity (m/s) in the x-direction as a function of time (s). F = female, M = male, 4 = 4 kmlh level, 8 = 8 kmlh level

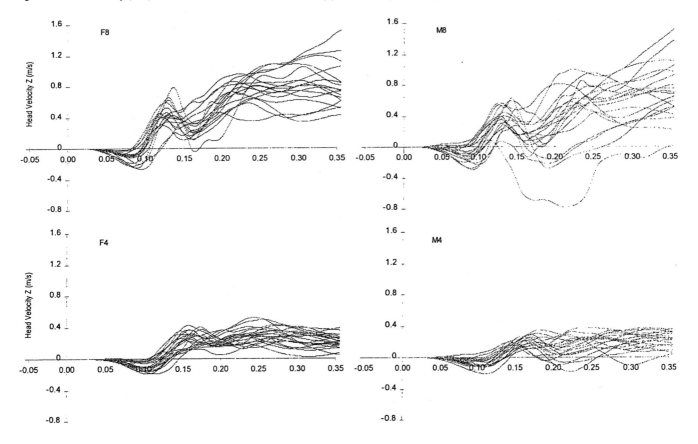

Figure A4. Head velocity (m/s) in the z-direction as a function of time (s). F = female, M = male, 4 = 4 kmlh level, 8 = 8 kmlh level

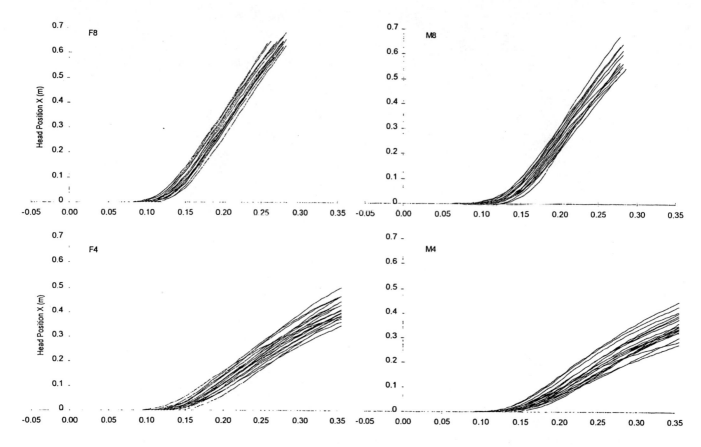

Figure A5. Head position (m) in the x-direction as a function of time (s). F = female, M = male, 4 = **4** km/h level, 8 = 8 kmlh level

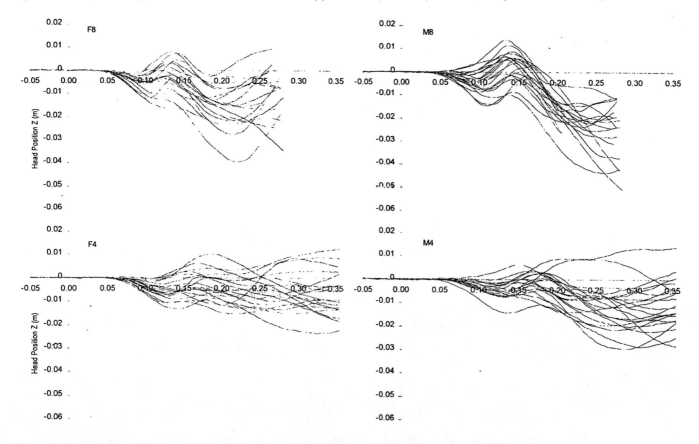

Figure A6. Head position (m) in the z-direction as a function of time **(s).** F = female, M = male, **4** = 4 kmlh level, 8 = 8 km/h level

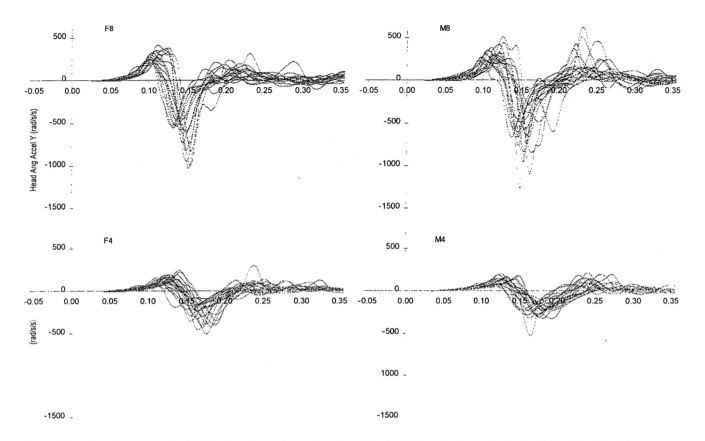

Figure A7. Head angular acceleration (radls²) about the y-axis as a function of time (s). F = female, M = male, **4** = 4 kmlh level, 8 = 8 kmlh level.

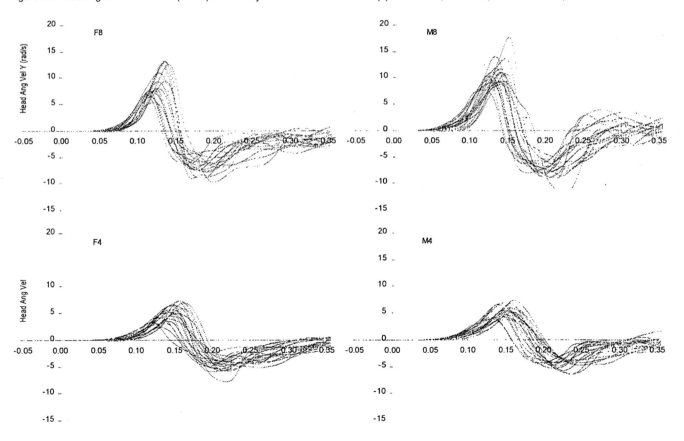

Figure A8. Head angular velocity (radls) about the y-axis as a function of time (s). F =female, M = male, **4** = **4** kmlh level, 8 = 8 kmlh level.

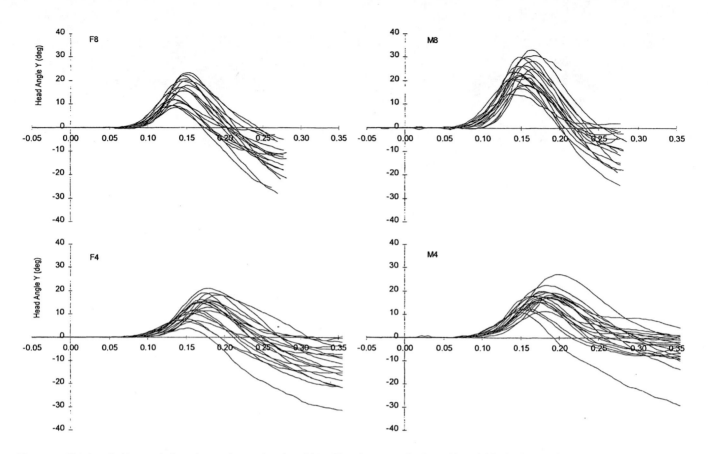

Figure A9. Head angle (degrees) about the y-axis as a function of time. Note that extension is positive and flexion is negative.

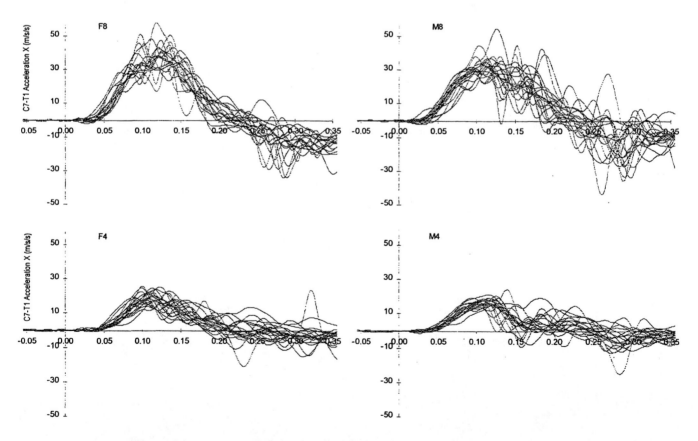

Figure A10. C7-T1 acceleration (m/s²) in the x-direction as a function of time (s). F = female, M = male, 4 = 4 kmlh level, 8 = 8 km/h level

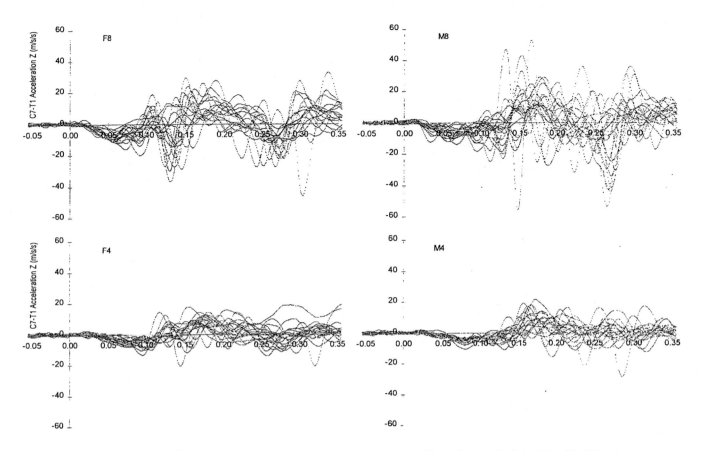

Figure A11. C7-T1 acceleration (m/s²) in the z-direction as a function of time *(s).* F = female, M = male, **4** = 4 kmlh level, 8 = 8 kmlh level.

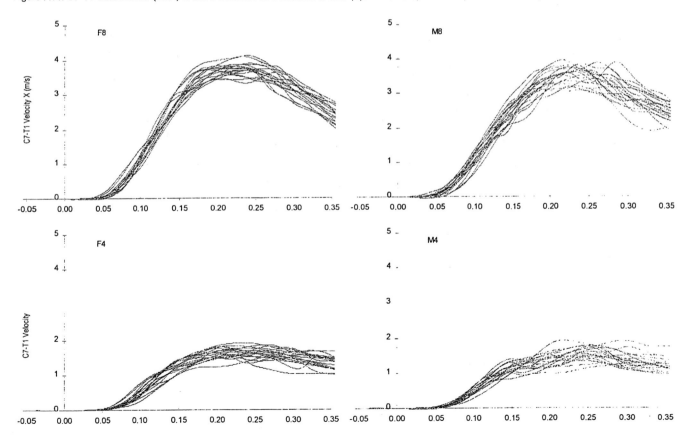

Figure A12. C7-T1 velocity (m/s) in the x-direction as a function of time (s). F =female, M = male, **4** = 4 kmlh level, 8 = 8 kmlh level.

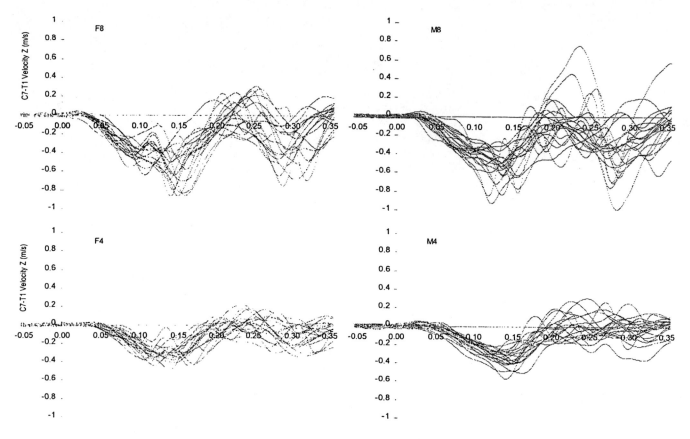

Figure A13. C7-T1 velocity (m/s) in the z-direction as a function of time (s). F = female, M = male, 4 = 4 km/h level. 8 = 8 kmlh level

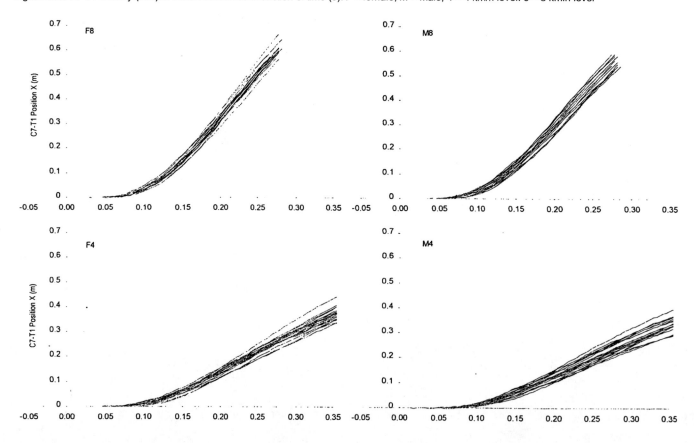

Figure A14. C7-T1 position (m) in the x-direction as a function of time (s). F = female, M = male, 4 = 4 kmlh level, 8 = 8 kmlh level.

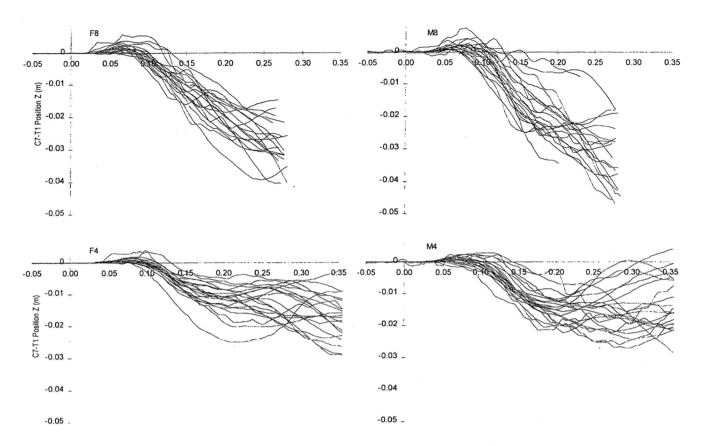

Figure A15. C7-T1 position (m) in the z-direction as a function of time (s). F = female, M = male, **4** = **4** km/h level, 8 = 8 kmlh level

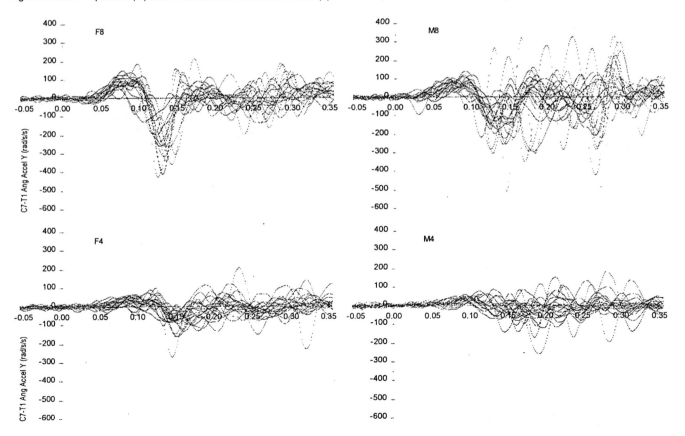

Figure A16. C7-T1 angular acceleration (rad/s²) about the y-axis as a function of time (s). F = female, M = male, 4 = 4 kmlh level, 8 = 8 km/h level.

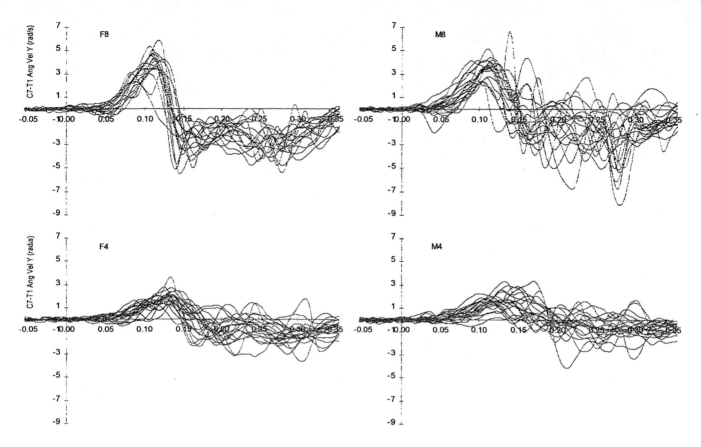

Figure A17. C7-T1 angular velocity (rad/s) about the y-axis as a function of time (s). F = female, M = male, 4 = 4 kmlh level, 8 = 8 kmlh level.

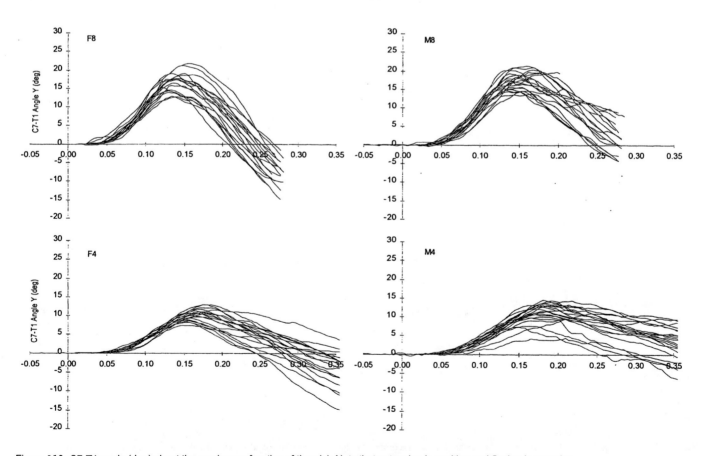

Figure A18. C7-T1 angle (deg) about the y-axis as a function of time (s). Note that extension is positive and flexion is negative

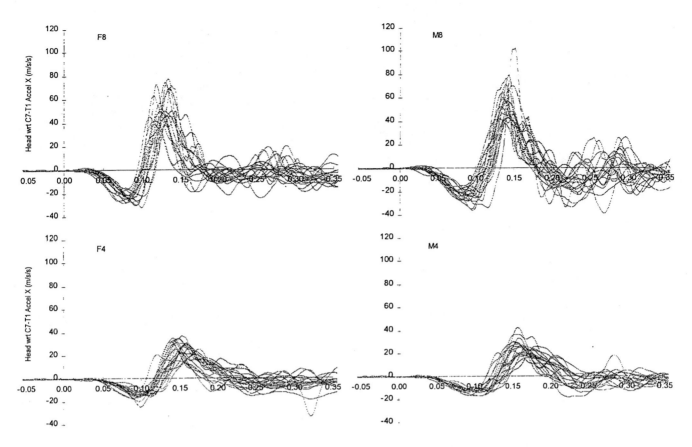

Figure A19. Head acceleration (m/s²) with respect to C7-T1 in the x-direction as a function of time (s). F = female, M = male, 4 = 4 kmlh, 8 = 8 km/h

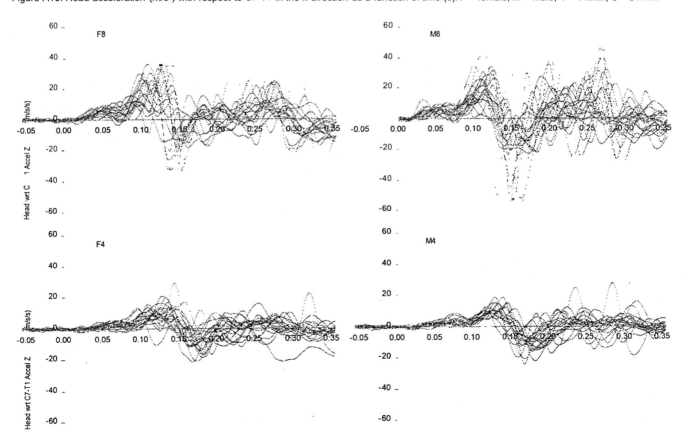

Figure A20. Head acceleration (mls²) with respect to C7-T1 in the z-direction as a function of time (s). F = female, M = male, 4 = 4 kmlh, 8 = 8 kmlh

461

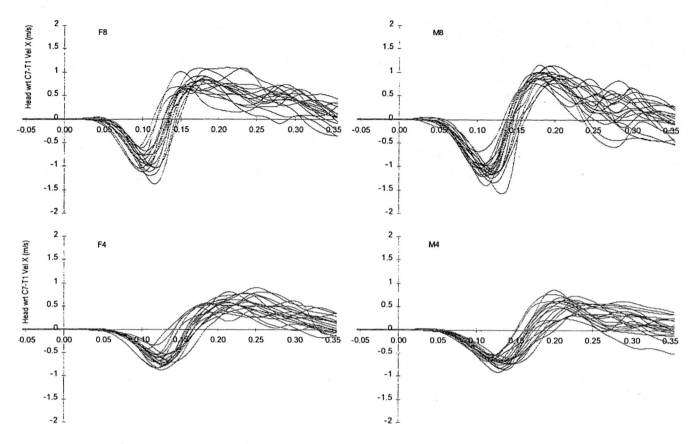

Figure A21. Head velocity (mls) with respect to C7-T1 in the x-direction as afunction of time (s). F = female, M = male, **4 = 4** kmlh, 8 = 8 kmlh.

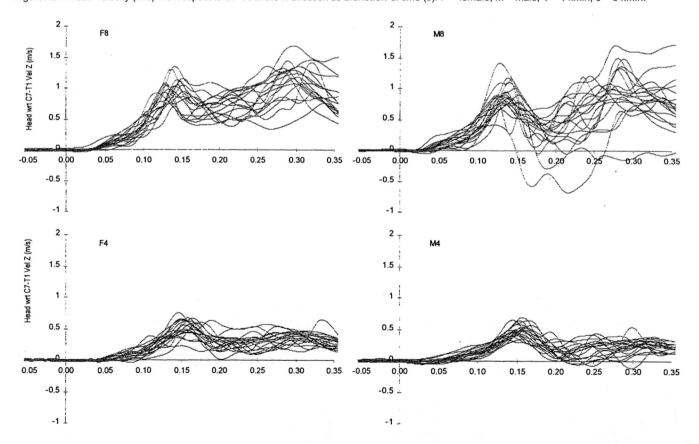

Figure A22. Head velocity (m/s) with respect to C7-T1 in the z-direction as **a** function of time (s). F = female, M = male, **4 = 4** kmlh, 8 = 8 kmlh.

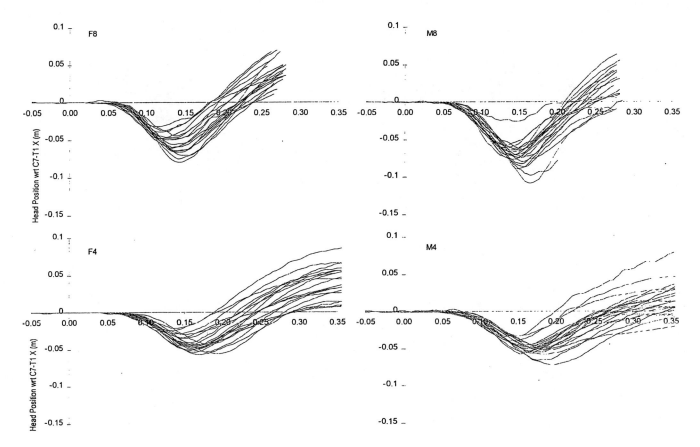

Figure A23. Head position (m) with respect to C7-T1 in the x-direction as a function of time (s). F = female, M = male, 4 = 4 km/h, 8 = 8 km/h

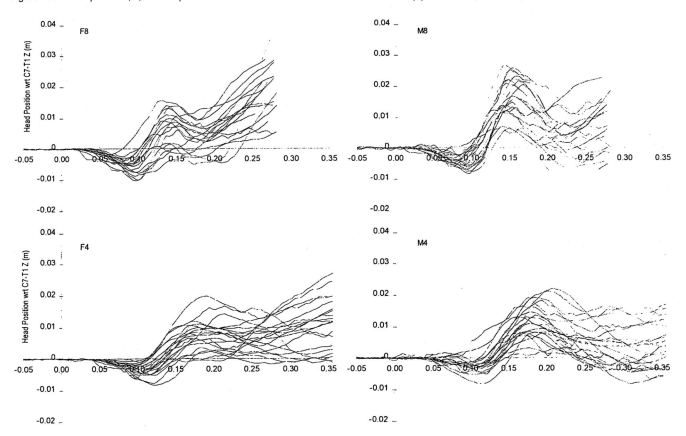

Figure A24. Head position (m) with respect to C7-T1 in the z-direction as a function of time (s). F = female, M = male, **4** = 4 km/h, 8 = 8 km/h

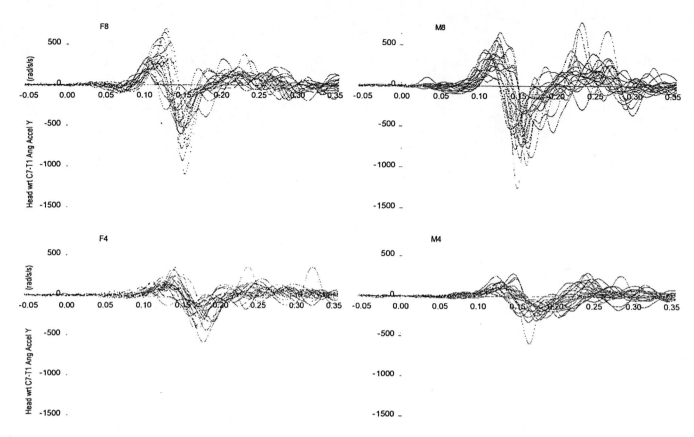

Figure **A25.** Head angular acceleration (rad/s²) with respect to C7-T1 about the y-axis as a function of time (s). F = female, M = male, **4** = **4** km/h, 8 = 8 km/h.

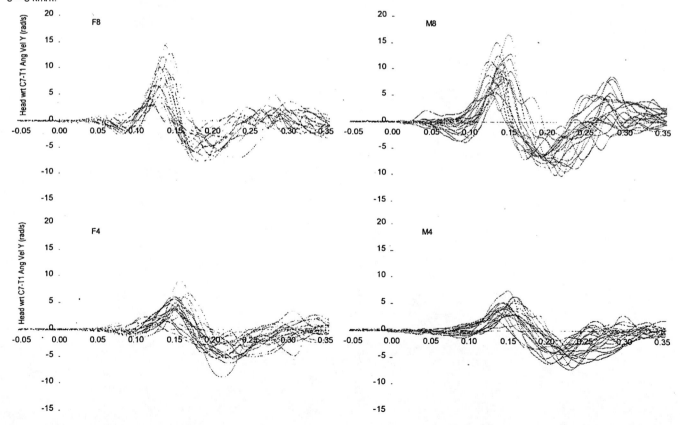

Figure A26. Head angular velocity (rad/s) with respect to C7-T1 in the x-direction as a function of time **(s).** F = female, M = male, **4** = **4** km/h, 8 = 8 kmlh.

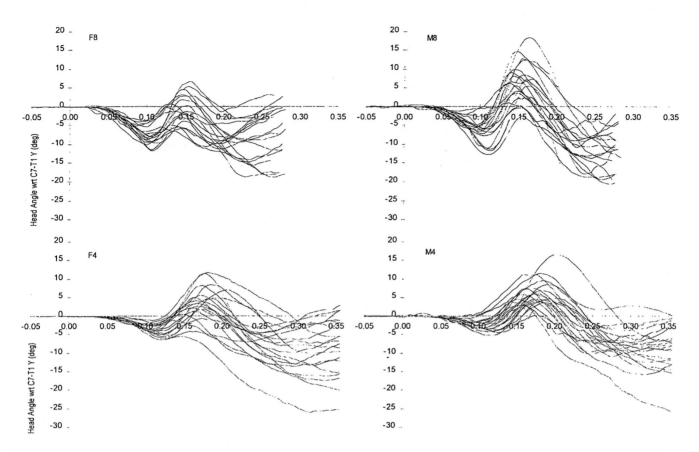

Figure A27. Head angle (deg) with respect to C7-T1 about the y-axis as a function of time (s). Note that extension is positive and flexion is negative

2006-01-0253

Assessment of 3 and 6-Year-Old Neck Injury Criteria Based on Field Investigation, Modeling, and Sled Testing

Mark R. Sochor, Daniel P. Faust and Kathryn F. Anderson
University of Michigan Program for Injury Research and Education

Skip Barnes and Stephen A. Ridella
TRW Automotive

Stewart C. Wang
University of Michigan Program for Injury Research and Education

ABSTRACT

The intent of this study was to compare the neck responses measured from the Hybrid III 3 and 6-year-old ATDs in laboratory testing to injuries sustained by three children in a field crash and investigate the appropriateness of recommended in-position neck injury assessment reference values (IARVs), and the regulated out-of-position (OOP) IARVs specified in FMVSS 208 for the Hybrid III 3 and 6-year-old ATDs. This paper principally reports on apparent artifacts associated with the Hybrid III 3 and 6-year-old ATDs, which complicated investigating the appropriateness of the in-position and out-of-position neck IARVs. In tests using 3-point belt restraints, these apparent artifacts included: 1) High neck extension moments, which produced the peak Nij values, without significant observed relative head-to-neck motion, 2) Neck tension forces well in excess of the IARVs that occurred when the ATD's chin contacted the chest. Because humanlike kinematics and force deflection properties for belt restrained chin-to-chest contact have not been defined, and ATD force deflection criteria are not specified for chin-to-chest contact; the biofidelity of the neck data measured by the ATD during chin-to-chest contact is doubtful. This paper also compares the measured neck injury criteria with 3 and 6-year-old Hybrid III ATDs to the neck injuries sustained by the three children in a previously reported frontal crash. To the extent the test environment simulated the field crash, the out-of-position neck tension IARVs seemed consistent with the injuries sustained by the 3 and 6-year-old children who were in 3-point restraints, but had placed the shoulder belts behind their backs. Excluding neck tension measurements during chin-to-chest contact, the neck tensions measured in the lap/shoulder belted sled tests seemed consistent with the injuries sustained by the lap/shoulder belted 7-year-old child in the field crash. While the Hybrid III 3 and 6-year-old ATDs are the best available tools for assessing child

neck injury potential, judgment is needed in interpreting the neck data from tests with belt restraint systems. It is inappropriate to only consider the peak neck measurements without examining the events surrounding those peak values.

INTRODUCTION

Neck responses in excess of the accepted injury assessment reference values have been measured from child anthropomorphic test devices (ATDs) during various sled testing of child restraints and 3-point belts [11, 29]. However, child neck injuries are seldom observed in the field under similar restraint conditions. Federal Motor Vehicle Safety Standard (FMVSS) 208 (Occupant Crash Protection) [5] and FMVSS 213 (Child Restraint Systems) [6] are intended, among other things, to help minimize the risk of injury to children in MVCs. While FMVSS 208 contains child neck injury criteria, it does not address in-position belt restrained children. While FMVSS 213 addresses dynamic, in-position child restraints, it does not regulate neck tolerance measurements.

The current FMVSS 208 standard utilizes a series of child-sized anthropomorphic test devices (ATDs). The neck structures in the Hybrid III 3-year-old and 6-year-old ATDs are designed to human response corridors that were scaled from 50th% adult male neck response corridors [9, 12]. FMVSS 208 specifies these ATDs to measure the injury potential to an out-of-position child that may be in extreme close proximity to a deploying passenger airbag. For child out-of-position occupant evaluations, FMVSS 208 specifies a set of injury assessment reference values for the Hybrid III 3 and 6-year-old ATDs that must not be exceeded if a vehicle manufacturer is attempting to comply an air bag system with the low risk deployment option of FMVSS 208. Mertz et al. [14] conducted a test program to understand the types of injuries that may be induced when a child is in close proximity to an inflating passenger air bag. The primary objective of this testing program was to understand the potential for injury to the head, neck and chest. Animal surrogates were utilized to determine the threshold of injury. The surrogates were exposed to deploying airbags and were evaluated for resulting injury. The most frequently observed neck injury was hemorrhage within the atlas occipital joint capsules and /or hemorrhage dorsal of the membrane covering the midsaggital-ventral aspect of the A-O junction. Such hemorrhage was described as a good indicator of the onset of damage to the apical odontoid ligament. This is in agreement with the neurosurgical medical literature [32] which states that involvement of the tectorial membrane was the critical threshold in the transition from a stable to an unstable neck injury. See Figure 1.

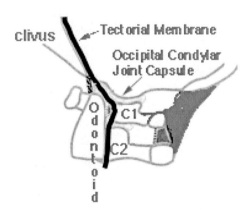

Figure 1: Occipital atlas interface

These same types of out-of-position tests were repeated with a 3-year-old sized ATD that was developed for airbag inflation induced injury assessment and utilized an upper neck load cell located in the area of the occipital condyles [34]. Neck injury thresholds and risk curves were then developed through research conducted by Mertz and Weber [13], Mertz et al. [15], Mertz and Prasad [16], and Mertz et al. [17]. Most recently, injury assessment reference values have been established for application to both in-position and out-of-position testing of ATDs [17]. The injury assessment reference values developed from this research were used as the basis for the neck injury values specified for the Hybrid III child ATDs in the FMVSS 208 out-of-position (OOP), air bag test requirements [2].

FMVSS 213, which regulates the performance and testing of child restraint systems, previously utilized only the Hybrid II ATD, which is not capable of recording neck loads. In addition to the Hybrid II, the Hybrid III ATD, which has the capability of measuring upper neck loads,

is allowed to be used for FMVSS 213 compliance [6]. FMVSS 213 calls for the Hybrid III to replace the Hybrid II ATD in the future. While FMVSS 208 specifies static tests with the Hybrid III child ATDs positioned very close to a deploying passenger air bag, FMVSS 213 is the only Federal regulation that addresses dynamic, in-position child restraints.

The FMVSS 208, Final Rule dated May 12, 2000 [19], specifies Nij neck injury criterion that account for combined loading conditions of neck flexion, extension, tension, and compression and their cumulative effects on injury risk. The final rule explains that neck tension and compression limits were added based on concerns expressed by the Alliance of Automobile Manufactures, the Insurance Institute for Highway Safety, and the National Transportation Safety Board that the peak neck tension and compression allowed by Nij when the moment was zero were too great. Several automobile manufacturers had concerns that the current Hybrid III ATD neck was inadequate to accurately assess flexion/extension injuries and would affect calculated Nij responses. They recommended delaying the extension related criteria and using neck tension as the only neck injury criteria.

The FMVSS 208 Final Rule further explains that the NHTSA and commenters observed high measured neck moments at the upper neck load cell within the first 20 ms, with initiation of large neck shear loads, without observing substantial angular deformation of the neck. The question was raised concerning whether these moments measured without substantial angular deformation of the neck were biomechanically realistic. However, because of the need to minimize the likelihood of neck injury and the lack of testing alternatives, the FMVSS 208 Final Rule specified the existing Hybrid III neck.

Kang et al. [10] found that moments at the occipital condyles of the Hybrid III 5th percentile dummy, in certain modes of air bag neck interaction, go out of the Mertz and Patrick human response corridors [12], with very little head rotation relative to the torso. Agaram et al. [1] discussed the occurrence of large moments at the occipital condyles with small angular deformation of the neck and that such responses are unlikely in humans because of greater laxity of the human atlanto occipital joint. They concluded that high extension/flexion moments were produced with little neck angular deformation when the Hybrid III 5th percentile neck was subjected to second mode bending. The Hybrid III neck was developed to respond in a manner representative of humans only in first mode beam bending. Similar neck structures, utilizing a pivoting OC joint and nodding blocks are used in both the Hybrid III 5th percentile and Hybrid III 6-year-old ATDs, so neck responses were expected to be similar. In addition, researchers have made observations regarding the biofidelity of the head kinematics as a result of the rigid thoracic spine in the

Hybrid III child ATDs and the resulting head, neck and chest responses in a dynamic environment [29, 11, 33, 3].

As the government and industry continue to develop solutions to enhance protection for children in an automotive crash environment, the responses of the Hybrid III family of child ATDs and the injury assessment reference values will come under increasing scrutiny. Based on several studies, NHTSA and other organizations have for several years emphasized the importance of proper child restraint usage. As government and industry strive to protect children, ATD measured responses need to appropriately predict child injury potential in a dynamic test environment. Assessing the appropriateness of the existing injury criteria or developing new ones is a difficult task as there is little child impact data from the biomechanical laboratories. This has led public health officials and bioengineers to utilize the sources of limited data, which are available to them. Because of these limitations, real world crash data and detailed injury reconstructions could be an important source of information in deciding the acceptance or rejection of proposed injury assessment reference values [23, 31, 32].

In a previous paper by Sochor et al. [30], a crash involving three seat belt restrained children was investigated as part of the University of Michigan CIREN (Crash Injury Research and Engineering Network) project. See Figure 2.

Figure 2: Case Vehicle.

The ages of the children involved in the crash were 3, 6 and 7 years old. This real world crash was studied because the children all suffered threshold-type (tectorial membrane hemorrhage and occipital condyle hemorrhage) neck injuries. The children were similar in height and weight to the Hybrid III 3-year-old ATD (ht = 945mm, wt = 16.2kg) and the Hybrid III 6-year-old ATD (ht = 1168, wt = 23.4kg).

The three children sustained the following injuries:

Left Rear Seated, Lap Belted Only, 3-year-old (ht=1000mm wt=15kg) ISS = 17

- synovial capsule with tectorial membrane hemorrhage (AIS = 2 (650208.2))(See Figure 3A and B)
- small bowel perforation (AIS = 3 (541424.3))
- bilateral iliac wing fractures (AIS = 2 (852602.2))
- Abdominal contusions (AIS = 1)

Right Rear Seated, Lap Belted Only, 6-year-old (ht=1220mm wt=23kg) ISS = 26

- tectorial membrane hemorrhage with occipital condyle ligamentous injury (AIS = 2 (650208.2))
- Right frontal bone depressed skull fracture with underlying subarachnoid hemorrhage (AIS = 3 (150404.3))
- small bowel devascularization x 2 (AIS = 4 (541426.4))
- colon perforation (AIS = 3 (540824.3)
- L2-3 spinous process avulsion fractures (AIS = 2 (650618.2))
- L4 vetebral body fracture (AIS = 2 (650632.2))
- Abdominal contusion (AIS = 1)

Right Front Seated, Lap/shoulder Belted, 7-year-old (ht=1240mm wt=35kg) ISS = 21

- 6th cranial nerve palsy (AIS = 2 (650208.2)) (The attending neurosurgeon opined that this injury resulted from stretching of the nerve by distracting the occiput from C1).
- Bilateral Pulmonary Contusions (AIS = 4 (441410.4))
- Left #2-6 rib fractures (AIS = 3 (450230.3))
- Lower Abdominal Contusions (AIS = 1)

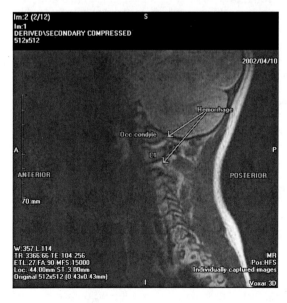

Figure 3A: MRI of tectorial membrane hemorrhage in 3-year-old victim.

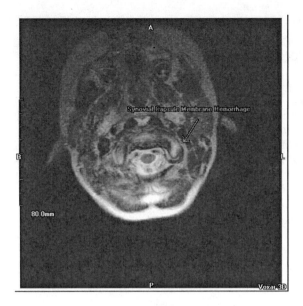

Figure 3B: Synovial Capsule Hemorrhage 3 year-old

In this crash, the case vehicle experienced a 12 o'clock, fully distributed, frontal impact when it T-boned a passenger car. An accident reconstruction was conducted, and WINDSMASH estimated the change in longitudinal velocity (delta V) of the case vehicle to be 45 kph (28mph). The 3 and 6-year-old children were both restrained in the rear outboard seating locations. Even though these seating locations contained a lap/shoulder belt system, both children were reportedly wearing the shoulder belt behind their back at the time of the crash. Therefore, for the purposes of this analysis, these two children were considered to be restrained by a lap belt only at the time of the crash. The 7-year-old child was seated in the right front passenger seat, and was restrained by the three-point vehicle lap/shoulder belt. Despite the differences in the children's sizes, restraint configurations, and seating locations, each child suffered a threshold A-O injury similar to those shown in the experiments run by Mertz et al. [14]. This field crash offered the unique opportunity to evaluate the neck IARVs using modeling and laboratory sled testing under conditions similar to this field event.

The intent of this study was to compare the neck responses measured from the Hybrid III 3-year-old and 6-year-old ATDs in laboratory testing to injuries sustained by the three children in the field crash and investigate the appropriateness of the neck injury assessment reference values. Apparent artifacts associated with the Hybrid III 3-year-old and 6-year-old ATDs were examined.

METHODS

SLED TESTING

A series of five sled tests were conducted on a rebound-type deceleration sled at the University of Michigan Transportation Research Institute (UMTRI) in Ann Arbor, Michigan. The Hybrid III 3-year-old and Hybrid III 6-year-old ATDs were used to simulate the sizes of the children involved in the field crash. The 3-year-old and 6-year-old ATDs, corresponding to the left and right rear occupants respectively, were very similar to the stature of the case vehicle occupants in those seating positions. The 7-year-old in the right front seating position is 72 mm taller and 11.6 kg heavier than the 6-year-old ATD.

Each sled test contained two ATDs, which were positioned in the outboard seating locations of a standard FMVSS 213 bench seat [6]. The ATDs were restrained by either a lap belt (three point restraint with shoulder belt portion behind child's back) or a lap/shoulder belt. The sled tests were recorded using high-speed digital video cameras (1000 frames per second) which provided side view and overhead coverage, and an 8-frame rapid-sequence Polaroid camera. See Table 1 for a description of the configuration for each sled test.

TABLE 1: Matrix of sled tests conducted at UMTRI utilizing a 25 mph, modified FMVSS 213 sled pulse.

Test No.	ATD	Belt System	Simulated Seating Location
Test 1	Hybrid III 3-yr-old	Lap Belt Only	Left Rear Belt Geometry
	Hybrid III 6-yr-old	Lap Belt Only*	Right Rear Belt Geometry
Test 2	Hybrid III 3-yr-old	Lap Belt Only	Left Rear Belt Geometry
	Hybrid III 6-yr-old	Lap Belt Only*	Right Rear Belt Geometry
Test 3	Hybrid III 3-yr-old	Lap/Shld Belt	Left Rear Belt Geometry
	Hybrid III 6-yr-old	Lap/Shld Belt	Right Front Belt Geometry
Test 4	Hybrid III 3-yr-old	Lap/Shld Belt	Left Rear Belt Geometry
	Hybrid III 6-yr-old	Lap/Shld Belt	Right Front Belt Geometry
Test 5	Hybrid III 3-yr-old	Lap/Shld Belt	Left Rear Belt Geometry
	Hybrid III 6-yr-old	Lap/Shld Belt	Right Front Belt Geometry

*The 6-yr-old lap belt only tests were conducted with 150 mm of slack in the belt system.

Due to lack of availability of a test fixture and seats like the SUV case vehicle, the FMVSS 213 test fixture and seat were used in the sled tests [6]. Differences between the SUV and FMVSS 213 seats are not expected to significantly affect the potential for neck injury because only the legs and buttocks of the ATD were contacting the seat when the peak neck loads were generated. The dummy H-point was used as the reference point for locating the belt anchorages, therefore the belt position on the ATD was the same for the 213 seat as it would have been with the SUV seat. Occupant compartment intrusion did not occur in the field crash which eliminated it as a variable when comparing the sled tests and the field crash.

The Hybrid III 3-year-old and 6-year-old ATDs were instrumented with tri-axial head and chest accelerometers; a chest potentiometer, and a six-axis upper neck load cell. The measured responses are filtered per SAE J211 [26] and reported per the SAE J1733 sign convention [25]. Belt loads were also measured during each test. Appendix A contains the external resultant head impact force equation, the moment transfer equation to obtain the upper neck moment about the y-axis at the occipital condyle for the Hybrid III 6-year-old ATD, and the neck injury predictor, Nij, equations used for the calculated dummy responses.

All test data were filtered according to the procedures in SAE J211 [26]. The Nij biomechanical neck injury predictors (N_{TE} = tension/extension and N_{TF} = tension/flexion) were calculated for the upper neck as specified in Appendix A [2, 19, 18, 15, 16, 17]. Table 2 summarizes the upper neck injury assessment reference values specified for in-position testing by Mertz et al. [17], and the upper neck limits specified for out-of-position testing by FMVSS 208 [5] for the Hybrid III 3-year-old and Hybrid III 6-year-old ATDs. The data generated from the sled testing was compared to both sets of injury values.

TABLE 2: Upper Neck Injury Criteria for the Hybrid III 3-year-old and Hybrid III 6-year-old ATDs.

Upper Neck Injury Criteria	In-Position IARVs Hybrid III 3-year-old ATD	FMVSS 208 OOP Limits Hybrid III 3-year-old ATD	In-Position IARVs Hybrid III 6-year-old ATD	FMVSS 208 OOP Limits Hybrid III 6-year-old ATD
Nij	1.0	1.0	1.0	1.0
F_{Tens} intercept (N)	2330	2120	3080	2800
F_{Comp} intercept (N)	2130	2120	2820	2800
M_{Flex} intercept (Nm)	67	68	96	93
M_{Ext} intercept (Nm)	29.3	27	42	37
Peak Axial Tension (N)	1430	1130	1890	1490
Peak Axial Compression (N)	1380	1380	1820	1820
Peak Extension Moment (Nm)	21	-	30	-
Peak Flexion Moment (Nm)	42	-	60	-

Figure 4 and 5 show the sled test setup and belt configurations for the lap belt only tests (Tests 1 and 2) and the lap/shoulder belt tests (Tests 3, 4, and 5), respectively. To simulate the restraint use of the left rear 3-year-old occupant in the field crash, a Hybrid III 3-year-old ATD was restrained only by the lap belt with the shoulder portion positioned behind the back for Tests 1 and 2. The rear seat belt geometry used in the sleds duplicated the rear seat belt geometry of the case vehicle. To simulate the restraint use of the right rear 6-year-old occupant in the field crash, a Hybrid III 6-year-old ATD was restrained only by the lap belt with the shoulder portion positioned behind the back for Tests 1 and 2. Case vehicle rear seat belt geometry was duplicated, but 150 millimeters of slack was introduced into this belt system. In the field crash, the right rear 6-year-old occupant's head contacted the back of the right front passenger seat in the case vehicle. Modeling studies indicated that 150 millimeters of slack was necessary in the lap portion of the belt system to reproduce the field crash kinematics that allowed the occupant's head to make contact with the back of the front seat back [30]. No front seat was present in the sled tests, so head to seat back contact could not occur. The sled tests did not attempt to fully re-create the right rear 6-year-old occupant's head contact with the front seat back in the field crash, so that camera coverage of the ATDs' kinematics could be optimized. Front seats were also not run in the sled tests so that the head, neck

and chest responses for all of the ATDs could be analyzed and compared directly without the added variable of head contact with other interior components.

Figure 4: Hybrid III 6-year-old (simulated right rear belt anchorages) and Hybrid III 3-year-old (simulated left rear belt anchorages) lap belted only sled test set-up (Tests 1 and 2).

Figure 5: Hybrid III 6-year-old (simulated right front belt anchorages) and Hybrid III 3-year-old (simulated left rear belt anchorages) LAP/SHOULDER belted sled test set-up (Tests 3, 4 and 5).

To simulate the restraint use of the right front 7-year-old occupant in the field crash, a Hybrid III 6-year-old ATD was restrained by a three-point lap/shoulder belt for Tests 3, 4, and 5. The right front seat belt geometry of the case vehicle was duplicated. To evaluate what effect a three-point restraint might have had on the 3-year-old case occupant, a Hybrid III 3-year-old ATD was restrained by the left rear three-point lap/shoulder belt in Tests 3, 4, and 5. The case vehicle left rear seat belt geometry was duplicated. New seat belts, similar in material properties and with the same geometry as the actual belts installed in the case vehicle (case vehicle belts out of production), were used for each test.

The 30 mph sled pulse specified in FMVSS 213 [6] was modified to a 25 mph, 80 millisecond (msec) pulse for use in these sled tests, in order to more closely simulate the case vehicle's response in the field crash. Also, due to constraints in the ability of the available rebound sled to produce longer duration pulses like the FMVSS 208 30 mph, 125 msec sled pulse [4], the shorter duration, 25 mph, modified FMVSS 213 sled pulse was used. All the sled tests were conducted using the same 25 mph, modified FMVSS 213 pulse, and the longitudinal acceleration and velocity associated with this sled pulse is shown in Figures 6 and 7.

Modeling

As previously described by Sochor et al., 2004, the occupant responses (kinematics and injury criteria) from the field crash were evaluated using rigid body vehicle and occupant modeling techniques. The MADYMO occupant simulation software (TNO Automotive, Delft, The Netherlands), was used for this simulation. A MADYMO model was created of a generic sport utility vehicle having internal geometry similar to the case vehicle. The seats, seat positions, belt mounting positions, and other relevant interior structures were derived from drawings of the vehicle. Interior structures were modeled with rigid bodies. Material properties for seats and other interior structures were derived from previously developed models. Body structures in certain areas (glass, sheetmetal and seatbelts) were modeled with finite elements to look more closely at potential interactions.

For this study, modeling was used to understand the effects of variation in crash pulse on the ATD responses. A model was created to simulate the FMVSS 213 fixture tests that were done for this study. A rigid body sled buck model was developed that had the proper seat geometry and foam force deflection properties (utilizing solid finite elements) that exist for a seat as designated in the 213 standard. The child dummies were modeled in the same position relative to the seat as the actual test set-up.

While the actual deceleration pulse of the field crash is not known, the impact of the front of the SUV case vehicle into the side of a passenger car suggests a pulse longer in duration than that of a rigid barrier test. WINDSMASH estimated the longitudinal delta V in the field crash to be 28 mph. The configuration of the crash and the estimated delta V suggest that the FMVSS 208 30 mph, 125 millisecond sled pulse would be reasonably representative of the field accident. Due to constraints in the ability of the available rebound sled to produce longer duration pulses like the FMVSS 208 30 mph, 125 millisecond sled pulse, the shorter duration FMVSS 213 sled pulse was modified to produce a 25 mph change in velocity.

MADYMO modeling was used to compare the sensitivity of the ATD responses to the modified 25 mph, FMVSS 213 pulse; the FMVSS 208 30 mph, 125 msec pulse; and a 30 mph rigid barrier crash pulse. The pulses evaluated were 1) a 30 mph barrier pulse of the exact same make and model vehicle from the NHTSA vehicle

crash database, 2) the FMVSS 208, 30 mph, 125 msec generic sled pulse first specified in the March 19, 1997, version of FMVSS 208 [4], and 3) a modified FMVSS 213 sled pulse [6] that took into account the pulse limitations of the sled system used for the 213 tests. These acceleration and velocity pulses are shown in Figures 6 and 7.

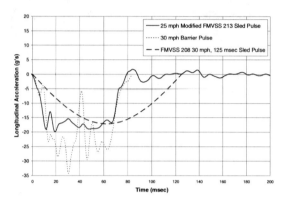

Figure 6: Comparison of longitudinal acceleration pulses (g's).

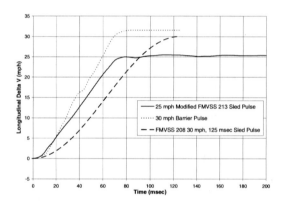

Figure 7: Comparison of longitudinal velocity pulses (mph).

Figures 8 and 9 compare the modeled upper neck force, F_z (tension/compression), and neck moment, M_y (flexion/extension), for the three pulses with a lap-belted 6-year-old Hybrid III ATD. Figures 10 and 11 compare the modeled upper neck force, F_z (tension/compression), and neck moment, M_y (flexion/extension), for the three pulses with a lap-shoulder-belted 6-year-old Hybrid III ATD. The model predicted that the modified 25 mph, FMVSS 213 pulse would produce neck tensions and moments similar in magnitude but offset earlier in the event than the FMVSS 208 30 mph, 125 ms pulse. The modeled barrier pulse produced neck tensions and moments similar in timing but greater in magnitude to the modified 25 mph, FMVSS 213 pulse.

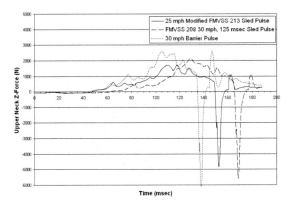

Figure 8: Modeling Results Comparing the Upper neck z-force (tension/compression) vs time for the Hybrid III six-year-old lap belted occupant.

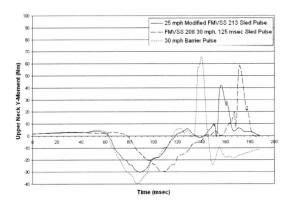

Figure 9: Modeling Results Comparing the Upper neck y-moment (flexion/extension) vs time for the Hybrid III six-year-old lap belted occupant.

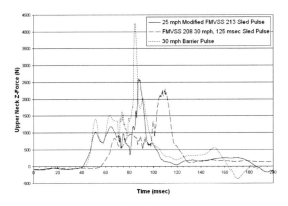

Figure 10: Modeling Results Comparing the Upper neck z-force (tension/compression) vs time for the Hybrid III six-year-old lap/shoulder belted occupant.

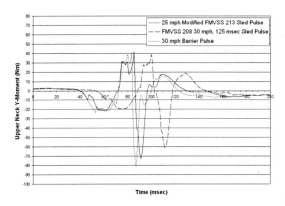

Figure 11: Modeling Results Comparing the Upper neck y-moment (flexion/extension) vs time for the Hybrid III six-year-old lap/shoulder belted occupant.

RESULTS

HEAD, NECK, AND CHEST DATA

Time history plots of several dummy responses for each of the sled tests are provided in Appendix B for the lap belted only tests and in Appendix C for the lap/shoulder belted tests. The same overlays are provided within each Appendix, and are presented in the same order.

NECK RESPONSES – LAP BELTED SLED TESTS

The highest Nij value recorded during the lap belted (shoulder belt behind back), Hybrid III 3-year-old tests, was N_{TE}, the combined neck tension/extension injury value. The N_{TE} time history curves for the lap belted Hybrid III 3-year-old in Tests 1 and 2 are displayed in Figure 12. The maximum N_{TE} and peak neck tension values for the lap belted Hybrid III 3-year-old tests are summarized in Table 3. These neck responses were compared to both the in-position and out-of-position neck injury assessment reference values [17, 5].

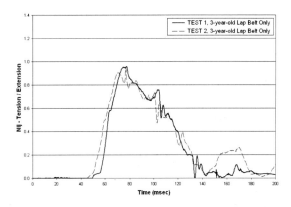

Figure 12: N_{TE} vs. time for the lap belted Hybrid III 3-year-old (using in-position intercepts).

TABLE 3: Lap belted Hybrid III 3-year-old peak N_{TE} and peak neck tension and the time at which the peak value occurred.

	Peak N_{TE}		Peak Neck Tension	
	In-Pos	OOP	In-Pos	OOP
IARV	1.0	1.0	1430 N	1130 N
Test 1	96%	104%	74%	93%
	74-78 msec		91 msec	
Test 2	95%	103%	77%	97%
	77 msec		85 msec	

The Hybrid III 3-year-old neck extension peaked near the same time as N_{TE} (75 msec, 20 Nm in the first test and at 70 msec, 19 Nm in the second test). The Hybrid III 3-year-old peak neck tensions occurred later than the peak extension and peak N_{TE}.

The highest Nij value recorded for the lap belted, Hybrid III 6-year-old tests, was also N_{TE}. The N_{TE} time history curves for the lap belted Hybrid III 6-year-old in Tests 1 and 2 are displayed in Figure 13. The maximum N_{TE} and peak neck tension values for the lap belted Hybrid III 6-year-old tests are summarized in Table 4.

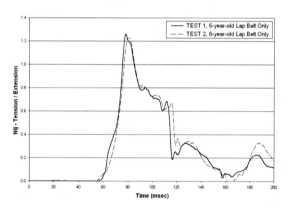

Figure 13: N_{TE} vs. time for the lap belted Hybrid III 6-year-old (using in-position intercepts).

TABLE 4: Lap belted Hybrid III 6-year-old peak N_{TE} and peak neck tension and the time at which the peak value occurred.

	Peak N_{TE}		Peak Neck Tension	
	In-Pos	OOP	In-Pos	OOP
IARV	1.0	1.0	1890	1490
Test 1	126%	142%	82%	104%
	78 msec		103 msec	
Test 2	123%	138%	86%	109%
	79-81 msec		105 msec	

The Hybrid III 6-year-old neck extension peaked at approximately the same time N_{TE} peaked (79 msec, 37 Nm in the first test and peaked at 81 msec, 38 Nm in the second test). The lap belted Hybrid III 6-year-old peak neck tensions also occurred later than the peak neck extensions and peak N_{TE}.

CHIN-TO-CHEST CONTACT ANALYSIS

The third, fourth, and fifth sled tests were run with lap/shoulder belted Hybrid III 3-year-old and 6-year-old ATDs. In each of these three tests, chin-to-chest contact was observed with both ATDs. The results from the lap/shoulder belted tests were analyzed while considering the neck responses in the period of time before chin-to-chest contact, and the responses resulting from chin-to-chest contact. In order to quantify the timing and duration of this contact, resultant external head impact forces were calculated per the procedure specified in SAE J2052 [27]. See Appendix A for the equations and head masses used in the calculations. Tables 5 and 6 summarize the time of chin-to-chest contact for all of the lap/shoulder belted tests, as determined by using the external head impact force calculation procedure.

TABLE 5: Chin-to-Chest contact times for the lap/shoulder belted Hybrid III 3-year-old.

Chin-to-Chest Contact	Hybrid III 3-year-old Lap/Shoulder Belted		
	Begin Contact (ms)	End Contact (ms)	Contact Duration (ms)
Test 3	71.9	95.5	23.6
Test 4	74.3	97.5	23.2
Test 5	71.4	93.1	21.7

TABLE 6: Chin-to-Chest contact times for the lap/shoulder belted Hybrid III 6-year-old.

Chin-to-Chest Contact	Hybrid III 6-year-old Lap/Shoulder Belted		
	Begin Contact (ms)	End Contact (ms)	Contact Duration (ms)
Test 3	82.7	116.8	34.1
Test 4	80.6	115.8	35.2
Test 5	80.6	115.6	35

Figures 14 and 15 overlay the resultant external head impact forces for the lap/shoulder belted Hybrid III 3-year-old and Hybrid III 6-year-old, respectively. The shaded area graphically represents the approximate duration of chin-to-chest contact, considering the earliest time of engagement and the latest time of disengagement detected in any of the three tests.

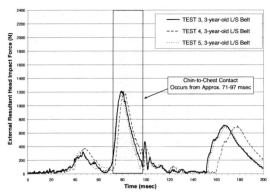

Figure 14: External Resultant Head Impact Force vs. time for the lap/shoulder belted Hybrid III 3-year-old ATD.

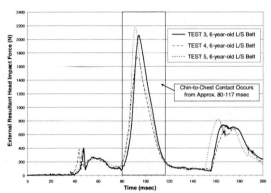

Figure 15: External Resultant Head Impact Force vs. time for the lap/shoulder belted Hybrid III 6-year-old ATD.

NECK RESPONSES – LAP/SHOULDER BELTED SLED TESTS

In the three lap/shoulder belted, Hybrid III 3-year-old tests, the highest calculated Nij values were N_{TE} and these maximum values occurred during the portion of the event before chin-to-chest contact. Figure 16 overlays the N_{TE} time history curves for the lap/shoulder belted Hybrid III 3-year-olds in Tests 3, 4, and 5. The maximum N_{TE} and peak neck tension values for the lap/shoulder belted Hybrid III 3-year-old tests are reported in Table 7 for the period of time before chin-to-chest contact.

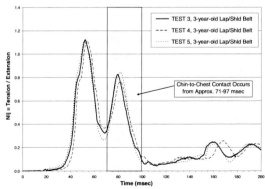

Figure 16: N_{TE} vs. time for the lap/shoulder belted Hybrid III 3-year-old (using in-position intercepts).

TABLE 7: Lap/shoulder belted Hybrid III 3-year-old peak N_{TE}, peak N_{TF}, and peak neck tension and the time at which the peak values occurred.

	Before Chin-to-Chest Contact				During Chin-to-Chest Contact			
	Peak N_{TE}		Peak Neck Tens.		Peak N_{TF}		Peak Neck Tens.	
	In-Pos	OOP	In-Pos	OOP	In-Pos	OOP	In-Pos	OOP
IARV	1.0	1.0	1430 N	1130 N	1.0	1.0	1430 N	1130 N
Test 3	112%	122%	78%	99%	98%	106%	135%	170%
	52 msec		57 msec		78 msec		80 msec	
Test 4	112%	122%	76%	96%	94%	101%	124%	158%
	54 msec		59 msec		80 msec		81 msec	
Test 5	111%	121%	71%	90%	82%	89%	128%	162%
	51 msec		56 msec		78 msec		79 msec	

The peak NTE responses occurred prior to chin-to-chest contact. After chin-to-chest contact occurred, the highest calculated Nij value was NTF for all three tests. Figure 17 overlays the NTF time history curves for the lap/shoulder belted Hybrid III 3-year-olds in Tests 3, 4, and 5. Table 7 reports the maximum NTF and peak neck tension values that resulted from chin-to-chest contact. These neck responses were compared to both the injury assessment reference values specified for in-position testing [17], and the upper neck limits specified for out-of-position testing by FMVSS 208 [5] (as listed in Table 2), and are reported as a percentage of those values.

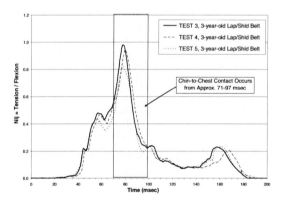

Figure 17: N_{TF} vs. time for the lap/shoulder belted Hybrid III 3-year-old (using in-position intercepts).

During the period of chin-to-chest contact, N_{TF} was the maximum Nij.

The Hybrid III 3-year-old peak neck tension values increased as a result of chin-to-chest contact in all three lap/shoulder belted tests. The upper neck z-force (tension/compression) time history responses can be found in Appendix C.

In the three lap/shoulder belted, Hybrid III 6-year-old tests, the highest Nij value occurring during the portion of the sled test before chin-to-chest contact was NTE. Figure 18 overlays the NTE time history curves for the lap/shoulder belted Hybrid III 6-year-olds in Tests 3, 4, and 5. The maximum NTE and peak neck tension values for the lap/shoulder belted Hybrid III 6-year-old tests are reported in Table 8 for the period of time before chin-to-chest contact. These neck injury values were compared to both the out-of-position and the in-position injury assessment reference values, and are reported as a percentage of those values.

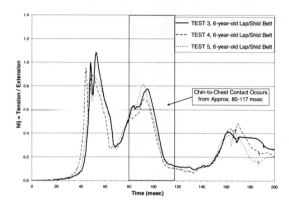

Figure 18: N_{TE} vs. time for the lap/shoulder belted Hybrid III 6-year-old (using in-position intercepts).

TABLE 8: Lap/shoulder belted Hybrid III 6-year-old peak N_{TE}, peak N_{TF} and peak neck tension and the time at which the peak values occurred.

	Before Chin-to-Chest Contact				During Chin-to-Chest Contact			
	Peak N_{TE}		Peak Neck Tens.		Peak N_{TF}		Peak Neck Tens.	
	In-Pos	OOP	In-Pos	OOP	In-Pos	OOP	In-Pos	OOP
IARV	1.0	1.0	1430 N	1130 N	1.0	1.0	1430 N	1130 N
Test 3	108%	122%	56%	71%	102%	111%	127%	161%
	52 msec		48 msec		95 msec		95 msec	
Test 4	95%	107%	65%	83%	92%	99%	112%	142%
	44 msec		44 msec		93 msec		93 msec	
Test 5	87%	98%	54%	68%	109%	118%	133%	169%
	52 msec		58 msec		91 msec		92 msec	

As reported in Table 8 the highest Nij value after the initial chin-to-chest contact occurred was N_{TF} for all three tests. Figure 19 overlays the N_{TF} time history curves for the lap/shoulder belted Hybrid III 6-year-old in Tests 3, 4 and 5.

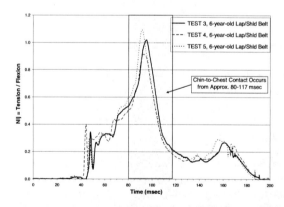

Figure 19: N_{TF} vs. time for the lap/shoulder belted Hybrid III 6-year-old (using in-position intercepts).

The Hybrid III 6-year-old peak neck tension values were highest during chin-to-chest contact. The upper neck z-force (tension/compression) time history responses can be found in Appendix C.

PERCENT RISK OF INJURY

The risk of injury as a result of the measured peak neck tensions was calculated for each of the lap only belted tests and lap/shoulder belted tests using updated injury risk curves specified by Mertz et al. [17]. The IARV for peak neck tension was based on a 3% risk of an AIS $\geq$ 3 injury. In the development of the neck tension IARV, it was assumed the neck muscles of an in-position occupant would be tensed to 80% of their maximum muscle tone [12], and the normalized values for use in applying the risk curves to in-position testing were adjusted to account for 80% muscle tone [17]. Out-of-position neck tension IARV is not adjusted for muscle tension. Tables 9 and 10 report the estimated percent risk of an AIS$\geq$3 neck injury due to neck tension for the lap belted Hybrid III 3-year-old and Hybrid III 6-year-old, respectively. The normalized values used in applying this injury risk curve to both in-position and OOP testing are also given in the tables, and percent risk is calculated for both conditions.

TABLE 9: Lap belted, Hybrid III 3-year-old estimate of % risk of an AIS > 3 injury due to neck tension

	Pk. Neck Tens. (N)	Normalized Tens.		% Risk of AIS>3	
		In-Pos	OOP	In-Pos	OOP
Test 1	1055	0.74	0.93	<0.1%	1%
Test 2	1098	0.77	0.97	<0.1%	2%

TABLE 10: Lap belted, Hybrid III 6-year-old estimate of % risk of an AIS $\geq$ 3 injury due to neck tension

	Pk. Neck Tens. (N)	Normalized Tens.		% Risk of AIS>3	
		In-Pos	OOP	In-Pos	OOP
Test 1	1553	0.82	1.04	<0.1%	7%
Test 2	1621	0.86	1.09	0.1%	13%

Tables 11 and 12 report the estimated percent risk of an AIS$\geq$3 neck injury due to neck tension for the lap/shoulder belted Hybrid III 3-year-old and Hybrid III 6-year-old, respectively. The percent risk of injury was determined using the peak neck tension values measured during the portion of the event before chin-to-chest contact occurred, and also using the peak neck tension values that resulted from chin-to-chest contact. The injury risk curve was again applied using both in-position and OOP normalizing values.

TABLE 11: Lap/Shoulder belted, Hybrid III 3-year-old estimate of % risk of an AIS ≥ 3 injury due to neck tension

| | Before Chin-to-Chest Contact | | | | During Chin-to-Chest Contact | | | | |
| | Pk. Neck | Normalized Tens. | | % Risk of AIS≥3 | | Pk. Neck | Normalized Tens. | | % Risk of AIS≥3 | |
	Tens. (N)	In-Pos	OOP	In-Pos	OOP	Tens. (N)	In-Pos	OOP	In-Pos	OOP
Test 3	1119	0.78	0.99	<0.1%	3%	1924	1.35	1.70	77%	100%
Test 4	1085	0.76	0.96	<0.1%	1.5%	1780	1.24	1.58	49%	100%
Test 5	1022	0.71	0.9	<0.1%	0.5%	1835	1.28	1.62	60%	100%

TABLE 12: Lap/Shoulder belted, Hybrid III 6-year-old estimate of % risk of an AIS > 3 injury due to neck tension

| | Before Chin-to-Chest Contact | | | | During Chin-to-Chest Contact | | | | |
| | Pk. Neck | Normalized Tens. | | % Risk of AIS≥3 | | Pk. Neck | Normalized Tens. | | % Risk of AIS≥3 | |
	Tens. (N)	In-Pos	OOP	In-Pos	OOP	Tens. (N)	In-Pos	OOP	In-Pos	OOP
Test 3	1064	0.56	0.71	<0.1%	<0.1%	2393	1.27	1.61	57%	100%
Test 4	1234	0.65	0.83	<0.1%	<0.1%	2122	1.12	1.42	17%	90%
Test 5	1021	0.54	0.69	<0.1%	<0.1%	2515	1.33	1.69	74%	100%

DISCUSSION

The intent of this study was to compare the neck responses measured from the Hybrid III 3 and 6-year-old ATDs in laboratory testing to injuries sustained by three children in a field crash and investigate the appropriateness of recommended in-position neck injury assessment reference values (IARVs), and the regulated out-of-position IARVs specified in FMVSS 208 for the Hybrid III 3 and 6-year-old ATDs. Such comparisons are relevant only to the extent the test environment simulates the field crash and to the extent the neck forces and moments measured by the Hybrid III 3 and 6-year-old ATDs are humanlike.

The in-position injury assessment reference values and the FMVSS 208 out-of-position limits for neck injury criteria that were approached or exceeded in the belted frontal crash environment simulated were the peak neck tension, and the N_{TE} and N_{TF}. The discussion focuses on these measures of injury potential from the sled tests and the extent to which they simulate the field crash. Peak upper neck compression, N_{CE}, and N_{CF} values were found to be insignificant in this study, and therefore are not further discussed in detail.

The parameters of the field crash are not precisely known, and the sled test environment could not exactly replicate the field crash. However, the MADYMO model indicates the field crash and the sled tests are similar in severity and would be expected to produce similar neck tensions and moments.

NECK RESPONSES - LAP BELTED SLED TESTS

The first and second sled tests used the same test conditions and produced essentially the same results. The shoulder belt was placed behind the backs of both the Hybrid III 3-year-old positioned on the left side of the FMVSS 213 bench seat and the Hybrid III 6-year-old positioned on the right side of the seat.

The head, neck, and torso kinematics of both the lap belted Hybrid III 3-year-old and Hybrid III 6-year-old were similar, except for belt slack in the Hybrid III 6-year-old's belt system that caused the measured dummy responses to shift later in time and delayed the time at which the ATD pitched forward into the restraint system. As the lap belt restrained the lower torso, the upper torso pitched forward before the head rotated significantly resulting in neck extension with relatively small angular bending of the neck. For example, Figures 20 and 21 show the position of the ATD during Test 1 at this time of measured peak neck extension moment.

Figure 20 – Hybrid III 3-year-old lap belted ATD position at the time of peak neck extension moment during Test 1 (75 ms).

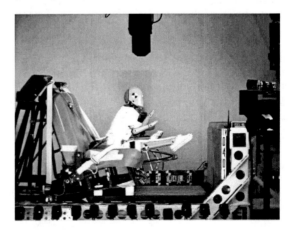

Figure 21 – Hybrid III 6-year-old lap belted ATD position at the time of peak neck extension moment during Test 1 (79 ms).

In all of the lap belted tests with the Hybrid III 3-year-old and 6-year-old ATDs, the neck extension and NTE peaked at the same time. For example, Figure 22 illustrates the relative timing of the neck extension, neck tension, and NTE for the lap belted Hybrid III 6-year-old in Test 1. The NTE curve shown in this figure was calculated using the in-position intercepts. For demonstration of the timing of the occurrence of extension and tension throughout the event, the extension and tension were divided by their respective in-position intercepts for comparison purposes to the timing of the NTE in order to graphically see the contribution of each of these two components to the combined tension-extension NTE. Early in the event the neck extension is more of a contributor to NTE than the neck tension, but neck tension becomes more of a contributor to NTE later in the event.

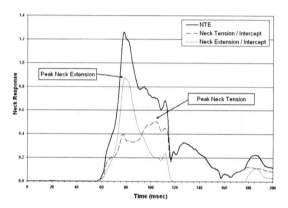

Figure 22 – Lap belted only, Hybrid III 6-year-old comparison of the contributions of extension and tension to N_{TE} for Test 1.

In the lap belted only tests, neck tension peaked later than neck extension for all ATDs, after the head rotated to a position approximately in line with the torso, and the torso was about 20 degrees above the horizontal. For example, Figures 23 and 24 show the position of the ATD during Test 1 at this time of measured peak neck tension. Time history graphs of the upper neck moment about the y-axis (flexion/extension) and upper neck z-force (tension/compression) are provided in Appendix B for the lap belted Hybrid III 3-year-old and Hybrid III 6-year-old tests. Head contact with the lower extremities and head contact with the base of the seat in the case of the Hybrid III 6-year-old, occurred later in the test sequence but did not affect the peak Nij values, neck forces, or neck moments. No chin-to-chest contact occurred in the lap belted tests.

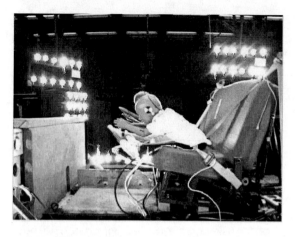

Figure 23 – Hybrid III 3-year-old lap belted ATD position at the time of peak neck tension (91 ms) during Test 1.

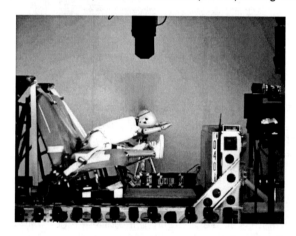

Figure 24 – Hybrid III 6-year-old lap belted ATD position at the time of peak neck tension during Test 1 (103 ms).

The high neck extension values, which produced the peak Nij values, were generated with relatively little head-to-neck motion. This response has been reported by Kang et al. [10] and Agaram et al. [1] during air bag loading of the Hybrid III 5th percentile ATD and was attributed to lack of laxity in the occipital condyle joint in the neck of the Hybrid III 5th percentile ATD. However, Agaram et al. [1] felt this was not an issue in belt restraint testing. The Hybrid III 6-year-old ATD has a similair OC joint to the Hybrid III 5th percentile ATD, and both of these ATD necks contain nodding blocks. Although the Hybrid III 3-year-old neck structure is different from the 6 year-old, both are designed to meet human response corridors, and both ATDs measured high neck extension moments without significant relative head motion in these sled tests. This measured response is likely due to the lack of laxity at the occipital condyle region.

Nusholtz et al. [24] found that no values used in the Nij criterion corresponding to compressive force, extension moment and flexion moment optimally separated the injury and non-injury cases in the biomechanical data set examined. Only the tensile force allowed identification of the injury risk. The peak neck tensions for the tests with the lap belted Hybrid III 3-year-old were 93% and 97% of the out-of-position IARV and 74% and 77% of the in-position IARV. The peak neck tensions for the tests with the lap belted Hybrid III 6-year-old were 104% and 109% of the out-of-position IARV and 82% and 86% of the in-position IARV.

The neck tension IARVs are intended to correspond to a low risk (3%) of AIS 3 injury [17]. The Hybrid III 3-year-old neck tension values from the lap belted Tests 1 and 2 correspond to a risk of AIS = 3 or greater neck injury of 1% and 2% respectively, using the out-of-position criteria and a risk level of less than 0.1% for both tests using the in-position criteria. The Hybrid III 6-year-old neck tension values from the lap belted Tests 1 and 2 correspond to a risk of AIS = 3 or greater neck injury of 7% and 13% respectively, using the out-of-position criteria and a risk level of 0.1% or less for both tests using the in-position criteria.

The two lap belted children in the field crash sustained AIS = 2 neck injuries indicating an incipient condition leading to potentially more severe neck injuries [32]. It is unlikely that the two lap belted children in the rear seat tensed their neck muscles in response to the impending crash. Absent tensing of the neck muscles, the out-of-position IARVs would be the most appropriate for comparison.

The atlanto-occipital dislocations that these children suffered were AIS = 2 injuries. The interpretation of the imaging studies and physical exams revealed no evidence of cord laceration or contusion and a lack of fracture present in the cervical vertebrae. Hemorrhage between the tectorial membrane and clivus does not represent an epidural hemorrhage thus these are considered AIS = 2 injuries. However, the neurosurgeons felt the 3-year-olds AIS =2 injury represented an unstable AO complex and recommended surgical intervention. To the extent the test environment simulates the field crash, the Hybrid III 3 and 6-year-old ATDs and the neck tension IARVs seem consistent with the injuries sustained by the lap belted 3 and 6-year-old children in the field crash.

NECK RESPONSES – LAP/SHOULDER BELTED SLED TESTS

The kinematics of both the lap/shoulder belted Hybrid III 3-year-old and Hybrid III 6-year-old were similar in all lap/shoulder belted Tests 3, 4 and 5. As the lap and shoulder belts restrained the upper torso and lower torso, the head initially translated forward resulting in neck tension and extension with little or no perceivable angular bending of the neck. The combination of tension and extension in the earlier stages of the sled tests produced the highest NTE values near the time the neck extension peaked. Figures 25 and 26 show the positions of the ATDs during Test 3 at the time of measured peak neck extension moment.

Figure 25 – Hybrid III 3-year-old lap/shoulder belted ATD position at the time of peak neck extension during Test 3 (52 ms).

Figure 26 – Hybrid III 6-year-old lap/shoulder belted ATD position at the time of peak neck extension during Test 3 (52 ms).

The head then rotated forward producing neck flexion. The chin of both the Hybrid III 3-year-old and the Hybrid III 6-year-old ATDs ultimately impacted the chest. The chin-to-chest contact increased the chest acceleration and deflection measures at the time of contact and resulted in the highest measured neck tensions and NTF values throughout the lap/shoulder belted sled tests. Figures 27 through 32 show NTF vs. time, resultant chest acceleration vs. time, and chest deflection vs. time for the Hybrid III 3-year-olds and 6-year-olds, respectively. The time the chin is in contact with the chest is indicated by the shaded region.

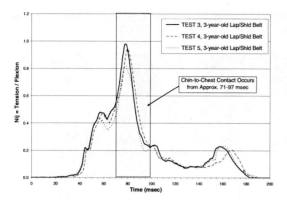

Figure 27: NTF vs. time for the lap/shoulder belted Hybrid III 3-year-old (using in-position intercepts).

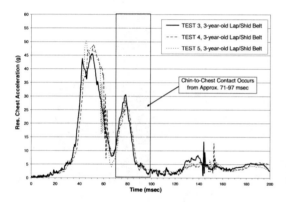

Figure 28: Resultant chest acceleration vs. time for the lap/shoulder belted Hybrid III 3-year-old.

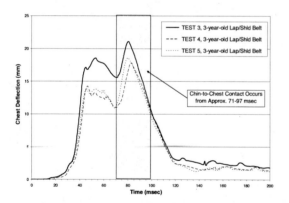

Figure 29: Chest deflection vs. time for the lap/shoulder belted Hybrid III 3-year-old.

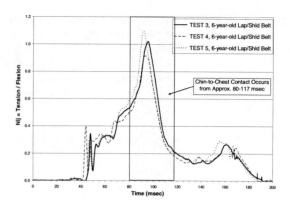

Figure 30: NTF vs. time for the lap/shoulder belted Hybrid III 6-year-old (using in-position intercepts).

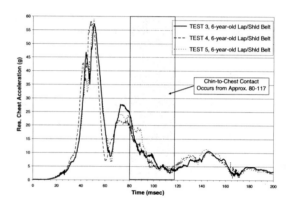

Figure 31: Resultant chest acceleration vs. time for the lap/shoulder belted Hybrid III 6-year-old.

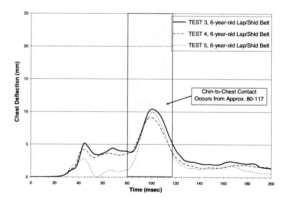

Figure 32: Chest deflection vs. time for the lap/shoulder belted Hybrid III 6-year-old.

Figure 33 shows the NTE for the lap/shoulder belted Hybrid III 6-year-old in Test 3, overlaid with the neck tension response (divided by the in-position neck tension Nij intercept value) and the neck extension response (divided by the in-position neck extension Nij intercept value). This comparison displays the different contributions of neck extension and neck tension throughout the duration of the sled test. Early in the event, neck extension is the primary contributor to NTE,

480

but between approximately 65 and 147 msec, neck tension is the sole contributor to NTE. Chin-to-chest contact occurred during this period of time. This comparison is similar for the other Hybrid III 6-year-olds in Tests 4 and 5, and the Hybrid III 3-year-olds in Tests 4, 5 and 6. The Hybrid III 6-year-old data from Test 3 are shown as a representative example.

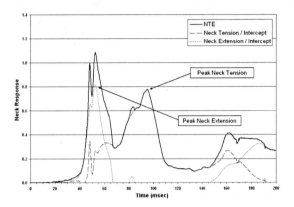

Figure 33 – Lap/shoulder belted, Hybrid III 6-year-old neck responses vs. N_{TE} for Test 3.

The high neck extension values, which produced the peak Nij values, were generated with relatively little or no head-to-neck motion. Figure 26 shows the position of the Hybrid III 6-year-old ATD during Test 3 at the time of maximum recorded neck extension. This ATD response was discussed previously relative to the lap belted tests, and the same response was observed in the lap/shoulder belted tests.

Figure 34 shows the N_{TF} for the lap/shoulder belted Hybrid III 6-year-old in Test 3, overlaid with the neck tension response (divided by the in-position neck tension Nij intercept value) and the neck flexion response (divided by the in-position neck flexion Nij intercept value). This comparison displays the different contributions of neck flexion and neck tension throughout the duration of the sled test. Neck tension is essentially the sole contributor to N_{TF} until approximately 65 msec, and remains a major contributor to N_{TF} throughout the rest of the event. This comparison is similar for the other Hybrid III 6-year-olds in Tests 4 and 5, and the Hybrid III 3-year-olds in Tests 4, 5 and 6. One of the comparisons is shown here as an example.

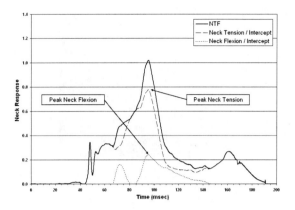

Figure 34 – Lap/shoulder belted, Hybrid III 6-year-old neck responses vs. NTF for Test 3.

In the lap/shoulder belted tests chin-to-chest contact produced the highest NTF values for the event, and a second peak in neck tension that was higher than the neck tension peak before chin-to-chest contact. Figures 35 and 36 below show the point in time during Test 3 where peak neck tension, peak neck flexion, and maximum chin-to-chest interaction occurred for the Hybrid III 3-year-old and Hybrid III 6-year-old, respectively.

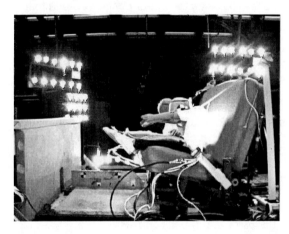

Figure 35 – Hybrid III 3-year-old lap/shoulder belted ATD position at the time of peak neck flexion, peak neck tension, and maximum chin-to-chest interaction during Test 3 (80 ms).

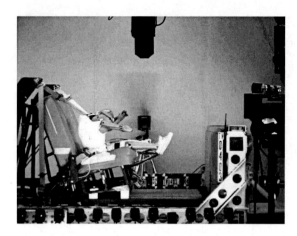

Figure 36 – Hybrid III 6-year-old lap/shoulder belted ATD position at the time of peak neck flexion, peak neck tension, and maximum chin-to-chest interaction during Test 3 (95 ms).

Hybrid III 3-year-old, lap/shoulder belted tests produced peak neck tensions prior to chin-to-chest contact of 99%, 96%, and 90% of the out-of-position IARV and 78%, 76%, and 71% of the in-position IARV. The 3 year-old neck tension values produced prior to chin-to-chest contact correspond to a risk of AIS = 3 or greater neck injury of 0.5% - 3% using the out-of-position criteria that don't incorporate muscle tone, and a risk level of less than 0.1% using the in-position criteria which do consider muscle tone. During chin-to-chest contact, the Hybrid III 3-year-old produced neck tensions of 170%, 158%, and 162% of the out-of-position IARV and 135%, 124%, and 128% of the in-position IARV. The 3-year-old neck tension values produced from chin-to-chest contact correspond to a risk of AIS = 3 or greater neck injury of 100% using the out-of-position criteria and a risk level of 49% - 77% using the in-position criteria.

Hybrid III 6-year-old, lap/shoulder belted tests produced peak neck tensions prior to chin-to-chest contact of 71%, 83%, and 68% of the out-of-position IARV and 56%, 65%, and 54% of the in-position IARV. The 6 year-old neck tension values produced prior to chin-to-chest contact correspond to a risk of AIS = 3 or greater neck injury of less than 0.1% using the out-of-position or in-position criteria. During chin-to-chest contact, the Hybrid III 6-year-old produced neck tensions of 161%, 142%, and 169% of the out-of-position IARV and 127%, 112%, and 133% of the in-position IARV. The 6-year-old neck tension values produced from chin-to-chest contact correspond to a risk of AIS = 3 or greater neck injury of 90% to 100% using the out-of-position criteria and a risk level of 17% - 74% using the in-position criteria.

When the MADYMO model was used to compare the sensitivity of the ATD responses to the sled pulse, it predicted peak neck tensions and N$_{TF}$ in the lap/shoulder belted tests would occur during chin-to-chest contact. Differences between what the model predicted during chin-to-chest contact and the experimental sled results

are probably due to lack of information concerning the actual force deflection properties of the ATD in the area of the chin and the upper chest when contacted by the chin. The MADYMO model uses two ellipsoids, a rigid one for the head and a deformable one for the face, including the chin. Uniform force/deflection properties are defined for the MADYMO face ellipse. The face ellipse force deflection properties were derived from the head drop test to the forehead specified in Part 572 [21, 22]. It is unlikely that the actual force/deflection properties of the chin are the same as the forehead. Furthermore, the model used the force/deflection properties from the pendulum tests to the center of the chest as specified in Part 572 for the entire chest region. Again, it is unlikely that the force/deflection properties of the upper chest, when impacted by the chin are the same as those described in Part 572.

To the extent the test environment simulates the field crash, the lap/shoulder belted, Hybrid III 6-year-old, neck tensions measured during chin-to-chest contact and the predicted risk of AIS = 3 or greater injury seem inconsistent with the AIS = 2 neck injuries sustained by the lap/shoulder belted 7-year-old child in the field crash. The biofidelity of chin-to-chest contact in the Hybrid III child ATDs and its effect on HIC measurements has been noted in Dorel Juvenile Group's comments to the FMVSS 213 docket [3], and by Tylko and Dalmotas [33]. After observing neck loads exceeding published injury thresholds in all tests with a Hybrid III 6-year-old ATD, even though non-contact cervical injuries are rare in correctly restrained children, Sherwood et al. [29], concluded that the stiff thoracic spine of the ATD resulted in non-humanlike kinematics, neck forces, and neck moments. Sherwood et al. [29] also noted that an area of concern was the biofidelity of the chin-to-chest interface. Possible reasons for the stiffer chin-to-chest contact were identified as a stiffer metal chin of dummy, lack of a temporomandibular joint in the dummy, lack of anterior-posterior translation in the dummy's cervical spine, the stiff chest of the dummy, and mass of the ribs in the dummy. Menon et al. [11] observed similar kinematics with the Hybrid III 6-year-old in a high-back-booster seat at a test speed of 56 kph. There was no indication that the chin of the lap-shoulder belted 7-year-old impacted the child's chest with significant force in the field crash. It is not known whether there was contact between the child's chin and chest or to what degree such contact may have contributed to the child's A-O injury.

CONCLUSIONS

Investigating the appropriateness of the in-position and OOP neck IARVs with sled testing was complicated by apparent artifacts associated with the Hybrid III 3 and 6-year-old ATDs. These apparent artifacts significantly affected neck injury values measured with the 3 and 6-year-old Hybrid III ATDs.

In both the lap belted tests and the lap/shoulder belted tests, high neck extension measurements were generated, which produced the peak N_{TE} values without significant relative head-to-neck motion of the Hybrid III neck. Although the neck configurations of the 3 and 6 year-old ATDs are different, the high neck extension phenomena without significant relative head motion observed in these sled tests is likely due to the lack of laxity in the OC joint.

In the lap/shoulder belted tests, high neck tension measurements were generated from the ATDs chin contacting the chest, which produced peak neck tension measurements well in excess of the neck IARVs. Because humanlike kinematics and force deflection properties for belt-restrained chin-to-chest contact have not been defined and ATD force deflection criteria are not specified for chin-to-chest contact [21, 22], the relevance of the neck data measured by the ATD after the point of chin-to-chest contact is in question.

The two lap belted children in the field crash sustained AIS = 2 neck injuries indicating an incipient condition leading to potentially more severe neck injuries. It is unlikely that the two lap belted children in the rear seat tensed their neck muscles in response to the impending crash. To the extent the test environment simulated the field crash, the out-of-position IARVs seemed consistent with the injuries sustained by the lap belted 3 and 6-year-old children in the field crash.

The injuries sustained by the lap/shoulder belted 7-year-old in the field crash, were not consistent with the sled test peak neck tension measurements that occurred during chin-to-chest contact. The neck tension measurements from the ATD that were recorded up to the point of chin-to-chest contact, were consistent with the neck injury sustained by the 7-year-old lap/shoulder belted right front passenger. It is not known whether there was contact between the chin and chest of the lap-shoulder belted 7-year-old in the field crash or to what degree such contact may have contributed to the child's neck injury.

While the Hybrid III 3-year-old and 6-year-old ATDs are the best available tools for assessing child neck injury potential, judgment is needed in interpreting the neck data from tests with belt restraint systems. It is inappropriate to only consider the peak neck measurements without examining the events surrounding those peak values.

ACKNOWLEDGMENTS

TRW Automotive for the financial support for both sled testing and modeling.

DOT/NHTSA for support of the research conducted under the Crash Injury Research and Engineering Network (CIREN) project funded by the U.S. DOT/NHTSA.

The published material represents the position of the authors and not necessarily that of DOT/NHTSA. This document is disseminated under the auspices of the Crash Injury Research and Engineering Network (CIREN). The United States Government assumes no responsibility for the contents or use thereof.

REFERENCES

1. Agaram, V., Kang, J., Nusholtz, G., and Kostyniuk, G., "Hybrid III Dummy Neck Response to Air Bag Loading." Proceedings of the 17th International Technical Conference on the Enhanced Safety of Vehicles, 2000: Paper No. 469. Amsterdam, Holland.

2. Alliance of Automobile Manufacturers. 1999. "Dummy Response Limit for FMVSS 208 Compliance Testing." Annex 2 of the Alliance of Automobile Manufacturers Submission to SNPRM, Docket No. 99-6407, December 23, 1999.

3. Dorel Juvenile Group, 2005. Response to Docket 21247 NPRM, Qualification of HIII-10C Dummy. September 12, 2005.

4. FMVSS 208. 1997. "Occupant Crash Protection." Docket of Federal Regulations 49, Part 571.208, US Government Printing Office, Washington, D.C.

5. FMVSS 208. 2004. "Occupant Crash Protection." Docket of Federal Regulations 49, Part 571.208, US Government Printing Office, Washington, D.C.

6. FMVSS 213. 2003. "Child Restraint Systems." Docket of Federal Regulations 49, Part 571.213. US Government Printing Office, Washington, D.C.

7. Huelke, D., Mackay, G., et al., "A Review of Cervical Fractures and Fracture-Dislocations without Head Impacts Sustained by Restrained Occupants." Accident Analysis and Prevention, 1991. 25(6): p. 731-743.

8. Huelke, D.F., Mackay, G.M., Morris, A., and Bradford, M. "Car Crashes and Non-head Impact Cervical Spine Injuries in Infants and Children." SAE 920562, 1992. Society of Automotive Engineers, Warrendale, PA.

9. Irwin, A.L. and Mertz, H.J., "Biomechanical Bases for the CRABI and Hybrid III Child Dummies." Proceedings of the 41st Stapp Car Crash Conference, 1997: 1-12. Society of Automotive Engineers, Warrendale, PA.

10. Kang, J., Agaram, V., Nusholtz, G.S., and Kostyniuk, G.W., "Air Bag Loading on In-Position Hybrid III Dummy Neck". SAE 2001-01-0179, 2001: p. 69-84.

11. Menon, R., Ghati, Y., Ridella, S., Roberts, D., and Winston, F., "Evaluation of restraint type and performance tested with 3- and 6-year-old Hybrid III dummies at a range of speeds". SAE 2004-01-0319, 2004: p. 59-69.

12. Mertz, H.J. and Patrick, L.M., "Strength and Response of the Human Neck." Proceedings of the 15th Stapp Car Crash Conference, 1971: 207-255. Society of Automotive Engineers, Warrendale, PA.

13. Mertz, H.J. and Weber, D.A., "Interpretations of the Impact Responses of a 3-Year-Old Child Dummy Relative to Child Injury Potential." Proceedings of the Ninth International Technical Conference on Experimental Safety Vehicles, 1982: 368-376. Kyoto, Japan. (Also published as SAE 826048, SP-736).

14. Mertz, H.J., Driscoll, G.D., Lenox, J.B., Nyquist, G.W., and Weber, D.A., "Responses of animals exposed to deployment of various passenger inflatable restraint system concepts for a variety of collision severities and animal positions." Proceedings of the Ninth International Technical Conference on Experimental Safety Vehicles, 1982: 352-368. Kyoto, Japan. (Also published as SAE 826047 PT31).

15. Mertz, H.J., Prasad, P., and Irwin, A.L., "Injury risk curves for children and adults in frontal and rear collisions." SAE 973318, 1997: p. 13-30.

16. Mertz, H.J. and Prasad, P., "Improved neck injury risk curves for tension and extension moment measurements of crash dummies." Stapp Car Crash Journal, 2000. 44: p. 59-75.

17. Mertz, H.J., Irwin, A.L., and Prasad, P., "Biomechanical and Scaling Bases for Frontal and Side Impact Injury Assessment Reference Values." Stapp Car Crash Journal, 2003. 47: p. 155-188.

18. National Highway Traffic Safety Administration. 1998. "Notice of Proposed Rule Making (NPRM) for Advanced Airbags." Federal Register (FR), Vol. 63, No. 181, September 18, 1998.

19. National Highway Traffic Safety Administration. 2000. "Final Rule on Advanced Airbags." Federal Register (FR), Vol. 65, 30680, May 12, 2000.

20. National Highway Traffic Safety Administration, "Proposed Amendment to FMVSS No 213, Frontal Test Procedure, Part III, Proposed Dummies, Injury Criteria and Other Changes". 2002, U.S. Department of Transportation: Washington, DC.

21. National Highway Traffic Safety Administration, "Laboratory Test Procedure for FMVSS 208 Occupant Crash Protection (TP208-12)", in Appendix C, Part 572 Subpart N (6-Year-Old) Dummy Performance Calibration Test Procedure. 2003, U.S. Department of Transportation: Washington, DC. p. C1-C45.

22. National Highway Traffic Safety Administration, "Laboratory Test Procedure for FMVSS 208 Occupant Crash Protection (TP208-12)", in Appendix D, Part 572, Subpart P (3 Year-Old) Dummy Performance Calibration Test Procedure. 2003, U.S. Department of Transportation: Washington, DC. p. D1-D40.

23. Newman, J.A. and Dalmotas, D., "Atlanto-occipital fracture dislocation in lap-belt restrained children". SAE 933099, 1993: p. 165-171.

24. Nusholtz, G.S., Di Domenico, L., Shi, Y., and Eagle, P., "Studies of Neck Injury Criteria Based on Existing Biomechanical Test Data." Accident Analysis & Prevention, 2003. 35: p. 777-786.

25. SAE J1733. 1994. "Sign Convention for Vehicle Crash Testing." Society of Automotive Engineers, Warrendale, PA.

26. SAE J211. 1995. "Instrumentation for Impact Test – Part 1 – Electronic Instrumentation." Society of Automotive Engineers, Warrendale, PA.

27. SAE J2052. 1997. "Test Device Head Contact Duration Analysis." Society of Automotive Engineers, Warrendale, PA.

28. SAE PT-44. 1994. "Hybrid III – The First Humanlike Crash Test Dummy." Society of Automotive Engineers, Warrendale, PA.

29. Sherwood, C.P., Shaw, C.G., Van Rooij, L., Kent, R.W., Gupta, P.K., Crandall, J.R., Orzechowski, K.M., Eichelberger, M.R., and Kallieris, D., "Prediction of cervical spine injury risk for the 6-year-old child in frontal crashes". 46th Annual Proceedings AAAM, 2002.

30. Sochor, M.R., Faust, D.P., Garton, H., Ridella, S.A., Barnes, S., Fischer, K., and Wang, S.C., "Simulation of Occipitoatlantoaxial Injury: Utilizing a MADYMO Model". SAE 2004-01-0326, 2004: p. 1-9.

31. Stalnaker, R.L., "Spinal Cord Injuries to Children in Real World Accidents". SAE 933100, 1993: p. 173-183.

32. Sun, P.P., Poffenbarger, G.J., Durham, S., and Zimmerman, R.A., "Spectrum of occipitoatlantoaxial injury in young children". J Neurosurg, 2000. 93(1 Suppl): p. 28-39.

33. Tylko, S. and Dalmotas, D., "Protection of Rear Seat Occupants in Frontal Crashes." Experimental Safety Vehicles Conference, 2005. Paper 05-258. Washington, D.C.

34. Wolanin, M.J., Mertz, H.J., Nyznyk, R.S., and Vincent, J.H., "Description and Basis of a Three-Year-Old Child Dummy for Evaluating Passenger Inflatable Restraint Concepts." Proceedings of the Ninth International Technical Conference on Experimental Safety Vehicles, 1982: 287-299. Kyoto, Japan.

APPENDIX A

1. Resultant external head impact force calculation [26]:

To calculate the inertia loads on the head*:

$$F_{x_{inertiaload}} = ma_{x_{head}}$$

$$F_{y_{inertiaload}} = ma_{y_{head}}$$

$$F_{z_{inertiaload}} = ma_{z_{head}}$$

*A Hybrid III 3-year-old ATD head mass of 2.63 kg, and a Hybrid III 6-year-old ATD head mass of 3.19 kg was used in the calculation, which represented the masses of the heads above the measurement strain gage in the upper neck load cell.

To calculate the head impact forces:

$$F_{x_{head}} = F_{x_{inertiaload}} - F_{x_{upperneck}}$$

$$F_{y_{head}} = F_{y_{inertiaload}} - F_{y_{upperneck}}$$

$$F_{z_{head}} = F_{z_{inertiaload}} - F_{z_{upperneck}}$$

To calculate the resultant head impact force:

$$F_{reshead} = \sqrt{(F_{xhead})^2 + (F_{yhead})^2 + (F_{zhead})^2}$$

2. Calculation of the translated moment about the y-axis of the occipital condyle for the Hybrid III 6-year-old ATD [24]:

The IARV for neck moments was based about the occipital condyle, however, for the Hybrid III 6-year-old ATD, the moment is not measured about the same location. The following equations allow for the moment to be translated from the point of measurement to where the IARV is applicable (the corrected moment). The measurements are at the correct location for the Hybrid III 3-year-old, so the moment does not need to be corrected for this ATD.

$$M_{OCy} = M_{MeasuredY} - (D * F_x)$$

Where,

M_{OCy} = the corrected moment about the y-axis (Nm)

F_x = the measured force output in the x-direction (N)

$M_{MeasuredY}$ = measured y-axis moment output (Nm)

D = vertical distance between the axis of the load cell and the axis of the condyle (0.01778 m)

3. Calculation of Nij – NTE and NTF [2, 18, 17, 14, 15, 16]:

$$N_{TE} = \frac{F_z}{F_{TENS}} + \frac{M_{OCy}}{M_{EXT}}$$

and

$$N_{TF} = \frac{F_z}{F_{TENS}} + \frac{M_{OCy}}{M_{FLEX}}$$

where F_{TENS}, M_{EXT} and M_{FLEX} intercept values are as listed in Table 2.

APPENDIX B

Lap Belted Sled Test Time History Overlays: Head x, y, z, and resultant accelerations; Upper neck Fx, Fz, and My; Chest x, y, z, and resultant accelerations; and Chest deflections (Test 1 is represented by a solid line and Test 2 by a dashed line).

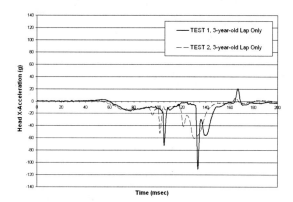

Figure B1: Head x-acceleration vs. time for the Hybrid III 3-year-old lap belted tests.

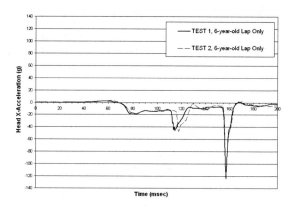

Figure B2: Head x-acceleration vs. time for the Hybrid III 6-year-old lap belted tests.

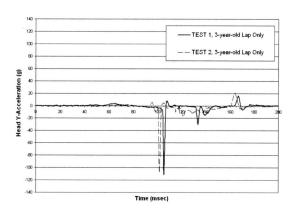

Figure B3: Head y-acceleration vs. time for the Hybrid III 3-year-old lap belted tests.

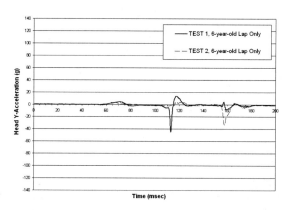

Figure B4: Head y-acceleration vs. time for the Hybrid III 6-year-old lap belted tests.

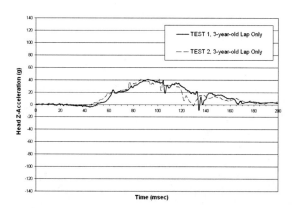

Figure B5: Head z-acceleration vs. time for the Hybrid III 3-year-old lap belted tests.

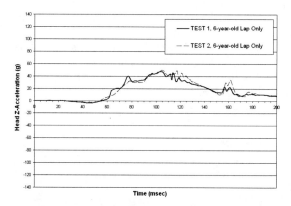

Figure B6: Head z-acceleration vs. time for the Hybrid III 6-year-old lap belted tests.

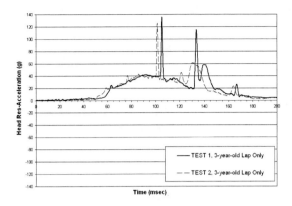

Figure B7: Head resultant acceleration vs. time for the Hybrid III 3-year-old lap belted tests.

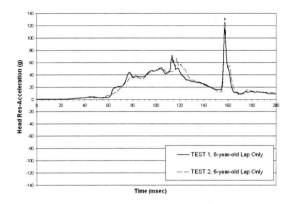

Figure B8: Head resultant acceleration vs. time for the Hybrid III 6-year-old lap belted tests.

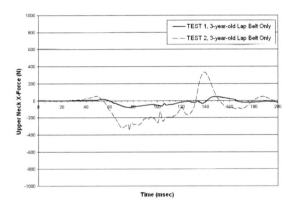

Figure B9: Upper neck x-force vs. time for the Hybrid III 3-year-old lap belted tests.

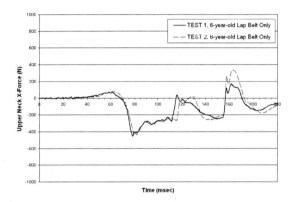

Figure B10: Upper neck x-force vs. time for the Hybrid III 6-year-old lap belted tests.

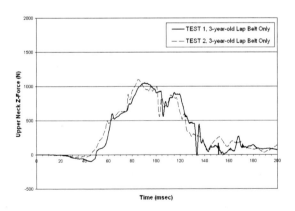

Figure B11: Upper neck z-force (tension/compression) vs. time for the Hybrid III 3-year-old lap belted tests.

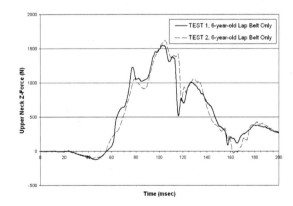

Figure B12: Upper neck z-force (tension/compression) vs. time for the Hybrid III 6-year-old lap belted tests.

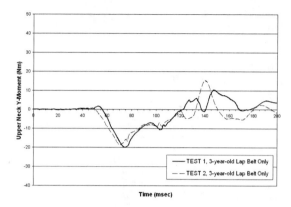

Figure B13: Upper neck y-moment (flexion/extension) vs. time for the Hybrid III 3-year-old lap belted tests.

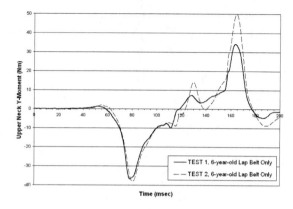

Figure B14: Upper neck y-moment (flexion/extension) vs. time for the Hybrid III 6-year-old lap belted tests.

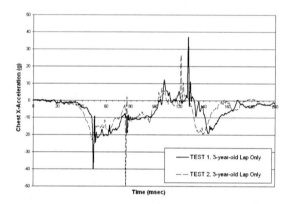

Figure B15: Chest x-acceleration vs. time for the Hybrid III 3-year-old lap belted tests.

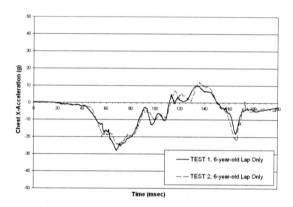

Figure B16: Chest x-acceleration vs. time for the Hybrid III 6-year-old lap belted tests.

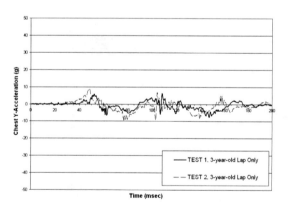

Figure B17: Chest y-acceleration vs. time for the Hybrid III 3-year-old lap belted tests.

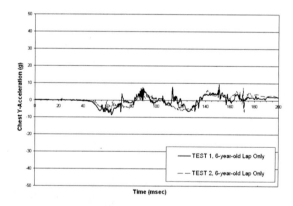

Figure B18: Chest y-acceleration vs. time for the Hybrid III 6-year-old lap belted tests.

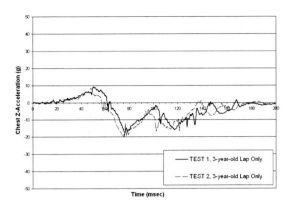

Figure B19: Chest z-acceleration vs. time for the Hybrid III 3-year-old lap belted tests.

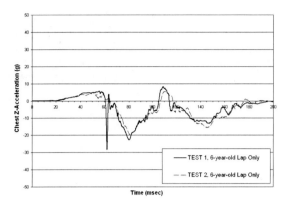

Figure B20: Chest z-acceleration vs. time for the Hybrid III 6-year-old lap belted tests.

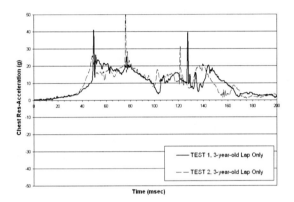

Figure B21: Chest resultant acceleration vs. time for the Hybrid III 3-year-old lap belted tests.

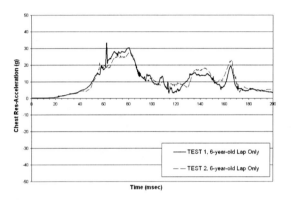

Figure B22: Chest resultant acceleration vs. time for the Hybrid III 6-year-old lap belted tests.

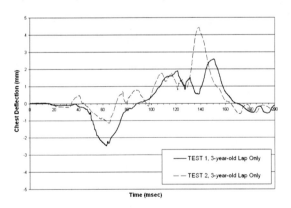

Figure B23: Chest Deflection vs. time for the Hybrid III 3-year-old lap belted tests.

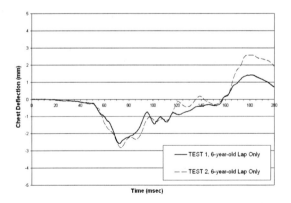

Figure B24: Chest Deflection vs. time for the Hybrid III 6-year-old lap belted tests.

APPENDIX C

Lap/Shoulder Belted Sled Test Time History Overlays: Head x, y, z, and resultant accelerations; Upper neck Fx, Fz, and My; Chest x, y, z, and resultant accelerations; and Chest deflections (Test 3 is represented by a solid line, Test 4 by a dashed line, and Test 5 by a dotted line).

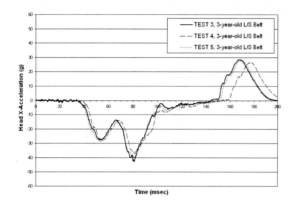

Figure C1: Head x-acceleration vs. time for the Hybrid III 3-year-old lap/shoulder belted tests.

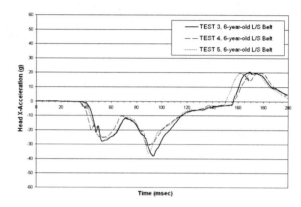

Figure C2: Head x-acceleration vs. time for the Hybrid III 6-year-old lap/shoulder belted tests.

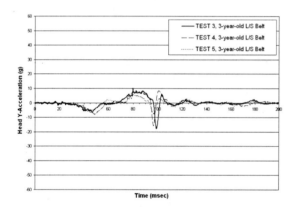

Figure C3: Head y-acceleration vs. time for the Hybrid III 3-year-old lap/shoulder belted tests.

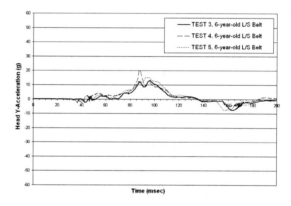

Figure C4: Head y-acceleration vs. time for the Hybrid III 6-year-old lap/shoulder belted tests.

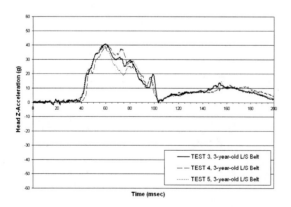

Figure C5: Head z-acceleration vs. time for the Hybrid III 3-year-old lap/shoulder belted tests.

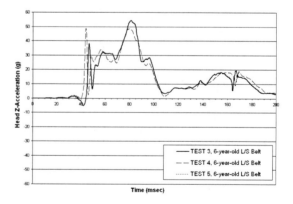

Figure C6: Head z-acceleration vs. time for the Hybrid III 6-year-old lap/shoulder belted tests.

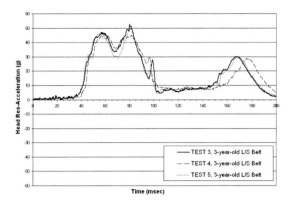

Figure C7: Head resultant acceleration vs. time for the Hybrid III 3-year-old lap/shoulder belted tests.

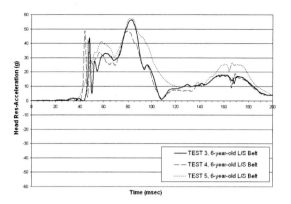

Figure C8: Head resultant acceleration vs. time for the Hybrid III 6-year-old lap/shoulder belted tests.

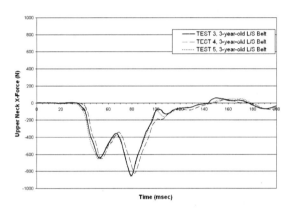

Figure C9: Upper neck x-force vs. time for the Hybrid III 3-year-old lap/shoulder belted tests.

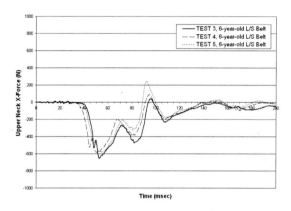

Figure C10: Upper neck x-force vs. time for the Hybrid III 6-year-old lap/shoulder belted tests.

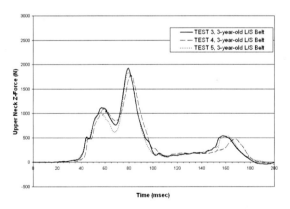

Figure C11: Upper neck z-force (tension/comp) vs. time for the Hybrid III 3-year-old lap/shoulder belted tests.

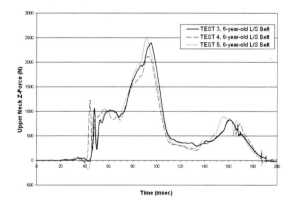

Figure C12: Upper neck z-force (tension/comp) vs. time for the Hybrid III 6-year-old lap/shoulder belted tests.

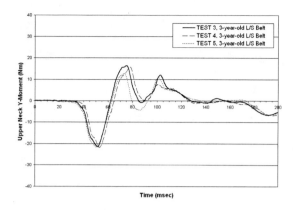

Figure C13: Upper neck y-moment (flexion/extension) vs. time for the Hybrid III 3-year-old lap/shoulder belted tests.

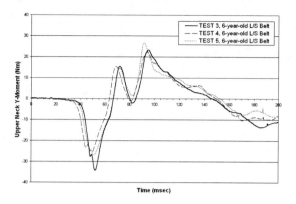

Figure C14: Upper neck y-moment (flexion/extension) vs. time for the Hybrid III 6-year-old lap/shoulder belted tests.

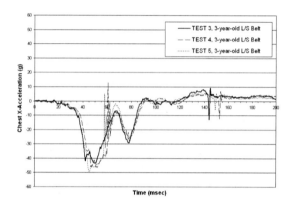

Figure C15: Chest x-acceleration vs. time for the Hybrid III 3-year-old lap/shoulder belted tests.

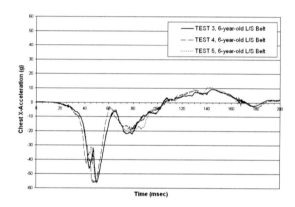

Figure C16: Chest x-acceleration vs. time for the Hybrid III 6-year-old lap/shoulder belted tests.

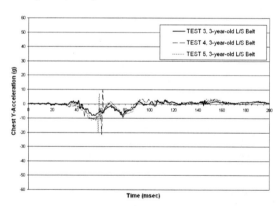

Figure C17: Chest y-acceleration vs. time for the Hybrid III 3-year-old lap/shoulder belted tests.

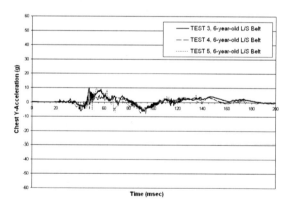

Figure C18: Chest y-acceleration vs. time for the Hybrid III 6-year-old lap/shoulder belted tests.

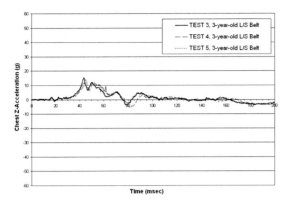

Figure C19: Chest z-acceleration vs. time for the Hybrid III 3-year-old lap/shoulder belted tests.

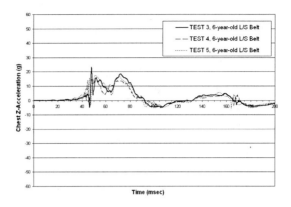

Figure C20: Chest z-acceleration vs. time for the Hybrid III 6-year-old lap/shoulder belted tests.

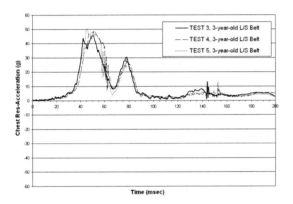

Figure C21: Chest resultant acceleration vs. time for the Hybrid III 3-year-old lap/shoulder belted tests.

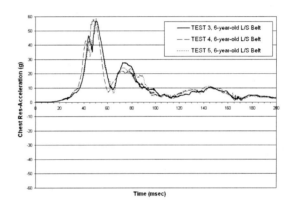

Figure C22: Chest resultant acceleration vs. time for the Hybrid III 6-year-old lap/shoulder belted tests.

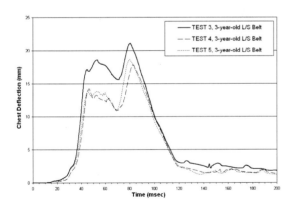

Figure C23: Chest Deflection vs. time for the Hybrid III 3-year-old lap/shoulder belted tests.

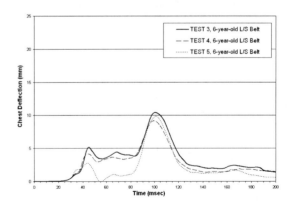

Figure C24: Chest Deflection vs. time for the Hybrid III 6-year-old lap/shoulder belted tests.

Responses of Human Surrogates to Simulated Rear Impact: Velocity and Level Dependent Facet Joint Kinematics

Brian D. Stemper, Narayan Yoganandan and Frank A. Pintar
Department of Neurosurgery
Medical College of Wisconsin and VA Medical Center

ABSTRACT

The objective of the present study was to determine the kinematics of the human head-neck complex with specific reference to posterior facet joints as a function of rear impact acceleration. Six intact human head-neck complexes were prepared by fixing the first thoracic vertebra in polymethylmethacrylate. The specimens were oriented such that the Frankfurt plane was horizontal and the cervico-thoracic disc was at an angle of 25 degrees to simulate the normal driving position. Retroreflective targets were inserted to the cervical vertebrae. The specimens were subjected to simulated rear impact accelerations using a minisled apparatus. A series of tests were conducted with velocities of 2.1, 4.6, 6.6, 9.3, and 12.4 km/h. In this study, to achieve the objective, results are presented on the facet joint motions at the C4-5, C5-6, and C6-7 levels as a function of change in velocity. Data were extracted from high-resolution high-speed video photography such that the compressive and sliding motions of the facet joints at the three levels were quantified and compared using statistical techniques. Results indicated that the cervical facet joints demonstrate compression of the most dorsal regions accompanied by distractions of the most ventral regions. Data demonstrated monotonically increasing variations of peak compression and peak sliding kinematics with increasing changes in velocity. However, C5-6 facet joints responded with higher magnitudes of compression at higher changes in velocity than their adjacent counterparts, but lower magnitudes at lower changes in velocity. These data may be of value in the determination of local changes in spinal component kinematics secondary to varying changes in rear impact acceleration.

INTRODUCTION

Biomechanics of the human head-neck complex due to rear end crashes has been studied using experimental models. They have included static loading of ligamentous columns, isolated cervical spine units, and whole-body specimens [10-15]. Although intact whole-body specimens provide information on the kinematics of the head-spine as an intact entity, it is difficult to discern the local motions of the various intervertebral joints that contribute to the mechanics of the human cervical spinal column. This is primarily due to the difficulties in resolving the motions with sufficient accuracy as the joints of the cervical spine have minimal dimensions (e.g., facet joints). Studies using segmented units of the cervical spine, on the other hand, are incapable of accommodating the curvature and segmental changes of the cervical column due to impact acceleration [4, 9]. Although these models have attempted to provide answers to questions such as the role of the thoracic spine on cervical kinetics (from whole-body specimens), it is important to delineate local spinal component kinematics (facet joint motions as a function of spinal level) as a function of change in velocity. Such evaluations may have implications in the analysis of head-neck kinematics secondary to rear impact as it is well known that motions indirectly determine the local integrity of spinal components [5, 14]. For example, excessive compressions of the facet joints may result in cartilage-to-cartilage contact that result in a compromise of its mechanical integrity, and these lead to changes in normal physiological responses [1-3, 6-8]. The quantification of the velocity-dependent kinematics of these posterior and well-innervated structural components of the spine could lead to the establishment of thresholds of spinal motions in rear impact.

Therefore, the objective of the present study was to determine the kinematics of the human head-neck complex with specific reference to posterior facet joints as a function of rear impact acceleration.

SPECIMENS

Six unembalmed human cadaver head-neck complexes were used in this study. Four males and two female specimens were used for achieving the objective of the study. Prior to head-neck isolation, each specimen was screened for HIV and Hepatitis A, B, and C. Specimens were also evaluated for pre-existing trauma or bone degeneration using radiography and medical records. The anatomical orientation, defined using the right hand coordinate system, was as follows.

+x direction = posterior to anterior
+y direction = right to left lateral
+z direction = inferior to superior

Positive rotations were determined using the right hand rule. Positive x, y, and z rotations represented right lateral bending, forward flexion, and left axial rotation, respectively.

Intact head-neck complexes were isolated and prepared for testing in the following manner. The head-neck complex was surgically removed at the second thoracic vertebral level. To enable rigid mounting of the complex to the minisled

apparatus and facilitate proper head-neck orientation during testing, each complex was mounted at the first thoracic vertebra in polymehtylmethacrylate (PMMA) and oriented at +25 degrees about the y-axis. The head-neck complex remained intact with the exception of skin and muscle partially transected on the right lateral side. This procedure was performed to insert retroreflective targets.

Targets were inserted to track relative rigid body displacements and rotations of the bony elements of the head-neck complex (Figure 1). Two targets were placed along the Frankfurt Plane (line passing through the auditory meatus and inferior orbit). Targets were placed in the anterior vertebral body and superior transverse process of each of the seven cervical vertebrae to track overall motions. Targets were also placed to outline the facet joints at C4-5, C5-6, and C6-7 levels. Two targets were placed along the inferior and superior facet joint surfaces to define relative facet joint motions. Insertion of these targets, which required minimal tissue exposure (approximately 25-40 mm width) on the lateral side, did not affect the integrity of the head-neck complex. In addition, four targets were placed on the PMMA to correspond to the positive x displacement of the thoracic vertebra.

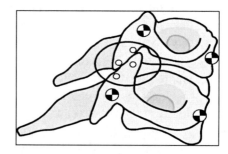

Figure 1: Placement of Retroreflective Targets on Cervical Vertebrae

TEST SETUP

All tests were performed implementing a pendulum/minisled apparatus as shown in Figure 2. The pendulum was suspended from the ceiling using four cables and allowed to accelerate in the global positive x direction [13]. Energy absorbing material was placed in front of the impacting edge of the pendulum to shape the impact pulse. The minisled apparatus consisted of a sled on two ground rails with a 2.5 m runoff on precision linear bearings.

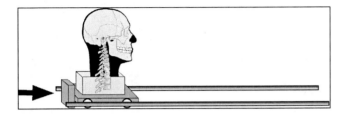

Figure 2: Test Setup

The input pulse to the specimen was modeled to mimic the positive x acceleration at the first thoracic vertebra from previous rear impact full-body cadaver tests performed in our laboratory. Modeling criteria were pulse shape and pulse width. Pulse amplitude was scaled to the specific values of this experiment because the previous rear impact study was run at 15 km/h. Initial tests were performed to obtain the desired input pulse to the first thoracic vertebra. It was desired to obtain consistent changes in positive x velocity of the minisled, which is equal to change in velocity of the first thoracic vertebra. A 50th percentile Hybrid III dummy head-neck complex was used for these initial pendulum tests. It was desired to obtain repeatable ΔV (change in velocity) values in the range of 2, 4, 6, 9, and 12 km/h. The initial height of the impactor (affecting pendulum velocity on impact) and energy absorbing material on the impactor (affecting pulse shape) were altered to obtain correct input pulses. Input ΔV pulses were obtained by integrating minisled x accelerations recorded from minisled accelerometers. Final input pulses, determined by change in velocity, pulse width, and initial acceleration of the sled, consisted of ΔV values of 2.1, 4.6, 6.6, 9.3, and 12.4 km/h.

PENDULUM TESTING

For each specimen, a series of tests was performed consisting of ten runs at different changes in velocity. A 2.1 km/h test was performed twice at the start of the experiment to describe baseline motions then once between each higher velocity test for the purpose of comparison. At the start of each run, the specimen was positioned with the Frankfurt plane in a horizontal position. The pendulum was then lifted to the test-specific height, determined from the tests using Hybrid III dummy, and released. Immediately after each test, the input pulses were analyzed for consistency between ΔV, pulse width, and maximum x acceleration. Between tests, the specimen was inspected for injury, which included visual inspection and radiography. Specifically, the anterior longitudinal ligament was evaluated for tears, facet and disc joints were evaluated for rupture, and vertebral bodies were examined for fracture. After the completion of 12.4 km/h tests, a final x-ray was taken.

DATA ACQUISITION

Sixteen channels of data were acquired at a sampling rate of 12500 Hertz. This data included local x force and acceleration of the impactor. Moments about the three axes and forces in the three directions were measured using a six-axis load cell attached to the base of the PMMA. Two accelerometers were placed on the minisled for the purpose of measuring positive x acceleration of the minisled. Local rotational velocities about the three axes and linear accelerations in the three directions were obtained from a dynacube (ATA Sensors) mounted to the left lateral side of the skull. High-speed video coverage was obtained using a Kodak high-resolution video system. Digital video data were obtained at 1000 Hertz and at a resolution of 512 X 512 pixels. Sigma Scan 5.0 software (Jandel Scientific Software) was used to digitize motion of the reflective targets.

DATA ANALYSIS

Motion analysis data were obtained at 1000 Hertz from the retroreflective targets photographed using the high-resolution camera.

The motion of each target was mapped in the sagittal plane using the kinematic software package. This software package allows the user to point and click on the reflective targets over a series of digital images. At each click, it obtains the pixel position from the 512X512 digital grid. Then a real space conversion factor, obtained by correlating the actual size of a landmark to its digitized size in pixels, was used to convert pixel position into real space position. A minimum of 140 msec of data were obtained for high velocity ($\Delta V=12.4$ km/h) runs, and a maximum of 400 msec of data were obtained for low velocity ($\Delta V=2.1$ km/h) runs. Data were then organized into x and z positions of each retro-reflective target versus time.

Since analysis of facet joint kinematics was the focus, a local coordinate system was established for each facet joint target with the positive x-direction in the posterior-anterior direction of the inferior facet joint surface, and the positive z-direction in the inferior-superior direction perpendicular to the positive x-direction. The lower facet joint surface was assumed to lie parallel to the line through the two inferior facet joint targets. This was confirmed by radiography. The local coordinate system changed with the lower facet joint surface at every facet joint level; the origin always remained at the posterior-inferior target, and the positive x-direction through the inferior facet joint targets. The local coordinate system is shown in Figure 3.

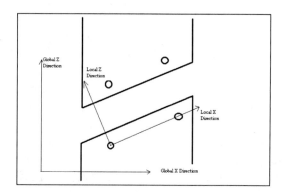

Figure 3: Local Coordinate System of the Facet Joint

The equations used to transform the coordinates from the global system to the local system are as follows.

$$X_L = X_G\cos(\alpha) + Z_G\cos(90 - \alpha)$$
$$Z_L = X_G\cos(90 + \alpha) + Z_G\cos(\alpha)$$

The local and global y directions remained the same in this transformation, although the origin moved to the posterior inferior facet joint target. The angle α accounts for the angle from the global coordinate system to the local coordinate system. Once data were transformed into the local coordinate system, motion analysis was performed. Motions of the targets were separated into compression/ distraction and sliding. All motions were recorded as the anterior/posterior superior target with respect to the anterior/posterior inferior target. For instance, posterior compression was obtained by posterior-superior target decreasing local z-axis distance to the posterior-inferior target. Likewise, posterior sliding was obtained by a change in local x distance between the posterior superior target and the posterior inferior target. Positive sliding was defined as superior target moving in the local positive x-direction with respect to the inferior target. For consistency between tests, facet joint motion data was collected at the time of the end of the input pulse. This time ranged from 103 to 110 msec after the impact.

STATISTICAL ANALYSIS

Statistical analysis was performed using Statview 5.0 software (Abacus Concepts). Compression/distraction and sliding at the three levels were analyzed using linear regression and analysis of variance (ANOVA) techniques. Linear regression analysis was used to evaluate trends in the data. Mean values of anterior and posterior compression/distraction and sliding kinematics were evaluated as a function of ΔV. Linear regressions were performed on the data with ΔV as the independent variable and the specific facet joint motion (posterior compression/ distraction, anterior sliding, etc.) as the dependent variable. The statistic of interest in the linear regression analysis was the R^2 value. The regression equation was forced through zero. ANOVA statistics was used to evaluate trends between facet joint motion and either vertebral level or change in velocity. The statistic of interest in the ANOVA was the p-value.

MOTION ANALYSIS

This discussion is divided into compression/distraction (local z) and sliding motions (local x) of the facet joints. For all three levels (C4-5, C5-6, and C6-7), anterior motions demonstrated distraction and posterior motions illustrated compression (Figures 4, 5). For the ensemble, mean compression/ distraction motion increased with increasing input ΔV with the exception of C5-6 anterior motion and C6-7 posterior motion. In both of these cases, the magnitude of motion reached a maximum at 6.6 km/h ΔV.

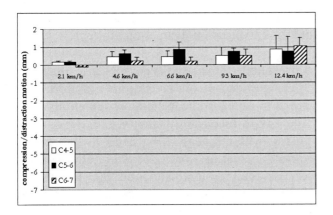

Figure 4: Mean Anterior Compression/Distraction Facet Joint Motion (Distraction is Positive)

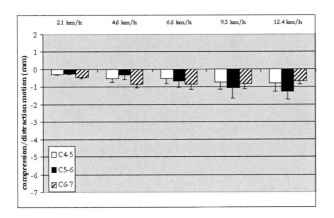

Figure 5: Mean Posterior Compression/Distraction Facet Joint Motion (Compression is Negative)

Sliding motion of the facet joints exhibited increasing motion with increasing input ΔV (Figures 6 and 7). Motion of the C4-5 facet joint had lower magnitude than that of C5-6 and C6-7 joints, which were approximately equal.

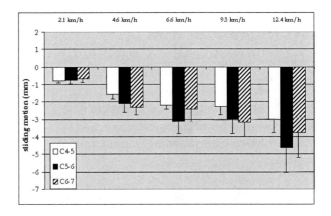

Figure 6: Mean Posterior Sliding Facet Joint Motion

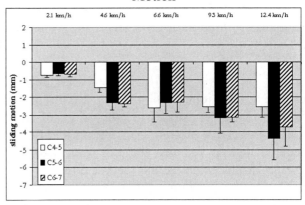

Figure 7: Mean Anterior Sliding Facet Joint Motion

REGRESSION ANALYSIS

Four regression analyses were performed on the data with input ΔV as the independent variable and facet joint motions as dependent variables. The data were split by spinal level. This resulted in R^2 values for each of the four motions at all three investigated levels. The results of this analysis are given in Table 1.

Table 1: Linear Regression Output

Motion	C4-5		C5-6		C6-7	
	R^2	slope	R^2	slope	R^2	slope
PX	0.978	-0.264	0.982	-0.381	0.977	-0.338
PZ	0.959	-0.074	0.992	-0.103	0.808	-0.086
AX	0.942	-0.261	0.988	-0.359	0.975	-0.331
AZ	0.974	0.069	0.898	0.083	0.871	0.063

PX = Posterior Sliding
PZ = Posterior Compression/Distraction
AX = Anterior Sliding
AZ = Anterior Compression/Distraction

The linear regression analysis illustrates that the facet joint motions increase linearly with change in velocity. All of the R^2 values are above 0.900 with the exception of AZ at C5-6, PZ at C6-7 and AZ at C6-7. In general, R^2 values for compressive motion at C4-5 and C5-6 are greater than that of C6-7.

The slope statistic indicates the rate of change of facet joint motion due to changes in ΔV. Sliding motion illustrates greater values of slope than compressive motion. This means that for the same change in velocity, the joint will undergo a greater displacement in the local x direction than in the local z direction. This is also illustrated by the slope data is a level dependence. Motion of the C5-6 facet joint is greater than that of C4-5 and C6-7 for all four motions. For all anterior compressive motion, C6-7 motion is greater than that of C4-5. In the figure, notice that anterior compressive motion is positive, indicating distraction, and that posterior compressive motion is negative, indicating compression.

ANOVA ANALYSIS

The first ANOVA test performed illustrated the effects of the two factors, spinal level and ΔV, on facet joint motions. This was conducted to determine whether the two factors were interrelated. This resulted in a p-value of 0.2150, indicating that the two factors were not interrelated. Therefore, independent ANOVA analyses between facet joint level and facet joint motions and ΔV and facet joint motions were performed.

In the analysis of the variability of facet joint motions based on facet joint level, the independent variable was facet joint level and the dependent variable was facet joint motion. Independent analyses were performed for each of the four facet joint motions; i.e., posterior compression/distraction, posterior sliding, anterior compression/distraction, and anterior sliding. The data was then split by ΔV, resulting in an analysis of the facet joint motion based on spinal level at a specific velocity. Results of this analysis illustrated little significance ($P<0.05$) between level and facet joint motion, with the exception of the lowest change in velocity for posterior and anterior compression/distraction motions.

Though the ANOVA analysis illustrated little significance of facet joint motion with respect to level, some general observations were made. With increasing velocity, relative motions of C5-6 (with respect to C4-5 and C6-7) increased for both anterior and posterior sliding, and posterior compression/distraction. For instance, at 2.1 km/h the magnitude of posterior compression/ distraction of C5-6 is smaller than C4-5 and C6-7. However at 12.4 km/h the posterior compression/ distraction motion of C5-6 is greater than C4-5 and C6-7. This phenomenon was discovered with both posterior sliding and anterior siding. This trend is illustrated in Figure 8 for posterior compression/distraction. The opposite trend was noted for anterior compression/ distraction. With increasing velocity, the anterior compression/ distraction of C5-6 became smaller with respect to the compression/ distraction of C4-5 and C6-7.

Results of the ANOVA test with velocity as the independent variable and facet joint motion as the dependent variable are illustrated in Table 2. In this table, P1 indicates a p-value less than 0.05 and P2 indicates a p-value less than 0.10.

The variability in posterior and anterior sliding was significantly affected by change in velocity. The compression/distraction data in both the anterior and posterior of the facet joint illustrated a less significant relationship between velocity and facet joint motion (Table 2). The only significant compressive motions occurred at C5-6 in the posterior and C6-7 in the anterior region of the cervical spine.

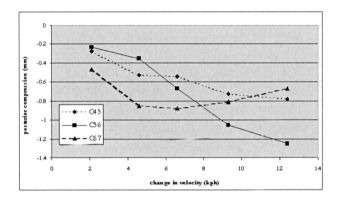

Figure 8: Posterior Compression/Distraction with Respect to Level

Table 2: ANOVA Analysis p-values
(Factor = Velocity)

Level	PX	PZ	AX	AZ
C4-5	P1	NS	P1	NS
C5-6	P1	P1	P1	P2
C6-7	P1	NS	P1	P1

PX = Posterior Sliding
PZ = Posterior Compression/Distraction
AX = Anterior Sliding
AZ = Anterior Compression/Distraction
P1 < 0.01; P2 < 0.05,
NS = not significant

CONCLUSIONS

1. With increasing input velocity, anterior facet joint distraction and posterior facet joint compression/distraction generally increased.
2. Facet joint sliding motion increased with increasing input velocity.
3. Regression analysis generally indicated linearity between change in velocity and facet joint motions.
4. For the same change in velocity, the sliding motion had a greater increase than that of compressive motion.
5. For all kinematic measurements, larger differences by ΔV occurred at C5-6 compared to C4-5 and C6-7 levels.
6. Magnitudes of facet joint motions were independent of spinal level. However, for lower velocities, C6-7 motions were highest ,and for higher velocities C5-6 motions were highest.
7. Variability in sliding facet joint motions was significantly affected by changes in input velocity. In contrast, variability in compressive facet joint motions was not significantly affected by changes in input velocity.

ACKNOWLEDGMENTS

This research was supported in part by PHS CDC Grant R49CCR 515433 and VA Medical Research.

REFERENCES

1. Aprill, C., A. Dwyer, and N. Bogduk, Cervical zygapophyseal joint pain patterns II: A clinical evaluation. Spine, 1990. 15(6): p. 458-461.
2. Barnsley, L., S. Lord, and N. Bogduk, Clinical Review - Whiplash Injury. Pain, 1994. 58: p. 283-307.
3. Bogduk, N., Pathonatomical assessment of whiplash, in Frontiers in Head and Neck Trauma: Clinical and Biomechanical, N. Yoganandan, et al., Editors. 1998, IOS Press: The Netherlands. p. 299-307.
4. Clark, C.R., et al., The Cervical Spine. Third ed. 1998, Philadelphia, PA: Lippincott-Raven. 1003.
5. Cusick, J.F., et al., Biomechanics of cervical spine facetectomy and fixation techniques. Spine, 1988. 13(7): p. 808-812.
6. Kaneoka, K., et al., Motion analysis of cervical vertebrae during whiplash loading. Spine, 1999. 24(8): p. 763-770.

7. Ono, K. and K. Kaneoka. Motion analysis of human cervical vertebrae during low speed rear impacts by the simulated sled. in IRCOBI Conference. 1997. Hannover, Germany.

8. Ono, K., et al. Cervical injury mechanism based on the analysis of human cervical vertebral motion and head-neck-torso kinematics during low speed rear impacts. in Proc 41st Stapp Car Crash Conference. 1997. Lake Buena Vista, FL: Society of Automotive Engineers, Inc.

9. Sherk, H.H., et al., The Cervical Spine. Second ed. 1989, Philadelphia, PA: JB Lippincott Co. 881.

10. Winkelstein, B.A., et al., Cervical facet joint mechanics: Its application to whiplash injury Spine, 2000. 25(10): p. 1238-1246.

11. Yoganadan, N et al, eds. Fronteriers in Head & Neck Trauma: Clinical & Biomechanical. 1998, IOS Press: The Netherlans. 743

12. Yoganandan, N. and F.A. Pintar, eds. Frontiers in Whiplash Trauma: Clinical and Biomechanical. 2000, IOS Press: The Netherlands. 590.

13. Yoganandan, N., et al. Head-neck biomechanics in simulated rear impact. in 42nd Association for the Advancement of Automotive Medicine. 1998. Charlottesville, VA.

14. Yoganandan, N., et al., Injury biomechanics of the human cervical column. Spine, 1990. 15(10): p. 1031-1039.

15. Yoganandan, N, Pintar, FA, Kleinberger, M. Whiplash injury biomechanical experimentations. Spine, 1999. 24)1): p. 83-85.

Rear-End Collisions - A Study of the Influence of Backrest Properties on Head-Neck Motion using a New Dummy Neck

Mats Y. Svensson and Per Lövsund
Chalmers University of Technology

Yngve Håland
Electrolux Autoliv AB

Stefan Larsson
SAAB Automobile AB

ABSTRACT

Neck injuries in rear-end collisions are usually caused by a swift extension-flexion motion of the neck and mostly occur at low impact velocities (typically less than 20 km/h). Although the injuries are classified as AIS 1, they often lead to permanent disability. The injury risk varies a great deal between different car models. Epidemiological studies show that the effectiveness of passenger-car head-restraints in rear-end collisions generally remains poor.

Rear-end collisions were simulated on a crash-sled by means of a Hybrid III dummy with a new neck (Rear Impact Dummy-neck). Seats were chosen from production car models. Differences in head-neck kinematics and kinetics between the different seats were observed at velocity changes of 5 and 12.5 km/h. Comparisons were made with an unmodified Hybrid III.

The results show that the head-neck motion is influenced by the stiffness and elasticity of the backrest as well as by the properties of the head-restraint. The elastic rebound of the backrest can aggravate the violence of the whiplash-motion and delay contact between the head and the head-restraint.

INTRODUCTION

Neck injuries in rear-end impacts usually occur at low impact-velocities, typically less than 20 km/h (Kahane, 1982; Romilly et al., 1989; Olsson et al., 1990). The protective effect of the head-restraints is small, typically 20% (O'Neill et al., 1972; Huelke and O'Day, 1975; Nygren et al., 1985).

The injury symptoms include pain, weakness or abnormal response in the parts of the body (mainly the neck, shoulders and upper back) that are connected to the central nervous system via the cervical nerve-roots (Nygren et al., 1985). The mechanisms causing these injuries are not fully understood and the exact injury sites are unknown. A

theory as to the site and causes of injury has been presented by Svensson et al. (1989). This theory predicts that the injuries occur in the nerve root region of the cervical spine due to a pressure gradient along the intervertebral foramen. The pressure gradient is the result of the dynamic flow conditions occuring during the swift neck motion in the inner and outer vein-plexa of the cervical spine. Thus the theory indicates that the risk of injury is related to the velocity and acceleration of the head-neck motion.

Nygren et al. (1985) found that the risk of neck injury in a rear-end collision was not reduced in newer cars. The study disclosed large differences in protective performance between different car models. Several authors have reported a considerably smaller risk of neck-injury in the rear-seat compared to the front seat for adult car occupants (Kihlberg, 1969; States et al., 1972; Carlsson et al., 1985; Lövsund et al., 1988; Otremski et al., 1989). This could be due to the fact that the seat-back of the rear seat is generally much stiffer than the seat-back of the front-seat.

Mertz and Patrick (1967) using a rigid seat and a rigid head-support simulated rear-end collisions with a volunteer. They found that the volunteer could withstand an impact velocity of 44 mph (70 km/h) with only little discomfort when the head was in contact with the head-support already at the beginning of the impact. A similar situation can be said to occur in frontal impacts for children in rearward-facing child-seats. These seats are generally very rigid, and since frontal impacts are usually preceded by the braking of the vehicle, the child's head is mostly in contact with the high seat-back at the start of the crash event. Data by Aldman et al. (1987) indicate that neck injuries to children in rearward-facing restraints are uncommon.

States et al. (1969) presented the following theory: The timing of the elastic rebound of the seat-back in a rear-end collision can be such that the torso is pushed forwards in the passenger compartment while the head is still in the process of moving backwards. This would increase the relative velocity between the head and the torso and thus increase the risk of neck injury. Later studies support this theory (McKenzie and Williams, 1971; Prasad et al., 1975; Rommily et al., 1989; Foret-Bruno et al., 1991).

Today, there is still no adequate method for simulating the head-neck motion of a car occupant in a rear-end collision. The best available dummy at present is the Hybrid III. The neck and spinal structure of this dummy are stiff and unlikely to interact with the seat-back in the same compliant way as would the human spine. Another problem is that the correlation between different kinematic and kinetic parameters and the risk of neck injury is not completely known.

As a first step in the construction of a new rear-impact dummy, Svensson and Lövsund (1992) developed a Rear Impact Dummy-neck (RID-neck) for the Hybrid III-dummy. The neck was especially designed for low-velocity rear-impact testing and resulted in more human-like head-neck kinematics in this type of impact than does the original Hybrid III dummy-neck. Thus, with this new neck, the Hybrid III attained a significantly improved bio-fidelity in low-velocity rear-end collision testing.

The aim of this study was to examine by means of a Hybrid III-dummy with a RID-neck the head-neck kinematics in low-velocity rear-end collisions for different production-car seats with differing stiffness properties. The influence of the backrest properties on the motion of the head and the torso as well as the relative motion and the contact between the head and the head-restraint were studied.

MATERIALS AND METHODS

A series of staged rear-end collisions was carried out involving five different front-seats (F1-F5) and three rear-seats (R1-R3) chosen from various car makes. The rear-seats R1 and R2 were identical except for the fact that R1 had no head-restraint. All the other seats had head-restraints. Figure 1 shows a schematic drawing of a seat-back frame of a front-seat and the dimensions are given in Table 1. The seat-backs of the rear-seats had a steel plate on the rear side and a cushion in front of this plate. This cushion was of a steel-spring design in R1 and R2 and of a polymer-foam design in R3.

The tests were done on a crash-sled at the Department of Injury Prevention. Two different pre-impact sled-velocities, 5 and 12.5 km/h (3.1 and 7.8 mph) were chosen. These sled velocities correspond to somewhat more than half the relative pre-impact velocity between the bullet vehicle and the impacted vehicle in a real rear-end collision involving two passenger cars. The sled deceleration was set to approximately a square pulse with an amplitude of 50 m/s^2 for the 5 km/h tests, and 70 m/s^2 for the 12.5 km/h tests. The seats were mounted in a rearward facing position on the sled. The front-seat seat-backs were adjusted to a 25 degree rearward angle. The front seats were mounted to the sled-floor in their standard attachment holes with their standard angular position of the lower seat-frame.

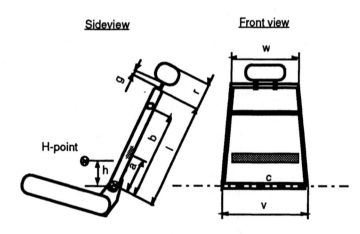

Sideview   **Front view**

l: Distance from joint to seat-back top.
r: Distance from top of seat-back to top of head-restraint.
g: Gap between seat-back and head-restraint.
h: Vertical distance between joint and H-point.
a: Distance between joint and lumbar supporting belt (if applicable).
b: Distance between joint and cross-member.
c: Presence of stiff axis between the two frame-joints (yes / no).
w: Upper seat-back width.
v: Lower seat-back width.

Figure 1: Schematic drawing of a front-seat inner frame seen from the left side and from the front.

Table 1: Dimensions of the five front-seats (as defined in Figure 1).

	F1	F2	F3	F4	F5
l (m)	0.68	0.68	0.63	0.55	0.50
r (m)	0.14	0.13	0.22	0.25	0.23
v (m)	0.46	0.50	0.50	0.50	0.45
w (m)	0.36	0.35	0.35	0.44	0.43
g (m)	0	0	0	0.05	0.07
h (m)	0.15	0.15	0.16	0.11	0.07
a (m)	0.16	-	-	0.14	-
b (m)	0.48	0.46	0.52	0.49	0.43
c	no	no	no	yes	yes

The sled-floor consisted of a 10 mm thick steel-plate. The rear-seats were firmly attached to the sled floor with the same seat-back angle as in the real vehicles. All tests were done twice to test the repeatability.

A 50th percentile Hybrid III-dummy equipped with a RID-neck was used. The RID-neck has been described in detail by Svensson and Lövsund (1992). The arms of the dummy were fixed with adhesive tape in a folded position in front of the chest. This to improve the repeatability of the tests and prevent the arms from obstructing the vision of the high-speed camera. The dummy was equipped with accelerometers in the head, chest and pelvis and with force-moment transducers at the upper neck (R.A.Denton, type:1716) and at the lower neck (R.A.Denton, type:1794). The sled acceleration was measured and all tests were filmed with a high-speed camera.

The seat-types, F1, F2 and F5 were equipped with vertically adjustable head-restraints. These were adjusted so that the upper edge of the head-restraint was at the level of the eyebrows.

RESULTS

The angular displacements between the head and the neck were generally larger for the rear-seats compared to the front-seats. The rear-seat without head-restraint, R1, showed the greatest displacement (Figs. 2 and 4). The linear displacements between the head and the upper torso were also largest for the rear seats but here the R3-seat showed the largest displacement (Figs. 3 and 5). Front-seat F4 showed the smallest displacements. (A more complete set of graphs of the results are shown in the Appendix, Figures A1 to A4.)

Table 2 shows the maximum values for accelerations and force-moment measurements.

Table 2: Maximum values for accelerations and force-moment measurements.

5 km/h:

Seat type	F1	F2	F3	F4	F5	R1	R2	R3
Head acc. (g)	8.5	9.1	9.7	6.8	11.1	4.4	4.2	3.6
Chest acc. (g)	3.6	3.4	2.9	3.5	4.2	3.2	3.4	3.1
Pelvic acc. (g)	2.6	2.0	1.9	3.0	2.3	1.8	2.2	3.2
Upper-neck x-force (kN)	0.15	0.14	0.17	0.12	0.22	0.15	0.16	0.15
Upper-neck z-force (kN)	0.25	0.27	0.22	0.17	0.28	0.12	0.14	0.12
Upper-neck y-mom. (Nm)	12	5	8	8	15	15	5	14
Lower-neck x-force (kN)	0.24	-	0.25	0.15	0.36	0.16	0.22	0.16
Lower-neck z-force (kN)	0.17	-	0.08	0.15	0.20	0.09	0.07	0.08
Lower-neck y-mom. (Nm)	41	-	42	24	61	42	43	46

12.5 km/h:

Seat type	F1	F2	F3	F4	F5	R1	R2	R3
Head acc. (g)	25	13	19	21	31	13	12	21
Chest acc. (g)	9	8	6	10	8	16	25	19
Pelvic acc. (g)	8	5	6	9	9	13	15	16
Upper-neck x-force (kN)	0.30	0.20	0.34	0.30	0.48	0.56	0.42	0.42
Upper-neck z-force (kN)	1.05	0.48	0.35	0.50	0.65	0.68	0.90	0.70
Upper-neck y-mom. (Nm)	45	28	-10	-20	-30	70	43	55
Lower-neck x-force (kN)	0.75	0.50	0.50	0.55	0.82	0.84	0.73	0.84
Lower-neck z-force (kN)	0.65	0.30	-0.30	0.40	0.30	0.48	0.70	0.60
Lower-neck y-mom. (Nm)	125	78	78	75	115	137	83	120

Figure 6 shows the contours of the dummy, the seat-back and the head-restraint of the 12.5 km/h impacts for three different time values. The three time values correspond to 1) the first moment of impact, 2) the time of maximum linear X-displacement at shoulder level and 3) the time of maximum angular displacement of the head relative to the torso. In all the tests the rebound of the torso starts when the head is still moving backwards relative to the sled.

DISCUSSION

Pilot tests were done with one of the seat-types (F1) to test the repeatability of the test set-up and to test the influence of the impact velocity. Four different velocities were tested. The pre-impact sled-velocities were set to 5; 10; 12.5 and 15 km/h (3.1; 6.2; 7.8 and 9.4 mph). The 10 km/h test was repeated three times in order to study the repeatability of the test set-up.

A Hybrid III dummy with a standard neck was also tested in this pilot series at 12.5 km/h and 15 km/h to enable a comparison between the RID-neck and the standard neck.

An assessment of the repeatability is given in Figure 7 which shows the angular displacement of the head relative to the torso for the three identical tests done with an F1 seat and with a sled velocity of 10 km/h. The deviation between the three curves is >10% of the maximum value.

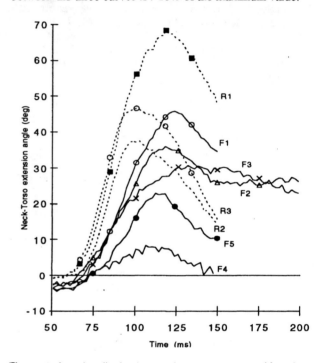

Figure 4: Angular displacements between torso and head at 12.5 km/h sled velocity:

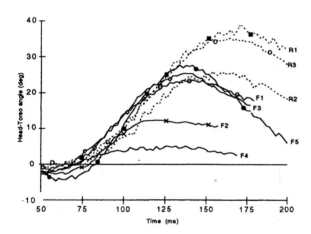

Figure 2: Angular displacements between torso and head 5 km/h sled velocity.

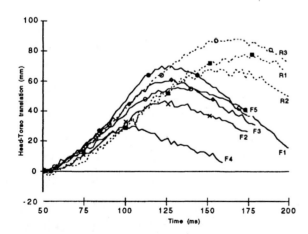

Figure 3: Linear X-displacements between torso and head at 5 km/h sled velocity.

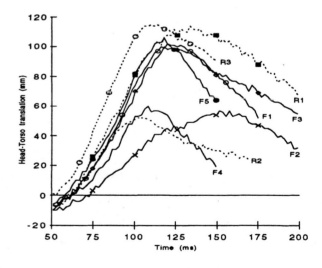

Figure 5: Linear X-displacements between head and torso at 12.5 km/h sled velocity.

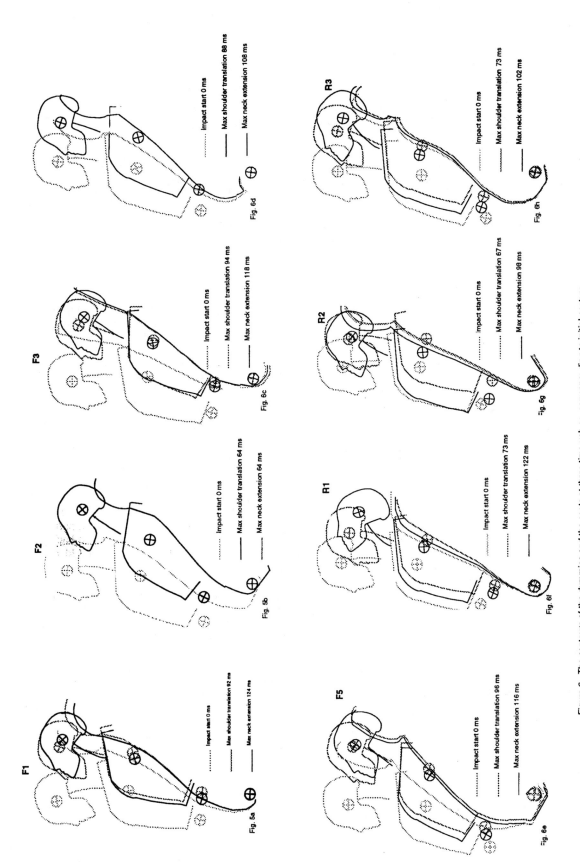

Figure 6: The contours of the dummies and the seats at three time-values, corresponding to initial posture, maximum rearward shoulder displacement and maximum neck extension-angle. In 6b the two last events occur simultaneously and in 6c the displacements between the two last events are very small and have been left out. The tests were done at 12.5 km/h sled velocity. In all tests exept 6b and 6c a rebound at shoulder level can be seen between the second last and the last time-value.

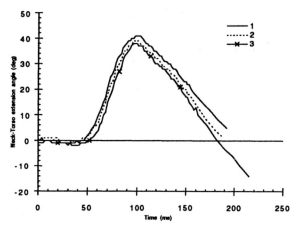

Figure 7: Angular displacement between head and torso for three identical consecutive runs in the same seat at 10 km/h sled velocity (seat F1).

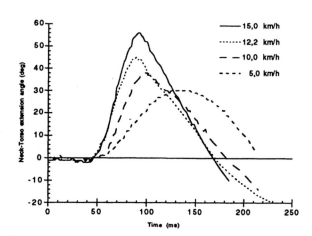

Figure 8: Angular displacement between head and torso at four different sled velocities (seat F1).

Figure 8 shows the angular displacement between the head and the torso for a sled velocity of 5, 10, 12.5 and 15 km/h. The maximum angular displacement between the head and the torso increases in magnitude and occurs earlier with increased sled velocity. The original Hybrid III-neck and the RID-neck are compared in Figure 9 which shows the angular displacement between head and torso for the two neck types at 12.5 km/h and 15 km/h.

The Hybrid III-dummy, fitted with the RID-neck, is believed to be a better instrument than the unmodified Hybrid III for studying the head-neck motion during a staged rear-end collision at low impact-velocity (Svensson and Lövsund, 1992). The resistance to rearward angular displacement of the RID-neck is closer to that of a human than is the original Hybrid III-neck.

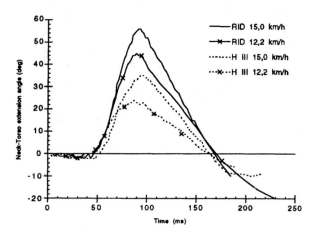

Figure 9: Angular displacements between torso and head at 12.5 km/h and 15 km/h, using booth the original Hybrid III-neck and the RID-neck. in seat F1.

The resistance to horizontal translational displacement between the torso and the head (without angular displacement) is much lower for the RID-neck, compared with the original Hybrid III-neck, but this motion has not yet been adequately validated and the RID-neck might still be too stiff. This might cause the rearward angular head-motion to start earlier for the RID-neck than what would have been the case for a real car occupant.

The spine of the upper torso of the Hybrid III is completely rigid and the lumbar part of the dummy spine is very stiff. This means that the dummy torso is insensitive to the variations in stiffness between different parts of the seat-back and does not interact with the seat-back in the same compliant way as a human torso would during a rear impact.

The lower-neck force and moment transducer protrudes about 50 mm at the upper posterior side of the torso and produces a 55 mm wide, horizontal crest that can penetrate into the soft seat-back cushion and come in contact with the harder structures below. A hard structure under the cushion at the same level as the transducer will thus stop the penetration of the upper dummy torso into the seat-back cushion about 50 mm earlier than if the transducer had not been used. This earlier and perhaps more abrupt stop of the upper torso will affect the head-neck motion of the dummy. The head-neck motion will be initiated at an earlier stage at which the distance between the head and the head-restraint is wider. This means that the lower-neck force and moment transducer decreases the bio-fidelity of the dummy at rear-end collision testing.

For the seats with adjustable head-restraints, F1, F2 and F5, the top of the head-restraint was placed at the level of the eyebrows. This height is often used as a recommended minimum height for correct head-restraint adjustment since it is situated slightly above the centre of gravity of the head. Thus the head-restraints of F1 and F2 were placed in the highest position and that of F5 in the lowest position.

During the crash event at a sled velocity of 12.5 km/h, the head-restraints of F1 and F2 were pushed down, by the head, to the lowest position. The head-restraint seemed to provide only a limited resistance to this downward motion. It is thus likely that the initial position of the head-restraint was of minor importance for F1 and F2.

The stiffness might differ between the real car floors and the sled floor and the floor stiffness is likely to influence the seat response to some extent.

The initial horizontal and vertical distance between the head and the head-restraint is important for the maximum allowable head-neck displacement. The initial distances between the head and the head-restraint for the different seats in this study can be seen in Figure 6. The largest distances can be observed for the rear-seats, which also display the largest maximum displacements (Figs. 2-5), and the smallest distance occured for F4 which also had relatively low displacements (Figs. 2-5).

Deformation or displacement of the head-restraint relative to the seat-back during contact with the head probably has the same influence on the head-neck motion as an increased initial distance between the head and the head-restraint would have. This phenomenon was particularly evident for tests F1, F2 and R2 at 12.5 km/h.

If the upper torso can compress the upper seat-back cushion early in the crash event, without deforming the seat-back frame, this will decrease the horizontal distance between the

head and the headrest before any larger displacements between the head and the torso take place. This type of movement (illustrated in Figure 10) appeared in seat F4.

The elasticity of the seat-back structure and of its pivoting joint at its lower end will influence the crash event. This elasticity permits the seat-back to yield at the initial stage of the crash event, when it is loaded by the mass of the torso due to the vehicle acceleration, and will cause the torso to rebound forwards in a later stage of the event. This, in turn, will give the torso a velocity change forwards that is greater than that of the impacted vehicle. If this rebound of the torso starts before the head has reached its rearmost position, the relative velocity between the head and the torso will increase. These rebound phenomena were evident in most of the tests in this test series and this is illustrated in Figure 6a-h for the 12.5 km/h sled velocity.

The stiffness of the seat-back frame is another important parameter. When the seat-back frame yields backwards, the head-restraint will follow thereby increasing the distance, and delaying the contact, between the head and the head-restraint. This is illustrated in Figure 11. This type of motion is evident for the seats F2, F3 and F5.

All five front-seats in this study have seat-backs that are built around a metal frame. As seen from the back of the seat, the frame consists of two vertical side members hinged to the bottom seat frame, and one horizontal cross-member connecting the upper ends of the side-members. The F4 and F5 seats had an axis connecting the two seat-back joints. Inside this frame was a metal-thread net that was attached to the frame with a number of coil springs.

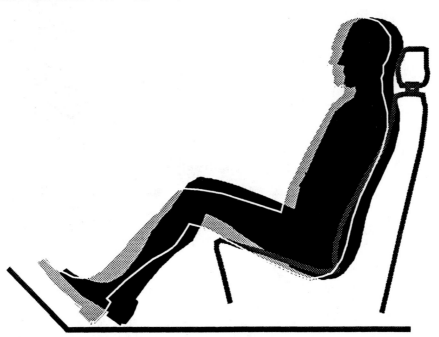

Figure 10: The torso sinks into the soft surface of the seat-back cushion early in the crash event, thus decreasing the distance between the head and the head-restraint before any larger motions between the head and the torso have commenced.

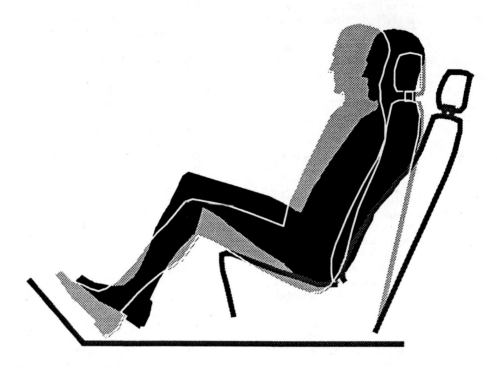

Figure 11: The seat-back frame yields backwards at the beginning of the crash event and this in turn distances the head-restraint further away from the head.

Figure 12: The upper part of this seat-back cushion is relatively stiff due to the upper crossbeam of the seat-back frame, and it does not allow the upper torso to penetrate very deeply. The lower part of the seat-back cushion is soft and allows deep penetration of the lower torso, and particularly of the pelvis which is loaded by the leg mass. This results in a large angular displacement of the torso relative to the seat-back frame.

Between this structure and the occupant there was polymer-foam upholstering. The seat-back frames of seats F1, F2 and F3 lacked an axis between the joints (Fig. 1) making the frames look something like a rectangular arc. This construction permitted the lower torso to penetrate deep into the lower seat-back which resulted in a significant angular displacement between the torso and the seat-back frame. Figure 12 is an illustration of this sort of movement. This difference in stiffness between the top and the lower part of the seat-back could cause the upper torso to rebound at a different time than the lower torso. The pelvic area is loaded by the mass of the legs and might penetrate deeper and rebound later than the mid and upper torso.

No particular parameter in the results of this study can be singled out as a predictor of the risk of neck injury in a low-velocity rear-end collision. There is, however, a clear indication that the maximum rearward displacement of the head relative to the torso does not correlate to the risk of injury since the largest displacements are found in the rear-seats in which the injury risk is the lowest according to accident data figures (Kihlberg, 1969; States et al., 1972; Carlsson et al., 1985; Lövsund et al., 1988; Otremski et al., 1989). In the 5 km/h tests (where the problems due to the hard contact between the lower-neck force and moment transducer and the hard seat structures is less pronounced) the maximum accelerations of the motion between the head and the torso are the lowest for the three rear-seats (Figures 2 and 3). This suggests that the acceleration of the head-neck motion is correlated to the risk of injury.

CONCLUSIVE REMARKS

A Hybrid III-dummy equipped with a RID-neck was used in a series of rear-end collision sled tests. Five types of front-seats and three types of rear-seats from passanger cars were tested at 5 km/h and 12.5 km/h pre-impact sled-velocity.

The stiffness and elastic properties of the seat-back and head-restraint as well as the initial distance between the head and the head-restraint were found to influence the head-neck kinematics. Elastic rebound of the seat-back causing the torso to move forwards at a stage when the head is still moving backwards increases the velocity difference between the head and the torso thus increasing the violence of the head-neck motion.

The lower-neck force and moment transducer was found to influence the outer contour of the back of the dummy torso and thereby to affect the validity of the test results.

There seems to be a correlation between the acceleration of the head-neck motion and the expected risk of injury. However, the maximum linear and angular displacement between the head and the torso did not correlate to the expected risk of injury.

ACKNOWLEDGEMENTS

This study was supported by the Swedish Transport Research Board (TFB) and the Swedish Road Safety Office (TSV). We thank Fredrik Lindh, M.Sc. and John Lindhe, M.Sc. for an excellent job in arranging the test set-up and running the tests.

REFERENCES

Aldman, B.; Gustafsson,H.; Nygren, Å.; Tingvall, C. Child restraints. A prospective study of children as car passengers in road traffic accidents with respect to restraint effectiveness. Acta Paediatr. Scand., Suppl .339: II:1-22; 1987

Carlsson, G.; Nilsson, S.; Nilsson-Ehle, A.; Norin, H.; Ysander, L.; Örtengren, R. Neck Injuries in Rear End Car Collisions. Biomechanical considerations to improve head restraints. Proc. Int. IRCOBI/AAAM Conf. Biomech. of Impacts, Göteborg, Sweden, 277-289; 1985

Foret-Bruno, J.Y.; Dauvilliers, F.; Tarriere, C. Influence of the Seat and Head Rest Stiffness on the Risk of Cervical Injuries in Rear Impact. Proc. 13th ESV Conf. in Paris, France, paper 91-S8-W-19, NHTSA, USA; 1991

Huelke, D.F.; O'Day, J. The Federal Motorvehicle Safety Standards: Recommendations for Increased Occupant Safety. Proc. Fourth Int. Cong. on Automotive Safety, NHTSA, USA, 275-292; 1975

Kahane, C.J. An Evaluation of Head Restraints - Federal Motor Vehicle Safety Standard 202. NHTSA Technical Report, DOT HS-806 108, National Technical Information Service, Springfield, Virginia 22161, USA; 1982

Kihlberg, J.K. Flexion-Torsion Neck Injury in Rear Impacts. Proc. 13th AAAM Ann. Conf., The Univ. of Minnesota, Minneapolis, USA, 1-17; 1969

Lövsund, P.; Nygren, Å.; Salen, B.; Tingvall, C. Neck Injuries in Rear End Collisions among Front and Rear Seat Occupants. Proc. Int. IRCOBI Conference Biomech. of Impacts, Bergisch-Gladbach, F.R.G., 319-325; 1988

McKenzie, J.A.; Williams, J.F. The Dynamic Behaviour of the Head and Cervical Spine during Whiplash. J. Biomech. 4:477-490; 1971

Mertz, H.J.; Patrick, L.M. Investigation of the Kinematics and Kinetics of Whiplash. Proc. 11th STAPP Car Crash Conf, SAE Inc., New York, USA, 267-317; 1967

Nygren, Å; Gustafsson, H.; Tingwall, C. Effects of Different Types of Headrests in Rear-End Collisions. 10th Int. Conference on Experimental Safety Vehicles, NHTSA, USA, 85-90; 1985

Olsson, I.; Bunketorp, O.; Carlsson, G.; Gustafsson, C.; Planath, I.; Norin, H.; Ysander, L. An In-Depth Study of Neck Injuries in Rear End Collisions. Proc. 1990 Int. IRCOBI Conf. on the Biomechanics of Impacts, Bron, Lyon, France, 269-282; 1990

O'Neill, B.; Haddon, W.; Kelley, A.B.; Sorenson, W.W.

Automobile Head Restraints: Frequency of Neck Injury Insurance Claims in Relation to the Presence of Head Restraints. Am J Publ Health, 62(3):399-406; 1972

Otremski, I.; Marsh, J.L.; Wilde, B.R.; McLardy Smith, P.D.; Newman, R.J. Soft Tissue Cervical Spinal Injuries in Motor Vehicle Accidents. Injury 20:349-351; 1989

Prasad, P.; Mital, N.; King, A.I.; Patrick, L.M. Dynamic Response of the Spine During + Gx Acceleration. Proc. Nineteenth STAPP Car Crash Conf., SAE Inc., USA, 869-897; 1975

Romilly, D.P.; Thomson, R.W.; Navin, F.P.D.; Macnabb, M.J. Low Speed Rear Impacts and the Elastic Properties of Automobiles. Proc. Twelfth Int. Techn. Conf. Experimental Safety Vehicles, US Dept. of Transp., NHTSA, USA, 1199-1205; 1989

Svensson, M. Y.; Örtengren, T.; Aldman, B.; Lövsund, P.; Seeman, T.; A Theoretical Model for and a Pilot Study Regarding Transient Pressure Changes in the Spinal Canal under Whip-lash Motion. Dept. of Injury Prevention, Chalmers University of Technology, Göteborg, Sweden, R 005, 1989

Svensson, M. Y.; Lövsund, P. A Dummy for Rear-End Collisions. Proc. 1992 Int. IRCOBI Conf. on the Biomechanics of Impacts, Verona, Italy, 299-310; 1992

States, J.D.; Korn, M.W.; Masengill, J.B. The Enigma of Whiplash Injuries. Proc. Thirteenth Ann. Conf. AAAM, Minnesota, USA, 83-108; 1969

States, J.D.; Balcerak, J.C.; Williams, J.S.; Morris, A.T.; Babcock, W.; Polvino, R.; Riger, P.; Dawley, R.E. Injury Frequency and Head Restraint Effectiveness in Rear-End Impact Accidents. Proc. 16th Stapp Car Crash Conf., SAE, New York, 228-245; 1972

APPENDIX

Figures A1 to A4 of this appendix show:

1) The horizontaldisplacements of the pelvis, shoulder, head and the upper seat-back. The relative horizontal displacement between the head and the shoulder is also shown.

2) The angular displacement of the head and the torso as well as the relative angular displacement between the head and the torso.

3) The X-accelerations of the head, chest and pelvis.

4) The Y-torque measured in the upper and lower neck transducers.

5) The X-shear-forces measured in the upper and lower neck transducers.

6) The Z-axial-forces measured in the upper and lower neck transducers.

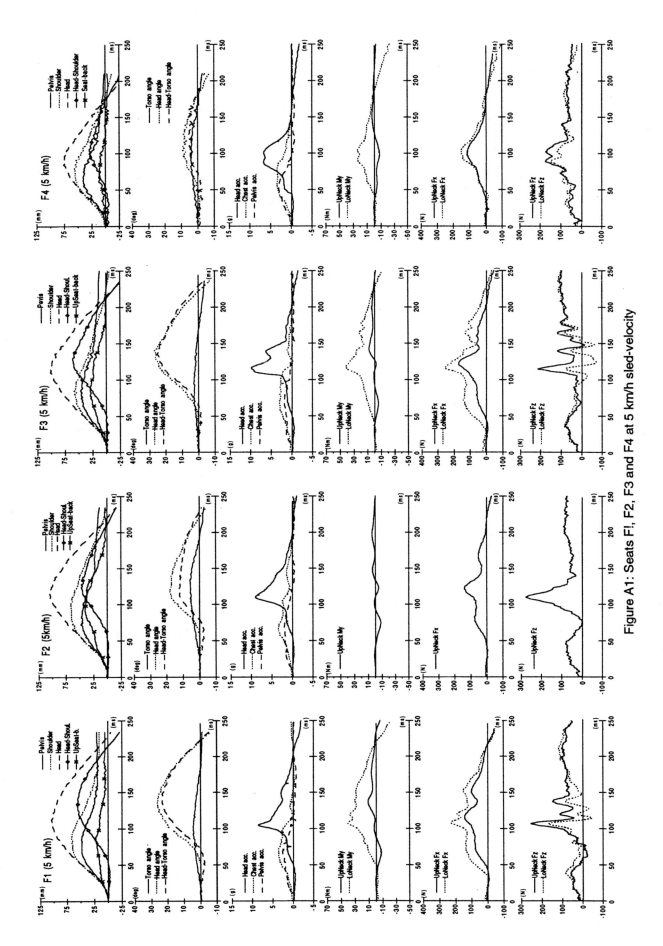

Figure A1: Seats F1, F2, F3 and F4 at 5 km/h sled-velocity

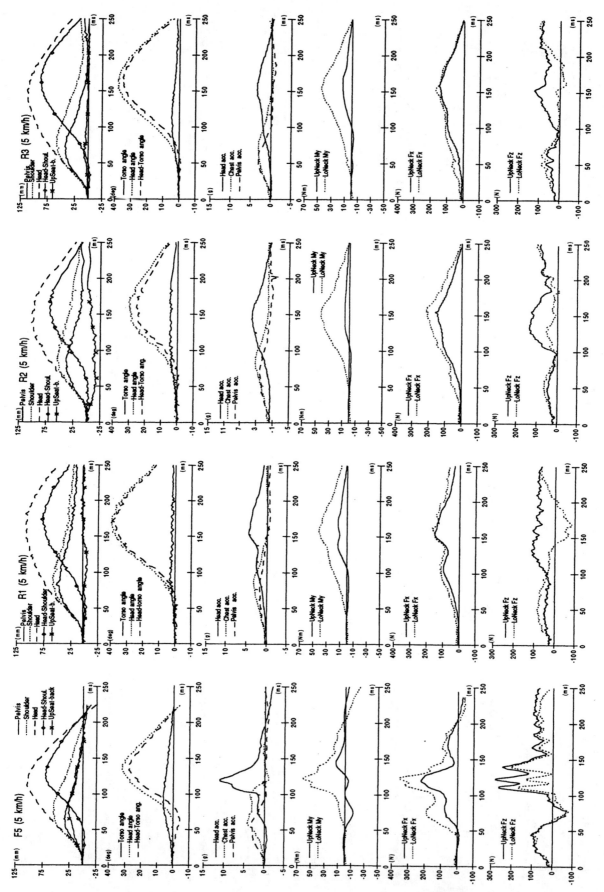

Figure A2: Seats F5, R1, R2 and R3 at 5 km/h sled-velocity

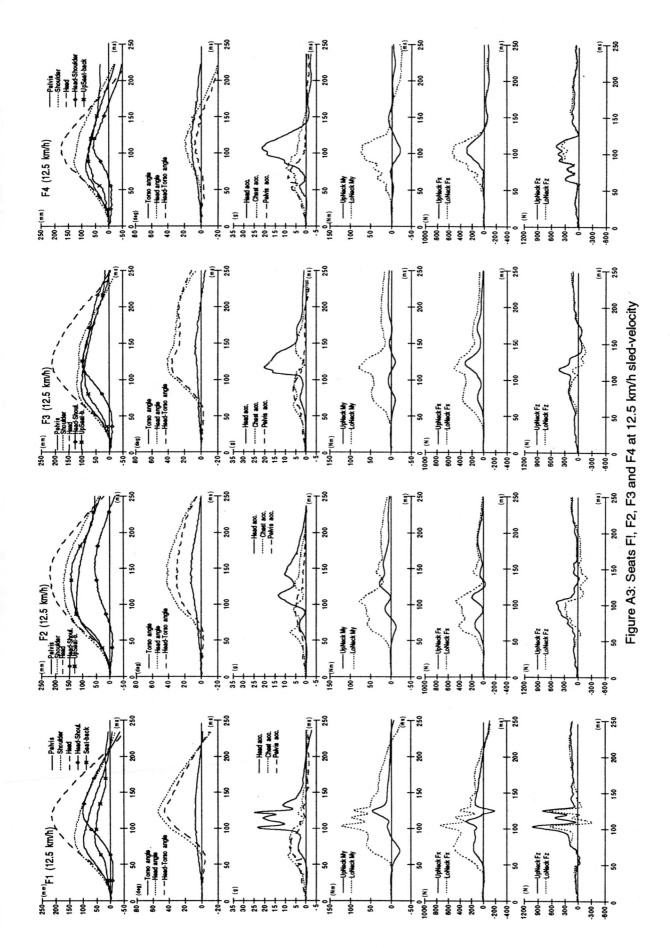

Figure A3: Seats F1, F2, F3 and F4 at 12.5 km/h sled-velocity

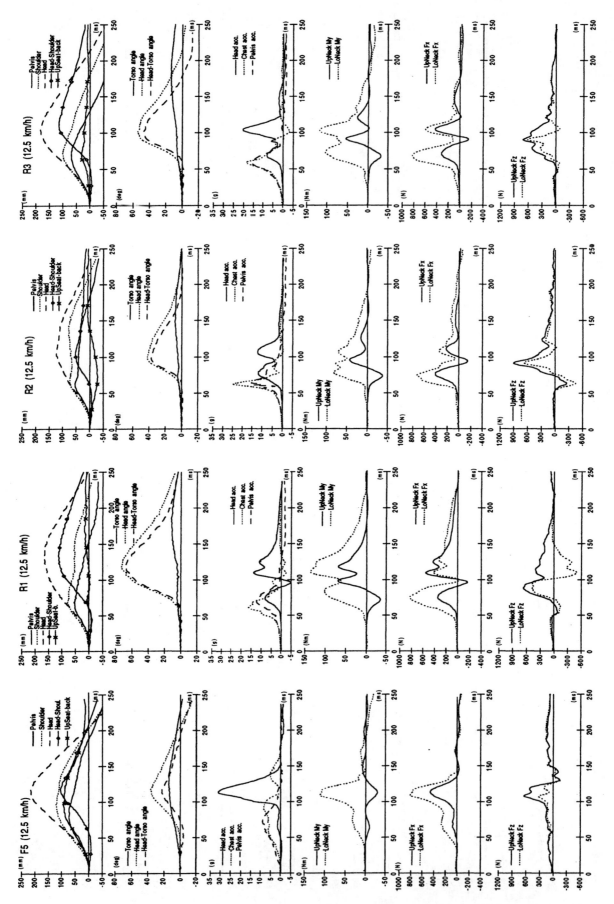

Figure A4: Seats F5, R1, R2 and R3 at 12.5 km/h sled-velocity

Human Subject Kinematics and Electromyographic Activity During Low Speed Rear Impacts

Thomas J. Szabo and Judson B. Welcher
Biomechanical Research & Testing, LLC

ABSTRACT

Research into the biomechanics of low speed rear impacts has focused primarily on the kinematic responses of anthropometric dummies and human subjects. Occupant muscular activity during low speed rear impacts remains largely unquantified however. The current study enhances the existing database of human subject test exposures with an emphasis on electromyographic activity before, during, and after low speed rear impact. This information may provide insight into injury mechanisms, occupant mathematical modeling, and aspects of seat and head restraint design.

Low speed rear impacts using instrumented human subjects were conducted. Ten nominal 16 km/h closing speed car-to-car impacts were conducted using male and female subjects aged 22-54 years, with struck vehicle velocity changes of up to 10 km/h. Two head restraint conditions were studied. One was a standard seat integrated head restraint. For the second condition the integrated head restraint was modified by adding 2 inches of padding to the existing head restraint, thus reducing the initial head-to-head restraint horizontal distance.

Accelerometers were affixed to the target vehicle static center of gravity and the occupant's head, cervical spine, and lumbar spine. Accelerations at the head static center of gravity were obtained via a 9-accelerometer headgear array and algorithm. Anterior paracervical, posterior paracervical, trapezius, and paralumbar muscle activity was monitored using surface mounted electrodes. Kinematics in the sagittal plane were obtained via high speed video.

No injuries were sustained by any occupant. In all cases the subjects exhibited pre-impact muscle activity commensurate with that of a relaxed seated posture, indicating that the test protocol inhibited the subjects from bracing in anticipation of the impacts. Initial muscle activity typically occurred approximately 100 to 125 milliseconds after the moment of bumper contact, during the initial phase of impact as the occupant's cervical spine was extending. Full muscle tension likely did not develop until the cervical spine was flexing. Cervical flexor, cervical extensor and lumbar paraspinal musculature demonstrated similar activity onset times. A centrally generated response was thus hypothesized for initial onset of muscle activity. This response was consistent with being triggered by lumbar spine acceleration, and typically occurred approximately 90-120 milliseconds after onset of lumbar spine acceleration.

No significant differences were noted between muscle response times for the two head restraint conditions. Decreases in rearward head displacement, cervical spine extension, and head acceleration were found for the modified head restraint.

INTRODUCTION

Historically, biomechanical research into automotive safety has relied heavily on human surrogate testing in order to enhance the understanding of injury mechanisms and improve vehicle safety systems. This includes cadaver tests, anthropometric dummy tests, and the use of mathematical models. Although the subject of some debate, the similarity of these surrogates to actual human occupants in high speed automotive impacts has been reasonably well established, and as a result the biomechanics of high energy impacts are relatively well understood.

Safety system design and the understanding of injury mechanisms in low speed rear impacts has not developed at the same rate as in higher energy impacts. This may be due in part to the relatively poor performance of many current surrogates in low speed impacts, which in turn may be associated with the surrogates' apparent inability to accurately simulate muscular inputs during low speed impacts.

Cadavers clearly lack muscle tone during any impact, and most dummies do not actively account for muscular effects. Some mathematical models intended for use in low speed impacts have either not taken into account the effects of musculature (McKenzie and Williams, 1971), or are unclear on the extent of muscular simulation used (Martinez and Garcia, 1968). Those models which have attempted to account for muscular input (Huston et al., 1978; Jakobsson et al.,1993; Soechting and Paslay, 1973) have done so largely without the benefit of human subject crash test data, particularly with respect to the onset of muscular activity.

Given the financial, ethical, and logistical hurdles associated with live human occupant testing, it is desirable to improve these other forms of surrogate testing so their benefits can be fully realized in the low speed area. Toward that end, a better understanding of muscular inputs during low speed rear impact may be beneficial, particularly with respect to the timing of any muscular activity during an unanticipated rear impact. Although it has been argued that muscles do not respond in sufficient time to manifest any changes in occupant motions during such an impact, this claim has been made largely without the benefit of actual testing. In addition, given the nature of many soft tissue injuries sustained during rear impacts, muscle activity patterns may provide some insight into potential injury mechanisms.

Another potential benefit of muscle activity analysis may be in evaluating the preparedness of human subjects during staged impact tests. Many low speed rear impact studies have described using "relaxed" or "unaware" human subjects, although no quantification of these descriptions were provided (Matsushita et al.,1994; Mertz and Patrick,1967; Severy et al.,1955; Szabo et al.,1994; West et al.,1993). Significant muscular activity immediately prior to impact is likely a function of occupant bracing for impact, and is a measure by which certain aspects of subject anticipation can be evaluated. EMG activity immediately prior to impact was thus monitored as part of the current study.

Head restraint geometry and position have long been implicated as factors in occupant injury mitigation during low speed rear impact. In particular, several studies have cited a decreased risk of injury with a decreased head-to-head restraint horizontal distance (Foret-Bruno et al.,1991; Geigl et al.,1994; Jakobsson et al.,1993; Olsson et al.,1990; Svensson et al.,1993; Weißner and Enßlen,1985). The current study examined occupant kinematics and muscular activity for a standard seat-integrated head restraint and for the same head restraint with 2 inches of padding added to decrease the initial head-to-head restraint horizontal distance.

REVIEW OF LITERATURE

In order to help establish an acceptable level of impact exposure for the current study, several studies in which live human subjects were exposed to actual or simulated rear impacts were reviewed. The number of exposures and symptoms incurred for rear impacts of various severities are summarized in Figure 1. Note that Figure 1 contains only those impacts for actual vehicle-to-vehicle impacts or vehicle-to-barrier impacts. Other forms of testing, such as sled tests have not been included.

Given this prior research, it was decided those exposures for which the change in velocity of the target vehicle was 9 km/h or less would be unlikely to result in significant injury (242 total exposures; 230 with negligible or no symptoms; 12 with minor symptoms). Since injury threshold levels were not under direct consideration, a change in velocity of 9 km/h was chosen as the target exposure level for the current study.

Relatively few studies have directly measured electromyographic activity during the sudden cervical spine extension-flexion associated with low speed rear impacts. No study was found which directly measured muscular activity during low speed vehicle-to-vehicle crashes with live human occupants. Matsushita et al. (1994) and Ono and Kanno (1993) conducted sled tests in which human subjects were exposed to simulated rear impacts. Although it was indicated electromyographic activity was monitored during the impacts, no EMG results were presented for either study.

Some studies have measured reflex times of various muscles to a sudden, unanticipated stimulus. Snyder et al. (1975) imposed controlled dynamic jerks in the fore-aft direction to human subjects' heads via a headband and rope-and-pulley arrangement. The "stretch reflex" was defined as the time lag between the onset of head acceleration and a distinct increase in muscle activity. Sternocleidomastoid activity was monitored during rearward head jerks, and splenius and semispinalis capitus activity monitored during forward head jerks.

Average reflex times for neck flexor and extensor muscles in male subjects were 77 and 66 milliseconds, respectively. Average female reflex times were 66 and 63 milliseconds for the cervical flexors and extensors, respectively. The authors also established a "muscle activation time" as the time from onset of muscular activity to the time of maximum head deceleration. The average muscle activation time for neck flexors and extensors was 61 and 69 milliseconds, respectively. Muscle reaction time was not influenced by sex, age or muscle location.

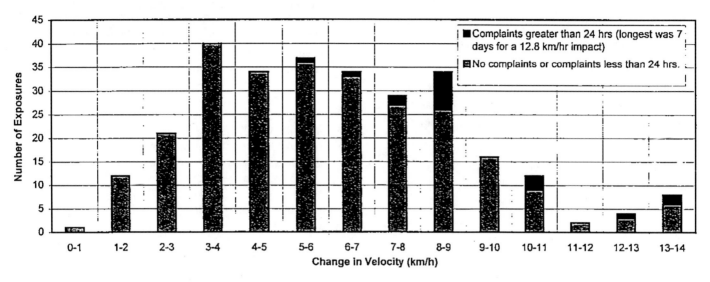

Baker Engineering unpublished data {55 exposures}	Szabo and Welcher unpublished data {34 exposures}
West et al. (1993) {45 exposures}	Szabo et al. (1994) {7 exposures}
McConnell et al. (1993) {9 exposures}	
McConnell et al. (1995) {18 exposures}	MacInnis Engineering unpublished data
Rosenbluth and Hicks (1994) {4 exposures}	King et al. (1993) {99 exposures}
SATAI tests (1995) {11 exposures}	King and Siegmund (1994)
Severy et al. (1955) {2 exposures}	Siegmund et al. (1994)

Figure 1: Human Subject Vehicle-to-Vehicle Impact Tests (284 total exposures)

Reid et al. (1981) subjected human subjects to similar rearward head jerks as in the Snyder study. Sternocleidomastoid activity was monitored. Reflex time was defined as the time from the initiation of head jerk to the onset of muscular activity, and was found to be approximately 90 milliseconds for an unaware test subject.

Forssberg and Hirschfeld (1994) exposed seated human subjects to sudden small translations and rotations, and monitored the activity of several muscles including the lumbar extensors, neck extensors and neck flexors. Only a weak correlation was found between muscle stretch and muscle activity, and as a result a centrally generated response (involving the central nervous system) was hypothesized, as opposed to a stretch reflex. Forward translation of the platform on which the subjects were seated elicited a response in the cervical and lumbar musculature approximately 75-120 milliseconds after onset of platform movement. Hip extension and pelvis rotation were induced early enough to be implicated in the initiation or triggering of this muscle response. Torso rotation, head displacement, and head rotation occurred simultaneous to or subsequent to muscle activation and were thus not considered as causative factors for the initial muscle response.

METHODOLOGY

Ten nominal 16 km/h closing speed car-to-car impacts were conducted using two mid 1970's Volvos. The impacts were aligned bumper-to-bumper impacts. Human subjects were in the driver's positions in both the bullet and target vehicles. Only the target vehicle occupant was instrumented. Each subject was exposed to 2 impacts over a 2 day period with the only difference between impacts being a modified head restraint for the second impact (2 inches of padding added). Tests were conducted in February of 1996 at Biomechanical Research & Testing LLC, in Los Angeles, CA.

519

Coordinate systems

Several coordinate systems were used for data analysis. System 1 originated at the approximate static vehicle center of gravity with the X axis parallel to the longitudinal axis of the vehicle. System 2 originated at the occupant's approximate head static center of gravity with the X axis in the sagittal plane and parallel to the Frankfort plane. System 3 originated at the approximate posterior base of the cervical spine with the X axis in the sagittal plane and perpendicular to the cervical spine at the origin. System 4 originated at the approximate posterior location of L5-S1 with the X axis in the sagittal plane, and perpendicular to the lumbar spine at the origin.

All axis systems were in accordance with SAE J211 recommended practice with the X, Y and Z axes forward, rightward and downward, respectively. A numerical designation after the axis letter was defined so that both the axis system and direction were identifiable (i.e. head Z acceleration was "Z2", lumbar spine X acceleration was "X4", etc.). Figure 2 shows the coordinate systems.

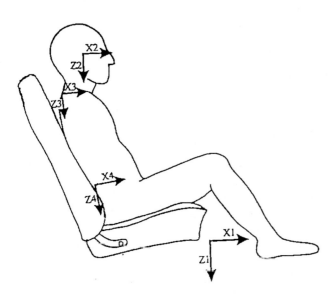

Figure 2: Coordinate Systems

Vehicles

The target vehicle was a 1976 Volvo 242 DL (VIN: VC24245E1094292). The bullet vehicle was a 1977 Volvo 244 (VIN: VC24445H1239613). Both vehicles were equipped with stock bumpers, seats, and restraint systems. The driver's door of the target vehicle was removed to facilitate high-speed video taping of entire body kinematics. The door was securely stored in the vehicle so as to maintain vehicle weight. Bullet and target vehicle weights were 1326 kg and 1304 kg, respectively.

Volvos were selected for several logistical reasons, as it was desirable to minimize any between-test vehicle repairs or adjustments and maximize the repeatability of all impacts. Insurance Institute for Highway Safety tests found the mid-70's Volvo bumper system withstood low speed impacts with no structural damage (IIHS, 1978), while King et al. (1993) noted piston-type energy absorbing bumper systems to have demonstrated a reasonable degree of repeatability in low speed impacts. It was also desirable to minimize any potential plastic seat back deformation which has been found to occur after repeated low speed impacts (observations from tests conducted by Szabo et al.,1994). The 1976 Volvo 242 DL seat exhibited high strength in static tests (Warner et al.,1991), and was deemed unlikely to yield during multiple impacts in the current study.

The standard Volvo seat with integrated head restraint was used in the first test for each occupant (tests T208A1 to T208E1). For each occupant's second test the head restraint was modified by securing an additional 2 inches of head restraint material (from another identical Volvo seat) to the front surface of the existing head restraint (tests T208A2 to T208E2). Table 1 contains descriptions of all tests.

Test	Date	Subject	Head Restraint
T208A1	03-Feb-96	A	Unmodified
T208B1	03-Feb-96	B	Unmodified
T208C1	03-Feb-96	C	Unmodified
T208D1	04-Feb-96	D	Unmodified
T208E1	04-Feb-96	E	Unmodified
T208A2	04-Feb-96	A	Modified
T208B2	04-Feb-96	B	Modified
T208C2	04-Feb-96	C	Modified
T208D2	04-Feb-96	D	Modified
T208E2	04-Feb-96	E	Modified

Table 1: Test Descriptions

Preliminary testing demonstrated the desired closing velocity could be attained within a reasonable degree of accuracy by idling the automatic transmission of the bullet vehicle. A time trap triggered by pressure sensitive tape switches recorded the bullet vehicle's velocity at impact. To avoid secondary impacts the bullet vehicle was equipped with an engine kill switch activated by a pressure sensitive tape switch affixed to the front bumper. In addition, the driver of the bullet vehicle applied the vehicle's brakes following the impact.

The target vehicle ignition was on (standard transmission in neutral) for all tests. Both bullet and target vehicles were inspected for any signs of damage after each test. Target vehicle post-impact roll out distance was measured after each test.

A triaxial block of IC Sensor 3031-050 (50 g) accelerometers was mounted at the target vehicle's approximate static center of gravity, and measured accelerations in the X1, Y1 and Z1 directions.

Occupants

One female and four males (22-54 years) served as test subjects. Informed consent forms and health questionnaires were filled out prior to testing. All subjects described themselves as being in good health on the pre-test questionnaires, although at least three of the subjects reported prior injuries including multiple knee surgeries, cervical injury, and mild closed head injury. Table 2 contains anthropometric data for the subjects.

For each test the subject was instructed to adjust the target vehicle seat and seat back as he/she normally would to drive. Subjects were instructed to sit in the same position they would adopt while stopped at a red light, and to apply brake pressure commensurate with that utilized while stopped for a light. Subjects were instructed to minimize the post impact movement of the target vehicle by braking after impact.

Since many occupants in vehicles struck from the rear report being unaware at the time of impact every effort was made to simulate an unanticipated impact. Subjects were free to look at whatever they chose during impact, but all potential visual cues were removed (i.e. rearview mirror removed, passenger side window covered, no test observers in occupant's field of view) to promote unanticipated impact. Subjects were required to wear portable stereo earphones and listen to an audio tape at high volume for several minutes prior to and during impact so that potential audio cues of impending impact could also be eliminated. Time from "last contact with the test subject" to the point of impact was varied randomly, and ranged from approximately 2 to 5 minutes.

Occupant head accelerations were obtained via 3 triaxial blocks of IC Sensors 3031-050 (50 g) accelerometers affixed to the head via a lightweight headband. The headband was made of rubber which, when tightly fastened to the subject's head, formed a secure bond. One triaxial block was located in the sagittal plane over the forehead while the other two were located approximately above the external auditory meatus, all in a plane parallel to the Frankfort plane.

Peripheral head acceleration measurements were resolved to the approximate head static center of gravity via an algorithm which utilized the locations of each triaxial block relative to known anatomical landmarks (based on Alem and Holstein, 1977). The ability of the headgear and algorithm to accurately predict head center of gravity accelerations was confirmed via testing with the Hybrid III anthropometric dummy. The headband measured accelerations in the X2, Y2 and Z2 directions.

A specially developed low profile (<1 cm) triaxial block of accelerometers was constructed using two Entran EGAXT-50 (50 g) accelerometers and one IC Sensors 3031-050 (50g) accelerometer. This was affixed to the occupant with medical adhesive and a tightly fitted belt at the approximate midline level of L5-S1, and measured accelerations in the X4, Y4, Z4 directions. A lightweight uniaxial IC Sensors 3031-050 (50 g) accelerometer was affixed with medical adhesive to the base of each subject's cervical spine at the approximate location of the C7 spinous process, and measured acceleration in the X3 direction.

Surface electromyography was obtained using standard single-use electrodes. The electrodes were pre-gelled and self-adhesive, and contained a silver-silver chloride electrode measuring 1 cm in diameter. Electrodes were configured in a bipolar arrangement with a spacing on each muscle group of approximately 2 cm center-to-center.

Subject	Age	Sex	WT (kg)	HT (cm)	KN (cm)	SH (cm)	SO (cm)	AA (cm)	KB (cm)	HC (cm)	NC (cm)	CL (cm)	ROMF (deg)	ROME (deg)
A	28	F	54.3	162.9	45.7	88.9	60.3	77.5	55.2	55.2	31.8	12.7	60	87
B	54	M	76.9	177.2	45.7	94.9	63.5	85.1	58.4	59.7	40.6	14.0	50	66
C	32	M	117.2	182.9	50.8	96.5	67.3	85.1	62.9	61.0	44.5	8.9	50	64
D	22	M	62.9	170.2	50.8	86.4	57.8	82.6	57.2	57.2	35.6	13.3	66	88
E	28	M	87.8	182.9	50.8	94.9	67.3	85.7	56.8	62.2	42.5	13.0	68	65

WT: Weight
HT: Height
KN: Knee-to-floor (standing)

SH: Sitting height
SO: Shoulder height
AA: Anterior reach

KB: Knee-to-buttock (sitting)
HC: Head circumference
NC: Neck circumference

CL: Cervical link (neck length)
ROMF: Range of motion (flexion)
ROME: Range of motion (extension)

Table 2: Subject Anthropometry

Electrodes were placed over the right and left sternocleidomastoid muscles (RSCM & LSCM), right and left suboccipital cervical extensor muscles (RCE & LCE), right and left superior trapezius muscles (RTRAP & LTRAP), and right and left paralumbar muscles (RLUM & LLUM). Figure 3 shows the electrode locations.

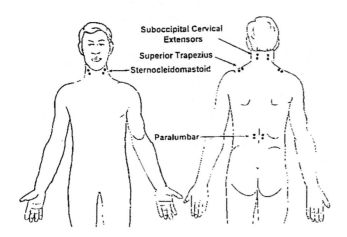

Figure 3: EMG Electrode Locations

EMG recordings were obtained during isometric maximal contractions over an hour prior to each test for all subjects, for purposes of comparison to the muscle activity before and during impact. Maximal contractions for the sternocleidomastoids, cervical extensors, and trapezius were obtained with the body in the seated position. Maximal contractions for the lumbar paraspinals were obtained by hyperextending the lumbar spine while lying prone. Quiescent EMG activity in all muscles was also recorded for each subject prior to each test.

EMG activity was obtained for each subject several minutes prior to testing while sitting relaxed in the vehicle seat, with the foot off the brake. This was later used as a basis for comparison with muscular activity levels recorded immediately prior to impact to assess subject bracing prior to impact.

Subjects completed post-test questionnaires immediately after each test to provide information regarding the subjective experience of the impact. Subjects retained the post-test questionnaire for 2 weeks post-impact, and were instructed to report any test related injuries or symptoms which arose over that period of time. Both pre and post-test questionnaires were based on forms provided by Spitzer et al.(1995).

Instrumentation and data processing-displacements

Occupant displacements were recorded at 500 frames per second with an NAC Memrecam-CI high speed color video camera attached to a lightweight, rigid camera boom mounted to the driver's side of the target vehicle. Standard 30 frame per second video cameras captured various external views of each test.

Targets were mounted on rigid portions of the seat back and head restraint, and on the occupant's approximate sagittal projections of the head center of gravity, shoulder, wrist, pelvis and knee. Displacements were obtained from the high speed video using the NAC Image Express motion analysis system. Two targets on the head were tracked and resolved so the Frankfort plane angle was obtained (defined as the "head angle"). The "upper torso angle" was defined as a line drawn through the approximate locations of the pelvis midline sagittal projection and the base of the cervical spine for each video frame.

In order to select an appropriate filter to remove higher frequency artifacts resulting from the digitization process, a power spectral density analysis of the raw digitized displacement data was conducted. Signal power above 10 Hz was found to be minimal, and deemed to be associated with digitization artifacts. Displacement data were thus low pass filtered at 10 Hz so as to obtain a smoother, more realistic depiction of occupant displacement.

Cervical flexion-extension angle was defined as the angle of the head relative to the upper torso (AMA,1993). Zero degrees of cervical flexion-extension was defined as the position at which the Frankfort plane and the upper torso angle formed a right angle. Standing anatomical position is thus 0 degrees of cervical flexion-extension (White and Panjabi,1990). For an occupant in an automotive seat however, the typical posture adopted is actually one of anatomical cervical flexion, the result of rearward inclination of the seat back and forward flexion of the neck to maintain a line of sight parallel with the ground.

This protocol was adopted so that cervical flexion-extension angle was reported relative to the anatomical position, as opposed to the occupant's initial position in the vehicle. This allowed for measurement of true anatomical cervical flexion-extension during the impact which is likely a more meaningful parameter than cervical flexion-extension angle relative to the initial seated position.

The SAE sign convention dictated cervical flexion be negative, while cervical extension be positive. This has been adopted by others investigating occupant kinematics in low speed impacts (Svensson et al.1993; Szabo et al.,1994).

Instrumentation and data processing-accelerations

Vehicle and occupant accelerometer data were collected with a self contained, on-board data acquisition system. Analog to digital conversion was performed by a 12 bit A/D converter operating with a maximum conversion rate of 330,000 samples per second. Data collection was initiated by a pressure sensitive tape switch on the target vehicle's rear bumper.

Acceleration data were collected at 10,000 samples per second in accordance with SAE J211 recommended practice (SAE,1996). Target vehicle accelerations were filtered using an SAE Class 60 filter (SAE Class 180 filter used for change in velocity calculations). Power spectral density analysis indicated that SAE Class 60 filtering of occupant acceleration data was appropriate for these low speed rear impacts.

Instrumentation and data processing-electromyography

Electromyographic data were collected at 500 samples per second using TELEMG (BTS, Milan, Italy), a 10 channel system consisting of integrated preamplified electrode leads connected to a secondary amplifier/multiplexor unit which was located within the vehicle. A fiber optic connection was used between the secondary amplifier/multiplexor unit and a demultiplexor/signal conditioning unit which was located outside the vehicle. Total system amplification was 1000. Analog to digital conversion was performed by a RTI-860 (Analog Devices) A/D converter. Data capture was manually triggered approximately 5 seconds prior to impact, with a total collection time of 8 seconds. Target vehicle bumper contact switch closure provided a time zero marker for the EMG data.

It has been recommended that EMG data be high pass filtered at 30 Hz in order to minimize any unwanted signal components due to motion artifact (Basmajian and DeLuca,1985; Gerleman and Cook,1992). Given the possibility of transient, potentially high frequency characteristics of occupant motion during vehicular impact, motion artifacts with frequencies above 30 Hz may be encountered during such an event. However, absent any recommendations in the literature in this regard, and in the interests of preserving the maximum amount of data, EMG signals were high pass filtered at 30 Hz.

RESULTS

Vehicle kinematics

Neither the bullet nor target Volvos sustained structural damage in any of the 10 tests. The target vehicle seat backs and vehicle interiors also remained undamaged. Table 3 contains information regarding the vehicle kinematics.

Test	Impact Velocity (km/h)	Target Vehicle Kinematics			
		Max Accn X1 (g)	Onset (msec)	Delta V X1 (km/h)	Delta X (cm)
T208A1	14.3	6.5	7	9.6	287
T208B1	14.6	6.2	6	9.3	158
T208C1	12.9	4.4	4	7.5	168
T208D1	15.0	7.0	5	10.0	491
T208E1	14.2	6.5	7	8.6	198
T208A2	15.0	7.0	3	10.0	238
T208B2	14.8	7.0	3	9.6	253
T208C2	14.0	6.0	5	8.8	207
T208D2	13.8	6.1	5	9.6	180
T208E2	14.0	6.5	5	9.0	162

Table 3: Vehicle Kinematics

Occupant injuries and subjective impressions of impact

All subjects described themselves as relaxed at the time of impact, and stated that they received no forewarning of impact. No subject complained of pain or injury at any time during testing, or over the 2 week period following the impacts. Although a physician was present during testing, no medical consultations were required at any time for any subject. In 7 of the 10 tests, the occupant described his/her motion as rearward-then-forward, while in the remainder a forward-then-rearward motion was described. Four of five subjects described the second test (with the decreased head-to-head restraint distance) as being less severe than the first, while the remaining subject indicated a similar severity for the two impacts.

Occupant displacements

The now familiar rearward then forward (rebound) occupant motion, as described by McConnell et al. (1993; 1995) and Szabo et al. (1994), was observed in all tests. Figure 4 contains still photographs obtained from the high speed video for a subject undergoing rear impact with the unmodified head restraint.

Figure 4: Occupant Kinematics during Rear Impact [T208D1]

220 msec

280 msec

200 msec

260 msec

180 msec

240 msec

300 msec

Figure 4: Occupant Kinematics during Rear Impact [T208D1]

Test	Rearward (Initial) Motion						Forward (Rebound) Motion					
	Head (cm)	Time (msec)	Shoulder (cm)	Time (msec)	Knee (cm)	Time (msec)	Head (cm)	Time (msec)	Shoulder (cm)	Time (msec)	Knee (cm)	Time (msec)
T208A1	dl		dl		dl		dl		dl		dl	
T208B1	dl		dl		dl		dl		dl		dl	
T208C1	-16.7	156	-10.8	196	-7.3	184	5.8	332	0.6	432	-0.4	348
T208D1	-13.5	116	-12.0	112	-8.8	120	8.4	300	-1.7	284	-3.5	264
T208E1	-16.4	140	-12.2	116	-9.1	120	11.6	328	2.0	332	-1.1	296
T208A2	-6.9	100	-8.2	104	-7.2	108	10.0	364	0.1	208	-0.2	204
T208B2	-9.9	116	-11.6	124	-7.9	116	8.9	196	0.4	236	-2.9	228
T208C2	-15.7	140	-11.9	128	-9.0	124	12.4	332	4.6	320	-0.1	292
T208D2	-8.6	108	-9.8	116	-6.5	112	7.3	596	-0.6	416	-1.3	212
T208E2	-12.3	124	-12.2	120	-10.0	120	11.0	308	0.6	316	-1.6	280

dl denotes data loss

All measurements are relative to the initial position of each body segment, with rearward displacement negative, and forward displacement positive

A negative value for maximum forward motion indicates that the body segment did not return through it's initial x position during impact

Table 4: Occupant Linear Displacements

Initial head-to-head restraint horizontal distances for the unmodified head restraint ranged from 7.6 to 11.4 cm. Rearward shoulder motion occurred within the first 200 milliseconds after bumper contact, followed by forward motion for an additional 200-250 milliseconds. Figure 5 contains representative displacement trajectories from an impact with the unmodified head restraint.

The shoulder belt locked during all tests, restraining the occupant during rebound, and the occupant's hands remained in the vicinity of the steering wheel (although a secure grip did not appear to be maintained). Table 4 contains displacement data for all tests. Note that a negative maximum displacement during the rebound motion indicates that the segment did not return through it's initial position during rebound.

In all cases cervical extension and flexion remained well within each subject's individual range of motion, consistent with the findings of McConnell et al.(1993,1995) and Szabo et al.(1994). No cervical spine hyperextension or hyperflexion was found for any test. Table 5 contains maximum angular displacements and the times at which they occurred.

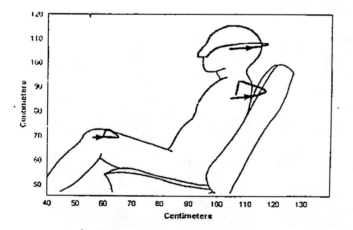

Figure 5: Typical Occupant Displacement
[T208D1]

	Rearward (Initial) Motion				Forward (Rebound) Motion		
Test	Initial (deg)	Ext. (deg)	Delta (deg)	Time(E) (msec)	Flx. (deg)	Delta (deg)	Time(F) (msec)
T208A1	dl		dl		dl		
T208B1	dl		dl		dl		
T208C1	-18	1	-19	172	-28	10	356
T208D1	-12	8	-20	124	-35	22	296
T208E1	-22	4	-26	160	-57	35	340
T208A2	-19	-12	-8	128	-43	24	280
T208B2	-18	-5	-13	128	-35	17	336
T208C2	-19	-12	-7	156	-37	18	256
T208D2	-16	-7	-9	124	-19	4	244
T208E2	-23	-11	-13	148	-57	33	324

Initial: Initial flexion angle at impact

Ext.: Maximum cervical spine extension angle

Flx.: Maximum cervical spine flexion angle

Time(E): Time at which maximum extension occurs

Time(F): Time at which maximum flexion occurs

Table 5: Occupant Angular Displacements

Figure 6 contains representative time histories of cervical flexion-extension angle for the standard and modified head restraints. Note that the sign convention dictated that cervical spine extension be positive.

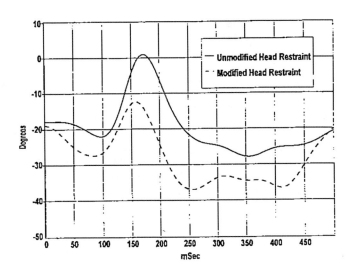

Figure 6: Cervical Flexion-Extension Angle
[T208C1 & T208C2]

Occupant accelerations

Figures 7 and 8 contain typical vehicle, head, and lumbar spine X direction accelerations for tests using the standard and modified head restraints, respectively.

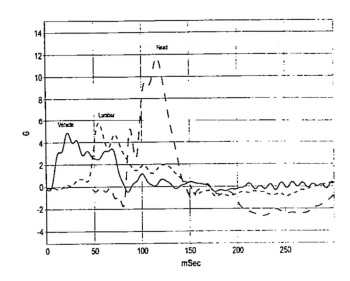

Figure 7: Typical Vehicle X1, Lumbar Spine X4, and Head X2 Accelerations (Unmodified Head Restraint)
[T208D1]

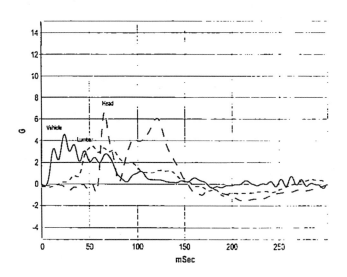

Figure 8: Typical Vehicle X1, Lumbar Spine X4, and Head X2 Accelerations (Modified Head Restraint)
[T208D2]

Table 6 contains peak values for the head center of gravity (X2, Z2, resultant) and lumbar spine (resultant) accelerations. Onset times for the head and lumbar resultant accelerations were defined as the time at which the acceleration reached 5% of peak, and are also included in Table 6. In some tests, particularly those using the modified head restraint, direct impacts between the head restraint and cervical spine accelerometer occurred. These impacts tended to artificially increase the accelerometer readings, and as a result peak cervical spine accelerations are not presented.

Test	Head Res. Max (g)	Onset (msec)	Head X2 Max (g)	Min (g)	Head Z2 Max (g)	Min (g)	Lumbar Res Max (g)	Onset (msec)
T208A1	14.8	71	14.8	-3.0	9.8	-1.9	5.2	9.0
T208B1	11.7	57	7.5	-2.2	10.0	-2.7	4.5	12.0
T208C1	16.6	79	7.5	-2.6	16.2	-3.0	4.3	10.0
T208D1	13.1	63	11.7	-2.6	6.2	-1.8	6.3	11.0
T208E1	17.2	66	6.5	-2.7	17.2	-2.1	5.1	12.0
T208A2	7.9	25	7.5	-2.0	3.2	-2.8	7.5	11.0
T208B2	7.9	41	5.2	-2.0	5.4	-2.7	6.6	10.0
T208C2	13.1	73	8.0	-2.8	12.2	-5.0	4.7	10.0
T208D2	6.6	54	6.6	-1.5	3.5	-1.3	3.9	10.0
T208E2	7.5	64	6.1	-2.2	5.5	-2.7	5.1	10.0

Table 6: Occupant Linear Accelerations

Electromyography

Figure 9 shows typical EMG traces during an impact with the unmodified head restraint. Note that only the right side activity for each muscle group is presented, as right and left sided muscle group activities were essentially similar. Bumper contact corresponds to "0" on the X-axis.

Table 7 contains the onset times for EMG activity in all tests. Onset times were defined as the first noticeable increase in EMG activity for the unfiltered signal.

Test	RSCM (msec)	LSCM (msec)	RCE (msec)	LCE (msec)	RTRA (msec)	LTRA (msec)	RLUM (msec)	LLUM (msec)
T208A1	114	120	106	100	112	126	110	104
T208B1	126	120	102	106	132	124	120	114
T208C1	124	126	134	110	102	128	104	116
T208D1	130	124	126	130	130	126	114	110
T208E1	116	124	126	122	122	120	136	108
Ave.	122	123	119	114	120	125	117	110
T208A2	116	108	104	114	120	116	98	120
T208B2	116	106	100	118	116	126	98	94
T208C2	124	114	138	132	104	124	96	102
T208D2	112	100	114	122	110	126	106	114
T208E2	120	124	116	116	126	126	94	104
Ave.	118	110	114	120	115	124	98	107

SCM: sternocleidomastoid   TRAP: upper trapezius
CE: cervical extensors   LUM: lumbar paraspinals

Table 7: Muscle Activity Onset Times

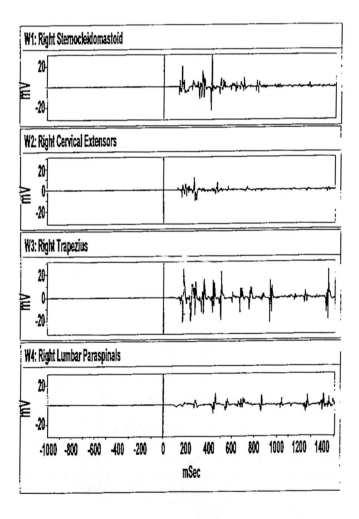

Figure 9: Typical Unfiltered EMG Histories
[T208D1]

DISCUSSION

No subject sustained injury or complained of pain during the two week period following the impacts. This is consistent with the results obtained in prior rear impact tests using human subjects (see Figure 1), and reinforces the notion that a reasonably healthy human occupant can withstand a rear impact with a change in velocity of 8 km/h without sustaining significant injury, assuming the presence of a head restraint and a reasonably "normal" initial seating position.

Quantification of EMG signal

In order to quantitatively assess EMG signals several analysis techniques are available. One such technique is the Root-Mean-Square (RMS). RMS is used for analysis of noisy, nonperiodic and nonsinusoidal signals in many engineering applications. The root-mean-square value for a signal is a measure of the magnitude of the AC component of the signal and is calculated according to the following:

$$RMS\{m(t)\} = \left(\frac{1}{T} \int_{t}^{t+T} m^2(t)dt \right)^{\frac{1}{2}}$$

Basmajian and DeLuca (1985), DeLuca (1979), Johnson (1978) and Lawrence and DeLuca (1983) noted the RMS value contains more relevant information than other signal analysis techniques such as the mean rectified or integrated EMG signal, and recommended this technique for EMG signal analysis.

Further, many studies have found a linear or near linear relationship between RMS values and muscle force (Basmajian and DeLuca,1985; DeVries,1968; Lawrence and DeLuca,1983; Leveau and Andersson,1992; Lind and Petrofsky,1979). A large intersubject variability in the absolute amplitude of the RMS has been reported (Lind and Petrofsky,1979), and as a result the RMS technique is limited to intrasubject analyses.

RMS was chosen as the EMG analysis technique in the current study. As RMS is time-period dependent for relatively transient signals, it was desirable to select a constant time period with which to calculate the RMS for all signals. Given that the majority of muscle activity in the vehicle impact occurred over a period of 500 milliseconds, this served as the RMS period for all signals analyzed. Electromyographic data were processed and analyzed using the DADiSP data analysis software package.

RMS values for each impact were calculated using the 500 millisecond period immediately after the onset of muscle activity. RMS values for maximal isometric contractions and quiescent signals used a 500 millisecond period over which a consistent level of activity was observed. Table 8 contains RMS values for all maximal isometric contractions, and for each impact (expressed as a percentage of maximal isometric RMS). EMG signals were processed with a 30 Hz high-pass cutoff filter prior to RMS calculations.

Test	RSCM MIC (RMS)	RSCM VI (%RMS)	LSCM MIC (RMS)	LSCM VI (%RMS)	RCE MIC (RMS)	RCE VI (%RMS)	LCE MIC (RMS)	LCE VI (%RMS)	RTRAP MIC (RMS)	RTRAP VI (%RMS)	LTRAP MIC (RMS)	LTRAP VI (%RMS)	RLUM MIC (RMS)	RLUM VI (%RMS)	LLUM MIC (RMS)	LLUM VI (%RMS)
T208A1	4.9	92.3%	4.0	85.8%	dl	dl	dl	dl	1.1	145.6%	1.7	174.6%	2.8	12.1%	3.4	6.1%
T208B1	3.8	43.3%	4.0	50.6%	dl	dl	dl	dl	0.7	122.8%	1.7	97.2%	1.6	11.6%	1.3	19.5%
T208C1	4.6	43.3%	5.3	44.2%	1.3	78.4%	1.0	78.4%	1.4	96.1%	1.2	135.8%	3.2	10.4%	3.2	10.9%
T208D1	7.3	59.8%	4.2	73.3%	5.3	37.3%	2.7	37.3%	16.7	31.4%	6.5	39.0%	4.5	23.7%	3.8	37.1%
T208E1	6.0	38.8%	4.6	35.1%	2.9	19.6%	1.9	19.6%	2.4	52.3%	2.0	49.0%	3.9	9.0%	5.3	6.6%
T208A2	6.4	40.9%	4.7	35.4%	2.3	61.9%	1.9	61.9%	7.6	17.5%	3.0	45.4%	3.1	27.5%	3.5	21.9%
T208B2	5.2	36.3%	4.6	44.6%	1.2	41.3%	1.5	41.3%	2.0	52.7%	3.5	33.8%	1.3	66.7%	1.1	47.9%
T208C2	4.6	55.4%	5.3	51.5%	1.5	117.8%	1.6	117.8%	1.1	254.9%	1.7	112.4%	3.2	9.2%	2.7	10.0%
T208D2	7.3	28.2%	4.2	59.4%	5.3	37.5%	2.7	37.5%	16.7	32.2%	6.5	27.1%	4.5	25.4%	3.8	32.0%
T208E2	6.0	42.3%	4.6	42.9%	2.7	20.1%	2.6	20.1%	2.4	74.3%	2.0	75.0%	3.9	8.3%	5.3	5.7%

MIC: Maximum isometric contraction
VI: Rear impact

RSCM: Right sternocleidomastoid
LSCM: Left sternocleidomastoid
RCE: Right cervical extensors
LCE: Left cervical extensors

RTRAP: Right trapezius
LTRAP: Left trapezius
RLUM: Right lumbar paraspinal
LLUM: Left lumbar paraspinal

Table 8: RMS Values for Impacts and Maximal Isometric Contractions

The RMS values during impact were relatively high when compared to those obtained during maximal isometric contractions. In fact Table 8 shows that some of the RMS values actually exceeded those obtained during the maximal isometric contraction (RMS >100% maximum) which is a seemingly problematic result.

One explanation lies in the potential presence of relatively high frequency motion artifacts. Such artifacts could contribute to the RMS of the signal during impact, without any actual underlying increase in muscle activity. The EMG signals were high-pass filtered at 30 Hz. However during the relatively rapid motion incurred by the human body during low speed rear impact some motion artifacts with frequencies higher than 30 Hz may have been encountered. These frequencies would be superimposed upon true muscle activity and become difficult to isolate and remove. A higher cutoff frequency for the high pass filter was considered, however given the frequency response for surface EMG (Winter, 1990) it was felt that such a filter may remove some of the actual muscle signal which is clearly undesirable.

It may be possible to decrease the likelihood of high frequency motion artifact with the use of fine wire electrodes. Given that electromyographic signals obtained with fine wire electrodes have a higher frequency content than those obtained with surface electrodes, it may also be appropriate to filter fine wire signals with a higher cutoff frequency. Future endeavors in the area of EMG and human subject impacts tests should weigh these potential benefits against the increased methodological complexities associated with the fine wire technique.

A second possible explanation for the relatively high RMS values found in the current study may be the contraction velocity dependence of EMG activity. Increased electromyographic activity has been reported for increased velocity of contraction (Heckathorne and Childress, 1981). Since most of the muscles in the current study likely underwent relatively high velocity contractions at some point it may not be surprising that some RMS values were near or above 100% of maximal isometric contraction RMS values.

Another consideration for muscle force calculation from EMG signals is summarized by Loeb and Gans (1986), who stated that "muscles frequently operate under conditions of imposed lengthening, for which they can generate more than the maximal isometric force, particularly for short excursions". This phenomenon was also noted by Basmajian and DeLuca (1985), Lamb and Hobart (1992) and Winter (1990).

Given that maximal muscle activity levels were obtained isometrically while the rear impact muscle activity levels were obtained dynamically, the possibility that the EMG signals contained some high frequency motion artifacts, and the likely existence of some eccentric contractions during the impacts, no attempt was made to correlate muscle RMS values with force output during the dynamic portions of the rear impact. This conservative approach is consistent with the recommendations of several researchers including Lieber (1992) who stated "in isometric or perfectly eccentric or concentric conditions, force might be estimated [using EMG], but otherwise, be very careful!"

Future work in the area may consider maximal activity tests which more closely resemble the event under study than isometric contractions in terms of joint angles and velocity of movement (Hawkins and Hull, 1992; Loeb and Gans, 1986). It is questionable whether sufficient joint velocity could ever be practically generated for application to human subject vehicular impacts, however.

Pre-impact muscle activity

Since virtually no body motion occurred prior to impact and motion artifact and velocity dependent contractions were not an issue for this time period, the pre-impact RMS values are likely an accurate reflection of muscle activity immediately prior to the impact.

As a means of evaluating the preparedness of the occupant for impact Table 9 contains the percent maximum RMS values for the 500 millisecond period immediately pre-impact, and for a 500 millisecond period obtained while the occupant was sitting relaxed in the vehicle during set-up, several minutes prior to impact. EMG signals were processed with a 30 Hz high-pass filter prior to RMS calculations.

RMS values recorded immediately prior to impact and those recorded while sitting relaxed in the car (several minutes prior to impact) were typically 1-5% of maximum isometric contraction RMS. No significant differences between percent maximal RMS values for the pre-impact ($\bar{X}$ = 3.7%) and "relaxed sitting" ($\bar{X}$ = 3.1%) conditions were found (P > 0.1). Isolated cases where the EMG activity in the trapezius or lumbar paraspinal musculature immediately prior to impact was greater than that sitting relaxed in the vehicle (see Table 9) were likely due to the subjects not braking when the "relaxed sitting" readings were obtained.

Test	RSCM PRE-VI (%RMS)	SIT (%RMS)	LSCM PRE-VI (%RMS)	SIT (%RMS)	RCE PRE-VI (%RMS)	SIT (%RMS)	LCE PRE-VI (%RMS)	SIT (%RMS)	RTRAP PRE-VI (%RMS)	SIT (%RMS)	LTRAP PRE-VI (%RMS)	SIT (%RMS)	RLUM PRE-VI (%RMS)	SIT (%RMS)	LLUM PRE-VI (%RMS)	SIT (%RMS)
T208A1	3.1%	1.6%	2.8%	2.4%	dl	dl	dl	dl	4.6%	3.6%	9.9%	10.6%	1.7%	1.5%	1.5%	1.2%
T208B1	2.0%	2.1%	1.8%	2.1%	dl	dl	dl	dl	12.9%	6.3%	16.5%	2.4%	2.5%	2.9%	4.0%	3.4%
T208C1	1.7%	1.1%	1.8%	1.0%	5.4%	5.4%	4.7%	5.1%	6.4%	12.5%	4.3%	7.9%	1.4%	1.4%	1.4%	1.5%
T208D1	1.6%	0.9%	3.5%	2.1%	1.5%	1.2%	2.5%	1.9%	0.5%	0.4%	1.0%	0.9%	1.2%	1.3%	2.8%	1.6%
T208E1	2.2%	1.1%	2.3%	1.8%	2.4%	2.4%	3.5%	3.2%	5.9%	2.4%	2.8%	3.1%	1.4%	1.3%	1.0%	1.1%
T208A2	2.1%	1.0%	2.0%	2.2%	5.1%	3.9%	5.3%	7.2%	0.8%	11.1%	2.9%	6.4%	3.8%	2.7%	1.7%	1.7%
T208B2	1.5%	1.4%	1.8%	1.6%	6.1%	6.2%	4.0%	4.0%	3.6%	2.9%	6.5%	2.4%	6.9%	4.6%	6.0%	6.1%
T208C2	2.9%	1.2%	2.7%	1.2%	6.6%	4.9%	4.7%	3.8%	22.1%	12.6%	4.7%	3.8%	1.5%	1.4%	2.1%	1.9%
T208D2	1.5%	0.8%	3.7%	1.6%	1.4%	1.7%	2.9%	2.7%	1.6%	0.4%	1.1%	0.7%	1.4%	1.0%	2.1%	3.0%
T208E2	2.4%	0.8%	2.5%	1.3%	2.8%	2.6%	3.9%	2.1%	2.4%	1.7%	3.0%	2.5%	1.4%	1.2%	0.9%	1.0%

PRE-VI: immediately prior to impact   RSCM: Right sternocleidomastoid   RTRAP: Right trapezius

SIT: Relaxed, sitting in car several minutes prior to impact   LSCM: Left sternocleidomastoid   LTRAP: Left trapezius

RCE: Right cervical extensors   RLUM: Right lumbar paraspinal

LCE: Left cervical extensors   LLUM: Left lumbar paraspinal

Table 9: Percent Maximum RMS for Pre-Impact and for "Relaxed Sitting"

The current study thus simulated rear impacts in which the occupant in the target vehicle was not braced for impact. This confirmed the subjective evaluations made by all subjects, who indicated they were relaxed and not anticipating the impending impact. The contention that human subject tests inherently incorporate an occupant who is aware and braced for impact is not supported by this study. Under the proper conditions, an unbraced state is relatively easily reproduced in the testing environment.

Note that although the pre-impact and "relaxed sitting" percent maximal RMS values are low, they still represent some baseline muscle activity. When compared to quiescent RMS levels, they were found to be significantly higher (P < 0.1). This is a reasonable result since occupants in the seated position exhibit some postural muscle activity (Schuldt, 1988).

Muscle Activity Onset Times

Muscle onsets in the current test series typically occurred during the initial phase of impact, while the cervical spine was extending. Figure 10 shows the EMG activity relative to cervical spine flexion-extension for a test with the unmodified head restraint. It should be noted that full muscle tension does not develop until approximately 60-70 milliseconds after the onset of muscle activity (Snyder et al., 1975), and thus likely did not develop until the cervical spine was undergoing flexion in the current study.

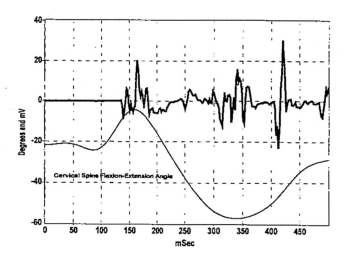

Figure 10: Typical Cervical Flexion-Extension History and Unfiltered EMG Activity
[T208E1]

The onset of muscle activity is a parameter potentially relevant to low speed occupant mathematical models, occupant protection strategies, and soft tissue injury mechanisms. Before any analysis of muscular activity onset times was conducted, the issue of potential motion artifact was evaluated.

Figure 11 shows the muscle activity for 3 different muscles in test T208A1. Vehicle X1, lumbar spine resultant, and head acceleration resultant are superimposed on the EMG traces for reference.

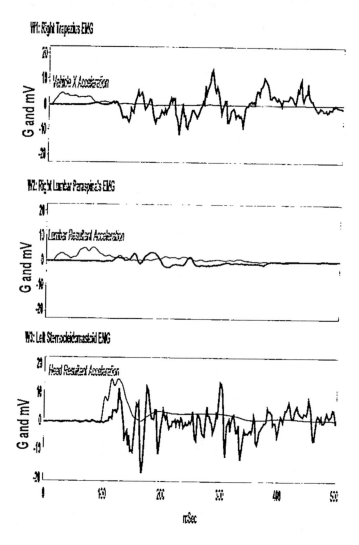

W1: Right Trapezius EMG

W2: Right Lumbar Paraspinals EMG

W3: Left Sternocleidomastoid EMG

Figure 11: Typical Unfiltered EMG Activity and Accelerations [T208A1]

The majority of lumbar spine acceleration occurred prior to any noticeable signal from the lumbar spine EMG electrodes. Lumbar movement clearly occurred prior to the onset of lumbar EMG signal in spite of the susceptibility of the lumbar spine electrodes to motion artifact (due to contact with the seat back). It is thus likely the onset times for muscular activity in the current study are an accurate representation of initial muscle activity, as opposed to motion artifact.

It appears the onset of muscle activity in different parts of the body occurred at approximately the same time and were not necessarily dependent on acceleration or movement in that area of the body. For example in test T208A1 (Figure 11), muscle onset times for RTRAP, LLUM and LSCM occurred within a 16 millisecond period.

Lumbar muscle activity in test T208A1 occurred about 80 milliseconds after the onset of lumbar acceleration, while sternocleidomastoid activity was initiated relatively close in time to the onset of head acceleration. Similar results were found for other tests. It is reasonable to hypothesize from these observations that some form of centrally generated muscle response may have been present during the initial stages of impact, as opposed to a localized stretch reflex. This is consistent with the findings of Forssberg and Hirschfeld (1994), who also hypothesized such a central response for low level anterior-posterior perturbations of a seated subject. Stretch reflexes may later contribute to the overall muscle activity, but did not appear to be significant in triggering initial muscle response for this test series.

At this time the physiological basis for a centrally generated response is unclear. Forssberg and Hirschfeld (1994) hypothesized that the central nervous system (CNS) shifts the equilibrium point of the limbs, resulting in an increased level of muscle activity in both the agonist and antagonist muscle group. Evaluation of this, or other hypotheses, is beyond the scope of the current study.

Trigger Mechanism for Centrally Generated Response
Muscle onset times relative to certain occupant kinematic parameters were calculated for the current test series in an effort to gain insight into the mechanism by which a central response may have been triggered during rear impact. Horak et al.(1990) indicated such a response comes from at least three sources: 1) the somatosensory system, 2) the vestibular system, and 3) vision.

Relative onset times were defined as "latency periods", for a given muscle and kinematic parameter. Figures 12 and 13 contain latency periods for all muscles relative to the onset of head resultant, lumbar spine resultant, and vehicle X1 acceleration for tests with the modified and unmodified head restraints, respectively.

No clear pattern of muscle recruitment was noted. There was no clear distinction between latency periods for impacts with the modified and unmodified head restraints. Overall representative latency periods were estimated for all muscles for each of three kinematic parameters (vehicle X1, lumbar spine resultant and head acceleration resultant onset), and are presented in Table 10.

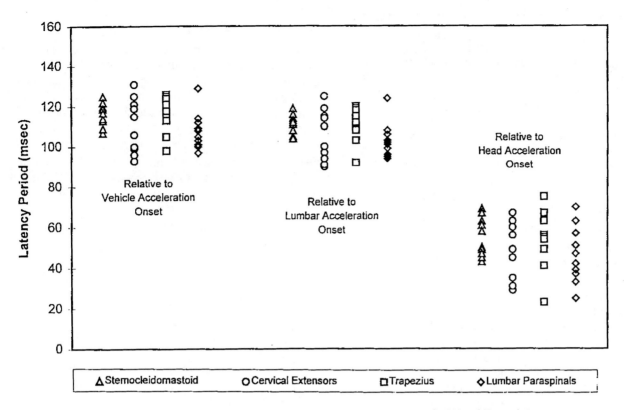

Figure 12: Muscle Activity Latency Periods for Unmodified Head Restraint

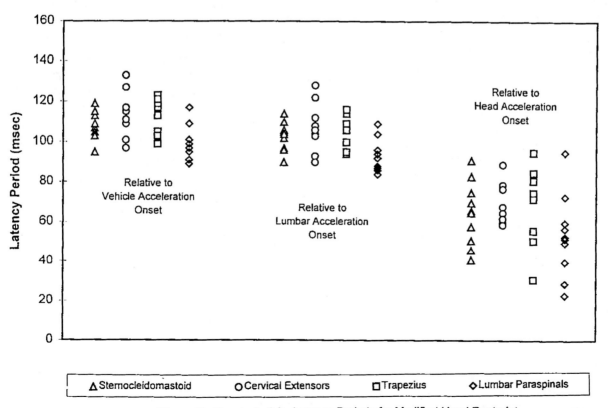

Figure 13: Muscle Activity Latency Periods for Modified Head Restraint

Parameter	Representative Latency Period (msec)
Onset of Vehicle X1 Acceleration	100-125
Onset of Lumbar Resultant Acceleration	90-120
Onset of Head Resultant Acceleration	30-90

Table 10: Representative Latency Periods (All Tests)

Latency periods relative to the onset of head acceleration were as low as 20-30 milliseconds in some tests, significantly lower than reflex times reported by Snyder et al. (1975) and Reid et al. (1981). It is not unreasonable to infer from this that the stimulus for the centrally generated response mechanism occurred prior to the onset of head acceleration.

The representative latency period for muscle activity relative to the onset of lumbar spine acceleration was 90-120 milliseconds. Forssberg and Hirschfeld (1994) hypothesized that pelvic movement likely triggered a centralized response in seated subjects, and found a latency period for cervical and lumbar muscles of approximately 75-120 milliseconds from the onset of pelvic motion. The latency periods for the cervical and lumbar spine muscles obtained in the current study compare favorably to those in the Forssberg study.

The centrally generated response may thus have been triggered by lumbar spine acceleration, occurring approximately 90-120 milliseconds after the onset of lumbar acceleration. Following this other somatosensory, vestibular and visual stimuli may have contributed to the overall muscle activity. Quantification of these other stimuli was beyond the scope of the current study.

Although all subjects in the current study indicated they were not anticipating the impacts, and insignificant pre-impact muscle activity was found, it is unknown whether some form of mental preparedness existed prior to impact, and whether any such preparedness could have an effect on muscle response and latency periods. Other parameters (such as the interaction between the hands or feet and the vehicle interior) were also not monitored in the current study. It is unknown whether any of these parameters influenced muscle response patterns or the triggering mechanism for the centrally generated response.

Jones and Kennedy (1951) noted that sudden sounds alone could induce a 'startle pattern' of muscle activity in the paracervical musculature, with a latency period of 25-50 milliseconds. It is therefore also possible that auditory cues may provide a stimulus for a centrally generated response during low speed rear impact. This phenomenon was not investigated in the current study since subjects were deprived of audio cues prior to and during impact.

Head restraint condition

All subjects but one reported that the impact with the modified head restraint felt less severe than that with the unmodified head restraint. One subject felt that the two impacts were similar. The modified head restraint resulted in decreased rearward head displacements (average decrease of 3.3 cm), decreased cervical extension (average decrease of 12.3 deg.), and decreased peak positive (average decrease of 2.9 G) and peak negative (average decrease of 0.5 G) head X2 accelerations. Head peak resultant accelerations were decreased an average of 6.1 G. No differences in occupant muscle activity onset were found between head restraint configurations.

CONCLUSIONS

1. No human subject sustained injury after undergoing two rear impacts with a change in velocity of 8-10 km/h over a span of two days. This lends support to the notion that a rear impact with a change in velocity of 8 km/h or less is within tolerance for a reasonably healthy occupant, assuming the presence of a head restraint and relatively "normal" initial seating position.

2. The surface EMG method used was found to be appropriate for determining muscle activity onset times and pre-impact muscular activity (bracing). Limitations in muscle force estimation during impact were encountered, possibly due to relatively high frequency motion artifacts and/or the contraction velocity dependence of the EMG signal.

3. Muscle activity immediately before impact was essentially similar to that obtained while the occupant was sitting relaxed in the vehicle several minutes prior to impact. The current study thus simulated "unbraced" impacts consistent with subject indications that they were not anticipating the impact.

4. Muscle onset typically occurred during the initial phase of impact while the cervical spine was extending. Full muscle tension likely did not develop until later while the cervical spine was flexing.

5. No clear differences were noted between different muscle group onset times. Muscle onsets appeared to occur independently of localized accelerations, indicating the likelihood of a centrally generated response mechanism for muscle recruitment.

6. Initial muscle activity was consistent with having been triggered by the onset of lumbar spine acceleration, occurring approximately 90-120 milliseconds after onset of lumbar resultant acceleration.

7. The addition of 2 inches of padding to the head restraint (thus reducing the initial head-to-head restraint horizontal distance) resulted in decreases in head acceleration, decreases in rearward head displacement, decreases in cervical extension, and a reduction in the subjects' perception of impact severity.

8. Future work in this area should consider fine wire electrodes as a potential measure to decrease the possibilities of high frequency motion artifacts contained in the EMG signals. Baseline maximal activity levels should also attempt to incorporate maximum practical joint velocities, although it is questionable as to whether sufficient velocity could be attained to allow meaningful muscle force estimation, particularly with respect to the cervical spine. If possible, the existence and/or significance of mental preparedness for impact should also be examined. Other potential stimuli such as the hands and feet interacting with the vehicle, or the initial sound of a rear impact should be evaluated for potential effects on muscle activity response patterns and/or latency periods. Monitoring EMG activity when the subject is fully aware and braced for the impending impact may also be beneficial.

ACKNOWLEDGEMENTS

The authors wish to express their appreciation to Dr. Kenton Kaufman and Steven Irby of Children's Hospital and Health Center, and to Richard Sutherland of Itronics for their professional assistance in data collection during the crash tests.

REFERENCES

Alem NM, Holstein, "Measurement of 3D Motion", HSRI Report No. UM-HSRI-77-46, October 1977.

American Medical Association (AMA), Guide to the Evaluation of Permanent Impairment, Edited by Theodore C. Doege, American Medical Association, Chicago, 1993.

Baker Engineering, Unpublished test data provided courtesy of Harvey West, 1996.

Basmajian JV, DeLuca CJ, Muscles Alive: Their Functions Revealed Through Electromyography, 5th edition, Williams & Wilkins, ISBN 0-683-00414-X, Baltimore, 1985.

DeLuca CJ, "Physiology and Mathematics of Myoelectric Signals", IEEE Transactions on Biomedical Engineering, BME-26(6), pp 313-325, 1979.

DeVries HA, "Efficiency of Electrical Activity as a Physiological Measure of the Functional State of Muscle Tissue", American Journal of Physical Medicine, 47, pp 10-22, 1968.

Foret-Bruno JY, Dauvilliers F, Tarriere C, Mack P, "Influence of the Seat and Head Rest Stiffness on the Risk of Cervical Injuries", 13th International Technical Conference on Experimental Safety Vehicles, pp 968-974, 1991.

Forssberg H, Hirschfeld H, "Postural Adjustments in Sitting Humans Following External Perturbations: Muscle Activity and Kinematics", Experimental Brain Research, 97(3), pp 515-527, 1994.

Foust DR, Chaffin DB, Snyder RG, Baum JK, "Cervical Range of Motion and Dynamic Response and Strength of Cervical Muscles", Proceedings of the 17th Stapp Car Crash Conference, (SAE No. 730975, pp 285-308, 1973.

Geigl BC, Leinzinger P, Roll, Muhlbauer M, Bauer G, "The Movement of the Head and Cervical Spine During Rearend Impact", 1994 International IRCOBI Conference on the Biomechanics of Impacts, pp 127-137, 1994.

Gerleman DG, Cook TM, "Instrumentation", in Selected Topics in Surface Electromyography for Use in the Occupational Setting: Expert Perspectives, National Institute for Occupational Safety and Health, U.S. Department of Health and Human Services, 1992.

Hawkins DA, Hull ML, "An Activation-Recruitment Scheme for Use in Muscle Modeling", Journal of Biomechanics, 25(12), pp 1467-1476, 1992.

Heckathorne CW, Childress DS, "Relationships of the Surface Electromyogram to the Force, Length, Velocity, and Contraction Rate of the Cineplastic Human Biceps", American Journal of Physical Medicine, 60(1), pp 1-9, 1981.

Horak FB, Nashner LM, Diener HC, "Postural Strategies Associated with Somatosensory and Vestibular Loss", Experimental Brain Research, 82(1), pp 167-177, 1990.

Huston RL, Huston JC, Harlow MW, "Comprehensive, Three-Dimensional Head-Neck Model for Impact and High Acceleration Studies", Aviation, Space, and Environmental Medicine, 49(1), pp 205-210, 1978.

IIHS (Insurance Institute for Highway Safety), "Report of 1978 Vehicle Low Speed Impact Testing", General Environments Corporation Report No. 6001, June 1978.

Jakobsson L, Norin H, Jernstrom C, Svensson SE, Isaksson-Hellman I, Svensson MY, "Analysis of Different Head and Neck Responses in Rear-End Car Collisions Using a New Humanlike Mathematical Model", 1993 International IRCOBI Conference on the Biomechanics of Impacts, pp 109-125, 1993.

Johnson JC, "Comparison of Analysis Techniques for Electromyographic Data", *Aviation, Space and Environmental Medicine*, 49(1), pp 14-18, 1978.

Jones FP, Kennedy JL, "An Electromyographic Technique for Recording the Startle Pattern", *Journal of Psychology*, 32, pp 63-68, 1951.

King DJ, Siegmund GP, "Staged Collisions: Roles of Bumpers, Estimating Impact Severity, Injury Potential", Low Speed Rear Impact Collision TOPTEC, Society of Automotive Engineers, Irvine CA, 1994.

King DJ, Siegmund GP, Bailey MN, "Automobile Bumper Behaviour in Low-Speed Impacts", Society of Automotive Engineers, SAE No. 930211, pp 1-18, 1993.

Lamb R, Hobart D, "Anatomic and Physiologic Basis for Surface Electromyography", in <u>Selected Topics in Surface Electromyography for Use in the Occupational Setting: Expert Perspectives</u>, National Institute for Occupational Safety and Health, U.S. Department of Health and Human Services, 1992.

Lawrence JH, DeLuca CJ, "Myoelectric Signal Versus Force Relationship in Different Human Muscles", *Journal of Applied Physiology*, 54(6), pp1653-1659, 1983.

LeVeau B, Andersson G, "Output Forms: Data Analysis and Applications", in <u>Selected Topics in Surface Electromyography for Use in the Occupational Setting: Expert Perspectives</u>, National Institute for Occupational Safety and Health, U.S. Department of Health and Human Services, 1992.

Lieber R, "<u>Skeletal Muscle Structure and Function: Implications for Rehabilitation and Sports Medicine</u>", Williams and Wilkins, Baltimore, 1992.

Lind AR, Petrofsky JS, "Amplitude of the Surface Electromyogram During Fatiguing Isometric Contractions", *Muscle & Nerve*, 2, pp 257-264, 1979.

Loeb GE, Gans C, <u>Electromyography for Experimentalists</u>, The University of Chicago Press, ISBN 0-226-49014-9, Chicago, 1986.

MacInnis Engineering Associates, Ltd., Unpublished test data provided courtesy of Mark Bailey, 1996.

Martinez JL, Garcia DJ, "A Model for Whiplash", *Journal of Biomechanics*, 1, pp 23-32, 1968.

Matsushita T, Sato TB, Hirabayashi K, Fujimura S, Asazuma T, Takatori T, "X-Ray Study of the Human Neck Motion Due to Head Inertia Loading", Proceedings of the 38th Stapp Car Crash Conference, SAE No. 942208, pp 55-64, 1994.

McConnell WE, Howard RP, Guzman HM, Bomar JB, Raddin JH, Benedict JV, Smith HL, Hatsell CP, "Analysis of Human Test Subject Kinematic Responses to Low Velocity Rear End Impacts", Society of Automotive Engineers, SAE No. 930889, pp 21-30, 1993.

McConnell WE, Howard RP, Krause R, Guzman HM, Bomar JB, Raddin JH, Benedict JV, Hatsell CP, "Human Head and Neck Kinematics After Low Velocity Rear-End Impacts - Understanding "Whiplash"", Proceedings of the 39th Stapp Car Crash Conference, SAE No. 952724, pp 215-238, 1995.

McKenzie JA, Williams JF, "The Dynamic Behaviour of the Head and Cervical Spine During "Whiplash"", *Journal of Biomechanics*, 4, pp 477-490, 1971.

Mertz HJ, Patrick LM, "Investigation of the Kinematics and Kinetics of Whiplash", Society of Automotive Engineers, SAE No. 670919, pp 43-71, 1967.

Morris F, "Do Head Restraints Protect the Neck from Whiplash Injuries?", *Archives of Emergency Medicine*, 6, pp 17-21, 1989.

Olsson I, Bunketorp O, Carlsson G, Planath I, Norin H, Ysander L, "An In-Depth Study of Neck Injuries in Rear End Collisions", 1990 International IRCOBI Conference on the Biomechanics of Impacts, pp 269-280, 1990.

Ono K, Kanno M, "Influences of the Physical Parameters on the Risk to Neck Injuries in Low Impact Speed Rear-end Collisions", 1993 International IRCOBI Conference on the Biomechanics of Impacts, pp 201-212, 1993.

Reid SE, Raviv G, Reid SE, "Neck Muscle Resistance to Head Impact", *Aviation, Space, and Environmental Medicine*, 52(2), pp 78-84, 1981.

Rosenbluth W, Hicks L, "Evaluating Low-Speed Rear-End Impact Severity and Resultant Occupant Stress Parameters", *Journal of Forensic Sciences*, 39(6), pp1393-1424, 1994.

SAE (Society of Automotive Engineers): "SAE Recommended Practice: Intrumentation for Impact Test -Part 1- Electronic Instrumentation - SAE J211/1 Mar95", <u>SAE Vehicle Occupant Restraint Systems and Components Standards Manual - 1996 Edition</u>, SAE HS-13, pp 83-92, 1996.

SATAI (Southwestern Association of Technical Accident Investigators), Tests conducted by MacInnis Engineering Associates, Ltd., Phoenix, AZ meeting, July 20-21, 1995.

Schuldt K, "On Neck Muscle Activity and Load Reduction in Sitting Postures: An Electromyographic and Biomechanical Study with Applications in Ergonomics and Rehabilitation", *Scandinavian Journal of Rehabilitation Medicine [Supplement]*, 19, pp 1-49, 1988.

Severy DM, Mathewson JH, Bechtol CO,: "Controlled Automobile Rear-End Collisions, an Investigation of Related Engineering and Medical Phenomena", *Canadian Services Medical Journal*, November, pp 727-759, 1955.

Siegmund GP, Bailey MN, King DJ, "Characteristics of Specific Automobile Bumpers in Low-Velocity Impacts", Society of Automotive Engineers, SAE No. 940916, pp 333-371, 1994.

Siegmund GP, Williamson PB, "Speed Change (Delta V) of Amusement Park Bumper Cars", Proceedings of the Canadian Multidisciplinary Road Safety Conference VIII, pp 299-308, 1993.

Snyder RG, Chaffin DB, Foust DR, "Bioengineering Study of Basic Physical Measurements Related to Susceptibility to Cervical Hyperextension-Hyperflexion Injury", Highway Safety Research Institute, UM-HSRI-BI-75-6, 1975.

Soechting JF, Paslay PR, "A Model for the Human Spine During Impact Including Musculature Influence", *Journal of Biomechanics*, 6(2), pp 195-203, 1973.

Spitzer WO, Skowron ML, Salmi LR, Cassidy JD, Duranceau J, Suissa S, Zeiss E, "Scientific Monograph of the Quebec Task Force on Whiplash-Associated Disorders: Redefining "Whiplash" and Its Management", *Spine* 20(8S), pp 1S-73S, 1995.

States JD, Korn MW, Masengill JB, "The Enigma of Whiplash Injuries", Proceedings of the 13th Conference of the American Association for Automotive Medicine, pp 83-108, 1969.

Svensson MY, Lovsund P, Haland Y, Larsson S, "The Influence of Seat-Back and Head Restraint Properties on the Head-Neck Motion During Rear Impact", 1993 International IRCOBI Conference on the Biomechanics of Impacts, pp 395-406, 1993.

Svensson MY, Lovsund P, Haland Y, Larsson S: "Rear-End Collisions - A Study of the Influence of Backrest Properties on Head-Neck Motion Using a New Dummy Neck", *Society of Automotive Engineers* SP-963 (SAE No. 930343), pp 129-142, 1993.

Szabo TJ, Welcher JB, Anderson RA, Rice MM, Ward JA, Paulo LR, Carpenter NJ, "Human Occupant Kinematic Response to Low Speed Rear-End Impacts", Society of Automotive Engineers, SAE No. 940532, 1994.

Warner CY, Strother CE, James MB, Decker RL, "Occupant Protection in Rear-End Collisions: II. The Role of Seat Back Deformation in Injury Reduction", Society of Automotive Engineers, SAE No. 912914, 1991.

Weißner R, Enßlen A, "The Head-Rest - A Necessary Safety Feature for Modern Passenger Cars", 1985 International IRCOBI Conference on the Biomechanics of Impacts, pp 269-276, 1985.

West DH, Gough JP, Harper GTK, "Low Speed Rear-End Collision Testing Using Human Subjects", *Accident Reconstruction Journal*, 5(3), pp 22-26, 1993.

White AA, Panjabi MM: Clinical Biomechanics of the Spine, 2nd ed., J.B. Lippincott Company, Philadelphia, 1990.

Winter DA, Biomechanics and Motor Control of Human Movement, 2nd edition, John Wiley & Sons, Inc., ISBN 0-471-50908-6, USA, 1990.

Traffic Injury Prevention, 9:144–152, 2008
Copyright © 2008 Taylor & Francis Group, LLC
ISSN: 1538-9588 print / 1538-957X online
DOI: 10.1080/15389580801894940

Taylor & Francis
Taylor & Francis Group

Analysis of Head Impacts Causing Neck Compression Injury

DAVID C. VIANO and CHANTAL S. PARENTEAU

ProBiomechanics LLC, Bloomfield Hills, Michigan, USA

Objective. Human cadavers have been subjected to inverted drop, linear, and pendulum impacts to the top of the head, causing neck compression injury. The data are not comparable on the basis of impact velocity because of differing impact masses and test conditions. This study analyzed the published biomechanical data and used peak head velocity to merge the datasets. Correlations were determined between biomechanical responses and serious injury (AIS 3+).

Methods. Three studies were found involving 33 inverted drop tests and three others involving 42 linear or pendulum impacts to the top of a cadaver's head. Various biomechanical responses were measured in the tests. The datasets could not be meaningfully merged on the basis of impact velocity. The coefficient of restitution (e) was determined and the peak head velocity calculated for tests with missing data. This allowed the datasets to be merged and statistically analyzed for relationships between head velocity, impact force, and serious injury. Power functions were fit to the biomechanical data, t-tests conducted for significant differences in injury, and logit risk functions determined.

Results. The coefficient of restitution was $e = 0.24 \pm 0.16$ $(n = 19)$ for the drop tests and $e = 0.21 \pm 0.12$ $(n = 20)$ for the impact tests. Peak head velocity was 22% higher than the impact velocity for the drop tests but -20% lower in the impact tests. Head velocity averaged 6.32 ± 1.29 m/s $(n = 51)$ causing serious injury and 3.75 ± 2.16 m/s $(n = 24)$ without injury $(t = 5.39, p = 0.00001, df = 31)$. Impact force was $7,382 \pm 3,632$ N with injury and $3,760 \pm 3,528$ N without $(t = 3.95, p = 0.0003, df = 42)$. A power function fit the impact force versus head velocity data $(F = 374V_h^{1.565}, R^2 = 0.758)$.

Conclusion. Peak head velocity was determined for inverted drop and impact tests as a means of merging and analyzing cadaver data on serious injury for impacts to the top of the head. There are relationships between head velocity, impact force, and serious injury. A 15% risk of serious injury is at 2.3 m/s (5.1 mph) head velocity and 50% risk at 4.2 m/s (9.4 mph); however, more data are needed in the 2-4 m/s head velocity range to clarify injury risks. In addition, many factors influence the risk of the neck injury, including the age and physical condition of the person; orientation of the head, neck, and torso; and location of impact and interface.

Keywords   Neck Injury; Fracture; Dislocation; Injury Risks; Biomechanics

INTRODUCTION

Head impacts causing neck compression injury occur in a number of situations. For example, diving into shallow water can involve head impact on the floor of the pool with inertia of the torso loading the neck in flexion-compression, resulting in cervical fracture-dislocations (McElhaney et al., 1979). Another example is in rollover crashes where occupants dive toward the ground and interact with the roof (Bahling et al., 1990; James et al., 2007). Injury typically occurs while the roof is in contact with the ground and the occupant is inverted. The diving kinematic is like the drop tests with inverted cadavers conducted by Nusholtz et al. (1983), Yoganandan et al. (1986),

Received 22 September 2007; accepted 5 January 2008.

Address correspondence to David C. Viano, ProBiomechanics LLC, 265 Warrington Rd., Bloomfield Hills, MI 48304-2952, USA. E-mail: dviano@comcast.net

and Sances et al. (1986). Thirty-three (33) cadaver drop tests were conducted in the series with free-fall heights of 0.1–1.8 m and impact velocities of 1.4–5.9 m/s. The drop tests caused fractures and ligament injury to the cervical spine associated with neck compression and deformation of cervical vertebrae.

Forty-two (42) linear and pendulum impact tests have been conducted by Culver et al. (1978), Nusholtz et al. (1981), and Alem et al. (1984). The cadaver was placed in a supine position and impacted on the top (vertex) of the head in-line with the spine. In most cases, the alignment of the head, neck and thoracic spine was controlled. Impact mass varied from 9.9 to 56.0 kg and impact velocity from 4.6 to 10.9 m/s. The vertex impacts caused compression of the spine and varying degrees of neck motion in flexion-extension, lateral bending, and rotation. The impact tests caused fractures and ligament injury to the cervical spine, and in some cases, thoracic spinal injury and basilar skull fractures depending on the orientation of the head, neck, and torso.

There have been a number of attempts to interpret the inverted drop and linear impact test data (CFIR, 2005; Xprts, 2005; Friedman and Nash 2005; Friedman et al., 2006; Nash and Friedman, 2007). However, these efforts have not identified statistically significant relationships among impact force, impact velocity, and injury. Based on an unpublished study, neither the individual datasets for the drop or impact tests nor the combined data could be interpreted by linear relationships constrained to pass through the origin. Consideration of the test methods and cadaver responses offered a possible means of analyzing and combining the data based on peak head velocity rather than impact velocity. The different masses used in the impacts and momentum exchange with the cadaver head make head velocity the appropriate response for analysis of the data. This study analyzes 75 cadaver impacts on the basis of peak head velocity.

Neck Injury Criteria

NHTSA reviewed biomechanical studies on neck injury and developed a neck injury criterion Nij (Kleinberger et al., 1998). The "ij" indices represent four injury mechanisms: tension-extension (Nte), tension-flexion (Ntf), compression-extension (Nce), and compression-flexion (Ncf). Nij is determined as a function of time by normalizing the flexion-extension (My) and adding it to the normalized axial load (Fz) to give Nij:

$$Nij = (Fz/Fzc) + (My/Myc) \qquad (1)$$

where Fzc is the critical intercept for axial neck loading and Myc the critical intercept for flexion-extension bending moment at the occipital condyles. The critical intercepts are Fzc = 6806 N for tension, Fzc = 6160 N for compression, Myc = 310 Nm for flexion, and Myc = 135 Nm for extension.

During an impact, all four combinations of neck response need to be below Nij = 1.0. In addition, peak neck tension cannot exceed 4170 N and compression 4000 N. The development of the Nij used information from experimental studies in the literature (Nightingale et al., 1997; Nusholtz et al., 1981; Pintar et al., 1990, 1995, 1998; Sances et al., 1981; Nuckley et al., 2002; Jarrod et al., 2002). However, the cadaver tests analyzed here do not have sufficient information to derive neck moments and forces. This study will focus on head velocity as a factor related to neck injury.

Neck tolerance for axial compression was estimated by using a Hybrid III dummy to measure neck loads when struck by a tackling block that had produced serious head and neck injuries in football players (Mertz et al., 1978). This biomechanical study was conducted to assess neck loads causing fracture-dislocations during head-down impacts in tackling-dummy practice (Mertz et al., 1978). Other studies evaluated game collisions and neck injury (Hodgson and Thomas 1980, 1983; Bishop and Wells 1986). Neck compression forces above 4000 N are considered sufficient to seriously injure a football player due to axial compression of the cervical spine.

Tolerance levels for neck flexion and extension were estimated using sled tests of volunteers and cadavers (Mertz and Patrick 1971; SAE J855, 1980). Volunteer tests provided data up to the pain threshold for extension bending moments and cadaver tests estimated tolerance limits for serious injuries at 57 Nm (Mertz and Patrick, 1971). The maximum voluntary flexion moment of 190 Nm was set as the tolerance limit. The bending moments were based on human responses, rather than dummy measurements.

Cadaver tests on neck tension showed failure at 3373 N (Yoganandan et al., 1996); however, lower forces were found for combined loading conditions of tension-extension (Shea et al., 1992). These and other studies have established injury tolerance criteria for neck loading that are used with test dummies to study injury risks in sports, automotive crashes and other impacts (McElhaney et al., 1979; King and Viano 1995). While the latest tolerances can be found in Mertz et al. (2003), additional analysis of biomechanical data is needed for compression and flexion-compression injury of the neck. The 4000 N neck compression limit is based on injury reconstructions with the Hybrid III dummy. Analysis of cadaver data would help evaluate the tolerance limit. This study addresses the need for that additional analysis.

MATERIALS AND METHODS

Inverted Drop Tests

Nusholtz et al. (1983), Yoganandan et al. (1986), and Sances et al. (1986) reported data on 33 inverted drop tests with cadavers. Figure 1 shows the drop test setups. After assembling the data, it was found that the reports of Yoganandan et al. (1986) and Sances et al. (1986) involve the same experiments. Some complementary results are presented in the two studies. The drop tests involved free-fall heights of 0.1–1.8 m and impact velocities of 1.4–5.9 m/s.

Table I summarizes the tests, biomechanical responses, and injuries. Sixteen (16) of the tests resulted in cervical fractures, dislocations, or ligament damage. Four tests had thoracic spine fractures and 5 had skull fractures. Two of the tests had both a skull and thoracic spine fracture. Ten (10) tests had no reported injury. Definitions of the individual parameters reported can be found in the original studies.

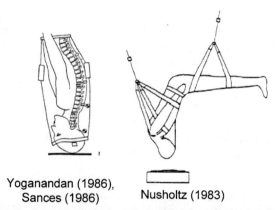

Yoganandan (1986), Sances (1986)   Nusholtz (1983)

Figure 1 Setups for the inverted drop tests of Yoganandan et al. (1986), Sances et al. (1986), and Nusholtz et al. (1983). The specimen was held in a fixture and dropped free-fall onto a load-measuring surface (modified from original figures).

Table I Inverted drop test data of Yoganandan et al. (1986), Sances et al. (1986), and Nusholtz et al. (1983)

Test no.	Cadaver no.	Padding (cm)	Impact location	Initial condition	Drop height (m)	Impact velocity (m/s)	Head velocity (m/s)	Impact force (N)	Neck force (N)	HIC	Serious injury	Some pathology information
Nusholtz et al. (1983)												
82L484	1	0.6		Constrained	1.0	4.4	5.4	6,700			Yes	T3fx, C4-C5 disc rupt.
82L485	2	2.5		Constrained	1.8	5.9	7.2				Yes	C1,C3-C4fx laminae, C2-C5 fx.
82L486	3	2.5		Constrained	1.5	5.4	6.6	5,900			Yes	C1-C3-C4fx.
82L487	4	2.5		Constrained	0.1	1.4	1.2	300		5	No	
82L488	4	2.5		Unconstrained	0.1	1.4	2.3	500		7	No	
82L489	4	2.5		Unconstrained	1.5	5.4	8.5	5,400		2240	Yes	C1-C2fx.
82L490	5	2.5		Constrained	0.1	1.4	1.3	600		2	No	
82L491	4	2.5		Unconstrained	0.1	1.4	1.6			4	No	
82L492	5	2.5		Unconstrained	0.1	1.4	1.5	500		3	No	
82L493	5	2.5		Unconstrained	0.1	1.4	1.7			3	No	
82L494	5	2.5		Unconstrained	1.1	4.6	6.0	5,200		477	Yes	C6-C7fx.
82L495	5	0.6		Unconstrained	0.1	1.4	1.4	900		14	No	
82L496	6	0.6		Unconstrained	0.1	1.4	1.7	900		12	No	
82L497	6	0.6		Unconstrained	0.1	1.4	1.8	1,000		27	No	
82L498	6	0.6		Unconstrained	0.1	1.4	1.6	900		17	No	
82L499	6	2.5		Unconstrained	0.9	4.2	4.3	3,200		210	Yes	C5-C7 disc rupture, T2 fx
82L500	7	2.5		Constrained	1.5	5.4	6.6	10,800			Yes	C1 fx, C4-C7 cervical disc ruptures, T3 chip fx.
82L501	8	2.5		Unconstrained	0.8	4.0	4.2	5,600		540	Yes	C7fx.
Yoganandan et al. (1986) & Sances et al. (1986)												
834212763	HS76	0.0	5 cm posterior of vertex	Free	0.9	4.2	5.2	4,687	1,100		Yes	Posterior ligament disruption C5-C6, C6 disc rupture, 7, T10 fx.
834217773	HS77	0.0	10 cm posterior of vertex	Free	0.9	4.2	5.2	3,129			Yes	Anterior subluxation of C5 on C6 with bilateral locked facets
834211753	HS75	0.0	2.5 cm posterior of vertex	Free	0.9	4.2	5.2	6,405	1,950		Yes	Linear parietal temporal skull fx.
834225804	HS80	0.0	Occipital proturbance	Free	1.2	4.8	5.9	5,684			Yes	T7 wedge compression fx T11-T12 fx.
834226814	HS81	0.0	10 cm posterior of vertex	Free	1.2	4.8	5.9	6,191			Yes	Disruption of posterior ligaments C6-C7; T7 fx.
844246835	HS83	0.0	2.5 cm posterior of vertex	Free	1.5	5.4	6.6	7,355	2,600		Yes	Bilateral temporal skull fx, C3 compression fx.
844247845	HS84	0.0	2.5 cm anterior of vertex	Free	1.5	5.4	6.6	7,185			Yes	Type II odontoid fx, avulsion of posterior ligaments C1-C2. T8 mid compression fx.
844248855	HS85	0.0	1.5 cm posterior of vertex	Restrained	1.5	5.4	6.6	14,922			Yes	Bilateral basilar skull fx. T3 fx.
844250865	HS86	0.0	on vertex	Restrained	1.5	5.4	6.6	14,329			Yes	Jefferson fx. C1, T4 burst fx.
844257875	HS87	1.2	2.5 cm posterior of vertex	Restrained	1.5	5.4	6.6		1,824		Yes	Occiput linear skull fx into base
844279884	HS88	1.2	2.5 cm left of sagittal plane &	Restrained	1.2	4.8	5.9				Yes	Right parietal skull fx into base, T1-T2 fx.
844285894	HS89	1.2	3 cm posterior of vertex	Restrained	1.2	4.8	5.9	9,786			Yes	T6fx linear skull fx.
844290914	HS91	1.2	4-6 cm posterior of vertex	Restrained	1.2	4.8	5.9	11,560	1,112		Yes	C6 spinous process fx. T7fx
844300924	HS92	1.2	on vertex	Restrained	1.2	4.8	5.9	12,840			Yes	C2 fx. T1 wedge fx
844314934	HS93	1.2	on vertex	Restrained	1.2	4.8	5.9	12,440	1,780		Yes	C3 spinous process fx. T8

Linear and Pendulum Impacts

Culver et al. (1978), Nusholtz et al. (1981), and Alem et al. (1984) reported data on 42 linear impact or pendulum tests with supine positioned cadavers. Figure 2 shows the impact test setups, which involved loading of the vertex of the head generally in-line with the spine. Impact mass varied from 9.9 to 56.0 kg and impact velocity from 4.6 to 10.9 m/s. Nusholtz et al.'s (1981) and Alem et al.'s (1984) reported the alignment of the head, neck, and thoracic spine.

Table II summarizes the tests, biomechanical responses, and injuries. Twenty-two (22) of the tests resulted in cervical fractures, dislocations, or ligament damage. Eight tests had thoracic spine fractures and 2 had skull fractures. Fourteen (14) tests had no reported injury. Definitions of the individual parameters

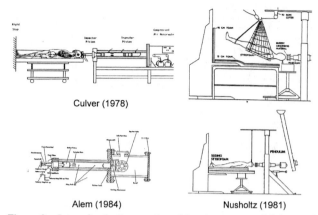

Figure 2 Setups for the linear and pendulum impact tests of Culver et al. (1978), Alem et al. (1984), and Nusholtz et al. (1981). The specimen was held supine on a table or in a net to align the head, neck, and torso with the impact (modified from original figures).

reported can be found in the original study. Definitions of the individual parameters reported can be found in the original studies.

Coefficient of Restitution (e)

Head impacts in these tests are not perfectly plastic collisions. The final velocity of the impact mass (V_f) is less than the peak head velocity (V_h), but V_f was not reported in any of the studies. Based on conservation of momentum and energy, the coefficient of restitution (e) in the impact can be determined from the following:

$$e = (m + m_h)V_h/mV - 1 \qquad (2)$$

where V is impact velocity, m is the impact mass, and m_h the mass of the head. The final velocity of the impact V_f is $V_f = V_h - eV$. For the inverted drop tests, the impact mass is infinite, so $m >> m_h$ and $e = V_h/V - 1$. Since $1 > e \geq 0$, head velocity is greater than the impact velocity.

The average coefficient of restitution was determined for the drop and impact tests and used to calculate peak head velocity for the tests with missing data:

$$V_h = (1 + e)mV/(m + m_h) \qquad (3)$$

For the inverted drop tests, $V_h = (1+e)V$. The average coefficient of restitution was used to calculate peak head velocity for the missing tests. For the inverted drop tests, 13 of 18 (72%) tests by Nusholtz et al. (1983) reported the peak head velocity; however, none of the tests by Yoganandan et al. (1986) or Sances et al. (1986) listed head velocity. Peak head velocity had to be calculated for two-thirds (66%) of the drop tests. In contrast, most (81%) of the impact tests reported peak head velocity. Only 8 tests needed a calculation of head velocity.

For this analysis, the reported head velocity was used, if available. It was determined by integration of triaxial accelerometers attached to the head and is a resultant value. The calculated head velocity was used only for the tests with missing data. The calculation assumes a linear, two-body impact, so forces in the neck

and off-axis effects are neglected in the calculation. Head mass was assumed to be $m_h = 4.54$ kg (Mertz et al., 1989).

For the inverted drop tests, 70% (16 out of 23) of the injuries were to the cervical spine. For the impact tests, 79% (22 out of 28) of the injuries were to the cervical spine. All serious injury (AIS 3+) in the impacts was considered neck compression related for the subsequent analysis. Some fractures were to the upper thoracic spine and others the basilar skull. Both are in close proximity to the neck connections to the torso and head.

Statistical Analysis

The significance of differences in biomechanical responses was determined using the t-test assuming unequal variance and two-sided distribution. Trendlines were fit to the data selecting the best fit among linear, exponential, polynomial, and power functions. A logit function was fit to the injury-noninjury data versus impact force and head velocity. The model is a sigmoidal relationship with two parameters. The probability (p) of injury is:

$$p(x) = 1/[1 + \exp(\alpha - \beta x)] \qquad (4)$$

where α and β are parameters defining injury risk as a function of a biomechanical response (x). An initial guess was made for α and β and the probability of injury determined for each test. The calculated risk was subtracted from the actual occurrence (1) or nonoccurrence (0) of injury to form an error. The error was squared and summed for all tests. The Solver algorithm was used to "best fit" α and β to achieve a minimum squared error. This gave a sigmoidal logit model for the risk of injury as a function of the biomechanical response. The t-test, trendlines, and Solver were performed using the analysis packages in Microsoft Excel 2007.

RESULTS

Inverted Drop Tests

Figure 3 shows the impact velocity and peak force on the head in the inverted drop tests with cases of serious injury and no injury shown. The data for all tests was reasonably fit by a power function relating peak force to impact velocity: $F = 334V_h^{1.96}$, $R^2 = 0.895$. For this function, a force of 4000 N occurs at an impact velocity of 3.55 m/s.

Linear and Pendulum Impacts

Figure 4 shows the impact velocity and peak force on the head in the linear and pendulum impact tests with the cases of serious injury and no injury shown. The data for all tests were poorly fit by a power function relating peak force to impact velocity: $F = 772V_h^{0.99}$, $R^2 = 0.223$. For this function, a force of 4000 N occurs at an impact velocity of 5.27 m/s. The poor fit is related to the range of impact masses and test conditions.

Merged Impact Velocity Data

The data from the inverted drop and the impact tests were merged. It showed clearly different trends for the two types of test. There was a broad scatter in the data. No statistically significant

Table II Linear and pendulum impact data from Culver et al. (1978), Alem et al. (1984), and Nusholtz et al. (1981)

Test No	Head angle (°)	Neck angle (°)	Torso angle (°)	Mass (kg)	Padding (cm)	Stroke (cm)	Impact velocity (m/s)	Head velocity (m/s)	Impact force (N)	Neck force (N)	Calculated peak resultant Acc. (g)	HIC	Serious injury	Some pathology information
Alem et al. (1984)														
H201				10		5.1	8.0	5.1	5,100	3,924	67	325	No	None
H202				10		5.1	8.0	4.7	5,210	3,136	63	286	No	None
H203				10		5.1	8.0	5.6	4,400	3,139	80	333	No	None
H204				10		5.1	8.0	4.7	3,930	2,904	72	249	No	None
H205				10		5.1	8.0	4.4	4,800	3,571	64	167	No	None
H401		30		10		5.1	8.4	8.4	4,200		130		No	None
H402		20		10		5.1	10.9	9.1	11,000				Yes	Bilateral fx. T2 lamina at base of spinous process
H403	100	25		10		5.1	10.9	8.1	10,500		160	1,031	Yes	Anerior-inferior chipfx. of C2. C3/C4 spinous process tip fx.
H404	95	25		10		5.1	7.8	6.5	4,000				Yes	Nearly complete tear an anterior longitudinal ligament at disk between C3-C4
H405	80	5		10		5.1	7.7	3.7	4.100		48	145	No	None
H406	80	5		10		5.1	8.0	5.8	4,000		70	288	Yes	Bilateral fx.of posterior C1 arch. Fx. of C2 dens
H407		5		10		5.1	9.2	6.9	4,500		99	503	Yes	Rupture of anterior longitudinal ligament and disk between C5&C6
H408	100	10		10		5.1	9.7	5.9	6,000		85	316	Yes	Bilateral fx. of C1 posterior arch, anterio-inferior C2 body fx. extending through C2-C3 disk.
H409		5		10		0.0	10.4	8.7	15,000				Yes	Circular depress fx. of apex of skull under impactor
H410		30		10		5.1	9.0	5.0	5,200		72	238	Yes	Fx. of anterior of C4
H411				10		5.1	7.2	3.5	4.100		48	76	No	None
H412		10		10		5.1	7.1	3.5	3,000		45	61	Yes	Teardrop fx. of lip of C5.
H413				10		0.5	9.0	7.6	17,000				Yes	Basilar skull fx.
H414				10		0.5	6.9	5.8	16,000				No	
Nusholtz et al. (1981)														
79L084	−45	5		56	2.5	17.8	5.6	6.0	3,300		238		Yes	Fx C3-C7
79L088	−30	5		56	2.5	3.0	5.6	3.0	2,200		57		No	Osteoporotic
79L092	−45	10		56	2.5	3.0	5.6	6.7	6,200		302		Yes	Fx C7, T1
80L096	−10	25		56	2.5	12.7	5.6	6.6	2,300		76		Yes	Fx C5-C6
80L101	−15	25		56	2.5	5.1	5.6	4.0	1,800		38		Yes	Fx C6-C7
80L108	−30	20		56	2.5	14.0	4.6	5.2	2,800		35		No	
80L113	−20	10		56	2.5	16.5	4.6	4.8	3,300		44		Yes	Fx C3-C4
80L117	−40	5		56	2.5	16.5	4.6	5.6	5,700		118		Yes	Fx C4
80L123	10	25	−22	56	2.5	0.0	5.7	6.3	6,000		116		Yes	Fx C5, C7, T2
80L128	−30	10	−22	56	2.5	11.9	5.6	6.4	7,100		133		Yes	Fx T4
80L134	−30	5	−15	56	2.5	17.8	5.6	6.3	11,100		217		Yes	Fx T3
80L139	−10	25	−25	56	2.5	15.2	5.6	6.3	10,300		228		Yes	Fx T1
Culver etal. (1978)														
77H101				9.9		15.2	7.9	6.6	6.700				No	
77H102				9.9		15.2	9.6	6.4	6,950		141		No	
77H103				9.9		15.2	8.8	6.7	7.200		165		Yes	Clavicle fx, C5 fx
77H104				9.9		20.3	10.0	8.4	8,850				Yes	C5 & T1 fx, T2 crushed
77H105				9.9		20.3	9.6	5.9	7.450		57		Yes	C2 fx
78H106				9.9		20.3	8.4	6.8	6,620		71		No	
78H107				9.9		10.2	10.2	8.3	8.450		106		Yes	C3-C4 fx
78H108				9.9		10.2	9.9	9.3	8,000		156		Yes	C1, C2, C4, C7, T1, T2 fx.
78H109				9.9		10.2	8.4	7.9	7.030		122		Yes	C7, T1 fx
78H110				9.9		10.2	6.8	4.1	4,710		60		Yes	C4, C5, C6 fx
78H111				9.9		10.2	7.6	6.9	6.050		100		Yes	C3, C4, C5 fx

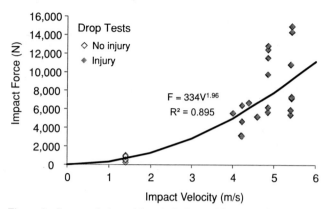

Figure 3 Impact velocity and force in the inverted drop tests of Yoganandan et al. (1986), Sances et al. (1986), and Nusholtz et al. (1983) showing injury and no injury data and trendline.

difference was found between the 6.48 ± 2.11 m/s impact velocity with serious injury and 5.01 ± 3.25 m/s without ($t = 2.02$, $p = 0.052$, $df = 32$). A trendline poorly fit a relationship between impact velocity and force: $F = 293V^{1.18}$, $R^2 = 0.58$.

In the drop tests, the head impacts involve an infinite mass of the ground. The head absorbs impact energy and rebounds. In this case, the head velocity is greater than the impact velocity. For the linear impact tests, the mass of the pendulum or impactor was greater than the head mass and a momentum exchange takes place influencing the amount of velocity transferred to the head. In this case, the head velocity is less than the impact velocity. Peak head velocity was a common response between the tests and a means of merging the datasets for analysis.

Coefficient of Restitution (e)

There were a suitable number of tests where the impact velocity and peak head velocity were reported. The coefficient of restitution was $e = 0.24 \pm 0.16$ ($n = 19$) for the drop tests and $e = 0.21 \pm 0.12$ ($n = 20$) for the impact tests. A t-test was run on the two datasets. It showed no statistical difference in the coefficient of restitution in the drop and impact tests ($t = 0.60$, $p = 0.55$, $df =$

33). The merged data had an average coefficient of restitution of $e = 0.22 \pm 0.14$ ($n = 39$), which was used to determine peak head velocity for the tests with missing data.

Peak Head Velocity (V_h)

For the inverted drop tests, the ground was assumed an infinite mass and the reduced formula was used to calculate peak head velocity: $V_h = (1 + e)V$. The coefficient of restitution was assumed $e = 0.22$. Table I shows the 25 calculated head velocities in cells with grey background. For all drop tests, head velocity averaged 22% higher than the impact velocity ($V_h = 4.7$ m/s versus $V = 3.8$ m/s).

For the linear and pendulum impact tests, the full equation was used with masses to determine peak head velocity: $V_h = (1 + e)mV/(m + m_h)$. Table II shows the 8 calculated head velocities in cells with grey background. For all impact tests, head velocity averaged -20% lower than the impact velocity ($V_h = 6.1$ m/s versus $V = 7.7$ m/s).

Merged Head Velocity Data

After calculating the peak head velocity for each test with a missing value, the datasets were merged. The full data had a peak head velocity of 6.32 ± 1.29 m/s for 51 impacts causing serious injury and 3.75 ± 2.16 m/s for 24 impacts without injury. The difference was statistically significant ($t = 5.39$, $p = 0.00001$, $df = 31$). Impact force was $7,382 \pm 3,632$ N with injury and $3,760 \pm 3,528$ N without; and the difference was also statistically significant ($t = 3.95$, $p = 0.0003$, $df = 42$).

Figure 5 shows the merged data from the inverted drop and impact tests. A power function fit the impact force versus head velocity data: $F = 374V_h^{1.565}$, $R^2 = 0.758$, although there is overlap in the injury and non-injury biomechanical data and a paucity of tests for 2–4 m/s head velocity. There is consistency in the merged data because head velocity was 22% higher than the impact velocity for the drop tests, whereas it was -20% lower in the impact tests. This reduced the scatter.

Injury risk functions were determined. Solver found that $\alpha = 3.839$ and $\beta = 0.917$ minimized the squared error. A 15% risk of serious injury is at 2.3 m/s (5.1 mph) head velocity and 50% at 4.2 m/s (9.4 mph). Ninety percent (90%) of the injury

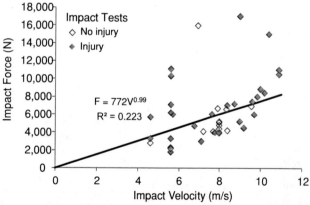

Figure 4 Impact velocity and force in the linear and pendulum impact tests of Culver et al. (1978), Alem et al. (1984), and Nusholtz et al. (1981) showing injury and no injury data and trendline.

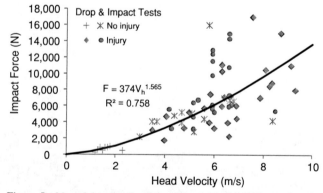

Figure 5 Merged data showing the peak head velocity and impact force in drop and linear impact tests. Injury and no injury data are shown with a trendline.

occurred at head velocities below 6.6 m/s (14.8 mph). However, the risk does not converge to zero indicating more data is needed in the 2–4 m/s head velocity range to clarify the injury risk function.

For an inverted drop on a rigid surface, the head velocity was 22% higher than the impact velocity because of rebound. A 15% risk of serious injury occurs at an impact velocity of 1.9 m/s (4.2 mph) and a 50% risk of injury at 3.4 m/s (7.7 mph) with 90% of injury occurring at impact velocities below 5.4 m/s (12.1 mph) for an inverted drop. Using the power function relationship between head velocity and peak force, a 15% risk of serious injury involves an impact force of 1377 N, a 50% risk at 3534 N, and 90% of injury is below a force of 7169 N. A logit analysis was conducted on impact force. Solver found $\alpha = 1.629$ and $\beta = 0.00048$ minimized the squared error. This function gives a 50% risk of serious injury at 3400 N, but crosses zero risk at 16% because of the limited data at low impact forces.

DISCUSSION

There are a number of factors influencing the risk of neck injury with impacts to the top of the head. Most important is the orientation of the head, neck, and torso with respect to the axis of impact. For the inverted drop tests, the mass of the ground is infinite so the head absorbs all of the impact momentum and rebounds during the continued dropping of the torso mass. This compresses the neck. The magnitude of head acceleration is determined by the type and thickness of padding. In these tests, head velocity is always greater than the impact velocity determined by the drop height.

For the impact and pendulum tests, the mass of the impactor is a factor as well as the impact alignment and orientation of the body. With a high velocity impact and light-weight impactor, the head can experience very high accelerations of short duration giving about the same head velocity as a heavier mass impact causing a longer duration and lower amplitude head acceleration. The interactions of head mass with the impactor cause the head velocity to be less than the impact velocity.

The use of head velocity in the drop and impact tests allowed the data to be merged. Figure 5 shows a clear increase in impact force with head velocity and a transition in the risk of serious injury. The data look reasonable and consistent on the basis of head velocity. While there are 75 tests in the combined datasets, many of them were at very high impact energy causing serious injury. Twenty-three (23) of the tests had more than 7000 N impact force with only one case without injury. Obviously, more data is needed in the 2–4 m/s head velocity range to clarify injury risk functions.

There was one injury in four tests (25%) with a head velocity of 3–4 m/s and five injuries in eight tests (63%) with a velocity of 4–5 m/s. The logit analysis gave a 15% risk of serious injury at 2.3 m/s (5.1 mph) head velocity, which is a 1.9 m/s (4.2 mph) impact velocity when converted for an inverted drop. These values are lower than, but consistent with, the data from

isolated cervical spine tests and other cadaver testing. Myers and Nightingale (1999) found that severe neck injury occurs at 3.1 m/s with a 40% effective dynamic body mass. They also found that buckling of the cervical spine causes flexion injuries in <20 ms without significant motion of the head and neck, following earlier studies (Myers et al., 1991). Large head motions occurred later in the impact at 20–100 ms. This velocity is consistent with earlier observations of significant neck injuries at 3.2 m/s, also without significant head motion (Nightingale et al., 1996). Other biomechanical data provide reasonably consistent tolerances for serious injury with head impacts causing neck deformation (McElhaney et al., 1995; Nightingale et al., 1991, 1996; Yoganandan et al., 1986).

The tests analyzed here had a range of initial orientations of the head, neck, and torso as well as differing impact conditions from an inverted drop to pendulum and linear impacts to the top of the head. The use of whole cadavers in these different test conditions resulted in various injury mechanism and head, neck and torso kinematics. Serious injuries were observed in the skull, cervical spine, and thoracic spine and various combinations of body regions. The inverted drop tests resulted in 16 cases of cervical spine fracture, dislocation, or ligament injury. This was 70% of the injury location with four cases of skull fracture and four cases of thoracic spine injury. The impact tests resulted in 21 cases of cervical spine fracture, dislocation, or ligament injury. This was 75% of the injury location with two cases of skull fracture and five cases of thoracic spine injury. For this analysis, the various injuries were treated as neck compression related and not analyzed separately. Furthermore, the most serious injury was rated only AIS 3, since no spinal cord injury was noted at autopsy. More serious functional injuries can only be interpreted by inference from the musculoskeletal injury, although even this is limited by a lack of quantitative measurement of vertebral dislocations and skeletal damage into the spinal canal.

Limitations

There is known error in the determination of velocity by integration linear accelerometers to determine velocity. An array of accelerometers was used to determine head cg acceleration in these tests. While the method was the best available, it raises a question about the reliability of the head velocity data report in the cadaver tests. Although there was spread in the values of the coefficient of restitution, the value of e = 0.22 was consistent with the different type of tests and laboratories involved in these studies. It could be argued that all of the head velocities should have been calculated rather than merely filling in the missing data.

The 75 tests were reanalyzed using only calculated head velocity for each experiment. The merged data had an average head velocity of 6.60 ± 1.10 m/s for impacts causing serious injury and 4.57 ± 2.52 m/s without (t = 3.79, p = 0.00076, df = 27). A power function was fit to the impact force versus head velocity data: $F = 293V_h^{1.62}$, $R^2 = 0.70$; and it provided a reasonably similar curve to that in Figure 5. Solver found that $\alpha = 2.621$ and $\beta = 0.615$ minimized the squared error.

There were enough tests where the peak head velocity was both measured and could be calculated. This allowed a determination of a percentage error between the measured and calculated head velocity. The average difference was $-7.6 \pm 19.7\%$ (n = 13) for the drop tests and $23.4 \pm 30.6\%$ (n = 34) for the impact test. This shows more variation in the impact tests, although only 8 values needed to be calculated.

The pathology findings summarized in Table I and II illustrate some of the most serious injuries observed. In some cases, the pathology was obtained from more than one publication and the data was not always consistent. The authors made an attempt to define the pathology by synthesizing information from all publications, but the reader is cautioned to consider the original studies.

This analysis aimed to study the risk of neck injury by the amount of head velocity in impact and drop tests. It showed that head impact velocity is not a consistent parameter to define risk among the various cadaver tests. Biomechanical responses to calculate Nij and other direct measures of neck loading, such as neck moments and forces, were not available, so these criteria could not be evaluated.

ACKNOWLEDGMENTS

Partial support for this research was provided by Chrysler LLC and Nissan North America. Also, the authors conducted an analysis of reports and literature on head impacts causing neck compression injury, which has been disclosed to the public (ProBiomechanics LLC Report, July 12, 2006). The data analysis presented here is an extension of that work.

REFERENCES

Alem NM, Nusholtz GS, Melvin JW. (1984) Head and Neck Response to Axial Impact. 28th Stapp Car Crash Conference, Society of Automotive Engineers, SAE 841667, Warrendale, PA.

Bahling GS, Bundorf RT, Kaspzyk GS, Moffatt EA, Orlowski KF, Stocke JE. (1990) Rollovers and Drop Tests—The Influence of Roof Strength on Injury Mechanics Using Belted Dummies. SAE 902314, 34th Stapp Car Crash Conference, Society of Automotive Engineers, Warrendale, PA, pp. 101—112.

Bishop PJ, Wells R. (1986) The Hybrid III Anthropometric Neck in the Evaluation of Head First Collisions. Society of Automotive Engineers, Warrendale PA, PT-44, SAE 860201, Warrendale PA.

CFIR Submission #2 to NPRM Roof Crush Docket 2005-22143, Subject: Injury Criteria for the Dynamic Evaluation of Rollover Occupant Protection, November 21, 2005.

Culver RH, Bender M, Melvin JW. (May 1978) Mechanisms, Tolerances and Responses Obtained Under Dynamic Superior-Inferior Head Impact. Final Report UM-HRSI-78-21.

Friedman D, Nash C. (June 2005) Reducing Rollover Occupant Injuries: How and How Soon. 19th International Technical Conference on Enhanced Safety of Vehicle Conference, 05-0417-W.

Friedman D, Nash CE, Bish J. (2006) Observations from Repeatable Dynamic Rollover Tests. ICrash Conference, Greece.

Hodgson VR, Thomas LM. (1980) Mechanisms of Cervical Spine Injury During Impact to the Protected Head. 24th Stapp Car Crash Conference, Society of Automotive Engineers, Warrendale PA, SAE 801300.

Hodgson VR, Thomas LM. (July 1983) The Biomechanics of Neck Injury from Direct Impact to the Head, Experimental Findings. in *Head and Neck Injury Criteria: A Consensus Workshop*, US Department of Transportation, DOT HS 806 434, pp. 110–120.

James MB, Nordhagen RP, Schneider DC, Koh SW. (2007) Occupant Injury in Rollover Crashes: A Reexamination of Malibu II. SAE-2007-01-0369, Society of Automotive Engineers, Warrendale PA.

Jarrod W, Carter JW, Ku GS, Nuckley DJ, Ching RP. (2002) Tolerance of the Cervical Spine to Eccentric Axial Compression. 46th Stapp Car Crash Conference, the Stapp Association, Ann Arbor, MI, SAE 2002-22-0022.

King AI, Viano DC. (1995) Mechanics of Head and Neck, In *The Biomedical Engineering Handbook*, Ed. J. D. Bronzino. Chapter 25, pp. 357–368. CRC Press, Inc. and IEEE Press, Boca Raton, FL.

Kleinberger M, Sun E, Eppinger R, Kuppa S, Saul R. (September 1998) Development of Improved Injury Criteria for the Assessment of Advanced Automotive Restraint Systems. NHTSA Docket No 1998-4405-9.

McElhaney JH, Hopper RH Jr, Nightingale RW, Myers BS. (1995) Mechanisms of Basilar Skull Fracture, *J. Neurotrauma*, Vol. 12, No. 4, pp. 669–678.

McElhaney J, Snyder RG, States JD, Gabrielsen MA. (1979) Biomechanical Analysis of Swimming Pool Injuries, In *Neck-Anatomy, Injury Mechanisms and Biomechanics*, SAE SP-438, Society of Automotive Engineers, Warrendale, PA, SAE 790137.

Mertz HJ, Hodgson VR, Thomas LM, Nyquist GW. (1978) An Assessment of Compressive Neck Loads Under Injury Producing Conditions, *Physician Sportsmedicine*, Vol. 6, No. 11, pp. 95–106.

Mertz HJ, Irwin AL, Melvin JW, Stalnaker RL, Beebe MS. (1989) Size, Weight and Biomechanical Impact Response Requirements for Adult Size Small Female and Large Male Dummies. SAE 890756, Society of Automotive Engineers, Warrendale, PA.

Mertz HJ, Irwin AL, Prasad P. (2003) Biomechanical and Scaling Bases for Frontal and Side Impact Injury Assessment Reference Values, SAE 2003-22-0009, *Stapp Car Crash Journal*, Vol. 47, pp. 155–188.

Mertz HJ, Patrick LM. (1971) Strength and Response of the Human Neck, 15th Stapp Car Crash Conference, Society of Automotive Engineers, Warrendale, PA, SAE 710855.

Myers BS, McElhaney JH, Richardson WJ, Nightingale RW and Doherty BJ. (1991) The Influence of End Condition on Human Cervical Spine Injury Mechanisms. SAE 912915, 35th Stapp Car Crash Conference, Society of Automotive Engineers, Warrendale, PA.

Myers BS, Nightingale R. (1999) Review: The Dynamics of Near Vertex Head Impact and Its Role in Injury Prevention and the Complex Clinical Presentation of Basicranial and Cervical Spine Injury, *J. Crash Prevention And Injury Control*, Vol. 1, No. 1, pp. 67–82.

Nash CE, Friedman D. (June 2007) A Rollover Human Dummy Head/Neck Injury Criteria, Paper #07-0357, 20th Enhanced Vehicle Safety (ESV) Conference, National Highway Safety Administration, Lyon, France.

Nightingale RW, McElhaney JH, Camacho DL, Kleinberger M, Winkelstein BA, Myers BS. (1997) The Dynamic Responses of the Cervical Spine: Buckling, End Conditions, and Tolerance in Compression Impacts. 41st Stapp Car Crash Conference, Society of Automotive Engineers, Warrendale, PA, SAE 973344, pp. 451–471.

Nightingale RW, McElhaney JH, Richardson WJ, Best TM, Myers BS. (1996) Experimental impact Injury to the Cervical Spine: Relating

Motion of the Head and the Mechanism of Injury, *J. Bone Joint Surg Am.*, Vol. 78, No. 3, pp. 412–421.

Nightingale RW, McElhaney JH, Richardson WJ, Myers BS. (1996) Dynamic Responses of the Head and Cervical Spine to Axial Impact Loading, *J. Biomech*, Vol. 29, No. 3, pp. 307–318.

Nightingale RW, Meyers BS, McElhaney J, Richardson WJ, Doherty BJ. (1991) The Influence of End Condition of Human Cervical Spine Injury Mechanisms. SAE 912915, Society of Automotive Engineers, Warrendale, PA.

Nuckley DJ, Hertsted SM, Ku GS, Eck MP, Ching RP. (2002) Compressive Tolerance of the Maturing Cervical spine, 46th Stapp Car Crash Conference, 2002-22-0021, the Stapp Association, Ann Arbor, MI.

Nusholtz GS, Huelke DE, Lux P, Alem NM, Montalvo F. (1983) Cervical Spine Injury Mechanisms. SAE 831616, 27th Stapp Car Crash Conference, Society of Automotive Engineers, Warrendale, PA.

Nusholtz GS, Melvin JW, Huelke DF, Alem NM, Blank JG. (1981) Response of the Cervical Spine to Superior-Inferior Head Impacts, SAE 81005, 25th Stapp Car Crash Conference, Society of Automotive Engineers, pp.197–237, Warrendale, PA.

Pintar FA, Sances A, Yoganandan N, Reinartz J et al., (1990) Biodynamics of the Total Human Cadaveric Cervical Spine, SAE 902309, 34th Stapp Car Crash Conference, Society of Automotive Engineers, Warrendale, PA.

Pintar FA, Yoganandan N, Voo L, Cusick JF, Maiman DJ, Sances A. (1995) Dynamic Characteristics of the Human Cervical Spine, SAE 952722, 39th Stapp Car Crash Conference, Society of Automotive Engineers, Warrendale, PA.

SAE J855. (Apr 1980) *Human Tolerance to Impact Conditions as Related to Motor Vehicle Design*, Society of Automotive Engineers, Warrendale, PA.

Sances A, Yoganandan N, Maiman D, Myklebust JB, Chilbert M, Larson SJ, Pech P, Pintar F, Myers T. (1986) Spinal Injuries with Vertical Impact, In *Mechanisms of Head and Spine Trauma*, ed. by A. Sances, D. Thomas, et al., Aloray Publishers, Goshen, NY, pp. 717–736.

Sances Jr A, Myklebust J, Cusick JF, Weber R, Houterman C, Larson SJ, Walsh P, Chilbert M, Prieto T, Zyvoloski M, Ewing C, Thomas D. (1981) Experimental Studies of Brain and Neck Injury. Proceedings of the 25th Stapp Car Crash Conference, Society of Automotive Engineers, Warrendale, PA, SAE 811032.

Shea M, Wittenberg RH, Edwards WT, White AA 3rd, Hayes WC. (1992) In Vitro Hyperextension Injuries in the Human Cadaveric Cervical Spine, *J. Orthop Res.*, Vol. 10, No. 6, pp. 911–916.

Xprts. LLC Submission #1 to NPRM Roof Crush Docket 2005-22143, Subject: Evaluation of Dynamic Criteria for Rollover Occupant Protection, November 21, 2005.

Yoganandan N, Pintar FA, Maiman DJ, Cusick JF, Sances A Jr, Walsh PR. (1996) Human Head-Neck Biomechanics Under Axial Tension, *Med Eng Phys.*, Vol. 18, No. 4, pp. 289–294.

Yoganandan N, Pintar FA, Sances A Jr and Maiman DJ. (1991) Strength and kinematic response of dynamic cervical spine injuries, *Spine*, Vol. 16, No. 10S, pp. 511–517.

Yoganandan N, Sances A, Maiman D, Myklebust JB, Pech P, Larson SJ. (1986) Experimental Spinal Injuries with Vertical Impact, *Spine*, Vol. 11, No. 9, pp. 855–860.

PERFORMANCE OF SEATS WITH ACTIVE HEAD RESTRAINTS IN REAR IMPACTS

Liming Voo
Bethany McGee
Andrew Merkle
Michael Kleinberger
Johns Hopkins University Applied Physics Laboratory
United States of America
Shashi Kuppa
National Highway Traffic Safety Administration
United States of America
Paper Number 07-0041

ABSTRACT

Seats with active head restraints may perform better dynamically than their static geometric characteristics would indicate. Farmer *et al.* found that active head restraints which moved higher and closer to the occupant's head during rear-end collisions reduced injury claim rates by 14-26 percent. The National Highway Traffic Safety Administration (NHTSA) recently upgraded their FMVSS No. 202 standard on head restraints in December 2004 to help reduce whiplash injury risk in rear impact collisions. This upgraded standard provides an optional dynamic test to encourage continued development of innovative technologies to mitigate whiplash injuries, including those that incorporate dynamic occupant-seat interactions. This study evaluates four original equipment manufacturer (OEM) seats with active head restraints in the FMVSS 202a dynamic test environment. The rear impact tests were conducted using a deceleration sled system with an instrumented 50th percentile Hybrid III male dummy. Seat performance was evaluated based on the FMVSS 202a neck injury criterion in addition to other biomechanical measures, and compared to the respective ratings by the Insurance Institute for Highway Safety (IIHS). Three of the four OEM seats tested were easily within the allowable FMVSS 202a optional dynamic test limits. The seat that was outside one of the allowable limits also received only an "acceptable" rating by IIHS while the other three seats were rated as "good." Results also suggest that the stiffness properties of the seat back and recliner influence the dynamic performance of the head restraint.

INTRODUCTION

Serious injuries and fatalities in low speed rear impacts are relatively few. However, the societal cost of whiplash injuries as a result of these collisions is quite high: the National Highway Traffic Safety Administration (NHTSA) estimates that the annual cost of these whiplash injuries is approximately $8.0 billion (NHTSA, 2004). Numerous scientific studies reported connection between the neck injury risk and seat design parameters during a rear impact (Olsson 1990, Svensson 1993, Eichberger 1996, Tencer 2002 and Kleinberger 2003). When sufficient height was achieved, the head restraint backset had the largest influence on the neck injury risk. In addition to its static position relative to the occupant head, the structural rigidity of the head restraint and its attachment to the seat back can have a significant impact on the neck injury risk in a rear impact (Voo 2004). Farmer et al. (2003) and IIHS (2005) examined automobile insurance claims and personal injury protection claims for passenger cars struck in the rear to determine the effects of changes in head restraint geometry and some new head restraint designs. Results from these studies indicated that cars with improved head restraint geometry reduced injury claims by 11-22 percent, while active head restraints that are designed to move higher and closer to occupants' heads during rear-end crashes were estimated to reduce claim rates by 14-26 percent.

In response to new evidence from epidemiological data and scientific research, NHTSA published the final rule that upgrades the FMVSS 202 head restraint standard (49 CFR Part 571) in 2004, and is participating in a Global Technical Regulation on head restraints. The new standard (FMVSS No. 202a) provides requirements that would make head restraints higher and closer to the head so as to engage the head early in the event of a rear impact. The rule also has provisions for a dynamic option to evaluate vehicle seats with a Hybrid III dummy in rear impact sled test that is intended in particular for active head restraints that may not meet the static head restraint position requirements such as height and backset. However, the dynamic option is not limited to active head restraints. By active head restraints we mean head restraints that move or deploy with respect to the seat back. These active head restraints might perform better in rear impact collisions than their static geometric measures may indicate. The neck injury criterion in this dynamic option uses the limit value of 12 degrees in the posterior head rotation relative to the torso of the dummy within the first 200 milliseconds of the rear impact event.

The Insurance Institute for Highway Safety (IIHS) has been publishing ratings of head restraint geometry since 1995 (IIHS, 2001). IIHS along with the International Insurance Whiplash Prevention Group (IIWPG) developed a dynamic test procedure (IIHS, 2006) to evaluate head restraints and have been rating head restraint systems since 2004 using a combination of their static measurement procedure and the newly developed dynamic test procedure. In this combined procedure, seat systems that obtain a "good" or "acceptable" rating according to the IIHS static head restraint measurement procedure, are put through a dynamic rear impact sled test with the BioRID II dummy, simulating a rear crash with a velocity change of 16 km/h. The dynamic evaluation is based on the time to head restraint contact, maximum forward T1 acceleration, and a vector sum of maximum upper neck tension and upper neck rearward shear force. This evaluation results in a dynamic rating of the seat ranging from "good" to "poor". As a consequence of this evaluation procedure by IIHS, head restraints that obtained a good or acceptable rating from the static head restraint measurements may obtain an overall poor rating from the dynamic test procedure. In addition, some active head restraint systems that obtain a marginal or poor static measurement rating are not even tested dynamically although their dynamic performance may actually be good.

This study evaluates the performance of a select group of automotive seats with active head restraints from original equipment manufacturers (OEM) under the environment of the optional FMVSS 202a dynamic test.

MATERIALS AND METHODS

Driver seats from four different passenger cars were evaluated: Saab 9-3, Honda Civic, Nissan Altima and Subaru Outback. The OEM driver seats were 2006 model year production stock, ordered directly from either the vehicle manufacturers or their suppliers, and included the seatbelt restraints. The seats were not modified in any way. Custom-designed rigid base brackets for each seat were used to anchor the seats to the impact sled such that the height and relative position of the seat to the B-pillar and floor pan would be similar to its position in the car. For each seat model, the corresponding OEM seatbelt was used as the restraining device during each test.

The seats were positioned nominally in accordance with sections S5.1 and S5.3 of FMVSS 202a. However, some aspects of the IIHS procedure (IIHS 2001) were implemented regarding the set up of the SAE J826 manikin and the seat back position. The procedure is briefly described below. Once fixed to the sled with its back toward the impact direction, the seat was positioned at the mid-track setting between the most forward and most rearward positions. Then the seat pan angle was set such that its front edge was at the lowest position relative to its rear edge. The vertical position of the seat was placed at the lowest position if a dedicated height adjustment mechanism existed independent of the seat pan incline adjustment. Once the seat pan angle and height were fixed, the seat back was reclined to a position such that the torso line of SAE J826 manikin (H-point machine) was at 25 degrees from the vertical, following a procedure similar to that used by IIHS (IIHS 2001). The head restraint height was measured at the highest and lowest adjustment settings using the head room probe of the H-point machine, and was then positioned midway between those two points or the next lower lockable setting. The head restraint backset and head-to-head-restraint height were measured using the Head Restraint Measurement Device (HRMD) in combination with the SAE J826 manikin with a procedure adopted by IIHS (IIHS 2001). The H-point of the seat as positioned was then recorded and marked to be used later in positioning the dummy.

A 50th percentile male Hybrid III dummy was used as the seat occupant for this study. The dummy was instrumented with triaxial accelerometers at the head CG and thorax CG, and a single accelerometer at T1. Angular rate sensors (IES 3100 series rate gyro) were mounted in the head and upper spine. The IES triaxial angular rate gyro was designed to meet the SAE J211/1 (rev. March 1995) CFC 600 frequency response requirement specified in FMVSS 202a and is capable of recording angular rates up to 4800 degree/second. The sensor weighs 22 grams and fits at the center of gravity of the Hybrid III dummy head on a custom mount. The Hybrid III head with the IES sensors was balanced so as to meet the mass specifications in Part 572. The upper neck and lower neck were instrumented with six-axis load cells, and the lumbar spine with a three-axis load cell.

The dummy was positioned in the test seat following the procedures outlined in S5.3.7 of FMVSS 202a (Figure 1) with the exception of the right foot and hands. The dummy was seated symmetric with respect to the seat centerline. Adjustments were made to align the hip joint with the seat H-point while keeping the head instrumentation platform level (± 0.5 degree). Both feet were positioned flat on the floor and the lower arms were positioned horizontally and parallel to each other with palms of the hands facing inward. The dummy was restrained using the OEM 3-point seatbelt harness for the corresponding seat during all tests. The position of the dummy head relative to the head restraint was measured in two ways: (1) the vertical distance from the top of the head to the top of the head restraint; and (2) the shortest horizontal distance between the head and the head restraint.

Video images were captured for these tests using two Phantom high-speed digital video cameras operating at 1000 frames per second. One camera was mounted on-board to provide a right lateral view of the dummy kinematics while the second camera was mounted overhead to provide a top view. Video collection was synchronized with the data acquisition system using a sled impact trigger with an optical flash that was visible within the field of view of both cameras to signal the time of initial sled impact.

Figure 1. Pre-impact setup of the dummy and seat for the FMVSS 202a rear impact sled tests.

The sled was accelerated to an impact velocity of approximately 17.3 km/h. Upon impact, the sled experienced a deceleration-time curve that conformed to the corridor described in the FMVSS 202a standard when filtered to channel class 60, as specified in the SAE Recommended Practice J211/1 (rev. Mar 95) (Figure 2). Upon sled impact, the sensor and video data were collected synchronously, including a head-to-head restraint contact sensor and

the sled linear accelerometer. All data were collected and processed in accordance with the procedures specified in SAE Recommended Practice J211/1 (rev. March 1995). Each seat was tested under FMVSS 202a dynamic conditions only once.

Angular displacements of the dummy head and torso were calculated through numerical integration of the angular velocity data obtained from the rate gyro sensors in the head and upper spine. The relative head-torso relative angular displacement values were calculated at each time step by subtracting the torso angular displacement value from the corresponding head angular displacement value. The maximum head-torso relative rotation value in the posterior direction was used to evaluate the relative whiplash injury risk associated with the different seats tested according to the FMVSS 202a dynamic option. Data from the load cells in the upper and lower neck were used to calculate the Nkm index (Schmitt, 2001). The positive shear (head moves posterior relative to the neck) was used in calculating Nkm and in comparing the upper neck and lower neck shear forces between tests. The moment measured at the lower neck load cell was corrected to represent the lower neck moment.

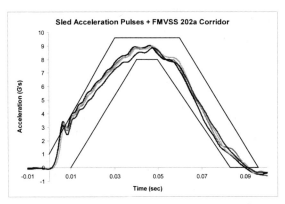

Figure 2. Sled impact deceleration pulses of rear impact testing of the four seats along with the FMVSS 202a corridor.

RESULTS

The head restraint height (vertical distance from the top of the head to the top of the head restraint) and backset (horizontal distance from the head restraint to the back of the head), as measured using the HRMD, ranged 15-45 mm and 25-70 mm respectively, as the OEM head restraint was in its mid-position (Table 1). The similar measurements representing the horizontal and vertical position of the head restraint relative to the Hybrid III dummy head are also presented in

Table 1 for comparison. In general, the head of the seated dummy was lower, but further away from the head restraint than the HRMD (Table 1). Note that among the four seats tested only the Nissan Altima had an independent seat height adjustment where the seat was set at the lowest position while the front edge of the seat pan was at the lowest position relative to its rear edge. For the other seats, the requirement of having the seat pan front edge to be at the lowest position relative to its rear edge forced the overall seat to be at the highest position.

Table 1: Head Restraint Geometric Measurements (Mid-Height Position)

2006 OEM Seat		Honda	Nissan	Saab	Subaru
HRMD	Backset (mm)*	40	25	70	48
HRMD	Head to HR Height (mm)*	45	39	29	15
Dummy	Horizontal Head to HR Distance(mm)	59	48	78	76
Dummy	Vertical Head to HR Distance(mm)	45	32	19	12

* IIHS procedure (IIHS 2001) was used to set up the SAE J826 manikin and the seat back position

Table 2 presents the results of the dummy responses in the FMVSS 202a optional dynamic test environment. The time that the dummy head made initial contact with the head restraint ranged from 56 to 74 milliseconds between the four seat tests, somewhat consistent with the horizontal head-to-head restraint distance values of the four seats (Table 1). The maximum posterior head-torso relative rotation of the Hybrid III dummy was less than 8 degrees for the Saab 9-3, Honda Civic, and the Subaru Outback, but exceeded the 12 degrees specified limit in FMVSS No. 202a for the Nissan Altima.

The performance of the seats, as measured by the peak posterior head-torso relative rotation (Figure 3), did not correlate with the initial relative position between the dummy head and head restraint. The greatest rotation occurred in the seat having the smallest horizontal dummy head to head restraint distance as well as the smallest backset and one of the seats with the smallest head-torso relative rotations occurred in a seat having the largest of these static dimensions (Table 1 and Figure 3). The head restraint height did not appear to be a strong factor in

seat performance as the head restraint at mid-position for all four seats were significantly higher than the head CG and were in the "Good" range for head restraint height as per the rating system by IIHS (IIHS 2001).

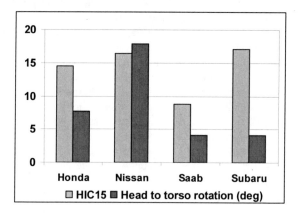

Figure 3. FMVSS 202a injury measures (Head-torso relative rotation in degrees and HIC15) for the four OEM seats in rear impact tests.

Table 2: Dynamic Test Results

2006 OEM Seat	Honda	Nissan	Saab	Subaru
Head Contact Time (ms)	69	56	69	74
Peak Head-Torso Rotation (deg)	7.7	17.9	4.1	4.1
Upper Neck Tension (N)	81	97	101	36
Upper Neck Shear (N)	110	160	87	98
Lower Neck Moment (Nm)	9	26	2	10
HIC 15msec	14.5	16.4	8.8	17.1
Nkm	0.07	0.24	0.13	0.06
Within FMVSS 202a Limits	Yes	No	Yes	Yes

The HIC15 injury measure for all seats was less than 20 (Table 2, Figure 3), which is significantly lower than the specified limit of 500 in FMVSS No. 202a. The relative performance of the seats measured by the head-torso relative posterior rotation was consistent with several other biomechanical measures such as the upper neck shear force (Figure 4), lower neck extension moment (Figure 5), and upper neck

Nkm index (Figure 6). Those measures all showed that the Altima seat, which had the smallest horizontal dummy head-to-head restraint distance and backset at mid-height position, sustained the highest relative motion and neck loads. The Saab had the lowest relative motion and neck loads, except for Nkm, and had the largest horizontal dummy head-to-head restraint distance and backset.

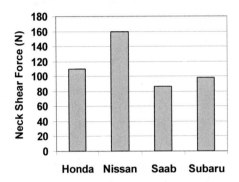

Figure 4. Upper neck positive shear forces for the four OEM seats in the FMVSS 202a dynamic test.

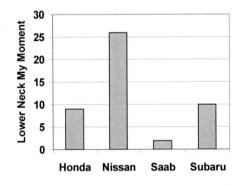

Figure 5. Lower neck extension moments for the four OEM seats in FMVSS 202a dynamic test.

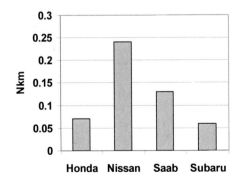

Figure 6. Shear-bending load index (Nkm) for the four OEM seats in FMVSS 202a dynamic test.

The time histories of the head, torso and head-torso relative rotation for the four OEM seats in the FMVSS 202a dynamic tests are presented in Figures 7-10. The maximum posterior head-torso relative rotation occurred before the maximum head or torso rearward rotation in all the seats. The maximum lower neck extension moment occurred approximately at the time of maximum head-torso rotation in all the seats except for the Saab seat where it had occurred somewhat earlier (Figure 9). The maximum shear force occurred after the maximum lower neck extension moment with all the seats.

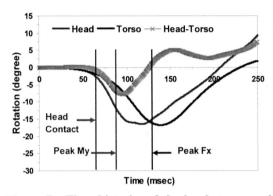

Figure 7. Time histories of the head, torso, and head-torso relative rotation in the Honda Civic seat in FMVSS 202a dynamic test.

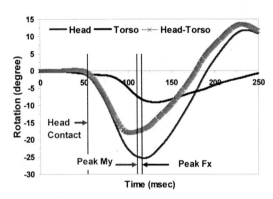

Figure 8. Time histories of the head, torso, and head-torso relative rotation in the Nissan Altima seat in FMVSS 202a dynamic test.

A detailed analysis of the dummy kinematics provided an understanding for the reasons why the Nissan Altima seat did not achieve the FMVSS 202a dynamic test requirements while the other three seats easily met the requirements. At the time of initial head contact with the head restraint, the head-torso rotation in the Altima seat was 0.9 degree (Figure 8) which was similar to that of the other three seats that ranged between 0.5 to 1.7 degrees (Figures 7, 9 and

10). However, after contact with the head restraint, the head continued to rotate up to a peak of 25 degrees in the Nissan Altima seat while the total torso rotation was only 9.1 degrees (Figure 8). The low torso rotation (lowest among all the seats tested) with respect to the head rotation (highest among all the four seats) resulted in high head-torso relative rotation with a peak of 17.9 degrees (Figure 8). On the other hand, the head restraints and the seat backs of the other three seats allowed the torso to undergo a similar total rotation as the head (Figures 7, 9, and 10). The seat-back stiffness, recliner stiffness, and the head restraint stiffness may have contributed to the different performances of the OEM seats.

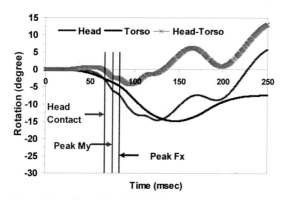

Figure 9. Time histories of the head, torso, and head-torso relative rotation in the Saab 9-3 seat in FMVSS 202a dynamic test.

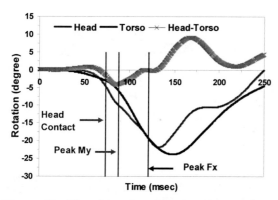

Figure 10. Time histories of the head, torso, and head-torso relative rotation in the Subaru Outback seat in FMVSS 202a dynamic test.

DISCUSSION

IIHS evaluated the 2006 Honda Civic, Nissan Altima, Saab 9-3, and the Subaru Outback using both their head restraint static measurement procedure as well as their dynamic test procedure (Table 3). The Honda Civic, Saab 9-3 and the Subaru Outback received "good" geometric and dynamic ratings, resulting in an overall "good" rating. The Nissan Altima received an "acceptable" geometric and dynamic rating, resulting in an overall "acceptable" rating. Note that the head restraint geometric rating by IIHS is based on height and backset measured in the lowest position or in the most favorably adjusted and locked position of the head restraint. The final static geometric rating is the better of the two, except that if the rating at an adjusted position is used, it is downgraded one category. The head restraint geometric measurements in this study were obtained with the head restraint at a locked position which is approximately mid-point of the highest and lowest position, since that is the position of the head restraint for the dynamic test.

Table 3: IIHS Seat Ratings and Dynamic Test Data using the BioRID Dummy

2006 OEM Seat	Honda	Nissan	Saab	Subaru
Geometric Rating	G	A	G	G
Peak T1 Accel.	13.7	9.7	16.2	11.2
Head Contact Time (ms)	62	64	64	67
Peak Neck Shear (N)	52	221	11	37
Peak Neck Tension (N)	677	660	287	308
Dynamic Rating	G	A	G	G
Overall Rating	G	A	G	G

The FMVSS 202a requirement of the 55 mm limit on the head restraint backset is more stringent than the IIHS backset limit of 75 mm for a "good" rating. This suggests that the seats that meet the FMVSS 202a static measurement requirement would likely receive a "good" geometric rating from IIHS unless the height dimension was insufficient. Comparison of the performance of the four OEM seats tested in the FMVSS 202a optional dynamic test procedure and the IIHS dynamic test procedure suggests that seats with active head restraints that are within the FMVSS No. 202a dynamic test limits are likely to obtain a "good" dynamic rating by IIHS. However, according to the IIHS procedure, if the seats with active head restraints do not obtain a "good" or "acceptable" geometric rating, they are not tested dynamically.

This study demonstrated that initial head restraint position relative to the head may not be a reliable indicator for the dynamic performance of seats with active head restraints. Real-world data and experimental studies have shown that a head restraint positioned closer to the head would provide more effective whiplash mitigation. Though the head restraints of all four OEM seats moved forward and closer to the head in a similar manner during the rear impact tests, their performance after the initial head contact differed (Figures 7-10). The Nissan Altima seat did not meet the optional dynamic test requirement of 12 degrees head-torso rotation, as a result of the large differential between the head and torso rotation after the initial head contact. This is evidenced in Figure 8, where the torso rotation is significantly smaller than that of the head.

Kinematic evaluation of the video data indicated that the seat back of the Altima was too stiff to allow sufficient torso movement into the seat back such that the torso and the head move together to minimize their relative motion. In contrast, the seat back stiffness, recliner stiffness, and the head restraint stiffness of the Honda Civic, Saab 9-3, and the Subaru Outback seats appeared to be optimized so that the head and torso rotated together and thereby minimized the relative rotation between the head and the torso at this test speed (Figures 7, 9, and 10). In addition, the head restraint of the Altima seat appeared to be too compliant, thus allowing too much posterior head rotation after the head made the initial contact with the head restraint. Previous research has found that a less rigid head restraint can increase the neck injury risk in rear impact (Voo 2004).

There are some seat positioning differences between the FMVSS 202a procedure and that of IIHS (NHTSA 2004, IIHS 2001):

- The FMVSS 202a seat positioning procedure, which this study attempted to follow, resulted in the seats of the Honda Civic, Saab 9-3 and Subaru Outback being at their highest position in order to obtain as shallow angle for the seat pan, which results in the highest H-point position relative to the seat back. The IIHS procedure would place those same seats at their lowest position regardless of the resulting seat pan angle (as per section 5.1.5 and 5.1.7 of IIHS 2001). This resulted in those same seat pans being adjusted to the most rearward tilted position (as per section 5.1.5 and 5.1.7 of IIHS 2001). On the other hand, both

procedures would set the Nissan Altima seat at its lowest position. The IIHS procedure would then place the seat pan at the mid-range of inclination.
- All the seats in this study were set at the mid-point between the most forward and most rearward positions of the seat track. The IIHS procedure would have set them at the most rearward position (as per section 5.1.6 of IIHS 2001).

Those seat positioning differences might have resulted in differences in head-restraint position measurement and/or dummy position relative to the head restraint. However, we do not believe that those differences have significantly altered the relative dynamic performance of the seats tested in this study and the similar ones by IIHS, even though different dummies (Hybrid III and BioRID) were used.

This study has demonstrated the complexity of designing a seat to mitigate whiplash injuries during a rear impact collision. Seats with active head restraints that have superior static (undeployed) geometry may not necessarily perform relatively well under dynamic conditions, whereas seats that do not have superior static (undeployed) geometry may still perform relatively well dynamically. The Saab 9-3 seat, for example, had an initial backset measurement of 70 mm (using the HRMD) but was still able to limit the head-torso relative rotation to approximately four degrees.

Results from this study demonstrated the importance of considering both the seat back and head restraint designs as a complete seating system to provide optimal protection to the occupants. Head restraint designs that are too compliant or seat-back designs that are too stiff may both result in excessive motion of the head relative to the torso.

ACKNOWLEDGEMENT

The authors would like to thank the National Highway Traffic Safety Administration for their support of this project under Cooperative Agreement No. DTNH22-05-H-01021.

REFERENCES

Eichberger A, Geigl BC, Moser A, Fachbach B, Steffan H, Hell W, Langwieder K. "Comparison of Different Car Seats Regarding Head-Neck Kinematics of Volunteers during Rear End Impact," International IRCOBI Conference on the Biomechanics of Impact, September, 1996, Dublin.

Farmer, C., Wells, J., Lund, A., "Effects of Head Restraint and Seat Redesign on Neck Injury Risk in Rear-End Crashes," Report of Insurance Institute for Highway Safety, October, 2002.

IIHS, (2005) "Insurance Special Report, Head Restraints and Personal Injury Protection Losses, "Highway Loss Data Institute, April 2005.

IIHS (2001) "A Procedure for Evaluating Motor Vehicle Head Restraints," http://www.iihs.org/ratings/protocols/pdf/head_restraint_procedure.pdf

IIHS (2006) RCAR-IIWPG Seat/Head Restraint, Evaluation Protocol, http://www.iihs.org/ratings/protocols/pdf/rcar_iiwpg_protocol.pdf

Kleinberger M, Voo LM, Merkle A, Bevan M, Chang S, "The Role of Seatback and Head Restraint Design Parameters on Rear Impact Occupant Dynamics," Proc 18th International Technical Conference on the Enhanced Safety of Vehicles, Paper #18ESV-000229, Nagoya, Japan, May 19-22, 2003.

NHTSA, "Federal Motor Vehicle Safety Standards; Head Restraints," (FMVSS 202a), Federal Register 49 CFR Part 571, Docket no. NHTSA-2004-19807, December 14, 2004.

Olsson, I., Bunketorp, O., Carlsson G.,Gustafsson, C., Planath, I., Norin, H., Ysander, L. "An In-Depth Study of Neck Injuries in Rear End Collisions", 1990 International Conference on the Biomechanics of Impacts, September, 1990, Lyon, France.

Schmitt, K., Muser, M., Niederer, P., "A New Neck Injury Criterion Candidate for Rear-End Collisions Taking into Account Shear Forces and Bending Moments," 17th ESV Conference, Paper No. 124, 2001.

Svensson, M., Lovsund, P., Haland, Y., Larsson, S. The Influence of Seat-Back and Head-Restraint Properties on the Head-Neck Motion during Rear-Impact, 1993 International Conference on the Biomechanics of Impacts, September, 1993, Eindhoven, Netherlands.

Tencer, A., Mirza, S., Bensel, K. Internal Loads in the Cervical Spine During Motor Vehicle Rear-End Impacts, SPINE, Vol. 27, No. 1 pp 34–42, 2002.

Voo LM, Merkle A, Wright J, and Kleinberger M: "Effect of Head-Restraint Rigidity on Whiplash Injury Risk," Proc Rollover, Side and Rear Impact (SP-1880), Paper #2004-01-0332, 2004_SAE World Congress, Detroit, MI, March 8 - 11, 2004.

Biofidelity of Rear Impact Dummies in Low-Speed Rear-end Impact
- Comparison of rigid seat and mass-production car seat
with human volunteers -

Kunio Yamazaki, Koshiro Ono, Mitsuru Ishii
Japan Automobile Research Institute (JARI)

ABSTRACT

Rear impact dynamic test methodologies for reducing minor neck injuries have recently been examined internationally by EEVC, GTR, NCAP (Europe, Japan), and others. In these test methodologies, the biofidelity of the dummies is of utmost importance, so this study evaluated the biofidelity of rear impact dummies (BioRID-II, RID3D and Hybrid-III) and compared the results with those of human volunteer tests.

To evaluate the biofidelity of rear impact dummies, two test series were conducted under the same conditions as with the human volunteer tests at the impact velocity of 8 km/h: the deceleration sled test and the acceleration sled test. In the deceleration sled test, a wooden rigid seat without headrest was mounted to a sled which was moved on ramped rails and collided into a damper. In the acceleration sled test, a mass-production car seat with headrest was mounted to a sled which was accelerated horizontally. In both test series, five tests were conducted for each dummy.

In the two test series, the behavior of BioRID-II was very similar to that of the volunteers. In addition, BioRID-II showed good biofidelity for the head angle with respect to T1 (the first thoracic spine), which is considered to be an important characteristic of a rear impact dummy. These results were possible due to the spine structure flexibility of BioRID-II. On the other hand, RID3D, which has flexibility in the neck, needs flexibility between the torso and the lower neck to improve its biofidelity. For Hybrid-III, the dummy showed a very different behavioral response compared to the human volunteers; its biofidelity was lower than that of the rear impact dummies.

Keywords: REAR IMPACT, BIOFIDELITY, MINI SLED, BACK IMPACT

WHIPLASH INJURY DURING A REAR END COLLISION causes serious damage and social cost, so multiple countries are collaborating to reduce neck injuries (Hynd et al. 2005). One method of evaluating the safety performance of the headrest, namely dynamic test methodologies using a rear impact dummy, is being examined by EEVC, GTR, and NCAP (Europe, Japan) (Avery et al. 2006, Bortenschlager et al. 2007). Although such examinations cover many topics like the testing conditions and evaluation method, the dummy is the most important measurement apparatus in the test and has attracted the most research attention.

Currently, BioRID-II and RID3D have been suggested as suitable candidates as rear impact dummies in addition to Hybrid-III for frontal crashes (Davidsson 1999a, Cappon et al. 2003, Foster et al. 2003). Several validations and improvements to these dummies have been made under various testing conditions. Philippens et al. (2002) compared the responses of BioRID-II, RID2 (the previous version of RID3D) and Hybrid-III in tests conducted under the same test conditions as the original tests used to develop and validate the rear impact dummies, and concluded that Hybrid-III should not be used for rear impact tests because its biofidelity was lower than that of BioRID-II and RID2. Willis et al. (2005) conducted sled tests using BioRID-II, RID3D and Thor-Alfa, and evaluated the biofidelity of these dummies by comparing the results with those of human volunteer tests. Their report concluded that the responses of BioRID-II and RID3D approximated the volunteers' responses but did not completely satisfy the range of their responses, and the response of Thor-Alfa was not close to that of human volunteers. Yaguchi et al. (2006) evaluated five types of dummies (BioRID-II, RID2, Hybrid-III, Thor-NT and Thor-FT) by comparing the results with those of mini-sled tests using human volunteers and the results of back impact tests using PMHS, and reported that BioRID-II was most similar to human volunteers and PMHS in terms of its response to low-speed rear impacts.

However, there is insufficient objective data to evaluate the biofidelity of the latest version of each dummy. Especially, there are few reports on the comparison between rear impact dummies and human volunteers for mass-production seats.

In this study, mini-sled tests using a rigid seat and a mass-production seat for three types of dummies (BioRID-II, RID3D and Hybrid-III) were conducted under the same conditions as human volunteer tests. By comparing the responses of the dummies with those of human volunteers, we evaluated the biofidelity of dummies in order to identify the most suitable dummy for use in low-speed rear impact tests.

METHODS AND MATERIALS

DUMMIES EVALUATED IN THIS STUDY: The dummies evaluated in this study were BioRID-II version g, RID3D version of December 2006 and Hybrid-III. BioRID-II and RID3D were developed for low-speed rear impacts. BioRID-II has a segmented spinal structure consisting of the same number of vertebrae as that of a human, i.e. 7 cervical, 12 thoracic and 5 lumbar vertebrae. The thoracic spine has a kyphosis and the lumbar spine is straight as in a human in the seated posture (Davidsson 1999b). RID3D basically has a rigid spinal structure, but it also has flexible elements in the thoracic and lumbar regions and a fully flexible neck design (Cappon et al. 2005). Hybrid-III was developed for high-speed frontal impacts, and basically has a rigid spinal structure except for the cervical part and the lumbar part (Foster et al. 1977).

MINI-SLED TEST: To evaluate the biofidelity of rear impact dummies, two test series were conducted under the same conditions as with the human volunteer tests: the deceleration sled test and the acceleration sled test. In both test series, five tests were conducted for each dummy. Dummies were positioned in the following states: (1) the bottom of the skull cup was kept horizontal, (2) The upper torso was pushed against the seat back as much as possible, and (3) hands were put on the handle (deceleration sled tests) or were put on the knees (acceleration sled tests) as well as the volunteer tests. The coordinates of the head and hip point were measured and checked to maintain the same seating posture in each test that used the same dummy.

In the deceleration sled test, a wooden rigid seat without headrest was mounted to a sled which was moved on ramped rails and collided into a damper at the impact velocity of 8 km/h (Figure 1). By using the rigid seat, the inherent impact response of the dummies can be evaluated without the influence of the mechanical properties of the seat.

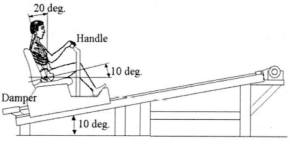

Figure 1. Ramp type sled

In the acceleration sled test, a mass-production car seat with headrest was mounted to a sled which was accelerated on horizontal rails by an electric motor. The change in velocity of the sled was 8 km/h (Figure 2). The objective of this test series was to evaluate the biofidelity of the dummies using a mass-production seat in consideration of accident conditions.

Figure 2. Horizontal type sled

DATA ANALYSIS: In each of the two test series, the motion of the head/neck/torso during low-speed rear impacts was recorded with a high-speed video camera. Based on the movement of target marks attached to the dummies, relative displacements and rotations of their head, neck, and torso were measured and analyzed. In addition, accelerations of head C.G. and T1 and the force/moment of the upper neck were measured by accelerometers and load cells attached to the dummy. The acceleration of the head C.G. (center of gravity) and upper neck loads of the volunteer were calculated using the six-degree freedom component measuring method (Ono and Kanno 1993). The T1 linear and angular displacements of the volunteer were calculated as a weighted average value from the film target data on the T1 and on the sternum. The method of measuring volunteer data is described in detail in the reference (Davidsson et al. 1999a).

Two categories of parameters were used to compare the dummy response with the human volunteers. One category of parameters concerned the external behavior obtained by video analysis, and the other concerned the impact response measured by sensors (Table 1). For parameters of the external behavior, changes of the head angle with respect to the neck (HA-NA), the neck angle with respect to T1 (NA-TA), the head angle with respect to T1 (HA-TA), the T1 x coordinate relative to the sled (T1-X-disp), the T1 z coordinate relative to the sled (T1-Z-disp) and the T1 angle with respect to the sled (Sled-TA) were used (Figure 3). For parameters of the impact response, accelerations of head C.G. in the x direction (Head Ax) and T1 in the x direction (T1 Ax), and shear force (Upper Neck Fx), axial force (Upper Neck Fz) and moment (Upper Neck My) were also used. Figure 4 shows the polarities of the impact response parameters.

Table 1. Evaluated parameters

Behavior	HA-NA	Change of head angle with respect to neck angle
	NA-TA	Change of neck angle with respect to T1 angle
	HA-TA	Change of head angle with respect to T1 angle
	T1-X-disp	Change of T1 X coordinate relative to sled
	T1-Z-disp	Change of T1 Z coordinate relative to sled
	Sled-TA	Change of T1 angle with respect to sled plane
Impact response	Head Ax	Head center of gravity acceleration in x direction
	T1 Ax	T1 (First thoracic vertebra) acceleration in x direction
	Upper Neck Fx	Upper neck shear force in x direction (Fx)
	Upper Neck Fz	Upper neck axial force in z direction (Fz)
	Upper Neck My	Upper neck moment about y axis (My)

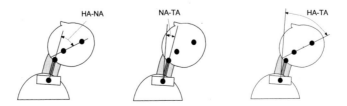

Figure 3. Explanation of evaluated parameters for behavior

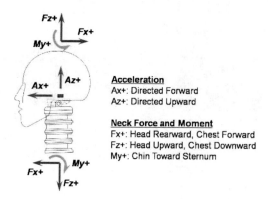

Acceleration
Ax+: Directed Forward
Az+: Directed Upward

Neck Force and Moment
Fx+: Head Rearward, Chest Forward
Fz+: Head Upward, Chest Downward
My+: Chin Toward Sternum

Figure 4. Polarities of responses

VOLUNTEER TEST DATA: The human volunteer test data of this study were obtained from the following two low-speed rear impact experiments:

(1) Deceleration sled test: Experiments on nine tests by seven volunteers using a ramp type sled with a rigid seat without headrest (Davidsson et al. 1999).

(2) Acceleration sled test: Experiments on six human volunteers using a horizontal type sled with a mass-production car seat with headrest (Pramudita et al. 2007).

From these experiments, human volunteers' corridors for parameters of external behavior and impact response were provided. These corridors were derived from the average value and standard deviation of volunteers' responses. In this study, these corridors were used to evaluate the biofidelity of the dummies.

RESULTS

DECELERATION SLED TEST: Figure 5 shows the sled acceleration pulse and the sled velocity pulse in the deceleration sled test. In this graph, each curve shows the average response of five tests conducted for each dummy. The shape of the acceleration pulse and duration showed a similar response to the volunteer test though some differences were seen in the peak acceleration of the installed dummy. Therefore, it is confirmed that the test conditions of each dummy were almost the same as those of the volunteer test.

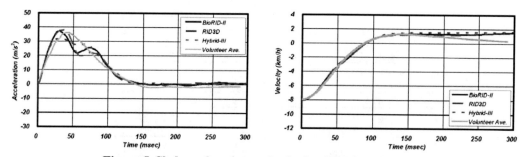

Figure 5. Sled acceleration and velocity (Deceleration sled tests)

External behavior of dummies – Figures 6 and 7 compare the external behavior of each part of the dummy (average response of five tests) with the volunteers' behavior. As for the sign of HA-NA, NA-TA and HA-TA, a plus value shows the flexion behavior and a minus value shows the extension behavior.

The response of HA-NA (the head rotation angle with respect to the neck) in human volunteers showed flexion before around 150 msec, and the angle change was comparatively small. The response of

HA-NA was quite different between the three dummies. RID3D showed flexion before 100 msec and showed extension afterwards while BioRID-II showed only flexion. Hybrid-III did not show flexion at all, but showed only extension.

As regards NA-TA (the neck rotation angle with respect to T1), the human volunteers' response showed slight flexion before around 80 msec, and showed gradual extension behavior that reached the range from –25 to –45 degrees at 250 msec. The response of NA-TA was different between the three dummies though extension behavior was indicated in each dummy. BioRID-II showed flexion at an early timing and showed extension after 100 msec. The extension behavior of the dummy was close to the lower limit of the volunteers' corridor. RID3D showed extension behavior after 50 msec and showed the same behavior as BioRID-II between 140 to 200 msec. Hybrid-III showed extension behavior that peaked at 130 msec, and returned forward quickly. Therefore, the extension behavior of Hybrid-III was obviously different from the response of the human volunteers.

HA-TA (the head rotation angle with respect to T1) means the sum of HA-NA and NA-TA. The response of HA-TA in the human volunteer tests showed flexion behavior at around 80 msec, and showed extension behavior that reached from –35 to –45 degrees at 250 msec. BioRID-II showed similar behavior to the human volunteers in terms of flexion at around 80 msec and extension until 250 msec. RID3D and Hybrid-III did not show flexion behavior at an early stage. RID3D reached a peak extension angle of –55 degree at 180 msec, and Hybrid-III reached a peak extension angle of –30 degree at 140 msec. Therefore, RID3D and Hybrid-III showed different behaviors to the human volunteers.

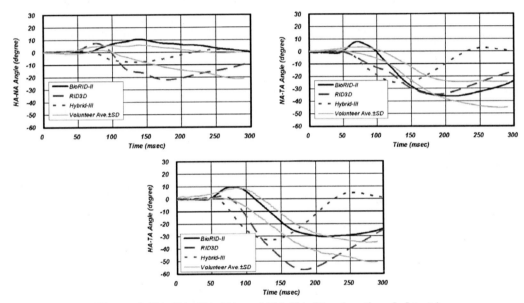

Figure 6. HA-NA, NA-TA and HA-TA (Deceleration sled tests)

Displacements and the angle change of T1 were influenced by the interaction between the back of the dummy and the seat back. For T1-X-disp and Sled-TA, the responses of BioRID-II were close to the lower limit of the volunteers' corridors, whereas those of RID3D were close to the upper limit of the corridors. As regards T1-Z-disp, the responses of both BioRID-II and RID3D were close to the lower limit of the corridor. The responses of Hybrid-III for T1-X-disp, T1-Z-disp and Sled-TA were different from volunteers' corridors.

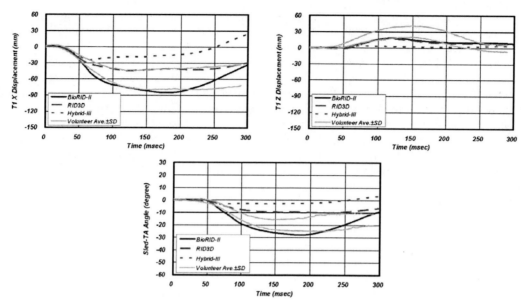

Figure 7. T1-X-disp, T1-Z-disp and Sled-TA (Deceleration sled tests)

Impact response of dummies – Figures 8 to 10 compare the impact responses of each part of the dummy (average response of five tests) with volunteers.

As regards Head Ax (anterior-posterior acceleration of the center of gravity of the head), the impact response of the human volunteers showed low posterior acceleration before around 80 msec, and showed anterior acceleration that reached from 10 to 50 m/s² at around 140 msec. Although the maximum anterior accelerations of Head Ax of the three dummies were almost the same as the upper limit of the volunteers' corridor, the timings of maximum accelerations differed among the dummies. The maximum acceleration of Hybrid-III occurred at 120 msec while those of BioRID-II and RID3D occurred from 170 to 190 msec. Only BioRID-II showed posterior acceleration before 80 msec, which was also seen in the volunteer's response.

Whereas T1 Ax of the human volunteers reached from 30 to 40 m/s² at around 70 msec, the three dummies showed around twice the acceleration compared with the human volunteers. T1 Ax of BioRID-II showed posterior acceleration at 40 msec, then showed anterior acceleration that reached a maximum at 70 msec. On the other hand, T1 Ax of RID3D and Hybrid-III showed a similar response, with anterior acceleration reaching a maximum at 60 msec without posterior acceleration initially.

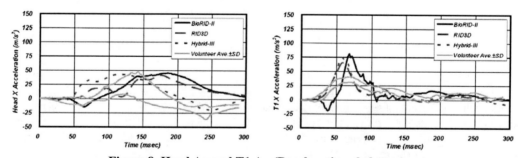

Figure 8. Head Ax and T1 Ax (Deceleration sled tests)

The response of upper neck Fx (upper neck shear force) in the volunteer tests showed a negative value before 140 ms, and changed to a positive value gradually. Only BioRID-II showed a negative value at an early timing in the three dummies. The tendency of the difference in the three dummies was the same as their Head Ax responses.

The response of upper neck Fz (upper neck axial force) in the volunteer tests showed compression before 110 ms, and then changed to tension. The three dummies were the same as the human volunteers in terms of the change in direction of force, although its peak value and timing were different. In addition, responses of BioRID-II and RID3D were comparatively similar to the volunteers' corridor.

The response of upper neck My (upper neck moment) in the volunteer tests showed flexion before 100 msec, and changed to extension gradually. The three dummies were the same as the human volunteers in terms of the change in direction of moment. However, the maximum values of the flexion of BioRID and RID3D were two or three times higher than that of the human volunteers. It was thought that the rigidity of the joint of the head and the neck in BioRID-II and RID3D was higher than that of the human volunteers though the dummies had flexibility in the neck. Although Hybrid-III indicated the same level of flexion as the volunteers, the change from flexion to extension was rapid and the maximum extension was three times higher than that of the volunteers. In Hybrid-III, it was thought that there was hardly any flexion behavior since the rigidity of the entire spine was higher than that of the other dummies, and the flexion moment was low.

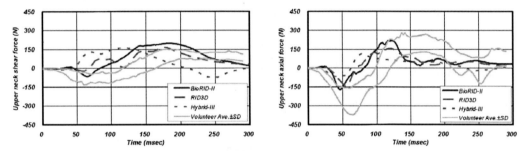

Figure 9. Upper neck shear force Fx and axial force Fz (Deceleration sled tests)

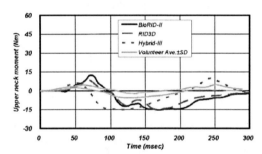

Figure 10. Upper neck moment My (Deceleration sled tests)

ACCELERATION SLED TEST: Figure 11 shows the sled acceleration pulse and the sled velocity pulse in the acceleration sled test. In this figure, each curve shows the average response of five tests conducted for each dummy. The shape of the acceleration pulse and duration showed a similar response to the volunteer test though some differences were seen after 120 msec depending on the installed dummy. Therefore, it is confirmed that the test conditions of each dummy were almost the same as those of the volunteer test.

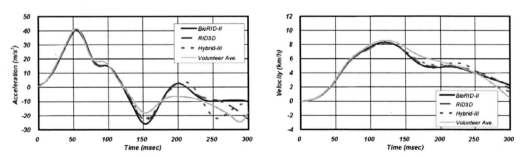

Figure 11. Sled acceleration and velocity (Acceleration sled tests)

External behavior of dummies – Table 2 lists the time at which the head made contact with the headrest for each test. Figures 12 to 13 compare the external behavior of each part of the dummy (average response of five tests) with the volunteers' behavior.

The response of HA-NA in the volunteer tests shows flexion between 100 and 150 msec. BioRID-II and RID3D showed flexion of around the lower limit of volunteers' corridor though the timings were around 30 msec later. The behavior of BioRID and RID3D regarding HA-NA was similar to that of the human volunteers, taking into consideration the fact that the headrest contact time of the dummies was later than that of the volunteers. On the other hand, Hybrid-III did not show flexion at all, unlike the other rear impact dummies.

The response of NA-TA in the volunteer tests showed flexion behavior in which the angle increased after the head made contact with the headrest. BioRID-II showed flexion except a small extension at 160 msec. RID3D showed extension before the head and headrest made contact, and changed from extension to flexion at 160 msec, then extension occurred again. Hybrid-III also showed extension behavior before the head and headrest made contact, and changed from extension to flexion after 240 msec when the dummy's back moved away from the seat back.

The response of HA-TA in the volunteer tests showed almost constant flexion after the head and headrest made contact. The response of BioRID-II was similar to that of the human volunteers, taking into consideration the fact that the head and headrest made contact 20 msec later than that of the volunteers, though the angle was small. The response of RID3D showed that flexion occurred later than BioRID-II, and changed to extension at 240 msec. Hybrid-III showed completely different behavior to the volunteers; the dummy showed extension only before the head and headrest made contact.

Table 2. Head restraint contact time

Unit ﹐msec

	Volunteer	BioRID-II	RID3D	Hybrid-III
	n=6	n=5	n=5	n=5
Earliest	74	112	114	124
Latest	112	116	120	128
Average	94	114	117	126
S.D.	14.2	1.7	2.3	2.0

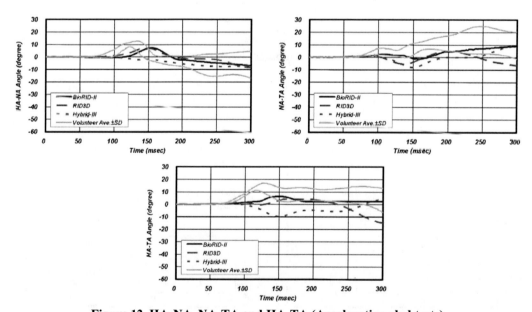

Figure 12. HA-NA, NA-TA and HA-TA (Acceleration sled tests)

For T1-X-disp, the responses of the three dummies were close to the upper limit of the volunteers' corridor. For T1-Z-disp, the response of BioRID-II was almost within the range of corridors before 150 msec. As regards Sled-TA, the response of BioRID-II was almost within the range of corridors.

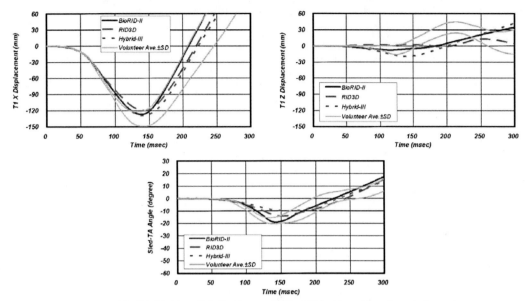

Figure 13. T1-X-disp, T1-Z-disp and Sled-TA (Acceleration sled tests)

Impact response of dummies – Figures 14 to 16 compare the impact responses of each part of the dummy (average response of five tests) with those of the volunteers.

As regards Head Ax, the impact response of the human volunteers showed anterior acceleration that reached from 40 to 80 m/s^2 between 100 and 150 msec. Although the three dummies showed around twice the acceleration compared with the human volunteers, the timing of the peak was similar to that of the volunteers taking into consideration the fact that the head and headrest made contact later.

The response of T1 Ax in the human volunteers showed anterior acceleration that reached from 35 to 50 m/s^2 between 70 and 170 msec. T1 Ax of the three dummies was close to the upper limit of the volunteers' corridor. For T1 Ax, RID3D was the most similar to the volunteers in terms of acceleration level, while BioRID-II was most similar in terms of the duration.

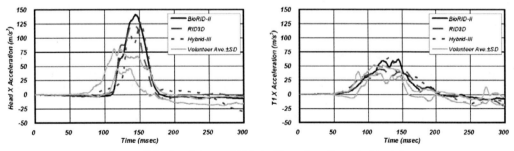

Figure 14. Head Ax and T1 Ax(Acceleration sled tests)

The response of upper neck Fx in the volunteer tests showed a negative value between 100 and 160 msec, which was the timing at which the head made contact with the headrest. The three dummies showed a different response to the volunteers. In BioRID-II, neck shear force did not occur. In RID3D, a slight shear force occurred, and the direction of the force changed at around 150 msec. In Hybrid-III, the shear force was positive while the head was in contact with the headrest.

The response of upper neck Fz in the volunteer tests showed tension between 100 and 160 msec and a similar tendency was seen in the three dummies. However, in BioRID-II the force was around twice that in the human volunteers. The responses of RID3D and Hybrid-III were almost the same as the upper limit of the volunteers' corridor.

As regards the upper neck My, the impact response of the volunteers showed flexion between 90 and 160 msec due to the contact between the head and headrest. The three dummies also showed flexion, though the peaks were small. The timing of the peak in Hybrid-III was later than in the other dummies, and this tendency was the same as upper neck Fz.

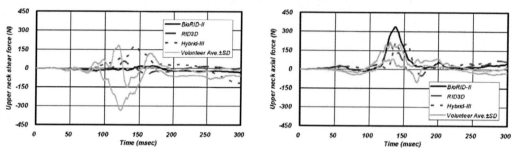

Figure 15. Upper neck shear force Fx and axial force Fz (Acceleration sled tests)

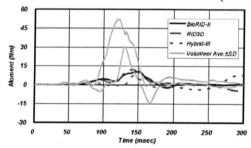

Figure 16. Upper neck moment My (Acceleration sled tests)

DISCUSSION

DECELERATION SLED TEST: Some differences were seen in the external behavior between two rear impact dummies. In BioRID-II, NA-TA showed flexion behavior first, and then shifted to extension behavior. The flexion behavior of HA-NA was caused by this NA-TA motion. On the other hand, in RID3D, the flexion behavior of HA-NA and the extension behavior of NA-TA were caused at the same time, and then HA-NA shifted to extension behavior.

These differences in behavior between BioRID-II and RID3D in the deceleration sled test were considered to be caused by the structural difference of their spines. The test results showed that RID3D has local flexibility in the upper part of the neck while BioRID-II has flexibility in the entire spine including the neck. These differences appeared clearly in HA-TA, which showed the entire change of angle between the head and the neck. As for HA-TA, only the behavior of BioRID-II was almost the same as the human volunteers' corridor. For Hybrid-III, the dummy showed a big difference in most of the parameters for evaluating external behavior compared to human volunteers.

The impact response of BioRID-II concerning Head Ax, Upper Neck Fx, and Fz was similar to that of the human volunteers. A response opposite to that of the human volunteers was seen in Head Ax and Upper Neck Fx of RID3D at an early time (before 100 msec). It was thought that these responses were due to the influence of the rigidity of the entire spine. The response of Head Ax and Upper Neck Fx of Hybrid-III was markedly earlier than that of the human volunteers.

The coefficient of variation (CV) for peak values of the evaluated parameters was calculated to evaluate the repeatability of the dummies. CV was expressed as a percentage after dividing the standard deviation of the peak measurement values of each dummy by the average value.

CV = standard deviation of the peak values / average value of the peak values × 100 [%]

Tables 3 to 6 show the CV in the deceleration sled test results. All three dummies showed good repeatability for HA-NA. For T1-Z-disp and Sled-TA, values of CV were large since average values were small. As for T1-X-disp, values of SD were almost the same in the three dummies. As regards impact response parameters, values of CV of the three dummies were within 10% except Fz of Hybrid-III.

Table 3. CV (Deceleration sled tests): BioRID-II

BioRID-II	HA-TA	T1-X-disp	T1-Z-disp	Sled-TA
	extension	rearward	upward	extension
Average of Max.	-30.4	-85.0	18.9	-27.8
S.D. of Max.	0.7	5.2	1.2	0.8
C.V. of Max.	2.3	6.1	6.3	2.7

BioRID-II	Head Ax	T1 Ax	Neck Up. Fx	Neck Up. Fz	Neck Up. My	
	forward	forward	positive	tension	flexion	extension
Average of Max.	48.7	84.7	199.0	231.2	12.5	-15.5
S.D. of Max.	1.8	7.1	1.3	5.0	0.6	0.1
C.V. of Max.	3.7	8.4	0.6	2.1	4.9	0.8

Table 4. CV (Deceleration sled tests): RID3D

RID3D	HA-TA	T1-X-disp	T1-Z-disp	Sled-TA
	extension	rearward	upward	extension
Average of Max.	-56.9	-44.9	17.0	-10.2
S.D. of Max.	1.2	5.7	2.0	0.8
C.V. of Max.	2.1	12.7	11.9	7.8

RID3D	Head Ax	T1 Ax	Neck Up. Fx	Neck Up. Fz	Neck Up. My	
	forward	forward	positive	tension	flexion	extension
Average of Max.	43.0	67.2	168.8	167.4	9.1	-15.2
S.D. of Max.	2.0	1.5	7.6	7.1	0.2	0.7
C.V. of Max.	4.7	2.3	4.5	4.2	2.6	4.5

Table 5. CV (Deceleration sled tests): Hybrid-III

Hybrid-III	HA-TA	T1-X-disp	T1-Z-disp	Sled-TA
	extension	rearward	upward	extension
Average of Max.	-32.9	-25.3	4.7	-3.3
S.D. of Max.	0.9	5.2	0.6	0.9
C.V. of Max.	2.7	20.6	13.0	27.0

Hybrid-III	Head Ax	T1 Ax	Neck Up. Fx	Neck Up. Fz	Neck Up. My	
	forward	forward	positive	tension	flexion	extension
Average of Max.	44.6	80.2	163.3	127.4	10.2	-15.2
S.D. of Max.	2.3	5.7	4.3	12.8	0.4	0.8
C.V. of Max.	5.1	7.1	2.6	10.1	3.7	5.5

ACCELERATION SLED TEST: A difference was seen between two rear impact dummies in NA-TA for external behavior. In BioRID-II, both HA-NA and NA-TA showed flexion behavior. On the other hand, in RID3D, HA-NA showed flexion behavior and NA-TA showed extension behavior. The difference in NA-TA caused the difference of the timing of the flexion behavior of HA-TA. HA-TA of BioRID-II showed better correlation with the volunteers than RID3D, though the angle was smaller and the time of contact with the headrest was around 20 msec slower than that of the human volunteers. In Hybrid-III, HA-TA showed extension behavior before the head came into contact with the headrest, and showed behavior opposite to that of the human volunteers.

As for the impact response, all three dummies showed about twice the peak value for Head Ax compared to the human volunteers though they showed a similar response for T1 Ax. There was a possibility that the impact velocity of the head with the headrest increased since the impact timing of the head and the headrest was slower than that of the human volunteers in the dummy. Hybrid-III showed a response of the opposite direction compared to the human volunteers in Upper Neck Fx before the head came into contact with the headrest.

Tables 6 to 8 show CV in the acceleration sled test results. For BioRID-II, the value of CV for HA-NA was large since the average value was small. For RID3D, CV for HA-TA was large because the maximum value for extension was measured in the rebound phase. As regards Head Ax and T1 Ax, all three dummies showed good repeatability.

Table 6. CV (acceleration sled tests): BioRID-II

BioRID-II	HA-TA	T1-X-disp	T1-Z-disp	Sled-TA
	extension	rearward	upward	extension
Average of Max.	-0.1	-127.0	34.5	-19.1
S.D. of Max.	0.1	2.6	1.6	0.6
C.V. of Max.	116.7	2.1	4.8	2.9

BioRID-II	Head Ax	T1 Ax	Neck Up. Fx	Neck Up. Fz	Neck Up. My	
	forward	forward	positive	tension	flexion	extension
Average of Max.	143.8	66.6	20.5	344.5	10.7	-2.4
S.D. of Max.	4.5	3.0	4.5	9.5	0.9	0.4
C.V. of Max.	3.1	4.6	22.0	2.8	8.0	15.6

Table 7. CV (acceleration sled tests): RID3D

RID3D	HA-TA	T1-X-disp	T1-Z-disp	Sled-TA
	extension	rearward	upward	extension
Average of Max.	-14.8	-120.0	13.5	-14.2
S.D. of Max.	3.8	2.1	3.2	0.2
C.V. of Max.	26.0	1.7	23.3	1.5

RID3D	Head Ax	T1 Ax	Neck Up. Fx	Neck Up. Fz	Neck Up. My	
	forward	forward	positive	tension	flexion	extension
Average of Max.	124.6	47.5	61.5	223.7	12.4	-0.1
S.D. of Max.	4.2	1.9	8.0	18.1	0.9	0.0
C.V. of Max.	3.4	4.0	13.0	8.1	7.2	30.7

Table 8. CV (acceleration sled tests): Hybrid-III

Hybrid-III	HA-TA	T1-X-disp	T1-Z-disp	Sled-TA
	extension	rearward	upward	extension
Average of Max.	-9.9	-129.6	42.4	-10.2
S.D. of Max.	1.1	5.0	2.6	0.6
C.V. of Max.	11.2	3.8	6.2	5.9

Hybrid-III	Head Ax	T1 Ax	Neck Up. Fx	Neck Up. Fz	Neck Up. My	
	forward	forward	positive	tension	flexion	extension
Average of Max.	131.5	67.4	189.2	249.9	11.3	-2.6
S.D. of Max.	4.1	4.2	2.6	10.4	1.0	0.3
C.V. of Max.	3.2	6.3	1.4	4.2	8.5	13.1

The cumulative variance ratio (CVR, Philippens et al. 2002) was calculated to evaluate the biofidelity of the dummies objectively. A lower CVR value indicates good biofidelity, as the variance of the dummy is more similar to corridor variance. A CVR value of over 5 means little or no correlation between the dummy response and the volunteer response. Therefore, CVR values of over 5 were set equal to 5 for calculating the average CVR in the same way as in previous literature.

$$CVR = \frac{\sum\limits_{t=Tstart}^{t=Tend}\left(dummy(t) - volunteer\_ave(t)\right)^2}{\sum\limits_{t=Tstart}^{t=Tend}\max\left(cor(t) - volunteer\_ave(t)\right)^2}$$

where, dummy(t) is the time response of the evaluated parameter of the dummy, volunteer_ave(t) is the average of the corresponding parameter of volunteers, and cor(t) is the volunteer corridor.

Since the CVR value is obtained by integration, it is affected by the end time of the calculation. Tables 9 and 10 show CVR values for external behavior parameters and impact response parameters until 150 msec (sled reached maximum velocity) and until 250 msec (end of response) respectively.

For external parameters, BioRID-II showed the smallest average value of CVR in the three kinds of dummies for all test and calculation conditions. As for impact response parameters, BioRID showed the smallest CVR in one case (deceleration sled test, calculation time of 150 msec), and RID3D showed the smallest CVR in the other three cases. However, the CVR value is calculated based on the difference from the average value of the volunteers. Therefore, to evaluate the biofidelity of the dummy, it is necessary to consider the shape of the time history of the response. For example, though values of CVR for HA-NA in the deceleration sled test were 2.8 in BioRID-II and Hybrid-III respectively, the shape of the response in BioRID-II was more similar to the volunteers' corridors (Figure 6).

Table 9. CVR (Deceleration sled tests)

[150ms]	BioRID-II	RID3D	Hybrid-III
HA-NA	2.8	5	2.8
NA-TA	2.9	5	5
HA-TA	0.4	5	5
T1-X-disp	1.5	0.8	5
T1-Z-disp	1.2	1.7	5
Sled-TA	3.0	4.8	5
Average	2.0	3.7	4.6

[250ms]	BioRID-II	RID3D	Hybrid-III
HA-NA	2.4	3.7	1.0
NA-TA	1.1	2.0	5
HA-TA	0.6	5	5
T1-X-disp	1.2	0.8	5
T1-Z-disp	1.1	1.5	4.3
Sled-TA	1.8	3.1	5
Average	1.4	2.7	4.2

[150ms]	BioRID-II	RID3D	Hybrid-III
HeadAx	0.4	0.8	3.2
T1Ax	5	5	5
Upper neck Fx	5	5	5
Upper neck Fz	2.8	2.7	4.0
Upper neck My	5	5	5
Average	3.6	3.7	4.4

[250ms]	BioRID-II	RID3D	Hybrid-III
HeadAx	5	4.6	2.4
T1Ax	5	4.1	5
Upper neck Fx	4.6	3.4	5
Upper neck Fz	2.5	2.5	3.3
Upper neck My	5	5	5
Average	4.4	3.9	4.1

Table 10. CVR (Acceleration sled tests)

[150ms]	BioRID-II	RID3D	Hybrid-III
HA-NA	2.8	1.5	5
NA-TA	1.0	5	5
HA-TA	5	5	5
T1-X-disp	1.2	3.1	1.0
T1-Z-disp	0.3	1.0	3.0
Sled-TA	1.6	5	5
Average	2.0	3.4	4.0

[250ms]	BioRID-II	RID3D	Hybrid-III
HA-NA	1.0	0.8	1.1
NA-TA	1.2	2.1	3.6
HA-TA	1.8	2.8	5
T1-X-disp	1.0	0.3	0.1
T1-Z-disp	3.1	4.3	5
Sled-TA	0.2	0.8	1.4
Average	1.4	1.9	2.7

[150ms]	BioRID-II	RID3D	Hybrid-III
HeadAx	5	5	4.0
T1Ax	1.8	0.5	1.7
Upper neck Fx	0.7	1.2	2.8
Upper neck Fz	4.6	0.6	2.2
Upper neck My	1.7	1.7	2.6
Average	2.8	1.8	2.6

[250ms]	BioRID-II	RID3D	Hybrid-III
HeadAx	5	4.1	5
T1Ax	2.1	1.7	3.1
Upper neck Fx	0.7	1.3	3.0
Upper neck Fz	3.3	0.9	2.9
Upper neck My	1.8	1.7	2.6
Average	2.6	1.9	3.3

In the above-mentioned result, the biofidelity of BioRID-II was slightly higher than that of RID3D in the study based on the results of two types of sled tests. Hybrid-III showed different behavior from the volunteers. Similar results were reported in the EEVC research report (EEVC 2007) and the presentation for UN ECE WP29 head restraint GTR informal group (Hynd 2007). Prasad et al. (1997) reported that Hybrid III was biofidelic when compared to the data from Mertz and Patrick (1967), however, the biofidelity evaluation was limited to head rotations only. The same studies also showed that Hybrid III was not biofidelic in consideration of seat interaction.

In this study, tests were conducted at 8 km/h in order to compare the response with that of human volunteers. Therefore, biofidelity evaluations of dummies were limited to low-speed impact. However, the biofidelity of BioRID at high-speed impact (16.5 km/h and 23.7 km/h) was confirmed by comparison with PMHS by back impact tests using a pendulum (Astrid el al. 2002, Yaguchi et al. 2006).

CONCLUSION

In the two test series, it was shown that the external behavior and the impact response of BioRID-II were very similar to those of the human volunteers. In the test using the rigid seat, the change of head angle with respect to T1, which is considered to be an important characteristic of a rear impact dummy, was almost the same as that of the human volunteers. BioRID-II also showed a similar response to the human volunteers in the test using a mass-production seat, considering the delay of the contact time of the head and the headrest. These results may have been due to the spine structure flexibility of BioRID-II.

The flexibility of the neck was observed from the change of angle in each part in the deceleration sled test for RID3D, but the flexibility was limited to the neck. To improve the biofidelity of RID3D, it is necessary to have flexibility between the torso and the lower neck.

For Hybrid-III, the dummy showed a very different behavioral response compared to the human volunteers. The biofidelity of Hybrid-III was lower than those of the other rear impact dummies; the dummy showed behavior different from that of the human volunteers in some evaluation parameters.

ACKNOWLEDGEMENTS

This research was conducted as part of a joint project under contract with the Ministry of Land, Infrastructure, Transport and Tourism.

REFERENCES

Astrid, L., Mats, S., Viano, D.: Evaluation of the BioRID P3 and the Hybrid III in Pendulum Impacts to the Back: A Comparison with Human Subject Test Data: Traffic Injury Prevention, Vol3(2),2002, p.159-166

Avery, M. and Weekes, A. M.: Dynamic testing of vehicles seats to reduce whiplash injury risk: an international protocol: International Conference of Crashworthiness, 2006

Bortenschlager, K., Hartlieb, M., Barnsteiner, K., Ferdinand, L., Kramberger, D., Siems, S., Muser, M. and Schmitt, K. U.: Review of Existing Injury Criteria and Their Tolerance Limits for Whiplash Injuries with Respect to Testing Experience and Rating Systems: 20th International Technical Conference on the Enhanced Safety of Vehicles, 2007

Cappon, H., van Ratingen, M., Wismans, J., Hell, W., Lang, D. and Svensson, M.: Whiplash Injuries, Not Only a Problem in Rear-End Impact: 18th International Technical Conference on the Enhanced Safety of Vehicles, 2003

Cappon, H., Hell, W., Hoschopf, H., Muser, M., Song, E. and Wismans, J.: Correlation of Accident Statistics to Whiplash Performance Parameters Using the RID3D and BioRID Dummy: International Research Council On the Biomechanics of Impact, 2005

Davidsson, J., Lovsund P., Ono, K. and Svensson, M. Y.: A comparison between volunteer, BioRID P3 and Hybrid III performance in rear impacts: International Research Council On the Biomechanics of Impact, 1999a

Davidsson, J.: BioRID-II Final Report: Chalmers University of Technology, 1999b

European Enhanced Vehicle-Safety Committee(EEVC): The use of Hybrid III Dummy in Low-speed Rear Impact Testing, WG12 report, September 2007

Foster, J. K., Kortge, J. O. and Wolanin, M. J.: Hybrid III – a biomechanically-based crash test dummy: 21st Stapp Car Crash Conference, 1977

Hynd, D. and van Ratingen, M.: Challenges in the Development of a Regulatory Test Procedure for Neck Protection in Rear Impacts: Status of the EEVC WG20 and WG12 Joint Activity, 19th International Technical Conference on the Enhanced Safety of Vehicles, 2005

Hynd, D.: EEVC WG12 Rear Impact Biofidelity Evaluation Programme (HR-10-09e), November 2007, (http://www.unece.org/trans/doc/2007/wp29grsp/HR-10-09e.pdf)

ISO: International Standard – ISO 17373; Road Vehicles – Sled Test Procedure for Evaluating Occupant Head and Neck Interactions with Seat/Head Restraint Designs in Low-Speed Rear-End Impact, 2005

Ono, K. and Kanno, M.: Influences of the Physical Parameters on the Risk to Neck Injuries in Low Impact Speed Rear-end Collisions: International Research Council On the Biomechanics of Impact, 1993

Philippens, M., Cappon, H., van Ratingen, M., Wismans, J., Svensson, M., Sirey, F., Ono, K., Nishimoto, N. and Matsuoka, F.: Comparison of the Rear Impact Biofidelity of BioRID II and RID2: 46th Stapp Car Crash Conference, 2002

Pramudita, J. A., Ono, K., Ejima, S., Kaneoka, K., Shiina, I. and Ujihashi, S.: Head/Neck/Torso Behavior and Cervical Vertebral Motion of Human Volunteer during Low Speed Rear Impact: Mini-sled Tests with Mass Production Car Seat: International Research Council On the Biomechanics of Impact, 2007

Prasad, P., Kim, A., and Weerappuli, D.: Biofidelity of anthropomorphic test devices for rear impact: 41st STAPP Car Crash Conference, 1997

Mertz, Jr. and Patrick, L.M.: Investigation of Kinematics and Kinetics of Whiplash: 11th STAPP Car Crash Conference, 1967

Willis, C., Carroll, J. and Roberts, A.: An Evaluation of a Current Rear Impact Dummy Against Human Response Corridors in both Pure and Oblique Rear Impact: 19th International Technical Conference on the Enhanced Safety of Vehicles, 2005

Yaguchi, M., Ono, K. and Kubota, M.: Comparison of Biofidelic Responses to Rear Impact of the Head/Neck/Torso among Human Volunteers, PMHS, and Dummies: International Research Council On the Biomechanics of Impact, 2006

970497

On the Role of Cervical Facet Joints in Rear End Impact Neck Injury Mechanisms

King H. Yang, and Paul C. Begeman
Wayne State Univ.

Marcus Muser, Peter Niederer, and Felix Walz
Swiss Federal Institute of Technology and University of Zürich

ABSTRACT

After a rear end impact, various clinical symptoms are often seen in car occupants (e.g. neck stiffness, strain, headache). Although many different injury mechanisms of the cervical spine have been identified thus far, the extent to which a single mechanism of injury is responsible remains uncertain. Apart from hyperextension or excessive shearing, a compression of the cervical spine can also be seen in the first phase of the impact due to ramping or other mechanical interactions between the seat back and the spine.

It is hypothesized that this axial compression, together with the shear force, are responsible for the higher observed frequency of neck injuries in rear end impacts versus frontal impacts of comparable severity. The axial compression first causes loosening of cervical ligaments making it easier for shear type soft tissue injuries to occur.

To test this hypothesis, an *in vitro* experiment was designed to investigate the theory that axial compression reduces the shear stiffness when the cervical spine is moved due to a rear-end impact. Specimens from C1-T1 were tested. Results showed that shear stiffness values were reduced significantly with increased axial compressions. More dynamic tests are needed to further test this hypothesis for the rear-end impact neck injury mechanism. Also, investigations of soft tissue injuries in rear-end crash victims are necessary to guarantee the merit of this hypothesis.

INTRODUCTION

The term whiplash has been widely used to describe neck injuries associated with rear-end impacts. Although some physicians disagreed with the term "whiplash injury," it is used because the injury mechanism is assumed to be similar to the whiplash mechanism (Walz, 1993). Patients who report whiplash pain typically experience a low to moderate speed rear-end impact with very little vehicular damage. Victims typically complain of neck stiffness, strain, headache, and pain in the neck and shoulder muscles. Other symptoms include anxiety, depression and insomnia (Balla, 1980 and Pearce, 1989). Symptoms can occur either immediately or are delayed hours or days before they become apparent.

In 1968, FMVSS 202 made head restraints mandatory for passenger cars sold in the U.S. Since then, neck injury cases still account for up to 20% of rear-end accidents. In one report, the introduction of head restraint even increased neck injury rates (O'Neill et al, 1972). These data indicate that either the head restraint was not used properly or the neck injury mechanism is not hyperextension and thus head restraint is ineffective.

In volunteer tests, McConnell et al (1993) found that a vertical acceleration can be measured during a low speed rear-end impact. This ramping up phenomenon was due to the straightening of the spine or the mechanical interactions between the seatback and the torso. This same phenomenon was also reported in a high-speed x-rays study of the neck for volunteers subjected to rear impact forces (Matsushita et al, 1994) and in Hybrid II dummy tests by Viano (1992). However, the measured vertical acceleration and movement were rather small that McConnell et al later reported to be insignificant compared to those measured horizontally (McConnell et al, 1995).

Although the vertical acceleration may seem small, it plays a significant role in the cervical spine biomechanics. The head generally possesses about 4.5 kg (10 lbs) of inertial mass. Even a small acceleration could generate a significant compressive force at the neck. In a rear-end impact, the car seat pushes (shears) the torso forward while the neck is subjected to this axial compression. Based on this observation, Yang and Begeman (1996) proposed a new hypothesis to explain the rear-end neck injury mechanism stating that axial compression can cause loosening of ligaments and make it easier for the facet joint capsule and other soft tissues to be injured. Because these injuries occur in soft tissues, this new theory explains why there is generally no objective evidence.

The facet joint geometry of the cervical spine also plays an important role. In frontal impacts, the upper vertebra will shear anteriorly, relative to the lower vertebra. By observing the anatomy of the facet joints, it is evident that contact of the facet joints can protect against excessive frontal shear. However, in a rear-end impact, the lower vertebra shears anteriorly, the facet joint offers no protection to such a motion. This can be the reason that the rate of neck injury is much lower in frontal impacts of the same or even higher severity.

The purpose of this study is to biomechanically test the hypothesis that combined axial compression and shear due to a rearend impact can cause soft tissue injuries in the neck and may be a cause of neck pain.

MATERIAL AND METHODS

Cervical spine specimens from C1-T1 were dissected from the entire spine. They were provided by the Department of Anatomy of Wayne State University through the willed body program. Prior to testing, the specimens were x-rayed and physically examined for any anatomic and pathologic abnormalities.

The C1 vertebra was fixed to an aluminum plate with screws. The other end (T1) was potted in epoxy and attached to a six-axis load cell. During the embedding process, the natural curvature of the specimen was maintained. The entire assembly was placed in a jig on an Instron testing machine. This jig limits the C1 vertebra from moving to simulate the inertial effect of the head. The T1 vertebra was attached to the actuator of the Instron testing machine. During the test, the actuator moves upward to simulate the seat back pushing from behind. Two pairs of bearing were used to release the constraint in two directions normal to the direction of the applied AP shear force.

Two LED markers were attached to each vertebral body from C2-C7. One additional LED marker was attached to the frame of the Instron as a reference. Figure 1 shows the testing setup. A 3D motion analysis system (OPTOTRAK) was used to track motions of the markers at 60Hz. . Force and moment data were recorded on a Tektronix Testlab 2510 analog-to-digital converter.

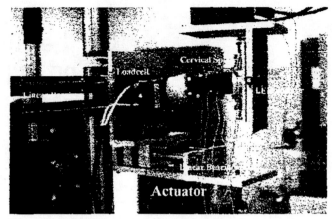

Figure 1: *In vitro* experimental test setup.

Five tests were done for each specimen. In the first test, the T1 was moved anteriorly to simulate a rear-end impact for 20 mm (0.79 Inch) displacement at a quasi-static speed of 0.04 m/s. In the next four tests, an axial compression of 44.5 N (10 lbs), 89.0 N (20 lbs), 133.5 N (30 lbs) and 178 N (40 lbs) of dead weight were applied through a cable-pulley system. The same procedure as in the first test was then repeated. Shear stiffness values were calculated from the load cell and motion data.

RESULTS

Figure 2 shows the total shear force vs. the shear deflection data for a typical C5-C6 motion segment. It can be clearly seen that the shear stiffness decreased as the applied axial compression increased.

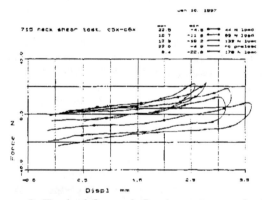

Figure 2: Typical force-deflection curves calculated from a C5-C6 motion segment subjected to a simulated rear-end impact.

The shear force vs deflection curves were nonlinear due to coupling rotations of vertebrae. The shear stiffness, defined as the final linear portion of the force-deflection curve, reduced significantly with increased axial compression. For example, for the C2-C3 portion of the specimen no. 715, the shear stiffness was only 50 % of that without axial preload (Table I).

Table I: Shear stiffness values calculated at each vertebra level for specimen no. 715

| | STIFFNESS (N/MM) | | | |
	C2-C3	C3-C4	C4-C5	C5-C6
NO PRE-LOAD	14.9	9.0	10.9	18.6
40 LBS PRE-LOAD	7.5	4.5	6.3	5.0

DISCUSSION AND CONCLUSIONS

In typical static tests, the shear stiffness is expected to increase as the axial compression increase. Our experimental data show the opposite trend. This explains why the neck injury rate is higher in a rear-end impact than that of a frontal impact. The axial compression presented in rear-end impacts reduce the shear stiffness of the cervical spine and make it easier to be injured. Dynamic tests can give us more insight into the neck injury mechanism. Those data are useful in the design of new equipment (head restraint) to protect the neck from rear-end injury.

This hypothesis is also neurophysiologically sound. Barnsley et al (1995) concluded that cervical facet joint pain was the most common source of chronic neck pain after whiplash. In our laboratory, Dr. Cavanaugh has studied the neurophysiology of the facet joint and has shown that in anesthetized rats and rabbits, facet joints can be a source of spinal pain (Cavanaugh, 1996). He and co-workers were able to demonstrate how the facet joint pain can be initiated or increased by stretching the facet joint capsule. This series of neurophysiological studies provide a logical physiological background for the mechanism of rear-end neck injury.

The calculated stiffness values were smaller compared with studies by Moroney et al (1988) and Panjabi et al (1986). We believe that coupling rotations during our tests were responsible for the difference. In addition, the shear stiffness values in detail because they were highly dependent on the initial test configuration. Nevertheless, the general trend remains the same for the shear stiffness to reduce as the axial compression increases.

REFERENCES

Balla, J.I. (1980) "The Late Whiplash Syndrome" Aust NZ J. Surg 50(6)610-614.

Barnsley, L.; Lord, S.M.; Wallis, B.J.; Bogduk, N. (1995) "The Presence of Cervical Zygapophyseal Joint Pain after Whiplash" Spine 20:20-26.

Cavanaugh, J.M. (1996) "A Proposed Role for Facet Joints in Neck Pain in Low to Moderate Speed Rear End Impacts Part II: Neuroanatomy and Neurophysiology" 6th Injury Prevention Through Biomechanics Symposium, May 9-10, 1996 at WSU, pp. 65-71.

Matsushita, T.; Sato, T.B.; Hirabayashi, K.; Fujimura, S. and Asaszuma, T. (1994) "X-ray Study of the Human Neck Motion Due to Head Inertia Loading" Proc. 38th Stapp Car Crash Conf. pp. 55-64, SAE paper no. 942208.

McConnell, W.E.; Howard, R.P.; Guzman, H.M.; Boniar, J.B.; Raddin, J.H.; Benedit, J.V.; Smith, H.L.; and Hatsell, C.P. (1993) "Analysis of Human Test Subject Kinematic Responses to Low Velocity Rear-End Impacts" SAE Paper No. 930889.

McConnell, W.E.; Howard, R.P.; Van Poppel, J.; Krause, R.; Guzman, H.M.; Boniar, J.B.; Raddin, J.H.; Benedit, J.V.; and Hatsell, C.P. (1995) "Human Head and Neck Kinematics after Low Velocity Rear-End Impacts - Understanding "Whiplash" Proc. 39th Stapp Car Crash Conf. pp 215-238, SAE Paper No. 952724.

Moroney, S.P.; Schultz, A.B.; Miller, J.A.A. and Andersson, G.B.J. (1988) "Load-displacement properties of Lower Cervical Spine Motion Segments" J. Biomechanics 21:769-779.

O'Neill, B.; Haddon, W. Jr.; Kelley, A.B.; and Sorenson, W.W. (1972) "Automobile Head Restraints: Frequency of Neck Injury Insurance Claims in Relation to the Presence of Head Restraints" Am. J. Pub. Health 62:399-406.

Panjabi, M.M.; Summers, D.J.; Pelker, R.R.; Videman, T.; Friedlaender, G.E. and Southwick, W.O. (1986) "Three Dimensional load-displacement Curves Due to Forces on the Cervical Spine" J. Orthopaedic Research 4:152-161.

Pearce, J.M.S. (1989) "Whiplash Injury: A Reappraisal" J. Neurosurg. Psychiatry 52:1329-1331.

Viano, D.C. (1992) "Influence of Seatback Angle on Occupant Dynamics in Simulated Rear-End Impacts" SAE Paper No. 922521.

Walz, F.H. (1993) Biomechanical Aspects of Cervical Spine Injuries" First European Congress of Orthopaedics, Paris, April 21-23.

Yang, K.H. and Begeman, P.C. (1996) "A Proposed Role for Facet Joints in Neck Pain in Low to Moderate Speed Rear End Impacts Part I: Biomechanics" 6th Injury Prevention Through Biomechanics Symposium, May 9-10, 1996 at WSU, pp. 59-63.

Biomechanical Assessment of Human Cervical Spine Ligaments

Narayan Yoganandan, Frank A. Pintar and Srirangam Kumaresan
Department of Neurosurgery
Medical College of Wisconsin and VA Medical Center

Ali Elhagediab
Aerotek Corp.

ABSTRACT

There is an increasing need to accurately define the soft tissue components of the human cervical spine in order to develop and exercise mathematical analogs such as the finite element model. Currently, a paucity of data exists in the literature and researchers have constantly underscored the need to obtain accurate data on cervical spine ligaments. Consequently, the objective of the study was to determine the geometrical and biomechanical properties of these ligaments from the axis to the first thoracic level. A total of thirty-three human cadavers were used in the study. Geometrical data included the length and cross-sectional area measurements; and the biomechanical properties included the force, deflection, stiffness, energy, stress, strain, and Young's modulus of elasticity data.

Data were obtained for the following ligaments: anterior and posterior longitudinal ligaments, joint capsules, ligamentum flavum, and interspinous ligament. Cryomicrotomy techniques were used to determine the geometrical characteristic. The length data were obtained using sagittal images and area data were determined using axial images. The biomechanical tests involved conducting failure tensile tests at a quasistatic rate of 10 mm/sec using in situ principles (ligaments were not tested in isolation) and the data were analyzed and synthesized as follows. The force-deformation relationships for each ligament type and at each spinal level were normalized. The mean force-deformation curves were obtained and grouped into clinically relevant mid and lower cervical regions.

The anterior and posterior longitudinal ligaments responded with the highest length measurements in both regions of the spine. In contrast, the ligamentum flavum and joint capsules exhibited the highest area of cross-section. All ligaments demonstrated increasing cross-sectional areas in the lower cervical group compared to the mid-cervical group. Detailed force-deflection characteristics are provided for all the five types of ligaments in both groups. The stiffness parameters were higher in the mid-cervical region than in the lower cervical region for the anterior longitudinal ligament, interspinous ligament, and ligamentum flavum, while the reverse was true for the other ligaments. The energy was higher in the lower cervical region than in the mid-cervical region for the joint capsules, ligamentum flavum, interspinous ligament, and anterior longitudinal ligament. The lowest values of energy to failure were observed for the interspinous ligament followed by the posterior longitudinal ligament, ligamentum flavum, anterior longitudinal ligament, and joint capsules in both regions. The anterior and posterior longitudinal ligaments responded with the highest stress followed by the joint capsules, interspinous ligament, and ligamentum flavum. While the joint capsules and ligamentum flavum demonstrated large strain to failure, the anterior and posterior longitudinal ligaments responded with the least percentage of strain to failure. The Young's modulus of elasticity based on a bilinear fit for each ligament type at each of the two regions is given. These studies provide important fundamental data on the properties of human cervical spine ligaments.

INTRODUCTION

To understand the biomechanical behavior of the human cervical spine, several approaches have been used in the literature. These include experimental investigations using in vivo animals, physical models, and isolated and intact human cadaver specimens [12, 14, 19, 22, 24]. Mathematical modeling studies have included lumped parameter and finite element approaches. In vivo experimental studies have the unique ability to determine the physiologic responses secondary to external mechanical forces. Experimental studies using physical models such as anthropomorphic test devices cannot identify physiologic responses. The advantages of these physical model studies however, include repeatability and reproducibility. Human cadaver experimentation on the other hand, has the advantage of reproducing clinically seen cervical spine trauma, and determining tissue tolerances.

Mathematical models are often used to supplement experimental research because of their unique capabilities of delineating intrinsic responses such as stress-strain distributions in addition to the experimental force-deformation responses. With the advent of high-speed computing technology and sophisticated imaging techniques, the idea of a complex anatomically accurate three-dimensional model of the human cervical spine is reaching realistic levels [2, 3, 5-9, 11, 18-22, 24]. However, for a mathematical analogue such as the finite element model of the human vertebral column to have biomechanical validity and to obtain realistic estimates of the intrinsic behavior of the spinal structure, it must incorporate appropriate properties for the components such as ligaments. The discrepancies in the correlation of the finite element output with experimental data have been primarily attributed to a lack of accurate characterization of the properties of the spinal components such as ligaments. Human cervical spine finite element studies have consistently underscored the need to obtain appropriate material data (e.g., [5, 19, 22, 24]). The present investigation is therefore focused on the determination of the characteristics of cervical spine ligaments to improve the accuracy of mathematical models. In particular, this study was conducted to provide experimentally determined geometrical and mechanical parameters of ligaments from the second cervical to the first thoracic spinal vertebral level representing the mid to lower human cervical spine column.

METHODS

A total of eight unembalmed human cadaver specimens were used to determine the geometry of the cervical spine ligaments from the axis to the first thoracic vertebra. The age of the cadavers ranged from 42 to 88 years (mean: 63 years), height ranged from 1.7 to 1.8 m (mean: 1.76 m), and weight ranged from 57 to 104 kg (mean: 77 kg). The selected specimens excluded history of spine disease or trauma. Pretest radiographs of the specimens were obtained. A cryomicrotome (Bromma, Sweden) was used to determine the initial length and cross-sectional area data for the following ligaments: anterior and posterior longitudinal ligaments, joint capsules, ligamentum flavum, and interspinous ligament. It should be noted that in both geometrical assessment and biomechanical testing, the joint capsule was evaluated as a whole joint. In other words, the capsular ligaments were not separated from the synovial joint, and therefore the strength and geometry data include contribution from both the ligament and the synovial capsule; thus the term "joint capsule". The specimens were initially frozen in the intact state. The normal spinal curvature was maintained during the freez-

ing process. The cervical spinal columns were sectioned after freezing and placed in Styrofoam boxes. The specimens were embedded in the Styrofoam box with carboxy methylcellulose. Computed tomography (CT) images of the specimens in the box were obtained. The alignment of the box was physically adjusted until a true sagittal (or axial) CT section was obtained; after which the built-in laser light of the CT scanner was used to mark the box in this final orientation. Lines were drawn on the outside of the box to facilitate alignment in the cryomicrotome. The tissue box was cut and sectioned parallel to the plane marked with the CT imager. The cut sections were fixed onto the stage of the cryomicrotome device such that anatomic sections were taken in the same plane as the CT sections. The table travel of the cryomicrotome device was adjusted to obtain sequential anatomic sections at intervals ranging from 20 to 40 microns. A standard 35 mm camera was used to photograph the frozen cadaver specimen following the table travel. When the individual spinal ligaments were observed, close-up photographs were taken to identify additional spinal characteristics. The normal anatomical features were documented using the cryomicrotome device. The geometrical properties were extracted from these photographs on a level-by-level basis from the axis to the first thoracic spinal vertebral level.

Using the sagittal sequential anatomic sections, the initial lengths of the cervical spinal ligaments were obtained. The following definitions for the ligaments were used in the analysis. The anterior and posterior longitudinal ligaments were defined from the mid-height of the superior vertebral body to the mid-height of the inferior body. The ligamentum flavum and interspinous ligaments were defined from the superior points of attachment to the corresponding inferior points of attachment. The joint capsules were defined from the superior tip of the caudal facet articulation to the inferior tip of the rostral facet articulation. These definitions are similar to those used in previous biomechanical studies of the other human vertebral column ligaments [13].

The cross-sectional areas of the above described ligaments were defined using sequential axial (transverse) anatomic images. The anterior and posterior longitudinal ligaments were distinguished based on the fiber-orientation anatomy between the intervertebral disc and ligament structures. The characteristic yellow color of the ligamentum flavum differentiated its anatomy. The synovial joint characteristics were used to distinguish the joint capsules. Since the joint capsule components include the synovial fluid surrounded by the rostral and caudal cartilages of the superior and inferior processes, and encapsulated by the synovial membrane, the outer boundary of the capsule spanning between the two opposing cartilages was defined to contribute to the cross-sectional area. The interspinous ligaments were distinguished based on the contrast in the surrounding tissues. The cross-sectional areas were based on the transverse sections approximately at the mid-height of the cervical spine intervertebral disc for the anterior and posterior longitudinal ligaments, and the ligamentum flavum. For the joint capsules, the measurements were taken at the mid-capsule height. For the interspinous ligament, the measurements were obtained midway between the two adjacent spinous processes. The photographs were projected and an outline of the ligament boundary was traced using the above definitions. A millimeter ruler was placed on the cryomicrotome just prior to obtaining a photograph and this process assisted in quantifying the geometrical characteristics. An independent blind examination by three members of our laboratory with differing experience was done and measurements were obtained to check the accuracy of the system. The technique was accurate to 100 microns. The computations were done using a computer-aided design software. Figures 1 and 2 illustrate the transverse and sagittal images indicating the human cervical spine ligaments.

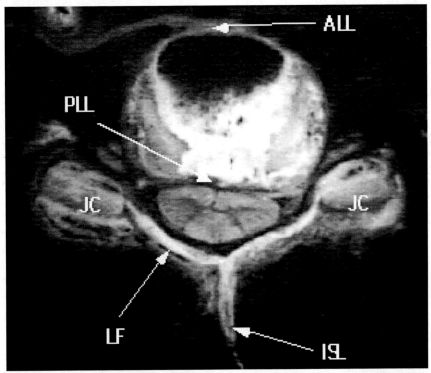

Figure 1. Transverse cryomicrotome section illustrating the various ligaments. ALL: Anterior longitudinal ligament; PLL: Posterior longitudinal ligament; JC: Joint capsules; LF: Ligamentum flavum; ISL: Interspinous ligament.

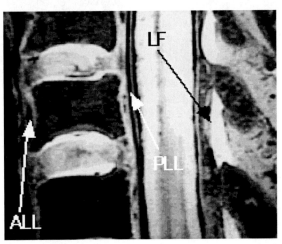

Figure 2. Sagittal cryomicrotome section illustrating the anterior (ALL) and posterior (PLL) longitudinal ligaments, and the ligamentum flavum (LF).

The biomechanical properties of the five cervical spine ligaments from the second cervical through the first thoracic vertebral level were determined from a separate group of twenty-five human cadavers. The mean age, height and weight were: 68 years, 176 cm and 63 kg, respectively. In situ mechanical testing procedures were adopted to subject each ligament under tensile loading at each level of the spinal column. The cadaver selection and initial condition followed the above described procedures. At each level, the intervertebral disc and all supporting structures except the ligament under test were transected. The vertebral bodies superior and inferior to the ligament under test were rigidly fixed in a custom designed frame with Steinmann pins. The specimen was pulled under tension at a loading rate of 10 mm/sec. The force was measured using a load cell attached in series with the electrohydraulic piston. The displacement was measured using a linear variable differential transformer built into the piston. The tensile force and distraction data were obtained as a function of time using a data acquisition system. Failure was defined as the point at which a further increase in the tensile deformation resulted in a decrease in the external tensile force. The energy to failure was defined as the area under the tensile force-deformation curve up to failure. The stiffness was defined as the slope of the least square fit line in the most linear portion of the curve as reported earlier [1, 4, 13]. The stress at failure was obtained as a ratio of the force at failure to the original cross-sectional area. Similarly, the strain at failure was obtained as a ratio of the deformation at failure to the original length. It should be noted that all geometric parameters were obtained from the previously described cryomicrotomy studies.

The mean tensile force-deformation responses for each ligament at each spinal level were determined in the following manner. Force-time and deformation-time signals were normalized with respect to the corresponding failure force and deformation magnitudes, respectively. Using the deformation as the governing variable for each force-deformation curve, at every normalized deformation data point, normalized force values were computed. This was done for 100 points on each curve. This procedure was adopted because the experiments were conducted by applying a linear displacement using the electrohydraulic testing device [16]. Further evaluations were done to group the spinal ligament responses by type and region. The mean of these values for each type of ligament at each cervical level resulted in the mean normalized force at the corresponding normalized deformation points. This procedure reduced the matrix into a five by six array (five ligaments at six levels). Further analysis was carried out to group the data into mid and lower cervical spinal regions. This resulted in ten curves (five under each region, one for each ligament type). These mean normalized force-deformation curves were transformed into engineering units by suitably multiplying the abscissa and ordinate with the mean failure deformation and mean failure force values, respectively, calculated for that particular ligament type and region. This procedure of grouping

the ligament responses into mid-cervical and lower cervical regions simplified the data and reduced the final number of curves for presentation. In addition, these regions represent the commonly used clinical classifications of the cervical spine from the axis to the first thoracic level. A bilinear fit was used to determine the Young's modulus of elasticity for each ligament at each region. The first region of the bilinear fit was taken to span four-tenths of the failure strain and this was based on an examination of all the biomechanical responses.

RESULTS

A summary of the geometrical properties of the human cervical spine ligaments from the second cervical to first thoracic vertebral level is included in Figures 3 and 4. The cross-sectional areas in the mid-cervical spine, increased from the longitudinal ligaments, interspinous ligament, joint capsules, ligamentum flavum. In the lower cervical spine the areas increased from the longitudinal ligaments, interspinous ligament, and ligamentum flavum to the joint capsules. Specifically, the anterior longitudinal ligament had the least cross-sectional area at the mid (mean: 11.12 square mm) and lower (mean: 12.09 square mm) cervical spinal regions. The ligamentum flavum exhibited the highest area of cross-section at the mid-cervical region (mean: 45.99 square mm). The joint capsules exhibited the highest area at the lower cervical region (mean: 49.53 square mm) Figure 3), while both the anterior and posterior longitudinal ligaments (mean: 18.76 mm for anterior, and 19.04 mm for posterior ligament at mid-cervical; and mean: 18.3 mm for anterior, and 17.95 for posterior ligament at lower cervical) responded with the highest length measurements (Figure 4). All ligaments demonstrated increasing cross-sectional areas in the lower cervical region compared to the mid-cervical region.

Although 150 spinal ligaments (25 subjects and 6 levels) were potentially available for inclusion in the study, technical difficulties in isolating and testing the ligaments in situ, sometimes resulted in non-usable samples. This reduced the sample size to 107, with 55 for the mid-cervical and 52 for the lower cervical regions. The biomechanical stiffness, energy, stress, and strain properties based on the above classification are given in Figures 5-8, respectively. The number of specimens used in these computations is given in table 1.

The stiffness parameters were higher in the mid-cervical region compared to the lower cervical region for the anterior longitudinal ligament, interspinous ligament, and ligamentum flavum, while the reverse was true for the other ligaments (Figure 5). The energy was higher in the lower cervical region than the mid-cervical region for all ligaments, except the posterior longitudinal ligament (Figure 6). For all ligaments, the mean values of the stiffnesses ranged from 7.6 to 33.6 N/mm at the mid-cervical region and from 6.4 to 34.4 N/mm at the lower cervical region. The lowest values of energy to failure were observed for the interspinous ligament (mean: 0.16 Nm at the mid-cervical, and mean: 0.18 Nm at the lower cervical) followed by the posterior longitudinal ligament (mean: 0.47 Nm at the mid-cervical, and mean: 0.4 Nm at the lower cervical), ligamentum flavum (mean: 0.51 Nm at the mid-cervical, and mean: 0.91 Nm at the lower cervical), anterior longitudinal ligament (mean: 0.61 Nm at the mid-cervical, and mean: 0.94 Nm at the lower cervical), and joint capsules (mean: 1.52 Nm at the mid-cervical, and mean: 1.69 Nm at the lower cervical). It should be re-emphasized that the joint capsules were tested together.

In both regions of the cervical spine, the anterior and posterior longitudinal ligaments responded with the highest stress followed by the joint capsules, interspinous ligament, and ligamentum flavum (Figure 7). While the joint capsules and ligamentum flavum demonstrated considerable strain to failure, the longitudinal ligaments responded with the least percentage of strain to failure (Figure 8) in both regions.

The derived mean force-deformation relationships for each of the five types of ligaments at the two regions of the cervical spine are illustrated (mid-cervical: Figures 9-13, and lower cervical: Figures 14-18). The values of the Young's modulus of elasticity based on a bilinear fit are given in Table 2 for all the ligaments in both regions of the cervical spine.

DISCUSSION

The use of the human cadaver material is an accepted way to describe the anatomy and biomechanics of the cervical spine. Similar procedures have been used in cervical and other regions of the human vertebral column [16, 25]. Although other methods such as visual observations of the anterior longitudinal ligament, joint capsule, and interspinous ligament are possible for these ligament structures, the cryomicrotome technique is superior because of the ability to obtain sequential anatomic specimens in an undeformed state. The cryomicrotome technique preserves the in situ anatomic features of the cervical spinal column and therefore measurements obtained from this method are reliable and superior to that estimated from gross-dissection procedures. The technique also delineates the characteristics of the internally apparent ligaments (i.e., the ligamentum flavum and posterior longitudinal ligament) of the human cervical

spine. This procedure was therefore used in the present investigation.

As indicated earlier, the rationale for grouping the geometrical and material properties is based on the common clinical classification of upper (occiput to axis), mid (axis to fifth cervical), and lower (fifth cervical to first thoracic) cervical regions. The present study is focused only on the mid and lower cervical regions. It should be emphasized that, to our best knowledge, this is the only study that has attempted to assimilate data from such a large group of samples to characterize the ligament properties of the mid and lower cervical spine regions.

The geometry and mechanical property information determined in the present investigation generally agrees with published literature [13, 15, 17]. The length of the cervical spinal ligaments obtained in this study, when compared with the measurements reported by others who have used gross-dissection and radiographic techniques, must be viewed based on the accuracy of the procedures. The biomechanical parameters such as stiffness and energy determined in this study for the cervical anterior longitudinal ligament and ligamentum flavum agree well with data from another earlier study [23]. For example, the mean stiffness of the anterior longitudinal ligament at the mid (16.5 N/mm) and lower cervical (15.1 N/mm) regions compares well with the mean stiffness of 14.9 N/mm (C2-T1 levels) reported by Yoganandan et al [23]. The mean stiffness of the ligamentum flavum at mid-cervical (23.2 N/mm) and lower cervical (21.6 N/mm) regions from the present study also agrees well with the mean stiffness of 21.9 N/mm (C2-T1 levels) reported earlier. These above cited stiffness data from the Yoganandan et al., research, were obtained from samples that were tested at approximately the same rate of loading used in the present study. The geometrical properties of these two ligaments were not determined in this earlier study and, in addition, these types of biomechanical data for other cervical spine ligaments were not reported. In fact, traditionally, studies have determined the force-deformation characteristics or the geometry of the ligaments. In contrast, the current research has presented information on geometrical measurements and mechanical testing to provide a comprehensive definition of the characteristics of the structure, representing the entire mid and lower cervical regions.

Table 1.   Sample Size Matrix

	ALL	PLL	JC	LF	ISL
C2-C3	5	3	3	4	3
C3-C4	3	4	4	4	3
C4-C5	4	4	3	5	3
C5-C6	3	3	3	4	3
C6-C7	3	4	4	3	2
C7-T1	5	3	5	4	3

ALL: Anterior Longitudinal Ligament
PLL: Posterior Longitudinal Ligament
JC: Joint Capsules
LF: Ligamentum Flavum
ISL: Interspinous Ligament

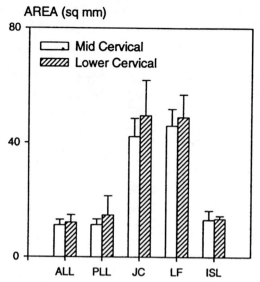

Figure 3. Cross-sectional area of the ligaments (mm^2) in the mid and lower cervical spine.

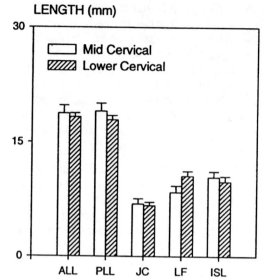

Figure 4. Length of the ligaments (mm) in the mid and lower cervical spine.

Table 2. Bilinear Young's Modulus (MPa) of Cervical Ligaments

Type	Mid-Cervical Region		Lower Cervical Region	
	$E\ (\varepsilon, \varepsilon_1)$	$E\ (\varepsilon_1, \varepsilon_m)$	$E\ (\varepsilon, \varepsilon_1)$	$E\ (\varepsilon_1, \varepsilon_m)$
Ant. long. ligament	43.8 (0, 12.9)	26.3 (12.9, 32.1)	28.2 (0, 14.8)	28.4 (14.8, 36.9)
Post. long. ligament	40.9 (0, 11.1)	22.2 (11.1, 27.7)	23.0 (0, 11.2)	24.6 (11.2, 28.1)
Joint capsules	5.0 (0, 56.8)	3.3 (56.8, 142.1)	4.8 (0, 57.0)	3.4 (57.0, 143.6)
Ligamentum flavum	3.1 (0, 40.7)	2.1 (40.7, 101.7)	3.5 (0, 35.3)	3.4 (35.3, 88.2)
Interspinous ligament	4.9 (0, 26.1)	3.1 (26.1, 65.3)	5.0 (0, 27.0)	3.3 (27.0, 67.9)

Note: $E\ (\varepsilon, \varepsilon_1)$ denotes the value of the Young's modulus of elasticity (E) corresponding to the first region of the bilinear fit with ε_1 denoting the upperbound for the strain. $E\ (\varepsilon_1, \varepsilon_m)$ denotes the modulus for the second region with ε_m denoting the failure/maximum strain.

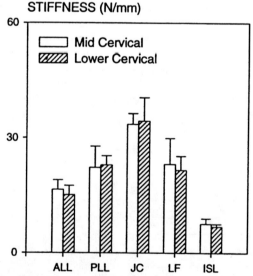

Figure 5. Stiffness of the ligaments in the mid and lower cervical spine.

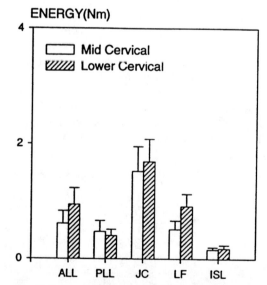

Figure 6. Failure energy of the ligaments in the mid and lower cervical spine.

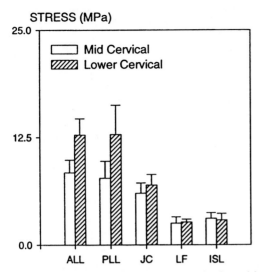

Figure 7. Failure stress of the ligaments in the mid and lower cervical spine.

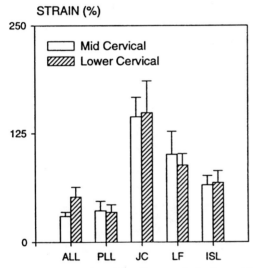

Figure 8. Failure strain of the ligaments in the mid and lower cervical spine.

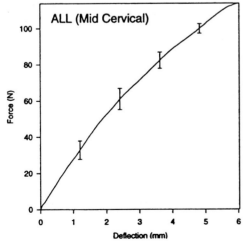

Figure 9. Force-deformation properties of the anterior longitudinal ligament for the mid-cervical region.

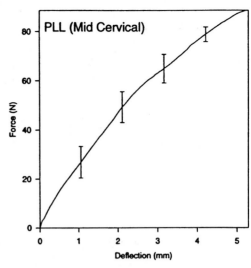

Figure 10. Force-deformation properties of the posterior longitudinal ligament for the mid-cervical region.

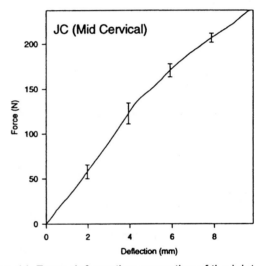

Figure 11. Force-deformation properties of the joint capsules for the mid-cervical region.

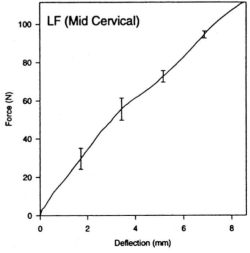

Figure 12. Force-deformation properties of the ligamentum flavum for the mid-cervical region.

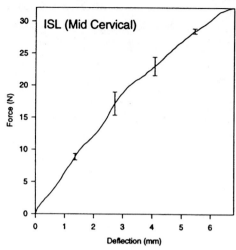

Figure 13. Force-deformation properties of the interspinous ligament for the mid-cervical region.

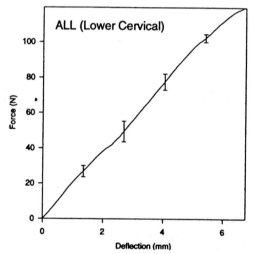

Figure 14. Force-deformation properties of the anterior longitudinal ligament for the lower cervical region.

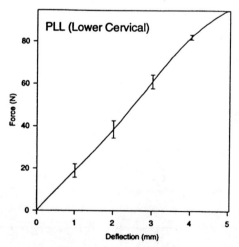

Figure 15. Force-deformation properties of the posterior longitudinal ligament for the lower cervical region.

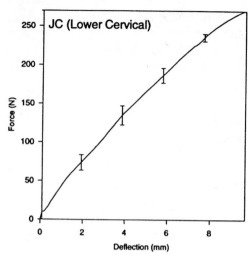

Figure 16. Force-deformation properties of the joint capsules for the lower cervical region.

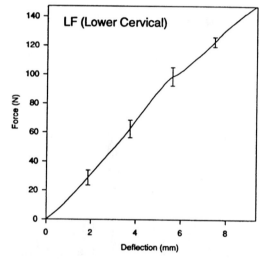

Figure 17. Force-deformation properties of the ligamentum flavum for the lower cervical region.

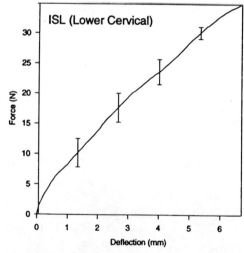

Figure 18. Force-deformation properties of the interspinous ligament for the lower cervical region.

Normalization procedures were adopted to derive the mean force-displacement curves. In general, the curves do not demonstrate the typical sigmoidal shape. This is due to several reasons. In the quest to derive the mean curves one cannot simply average the dimensional values of force and displacement, because of the variation in the total response up to failure. Therefore, values of the force and deformation at failure were used to normalize each curve to unity. The normalized curves could then be averaged together with the nondimensionalized values. This process tends to remove some of the nonlinearities because every ligament (even for one type of ligament at one level) will not have the same degree of nonlinearity. The toe region (the first portion of the sigmoidal curve, where a greater amount of deformation leads to a relatively small amount of force) is susceptible to the averaging process because some ligaments may have had little toe region and some may have had a longer toe region. This is due to biological variations. Inclusion of the error bars in the mean response provides the modeler with sufficient information to account for data variations. The results were accomplished using the force-deformation responses from one group of specimens and combining with the geometrical characteristics obtained from another ensemble. While this may appear to be a limitation, the closeness in the anthropomorphic data between the two groups of specimens (mean age of 63 years for the geometrical group, and 68 years for the mechanical testing group) tends to minimize the variability in the output. The present synthesis of data, which has provided mean force-deformation curves or Young's modulus values for human cervical spine ligaments can be used in a finite element model [7-9]. It should be emphasized that the definitions adopted to describe and quantify the ligament characteristics must be followed in order to use the values reported in Table 2.

The purpose of the present study was to generate appropriate geometrical and mechanical parameters which can be used in mathematical models. Complex computer models often reproduce the in situ geometry of the musculoskeletal architecture [10]. The nature of these mathematical models was taken into account while defining the characteristics of the ligaments. Thus, the ligaments were tested in their anatomical in situ conditions to allow for direct usage in such models. In addition, the geometry of the ligaments was defined so that realistic anatomical functional relationships and the practical needs of developing an accurate mathematical model can be simultaneously achieved. For example, the anterior longitudinal ligament was defined from the mid height of one vertebral body to the mid height of the adjacent body. This definition allows for the functional anatomical nature of the continuous ligament and it also permits a straightforward attachment into the model geometry. In contrast, if one were to define this ligament to span between the two endplates or the edges of the intervertebral disc (as has been done in other studies), then in a mathematical model, it would be a difficult task to define the ligament structure that overlies the two vertebral bodies. These issues were given due consideration while defining and describing the characteristics of the ligaments in this investigation.

Furthermore, since the definitions for delineating the ligament geometry and mechanical property are obtained from a single laboratory source, these data can be immediately used by modelers to describe the biomechanical responses of the human cervical spine. It must be noted that the ligaments are rate sensitive viscoelastic structures. Data reported in this study are derived at a quasistatic rate, i.e., at 10 mm/sec. Because the present study design allowed for quasistatic testing of ligaments, it may be appropriate to apply a scaling factor for the dynamic injury scenario. Previous studies by Yoganandan et al., [23] evaluating the dynamic characteristics of cervical spine anterior longitudinal ligaments and ligamentum flavum demonstrated a rate dependence for failure force but not for failure deformation (Figure 19). The failure force parameter increased by a factor of 2.6 for ligamentum flavum and 2.9 for anterior longitudinal ligament between 0.01 m/sec and 2.5 m/sec rate of loading. Understanding that no similar rate dependency was found for the deformation variable, these dynamic factors may be applied to the elastic modulus values appropriately.

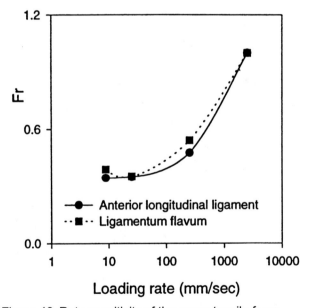

Figure 19. Rate sensitivity of the mean tensile force expressed as a ratio (Fr) of the failure force at a given loading rate to the failure force at the highest loading rate (2.5 m/s) for the cervical spine anterior longitudinal ligament and ligamentum flavum. Data from Yoganandan et al [23].

CONCLUSION

In summary, this study has provided geometrical and biomechanical properties of the human cervical spinal ligaments from the second cervical level to the first thoracic vertebral level. Data are grouped into mid and lower cervical regions. Tensile force-deformation curves for the anterior and posterior longitudinal ligaments, ligamentum flavum, joint capsules, and interspinous ligaments are provided for these two regions of the cervical spine. In addition, bilinear Young's modulus of elasticity, stress, strain, stiffness, and energy parameters are reported along with the geometrical length and cross-sectional area data. These data are of value to better develop and exercise mathematical analogs such as finite element and multibody dynamic models which can be used to more realistically predict the internal and external biomechanical responses of the human cervical spine.

ACKNOWLEDGMENT

This study was supported in part by DOT NHTSA Grant DTNH22-93-Y-17028, and the Department of Veterans Affairs Medical Research.

REFERENCES

1. 1.Chazal J, Tanguy A, Bourges M. Biomechanical properties of spinal ligaments and a histological study of the supraspinal ligament in traction. *J Biomech* 18: 167-176, 1985.
2. de Jager M, Sauren A, Thunnissen J, Wismans J. A global and a detailed mathematical model for head-neck dynamics. In: Proc 40th Stapp Car Crash Conf, Albuquerque, New Mexico, 1996, 269-281.
3. de Jager M, Sauren A, Thunnissen J, Wismans J. A three-dimensional head-neck model: Validation for frontal and lateral impacts. In: Proc 38th Stapp Car Crash Conf, Ft. Lauderdale, FL, 1994, 93-109.
4. Dumas GA, Beaudoin L, Drouin G. In situ mechanical behavior of posterior spinal ligaments in the lumbar region. An in vitro study. *J Biomech* 20: 301-310, 1987.
5. Khalil TB, Lin TC. Simulation of the Hybrid III dummy response to impact by nonlinear finite element analysis. In: Proc 38th Stapp Car Crash Conf, Ft. Lauderdale, FL, 1994, 325-343.
6. Kumaresan S, Yoganandan N, Pintar F. Age-specific pediatric cervical spine biomechanical responses: Three-dimensional nonlinear finite element models. In: 41st Stapp Car Crash Conf, Nov 12-14, Orlando, FL, 1997, 31-61.
7. Kumaresan S, Yoganandan N, Pintar F. Finite element analysis of anterior cervical spine interbody fusion. *Biomed. Mat. & Eng.* 7 (4): 221-230, 1997.
8. Kumaresan S, Yoganandan N, Pintar F, Voo L, Cusick J, Larson S. Finite element modeling of cervical laminectomy with graded facetectomy. *J Spinal Disord* 10 (1): 40-47, 1997.
9. Kumaresan S, Yoganandan N, Pintar FA. Finite element modeling approaches of human cervical spine facet joint capsule. *J Biomech* 31: 371-376, 1998.
10. Kumaresan S, Yoganandan N, Pintar FA. Methodology to quantify the uncovertebral joint in the human cervical spine. *J Musculoskeletal Research* 1 (2): 1-9, 1997.
11. Kumaresan S, Yoganandan N, Pintar FA. Pediatric neck modeling using finite element analysis. *International J Crashworthiness* 2 (4): 367-377, 1998.
12. Myers BS, Winkelstein B. Epidemiology, classification, mechanism, and tolerance of human cervical spine injuries. *Crit Rev Biomed Eng* 23 (5): 307-409, 1995.
13. Myklebust JB, Pintar FA, Yoganandan N, Cusick JF, Maiman DJ, Myers T, Sances A, Jr. Tensile strength of spinal ligaments. *Spine* 13 (5): 526-531, 1988.
14. Nitsche S, Krabbel G, Appel H, Haug E. Validation of a finite-element-model of the human neck. In: International IRCOBI Conference on the Biomechanics of Impact, Dublin, Ireland, 1996, 107-108.
15. Panjabi MM, Oxland TR, Parks EH. Quantitative anatomy of cervical ligaments -- Part II. middle and lower cervical spine. *J Spinal Disord* 4 (3): 276-285, 1991.
16. Pintar FA, Yoganandan N, Myers T, Elhagediab A, Sances A, Jr. Biomechanical properties of human lumbar spine ligaments. *J Biomech* 25 (11): 1351-1356, 1992.
17. Przybylski GJ, Patel PR, Carlin GJ, Woo SLY. Quantitative anthropometry of the subatlantal cervical longitudinal ligaments. *Spine* 23 (8): 893-898, 1998.
18. van der Horst MJ, Thunnissen JG, Happee R, van Haaster RM, Wismans JS. The influence of muscle activity on head-neck response during impact. In: Proc 41st Stapp Car Crash Conf, Lake Buena Vista, FL, 1997, 487-507.
19. Yoganandan N, Kumaresan S, Voo L, Pintar F. Finite element applications in human cervical spine modeling. *Spine* 21 (15): 1824-1834, 1996.
20. Yoganandan N, Kumaresan S, Voo L, Pintar F. Finite element model of the human lower cervical spine. *J Biomech Eng* 119 (1): 87-92, 1997.
21. Yoganandan N, Kumaresan S, Voo L, Pintar F, Larson S. Finite element modeling of the C4-C6 cervical spine unit. *Med Eng Phy* 18 (7): 569-574, 1996.
22. Yoganandan N, Myklebust JB, Ray G, Sances A, Jr. Mathematical and finite element analysis of spinal injuries. *CRC Review Biomed Eng* 15 (1): 29-93, 1987.
23. Yoganandan N, Pintar FA, Butler J, Reinartz J, Sances A, Jr, Larson SJ. Dynamic response of human cervical spine ligaments. *Spine* 14 (10): 1102-1110, 1989.
24. Yoganandan N, Pintar FA, Larson SJ, Sances A, Jr, eds. Frontiers in Head and Neck Trauma: Clinical and Biomechanical. Amsterdam, Netherlands: IOS Press, p. 740, 1998.
25. Yoganandan N, Sances A, Jr, Maiman DJ, Myklebust JB, Pech P, Larson SJ. Experimental spinal injuries with vertical impact. *Spine* 11 (9): 855-860, 1986.

Table I. Reprints by Category

No.	Author 1	Year	Impact Direction	Loading/ Moment
1	Agaram	2001	FR	Ex
2	Benson	1996	RE	Ex
3	Bostrom	1998	RE	Ex
4	Carlsson	2008	RE	Ex
5	Croft	2006	RE	DU
6	Deng	1998		FI
7	Eriksson	2006	RE	Ex
8	Gibson	2005	RE	Ex
9	Hassan	2000	FR	FR
10	Huelke	1979	GE	GE
11	Kitagawa	2008	RE	Ex
12	Kullgren	2007	RE	Ex
13	Laituri	2009	FR	GE
14	Linder	2003	RE	
15	McConnell	1995	RE	Ex
16	Mertz	1997	SI, FR	
17	Mertz	1971		FI, Ex
18	Millington	2005	FR	FI
19	Moore	2009	FR	GE
20	Morris	1996	RE, FR	
21	Muñoz	2005	RE	DU
22	Myers	1995	GE	
23	Nightingale	1997		FI, Ex
24	Ono	2007	SI	LF
25	Ono	1997	RE	Ex
26	Park	1998	FR	
27	Pintar	1995		Co
28	Pramudita	2007	RE	Ex
29	Prasad	1997	RE	DU
30	Ridella	2008	RO	
31	Siegmund	1997	RE	Ex
32	Sochor	2006	FR	DU
33	Stemper	2001	RE	Ex
34	Svensson	1993	RE	DU
35	Szabo	1996	RE	Ex
36	Viano	2008	Co	
37	Voo	2007	RE	Ex
38	Yamazaki	2008	RE	DU
39	Yang	1997	RE	Ex
40	Yoganandan	1998	SI	

Abbreviation	Category
Co	Compression
DU	Test Dummy
Ex	Extension
FI	Flexion
FR	Frontal
GE	General
RE	Rear
RO	Rollover
SI	Side Impact
Te	Tension

BIBLIOGRAPHY

1. Agaram, V., Kang, J., Nusholtz, G., and Kostyniuk, G. (2001): "Hybrid III Dummy Neck Response to Air Bag Loading," Proceedings of the 17th International Experimental Safety Vehicle Conference, National Highway Traffic Safety Administration, Washington, D.C.

2. Althoff, B. (1979): "Fractures of the Odontoid Process: An Experimental and Clinical Study," *Acta Orthop Scand(S)*, 1979, 177:1-95.

3. Anderson, J.S., Hsu, A.W., and Vasavada, A.N. (2005): "Morphology, Architecture, and Biomechanics of Human Cervical Multifidus," *Spine*, 30:E86-91.

4. Aprill, C., Dwyer, A., and Bogduk, N. (1990): "Cervical Zygapophyseal Joint Pain Patterns II: A Clinical Evaluation," *Spine*, 15(6):458-461.

5. Barnsley, L., Lord, S., and Bogduk, N. (1998): "The Pathophysiology of Whiplash," *Spine: State of the Art Reviews*, Vol. 12, No. 2, May 1998.

6. Been, B., Philippens, M., de Lange, R., and van Ratingen, M. (2004): "WorldSID Dummy Head-Neck Biofidelity Response," Paper No. 2004-22-0019, *Stapp Car Crash J.*, 48:431-454.

7. Bohman, K., Bostrom, O., Haland, Y., and Kullgren, A. (2000): "A Study of AIS1 Neck Injury Parameters in 168 Frontal Collisions Using a Restrained Hybrid III Dummy," Paper No. 2000-01-SC08, *Stapp Car Crash J.*, 44:103-116.

8. Bostrom, O., Svensson, M.Y., Aldman, B., Hansson, H.A., Haland, Y., Lovsund, P., Seeman, T., Suneson, A., Saljo, A., and Ortengren, T. (1996): "A New Neck Injury Criterion Candidate—Based on Injury Findings in the Cervical Spinal Ganglia After Experimental Neck Extension Trauma," Proceedings of the 1996 International Conference on the Biomechanics of Impact (IRCOBI, Zurich), 123-136.

9. Brault, J.R., Siegmund, G.P., and Wheeler, J.B. (2000): "Cervical Muscle Response During Whiplash: Evidence of a Lengthening Muscle Contraction," *Clinical Biomechanics*, 15(6):426-435.

10. Cappon, H., van Ritingen, M., Wismans, J., Hell, W., Lang, D., and Svensson, M. (2003): "Whiplash Injuries, Not Only a Problem in Rear-End Impact," Proceedings of the 18th International Experimental Safety Vehicle Conference, National Highway Traffic Safety Administration, Washington, D.C.

11. Cappon, H., Hell, W., Hoschopf, H., Muser, M., Song, E., and Wismans, J. (2005): "Correlation of Accident Statistics to Whiplash Performance Parameters Using the RID and BIORID Dummy," Proceedings of the 1995 International Conference on the Biomechanics of Impact (IRCOBI, Zurich), 229-244.

12. Carlson, E.J., Tominaga, Y., Ivancic, P.C., and Panjabi, M.M. (2007): "Dynamic Vertebral Artery Elongation During Frontal and Side Impacts," *Spine*, 7(2):222-228.

13. Cavanaugh, J.M. (2000): "Neurophysiology and Neuroanatomy of Neck Pain," in *Frontiers in Whiplash Trauma*, Yoganandan, N. and Pintar, F.A., Eds., IOS Press, The Netherlands.

14. Chancey, V.C., Nightingale, R.W., Van Ee, C.A., Knaub, K.E., and Myers, B.S. (2003): "Improved Estimation of Human Neck Tensile Tolerance: Reducing the Range of Reported Tolerance Using Anthropometrically Correct Muscles and Optimized Physiologic Initial Conditions," Paper No. 2003-22-0008, *Stapp Car Crash J.*, 47:135-153.

15. Crawford, N.R., Duggal, N., Chamberlain, R.H., and Park, S.C. (2002): "Unilateral Cervical Facet Dislocation: Injury Mechanism and Biomechanical Consequences," *Spine*, 27(17):1858-1863.

16. Davidsson, J., Lövsund, P., Ono, K., Svensson, M., and Inami, S. (1999): "A Comparison Between Volunteer, BioRID P3, and Hybrid III Performance in Rear Impacts," Proceedings of the 1999 International Conference on the Biomechanics of Impact (IRCOBI, Zurich).

17. Deng, B., Begeman, P.C., Yang, K.H., Tashman, S., and King, A.I. (2000): "Kinematics of Human Cadaver Cervical Spine During Low Speed Rear-End Impacts," Paper No. 2000-01-SC13, *Stapp Car Crash J.*, 44:171-188.

18. Dibb, A.T., Nightingale, R.W., Chancey, V.C., Fronheiser, L.E., Tran, L., Ottaviano, D., and Myers, B.S. (2006): "Comparative Structural Neck Responses of the THOR-NT, Hybrid III, and Human in Combined Tension-Bending and Pure-Bending," Paper No. 2006-22-0021, *Stapp Car Crash J.*, 50:567-581.

19. Doherty, B.J. and Heggeness, M.H. (1997): "The Role of Facet Angle Asymmetry in Fractures of the First Cervical Vertebra," Paper No. 970496, SAE International, Warrendale, PA.

20. Duma, S.M., Crandall, J.R., Rudd, R.W., Funk, J.R., and Pilkey, W.D. (1999): "Small Female Head and Neck Interaction with a Deploying Side Air Bag," Proceedings of the 1999 International Conference on the Biomechanics of Impact (IRCOBI, Zurich), 191-200.

21. Eichberger, A., Darok, M., Steffan, H., Leinzinger, P.E., Bostrom, O., and Svensson, M.Y. (2000): "Pressure Measurements in the Spinal Canal of Post-Mortem Human Subjects During Rear-End Impact and Correlation of Results to the Neck Injury Criterion," *Accid. Anal. Prev.*, 2:251-260.

22. Ejima, S., Ono, K., Kaneoka, K., and Fukushima, M. (2005): "Development and Validation of a Human Neck Muscle Model Under Impact Loading," Proceedings of the 2005 International Conference on the Biomechanics of Impact (IRCOBI, Zurich), 245-256.

23. Farmer, C.M., Wells, J.K., and Lund, A.K. (2003): "Effects of Head Restraint and Seat Redesign on Neck Injury Risk in Rear-End Crashes," *Traffic Injury Prevention*, 4:83-90.

24. Guez, M., Hildingsson, C., Stegmayr, B., and Toolanen, G. (2003): "Chronic Neck Pain of Traumatic and Non-Traumatic Origin," *Acta Orthop Scand*, 74(5):576-579.

25. Hodgson, V.R. and Thomas, L.M. (1980): "Mechanisms of Cervical Spine Injury During Impact to the Protected Head," Paper No. 801300, Proceedings of the 24th Stapp Car Crash Conference, SAE International, Warrendale, PA.

26. Hubbard, R.P. and Begeman, P.C. (1990): "Biomechanical Performance of a New Head and Neck Support," Paper No. 902312, Proceedings of the 34th Stapp Car Crash Conference, SAE International, Warrendale, PA.

27. Hubbard, R.P., Quinn, K.P., Martinez, J.J., and Winkelstein, B.A. (2008): "The Role of Graded Nerve Root Compression on Axonal Damage, Neuropeptide Changes, and Pain-Related Behaviors," Paper No. 2008-01-SC02, *Stapp Car Crash J.*, 52:33-58.

28. Ivancic, P.C., Panjabi, M.M., Tominaga, Y., and Malcolmson, G.F. (2006): "Predicting Multiplanar Cervical Spine Injury Due to Head-Turned Rear Impacts Using IV-NIC," *Traffic Injury Prevention*, 7(3):264-275.

29. Ivancic, P.C., Ito, S., Tominaga, Y., Rubin, W., Coe, M.P., Ndu, A.B., Carlson, E.J., and Panjabi, M.M. (2008): "Whiplash Causes Increased Laxity of Cervical Capsular Ligament," *Clin. Biomech.*, 23(2):159-165.

30. Jakobsson, L. and Norin, H. (2004): "Parameters Influencing AIS 1 Neck Injury Outcome in Frontal Impacts," *Traffic Injury Prevention*, 5:156-163.

31. James, M.B., Nordhagen, R.P., Schneider, D.C., and Koh, S.-W. (2007): "Occupant Injury in Rollover Crashes: A Reexamination of Malibu II," Paper No. 2007-01-0369, SAE International, Warrendale, PA.

32. Kaneoka, K., Ono, K., Inami, S., and Hayashi, K. (1999): "Motion Analysis of Cervical Vertebrae During Whiplash Loading," *Spine*, 24(8):763-769.

33. Kang, J., Agaram, V., Nusholtz, G.S., and Kostyniuk, G.W. (2001): Air Bag Loading on In-Position Hybrid III Dummy Neck," Paper No. 2001-01-0179, SAE International, Warrendale, PA.

34. Kang, J., Nusholtz, G., and Agaram, V. (2005): "Hybrid-III Dummy Neck Issues," Paper No. 2005-01-1704, SAE International, Warrendale, PA.

35. Kim, A., Anderson, K.F., Berliner, J., Bryzik, C., Hassan, J., Jensen, J., Kendall, M., Mertz, H.J., Morrow, T., Rao, A., and Wozniak, J.A. (2001): "A Comparison of the Hybrid III and BioRID II Dummies in Low-Severity, Rear-Impact Sled Tests," Paper No. 2001-22-0012, *Stapp Car Crash J.*, 45:257-284.

36. Klinich, K.D., Ebert, S.M., Van Ee, C.A., Flannagan, C.A.C., Prasad, M., Reed, M.P., and Schneider, L.W. (2004): "Cervical Spine Geometry in the Automotive Seated Posture: Variations with Age, Stature, and Gender," Paper No. 2004-22-0014, *Stapp Car Crash J.*, 48:301-330.

37. Kullgren, A., Krafft, M., Malm, S., Ydenius, A., and Tingvall, C. (2000): "Influence of Airbags and Seatbelt Pretensioners on AIS1 Neck Injuries for Belted Occupants in Frontal Impacts," Paper No. 2000-01-SC09, *Stapp Car Crash J.*, 44:117-125.

38. Lee, K.E., Davis, M.B., Mejilla, R.M., and Winkelstein, B.A. (2004): "In Vivo Cervical Facet Capsule Distraction: Mechanical Implications for Whiplash and Neck Pain," Paper No. 2004-22-0016, *Stapp Car Crash J.,* 48:373-395.

39. Lee, K.E., Franklin, A.N., Davis, M.B., and Winkelstein, B.A. (2006): "Tensile Cervical Facet Capsule Ligament Mechanics," *J. Biomechanics,* 39(7):1256-1264.

40. Linder, A., Avery, M., Krafft, M., Kullgren, A., and Svensson, M.Y. (2001): "Acceleration Pulses and Crash Severity in Low Velocity Rear Impacts—Real-World Data And Barrier Tests," Paper No. 2001-06-0253, SAE International, Warrendale, PA.

41. Lord, S.M., Barnsley, L., Wallis, B.J., and Bogduk, N. (1996): "Chronic Cervical Zygapophysial Joint Pain After Whiplash: A Placebo-Controlled Prevalence Study," *Spine,* 21(15):1737-1744.

42. Lu, Y., Chen, C., Kallakuri, S., Patwardhan, A., and Cavanaugh, J.M. (2005): "Neural Response of Cervical Facet Joint Capsule to Stretch: A Study of Whiplash Pain Mechanism," Paper No. 2005-22-0003, *Stapp Car Crash J.,* 49:49-65.

43. Maak, T.G., Ivancic, P.C., Tominaga, Y., and Panjabi, M.M. (2007): "Side Impact Causes Multiplanar Cervical Spine Injuries," *J. Trauma Injury Infection Crit. Care,* 63(6):1296-1307.

44. McElhaney, J.H., Hopper, R.H. Jr., Nightingale, R.W., and Myers, B.S. (1995): "Mechanisms of Basilar Skull Fracture," *J. Neurotrauma,* 12(4):669-678.

45. McLain, R.F. (1994): "Mechanoreceptor Endings in Human Cervical Facet Joints," *Spine,* 19:495-501.

46. Melvin, J.W. and Begeman, P.C. (2002): "Sled Test Evaluation of Racecar Head/Neck Restraints," Paper No. 2002-01-3304, SAE International, Warrendale, PA.

47. Mercer, S. and Bogduk, N. (1999): "The Ligaments and Annulus Fibrosus of Human Adult Cervical Intervertebral Discs," *Spine,* 24(7):619-626.

48. Mertz, H.J. and Patrick, L.M. (1967): "Investigation of the Kinematics and Kinetics of Whiplash During Vehicle Rear-End Collisions," Paper No. 670919, Proceedings of the 11th Stapp Car Crash Conference, SAE International, Warrendale, PA.

49. Mertz, H.J., Hodgson, V.R., Murray, T.L., and Nyquist, G.W. (1978): "An Assessment of Compressive Neck Loads Under Injury-Producing Conditions," *Phys. Sportsmed.,* 6:95-106.

50. Mertz, H.J. and Prasad, P. (2000): "Improved Neck Injury Risk Curves for Tension and Extension Moment Measurements of Crash Dummies," Paper No. 2000-01-SC05, *Stapp Car Crash J.,* 44:59-75.

51. Mertz, H.J., Irwin, A.L., and Prasad, P. (2003): "Biomechanical and Scaling Bases for Frontal and Side Impact Injury Assessment Reference Values," Paper No. 2003-22-0009, *Stapp Car Crash J.,* 47:155-188.

52. Moss, R.T., Bardas, A.M., Hughes, M.C., and Happer, A.J. (2005): "Injury Symptom Risk Curves for Occupants Involved in Rear-End, Low- Speed Motor Vehicle Collisions," Paper No. 2005-01-0296, SAE International, Warrendale, PA.

53. Nightingale, R.W., Richardson, W.J., and Myers, B.S. (1997): "The Effects of Padded Surfaces on the Risk for Cervical Spine Injury," *Spine,* 22(20):2380-2387.

54. Nightingale, R.W., Winkelstein, B.A., Knaub, K.E., Richardson, W.J., Luck, J.F., and Myers, B.S. (2002): "Comparative Strengths and Structural Properties of the Upper and Lower Cervical Spine in Flexion and Extension," *J. Biomechanics,* 35(6):725-732.

55. Nightingale, R.W., Chancey, V.C., Ottaviano, D., Luck, J.F., Tran, L., Prange, M., and Myers, B.S. (2007): "Flexion and Extension Structural Properties and Strengths for Male Cervical Spine Segments," *J. Biomechanics,* 40(3):535-542.

56. Nuckley, D.J., Hertsted, S.M., Ku, G.S., Eck, M.P., and Ching, R.P. (2002): "Compressive Tolerance of the Maturing Cervical Spine," Paper No. 2002-22-0021, *46th Stapp Car Crash J.,* 46:431-440.

57. Nusholtz, G.S., Di Domenico, L., Shi, Y., and Eagle, P. (2003): "Studies of Neck Injury Criteria Based on Existing Biomechanical Test Data," *Accid. Anal. Prev.,* 35(5):777-786.

58.   Oi, N., Pandy, M.G., Myers, B.S., Nightingale, R.W., and Chancey, V.C. (2004): "Variation of Neck Muscle Strength Along the Human Cervical Spine," Paper No. 2004-22-0017, *Stapp Car Crash J.,* 48:397-417.

59.   Ono, K., Kaneoka, K., and Inami, S. (1998): "Influence of Seat Properties on Human Cervical Vertebral Motion and Head/Neck/Torso Kinematics," Proceedings of the 1998 International Conference on the Biomechanics of Impact (IRCOBI, Zurich).

60.   Ono, K., Uwai, H., Kaneoka, K., Fukushima, M., and Ujihashi, S. (2003): "Influences of Neck Muscle Tension on Cervical Vertebral Motions During Direct Loading on Human Head," Proceedings of the 2003 International Conference on the Biomechanics of Impact (IRCOBI, Zurich).

61.   Ono, K., Ejima, S., Suzuki, Y., Kaneoka, K., Fukushima, M., and Ujihashi, S. (2006): "Prediction of Neck Injury Risk Based on the Analysis of Localized Cervical Vertebral Motion of Human Volunteers During Low-Speed Rear Impacts," Proceedings of the 2006 International Conference on the Biomechanics of Impact (IRCOBI, Zurich), 103-114.

62.   Panjabi, M.M., Oda, T., Crisco, J.J. III, Oxland, T.R., Katz, L., and Nolte, L.-P. (1991): "Experimental Study of Atlas Injuries I: Biomechanical Analysis of Their Mechanisms and Fracture Patterns," *Spine,* 16:S460-S465.

63.   Panjabi, M.M., Wang, J.-L., and Delson, N. (1999): "Neck Injury Criterion Based on Intervertebral Motions and Its Evaluation Using an Instrumented Neck Dummy," Proceedings of the 1999 International Conference on the Biomechanics of Impact (IRCOBI, Zurich).

64.   Pearson, A.M., Ivancic, P.C., Ito, S., and Panjabi, M.M. (2004): "Facet Joint Kinematics and Injury Mechanisms During Simulated Whiplash," *Spine,* 29(4):390-397.

65.   Philippens, M., Cappon, H., van Ratingen, M., Wismans, J., Svensson, M., Sirey, F., Ono, K., Nishimoto, N., and Matsuoka, F. (2002): "Comparison of the Rear Impact Biofidelity of BioRID II and RID2," Paper No. 2002-22-0023, *Stapp Car Crash J.,* 46:383-399.

66.   Pike, J.A. (1989): "The Quantitative Effect of Age on Injury Outcome," Paper No. 89-1A-W-020, Proceedings of the 12th International Experimental Safety Vehicle Conference, National Highway Traffic Safety Administration. Washington, D.C.

67.   Pike, J.A. (1997): *Automotive Safety: Anatomy, Injury, Testing, and Regulation* (R-171), SAE International, Warrendale, PA.

68.   Pike, J.A. (2000): "Whiplash Injury and Vehicle Design Concepts," in *Frontiers in Whiplash Trauma,* Yoganandan, N. and Pintar, F.A., Eds., IOS Press, The Netherlands.

69.   Pike, J.A. (2002): *Neck Injury: The Use of X-Rays, CTs And MRIs to Study Crash-Related Injury Mechanisms* (R-268), SAE International, Warrendale, PA.

70.   Pike, J.A. (2008): *Forensic Biomechanics: Using Medical Records to Study Injury Mechanisms* (R-379), SAE International, Warrendale, PA.

71.   Pintar, F.A., Yoganandan, N., Sances, A. Jr., Reinartz, J., Harris, G., and Larson, S.J. (1989): "Kinematic and Anatomical Analysis of the Human Cervical Spinal Column Under Axial Loading," Paper No. 892436, Proceedings of the 33rd Stapp Car Crash Conference, SAE International, Warrendale, PA.

72.   Pintar, F.A., Yoganandan, N., and Voo, L. (1998): "Effect of Age and Loading Rate on Human Cervical Spine Injury Threshold," *Spine,* 23(18):1957-1962.

73.   Pintar, F.A., Mayer, R.G., Yoganandan, N., and Sun, E. (2000): "Child Neck Strength Characteristics Using an Animal Model," Paper No. 2000-01-SC06, *Stapp Car Crash J.,* 44:77-83.

74.   Pintar, F.A., Yoganandan, N., and Baisden, J. (2002): "Characterizing Occipital Condyle Loads Under High-Speed Head Rotation," Paper No. 2005-22-0002, *Stapp Car Crash J.,* 49:33-47.

75.   Prasad, P., Kim, A.S., Weerappuli, D., Roberts, V.L., and Schneider, D.C. (1997): "Relationships Between Passenger-Car Seat Back Strength and Occupant Injury Severity in Rear-End Collisions: Field and Laboratory Studies," Paper No. 973343, Proceedings of the 41st Stapp Car Crash Conference, SAE International, Warrendale, PA.

76.   SAE International (2003): "Human Tolerance to Impact Conditions as Related to Motor Vehicle Design," Information Report J885, SAE International, Warrendale, PA.

77.   Shea, M., Wittenberg, R.H., Edwards, W.T., White, A.A. III, and Hayes, W.C. (1992): "In Vitro Hyperextension Injuries in the Human Cadaveric Cervical Spine," *J. Ortho. Res.,* 10(6):911-916.

78. Siegmund, G.P., Myers, B.S., Davis, M.B., Bohnet, H.F., and Winkelstein, B.A. (2000): "Human Cervical Motion Segment Flexibility and Facet Capsular Ligament Strain Under Combined Posterior Shear, Extension and Axial Compression," Paper No. 2000-01-SC12, Proceedings of the 44th Stapp Car Crash Conference, *Stapp Car Crash J.,* 44:159-170.

79. Siegmund, G.P., Heinrichs, B.E., Lawrence, J.M., and Philippens, M.M.G.M. (2001a): "Kinetic and Kinematic Responses of the RID2a, Hybrid III and Human Volunteers in Low-Speed Rear-End Collisions," Paper No. 2001-22-0011, Proceedings of the 45th Stapp Car Crash Conference, *Stapp Car Crash J.,* 45:239-256.

80. Siegmund, G.P., Myers, B.S., Davis, M.B., Bohnet, H.F., and Winkelstein, B.A. (2001b): "Mechanical Evidence of Cervical Facet Capsule Injury During Whiplash: A Cadaveric Study Using Combined Shear, Compression, and Extension Loading," *Spine,* 26: 2095-2101.

81. Siegmund, G.P., Blouin, J.S., Brault, J.R., Hedenstierna, S., and Inglis, J.T. (2007): "Electromyography of Superficial and Deep Neck Muscles During Isometric, Voluntary, and Reflex Contractions," *J. Biomech. Eng.,* 29(1):66-77.

82. Stemper, B.D., Stineman, M.R., Yoganandan, N., Pintar, F.A., Sinson, G.P., and Gennarelli, T.A. (2005): "Mechanical Characterization of Internal Layer Failure in the Human Carotid Artery," *Ann. Biomed. Eng.,* 35(2):285-289.

83. Sundararajan, S., Prasad, P., Demetropoulos, C.K., Tashman, S., Begeman, P.C., Yang, K.H., and King, A.I. (2004): "Effect of Head-Neck Position on Cervical Facet Stretch of Post Mortem Human Subjects During Low Speed Rear-End Impacts," Paper No. 2004-22-0015, *Stapp Car Crash J.,* 48:331-372.

84. Svensson, M.Y., Lövsund, P., Håland, Y., Larsson, S., and Svensson, M.Y. (1996): "The Influence of Seat-Back and Head-Restraint Properties on the Head-Neck Motion During Rear-Impact," *Accid. Anal. Prev.,* 28(2):221-227.

85. Svensson, M.Y., Boström, O., Davidsson, J., Hansson, H.A., Håland, Y., Lövsund, P., Suneson, A., and Säljö, A. (2000): "Neck Injuries in Car Collisions—A Review Covering a Possible Injury Mechanism and the Development of a New Rear-Impact Dummy," *Accid. Anal. Prev.,* 32(2):167-75.

86. Van Ee, C.A., Nightingale, R.W., Camacho, D.L.A., Chancey, V.C., Knaub, K.E., Sun, E.A., and Myers, B.S. (2000): "Tensile Properties of the Human Muscular and Ligamentous Cervical Spine," Paper No. 2000-01-SC07, *Stapp Car Crash J.,* 44:85-102.

87. Viano, D.C., Olsen, S., Locke, G.S., and Humer, M. (2002): "Neck Biomechanical Responses with Active Head Restraints: Rear Barrier Tests with BioRID and Sled Tests with Hybrid III," Paper No. 2002-01-0030, SAE International, Warrendale, PA.

88. Viano, David C. and Pellman, Elliott J. (2005): "Concussion in Professional Football: Biomechanics of the Striking Player—Part 8," *Neurosurgery,* 56(2):266-280.

89. Viano, D.C. and Parenteau, C.S. (2008): "Analysis of Head Impacts Causing Neck Compression Injury," *Traffic Injury Prevention,* 9(2):144-152.

90. Watanabe, Y., Ichikawa, H., Kayama, O., Ono, K., Kaneoka, K., and Inami, S. (1999): "Relationships Between Occupant Motion and Seat Characteristics in Low-Speed Rear Impacts," Paper No. 1999-01-0635, SAE International, Warrendale, PA.

91. Winkelstein, B.A., Nightingale, R.W., Richardson, W.J., and Myers, B.S. (1999): "Cervical Facet Joint Mechanics: Its Application to Whiplash Injury," Paper No. 99SC15, Proceedings of the 43rd Stapp Car Crash Conference.

92. Winkelstein, B.A. and Myers, B.S. (2000): "Experimental and Computational Characterization of Three-Dimensional Cervical Spine Flexibility," Paper No. 2000-01-SC11, Proceedings of the 44th Stapp Crash Conference.

93. Winkelstein, B.A., McLendon, R.E., Barbir, A., and Myers, B.S. (2001): "An Anatomical Investigation of the Human Cervical Facet Capsule, Quantifying Muscle Insertion Area," *J. of Anatomy,* 198(4):455-461.

94. Winston, F.K. and Reed, R. (1996): "Air Bags and Children: Results of a National Highway Traffic Safety Administration Special Investigation into Actual Crashes," Paper No. 962438, Proceedings of the 40th Stapp Car Crash Conference, SAE International, Warrendale, PA.

95. Yamaguchi, G.T., Carhart, M.R., Larson, R., Richards, D., Pierce, J., Raasch, C.C., Scher, I., and Corrigan, C.F. (2005): "Electromyographic Activity and Posturing of the Human Neck During Rollover Tests," Paper No. 2005-01-0302, SAE International, Warrendale, PA.

96. Yoganandan, N., Pintar, F.A., Maiman, D.J., and Cusick, J.F. (1996): "Human Head-Neck Biomechanics Under Axial Tension," *Med. Eng. Phys.,* 1(4):289-294.

97. Yoganandan, N., Pintar, F.A., Kumaresan, S., and Elhagediab, A. (1998): "Biomechanical Assessment of Human Cervical Spine Ligaments," Paper No. 983159, Proceedings of the 42nd Stapp Car Crash Conference.

98. Yoganandan, N., Pintar, F.A., Stemper, B.D., Schlick, M.B., Philippens, M., and Wismans, J. (2000): "Biomechanics of Human Occupants in Simulated Rear Crashes: Documentation of Neck Injuries and Comparison of Injury Criteria," Paper No. 2000-01-SC14, Proceedings of the 44th Stapp Car Crash Conference.

99. Yoganandan, N., Stemper, B.D., Pintar, F.A., Baisden, J.L., Shender, B.S., and Paskoff, G. (2008): "Normative Segment-Specific Axial and Coronal Angulation Corridors of Subaxial Cervical Column in Axial Rotation," *Spine*, 33(5):490-496.

100. Yoganandan, N., Pintar, F.A., Maiman, D.J, Philippens, M., and Wismans, J. (2009): "Neck Forces and Moments and Head Accelerations in Side Impact," *Traffic Injury Prevention,* 10(1):51-57.

Table II. Bibliography Listing by Year

Mertz	1967		Melvin	2002
Mertz	1978		Nightingale	2002
Althoff	1979		Nuckley	2002
Hodgson	1980		Philippens	2002
Pike	1989		Pike	2002
Pintar	1989		Pintar	2002
Aprill	1990		Viano	2002
Hubbard	1990		Cappon	2003
Panjabi	1991		Chancey	2003
Shea	1992		Farmer	2003
McLain	1994		Guez	2003
McElhaney	1995		Mertz	2003
Bostrom	1996		Nusholtz	2003
Lord	1996		Ono	2003
Svensson	1996		SAE	2003
Winston	1996		Been	2004
Yoganandan	1996		Jakobsson	2004
Doherty	1997		Klinich	2004
Nightingale	1997		Lee	2004
Pike	1997		Oi	2004
Prasad	1997		Pearson	2004
Barnsley	1998		Sundarajan	2004
Ono	1998		Anderson	2005
Pintar	1998		Cappon	2005
Yoganandan	1998		Ejima	2005
Davidsson	1999		Kang	2005
Duma	1999		Lu	2005
Kaneoka	1999		Moss	2005
Mercer	1999		Stemper	2005
Panjabi	1999		Viano	2005
Watanabe	1999		Yamaguchi	2005
Winkelstein	1999		Dibb	2006
Bohman	2000		Ivancic	2006
Brault	2000		Lee	2006
Cavanaugh	2000		Ono	2006
Deng	2000		Carlsson	2007
Eichberger	2000		James	2007
Kullgren	2000		Maak	2007
Mertz	2000		Nightingale	2007
Pike	2000		Siegmund	2007
Pintar	2000		Hubbard	2008
Siegmund	2000		Ivancic	2008
Svensson	2000		Pike	2008
Van Ee	2000		Viano	2008
Winkelstein	2000		Yoganandan	2008
Yoganandan	2000		Yoganandan	2009
Agaram	2001			
Kang	2001			
Kim	2001			
Linder	2001			
Winkelstein	2001			
Siegmund	2001a			
Siegmund	2001b			
Crawford	2002			